This book provides an introduction to the major mathematical structures used in physics today. It covers the concepts and techniques needed for topics such as group theory, Lie algebras, topology, Hilbert spaces and differential geometry. Important theories of physics such as classical and quantum mechanics, thermodynamics, and special and general relativity are also developed in detail, and presented in the appropriate mathematical language.

The book is suitable for advanced undergraduate and beginning graduate students in mathematical and theoretical physics. It includes numerous exercises and worked examples to test the reader's understanding of the various concepts, as well as extending the themes covered in the main text. The only prerequisites are elementary calculus and linear algebra. No prior knowledge of group theory, abstract vector spaces or topology is required.

PETER SZEKERES received his Ph.D. from King's College London in 1964, in the area of general relativity. He subsequently held research and teaching positions at Cornell University, King's College and the University of Adelaide, where he stayed from 1971 till his recent retirement. Currently he is a visiting research fellow at that institution. He is well known internationally for his research in general relativity and cosmology, and has an excellent reputation for his teaching and lecturing.

A Course in Modern Mathematical Physics

Groups, Hilbert Space and Differential Geometry

Peter Szekeres
Formerly of University of Adelaide

CAMBRIDGE UNIVERSITY PRESS
Cambridge, New York, Melbourne, Madrid, Cape Town,
Singapore, São Paulo, Delhi, Mexico City

Cambridge University Press
The Edinburgh Building, Cambridge CB2 8RU, UK

Published in the United States of America by Cambridge University Press, New York

www.cambridge.org
Information on this title: www.cambridge.org/9780521829601

© P. Szekeres 2004

This publication is in copyright. Subject to statutory exception
and to the provisions of relevant collective licensing agreements,
no reproduction of any part may take place without the written
permission of Cambridge University Press.

First published 2004
4th printing 2012

A catalogue record for this publication is available from the British Library

Library of Congress Cataloguing in Publication Data

Szekeres, Peter, 1940–
  A course in modern mathematical physics: groups, Hilbert space, and differential
  geometry / Peter Szekeres.
    p.   cm.
  Includes bibliographical references and index.
  ISBN 0 521 82960 7 – ISBN 0 521 53645 6 (pb.)
  1. Mathematical physics.   I. Title.

QC20 S965 2004
530.15–dc22   2004045675

ISBN 978-0-521-82960-1 Hardback

Cambridge University Press has no responsibility for the persistence or
accuracy of URLs for external or third-party internet websites referred to in
this publication, and does not guarantee that any content on such websites is,
or will remain, accurate or appropriate. Information regarding prices, travel
timetables, and other factual information given in this work is correct at
the time of first printing but Cambridge University Press does not guarantee
the accuracy of such information thereafter.

Contents

Preface		*page* ix
Acknowledgements		xiii
1	**Sets and structures**	**1**
	1.1 Sets and logic	2
	1.2 Subsets, unions and intersections of sets	5
	1.3 Cartesian products and relations	7
	1.4 Mappings	10
	1.5 Infinite sets	13
	1.6 Structures	17
	1.7 Category theory	23
2	**Groups**	**27**
	2.1 Elements of group theory	27
	2.2 Transformation and permutation groups	30
	2.3 Matrix groups	35
	2.4 Homomorphisms and isomorphisms	40
	2.5 Normal subgroups and factor groups	45
	2.6 Group actions	49
	2.7 Symmetry groups	52
3	**Vector spaces**	**59**
	3.1 Rings and fields	59
	3.2 Vector spaces	60
	3.3 Vector space homomorphisms	63
	3.4 Vector subspaces and quotient spaces	66
	3.5 Bases of a vector space	72
	3.6 Summation convention and transformation of bases	81
	3.7 Dual spaces	88
4	**Linear operators and matrices**	**98**
	4.1 Eigenspaces and characteristic equations	99
	4.2 Jordan canonical form	107

	4.3	Linear ordinary differential equations	116
	4.4	Introduction to group representation theory	120

5 Inner product spaces — 126
- 5.1 Real inner product spaces — 126
- 5.2 Complex inner product spaces — 133
- 5.3 Representations of finite groups — 141

6 Algebras — 149
- 6.1 Algebras and ideals — 149
- 6.2 Complex numbers and complex structures — 152
- 6.3 Quaternions and Clifford algebras — 157
- 6.4 Grassmann algebras — 160
- 6.5 Lie algebras and Lie groups — 166

7 Tensors — 178
- 7.1 Free vector spaces and tensor spaces — 178
- 7.2 Multilinear maps and tensors — 186
- 7.3 Basis representation of tensors — 193
- 7.4 Operations on tensors — 198

8 Exterior algebra — 204
- 8.1 r-Vectors and r-forms — 204
- 8.2 Basis representation of r-vectors — 206
- 8.3 Exterior product — 208
- 8.4 Interior product — 213
- 8.5 Oriented vector spaces — 215
- 8.6 The Hodge dual — 220

9 Special relativity — 228
- 9.1 Minkowski space-time — 228
- 9.2 Relativistic kinematics — 235
- 9.3 Particle dynamics — 239
- 9.4 Electrodynamics — 244
- 9.5 Conservation laws and energy–stress tensors — 251

10 Topology — 255
- 10.1 Euclidean topology — 255
- 10.2 General topological spaces — 257
- 10.3 Metric spaces — 264
- 10.4 Induced topologies — 265
- 10.5 Hausdorff spaces — 269
- 10.6 Compact spaces — 271

	10.7	Connected spaces	273
	10.8	Topological groups	276
	10.9	Topological vector spaces	279
11	**Measure theory and integration**		**287**
	11.1	Measurable spaces and functions	287
	11.2	Measure spaces	292
	11.3	Lebesgue integration	301
12	**Distributions**		**308**
	12.1	Test functions and distributions	309
	12.2	Operations on distributions	314
	12.3	Fourier transforms	320
	12.4	Green's functions	323
13	**Hilbert spaces**		**330**
	13.1	Definitions and examples	330
	13.2	Expansion theorems	335
	13.3	Linear functionals	341
	13.4	Bounded linear operators	344
	13.5	Spectral theory	351
	13.6	Unbounded operators	357
14	**Quantum mechanics**		**366**
	14.1	Basic concepts	366
	14.2	Quantum dynamics	379
	14.3	Symmetry transformations	387
	14.4	Quantum statistical mechanics	397
15	**Differential geometry**		**410**
	15.1	Differentiable manifolds	411
	15.2	Differentiable maps and curves	415
	15.3	Tangent, cotangent and tensor spaces	417
	15.4	Tangent map and submanifolds	426
	15.5	Commutators, flows and Lie derivatives	432
	15.6	Distributions and Frobenius theorem	440
16	**Differentiable forms**		**447**
	16.1	Differential forms and exterior derivative	447
	16.2	Properties of exterior derivative	451
	16.3	Frobenius theorem: dual form	454
	16.4	Thermodynamics	457
	16.5	Classical mechanics	464

17 Integration on manifolds — 481
- 17.1 Partitions of unity — 482
- 17.2 Integration of n-forms — 484
- 17.3 Stokes' theorem — 486
- 17.4 Homology and cohomology — 493
- 17.5 The Poincaré lemma — 500

18 Connections and curvature — 506
- 18.1 Linear connections and geodesics — 506
- 18.2 Covariant derivative of tensor fields — 510
- 18.3 Curvature and torsion — 512
- 18.4 Pseudo-Riemannian manifolds — 516
- 18.5 Equation of geodesic deviation — 522
- 18.6 The Riemann tensor and its symmetries — 524
- 18.7 Cartan formalism — 527
- 18.8 General relativity — 534
- 18.9 Cosmology — 548
- 18.10 Variation principles in space-time — 553

19 Lie groups and Lie algebras — 559
- 19.1 Lie groups — 559
- 19.2 The exponential map — 564
- 19.3 Lie subgroups — 569
- 19.4 Lie groups of transformations — 572
- 19.5 Groups of isometries — 578

Bibliography — 587
Index — 589

Preface

After some twenty years of teaching different topics in the Department of Mathematical Physics at the University of Adelaide I conceived the rather foolhardy project of putting all my undergraduate notes together in one single volume under the title *Mathematical Physics*. This undertaking turned out to be considerably more ambitious than I had originally expected, and it was not until my recent retirement that I found the time to complete it.

Over the years I have sometimes found myself in the midst of a vigorous and at times quite acrimonious debate on the difference between theoretical and mathematical physics. This book is symptomatic of the difference. I believe that mathematical physicists put the mathematics first, while for theoretical physicists it is the physics which is uppermost. The latter seek out those areas of mathematics for the use they may be put to, while the former have a more unified view of the two disciplines. I don't want to say one is better than the other – it is simply a different outlook. In the big scheme of things both have their place but, as this book no doubt demonstrates, my personal preference is to view mathematical physics as a branch of mathematics.

The classical texts on mathematical physics which I was originally brought up on, such as Morse and Feshbach [7], Courant and Hilbert [1], and Jeffreys and Jeffreys [6] are essentially books on differential equations and linear algebra. The flavour of the present book is quite different. It follows much more the lines of Choquet-Bruhat, de Witt-Morette and Dillard-Bleick [14] and Geroch [3], in which mathematical structures rather than mathematical analysis is the main thrust. Of these two books, the former is possibly a little daunting as an introductory undergraduate text, while Geroch's book, written in the author's inimitably delightful lecturing style, has occasional tendencies to overabstraction. I resolved therefore to write a book which covers the material of these texts, assumes no more mathematical knowledge than elementary calculus and linear algebra, and demonstrates clearly how theories of modern physics fit into various mathematical structures. How well I have succeeded must be left to the reader to judge.

At times I have been caught by surprise at the natural development of ideas in this book. For example, how is it that quantum mechanics appears before classical mechanics? The reason is certainly not on historical grounds. In the natural organization of mathematical ideas, algebraic structures appear before geometrical or topological structures, and linear structures are evidently simpler than non-linear. From the point of view of mathematical simplicity quantum mechanics, being a purely linear theory in a quasi-algebraic space (Hilbert space), is more elementary than classical mechanics, which can be expressed in

Preface

terms of non-linear dynamical systems in differential geometry. Yet, there is something of a paradox here, for as Niels Bohr remarked: 'Anyone who is not shocked by quantum mechanics does not understand it'. Quantum mechanics is not a difficult theory to express mathematically, but it is almost impossible to make epistomological sense of it. I will not even attempt to answer these sorts of questions, and the reader must look elsewhere for a discussion of quantum measurement theory [5].

Every book has its limitations. At some point the author must call it a day, and the omissions in this book may prove a disappointment to some readers. Some of them are a disappointment to me. Those wanting to go further might explore the theory of fibre bundles and gauge theories [2, 8, 13], as the stage is perfectly set for this subject by the end of the book. To many, the biggest omission may be the lack of any discussion of quantum field theory. This, however, is an area that seems to have an entirely different flavour to the rest of physics as its mathematics is difficult if nigh on impossible to make rigorous. Even quantum mechanics has a 'classical' flavour by comparison. It is such a huge subject that I felt daunted to even begin it. The reader can only be directed to a number of suitable books to introduce them to this field [10–14].

Structure of the book

This book is essentially in two parts, modern algebra and geometry (including topology). The early chapters begin with set theory, group theory and vector spaces, then move to more advanced topics such as Lie algebras, tensors and exterior algebra. Occasionally ideas from group representation theory are discussed. If calculus appears in these chapters it is of an elementary kind. At the end of this algebraic part of the book, there is included a chapter on special relativity (Chapter 9), as it seems a nice example of much of the algebra that has gone before while introducing some notions from topology and calculus to be developed in the remaining chapters. I have treated it as a kind of crossroads: Minkowski space acts as a link between algebraic and geometric structures, while at the same time it is the first place where physics and mathematics are seen to interact in a significant way.

In the second part of the book, we discuss structures that are essentially geometrical in character, but generally have an algebraic component as well. Beginning with topology (Chapter 10), structures are created that combine both algebra and the concept of continuity. The first of these is Hilbert space (Chapter 13), which is followed by a chapter on quantum mechanics. Chapters on measure theory (Chapter 11) and distribution theory (Chapter 12) precede these two. The final chapters (15–19) deal with differential geometry and examples of physical theories using manifold theory as their setting – thermodynamics, classical mechanics, general relativity and cosmology. A flow diagram showing roughly how the chapters interlink is given below.

Exercises and problems are interspersed throughout the text. The exercises are not designed to be difficult – their aim is either to test the reader's understanding of a concept just defined or to complete a proof needing one or two more steps. The problems at ends of sections are more challenging. Frequently they are in many parts, taking up a thread

Preface

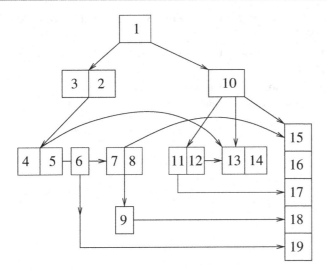

of thought and running with it. This way most closely resembles true research, and is my preferred way of presenting problems rather than the short one-liners often found in text books. Throughout the book, newly defined concepts are written in bold type. If a concept is written in italics, it has been introduced in name only and has yet to be defined properly.

References

[1] R. Courant and D. Hilbert. *Methods of Mathematical Physics*, vols 1 and 2. New York, Interscience, 1953.
[2] T. Frankel. *The Geometry of Physics*. New York, Cambridge University Press, 1997.
[3] R. Geroch. *Mathematical Physics*. Chicago, The University of Chicago Press, 1985.
[4] J. Glimm and A. Jaffe. *Quantum Physics: A Functional Integral Point of View*. New York, Springer-Verlag, 1981.
[5] J. M. Jauch. *Foundations of Quantum Mechanics*. Reading, Mass., Addison-Wesley, 1968.
[6] H. J. Jeffreys and B. S. Jeffreys. *Methods of Mathematical Physics*. Cambridge, Cambridge University Press, 1946.
[7] P. M. Morse and H. Feshbach. *Methods of Theoretical Physics*, vols 1 and 2. New York, McGraw-Hill, 1953.
[8] C. Nash and S. Sen. *Topology and Geometry for Physicists*. London, Academic Press, 1983.
[9] P. Ramond. *Field Theory: A Modern Primer*. Reading, Mass., Benjamin/Cummings, 1981.
[10] L. H. Ryder, *Quantum Field Theory*. Cambridge, Cambridge University Press, 1985.
[11] S. S. Schweber. *An Introduction to Relativistic Quantum Field Theory*. New York, Harper and Row, 1961.

[12] R. F. Streater and A. S. Wightman. *PCT, Spin and Statistics, and All That*. New York, W. A. Benjamin, 1964.

[13] A. Trautman. Fibre bundles associated with space-time. *Reports on Mathematical Physics*, **1**:29–62, 1970.

[14] C. de Witt-Morette, Y. Choquet-Bruhat and M. Dillard-Bleick. *Analysis, Manifolds and Physics*. Amsterdam, North-Holland, 1977.

Acknowledgements

There are an enormous number of people I would like to express my gratitude to, but I will single out just a few of the most significant. Firstly, my father George Szekeres, who introduced me at an early age to the wonderful world of mathematics and has continued to challenge me throughout my life with his doubts and criticisms of the way physics (particularly quantum theory) is structured. My Ph.D. supervisor Felix Pirani was the first to give me an inkling of the importance of differential geometry in mathematical physics, while others who had an enormous influence on my education and outlook were Roger Penrose, Bob Geroch, Brandon Carter, Andrzej Trautman, Ray McLenaghan, George Ellis, Bert Green, Angas Hurst, Sue Scott, David Wiltshire, David Hartley, Paul Davies, Robin Tucker, Alan Carey, and Michael Eastwood. Finally, my wife Angela has not only been an endless source of encouragement and support, but often applied her much valued critical faculties to my manner of expression. I would also like to pay a special tribute to Patrick Fitzhenry for his invaluable assistance in preparing diagrams and guiding me through some of the nightmare that is today's computer technology.

To my mother, Esther

1 Sets and structures

The object of mathematical physics is to describe the physical world in purely mathematical terms. Although it had its origins in the science of ancient Greece, with the work of Archimedes, Euclid and Aristotle, it was not until the discoveries of Galileo and Newton that mathematical physics as we know it today had its true beginnings. Newton's discovery of the calculus and its application to physics was undoubtedly the defining moment. This was built upon by generations of brilliant mathematicians such as Euler, Lagrange, Hamilton and Gauss, who essentially formulated physical law in terms of differential equations. With the advent of new and unintuitive theories such as relativity and quantum mechanics in the twentieth century, the reliance on mathematics moved to increasingly recondite areas such as abstract algebra, topology, functional analysis and differential geometry. Even classical areas such as the mechanics of Lagrange and Hamilton, as well as classical thermodynamics, can be lifted almost directly into the language of modern differential geometry. Today, the emphasis is often more structural than analytical, and it is commonly believed that finding the right mathematical structure is the most important aspect of any physical theory. Analysis, or the consequences of theories, still has a part to play in mathematical physics – indeed, most research is of this nature – but it is possibly less fundamental in the total overview of the subject.

When we consider the significant achievements of mathematical physics, one cannot help but wonder why the workings of the universe are expressable at all by rigid mathematical 'laws'. Furthermore, how is it that purely human constructs, in the form of deep and subtle mathematical structures refined over centuries of thought, have any relevance at all? The nineteenth century view of a clockwork universe regulated deterministically by differential equations seems now to have been banished for ever, both through the fundamental appearance of probabilities in quantum mechanics and the indeterminism associated with chaotic systems. These two aspects of physical law, the deterministic and indeterministic, seem to interplay in some astonishing ways, the impact of which has yet to be fully appreciated. It is this interplay, however, that almost certainly gives our world its richness and variety. Some of these questions and challenges may be fundamentally unanswerable, but the fact remains that mathematics seems to be the correct path to understanding the physical world.

The aim of this book is to present the basic mathematical structures used in our subject, and to express some of the most important theories of physics in their appropriate mathematical setting. It is a book designed chiefly for students of physics who have the need for a more rigorous mathematical education. A basic knowledge of calculus and linear algebra, including matrix theory, is assumed throughout, but little else. While different students will

Sets and structures

of course come to this book with different levels of mathematical sophistication, the reader should be able to determine exactly what they can skip and where they must take pause. Mathematicians, for example, may be interested only in the later chapters, where various theories of physics are expressed in mathematical terms. These theories will not, however, be developed at great length, and their consequences will only be dealt with by way of a few examples.

The most fundamental notion in mathematics is that of a *set*, or 'collection of objects'. The subject of this chapter is *set theory* – the branch of mathematics devoted to the study of sets as abstract objects in their own right. It turns out that every mathematical structure consists of a collection of sets together with some *defining relations*. Furthermore, as we shall see in Section 1.3, such relations are themselves defined in terms of sets. It is thus a commonly adopted viewpoint that all of mathematics reduces essentially to statements in set theory, and this is the motivation for starting with a chapter on such a basic topic.

The idea of sets as collections of objects has a non-rigorous, or 'naive' quality, although it is the form in which most students are introduced to the subject [1–4]. Early in the twentieth century, it was discovered by Bertrand Russell that there are inherent self-contradictions and paradoxes in overly simple versions of set theory. Although of concern to logicians and those mathematicians demanding a totally rigorous basis to their subject, these paradoxes usually involve inordinately large self-referential sets – not the sort of constructs likely to occur in physical contexts. Thus, while special models of set theory have been designed to avoid contradictions, they generally have somewhat artificial attributes and naive set theory should suffice for our purposes. The reader's attention should be drawn, however, to the remarks at the end of Section 1.5 concerning the possible relevance of fundamental problems of set theory to physics. These problems, while not of overwhelming concern, may at least provide some food for thought.

While a basic familiarity with set theory will be assumed throughout this book, it nevertheless seems worthwhile to go over the fundamentals, if only for the sake of completeness and to establish a few conventions. Many physicists do not have a good grounding in set theory, and should find this chapter a useful exercise in developing the kind of rigorous thinking needed for mathematical physics. For mathematicians this is all bread and butter, and if you feel the material of this chapter is well-worn ground, please feel free to pass on quickly.

1.1 Sets and logic

There are essentially two ways in which we can think of a **set** S. Firstly, it can be regarded as a collection of mathematical objects $a, b, \ldots$, called **constants**, written

$$S = \{a, b, \ldots\}.$$

The constants $a, b, \ldots$ may themselves be sets and, indeed, some formulations of set theory *require* them to be sets. Physicists in general prefer to avoid this formal nicety, and find it much more natural to allow for 'atomic' objects, as it is hard to think of quantities such as *temperature* or *velocity* as being 'sets'. However, to think of sets as consisting of lists of

1.1 Sets and logic

objects is only suitable for finite or at most countably infinite sets. If we try putting the real numbers into a list we encounter the Cantor diagonalization problem – see Theorems 1.4 and 1.5 of Section 1.5.

The second approach to set theory is much more general in character. Let $P(x)$ be a *logical proposition* involving a **variable** x. Any such proposition symbolically defines a set

$$S = \{x \mid P(x)\},$$

which can be thought of as symbolically representing the collection of all x for which the proposition $P(x)$ is true. We will not attempt a full definition of the concept of logical proposition here – this is the business of formal logic and is only of peripheral interest to theoretical physicists – but some comments are in order. Essentially, logical propositions are statements made up from an alphabet of symbols, some of which are termed **constants** and some of which are called **variables**, together with logical connectives such as **not**, **and**, **or** and **implies**, to be manipulated according to rules of standard logic. Instead of 'P implies Q' we frequently use the words '**if** P **then** Q' or the symbolic representation $P \Rightarrow Q$. The statement 'P **if and only if** Q', or 'P **iff** Q', symbolically written $P \Leftrightarrow Q$, is a shorthand for

$$(P \Rightarrow Q) \text{ and } (Q \Rightarrow P),$$

and signifies logical equivalence of the propositions P and Q. The two **quantifiers** $\forall$ and $\exists$, said **for all** and **there exists**, respectively, make their appearance in the following way: if $P(x)$ is a proposition involving a variable x, then

$$\forall x(P(x)) \text{ and } \exists x(P(x))$$

are propositions.

Mathematical theories such as *set theory*, *group theory*, etc. traditionally involve the introduction of some new symbols with which to generate further logical propositions. The theory must be complemented by a collection of logical propositions called **axioms** for the theory – statements that are taken to be automatically **true** in the theory. All other true statements should in principle follow by the rules of logic.

Set theory involves the introduction of the new phrase **is a set** and new symbols $\{\ldots \mid \ldots\}$ and $\in$ defined by:

(Set1) If S is any constant or variable then 'S is a set' is a logical proposition.
(Set2) If $P(x)$ is a logical proposition involving a variable x then $\{x \mid P(x)\}$ is a set.
(Set3) If S is a set and a is any constant or variable then $a \in S$ is a logical proposition, for which we say a **belongs to** S or a **is a member of** S, or simply a **is in** S. The negative of this proposition is denoted $a \notin S$ – said a **is not in** S.

These statements say nothing about whether the various propositions are true or false – they merely assert what are 'grammatically correct' propositions in set theory. They merely tell us how the new symbols and phrases are to be used in a grammatically correct fashion. The main axiom of set theory is: if $P(x)$ is any logical proposition depending on a variable x,

Sets and structures

then for any constant or variable a

$$a \in \{x \mid P(x)\} \Leftrightarrow P(a).$$

Every mathematical theory uses the equality symbol $=$ to express the identity of mathematical objects in the theory. In some cases the concept of mathematical identity needs a separate definition. For example **equality of sets** $A = B$ is defined through the *axiom of extensionality*:

Two sets A and B are equal if and only if they contain the same members. Expressed symbolically,

$$A = B \Leftrightarrow \forall a (a \in A \Leftrightarrow a \in B).$$

A **finite set** $A = \{a_1, a_2, \ldots, a_n\}$ is equivalent to

$$A = \{x \mid (x = a_1) \text{ or } (x = a_2) \text{ or } \ldots \text{ or } (x = a_n)\}.$$

A set consisting of just one element a is called a **singleton** and should be written as $\{a\}$ to distinguish it from the element a which belongs to it: $\{a\} = \{x \mid x = a\}$.

As remarked above, sets can be members of other sets. A set whose elements are all sets themselves will often be called a **collection** or **family** of sets. Such collections are often denoted by script letters such as $\mathcal{A}, \mathcal{U}$, etc. Frequently a family of sets $\mathcal{U}$ has its members **indexed** by another set I, called the **indexing set**, and is written

$$\mathcal{U} = \{U_i \mid i \in I\}.$$

For a finite family we usually take the indexing set to be the first n natural numbers, $I = \{1, 2, \ldots, n\}$. Strictly speaking, this set must also be given an axiomatic definition such as *Peano's axioms*. We refer the interested reader to texts such as [4] for a discussion of these matters.

Although the finer details of logic have been omitted here, essentially all concepts of set theory can be constructed from these basics. The implication is that all of mathematics can be built out of an alphabet for constants and variables, parentheses ($\ldots$), logical connectives and quantifiers together with the rules of propositional logic, and the symbols $\{\ldots \mid \ldots\}$ and $\in$. Since mathematical physics is an attempt to express physics in purely mathematical language, we have the somewhat astonishing implication that all of physics should also be reducible to these simple terms. Eugene Wigner has expressed wonderment at this idea in a famous paper entitled *The unreasonable effectiveness of mathematics in the natural sciences* [5].

The presentation of set theory given here should suffice for all practical purposes, but it is not without logical difficulties. The most famous is *Russell's paradox*: consider the set of all sets which are not members of themselves. According to the above rules this set can be written $R = \{A \mid A \notin A\}$. Is R a member of itself? This question does not appear to have an answer. For, if $R \in R$ then by definition $R \notin R$, which is a contradiction. On the other hand, if $R \notin R$ then it satisfies the criterion required for membership of R; that is, $R \in R$.

To avoid such vicious arguments, logicians have been forced to reformulate the axioms of set theory in a very careful way. The most frequently used system is the axiomatic scheme of *Zermelo and Fraenkel* – see, for example, [2] or the Appendix of [6]. We will adopt the 'naive' position and simply assume that the sets dealt with in this book do not exhibit the self-contradictions of Russell's monster.

1.2 Subsets, unions and intersections of sets

A set T is said to be a **subset** of S, or T is **contained in** S, if every member of T belongs to S. Symbolically, this is written $T \subseteq S$,

$$T \subseteq S \quad \text{iff} \quad a \in T \Rightarrow a \in S.$$

We may also say S is a **superset** of T and write $S \supset T$. Of particular importance is the **empty set** $\emptyset$, to which no object belongs,

$$\forall a \, (a \notin \emptyset).$$

The empty set is assumed to be a subset of any set whatsoever,

$$\forall S (\emptyset \subseteq S).$$

This is the default position, consistent with the fact that $a \in \emptyset \Rightarrow a \in S$, since there are no a such that $a \in \emptyset$ and the left-hand side of the implication is never true. We have here an example of the logical dictum that 'a false statement implies the truth of any statement'.

A common criterion for showing the equality of two sets, $T = S$, is to show that $T \subseteq S$ and $S \subseteq T$. The proof follows from the axiom of extensionality:

$$\begin{aligned} T = S &\iff (a \in T \iff a \in S) \\ &\iff (a \in T \Rightarrow a \in S) \text{ and } (a \in S \Rightarrow a \in T) \\ &\iff (T \subseteq S) \text{ and } (S \subseteq T). \end{aligned}$$

Exercise: Show that the empty set is unique; i.e., if $\emptyset'$ is an empty set then $\emptyset' = \emptyset$.

The collection of all subsets of a set S forms a set in its own right, called the **power set** of S, denoted 2^S.

Example 1.1 If S is a finite set consisting of n elements, then 2^S consists of one empty set $\emptyset$ having no elements, n singleton sets having just one member, $\binom{n}{2}$ sets having two elements, etc. Hence the total number of sets belonging to 2^S is, by the binomial theorem,

$$1 + \binom{n}{1} + \binom{n}{2} + \cdots + \binom{n}{n} = (1+1)^n = 2^n.$$

This motivates the symbolic representation of the power set.

Unions and intersections

The **union** of two sets S and T, denoted $S \cup T$, is defined as

$$S \cup T = \{x \mid x \in S \text{ or } x \in T\}.$$

Sets and structures

The **intersection** of two sets S and T, denoted $S \cap T$, is defined as

$$S \cap T = \{x \mid x \in S \text{ and } x \in T\}.$$

Two sets S and T are called **disjoint** if no element belongs simultaneously to both sets, $S \cap T = \emptyset$. The **difference** of two sets S and T is defined as

$$S - T = \{x \mid x \in S \text{ and } x \notin T\}.$$

Exercise: If S and T are disjoint, show that $S - T = S$.

The union of an arbitrary (possibly infinite) family of sets $\mathcal{A}$ is defined as the set of all elements x that belong to some member of the family,

$$\bigcup \mathcal{A} = \{x \mid \exists S \text{ such that } (S \in \mathcal{A}) \text{ and } (x \in S)\}.$$

Similarly we define the **intersection** of S to be the set of all elements that belong to *every* set of the collection,

$$\bigcap \mathcal{A} = \{x \mid x \in S \text{ for all } S \in \mathcal{A}\}.$$

When $\mathcal{A}$ consists of a family of sets S_i indexed by a set I, the union and intersection are frequently written

$$\bigcup_{i \in I} \{S_i\} \quad \text{and} \quad \bigcap_{i \in I} \{S_i\}.$$

Problems

Problem 1.1 Show the distributive laws

$$A \cap (B \cup C) = (A \cap B) \cup (A \cap C), \quad A \cup (B \cap C) = (A \cup B) \cap (A \cup C).$$

Problem 1.2 If $\mathcal{B} = \{B_i \mid i \in I\}$ is any family of sets, show that

$$A \cap \bigcup \mathcal{B} = \bigcup \{A \cap B_i \mid i \in I\}, \quad A \cup \bigcap \mathcal{B} = \bigcap \{A \cup B_i \mid i \in I\}.$$

Problem 1.3 Let B be any set. Show that $(A \cap B) \cup C = A \cap (B \cup C)$ if and only if $C \subseteq A$.

Problem 1.4 Show that

$$A - (B \cup C) = (A - B) \cap (A - C), \quad A - (B \cap C) = (A - B) \cup (A - C).$$

Problem 1.5 If $\mathcal{B} = \{B_i \mid i \in I\}$ is any family of sets, show that

$$A - \bigcup \mathcal{B} = \bigcup \{A - B_i \mid i \in I\}.$$

Problem 1.6 If E and F are any sets, prove the identities

$$2^E \cap 2^F = 2^{E \cap F}, \quad 2^E \cup 2^F \subseteq 2^{E \cup F}.$$

Problem 1.7 Show that if $\mathcal{C}$ is any family of sets then

$$\bigcap_{X \in \mathcal{C}} 2^X = 2^{\cap \mathcal{C}}, \quad \bigcup_{X \in \mathcal{C}} 2^X \subseteq 2^{\cup \mathcal{C}}.$$

1.3 Cartesian products and relations

Ordered pairs and cartesian products

As it stands, there is no concept of order in a set consisting of two elements, since $\{a, b\} = \{b, a\}$. Frequently we wish to refer to an **ordered pair** (a, b). Essentially this is a set of two elements $\{a, b\}$ where we specify the *order* in which the two elements are to be written. A purely set-theoretical way of expressing this idea is to adjoin the element a that is to be regarded as the 'first' member. An ordered pair (a, b) can thus be thought of as a set consisting of $\{a, b\}$ together with the element a singled out as being the *first*,

$$(a, b) = \{\{a, b\}, a\}. \tag{1.1}$$

While this looks a little artificial at first, it does demonstrate how the concept of 'order' can be defined in purely set-theoretical terms. Thankfully, we only give this definition for illustrative purposes – there is essentially no need to refer again to the formal representation (1.1).

Exercise: From the definition (1.1) show that $(a, b) = (a', b')$ iff $a = a'$ and $b = b'$.

Similarly, an **ordered n-tuple** $(a_1, a_2, \ldots, a_n)$ is a set in which the order of the elements must be specified. This can be defined inductively as

$$(a_1, a_2, \ldots, a_n) = (a_1, (a_2, a_3, \ldots, a_n)).$$

Exercise: Write out the ordered triple (a, b, c) as a set.

The **(cartesian) product** of two sets, $S \times T$, is the set of all ordered pairs (s, t) where s belongs to S and t belongs to T,

$$S \times T = \{(s, t) \mid s \in S \text{ and } t \in T\}.$$

The product of n sets is defined as

$$S_1 \times S_2 \times \cdots \times S_n = \{(s_1, s_2, \ldots, s_n) \mid s_1 \in S_1, s_2 \in S_2, \ldots, s_n \in S_n\}.$$

If the n sets are equal, $S_1 = S_2 = \cdots = S_n = S$, then their product is denoted S^n.

Exercise: Show that $S \times T = \emptyset$ iff $S = \emptyset$ or $T = \emptyset$.

Relations

Any subset of S^n is called an *n-ary* **relation** on a set S. For example,

$$\begin{aligned}\textbf{unary relation} &\equiv \text{1-ary relation} = \text{subset of } S\\ \textbf{binary relation} &\equiv \text{2-ary relation} = \text{subset of } S^2 = S \times S\\ \textbf{ternary relation} &\equiv \text{3-ary relation} = \text{subset of } S^3 = S \times S \times S, \text{ etc.}\end{aligned}$$

We will focus attention on binary relations as these are by far the most important. If $R \subseteq S \times S$ is a binary relation on S, it is common to use the notation aRb in place of $(a, b) \in R$.

Sets and structures

Some commonly used terms describing relations are the following:

R is said to be a **reflexive** relation if aRa for all $a \in S$.
R is called **symmetric** if $aRb \Rightarrow bRa$ for all $a, b \in S$.
R is **transitive** if $(aRb$ and $bRc) \Rightarrow aRc$ for all $a, b, c \in S$.

Example 1.2 Let $\mathbb{R}$ be the set of all real numbers. The usual ordering of real numbers is a relation on $\mathbb{R}$, denoted $x \le y$, which is both reflexive and transitive but not symmetric. The relation of strict ordering $x < y$ is transitive, but is neither reflexive nor symmetric. Similar statements apply for the ordering on subsets of $\mathbb{R}$, such as the integers or rational numbers. The notation $x \le y$ is invariably used for this relation in place of the rather odd-looking $(x, y) \in \le$ where $\le \subseteq \mathbb{R}^2$.

Equivalence relations

A relation that is reflexive, symmetric and transitive is called an **equivalence relation**. For example, equality $a = b$ is always an equivalence relation. If R is an equivalence relation on a set S and a is an arbitrary element of S, then we define the **equivalence class corresponding to** a to be the subset

$$[a]_R = \{b \in S \mid aRb\}.$$

The equivalence class is frequently denoted simply by $[a]$ if the equivalence relation R is understood. By the reflexive property $a \in [a]$ – that is, equivalence classes 'cover' the set S in the sense that every element belongs to at least one class. Furthermore, if aRb then $[a] = [b]$. For, let $c \in [a]$ so that aRc. By symmetry, we have bRa, and the transitive property implies that bRc. Hence $c \in [b]$, showing that $[a] \subseteq [b]$. Similarly $[b] \subseteq [a]$, from which it follows that $[a] = [b]$.

Furthermore, if $[a]$ and $[b]$ are any pair of equivalence classes having non-empty intersection, $[a] \cap [b] \ne \emptyset$, then $[a] = [b]$. For, if $c \in [a] \cap [b]$ then aRc and cRb. By transitivity, aRb, or equivalently $[a] = [b]$. Thus any pair of equivalence classes are either disjoint, $[a] \cap [b] = \emptyset$, or else they are equal, $[a] = [b]$. The equivalence relation R is therefore said to **partition** the set S into disjoint equivalence classes.

It is sometimes useful to think of elements of S belonging to the same equivalence class as being 'identified' with each other through the equivalence relation R. The set whose elements are the equivalence classes defined by the equivalence relation R is called the **factor space**, denoted S/R,

$$S/R = \{[a]_R \mid a \in S\} \equiv \{x \mid x = [a]_R, a \in S\}.$$

Example 1.3 Let p be a positive integer. On the set of all integers $\mathbb{Z}$, define the equivalence relation R by mRn if and only if there exists $k \in \mathbb{Z}$ such that $m - n = kp$, denoted

$$m \equiv n \pmod{p}.$$

This relation is easily seen to be an equivalence relation. For example, to show it is transitive, simply observe that if $m - n = kp$ and $n - j = lp$ then $m - j = (k + l)p$. The equivalence class $[m]$ consists of the set of integers of the form $m + kp, (k = 0, \pm 1, \pm 2, \dots)$. It follows

1.3 Cartesian products and relations

that there are precisely p such equivalence classes, $[0], [1], \ldots, [p-1]$, called the **residue classes modulo** p. Their union spans all of $\mathbb{Z}$.

Example 1.4   Let $\mathbb{R}^2 = \mathbb{R} \times \mathbb{R}$ be the cartesian plane and define an equivalence relation $\equiv$ on $\mathbb{R}^2$ by

$$(x, y) \equiv (x', y') \quad \text{iff} \quad \exists n, m \in \mathbb{Z} \text{ such that } x' = x + m, \ y' = y + n.$$

Each equivalence class $[(x, y)]$ has one representative such that $0 \le x < 1, \ 0 \le y < 1$. The factor space

$$T^2 = \mathbb{R}^2 / \equiv \ = \{[(x, y)] \, | \, 0 \le x, \ y < 1\}$$

is called the **2-torus**. The geometrical motivation for this name will become apparent in Chapter 10.

Order relations and posets

The characteristic features of an 'order relation' have been discussed in Example 1.2, specifically for the case of the real numbers. More generally, a relation R on a set S is said to be a **partial order** on S if it is reflexive and transitive, and in place of the symmetric property it satisfies the 'antisymmetric' property

$$a R b \text{ and } b R a \implies a = b.$$

The ordering $\le$ on real numbers has the further special property of being a **total order**, by which it is meant that for every pair of real numbers x and y, we have either $x \le y$ or $y \le x$.

Example 1.5   The power set 2^S of a set S is partially ordered by the relation of set inclusion $\subseteq$,

$$U \subseteq U \text{ for all } U \in S,$$
$$U \subseteq V \text{ and } V \subseteq W \implies U \subseteq W,$$
$$U \subseteq V \text{ and } V \subseteq U \implies U = V.$$

Unlike the ordering of real numbers, this ordering is *not* in general a total order.

A set S together with a partial order $\le$ is called a **partially ordered set** or more briefly a **poset**. This is an example of a *structured set*. The words 'together with' used here are a rather casual type of mathspeak commonly used to describe a set with an imposed structure. Technically more correct is the definition of a poset as an ordered pair,

$$\text{poset } S \equiv (S, \le)$$

where $\le \, \subseteq S \times S$ satisfies the axioms of a partial order. The concept of a poset could be totally reduced to its set-theoretical elements by writing ordered pairs (s, t) as sets of the form $\{\{s, t\}, s\}$, etc., but this uninstructive task would only serve to demonstrate how simple mathematical concepts can be made totally obscure by overzealous use of abstract definitions.

Sets and structures

Problems

Problem 1.8 Show the following identities:

$$(A \cup B) \times P = (A \times P) \cup (B \times P),$$
$$(A \cap B) \times (P \cap Q) = (A \times P) \cap (B \times P),$$
$$(A - B) \times P = (A \times P) - (B \times P).$$

Problem 1.9 If $\mathcal{A} = \{A_i \mid i \in I\}$ and $\mathcal{B} = \{B_j \mid j \in J\}$ are any two families of sets then

$$\bigcup \mathcal{A} \times \bigcup \mathcal{B} = \bigcup_{i \in I, j \in J} A_i \times B_j,$$

$$\bigcap \mathcal{A} \times \bigcap \mathcal{B} = \bigcap_{i \in I, j \in J} A_i \times B_j.$$

Problem 1.10 Show that both the following two relations:

$$(a, b) \le (x, y) \quad \text{iff} \quad a < x \text{ or } (a = x \text{ and } b \le y)$$
$$(a, b) \preceq (x, y) \quad \text{iff} \quad a \le x \text{ and } b \le y$$

are partial orders on $\mathbb{R} \times \mathbb{R}$. For any pair of partial orders $\le$ and $\preceq$ defined on an arbitrary set A, let us say that $\le$ is *stronger* than $\preceq$ if $a \le b \to a \preceq b$. Is $\le$ stronger than, weaker than or incomparable with $\preceq$?

1.4 Mappings

Let X and Y be any two sets. A **mapping** φ from X to Y, often written $\varphi : X \to Y$, is a subset of $X \times Y$ such that for every $x \in X$ there is a *unique* $y \in Y$ for which $(x, y) \in \varphi$. By **unique** we mean

$$(x, y) \in \varphi \text{ and } (x, y') \in \varphi \implies y = y'.$$

Mappings are also called **functions** or **maps**. It is most common to write $y = \varphi(x)$ for $(x, y) \in \varphi$. Whenever $y = \varphi(x)$ it is said that x **is mapped to** y, written $\varphi : x \mapsto y$.

In elementary mathematics it is common to refer to the subset $\varphi \subseteq X \times Y$ as representing the *graph* of the function φ. Our definition essentially identifies a function with its graph. The set X is called the **domain** of the mapping φ, and the subset $\varphi(X) \subseteq Y$ defined by

$$\varphi(X) = \{y \in Y \mid y = \varphi(x), x \in X\}$$

is called its **range**.

Let U be any subset of Y. The **inverse image of** U is defined to be the set of all points of X that are mapped by φ into U, denoted

$$\varphi^{-1}(U) = \{x \in X \mid \varphi(x) \in U\}.$$

This concept makes sense even when the *inverse map* φ^{-1} does not exist. The notation $\varphi^{-1}(U)$ is to be regarded as one entire symbol for the inverse image set, and should not be broken into component parts.

1.4 Mappings

Example 1.6 Let $\sin: \mathbb{R} \to \mathbb{R}$ be the standard sine function on the real numbers $\mathbb{R}$. The inverse image of 0 is $\sin^{-1}(0) = \{0, \pm\pi, \pm 2\pi, \pm 3\pi, \dots\}$, while the inverse image of 2 is the empty set, $\sin^{-1}(2) = \emptyset$.

An **n-ary function** from X to Y is a function $\varphi: X^n \to Y$. In this case we write $y = \varphi(x_1, x_2, \dots, x_n)$ for $((x_1, x_2, \dots, x_n), y) \in \varphi$ and say that φ has n **arguments** in the set S, although strictly speaking it has just one argument from the product set $X^n = X \times \cdots \times X$.

It is possible to generalize this concept even further and consider maps whose domain is a product of n possibly different sets,

$$\varphi: X_1 \times X_2 \times \cdots \times X_n \to Y.$$

Important maps of this type are the **projection maps**

$$\mathrm{pr}_i: X_1 \times X_2 \times \cdots X_n \to X_i$$

defined by

$$\mathrm{pr}_i: (x_1, x_2, \dots, x_n) \mapsto x_i.$$

If $\varphi: X \to Y$ and $\psi: Y \to Z$, the **composition** map $\psi \circ \varphi: X \to Z$ is defined by

$$\psi \circ \varphi(x) = \psi(\varphi(x)).$$

Composition of maps satisfies the **associative law**

$$\alpha \circ (\psi \circ \varphi) = (\alpha \circ \psi) \circ \varphi$$

where $\alpha: Z \to W$, since for any $x \in X$

$$\alpha \circ (\psi \circ \varphi)(x) = \alpha(\psi(\varphi(x))) = (\alpha \circ \psi)(\varphi(x)) = (\alpha \circ \psi) \circ \varphi(x).$$

Hence, there is no ambiguity in writing $\alpha \circ \psi \circ \varphi$ for the composition of three maps.

Surjective, injective and bijective maps

A mapping $\varphi: X \to Y$ is said to be **surjective** or **a surjection** if its range is all of T. More simply, we say φ is a mapping of X **onto** Y if $\varphi(X) = Y$. It is said to be **one-to-one** or **injective**, or **an injection**, if for every $y \in Y$ there is a *unique* $x \in X$ such that $y = \varphi(x)$; that is,

$$\varphi(x) = \varphi(x') \implies x = x'.$$

A map φ that is injective and surjective, or equivalently one-to-one and onto, is called **bijective** or **a bijection**. In this and only this case can one define the **inverse map** $\varphi^{-1}: Y \to X$ having the property

$$\varphi^{-1}(\varphi(x)) = x, \quad \forall x \in X.$$

Two sets X and Y are said to be in **one-to-one correspondence** with each other if there exists a bijection $\varphi: X \to Y$.

Exercise: Show that if $\varphi: X \to Y$ is a bijection, then so is φ^{-1}, and that $\varphi(\varphi^{-1}(x)) = x$, $\forall x \in X$.

Sets and structures

A bijective map $\varphi : X \to X$ from X onto itself is called a **transformation** of X. The most trivial transformation of all is the **identity map** id_X defined by

$$\mathrm{id}_X(x) = x, \quad \forall x \in X.$$

Note that this map can also be described as having a 'diagonal graph',

$$\mathrm{id}_X = \{(x,x) \mid x \in X\} \subseteq X \times X.$$

Exercise: Show that for any map $\varphi : X \to Y$, $\mathrm{id}_Y \circ \varphi = \varphi \circ \mathrm{id}_X = \varphi$.

When $\varphi : X \to Y$ is a bijection with inverse φ^{-1}, then we can write

$$\varphi^{-1} \circ \varphi = \mathrm{id}_X, \qquad \varphi \circ \varphi^{-1} = \mathrm{id}_Y.$$

If both φ and ψ are bijections then so is $\psi \circ \varphi$, and its inverse is given by

$$(\psi \circ \varphi)^{-1} = \varphi^{-1} \circ \psi^{-1}$$

since

$$\varphi^{-1} \circ \psi^{-1} \circ \psi \circ \varphi = \varphi^{-1} \circ \mathrm{id}_Y \circ \varphi = \varphi^{-1} \circ \varphi = \mathrm{id}_X.$$

If U is any subset of X and $\varphi : X \to Y$ is any map having domain X, then we define the **restriction** of φ to U as the map $\varphi|_U : U \to Y$ by $\varphi|_U(x) = \varphi(x)$ for all $x \in U$. The restriction of the identity map

$$i_U = \mathrm{id}_X\big|_U : U \to X$$

is referred to as the **inclusion map** for the subset U. The restriction of an arbitrary map φ to U is then its composition with the inclusion map,

$$\varphi|_U = \varphi \circ i_U.$$

Example 1.7 If U is a subset of X, define a function $\chi_U : X \to \{0, 1\}$, called the **characteristic function** of U, by

$$\chi_U(x) = \begin{cases} 0 & \text{if } x \notin U, \\ 1 & \text{if } x \in U. \end{cases}$$

Any function $\varphi : X \to \{0, 1\}$ is evidently the characteristic function of the subset $U \subseteq X$ consisting of those points that are mapped to the value 1,

$$\varphi = \chi_U \quad \text{where} \quad U = \varphi^{-1}(\{1\}).$$

Thus the power set 2^X and the set of all maps $\varphi : X \to \{0, 1\}$ are in one-to-one correspondence.

Example 1.8 Let R be an equivalence relation on a set X. Define the **canonical map** $\varphi : X \to X/R$ from X onto the factor space by

$$\varphi(x) = [x]_R, \quad \forall x \in X.$$

It is easy to verify that this map is onto.

More generally, any map $\varphi : X \to Y$ defines an equivalence relation R on X by $a\,R\,b$ iff $\varphi(a) = \varphi(b)$. The equivalence classes defined by R are precisely the inverse images of the singleton subsets of Y,

$$X/R = \{\varphi^{-1}(\{y\}) \mid y \in T\},$$

and the map $\psi : Y \to X/R$ defined by $\psi(y) = \varphi^{-1}(\{y\})$ is one-to-one, for if $\psi(y) = \psi(y')$ then $y = y'$ – pick any element $x \in \psi(y) = \psi(y')$ and we must have $\varphi(x) = y = y'$.

1.5 Infinite sets

A set S is said to be **finite** if there is a natural number n such that S is in one-to-one correspondence with the set $N = \{1, 2, 3, \ldots, n\}$ consisting of the first n natural numbers. We call n the **cardinality** of the set S, written $n = \mathrm{Card}(S)$.

Example 1.9 For any two sets S and T the set of all maps $\varphi : S \to T$ will be denoted by T^S. Justification for this notation is provided by the fact that if S and T are both finite and $s = \mathrm{Card}(S)$, $t = \mathrm{Card}(T)$ then $\mathrm{Card}(T^S) = t^s$. In Example 1.7 it was shown that for any set S, the power set 2^S is in one-to-one correspondence with the set of characteristic functions on $\{1, 2\}^S$. As shown in Example 1.1, for a finite set S both sets have cardinality 2^s.

A set is said to be **infinite** if it is not finite. The concept of infinity is intuitively quite difficult to grasp, but the mathematician Georg Cantor (1845–1918) showed that infinite sets could be dealt with in a completely rigorous manner. He even succeeded in defining different 'orders of infinity' having a *transfinite arithmetic* that extended the ordinary arithmetic of the natural numbers.

Countable sets

The lowest order of infinity is that belonging to the natural numbers. Any set S that is in one-to-one correspondence with the set of natural numbers $\mathbb{N} = \{1, 2, 3, \ldots\}$ is said to be **countably infinite**, or simply **countable**. The elements of S can then be displayed as a **sequence**, $s_1, s_2, s_3, \ldots$ on setting $s_i = f^{-1}(i)$.

Example 1.10 The set of all integers $\mathbb{Z} = \{0, \pm 1, \pm 2, \ldots\}$ is countable, for the map $f : \mathbb{Z} \to \mathbb{N}$ defined by $f(0) = 1$ and $f(n) = 2n$, $f(-n) = 2n + 1$ for all $n > 0$ is clearly a bijection,

$$f(0) = 1, \ f(1) = 2, \ f(-1) = 3, \ f(2) = 4, \ f(-2) = 5, \ldots$$

Theorem 1.1 *Every subset of a countable set is either finite or countable.*

Proof: Let S be a countable set and $f : S \to \mathbb{N}$ a bijection, such that $f(s_1) = 1$, $f(s_2) = 2, \ldots$ Suppose S' is an infinite subset of S. Let s'_1 be the first member of the sequence $s_1, s_2, \ldots$ that belongs to S'. Set s'_2 to be the next member, etc. The map $f' : S' \to \mathbb{N}$

Sets and structures

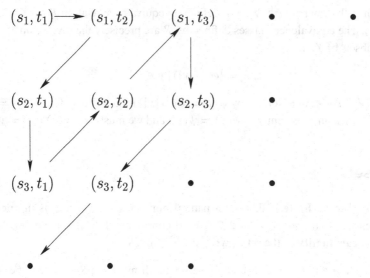

Figure 1.1 Product of two countable sets is countable

defined by

$$f'(s'_1) = 1, \ f'(s'_2) = 2, \ldots$$

is a bijection from S' to $\mathbb{N}$. ∎

Theorem 1.2 *The cartesian product of any pair of countable sets is countable.*

Proof: Let S and T be countable sets. Arrange the ordered pairs (s_i, t_j) that make up the elements of $S \times T$ in an infinite rectangular array and then trace a path through the array as depicted in Fig. 1.1, converting it to a sequence that includes every ordered pair. ∎

Corollary 1.3 *The rational numbers $\mathbb{Q}$ form a countable set.*

Proof: A rational number is a fraction n/m where m is a natural number (positive integer) and n is an integer having no common factor with m. The rationals are therefore in one-to-one correspondence with a subset of the product set $\mathbb{Z} \times \mathbb{N}$. By Example 1.10 and Theorem 1.2, $\mathbb{Z} \times \mathbb{N}$ is a countable set. Hence the rational numbers $\mathbb{Q}$ are countable. ∎

In the set of real numbers ordered by the usual $\leq$, the rationals have the property that for any pair of real numbers x and y such that $x < y$, there exists a rational number q such that $x < q < y$. Any subset, such as the rationals $\mathbb{Q}$, having this property is called a **dense set** in $\mathbb{R}$. The real numbers thus have a *countable* dense subset; yet, as we will now show, the entire set of real numbers turns out to be uncountable.

Uncountable sets

A set is said to be **uncountable** if it is neither finite nor countable; that is, it cannot be set in one-to-one correspondence with any subset of the natural numbers.

1.5 Infinite sets

Theorem 1.4 *The power set 2^S of any countable set S is uncountable.*

Proof: We use **Cantor's diagonal argument** to demonstrate this theorem. Let the elements of S be arranged in a sequence $S = \{s_1, s_2, \ldots\}$. Every subset $U \subseteq S$ defines a unique sequence of 0's and 1's

$$x = \{\epsilon_1, \epsilon_2, \epsilon_3, \ldots\}$$

where

$$\epsilon_i = \begin{cases} 0 & \text{if } s_i \notin U, \\ 1 & \text{if } s_i \in U. \end{cases}$$

The sequence x is essentially the characteristic function of the subset U, discussed in Example 1.7. If 2^S is countable then its elements, the subsets of S, can be arranged in sequential form, $U_1, U_2, U_3, \ldots$, and so can their set-defining sequences,

$$x_1 = \epsilon_{11}, \epsilon_{12}, \epsilon_{13}, \ldots$$
$$x_2 = \epsilon_{21}, \epsilon_{22}, \epsilon_{23}, \ldots$$
$$x_3 = \epsilon_{31}, \epsilon_{32}, \epsilon_{33}, \ldots$$
etc.

Let x' be the sequence of 0's and 1's defined by

$$x' = \epsilon'_1, \epsilon'_2, \epsilon'_3, \ldots$$

where

$$\epsilon'_i = \begin{cases} 0 & \text{if } \epsilon_{ii} = 1, \\ 1 & \text{if } \epsilon_{ii} = 0. \end{cases}$$

The sequence x' cannot be equal to any of the sequences x_i above since, by definition, it differs from x_i in the ith place, $\epsilon'_i \neq \epsilon_{ii}$. Hence the set of all subsets of S cannot be arranged in a sequence, since their characteristic sequences cannot be so arranged. The power set 2^S cannot, therefore, be countable. ∎

Theorem 1.5 *The set of all real numbers $\mathbb{R}$ is uncountable.*

Proof: Each real number in the interval [0, 1] can be expressed as a binary decimal

$$0.\epsilon_1\epsilon_2\epsilon_3\ldots \quad \text{where} \quad \text{each } \epsilon_i = 0 \text{ or } 1 \ (i = 1, 2, 3, \ldots).$$

The set [0, 1] is therefore uncountable since it is in one-to-one correspondence with the power set $2^\mathbb{N}$. Since this set is a subset of $\mathbb{R}$, the theorem follows at once from Theorem 1.1. ∎

Example 1.11 We have seen that the rational numbers form a countable dense subset of the set of real numbers. A set is called *nowhere dense* if it is not dense in any open interval (a, b). Surprisingly, there exists a nowhere dense subset of $\mathbb{R}$ called the **Cantor set**, which is uncountable – the surprise lies in the fact that one would intuitively expect such a set to

Figure 1.2 The Cantor set (after the four subdivisions)

be even sparser than the rationals. To define the Cantor set, express the real numbers in the interval [0, 1] as ternary decimals, to the base 3,

$$x = 0.\epsilon_1\epsilon_2\epsilon_3\ldots \quad \text{where} \quad \epsilon_i = 0, 1 \text{ or } 2, \quad \forall i.$$

Consider those real numbers whose ternary expansion contains only 0's and 2's. These are clearly in one-to-one correspondence with the real numbers expressed as binary expansions by replacing every 2 with 1.

Geometrically one can picture this set in the following way. From the closed real interval [0, 1] remove the middle third (1/3, 2/3), then remove the middle thirds of the two pieces left over, then of the four pieces left after doing that, and continue this process *ad infinitum*. The resulting set can be visualized in Fig. 1.2.

This set may appear to be little more than a mathematical curiosity, but sets displaying a similar structure to the Cantor set can arise quite naturally in non-linear maps relevant to physics.

The continuum hypothesis and axiom of choice

All infinite subsets of $\mathbb{R}$ described above are either countable or in one-to-one correspondence with the real numbers themselves, of cardinality $2^{\aleph}$. Cantor conjectured that this was true of all infinite subsets of the real numbers. This famous **continuum hypothesis** proved to be one of the most challenging problems ever postulated in mathematics. In 1938 the famous logician Kurt Gödel (1906–1978) showed that it would never be possible to prove the converse of the continuum hypothesis – that is, no mathematical inconsistency could arise by assuming Cantor's hypothesis to be true. While not proving the continuum hypothesis, this meant that it could never be proved using the time-honoured method of *reductio ad absurdum*. The most definitive result concerning the continuum hypothesis was achieved by Cohen [7], who demonstrated that it was a genuinely independent axiom, neither provable, nor demonstrably false.

In many mathematical arguments, it is assumed that from any family of sets it is always possible to create a set consisting of a representative element from each set. To justify this seemingly obvious procedure it is necessary to postulate the following proposition:

AXIOM OF CHOICE Given a family of sets $\mathcal{S} = \{S_i \mid i \in I\}$ labelled by an indexing set I, there exists a *choice function* $f : I \to \bigcup \mathcal{S}$ such that $f(i) \in S_i$ for all $i \in I$.

While correct for finite and countably infinite families of sets, the status of this axiom is much less clear for uncountable families. Cohen in fact showed that the axiom of choice was

1.6 Structures

an independent axiom and was independent of the continuum hypothesis. It thus appears that there are a variety of alternative set theories with differing axiom schemes, and the real numbers have different properties in these alternative theories. Even though the real numbers are at the heart of most physical theories, no truly challenging problem for mathematical physics has arisen from these results. While the axiom of choice is certainly useful, its availability is probably not critical in physical applications. When used, it is often invoked in a slightly different form:

Theorem 1.6 (Zorn's lemma) *Let $\{P, \leq\}$ be a partially ordered set (poset) with the property that every totally ordered subset is bounded above. Then P has a maximal element.*

Some words of explanation are in order here. Recall that a subset Q is *totally ordered* if for every pair of elements $x, y \in Q$ either $x \leq y$ or $y \leq x$. A subset Q is said to be **bounded above** if there exists an element $x \in P$ such that $y \leq x$ for all $y \in Q$. A **maximal element** of P is an element z such that there is no $y \neq z$ such that $z \leq y$. The proof that Zorn's lemma is equivalent to the axiom of choice is technical though not difficult; the interested reader is referred to Halmos [4] or Kelley [6].

Problems

Problem 1.11 There is a technical flaw in the proof of Theorem 1.5, since a decimal number ending in an endless sequence of 1's is identified with a decimal number ending with a sequence of 0's, for example,

$$.011011111\ldots = .0111000000\ldots$$

Remove this hitch in the proof.

Problem 1.12 Prove the assertion that the Cantor set is nowhere dense.

Problem 1.13 Prove that the set of all real functions $f : \mathbb{R} \to \mathbb{R}$ has a higher cardinality than that of the real numbers by using a Cantor diagonal argument to show it cannot be put in one-to-one correspondence with $\mathbb{R}$.

Problem 1.14 If $f : [0, 1] \to \mathbb{R}$ is a non-decreasing function such that $f(0) = 0$, $f(1) = 1$, show that the places at which f is not continuous form a countable subset of $[0, 1]$.

1.6 Structures

Physical theories have two aspects, the *static* and the *dynamic*. The former refers to the general background in which the theory is set. For example, special relativity takes place in Minkowski space while quantum mechanics is set in Hilbert space. These mathematical structures are, to use J. A. Wheeler's term, the 'arena' in which a physical system evolves; they are of two basic kinds, algebraic and geometric.

In very broad terms, an *algebraic structure* is a set of binary relations imposed on a set, and 'algebra' consists of those results that can be achieved by formal manipulations using the rules of the given relations. By contrast, a *geometric structure* is postulated as a set of

Sets and structures

relations on the power set of a set. The objects in a geometric structure can in some sense be 'visualized' as opposed to being formally manipulated. Although mathematicians frequently divide themselves into 'algebraists' and 'geometers', these two kinds of structure interrelate in all kinds of interesting ways, and the distinction is generally difficult to maintain.

Algebraic structures

A **(binary) law of composition** on a set S is a binary map

$$\varphi : S \times S \to S.$$

For any pair of elements $a, b \in S$ there thus exists a new element $\varphi(a, b) \in S$ called their **product**. The product is often simply denoted by ab, while at other times symbols such as $a \cdot b$, $a \circ b$, $a + b$, $a \times b$, $a \wedge b$, $[a, b]$, etc. may be used, depending on the context.

Most algebraic structures consist of a set S together with one or more laws of composition defined on S. Sometimes more than one set is involved and the law of composition may take a form such as $\varphi : S \times T \to S$. A typical example is the case of a vector space, where there are two sets involved consisting of vectors and scalars respectively, and the law of composition is *scalar multiplication* (see Chapter 3). In principle we could allow laws of composition that are n-ary maps ($n > 2$), but such laws can always be thought of as families of binary maps. For example, a ternary map $\phi : S^3 \to S$ is equivalent to an indexed family of binary maps $\{\phi_a \mid a \in S\}$ where $\phi_a : S^2 \to S$ is defined by $\phi_a(b, c) = \phi(a, b, c)$.

A law of composition is said to be **commutative** if $ab = ba$. This is *always* assumed to be true for a composition denoted by the symbol $+$; that is, $a + b = b + a$. The law of composition is **associative** if $a(bc) = (ab)c$. This is true, for example, of matrix multiplication or functional composition $f \circ (g \circ h) = (f \circ g) \circ h$, but is not true of vector product $\mathbf{a} \times \mathbf{b}$ in ordinary three-dimensional vector calculus,

$$\mathbf{a} \times (\mathbf{b} \times \mathbf{c}) = (\mathbf{a}.\mathbf{c})\mathbf{b} - (\mathbf{a}.\mathbf{b})\mathbf{c} \ne (\mathbf{a} \times \mathbf{b}) \times \mathbf{c}.$$

Example 1.12 A **semigroup** is a set S with an associative law of composition defined on it. It is said to have an **identity element** if there exists an element $e \in S$ such that

$$ea = ae = a, \quad \forall a \in S.$$

Semigroups are one of the simplest possible examples of an algebraic structure. The theory of semigroups is not particularly rich, and there is little written on their general theory, but particular examples have proved interesting.

(1) The positive integers $\mathbb{N}$ form a commutative semigroup under the operation of addition. If the number 0 is adjoined to this set it becomes a semigroup with identity $e = 0$, denoted $\hat{\mathbb{N}}$.

(2) A map $f : S \to S$ of a set S into itself is frequently called a **discrete dynamical system**. The successive iterates of the function f, namely $F = \{f, f^2, \ldots, f^n = f \circ (f^{n-1}), \ldots\}$, form a commutative semigroup with functional iteration as the law of composition. If we include the identity map and set $f^0 = \text{id}_S$, the semigroup is called the **evolution semigroup generated by the function** f, denoted E_f.

1.6 Structures

The map $\phi : \hat{\mathbb{N}} \to E_f$ defined by $\phi(n) = f^n$ preserves semigroup products,

$$\phi(n+m) = f^{n+m}.$$

Such a product-preserving map between two semigroups is called a **homomorphism**. If the homomorphism is a one-to-one map it is called a **semigroup isomorphism**. Two semigroups that have an isomorphism between them are called **isomorphic**; to all intents and purposes they have the same semigroup structure. The map ϕ defined above need not be an isomorphism. For example on the set $S = \mathbb{R} - \{2\}$, the real numbers excluding the number 2, define the function $f : S \to S$ by

$$f(x) = \frac{2x-3}{x-2}.$$

Simple algebra reveals that $f(f(x)) = x$, so that $f^2 = \text{id}_S$. In this case E_f is isomorphic with the residue class of integers modulo 2, defined in Example 1.3.

(3) All of mathematics can be expressed as a semigroup. For example, set theory is made up of finite strings of symbols such as $\{\ldots \mid \ldots\}$, and, not, $\in, \forall$, etc. and a countable collection of symbols for variables and constants, which may be denoted $x_1, x_2, \ldots$ Given two strings σ_1 and σ_2 made up of these symbols, it is possible to construct a new string $\sigma_1\sigma_2$, formed by concatenating the strings. The set of all possible such strings is a semigroup, where 'product' is defined as string concatenation. Of course only some strings are logically meaningful, and are said to be *well-formed*. The rules for a well-formed string are straightforward to list, as are the rules for 'universally valid statements' and the rules of inference. Gödel's famous *incompleteness theorem* states that if we include statements of ordinary arithmetic in the semigroup then there are propositions P such that neither P nor its negation, not P, can be reached from the axioms by any sequence of logically allowable operations. In a sense, the truth of such statements is unknowable. Whether this remarkable theorem has any bearing on theoretical physics has still to be determined.

Geometric structures

In its broadest terms, a geometric structure defines certain classes of subsets of S as in some sense 'acceptable', together with rules concerning their intersections and unions. Alternatively, we can think of a geometric structure $\mathcal{G}$ on a set S as consisting of one or more subsets of 2^S, satisfying certain properties. In this section we briefly discuss two examples: *Euclidean geometry* and *topology*.

Example 1.13 **Euclidean geometry** concerns points (singletons), straight lines, triangles, circles, etc., all of which are subsets of the plane. There is a 'visual' quality of these concepts, even though they are idealizations of the 'physical' concepts of points and lines that must have size or thickness to be visible. The original formulation of plane geometry as set out in Book 1 of *Euclid's Elements* would hardly pass muster by today's criteria as a rigorous axiomatic system. For example, there is considerable confusion between definitions and undefined terms. Historically, however, it is the first systematic approach to an area of mathematics that turns out to be both axiomatic and interesting.

The undefined terms are *point*, *line segment*, *line*, *angle*, *circle* and relations such as *incidence on*, *endpoint*, *length* and *congruence*. Euclid's five postulates are:

1. Every pair of points are on a unique line segment for which they are end points.
2. Every line segment can be extended to a unique line.
3. For every point A and positive number r there exists a unique circle having A as its centre and radius r, such that the line connecting every other point on the circle to A has length r.
4. All right angles are equal to one another.
5. *Playfair's axiom*: given any line ℓ and a point A not on ℓ, there exists a unique line through A that does not intersect ℓ – said to be *parallel* to ℓ.

The undefined terms can be defined as subsets of some basic set known as the Euclidean plane. Points are singletons, line segments and lines are subsets subject to Axioms 1 and 2, while the relation *incidence on* is interpreted as the relation of set-membership $\in$. An angle would be defined as a set $\{A, \ell_1, \ell_2\}$ consisting of a point and two lines on which it is incident. Postulates 1–3 and 5 seem fairly straightforward, but what are we to make of Postulate 4? Such inadequacies were tidied up by Hilbert in 1921.

The least 'obvious' of Euclid's axioms is Postulate 5, which is not manifestly independent of the other axioms. The challenge posed by this axiom was met in the nineteenth century by the mathematicians Bolyai (1802–1860), Lobachevsky (1793–1856), Gauss (1777–1855) and Riemann (1826–1866). With their work arose the concept of *non-Euclidean geometry*, which was eventually to be of crucial importance in Einstein's theory of gravitation known as *general relativity*; see Chapter 18. Although often regarded as a product of pure thought, Euclidean geometry was in fact an attempt to classify logically the geometrical relations in the world around us. It can be regarded as one of the earliest exercises in mathematical physics. Einstein's general theory of relativity carried on this ancient tradition of unifying geometry and physics, a tradition that lives on today in other forms such as gauge theories and string theory.

The discovery of analytic geometry by René Descartes (1596–1650) converted Euclidean geometry into algebraic language. The cartesian method is simply to define the Euclidean plane as $\mathbb{R}^2 = \mathbb{R} \times \mathbb{R}$ with a distance function $d : \mathbb{R}^2 \times \mathbb{R}^2 \to \mathbb{R}$ given by the Pythagorean formula

$$d((x, y), (u, v)) = \sqrt{(x - u)^2 + (y - v)^2}. \tag{1.2}$$

This theorem is central to the analytic version of Euclidean geometry – it underpins the whole Euclidean edifice. The generalization of Euclidean geometry to a space of arbitrary dimensions $\mathbb{R}^n$ is immediate, by setting

$$d(\mathbf{x}, \mathbf{y}) = \sqrt{\sum_{i=1}^{n} (x_i - y_i)^2} \quad \text{where} \quad \mathbf{x} = (x_1, x_2, \ldots, x_n), \text{ etc.}$$

The ramifications of Pythagoras' theorem have revolutionized twentieth century physics in many ways. For example, Minkowski discovered that Einstein's special theory of relativity could be represented by a four-dimensional *pseudo-Euclidean* geometry where time is

1.6 Structures

interpreted as the fourth dimension and a minus sign is introduced into Pythagoras' law. When gravitation is present, Einstein proposed that Minkowski's geometry must be 'curved', the pseudo-Euclidean structure holding only *locally* at each point. A *complex* vector space having a natural generalization of the Pythagorean structure is known as a *Hilbert space* and forms the basis of quantum mechanics (see Chapters 13 and 14). It is remarkable to think that the two pillars of twentieth century physics, relativity and quantum theory, both have their basis in mathematical structures based on a theorem formulated by an eccentric mathematician over two and a half thousand years ago.

Example 1.14 In Chapter 10 we will meet the concept of a **topology** on a set S, defined as a subset $\mathcal{O}$ of 2^S whose elements (subsets of S) are called *open sets*. To qualify as a topology, the open sets must satisfy the following properties:

1. The empty set and the whole space are open sets, $\emptyset \in \mathcal{O}$ and $S \in \mathcal{O}$.
2. If $U \in \mathcal{O}$ and $V \in \mathcal{O}$ then $U \cap V \in \mathcal{O}$.
3. If $\mathcal{U}$ is any subset of $\mathcal{O}$ then $\bigcup \mathcal{U} \in \mathcal{O}$.

The second axiom says that the intersection of any pair of open sets, and therefore of any finite collection of open sets, is open. The third axiom says that an arbitrary, possibly infinite, union of open sets is open. According to our criterion, a topology is clearly a geometrical structure on S.

The basic view presented here is that the key feature distinguishing an algebraic structure from a geometric structure on a set S is

$$\text{algebraic structure} = \text{a map } S \times S \to S = \text{a subset of } S^3,$$

while

$$\text{geometric structure} = \text{a subset of } 2^S.$$

This may look to be a clean distinction, but it is only intended as a guide, for in reality many structures exhibit both algebraic and geometric aspects. For example, Euclidean geometry as originally expressed in terms of relations between subsets of the plane such as points, lines and circles is the geometric or 'visual' approach. On the other hand, cartesian geometry is the algebraic or analytic approach to plane geometry, in which points are represented as elements of $\mathbb{R}^2$. In the latter approach we have two basic maps: the *difference map* $- : \mathbb{R}^2 \times \mathbb{R}^2 \to \mathbb{R}^2$ defined as $(x, y) - (u, v) = (x - u, y - v)$, and the *distance map* $d : \mathbb{R}^2 \times \mathbb{R}^2 \to \mathbb{R}$ defined by Eq. (1.2). The emphasis on maps places this method much more definitely in the algebraic camp, but the two representations of Euclidean geometry are essentially interchangeable and may indeed be used simultaneously to best understand a problem in plane geometry.

Dynamical systems

The evolution of a system with respect to its algebraic/geometric background invokes what is commonly known as 'laws of physics'. In most cases, particularly when describing

Sets and structures

a continuous evolution, these laws are expressed in the form of differential equations. Providing they have a well-posed initial value problem, such equations generally give rise to a unique evolution for the system, wherein lies the predictive power of physics. However, exact solutions of differential equations are only available in some very specific cases, and it is frequently necessary to resort to numerical methods designed for digital computers with the time parameter appearing in discrete packets. Discrete time models can also serve as a useful technique for formulating 'toy models' exhibiting features similar to those of a continuum theory, which may be too difficult to prove analytically.

There is an even more fundamental reason for considering discretely evolving systems. We have good reason to believe that on time scales less than the *Planck time*, given by

$$T_{\text{Planck}} = \sqrt{\frac{G\hbar}{c^5}},$$

the continuum fabric of space-time is probably invalid and a quantum theory of gravity becomes operative. It is highly likely that differential equations have little or no physical relevance at or below the Planck scale.

As already discussed in Example 1.12, a *discrete dynamical system* is a set S together with a map $f : S \to S$. The map $f : S \to S$ is called a **discrete dynamical structure** on S. The complexities generated by such a simple structure on a single set S can be enormous. A well-known example is the *logistic map* $f : [0, 1] \to [0, 1]$ defined by

$$f(x) = Cx(1 - x) \quad \text{where} \quad 0 < C \le 4,$$

and used to model population growth with limited resources or predator–prey systems in ecology. Successive iterates give rise to the phenomena of chaos and *strange attractors* – limiting sets having a Cantor-like structure. The details of this and other maps such as the **Hénon map** [8], $f : \mathbb{R}^2 \to \mathbb{R}^2$ defined by

$$f(x, y) = (y + 1 - ax^2, bx)$$

can be found in several books on non-linear phenomena, such as [9].

Discrete dynamical structures are often described on the set of states on a given set S, where a *state* on S is a function $\phi : S \to \{0, 1\}$. As each state is the characteristic function of some subset of S (see Example 1.7), the set of states on S can be identified with 2^S. A discrete dynamical structure on the set of all states on S is called a **cellular automaton** on S.

Any discrete dynamical system (S, f) induces a cellular automaton $(2^S, f^* : 2^S \to 2^S)$, by setting $f^* : \phi \mapsto \phi \circ f$ for any state $\phi : S \to \{0, 1\}$. This can be pictured in the following way. Every state ϕ on S attaches a 1 or 0 to every point p on S. Assign to p the new value $\phi(f(p))$, which is the value 0 or 1 assigned by the original state ϕ to the *mapped point* $f(p)$. This process is sometimes called a *pullback* – it carries state values 'backwards' rather than forwards. We will frequently meet this idea that a mapping operates on functions, states in this case, in the opposite direction to the mapping.

Not all dynamical structures defined on 2^S, however, can be obtained in the way just described. For example, if S has n elements, then the number of dynamical systems on S is n^n. However, the number of discrete dynamical structures on 2^S is the much larger number $(2^n)^{2^n} = 2^{n2^n}$. Even for small initial sets this number is huge; for example, for $n = 4$ it is $2^{64} \approx 2 \times 10^{19}$, while for slightly larger n it easily surpasses all numbers normally encountered in physics. One of the most intriguing cellular automata is Conway's **game of life**, which exhibits complex behaviour such as the existence of stable structures with the capacity for self-reproducibility, all from three simple rules (see [9, 10]). Graphical versions for personal computers are readily available for experimentation.

1.7 Category theory

Mathematical structures generally fall into 'categories', such as sets, semigroups, groups, vector spaces, topological spaces, differential manifolds, etc. The mathematical theory devoted to this categorizing process can have enormous benefits in the hands of skilled practioners of this abstract art. We will not be making extensive use of category theory, but in this section we provide a flavour of the subject. Those who find the subject too obscure for their taste are urged to move quickly on, as little will be lost in understanding the rest of this book.

A **category** consists of:

(Cat1) A class $\mathcal{O}$ whose elements are called **objects**. Note the use of the word 'class' rather than 'set' here. This is necessary since the objects to be considered are generally themselves sets and the collection of all possible sets with a given type of structure is too vast to be considered as a set without getting into difficulties such as those presented by Russell's paradox discussed in Section 1.1.

(Cat2) For each pair of objects A, B of $\mathcal{O}$ there is a set $\mathrm{Mor}(A, B)$ whose elements are called **morphisms** from A to B, usually denoted $A \xrightarrow{\phi} B$.

(Cat3) For any pair of morphisms $A \xrightarrow{\phi} B$, $B \xrightarrow{\psi} C$ there is a morphism $A \xrightarrow{\psi \circ \phi} C$, called the **composition** of ϕ and ψ such that

1. Composition is associative: for any three morphisms $A \xrightarrow{\phi} B$, $B \xrightarrow{\psi} C$, $C \xrightarrow{\rho} D$,

$$(\rho \circ \psi) \circ \phi = \rho \circ (\psi \circ \phi).$$

2. For each object A there is a morphism $A \xrightarrow{\iota_A} A$ called the **identity morphism** on A, such that for any morphism $A \xrightarrow{\phi} B$ we have

$$\phi \circ \iota_A = \phi,$$

and for any morphism $C \xrightarrow{\psi} A$ we have

$$\iota_A \circ \psi = \psi.$$

Example 1.15 The simplest example of a category is the **category of sets**, in which the objects are all possible sets, while morphisms are mappings from a set A to a set B. In this case the set $\mathrm{Mor}(A, B)$ consists of all possible mappings from A to B. Composition of morphisms is simply composition of mappings, while the identity morphism on

Sets and structures

an object A is the identity map id_A on A. Properties (Cat1) and (Cat2) were shown in Section 1.4.

Exercise: Show that the class of all semigroups, Example 1.12, forms a category, where morphisms are defined as semigroup homomorphisms.

The following are some other important examples of categories of structures to appear in later chapters:

Objects	Morphisms	Refer to
Groups	Homomorphisms	Chapter 2
Vector spaces	Linear maps	Chapter 3
Algebras	Algebra homomorphisms	Chapter 6
Topological spaces	Continuous maps	Chapter 10
Differential manifolds	Differentiable maps	Chapter 15
Lie groups	Lie group homomorphisms	Chapter 19

Two important types of morphisms are defined as follows. A morphism $A \xrightarrow{\varphi} B$ is called a **monomorphism** if for any object X and morphisms $X \xrightarrow{\alpha} A$ and $X \xrightarrow{\alpha'} A$ we have that

$$\varphi \circ \alpha = \varphi \circ \alpha' \implies \alpha = \alpha'.$$

The morphism φ is called an **epimorphism** if for any object X and morphisms $B \xrightarrow{\beta} Y$ and $B \xrightarrow{\beta'} Y$

$$\beta \circ \varphi = \beta' \circ \varphi \implies \beta = \beta'.$$

These requirements are often depicted in the form of **commutative diagrams**. For example, φ is a monomorphism if the morphism α is uniquely defined by the diagram shown in Fig. 1.3. The word 'commutative' here means that chasing arrows results in composition of morphisms, $\psi = (\varphi \circ \alpha)$.

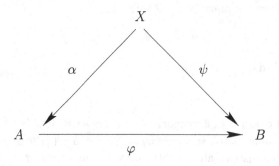

Figure 1.3 Monomorphism φ

1.7 Category theory

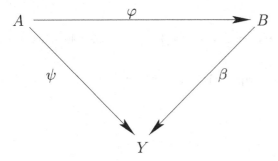

Figure 1.4 Epimorphism φ

On the other hand, φ is an epimorphism if the morphism β is uniquely defined in the commutative diagram shown on Fig. 1.4.

In the case of the category of sets a morphism $A \xrightarrow{\varphi} B$ is a monomorphism if and only if it is a one-to-one mapping.

Proof: 1. If $\varphi : A \to B$ is one-to-one then for any pair of maps $\alpha : X \to A$ and $\alpha' : X \to A$,

$$\varphi(\alpha(x)) = \varphi(\alpha'(x)) \implies \alpha(x) = \alpha'(x)$$

for all $x \in X$. This is simply another way of stating the monomorphism property $\varphi \circ \alpha = \varphi \circ \alpha' \implies \alpha = \alpha'$.

2. Conversely, suppose φ is a monomorphism. Since X is an arbitrary set, in the definition of the monomorphism property, we may choose it to be a singleton $X = \{x\}$. For any pair of points $a, a' \in A$ define the maps $\alpha, \alpha' : X \to A$ by setting $\alpha(x) = a$ and $\alpha'(x) = a'$. Then

$$\begin{aligned} \varphi(a) = \varphi(a') &\implies \varphi \circ \alpha(x) = \varphi \circ \alpha'(x) \\ &\implies \varphi \circ \alpha = \varphi \circ \alpha' \\ &\implies \alpha = \alpha' \\ &\implies a = \alpha(x) = \alpha'(x) = a'. \end{aligned}$$

Hence φ is one-to-one. ∎

It is left as a problem to show that in the category of sets a morphism is an epimorphism if and only if it is surjective. A morphism $A \xrightarrow{\varphi} B$ is called an **isomorphism** if there exists a morphism $B \xrightarrow{\varphi'} A$ such that

$$\varphi' \circ \varphi = \iota_A \quad \text{and} \quad \varphi \circ \varphi' = \iota_B.$$

In the category of sets a mapping is an isomorphism if and only if it is bijective; that is, it is both an epimorphism and a monomorphism. There can, however, be a trap for the unwary here. While every isomorphism is readily shown to be both a monomorphism and an epimorphism, the converse is not always true. A classic case is the category of Hausdorff topological spaces in which there exist continuous maps that are epimorphisms and monomorphisms but are not invertible. The interested reader is referred to [11] for further development of this subject.

Problems

Problem 1.15 Show that in the category of sets a morphism is an epimorphism if and only if it is onto (surjective).

Problem 1.16 Show that every isomorphism is both a monomorphism and an epimorphism.

References

[1] T. Apostol. *Mathematical Analysis*. Reading, Mass., Addison-Wesley, 1957.
[2] K. Devlin. *The Joy of Sets*. New York, Springer-Verlag, 1979.
[3] N. B. Haaser and J. A. Sullivan. *Real Analysis*. New York, Van Nostrand Reinhold Company, 1971.
[4] P. R. Halmos. *Naive Set Theory*. New York, Springer-Verlag, 1960.
[5] E. Wigner. The unreasonable effectiveness of mathematics in the natural sciences. *Communications in Pure and Applied Mathematics*, **13**:1–14, 1960.
[6] J. Kelley. *General Topology*. New York, D. Van Nostrand Company, 1955.
[7] P. J. Cohen. *Set Theory and the Continuum Hypothesis*. New York, W. A. Benjamin, 1966.
[8] M. Hénon. A two-dimensional map with a strange attractor. *Communications in Mathematical Physics*, **50**:69–77, 1976.
[9] M. Schroeder. *Fractals, Chaos, Power Laws*. New York, W. H. Freeman and Company, 1991.
[10] W. Poundstone. *The Recursive Universe*. Oxford, Oxford University Press, 1987.
[11] R. Geroch. *Mathematical Physics*. Chicago, The University of Chicago Press, 1985.

2 Groups

The cornerstone of modern algebra is the concept of a *group*. Groups are one of the simplest algebraic structures to possess a rich and interesting theory, and they are found embedded in almost all algebraic structures that occur in mathematics [1–3]. Furthermore, they are important for our understanding of some fundamental notions in mathematical physics, particularly those relating to *symmetries* [4].

The concept of a group has its origins in the work of Evariste Galois (1811–1832) and Niels Henrik Abel (1802–1829) on the solution of algebraic equations by radicals. The latter mathematician is honoured with the name of a special class of groups, known as *abelian*, which satisfy the commutative law. In more recent times, Emmy Noether (1888–1935) discovered that every group of symmetries of a set of equations arising from an action principle gives rise to conserved quantities. For example, energy, momentum and angular momentum arise from the symmetries of time translations, spatial translations and rotations, respectively. In elementary particle physics there are further conservation laws related to exotic groups such as $SU(3)$, and their understanding has led to the discovery of new particles. This chapter presents the fundamental ideas of group theory and some examples of how they arise in physical contexts.

2.1 Elements of group theory

A **group** is a set G together with a law of composition that assigns to any pair of elements $g, h \in G$ an element $gh \in G$, called their **product**, satisfying the following three conditions:

(Gp1) The **associative law** holds: $g(hk) = (gh)k$, for all $g, h, k \in G$.
(Gp2) There exists an **identity element** $e \in G$, such that

$$eg = ge = g \quad \text{for all } g \in G.$$

(Gp3) Each element $g \in G$ has an **inverse** $g^{-1} \in G$ such that

$$g^{-1}g = gg^{-1} = e.$$

More concisely, a group is a semigroup with identity in which every element has an inverse.

Sometimes the fact that the product of two elements is another element of G is worth noting as a separate condition, called the **closure property**. This is particularly relevant when G is a subset of a larger set with a law of composition defined. In such cases it is always necessary to verify that G is **closed** with respect to this law of composition; that is, for every pair $g, h \in G$, their product $gh \in G$. Examples will soon clarify this point.

Condition (Gp1) means that *all* parentheses in products may be omitted. For example, $a((bc)d) = a(b(cd)) = (ab)(cd) = ((ab)c)d$. It is a tedious but straightforward matter to show that all possible ways of bracketing a product of any number of elements are equal. There is therefore no ambiguity in omitting all parentheses in expressions such as $abcd$. However, it is generally important to specify the *order* in which the elements appear in a product.

The identity element e is easily shown to be unique. For, if e' is a second identity such that $e'g = ge' = g$ for all $g \in G$ then, setting $g = e$, we have $e = e'e = e'$ by (Gp2).

Exercise: By a similar argument, show that every $g \in G$ has a *unique* inverse g^{-1}.

Exercise: Show that $(gh)^{-1} = h^{-1}g^{-1}$.

A group G is called **abelian** if the law of composition is commutative,

$$gh = hg \qquad \text{for all } g, h \in G.$$

The notation gh for the product of two elements is the default notation. Other possibilities are $a \cdot b$, $a \times b$, $a + b$, $a \circ b$, etc. When the law of composition is written as an *addition* $g + h$, we will always assume that the commutative law holds, $g + h = h + g$. In this case the identity element is usually written as 0, so that (Gp2) reads $g + 0 = 0 + g = g$. The inverse is then written $-g$, with (Gp3) reading $g + (-g) = 0$ or, more simply, $g - g = 0$. Again, the associative law means we never have to worry about parentheses in expressions such as $a + b + c + \cdots + f$.

A **subgroup** H of a group G is a subset that is a group in its own right. A subset $H \subseteq G$ is thus a subgroup if it contains the identity element of G and is closed under the operations of taking products and inverses:

(a) $h, k \in H \implies hk \in H$ (closure with respect to taking products);
(b) the identity $e \in H$;
(c) $h \in H \implies h^{-1} \in H$ (closure with respect to taking inverses).

It is not necessary to verify the associative law since H automatically inherits this property from the larger group G. Every group has two **trivial subgroups** $\{e\}$ and G, consisting of the identity alone and the whole group respectively.

Example 2.1 The integers $\mathbb{Z}$ with addition as the law of composition form a group, called the *additive group of integers*. Strictly speaking one should write this group as $(\mathbb{Z}, +)$, but the law of composition is implied by the word 'additive'. The identity element is the integer 0, and the inverse of any integer n is $-n$. The *even integers* $\{0, \pm 2, \pm 4, \dots\}$ form a subgroup of the additive group of integers.

Example 2.2 The real numbers $\mathbb{R}$ form a group with addition $x + y$ as the law of composition, called the *additive group of reals*. Again the identity is 0 and the inverse of x is $-x$. The additive group of integers is clearly a subgroup of $\mathbb{R}$. The rational numbers $\mathbb{Q}$ are

2.1 Elements of group theory

closed with respect to addition and also form a subgroup of the additive reals $\mathbb{R}$, since the number 0 is rational and if p/q is a rational number then so is $-p/q$.

Example 2.3 The non-zero real numbers $\dot{\mathbb{R}} = \mathbb{R} - \{0\}$ form a group called the *multiplicative group of reals*. In this case the product is taken to be ordinary multiplication xy, the identity is the number 1 and the inverse of x is $x^{-1} = 1/x$. The number 0 must be excluded since it has no inverse.

Exercise: Show that the non-zero rational numbers $\dot{\mathbb{Q}}$ form a multiplicative subgroup of $\dot{\mathbb{R}}$.

Exercise: Show that the complex numbers $\mathbb{C}$ form a group with respect to addition, and $\dot{\mathbb{C}} = \mathbb{C} - \{0\}$ is a group with respect to multiplication of complex numbers.

Exercise: Which of the following sets form a group with respect to addition: (i) the rational numbers, (ii) the irrational numbers, (iii) the complex numbers of modulus 1? Which of them is a group with respect to multiplication?

A group G consisting of only a finite number of elements is known as a **finite group**. The number of elements in G is called its **order**, denoted $|G|$.

Example 2.4 Let k be any natural number and $\mathbb{Z}_k = \{[0], [1], \ldots, [k-1]\}$ the integers modulo k, defined in Example 1.3, with **addition modulo** k as the law of composition

$$[a] + [b] = [a+b].$$

$\mathbb{Z}_k$ is called the **additive group of integers modulo** k. It is a finite group of order k, written $|\mathbb{Z}_k| = k$. There is little ambiguity in writing the elements of $\mathbb{Z}_k$ as $0, 1, \ldots, k-1$ and $[a+b]$ is often replaced by the notation $a + b \mod k$.

Exercise: Show that the definition of addition modulo k is independent of the choice of representatives from the residue classes $[a]$ and $[b]$.

Example 2.5 If a group G has an element a such that its powers $\{a, a^2, a^3, \ldots\}$ run through all of its elements, then G is said to be a **cyclic group** and a is called a **generator** of the group. If G is a finite cyclic group and a is a generator, then there exists a positive integer m such that $a^m = e$. If m is the lowest such integer then every element $g \in G$ can be uniquely written $g = a^i$ where $1 \le i \le m$, for if $g = a^i = a^j$ and $1 \le i < j \le m$ then we have the contradiction $a^{j-i} = e$ with $1 \le (j-i) < m$. In this case the group is denoted C_m and its order is $|C_m| = m$. The additive group of integers modulo k is a cyclic group of order k, but in this case the notation a^n is replaced by

$$na = \underbrace{a + a + \cdots + a}_{n} \mod k.$$

Example 2.6 Let $p > 2$ be any prime number. The non-zero integers modulo p form a group of order $p - 1$ with respect to multiplication modulo p,

$$[a][b] = [ab] \equiv ab \mod p,$$

denoted G_p. The identity is obviously the residue class $[1]$, but in order to prove the existence of inverses one needs the following result from number theory: if p and q are relatively prime numbers then there exist integers k and m such that $kp + mq = 1$. Since p is a prime number, if $[q] \neq [0]$ then q is relatively prime to p and for some k and m

$$[m][q] = [1] - [k][p] = [1].$$

Hence $[q]$ has an inverse $[q]^{-1} = [m]$.

For finite groups of small order the law of composition may be displayed in the form of a multiplication table, where the (i, j)th entry specifies the product of the ith element and the jth element. For example, here is the multiplication table of G_7:

G_7	1	2	3	4	5	6
1	1	2	3	4	5	6
2	2	4	6	1	3	5
3	3	6	2	5	1	4
4	4	1	5	2	6	3
5	5	3	1	6	4	2
6	6	5	4	3	2	1

2.2 Transformation and permutation groups

All groups in the above examples are abelian. The most common examples of non-commutative groups are found in a class called *transformation groups*. We recall from Section 1.4 that a *transformation* of a set X is a map $g : X \to X$ that is one-to-one and onto. The map g then has an inverse $g^{-1} : X \to X$ such that $g^{-1} \circ g = g \circ g^{-1} = \mathrm{id}_X$. Let the product of two transformations g and h be defined as their functional composition $gh = g \circ h$,

$$(gh)(x) = g \circ h(x) = g(h(x)).$$

The set of all transformations of X forms a group, denoted $\mathrm{Transf}(X)$:

Closure: if g and h are transformations of X then so is gh;
Associative law: $f(gh) = (fg)h$;
Identity: $e = \mathrm{id}_X \in \mathrm{Transf}(X)$;
Inverse: if g is a transformation of X then so is g^{-1}.

Closure follows from the fact that the composition of two transformations (invertible maps) results in another invertible map, since $(f \circ g)^{-1} = g^{-1} \circ f^{-1}$. The associative law holds automatically for composition of maps, while the identity and inverse are trivial. By a **transformation group** of X is meant any subgroup of $\mathrm{Transf}(X)$.

If X is a finite set of cardinality n then the transformations of X are called **permutations** of the elements of X. The group of permutations of $X = \{1, 2, \ldots, n\}$ is called the **symmetric group of order** n, denoted S_n. Any subgroup of S_n is called a **permutation group**. A permutation π on n elements can be represented by the permutation symbol

$$\pi = \begin{pmatrix} 1 & 2 & \ldots & n \\ a_1 & a_2 & \ldots & a_n \end{pmatrix}$$

2.2 Transformation and permutation groups

where $a_1 = \pi(1)$, $a_2 = \pi(2)$, etc. The same permutation can also be written as

$$\pi = \begin{pmatrix} b_1 & b_2 & \cdots & b_n \\ c_1 & c_2 & \cdots & c_n \end{pmatrix}$$

where $b_1, b_2, \ldots, b_n$ are the numbers $1, 2, \ldots, n$ in an arbitrary order and $c_1 = \pi(b_1)$, $c_2 = \pi(b_2), \ldots, c_n = \pi(b_n)$. For example, the permutation π that interchanges the elements 2 and 4 from a four-element set can be written in several ways,

$$\pi = \begin{pmatrix} 1 & 2 & 3 & 4 \\ 1 & 4 & 3 & 2 \end{pmatrix} = \begin{pmatrix} 2 & 3 & 1 & 4 \\ 4 & 3 & 1 & 2 \end{pmatrix} = \begin{pmatrix} 4 & 1 & 2 & 3 \\ 2 & 1 & 4 & 3 \end{pmatrix}, \text{etc.}$$

In terms of permutation symbols, if

$$\sigma = \begin{pmatrix} 1 & 2 & \cdots & n \\ a_1 & a_2 & \cdots & a_n \end{pmatrix} \quad \text{and} \quad \pi = \begin{pmatrix} a_1 & a_2 & \cdots & a_n \\ b_1 & b_2 & \cdots & b_n \end{pmatrix}$$

then their product is the permutation $\pi\sigma = \pi \circ \sigma$,

$$\pi\sigma = \begin{pmatrix} 1 & 2 & \cdots & n \\ b_1 & b_2 & \cdots & b_n \end{pmatrix}.$$

Note that this product involves first performing the permutation σ followed by π, which is opposite to the order in which they are written; conventions can vary on this point. Since the product is a functional composition, the associative law is guaranteed. The identity permutation is

$$\text{id}_n = \begin{pmatrix} 1 & 2 & \cdots & n \\ 1 & 2 & \cdots & n \end{pmatrix},$$

while the inverse of any permutation is given by

$$\pi^{-1} = \begin{pmatrix} 1 & 2 & \cdots & n \\ a_1 & a_2 & \cdots & a_n \end{pmatrix}^{-1} = \begin{pmatrix} a_1 & a_2 & \cdots & a_n \\ 1 & 2 & \cdots & n \end{pmatrix}.$$

The symmetric group S_n is a finite group of order $n!$, the total number of ways n objects may be permuted. It is not abelian in general. For example, in S_3

$$\begin{pmatrix} 1 & 2 & 3 \\ 1 & 3 & 2 \end{pmatrix}\begin{pmatrix} 1 & 2 & 3 \\ 2 & 1 & 3 \end{pmatrix} = \begin{pmatrix} 2 & 1 & 3 \\ 3 & 1 & 2 \end{pmatrix}\begin{pmatrix} 1 & 2 & 3 \\ 2 & 1 & 3 \end{pmatrix} = \begin{pmatrix} 1 & 2 & 3 \\ 3 & 1 & 2 \end{pmatrix}$$

while

$$\begin{pmatrix} 1 & 2 & 3 \\ 2 & 1 & 3 \end{pmatrix}\begin{pmatrix} 1 & 2 & 3 \\ 1 & 3 & 2 \end{pmatrix} = \begin{pmatrix} 1 & 3 & 2 \\ 2 & 3 & 1 \end{pmatrix}\begin{pmatrix} 1 & 2 & 3 \\ 1 & 3 & 2 \end{pmatrix} = \begin{pmatrix} 1 & 2 & 3 \\ 2 & 3 & 1 \end{pmatrix}.$$

A more compact notation for permutations is the **cyclic notation**. Begin with any element to be permuted, say a_1. Let a_2 be the result of applying the permutation π to a_1, and let a_3 be the result of applying it to a_2, etc. Eventually the first element a_1 must reappear, say as $a_{m+1} = a_1$. This defines a **cycle**, written $(a_1 \, a_2 \ldots a_m)$. If $m = n$, then π is said to be a **cyclic permutation**. If $m < n$ then take any element b_1 not appearing in the cycle generated by a_1 and create a new cycle $(b_1 \, b_2 \ldots b_m)$ of successive images of b_1 under π. Continue until all the elements $1, 2, \ldots, n$ are exhausted. The permutation π may be written as the product of its cycles; for example,

$$\begin{pmatrix} 1 & 2 & 3 & 4 & 5 & 6 & 7 \\ 4 & 5 & 3 & 7 & 2 & 1 & 6 \end{pmatrix} = (1\,4\,7\,6)(2\,5)(3).$$

Note that it does not matter which element of a cycle is chosen as the first member, so that $(1\,4\,7\,6) = (7\,6\,1\,4)$ and $(2\,5) = (5\,2)$.

Cycles of length 1 such as (3) merely signify that the permutation π leaves the element 3 unchanged. Nothing is lost if we totally ignore such 1-cycles from the notation, writing

$$(1\,4\,7\,6)(2\,5)(3) = (1\,4\,7\,6)(2\,5).$$

The order in which cycles that have no common elements is written is also immaterial,

$$(1\,4\,7\,6)(2\,5) = (2\,5)(1\,4\,7\,6).$$

Products of permutations are easily carried out by following the effect of the cycles on each element in succession, taken in order from right to left. For example,

$$(1\,3\,7)(5\,4\,2)(1\,2)(3\,4\,6\,7)(1\,4\,6) = (1\,6\,5\,4)(2\,3)(7)$$

follows from $1 \to 4 \to 6$, $6 \to 1 \to 2 \to 5$, $5 \to 4$, $4 \to 6 \to 7 \to 1$, etc.

Exercise: Express each permutation on $\{1, 2, 3\}$ in cyclic notation and write out the 6×6 multiplication table for S_3.

Cycles of length 2 are called **interchanges**. Every cycle can be written as a product of interchanges,

$$(a_1\, a_2\, a_3\, \ldots\, a_n) = (a_2\, a_3)(a_3\, a_4) \ldots (a_{n-1}\, a_n)(a_n\, a_1),$$

and since every permutation π is a product of cycles, it is in turn a product of interchanges. The representation of a permutation as a product of interchanges is not in general unique, but the number of interchanges needed is either always odd or always even. To prove this, consider the homogeneous polynomial

$$f(x_1, x_2, \ldots, x_n) = \prod_{i<j}(x_i - x_j)$$

$$= (x_1 - x_2)(x_1 - x_3) \ldots (x_1 - x_n)(x_2 - x_3) \ldots (x_{n-1} - x_n).$$

If any pair of variables x_i and x_j are interchanged then the factor $(x_i - x_j)$ changes sign and the factor $(x_i - x_k)$ is interchanged with $(x_j - x_k)$ for all $k \neq i, j$. When $k < i < j$ or $i < j < k$ neither factor changes sign in the latter process, while if $i < k < j$ each factor suffers a sign change and again there is no overall sign change in the product of these two factors. The net result of the interchange of x_i and x_j is a change of sign in the polynomial $f(x_1, x_2, \ldots, x_n)$. Hence permutations may be called **even** or **odd** according to whether f is left unchanged, or changes its sign. In the first case they can be written as an even, and

2.2 Transformation and permutation groups

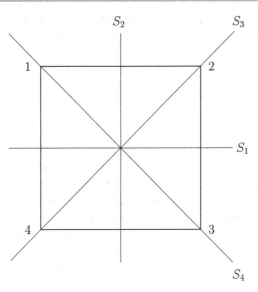

Figure 2.1 Symmetries of the square

only an even, number of interchanges, while in the second case they can only be written as an odd number. This quality is called the **parity** of the permutation and the quantity

$$(-1)^\pi = \begin{cases} +1 & \text{if } \pi \text{ is even,} \\ -1 & \text{if } \pi \text{ is odd,} \end{cases}$$

is called the **sign** of the permutation. Sometimes it is denoted sign π.

Exercise: Show that

$$(-1)^{\pi\sigma} = (-1)^{\sigma\pi} = (-1)^\pi (-1)^\sigma. \tag{2.1}$$

Example 2.7 In the Euclidean plane consider a square whose corners are labelled 1, 2, 3 and 4. The *group of symmetries of the square* consists of four rotations (clockwise by 0°, 90°, 180° and 270°), denoted R_0, R_1, R_2 and R_3 respectively, and four reflections S_1, S_2, S_3 and S_4 about the axes in Fig. 2.1.

This group is not commutative since, for example, $R_1 S_1 = S_4 \neq S_1 R_1 = S_3$ – remember, the rightmost operation is performed first in any such product! A good way to do these calculations is to treat each of the transformations as a permutation of the vertices; for example, in cyclic notation $R_1 = (1\,2\,3\,4)$, $R_2 = (1\,3)(2\,4)$, $S_1 = (1\,4)(2\,3)$, $S_3 = (1\,3)$, etc. Thus the symmetry group of the square is a subgroup of order 8 of the symmetric group S_4.

Exercise: Show that the whole group can be generated by repeated applications of R_1 and S_1.

Example 2.8 An important subgroup of S_n is the set of all even permutations, $(-1)^\pi = 1$, known as the **alternating group**, denoted A_n. The closure property, that the product of two

Groups

even permutations is always even, follows immediately from Eq. (2.1). Furthermore, the identity permutation id_n is clearly even and the inverse of an even permutation π must be even since

$$1 = (-1)^{\mathrm{id}_n} = (-1)^{\pi \pi^{-1}} = (-1)^{\pi}(-1)^{\pi^{-1}} = (-1)^{\pi^{-1}}.$$

Hence A_n is a subgroup of S_n. Its order is $n!/2$.

Example 2.9 Let π be any permutation of $1, 2, \ldots, n$. Since there are a total of $n!$ permutations of n objects, successive iterations $\pi^2, \pi^3, \ldots$ must eventually arrive at repetitions, say $\pi^k = \pi^l$, whence $\pi^{l-k} = \mathrm{id}_n$. The smallest m with the property $\pi^m = \mathrm{id}_n$ is called the **order of the permutation** π. Any cycle of length k evidently has order k, and since every permutation can be written as a product of cycles, the order of a permutation is the lowest common multiple of its cycles. For example, the order of $(1\,2\,3)(4\,5)$ is the lowest common multiple of 3 and 2, which is 6. The set of elements $\{\mathrm{id}_n, \pi, \pi^2, \ldots, \pi^{m-1} = \pi^{-1}\}$ form a subgroup of S_n, called the *subgroup generated by* π. It is clearly a cyclic group.

Problems

Problem 2.1 Show that the only finite subgroup of the additive reals is the singleton $\{0\}$, while the only finite subgroups of the multiplicative reals are the sets $\{1\}$ and $\{1, -1\}$.

Find all finite subgroups of the multiplicative complex numbers $\dot{\mathbb{C}}$.

Problem 2.2 Write out the complete 8×8 multiplication table for the group of symmetries of the square D_4 described in Example 2.7. Show that R_2 and S_1 generate an abelian subgroup and write out its multiplication table.

Problem 2.3 (a) Find the symmetries of the cube, Fig. 2.2(a), which keep the vertex 1 fixed. Write these symmetries as permutations of the vertices in cycle notation.
(b) Find the group of rotational symmetries of the regular tetrahedron depicted in Fig. 2.2(b).
(c) Do the same for the regular octahedron, Fig. 2.2(c).

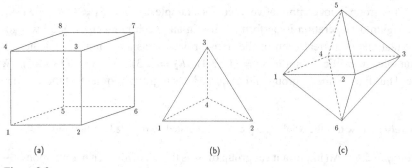

Figure 2.2

Problem 2.4 Show that the multiplicative groups modulo a prime G_7, G_{11}, G_{17} and G_{23} are cyclic. In each case find a generator of the group.

Problem 2.5 Show that the order of any cyclic subgroup of S_n is a divisor of $n!$.

2.3 Matrix groups

Linear transformations

Let $\mathbb{R}^n$ be the space of $n \times 1$ real column vectors

$$\mathbf{x} = \begin{pmatrix} x_1 \\ x_2 \\ \vdots \\ x_n \end{pmatrix}.$$

A mapping $A : \mathbb{R}^n \to \mathbb{R}^n$ is said to be **linear** if

$$A(a\mathbf{x} + b\mathbf{y}) = aA(\mathbf{x}) + bA(\mathbf{y})$$

for all vectors $\mathbf{x}, \mathbf{y} \in \mathbb{R}^n$ and all real numbers $a, b \in \mathbb{R}$. Writing

$$\mathbf{x} = \sum_{i=1}^{n} x_i \mathbf{e}_i \quad \text{where} \quad \mathbf{e}_1 = \begin{pmatrix} 1 \\ 0 \\ \vdots \\ 0 \end{pmatrix}, \mathbf{e}_2 = \begin{pmatrix} 0 \\ 1 \\ \vdots \\ 0 \end{pmatrix}, \ldots, \mathbf{e}_n = \begin{pmatrix} 0 \\ 0 \\ \vdots \\ 1 \end{pmatrix},$$

we have

$$A(\mathbf{x}) = \sum_{i=1}^{n} x_i A(\mathbf{e}_i).$$

If we set

$$A(\mathbf{e}_i) = \sum_{j=1}^{n} a_{ji} \mathbf{e}_j$$

the components x_i of the vector $\mathbf{x}$ transform according to the formula

$$\mathbf{x} \mapsto \mathbf{x}' = A(\mathbf{x}) \quad \text{where} \quad x_i' = \sum_{j=1}^{n} a_{ij} x_j \, (i = 1, \ldots, n).$$

It is common to write this mapping in the form

$$\mathbf{x}' = A\mathbf{x}$$

where $\mathsf{A} = [a_{ij}]$ is the $n \times n$ array

$$A = \begin{pmatrix} a_{11} & a_{12} & a_{13} & \cdots & a_{1n} \\ a_{21} & a_{22} & a_{23} & \cdots & a_{2n} \\ \cdots & \cdots & \cdots & \cdots & \cdots \\ a_{n1} & a_{n2} & a_{n3} & \cdots & a_{nn} \end{pmatrix}.$$

A is called the **matrix of the linear mapping** A, and a_{ij} are its **components**. The matrix AB of the product transformation AB is then given by the matrix multiplication rule,

$$(\mathsf{AB})_{ij} = \sum_{k=1}^{n} a_{ik} b_{kj}.$$

Exercise: Prove this formula.

Linear maps on $\mathbb{R}^n$ and $n \times n$ matrices as essentially identical concepts, the latter being little more than a notational device for the former. Be warned, however, when we come to *general* vector spaces in Chapter 3 such an identification cannot be made in a natural way. In later chapters we will often adopt a different notation for matrix components in order to take account of this difficulty, but for the time being it is possible to use standard matrix notation as we are only concerned with the particular vector space $\mathbb{R}^n$ for the rest of this chapter.

A **linear transformation** A is a one-to-one linear mapping from $\mathbb{R}^n$ onto itself. Such a map is invertible, and its matrix A has non-zero determinant, $\det \mathsf{A} \neq 0$. Such a matrix is said to be **non-singular** and have an inverse matrix A^{-1} given by

$$(\mathsf{A}^{-1})_{ij} = \frac{A_{ji}}{\det \mathsf{A}},$$

where A_{ji} is the (j, i) cofactor of the matrix A, defined as the determinant of the submatrix of A formed by removing its jth row and ith column and multiplied by the factor $(-1)^{i+j}$. The inverse of a matrix acts as both right and left inverse:

$$\mathsf{A}\mathsf{A}^{-1} = \mathsf{A}^{-1}\mathsf{A} = \mathsf{I}, \tag{2.2}$$

where I is the $n \times n$ **unit matrix**

$$\mathsf{I} = \begin{pmatrix} 1 & 0 & \cdots & 0 \\ 0 & 1 & \cdots & 0 \\ \vdots & & \ddots & \\ 0 & 0 & \cdots & 1 \end{pmatrix}.$$

The components of the unit matrix are frequently written as the **Kronecker delta**

$$\delta_{ij} = \begin{cases} 1 & \text{if } i = j, \\ 0 & \text{if } i \neq j. \end{cases} \tag{2.3}$$

The inverse of AB is given by the matrix identity

$$(\mathsf{AB})^{-1} = \mathsf{B}^{-1}\mathsf{A}^{-1}. \tag{2.4}$$

2.3 Matrix groups

Matrix groups

The set of all $n \times n$ non-singular real matrices is a group, denoted $GL(n, \mathbb{R})$. The key to this result is the product law of determinants

$$\det(AB) = \det(A)\det(B). \tag{2.5}$$

Closure: this follows from the fact that $\det A \neq 0$ and $\det B \neq 0$ implies that $\det(AB) = \det A \det B \neq 0$.
Associative law: $(AB)C = A(BC)$ is true of *all* matrices, singular or not.
Identity: the $n \times n$ unit matrix I is an identity element since $IA = AI = A$ for all $n \times n$ matrices A.
Inverse: from Eq. (2.2) A^{-1} clearly acts as an inverse element to A. Equation (2.5) ensures that A^{-1} is non-singular and also belongs to $GL(n, \mathbb{R})$, since

$$\det A^{-1} = \frac{1}{\det A}.$$

A similar discussion shows that the set of $n \times n$ non-singular matrices with complex components, denoted $GL(n, \mathbb{C})$, also forms a group. Except for the case $n = 1$, these groups are non-abelian since matrices do not in general commute, $AB \neq BA$. The groups $GL(n, \mathbb{R})$ and $GL(n, \mathbb{C})$ are called the **general linear groups of order** n. Subgroups of these groups, whose elements are matrices with the law of composition being matrix multiplication, are generically called **matrix groups** [5].

In the following examples the associative law may be assumed, since the law of composition is matrix multiplication. Frequent use will be made of the concept of the **transpose** A^T of a matrix A, defined as the matrix formed by reflecting A about its diagonal,

$$(A^T)_{ij} = a_{ji} \quad \text{where} \quad A = [a_{ij}].$$

The following identities can be found in many standard references such as Hildebrand [6], and should be known to the reader:

$$(AB)^T = B^T A^T, \tag{2.6}$$

$$\det A^T = \det A, \tag{2.7}$$

and if A is non-singular then the inverse of its transpose is the transpose of its inverse,

$$(A^{-1})^T = (A^T)^{-1}. \tag{2.8}$$

Example 2.10 The **special linear group** or **unimodular group** of degree n, denoted $SL(n, \mathbb{R})$, is defined as the set of $n \times n$ **unimodular** matrices, real matrices having determinant 1. Closure with respect to matrix multiplication follows from Eq. (2.5),

$$\det A = \det B = 1 \implies \det(AB) = \det A \det B = 1.$$

The identity $I \in SL(n, \mathbb{R})$ since $\det I = 1$, and closure with respect to inverses follows from

$$\det A = 1 \implies \det A^{-1} = \frac{1}{\det A} = 1.$$

Groups

Example 2.11 A matrix A is called **orthogonal** if its inverse is equal to its transpose,

$$AA^T = A^T A = I. \tag{2.9}$$

The set of real orthogonal $n \times n$ matrices, denoted $O(n)$, forms a group known as the **orthogonal group of order** n:

Closure: if A and B are orthogonal matrices, $AA^T = BB^T = I$, then so is their product AB,

$$(AB)(AB)^T = ABB^T A^T = AIA^T = AA^T = I.$$

Identity: the unit matrix I is clearly orthogonal since $I^T I = I^2 = I$.

Inverse: if A is an orthogonal matrix then A^{-1} is also orthogonal for, using (2.8) and (2.4),

$$A^{-1}(A^{-1})^T = A^{-1}(A^T)^{-1} = (A^T A)^{-1} = I^{-1} = I.$$

The determinant of an orthogonal matrix is always ± 1 since

$$AA^T = I \implies \det A \det A^T = \det(AA^T) = \det I = 1.$$

Hence $(\det A)^2 = 1$ by (2.7) and the result $\det A = \pm 1$ follows at once. The orthogonal matrices with determinant 1 are called **proper orthogonal matrices**, while those with determinant -1 are called **improper**. The proper orthogonal matrices, denoted $SO(n)$, form a group themselves called the **proper orthogonal group of order** n. This group is often known as the **rotation group in** n **dimensions** – see Section 2.7. It is clearly a subgroup of the special linear group $SL(n, \mathbb{R})$.

Example 2.12 Let p and q be non-negative integers such that $p + q = n$, and define G_p to be the $n \times n$ matrix whose components $G_p = [g_{ij}]$ are defined by

$$g_{ij} = \begin{cases} 1 & \text{if } i = j \leq p, \\ -1 & \text{if } i = j > p, \\ 0 & \text{if } i \neq j. \end{cases}$$

We use $O(p, q)$ to denote the set of matrices A such that

$$A^T G_p A = G_p. \tag{2.10}$$

It follows from this equation that any matrix belonging to $O(p, q)$ is non-singular, for on taking determinants,

$$\det A^T \det G_p \det A = \det G_p.$$

Since $\det G_p = \pm 1 \neq 0$ we have $\det A^T \det A = (\det A)^2 = 1$, and consequently

$$\det A = \pm 1.$$

The group properties of $O(p, q)$ follow:

2.3 Matrix groups

Closure: if **A** and **B** both satisfy Eq. (2.10), then so does their product **AB**, for
$$(AB)^T G_p (AB) = B^T A^T G_p AB = B^T G_p B = G_p.$$

Identity: the unit matrix $A = I$ clearly satisfies Eq. (2.10).

Inverse: if Eq. (2.10) is multiplied on the right by A^{-1} and on the left by $(A^{-1})^T$, we have from (2.8)
$$G_p = (A^{-1})^T G_p A^{-1}.$$

Hence A^{-1} satisfies (2.10) and belongs to $O(p, q)$.

The group $O(p, q)$ is known as the **pseudo-orthogonal group of type** (p, q). The case $q = 0$, $p = n$ reduces to the orthogonal group $O(n)$. As for the orthogonal group, those elements of $O(p, q)$ having determinant 1 form a subgroup denoted $SO(p, q)$.

Example 2.13 Let J be the $2n \times 2n$ matrix
$$J = \begin{pmatrix} O & I \\ -I & O \end{pmatrix},$$
where O is the $n \times n$ zero matrix and I is the $n \times n$ unit matrix. A $2n \times 2n$ matrix A is said to be **symplectic** if it satisfies the equation
$$A^T J A = J. \tag{2.11}$$

The argument needed to show that these matrices form a group is essentially identical to that just given for $O(p, q)$. Again, since $\det J = 1$, it follows immediately from (2.11) that $\det A = \pm 1$, and A is non-singular. The group is denoted $Sp(2n)$, called the **symplectic group of order** $2n$.

Exercise: Show that the symplectic matrices of order 2 are precisely the unimodular matrices of order 2. Hence for $n = 2$ all symplectic matrices have determinant 1. It turns out that symplectic matrices of *any* order have determinant 1, but the proof of this is more complicated.

Example 2.14 The **general complex linear group**, $GL(n, \mathbb{C})$, is defined exactly as for the reals. It is the set of non-singular complex $n \times n$ matrices, where the law of composition is matrix product using multiplication of complex numbers. We define special subgroups of this group the same way as for the reals:

$SL(n, \mathbb{C})$ is the **complex unimodular group of degree** n, consisting of complex $n \times n$ matrices having determinant 1;

$O(n, \mathbb{C})$ is the **complex orthogonal group of degree** n, whose elements are complex $n \times n$ matrices A satisfying $A^T A = I$;

$SO(n, \mathbb{C})$ is the **complex proper orthogonal group**, which is the intersection of the above two groups.

There is no complex equivalent of the pseudo-orthogonal groups since these are all isomorphic to $O(n, \mathbb{C})$ – see Problem 2.7.

Groups

Example 2.15 The **adjoint** of a complex matrix A is defined as its complex conjugate transpose $A^\dagger = \overline{A^T}$, whose components are $a^\dagger_{ij} = \overline{a_{ji}}$ where a bar over a complex number refers to its complex conjugate. An $n \times n$ complex matrix U is called **unitary** if

$$UU^\dagger = I. \qquad (2.12)$$

It follows immediately that

$$\det U \, \overline{\det U} = |\det U|^2 = 1,$$

and there exists a real number ϕ with $0 \le \phi < 2\pi$ such that $\det U = e^{i\phi}$. Hence all unitary matrices are non-singular and the group properties are straightforward to verify. The group of all $n \times n$ unitary matrices is called the **unitary group of order** n, denoted $U(n)$. The subgroup of unitary matrices having $\det U = 1$ is called the **special unitary group of order** n, denoted $SU(n)$.

Problems

Problem 2.6 Show that the following sets of matrices form groups with respect to addition of matrices, but that none of them is a group with respect to matrix multiplication: (i) real antisymmetric $n \times n$ matrices ($A^T = -A$), (ii) real $n \times n$ matrices having vanishing trace ($\operatorname{tr} A = \sum_{i=1}^n a_{ii} = 0$), (iii) complex hermitian $n \times n$ matrices ($H^\dagger = H$).

Problem 2.7 Find a diagonal complex matrix S such that

$$I = S^T G_p S$$

where G_p is defined in Example 2.12. Show that:

(a) Every complex matrix A satisfying Eq. (2.10) can be written in the form

$$A = SBS^{-1}$$

where B is a complex orthogonal matrix (i.e. a member of $O(n, \mathbb{C})$).

(b) The complex versions of the pseudo-orthogonal groups, $O(p, q, \mathbb{C})$, are all isomorphic to each other if they have the same dimension,

$$O(p, q, \mathbb{C}) \cong O(n, \mathbb{C}) \quad \text{where} \quad n = p + q.$$

Problem 2.8 Show that every element of $SU(2)$ has the form

$$U = \begin{pmatrix} a & b \\ c & d \end{pmatrix} \quad \text{where} \quad a = \bar{d} \text{ and } b = -\bar{c}.$$

2.4 Homomorphisms and isomorphisms

Homomorphisms

Let G and G' be groups. A **homomorphism** $\varphi : G \to G'$ is a map from G to G' that preserves products,

$$\varphi(ab) = \varphi(a)\varphi(b) \quad \text{for all } a, b \in G.$$

2.4 Homomorphisms and isomorphisms

Theorem 2.1 *Under a homomorphism $\varphi : G \to G'$ the identity e of G is mapped to the identity e' of G', and the inverse of any element g of G to the inverse of its image $\varphi(g)$.*

Proof: For any $g \in G$

$$\varphi(g) = \varphi(ge) = \varphi(g)\varphi(e).$$

Multiplying both sides of this equation on the left by $(\varphi(g))^{-1}$ gives the desired result,

$$e' = e'\varphi(e) = \varphi(e).$$

If $g \in G$ then $\varphi(g^{-1})\varphi(g) = \varphi(g^{-1}g) = \varphi(e) = e'$. Hence $\varphi(g^{-1}) = (\varphi(g))^{-1}$ as required. ∎

Exercise: If $\varphi : G \to G'$ is a homomorphism, show that the image set

$$\text{im}(\varphi) = \varphi(G) = \{g' \in G' \,|\, g' = \varphi(g),\ g \in G\} \tag{2.13}$$

is a subgroup of G'.

Example 2.16 For any real number $x \in \mathbb{R}$, define its *integral part* $[x]$ to be the largest integer that is less than or equal to x, and its *fractional part* to be $(x) = x - [x]$. Evidently $0 \le (x) < 1$. On the half-open interval $[0, 1)$ of the real line define *addition modulo 1* by

$$a + b \bmod 1 = (a+b) = \text{fractional part of } a + b.$$

This defines an abelian group, *the group of real numbers modulo 1*. To verify the group axioms we note that 0 is the identity element and the inverse of any $a > 0$ is $1 - a$. The inverse of $a = 0$ is 0.

The map $\varphi_1 : \mathbb{R} \to [0, 1)$ from the additive group of real numbers to the group of real numbers modulo 1 defined by $\varphi_1(x) = (x)$ is a homomorphism since $(x) + (y) = (x+y)$.

Exercise: Show that the *circle map* or *phase map* $C : \dot{\mathbb{C}} \to [0, 2\pi)$ defined by

$$C(z) = \theta \quad \text{where} \quad z = |z|e^{i\theta}, 0 \le \theta < 2\pi$$

is a homomorphism from the multiplicative group of complex numbers to the additive group of reals modulo 2π, defined in a similar way to the reals modulo 1 in the previous example.

Example 2.17 Let sign $: S_n \to \{+1, -1\}$ be the map that assigns to every permutation π its parity,

$$\text{sign}(\pi) = (-1)^\pi = \begin{cases} +1 & \text{if } \pi \text{ is even,} \\ -1 & \text{if } \pi \text{ is odd.} \end{cases}$$

From (2.1), sign is a homomorphism from S_n to the multiplicative group of reals

$$\text{sign}(\pi\sigma) = \text{sign}(\pi)\text{sign}(\sigma).$$

Exercise: From Eq. (2.5) show that the determinant map $\det : GL(n, \mathbb{R}) \to \dot{\mathbb{R}}$ from the general linear group of order n to the multiplicative group of reals is a homomorphism.

Isomorphisms

An **isomorphism** is a homomorphism that is one-to-one and onto. If an isomorphism exists between two groups G and G' they are said to be **isomorphic**, written $G \cong G'$. The two groups are then essentially identical in all their group properties.

Exercise: Show that if $\varphi : G \to G'$ is an isomorphism, then so is the inverse map $\varphi^{-1} : G' \to G$.

Exercise: If $\varphi : G \to G'$ and $\psi : G' \to G''$ are isomorphisms then so is $\psi \circ \varphi : G \to G''$.

These two statements show that isomorphism is a symmetric and transitive relation on the class of all groups. Hence it is an equivalence relation on the class of groups, since the reflexive property follows from the fact that the identity map $\mathrm{id}_G : G \to G$ is trivially an isomorphism. Note that the word 'class' must be used in this context because the 'set of all groups' is too large to be acceptable. *Group theory* is the study of equivalence classes of isomorphic groups. Frequently it is good to single out a special representative of an equivalence class. Consider, for example, the following useful theorem for finite groups:

Theorem 2.2 (Cayley) *Every finite group G of order n is isomorphic to a permutation group.*

Proof: For every $g \in G$ define the map $L_g : G \to G$ to be left multiplication by g,

$$L_g(x) = gx \quad \text{where} \quad x \in G.$$

This map is one-to-one and onto since

$$gx = gx' \implies x = x' \quad \text{and} \quad x = L_g(g^{-1}x) \quad \text{for all } x \in G.$$

The map L_g therefore permutes the elements of $G = \{g_1 = e, g_2, \ldots, g_n\}$ and may be identified with a member of S_n. It has the property $L_g \circ L_h = L_{gh}$, since

$$L_g \circ L_h(x) = g(hx) = (gh)x = L_{gh}(x), \quad \forall x \in G.$$

Hence the map $\varphi : G \to S_n$ defined by $\varphi(g) = L_g$ is a homomorphism,

$$\varphi(g)\varphi(h) = L_g \circ L_h = L_{gh} = \varphi(gh).$$

Furthermore, φ is one-to-one, for if $\varphi(g) = \varphi(h)$ then $g = L_g(e) = L_h(e) = h$. Thus G is isomorphic to the subgroup of $\varphi(G) \subseteq S_n$. ∎

From the abstract point of view there is nothing to distinguish two isomorphic groups, but different 'concrete' versions of the same group may have different applications. The particular concretization as linear groups of transformations or matrix groups is known as *group representation theory* and plays a major part in mathematical physics.

Automorphisms and conjugacy classes

An **automorphism** is an isomorphism $\varphi : G \to G$ of a group onto itself. A trivial example is the identity map $\mathrm{id}_G : G \to G$. Since the composition of any pair of automorphisms is

2.4 Homomorphisms and isomorphisms

an automorphism and the inverse of any automorphism φ^{-1} is an automorphism, it follows that the set of all automorphisms of a group G is itself a group, denoted Aut(G).

If g is an arbitrary element of G, the map $C_g : G \to G$ defined by

$$C_g(a) = gag^{-1} \tag{2.14}$$

is called **conjugation** by the element g. This map is a homomorphism, for

$$C_g(ab) = gabg^{-1} = gag^{-1}gbg^{-1} = C_g(a)C_g(b),$$

and $C_{g^{-1}}$ is its inverse since

$$C_{g^{-1}} \circ C_g(a) = g^{-1}(gag^{-1})g = a, \quad \forall a \in G.$$

Hence every conjugation C_g is an automorphism of G. Automorphisms that are a conjugation by some element g of G are called **inner automorphisms**. The identity $C_{gh} = C_g \circ C_h$ holds, since for any $a \in G$

$$C_{gh}(a) = gha(gh)^{-1} = ghah^{-1}g^{-1} = C_g(C_h(a)).$$

Hence the map $\psi : G \to \text{Aut}(G)$, defined by $\psi(g) = C_g$, is a homomorphism. The inner automorphisms, being the image of G under ψ, form a subgroup of Aut(G). Two subgroups H and H' of G that can be transformed to each other by an inner automorphism of G are called **conjugate subgroups**. In this case there exists an element $g \in G$ such that

$$H' = gHg^{-1} = \{ghg^{-1} \mid h \in H\}.$$

Exercise: Show that conjugacy is an equivalence relation on the set of all subgroups of a group G. What is the equivalence class containing the trivial subgroup $\{e\}$?

Conjugation also induces an equivalence relation on the original group G by $a \equiv b$ if and only if there exists $g \in G$ such that $b = C_g(a)$. The three requirements for an equivalence relation are easily verified: (i) reflexivity, $a = C_e(a)$ for all $a \in G$; (ii) symmetry, if $b = C_g(a)$ then $a = C_{g^{-1}}(b)$; (iii) transitivity, if $b = C_g(a)$ and $c = C_h(b)$ then $c = C_{hg}(a)$. The equivalence classes with respect to this relation are called **conjugacy classes**. The conjugacy class of an element $a \in G$ is denoted G_a. For example, the conjugacy class of the identity is always the singleton $G_e = \{e\}$, since $C_g e = geg^{-1} = e$ for all $g \in G$.

Exercise: What are the conjugacy classes of an abelian group?

Example 2.18 For a matrix group, matrices **A** and **B** in the same conjugacy class are related by a **similarity transformation**

$$\mathbf{B} = \mathbf{SAS}^{-1}.$$

Matrices related by a similarity transformation have identical invariants such as determinant, trace (sum of the diagonal elements) and eigenvalues. To show determinant is an invariant use Eq. (2.5),

$$\det \mathbf{B} = \det \mathbf{S} \det \mathbf{A} (\det \mathbf{S})^{-1} = \det \mathbf{A}.$$

Groups

For the invariance of trace we need the identity

$$\text{tr}(AB) = \text{tr}(BA), \tag{2.15}$$

which is proved by setting $A = [a_{ij}]$ and $B = [b_{ij}]$ and using the multiplication law of matrices,

$$\text{tr}(AB) = \sum_{i=1}^{n}\left(\sum_{j=1}^{n} a_{ij}b_{ji}\right) = \sum_{j=1}^{n}\left(\sum_{i=1}^{n} b_{ji}a_{ij}\right) = \text{tr}(BA).$$

Hence

$$\text{tr}\, B = \text{tr}(SAS^{-1}) = \text{tr}(S^{-1}SA) = \text{tr}(IA) = \text{tr}\, A,$$

as required. Finally, if λ is an eigenvalue of A corresponding to eigenvector $\mathbf{v}$, then $S\mathbf{v}$ is an eigenvector of B with the same eigenvalue,

$$A\mathbf{v} = \lambda\mathbf{v} \implies B(S\mathbf{v}) = SAS^{-1}S\mathbf{v} = SA\mathbf{v} = \lambda S\mathbf{v}.$$

Example 2.19 The conjugacy classes of the permutation group S_3 are, in cyclic notation,

$$\{e\}; \quad \{(1\,2), (1\,3), (2\,3)\}; \quad \text{and} \quad \{(1\,2\,3), (1\,3\,2)\}.$$

These are easily checked by noting that $(1\,2)^{-1}(1\,2\,3)(1\,2) = (1\,3\,2)$ and

$$(1\,2\,3)^{-1}(1\,2)(1\,2\,3) = (1\,3), \text{ etc.}$$

It is a general feature of permutation groups that conjugacy classes consist of permutations having identical cycle structure (see Problem 2.11).

Problems

Problem 2.9 Show that Theorem 2.2 may be extended to infinite groups as well. That is, any group G is isomorphic to a subgroup of Transf(G), the transformation group of the set G.

Problem 2.10 Find the group multiplication tables for all possible groups on four symbols e, a, b and c, and show that any group of order 4 is either isomorphic to the cyclic group $\mathbb{Z}_4$ or the product group $\mathbb{Z}_2 \times \mathbb{Z}_2$.

Problem 2.11 Show that every cyclic permutation $(a_1\, a_2 \ldots a_n)$ has the property that for any permutation π,

$$\pi(a_1\, a_2 \ldots a_n)\pi^{-1}$$

is also a cycle of length n. [*Hint*: It is only necessary to show this for interchanges $\pi = (b_1\, b_2)$ as every permutation is a product of such interchanges.]

(a) Show that the conjugacy classes of S_n consist of those permutations having the same cycle structure, e.g. $(1\,2\,3)(4\,5)$ and $(1\,4\,6)(2\,3)$ belong to the same conjugacy class.
(b) Write out all conjugacy classes of S_4 and calculate the number of elements in each class.

Problem 2.12 Show that the class of groups as objects with homomorphisms between groups as morphisms forms a category – the *category of groups* (see Section 1.7). What are the monomorphisms, epimorphisms and isomorphisms of this category?

2.5 Normal subgroups and factor groups

Cosets

For any pair of subsets A and B of a group G, define AB to be the set

$$AB = \{ab \mid a \in A \text{ and } b \in B\}.$$

If H is a subgroup of G then $HH = H$.

When A is a singleton set, say $A = \{a\}$, we usually write aB instead of $\{a\}B$. If H is a subgroup of G, then each subset aH where $a \in G$ is called a **(left) coset of** H. Two cosets of a given subgroup H are either identical or non-intersecting. For, suppose there exists an element $g \in aH \cap bH$. Setting

$$g = ah_1 = bh_2 \quad (h_1, h_2 \in H)$$

we have for any $h \in H$

$$ah = bh_2 h_1^{-1} h \in bH,$$

so that $aH \subseteq bH$. Equally, it can be argued that $bH \subseteq aH$, whence either $aH \cap bH = \emptyset$ or $aH = bH$. Since $g = ge$ and $e \in H$, any element $g \in G$ always belongs to the coset gH. Thus the cosets of H form a family of disjoint subsets covering all of G. There is an alternative way of demonstrating this partitioning property. The relation $a \equiv b$ on G, defined by

$$a \equiv b \quad \text{iff} \quad b^{-1}a \in H,$$

is an equivalence relation since it is (i) reflexive, $a^{-1}a = e \in H$; (ii) symmetric, $a^{-1}b = (b^{-1}a)^{-1} \in H$ if $b^{-1}a \in H$; and (iii) transitive, $a^{-1}b \in H$, $b^{-1}c \in H$ implies $a^{-1}c = a^{-1}bb^{-1}c \in H$. The equivalence classes defined by this relation are precisely the left cosets of the subgroup H, for $b \equiv a$ if and only if $b \in aH$.

Theorem 2.3 (Lagrange) *If G is a finite group of order n, then the order of every subgroup H is a divisor of n.*

Proof: Every coset gH is in one-to-one correspondence with H, for if $gh_1 = gh_2$ then $h_1 = g^{-1}gh_2 = h_2$. Hence every coset gH must have exactly $|H|$ elements, and since the cosets partition the group G it follows that n is a multiple of $|H|$. ∎

Corollary 2.4 *The order of any element is a divisor of $|G|$.*

Proof: Let g be any element of G and let m be its order. As shown in Example 2.5 the elements $\{g, g^2, \ldots, g^m = e\}$ are then all unequal to each other and form a cyclic subgroup of order m. By Lagrange's theorem m divides the order of the group, $|G|$. ∎

Exercise: If G has prime order p all subgroups are trivial – they are either the identity subgroup $\{e\}$ or G itself. Show that G is a cyclic group.

Normal subgroups

The **right cosets** Hg of a subgroup H are defined in a completely analogous way to the left cosets. While in general there is no obvious relationship between right and left cosets, there is an important class of subgroups for which they coincide. A subgroup N of a group G is called **normal** if

$$gNg^{-1} = N, \quad \forall g \in G.$$

Such subgroups are invariant under inner automorphisms; they are sometimes referred to as **invariant** or **self-conjugate** subgroups. The key feature of normal subgroups is that the systems of left and right cosets are identical, for

$$gN = gNg^{-1}g = Ng, \quad \forall g \in G.$$

This argument may give the misleading impression that every element of N commutes with every element of G, but what it actually demonstrates is that for every $n \in N$ and every $g \in G$ there exists an element $n' \in N$ such that $gn = n'g$. There is no reason, in general, to expect that $n' = n$.

For any group G the trivial subgroups $\{e\}$ and G are always normal. A group is called **simple** if it has no normal subgroups other than these trivial subgroups.

Example 2.20 The **centre** Z of a group G is defined as the set of elements that commute with all elements of G,

$$Z = \{z \in G \,|\, zg = gz \text{ for all } g \in G\}.$$

This set forms a subgroup of G since the three essential requirements hold:

Closure: if $z, z' \in Z$ then $zz' \in Z$ since

$$(zz')g = z(z'g) = z(gz') = (zg)z' = (gz)z' = g(zz').$$

Identity: $e \in Z$, as $eg = ge = g$ for all $g \in G$.

Inverse: if $z \in Z$ then $z^{-1} \in Z$ since

$$z^{-1}g = z^{-1}ge = z^{-1}gzz^{-1} = z^{-1}zgz^{-1} = gz^{-1}.$$

This subgroup is clearly normal since $gZ = Zg$ for all $g \in G$.

Factor groups

When we multiply left cosets of a subgroup H together, for example

$$gHg'H = \{ghg'h' \,|\, h, h' \in H\},$$

the result is not in general another coset. On the other hand, the product of cosets of a *normal* subgroup N is always another coset,

$$gNg'N = gg'NN = (gg')N,$$

2.5 Normal subgroups and factor groups

and satisfies the associative law,

$$(gNg'N)g''N = (gg'g'')N = gN(g'Ng''N).$$

Furthermore, the coset $eN = N$ plays the role of an identity element, while every coset has an inverse $(gN)^{-1} = g^{-1}N$. Hence the cosets of a *normal* subgroup N form a group called the **factor group** of G by N, denoted G/N.

Example 2.21 The even integers $2\mathbb{Z}$ form a normal subgroup of the additive group of integers $\mathbb{Z}$, since this is an abelian group. The factor group $\mathbb{Z}/2\mathbb{Z}$ has just two cosets $[0] = 0 + 2\mathbb{Z}$ and $[1] = 1 + 2\mathbb{Z}$, and is isomorphic to the additive group of integers modulo 2, denoted by $\mathbb{Z}_2$ (see Example 2.4).

Kernel of a homomorphism

Let $\varphi : G \to G'$ be a homomorphism between two groups G and G'. The **kernel** of φ, denoted $\ker(\varphi)$, is the subset of G consisting of those elements that map onto the identity e' of G',

$$\ker(\varphi) = \varphi^{-1}(e') = \{k \in G \mid \varphi(k) = e'\}.$$

The kernel $K = \ker(\varphi)$ of any homomorphism φ is a subgroup of G:

Closure: if k_1 and k_2 belong to K then so does $k_1 k_2$, since

$$\varphi(k_1 k_2) = \varphi(k_1)\varphi(k_2) = e'e' = e'.$$

Identity: $e \in K$ as $\varphi(e) = e'$.
Inverse: if $k \in K$ then $k^{-1} \in K$, for

$$\varphi(k^{-1}) = (\varphi(k))^{-1} = (e')^{-1} = e'.$$

Furthermore, K is a normal subgroup since, for all $k \in K$ and $g \in G$,

$$\varphi(gkg^{-1}) = \varphi(g)\varphi(k)\varphi(g^{-1}) = \varphi(g)e'(\varphi(g))^{-1} = e'.$$

The following theorem will show that the converse of this result also holds, namely that *every* normal subgroup is the kernel of a homomorphism.

Theorem 2.5 *Let G be a group. Then the following two properties hold:*

1. *If N is a normal subgroup of G then there is a homomorphism $\mu : G \to G/N$.*
2. *If $\varphi : G \to G'$ is a homomorphism then the factor group $G/\ker(\varphi)$ is isomorphic with the image subgroup $\mathrm{im}(\varphi) \subseteq G'$ defined in Eq. (2.13),*

$$\mathrm{im}(\varphi) \cong G/\ker(\varphi).$$

Proof: 1. The map $\mu : G \to G/N$ defined by $\mu(g) = gN$ is a homomorphism, since

$$\mu(g)\mu(h) = gNhN = ghNN = ghN = \mu(gh).$$

2. Let $K = \ker(\varphi)$ and $H' = \mathrm{im}(\varphi)$. The map φ is constant on each coset gK, for

$$k \in K \quad \Longrightarrow \quad \varphi(gk) = \varphi(g)\varphi(k) = \varphi(g)e' = \varphi(g).$$

47

Groups

Hence the map φ defines a map $\psi : G/K \to H'$ by setting

$$\psi(gK) = \varphi(g),$$

and this map is a homomorphism since

$$\psi(gKhK) = \psi(ghK) = \varphi(gh) = \varphi(g)\varphi(h) = \psi(gK)\psi(hK).$$

Furthermore ψ is one-to-one, for

$$\begin{aligned}\psi(gK) = \psi(hK) &\implies \varphi(g) = \varphi(h) \\ &\implies \varphi(gh^{-1}) = \varphi(g)(\varphi(h))^{-1} = e' \\ &\implies gh^{-1} \in K \\ &\implies g \in hK.\end{aligned}$$

Since every element h' of the image set H' is of the form $h' = \varphi(g) = \psi(gK)$, the map ψ is an isomorphism between the groups G/K and H'. ∎

Example 2.22 Let G and H be two groups with respective identity elements e_G and e_H. A law of composition can be defined on the cartesian product $G \times H$ by

$$(g, h)(g', h') = (gg', hh').$$

This product clearly satisfies the associative law and has identity element (e_G, e_H). Furthermore, every element has a unique inverse $(g, h)^{-1} = (g^{-1}, h^{-1})$. Hence, with this law of composition, $G \times H$ is a group called the **direct product** of G and H. The group G is clearly isomorphic to the subgroup $(G, e_H) = \{(g, e_H) \mid g \in G\}$. The latter is a normal subgroup, since

$$(a, b)(G, e_H)(a^{-1}, b^{-1}) = (aGa^{-1}, be_H b^{-1}) = (G, e_H).$$

It is common to *identify* the subgroup of elements (G, e_H) with the group G. In a similar way H is identified with the normal subgroup (e_G, H).

Problems

Problem 2.13 (a) Show that if H and K are subgroups of G then their intersection $H \cap K$ is always a subgroup of G.
(b) Show that the product $HK = \{hk \mid h \in H, k \in K\}$ is a subgroup if and only if $HK = KH$.

Problem 2.14 Find all the normal subgroups of the group of symmetries of the square D_4 described in Example 2.7.

Problem 2.15 The *quaternion group* G consists of eight elements denoted

$$\{1, -1, i, -i, j, -j, k, -k\},$$

subject to the following law of composition:

$$1g = g1 = g, \text{ for all } g \in Q,$$
$$-1g = -g, \text{ for } g = i, j, k,$$
$$i^2 = j^2 = k^2 = -1,$$
$$ij = k, jk = i, ki = j.$$

(a) Write down the full multiplication table for Q, justifying all products not included in the above list.
(b) Find all subgroups of Q and show that all subgroups of Q are normal.
(c) Show that the subgroup consisting of $\{1, -1, i, -i\}$ is the kernel of a homomorphism $Q \to \{1, -1\}$.
(d) Find a subgroup H of S_4, the symmetric group of order 4, such that there is a homomorphism $Q \to H$ whose kernel is the subgroup $\{1, -1\}$.

Problem 2.16 A *Möbius transformation* is a complex map,

$$z \mapsto z' = \frac{az+b}{cz+d} \quad \text{where} \quad a, b, c, d \in \mathbb{C}, ad - bc = 1.$$

(a) Show that these are one-to-one and onto transformations of the extended complex plane, which includes the point $z = \infty$, and write out the composition of an arbitrary pair of transformations given by constants (a, b, c, d) and (a', b', c', d').
(b) Show that they form a group, called the *Möbius group*.
(c) Show that the map μ from $SL(2, \mathbb{C})$ to the Möbius group, which takes the unimodular matrix $\begin{pmatrix} a & b \\ c & d \end{pmatrix}$ to the above Möbius transformation, is a homomorphism, and that the kernel of this homomorphism is $\{1, -1\}$; i.e. the Möbius group is isomorphic to $SL(2, \mathbb{C})/\mathbb{Z}_2$.

Problem 2.17 Assuming the identification of G with (G, e_H) and H with (e_G, H), show that $G \cong (G \times H)/H$ and $H \cong (G \times H)/G$.

Problem 2.18 Show that the conjugacy classes of the direct product $G \times H$ of two groups G and H consist precisely of products of conjugacy classes from the groups

$$(C_i, D_j) = \{(g_i, h_j) \mid g_i \in C_i, h_j \in D_j\}$$

where C_i is a conjugacy class of G and D_j a conjugacy class of H.

2.6 Group actions

A **left action** of a group G on a set X is a homomorphism φ of G into the group of transformations of X,

$$\varphi : G \to \text{Transf}(X).$$

It is common to write $\varphi(g)(x)$ simply as gx, a notation that makes it possible to write ghx in place of $(gh)x$, since

$$(gh)x = \varphi(gh)(x) = \varphi(g)\varphi(h)(x) = \varphi(g)(hx) = g(hx).$$

A left action ϕ of G on $\mathbb{R}^n$ all of whose images are linear transformations is a homomorphism

$$\phi : G \to GL(n, \mathbb{R}),$$

and is called an *n*-**dimensional representation** of G. Similarly, a homomorphism $\phi : G \to GL(n, \mathbb{C})$ is called a **complex *n*-dimensional representation** of G.

An **anti-homomorphism** is defined as a map $\rho : G \to \text{Transf}(X)$ with the property

$$\rho(gh) = \rho(h)\rho(g).$$

It can give rise to a **right action** $xg = \rho(g)(x)$, a notation that is consistent with writing xgh in place of $x(gh) = (xg)h$.

Exercise: If $\varphi : G \to H$ is a homomorphism show that the map $\rho : G \to H$ defined by $\rho(g) = \varphi(g^{-1})$ is an anti-homomorphism.

Let G be a group having a left action on X. The **orbit** Gx of a point $x \in X$ is the set of all points that can be reached from x by this action,

$$Gx = \{gx \mid g \in G\}.$$

We say the action of G on X is **transitive** if the whole of X is the orbit of some point in X,

$$\exists x \in X \text{ such that } X = Gx.$$

In this case any pair of elements $y, z \in X$ can be connected by the action of a group element, for if $y = gx$ and $z = hx$ then $z = g'y$ where $g' = hg^{-1}$. Hence $X = Gy$ for all $y \in X$.

If x is any point of X, define the **isotropy group** of x to be

$$G_x = \{g \mid gx = x\}.$$

If $gx = x \implies g = \text{id}_X$ the action of G on X is said to be **free**. In this case the isotropy group is trivial, $G_x = \{\text{id}_X\}$, for every point $x \in X$.

Exercise: Show that G_x forms a subgroup of G.

If $x \in X$ and $h, h' \in G$ then

$$hx = h'x \implies h^{-1}h' \in G_x \implies h' \in hG_x.$$

If G is a finite group, we denote the number of points in any subset S by $|S|$. Since hG_x is a left coset of the subgroup G_x and from the proof of Lagrange's theorem 2.3 all cosets have the same number of elements, there must be precisely $|G_x|$ group elements that map x to any point y of its orbit Gx. Hence

$$|G| = |Gx| |G_x|. \tag{2.16}$$

Example 2.23 The cyclic group of order 2, $\mathbb{Z}_2 = \{e, a\}$ where $a^2 = e$, acts on the real numbers $\mathbb{R}$ by

$$ex = x, \quad ax = -x.$$

2.6 Group actions

The orbit of any point $x \neq 0$ is $\mathbb{Z}_2 x = \{x, -x\}$, while $\mathbb{Z}_2 0 = \{0\}$. This action is not transitive. The isotropy group of the origin is the whole of $\mathbb{Z}_2$, while for any other point it is $\{e\}$. It is a simple matter to check (2.16) separately for $x = 0$ and $x \neq 0$.

Example 2.24 The additive group of reals $\mathbb{R}$ acts on the complex plane $\mathbb{C}$ by

$$\theta : z \mapsto z e^{i\theta}.$$

The orbit of any $z \neq 0$ is the circle centred 0, radius $r = |z|$. The action is not transitive since circles of different radius are disjoint. The isotropy group of any $z \neq 0$ is the set of real numbers of the form $\theta = 2\pi n$ where $n \in \mathbb{Z}$. Hence the isotropy group $\mathbb{R}_z$ for $z \neq 0$ is isomorphic to $\mathbb{Z}$, the additive group of integers. On the other hand the isotropy group of $z = 0$ is all of $\mathbb{R}$.

Example 2.25 A group G acts on itself by **left translation**

$$g : h \mapsto L_g h = gh.$$

This action is clearly transitive since any element g' can be reached from any other g by a left translation,

$$g' = L_{g'g^{-1}} g.$$

Any subgroup $H \subseteq G$ also acts on G by left translation. The orbit of any group element g under this action is the *right coset Hg* containing g. Similarly, under the right action of H on G defined by **right translation** $R_h : g \mapsto gh$, the orbits are the *left cosets gH*. These actions are not transitive in general.

Example 2.26 The process of conjugation by an element g, defined in Eq. (2.14), is a left action of the group G on itself since the map $g \mapsto C_g$ is a homomorphism,

$$C_{gh} a = (gh)a(gh)^{-1} = ghah^{-1}g^{-1} = C_g C_h a,$$

where we have written $C_g a$ for $C_g(a)$. The orbits under the action of conjugation are precisely the *conjugacy classes*. By Eq. (2.16) it follows that if G is a finite group then the number of elements in any conjugacy class, being an orbit under an action of G, is a divisor of the order of the group $|G|$.

If G has a left action on a set X and if x and y are any pair of points in X in the same orbit, such that $y = hx$ for some $h \in G$, then their isotropy groups are conjugate to each other,

$$G_y = G_{hx} = h G_x h^{-1}. \qquad (2.17)$$

For, let $g \in G_y$, so that $gy = y$. Since $y = hx$ it follows on applying h^{-1} that $h^{-1}ghx = x$. Hence $h^{-1}gh \in G_x$, or equivalently $g \in h G_x h^{-1}$. The converse, that $h G_x h^{-1} \subseteq G_y$, is straightforward: for any $g \in h G_x h^{-1}$, we have that

$$gy = ghx = hg'h^{-1}hx \quad \text{where} \quad g' \in G_x,$$

whence $gy = hx = y$ and $g \in G_y$. Thus the isotropy groups of x and y are isomorphic since they are conjugate to each other, and are related by an inner automorphism. If the

Groups

action of G on X is transitive it follows that the isotropy groups of any pair of points x and y are isomorphic to each other.

Exercise: Under what circumstances is the action of conjugation by an element g on a group G transitive?

Problem

Problem 2.19 If H is any subgroup of a group G define the action of G on the set of left cosets G/H by $g : g'H \mapsto gg'H$.

(a) Show that this is always a transitive action of H on G.
(b) Let G have a transitive left action on a set X, and set $H = G_x$ to be the isotropy group of any point x. Show that the map $i : G/H \to X$ defined by $i(gH) = gx$ is well-defined, one-to-one and onto.
(c) Show that the left action of G on X can be identified with the action of G on G/H defined in (a).
(d) Show that the group of proper orthogonal transformations $SO(3)$ acts transitively on the 2-sphere S^2,

$$S^2 = \{(x, y, z) \mid r^2 = x^2 + y^2 + z^2 = 1\} = \{\mathbf{r} \mid r^2 = \mathbf{r}^T \mathbf{r} = 1\},$$

where $\mathbf{r}$ is a column vector having real components x, y, z. Show that the isotropy group of any point $\mathbf{r}$ is isomorphic to $SO(2)$, and find a bijective correspondence between the factor space $SO(3)/SO(2)$ and the 2-sphere S^2 such that $SO(3)$ has identical left action on these two spaces.

2.7 Symmetry groups

For physicists, the real interest in groups lies in their connection with the symmetries of a space of interest or some important function such as the Lagrangian. Here the concept of a *space* X will be taken in its broadest terms to mean a set X with a 'structure' imposed on it, as discussed in Section 1.6. The definitions of such spaces may involve combinations of algebraic and geometric structures, but the key thing is that their definitions invariably involve the specification of certain functions on the space. For example, algebraic structures such as groups require laws of composition, which are functions defined on cartesian products of the underlying sets. Geometric structures such as topology usually involve a selection of subsets of X – this can also be defined as a characteristic function on the power set of X. For the present purposes let us simply regard a *space* as being a set X together with one or more functions $F : X \to Y$ to another set Y defined on it. This concept will be general enough to encapsulate the basic idea of a 'space'.

If F is a Y-valued function on X, we say a transformation $g : X \to X$ leaves F **invariant** if

$$F(x) = F(gx) \quad \text{for all } x \in X,$$

where, as in Section 2.6, we denote the left action by $gx \equiv g(x)$.

Theorem 2.6 *The set of all transformations of X leaving F invariant form a group.*

2.7 Symmetry groups

Proof: We show the usual three things:

Closure: if g and h leave F invariant then $F(x) = F(hx)$ for all $x \in X$ and $F(y) = F(gy)$ for all $y \in X$. Hence $gh \equiv g \circ h$ leaves F invariant since $F(ghx) = F(g(hx)) = F(hx) = F(x)$.

Identity: obviously $F(x) = F(\mathrm{id}_X(x))$; that is, id_X leaves F invariant.

Inverse: if g is a transformation then there exists an inverse map g^{-1} such that $gg^{-1} = \mathrm{id}_X$. The map g^{-1} leaves F invariant if g does, since

$$F(g^{-1}x) = F(g(g^{-1}x)) = F(x).$$

∎

It is a straightforward matter to extend the above theorem to an arbitrary set $\mathcal{F}$ of functions on X. The group of transformations leaving all functions $F \in \mathcal{F}$ invariant will be called the **invariance group** or **symmetry group** of $\mathcal{F}$. The following are some important examples of symmetry groups in mathematical physics.

Example 2.27 *The rotation group $SO(3)$.* As in Example 2.11, let $\mathbb{R}^3$ be the set of all 3×1 column vectors

$$\mathbf{r} = \begin{pmatrix} x \\ y \\ z \end{pmatrix} \quad \text{such that} \quad x, y, z \in \mathbb{R}.$$

Consider the set of all linear transformations $\mathbf{r} \mapsto \mathbf{r}' = \mathbf{Ar}$ on $\mathbb{R}^3$, where $\mathbf{A}$ is a 3×3 matrix, which leave the distance of points from the origin $r = |\mathbf{r}| = \sqrt{x^2 + y^2 + z^2}$ invariant. Since $r^2 = \mathbf{r}^T \mathbf{r}$, we have

$$r'^2 = \mathbf{r}'^T \mathbf{r}' = \mathbf{r}^T \mathbf{A}^T \mathbf{A} \mathbf{r} = r^2 = \mathbf{r}^T \mathbf{r},$$

which holds for *arbitrary* vectors $\mathbf{r}$ if and only if $\mathbf{A}$ is an orthogonal matrix, $\mathbf{A}\mathbf{A}^T = \mathbf{I}$. As shown in Example 2.11, orthogonal transformations all have determinant ± 1. Those with determinant $+1$ are called **rotations**, while transformations of determinant -1 must involve a reflection with respect to some plane; for example, the transformation $x' = x$, $y' = y$, $z' = -z$.

In a similar manner $O(n)$ is the group of symmetries of the distance function in n-dimensions,

$$r = \sqrt{x_1^2 + x_2^2 + \cdots + x_n^2}$$

and those with positive determinant are denoted $SO(n)$, called the **group of rotations in n-dimensions**. There is no loss of generality in our assumption of *linear* transformations for this group since it can be shown that any transformation of $\mathbb{R}^n$ leaving r^2 invariant must be linear (see Chapter 18).

Example 2.28 *The Euclidean group.* The **Euclidean space** $\mathbb{E}^3$ is defined as the cartesian space $\mathbb{R}^3$ with a distance function between any pair of points given by

$$\Delta s^2 = (\mathbf{r}_2 - \mathbf{r}_1)^2 = \Delta \mathbf{r}^T \Delta \mathbf{r}.$$

A transformation of $\mathbb{E}^3$ that leaves the distance between any pair of points invariant will be called a **Euclidean transformation**. As for the rotation group, a Euclidean transformation $\mathbf{r} \to \mathbf{r}'$ has

$$\Delta\mathbf{r}' = A\Delta\mathbf{r}, \qquad AA^T = I.$$

For any pair of points $\mathbf{r}'_2 - \mathbf{r}'_1 = A(\mathbf{r}_2 - \mathbf{r}_1)$, and if we set $\mathbf{r}_1 = 0$ to be the origin and $0' = \mathbf{a}$, then $\mathbf{r}'_2 - \mathbf{a} = A\mathbf{r}_2$. Since $\mathbf{r}_2$ is an arbitrary point in $\mathbb{E}^3$, the general Euclidean transformations have the form

$$\mathbf{r}' = A\mathbf{r} + \mathbf{a} \quad \text{where} \quad A^T A = I, \mathbf{a} = \text{const.} \tag{2.18}$$

Transformations of this form are frequently called **affine** or **inhomogeneous linear** transformations.

Exercise: Check directly that these transformations form a group – do not use Theorem 2.6.

The group of Euclidean transformations, called the **Euclidean group**, can also be written as a matrix group by replacing $\mathbf{r}$ with the 4×1 column matrix $(x, y, z, 1)^T$ and writing

$$\begin{pmatrix} \mathbf{r}' \\ 1 \end{pmatrix} = \begin{pmatrix} A & \mathbf{a} \\ 0^T & 1 \end{pmatrix} \begin{pmatrix} \mathbf{r} \\ 1 \end{pmatrix} = \begin{pmatrix} A\mathbf{r} + \mathbf{a} \\ 1 \end{pmatrix}.$$

This may seem an odd trick, but its value lies in demonstrating that the Euclidean group is isomorphic to a matrix group – the Euclidean transformations are affine, not linear, on $\mathbb{R}^3$, and thus cannot be written as 3×3 matrices.

Example 2.29 *The Galilean group.* To find the set of transformations of space and time that preserve the laws of Newtonian mechanics we follow the lead of special relativity (see Chapter 9) and define an **event** to be a point of $\mathbb{R}^4$ characterized by four coordinates (x, y, z, t). Define **Galilean space** $\mathbb{G}^4$ to be the space of events with a structure consisting of three elements:

1. Time intervals $\Delta t = t_2 - t_1$.
2. The spatial distance $\Delta s = |\mathbf{r}_2 - \mathbf{r}_1|$ between any pair of **simultaneous events** (events having the same time coordinate, $t_1 = t_2$).
3. Motions of inertial (free) particles, otherwise known as **rectilinear motions**,

$$\mathbf{r}(t) = \mathbf{u}t + \mathbf{r}_0, \tag{2.19}$$

where $\mathbf{u}$ and $\mathbf{r}_0$ are arbitrary constant vectors.

Note that only the distance between *simultaneous* events is relevant. A simple example should make this clear. Consider a train travelling with uniform velocity v between two stations A and B. In the frame of an observer who stays at A the distance between the (non-simultaneous) events $E_1 = $ 'train leaving A' and $E_2 = $ 'train arriving at B' is clearly $d = vt$, where t is the time of the journey. However, in the rest frame of the train it hasn't moved at all and the distance between these two events is zero! Assuming no accelerations at the start and end of the journey, both frames are equally valid Galilean frames of reference.

2.7 Symmetry groups

Note that Δt is a function on all of $\mathbb{G}^4 \times \mathbb{G}^4$, while Δs is a function on the subset of $\mathbb{G}^4 \times \mathbb{G}^4$ consisting of simultaneous pairs of events, $\{((\mathbf{r}, t), (\mathbf{r}', t')) \mid \Delta t = t' - t = 0\}$. We define a **Galilean transformation** as a transformation $\varphi : \mathbb{G}^4 \to \mathbb{G}^4$ that preserves the three given structural elements. All Galilean transformations have the form

$$t' = t + a \quad (a = \text{const}), \tag{2.20}$$
$$\mathbf{r}' = A\mathbf{r} - \mathbf{v}t + \mathbf{b} \quad (A^T A = I,\, \mathbf{v},\, \mathbf{b} = \text{consts}). \tag{2.21}$$

Proof: From the time difference equation $t' - 0' = t - 0$ we obtain (2.20) where $a = 0'$. Invariance of Property 2. gives, by a similar argument to that used to deduce Euclidean transformations,

$$\mathbf{r}' = A(t)\mathbf{r} + \mathbf{a}(t), \qquad A^T A = I \tag{2.22}$$

where $A(t)$ is a time-dependent orthogonal matrix and $\mathbf{a}(t)$ is an arbitrary vector function of time. These transformations allow for rotating and accelerating frames of reference and are certainly too general to preserve Newton's laws.

Property 3. is essentially the invariance of Newton's first law of motion, or equivalently Galileo's principle of inertia. Consider a particle in uniform motion given by Eq. (2.19). This equation must be transformed into an equation of the form $\mathbf{r}'(t) = \mathbf{u}'t + \mathbf{r}'_0$ under a Galilean transformation. From the transformation law (2.22)

$$\mathbf{u}'t + \mathbf{r}'_0 = A(t)(\mathbf{u}t + \mathbf{r}_0) + \mathbf{a}(t),$$

and taking twice time derivatives of both sides of this equation gives

$$0 = (\ddot{A}t + 2\dot{A})\mathbf{u} + \ddot{A}\mathbf{r}_0 + \ddot{\mathbf{a}}.$$

Since $\mathbf{u}$ and $\mathbf{r}_0$ are arbitrary constant vectors it follows that

$$0 = \ddot{A}, \quad 0 = \ddot{A}t + 2\dot{A} \quad \text{and} \quad 0 = \ddot{\mathbf{a}}.$$

Hence $\dot{A} = 0$, so that A is a constant orthogonal matrix, and $\mathbf{a} = -\mathbf{v}t + \mathbf{b}$ for some constant vectors $\mathbf{v}$ and $\mathbf{b}$. ∎

Exercise: Exhibit the Galilean group as a matrix group, as was done for the Euclidean group in (2.18).

Example 2.30 *The Lorentz group.* The Galilean transformations do not preserve the *light cone* at the origin

$$\Delta x^2 + \Delta y^2 + \Delta z^2 = c^2 \Delta t^2 \quad (\Delta x = x_2 - x_1,\ \text{etc.}).$$

The correct transformations that achieve this important property preserve the metric of **Minkowski space**,

$$\Delta s^2 = \Delta x^2 + \Delta y^2 + \Delta z^2 - c^2 \Delta t^2$$
$$= \Delta \mathbf{x}^T G \Delta \mathbf{x},$$

where

$$\mathbf{x} = \begin{pmatrix} x \\ y \\ z \\ ct \end{pmatrix}, \quad \Delta\mathbf{x} = \begin{pmatrix} \Delta x \\ \Delta y \\ \Delta z \\ c\Delta t \end{pmatrix} \quad \text{and} \quad \mathbf{G} = [g_{\mu\nu}] = \begin{pmatrix} 1 & 0 & 0 & 0 \\ 0 & 1 & 0 & 0 \\ 0 & 0 & 1 & 0 \\ 0 & 0 & 0 & -1 \end{pmatrix}.$$

The transformations in question must have the form

$$\mathbf{x}' = \mathbf{L}\mathbf{x} + \mathbf{a},$$

and the invariance law $\Delta s'^2 = \Delta s^2$ implies

$$\Delta\mathbf{x}'^T \mathbf{G} \Delta\mathbf{x}' = \Delta\mathbf{x}^T \mathbf{L}^T \mathbf{G} \mathbf{L} \Delta\mathbf{x} = \Delta\mathbf{x}^T \mathbf{G} \Delta\mathbf{x}.$$

Since this equation holds for arbitrary $\Delta\mathbf{x}$, the 4×4 matrix $\mathbf{L}$ must satisfy the equation

$$\mathbf{G} = \mathbf{L}^T \mathbf{G} \mathbf{L}. \tag{2.23}$$

The *linear* transformations, having $\mathbf{a} = 0$, are called **Lorentz transformations** while the general transformations with arbitrary $\mathbf{a}$ are called **Poincaré transformations**. The corresponding groups are called the **Lorentz group** and **Poincaré group**, respectively. The essence of the special theory of relativity is that all laws of physics are Poincaré invariant.

Problems

Problem 2.20 The *projective transformations* of the line are defined by

$$x' = \frac{ax + b}{cx + d} \quad \text{where} \quad ad - bc = 1.$$

Show that projective transformations preserve the cross-ratio

$$\frac{(x_1 - x_2)(x_3 - x_4)}{(x_3 - x_2)(x_1 - x_4)}$$

between any four points x_1, x_2, x_3 and x_4. Is every analytic transformation that preserves the cross-ratio between any four points on the line necessarily a projective transformation? Do the projective transformations form a group?

Problem 2.21 Show that a matrix $\mathbf{U}$ is unitary, satisfying Eq. (2.12), if and only if it preserves the 'norm'

$$\|\mathbf{z}\|^2 = \sum_{i=1}^{n} z_i \bar{z}_i$$

defined on column vectors $(z_1, z_2, \ldots, z_n)^T$ in $\mathbb{C}^n$. Verify that the set of $n \times n$ complex unitary matrices $U(n)$ forms a group.

Problem 2.22 Show that two rotations belong to the same conjugacy classes of the rotation group $SO(3)$ if and only if they have the same magnitude; that is, they have the same angle of rotation but possibly a different axis of rotation.

Problem 2.23 The general Galilean transformation

$$t' = t + a, \quad \mathbf{r}' = \mathbf{A}\mathbf{r} - \mathbf{v}t + \mathbf{b} \quad \text{where} \quad \mathbf{A}^T \mathbf{A} = \mathbf{I}$$

2.7 Symmetry groups

may be denoted by the abstract symbol $(a, \mathbf{v}, \mathbf{b}, \mathsf{A})$. Show that the result of performing two Galilean transformations

$$G_1 = (a_1, \mathbf{v}_1, \mathbf{b}_1, \mathsf{A}_1) \quad \text{and} \quad G_2 = (a_2, \mathbf{v}_2, \mathbf{b}_2, \mathsf{A}_2)$$

in succession is

$$G = G_2 G_1 = (a, \mathbf{v}, \mathbf{b}, \mathsf{A})$$

where

$$a = a_1 + a_2, \quad \mathbf{v} = \mathsf{A}_2 \mathbf{v}_1 + \mathbf{v}_2, \quad \mathbf{b} = \mathbf{b}_2 - a_1 \mathbf{v}_2 + \mathsf{A}_2 \mathbf{b}_1 \quad \text{and} \quad \mathsf{A} = \mathsf{A}_2 \mathsf{A}_1.$$

Show from this rule of composition that the Galilean transformations form a group. In particular verify explicitly that the associative law holds.

Problem 2.24 (a) From the matrix relation defining a Lorentz transformation L,

$$\mathsf{G} = \mathsf{L}^T \mathsf{G} \mathsf{L},$$

where G is the 4×4 diagonal matrix whose diagonal components are $(1, 1, 1, -1)$; show that Lorentz transformations form a group.
(b) Denote the Poincaré transformation

$$\mathbf{x}' = \mathsf{L}\mathbf{x} + \mathbf{a}$$

by $(\mathsf{L}, \mathbf{a})$, and show that two Poincaré transformations $(\mathsf{L}_1, \mathbf{a})$ and $(\mathsf{L}_2, \mathbf{b})$ performed in succession is equivalent to the Poincaré transformation

$$(\mathsf{L}_2 \mathsf{L}_1, \mathbf{b} + \mathsf{L}_2 \mathbf{a}).$$

(c) From this law of composition show that the Poincaré transformations form a group. As in the previous problem the associative law should be shown explicitly.

Problem 2.25 Let V be an abelian group with law of composition $+$, and G any group with a left action on V, denoted as usual by $g : v \mapsto gv$. Assume further that this action is a homomorphism of V,

$$g(v + w) = gv + gw.$$

(a) Show that $G \times V$ is a group with respect to the law of composition

$$(g, v)(g', v') = (gg', v + gv').$$

This group is known as the **semi-direct** product of G and V, and is denoted $G \circledS V$.
(b) Show that the elements of type $(g, 0)$ form a subgroup of $G \circledS V$ that is isomorphic with G and that V is isomorphic with the subgroup (e, V). Show that the latter is a normal subgroup.
(c) Show that every element of $G \circledS V$ has a unique decomposition of the form vg, where $g \equiv (g, 0) \in G$ and $v \equiv (e, v) \in V$.

Problem 2.26 The following provide examples of the concept of semi-direct product defined in Problem 2.25:

(a) Show that the Euclidean group is the semi-direct product of the rotation group $SO(3, \mathbb{R})$ and $\mathbb{R}^3$, the space of column 3-vectors.
(b) Show that the Poincaré group is the semi-direct product of the Lorentz group $O(3, 1)$ and the abelian group of four-dimensional vectors $\mathbb{R}^4$ under vector addition (see Problem 2.24).
(c) Display the Galilean group as the semi-direct product of two groups.

Problem 2.27 The group A of **affine transformations** of the line consists of transformations of the form

$$x' = ax + b, \quad a \neq 0.$$

Show that these form a semi-direct product on $\dot{\mathbb{R}} \times \mathbb{R}$. Although the multiplicative group of reals $\dot{\mathbb{R}}$ and the additive group $\mathbb{R}$ are both abelian, demonstrate that their semi-direct product is not.

References

[1] G. Birkhoff and S. MacLane. *A Survey of Modern Algebra*. New York, MacMillan, 1953.
[2] R. Geroch. *Mathematical Physics*. Chicago, The University of Chicago Press, 1985.
[3] S. Lang. *Algebra*. Reading, Mass., Addison-Wesley, 1965.
[4] M. Hammermesh. *Group Theory and its Applications to Physical Problems*. Reading, Mass., Addison-Wesley, 1962.
[5] C. Chevalley. *Theory of Lie Groups*. Princeton, N.J., Princeton University Press, 1946.
[6] F. P. Hildebrand. *Methods of Applied Mathematics*. Englewood Cliffs, N. J., Prentice-Hall, 1965.

3 Vector spaces

Some algebraic structures have more than one law of composition. These must be connected by some kind of *distributive laws*, else the separate laws of composition are simply independent structures on the same set. The most elementary algebraic structures of this kind are known as *rings* and *fields*, and by combining fields and abelian groups we create *vector spaces* [1–7].

For the rest of this book, vector spaces will never be far away. For example, *Hilbert spaces* are structured vector spaces that form the basis of *quantum mechanics*. Even in non-linear theories such as *classical mechanics* and *general relativity* there exist local vector spaces known as the *tangent space* at each point, which are needed to formulate the dynamical equations. It is hard to think of a branch of physics that does not use vector spaces in some aspect of its formulation.

3.1 Rings and fields

A **ring** R is a set with two laws of composition called *addition* and *multiplication*, denoted $a+b$ and ab respectively. It is required that R is an abelian group with respect to $+$, with identity element 0 and inverses denoted $-a$. With respect to multiplication R is to be a commutative semigroup, so that the identity and inverses are not necessarily present. In detail, the requirements of a ring are:

(R1) Addition is associative, $(a+b)+c = a+(b+c)$.
(R2) Addition is commutative, $a+b = b+a$.
(R3) There is an element 0 such that $a+0 = a$ for all $a \in R$.
(R4) For each $a \in R$ there exists an element $-a$ such that $a - a \equiv a + (-a) = 0$.
(R5) Multiplication is associative, $(ab)c = a(bc)$.
(R6) Multiplication is commutative, $ab = ba$.
(R7) The **distributive law** holds, $a(b+c) = ab + ac$. By (R6) this also implies $(a+b)c = ac + bc$. It is the key relation linking the two laws of composition, addition and multiplication.

As shown in Chapter 2, the additive identity 0 is unique. From these axioms we also have that $0a = 0$ for all $a \in R$, for by (R1), R(3), (R4) and (R7)

$$0a = 0a + 0 = 0a + 0a - 0a = (0+0)a - 0a = 0a - 0a = 0.$$

59

Vector spaces

Example 3.1 The integers $\mathbb{Z}$ form a ring with respect to the usual operation of addition and multiplication. This ring has a (multiplicative) identity 1, having the property $1a = a1 = a$ for all $a \in \mathbb{Z}$. The set $2\mathbb{Z}$ consisting of all even integers also forms a ring, but now there is no identity.

Example 3.2 The set M_n of all $n \times n$ real matrices forms a ring with addition of matrices $\mathbf{A} + \mathbf{B}$ and matrix product $\mathbf{AB}$ defined in the usual way. This is a ring with identity $\mathbf{I}$, the unit matrix.

Example 3.3 The set of all real-valued functions on a set S, denoted $\mathcal{F}(S)$, forms a ring with identity. Addition and multiplication of functions $f + g$, fg are defined in the usual way,

$$(f+g)(x) = f(x) + g(x), \qquad (fg)(x) = f(x)g(x).$$

The 0 element is the zero function whose value on every $x \in S$ is the number zero, while the identity is the function having the value 1 at each $x \in S$.

These examples of rings all fail to be groups with respect to multiplication, for even when they have a multiplicative identity 1, it is almost never true that the zero element 0 has an inverse.

Exercise: Show that if 0^{-1} exists in a ring R with identity then $0 = 1$ and R must be the trivial ring consisting of just one element 0.

A **field** $\mathbb{K}$ is a ring with a multiplicative identity 1, in which every element $a \ne 0$ has an inverse $a^{-1} \in \mathbb{K}$ such that $aa^{-1} = 1$. It not totally clear why the words 'rings' and 'fields' are used to describe these algebraic entities. However, the word 'field' is a perhaps a little unfortunate as it has nothing whatsoever to do with expressions such as 'electromagnetic field', commonly used in physics.

Example 3.4 The real numbers $\mathbb{R}$ and complex numbers $\mathbb{C}$ both form fields with respect to the usual rules of addition and multiplication. These are essentially the only fields of interest in this book. We will frequently use the symbol $\mathbb{K}$ to refer to a field which could be *either* $\mathbb{R}$ or $\mathbb{C}$.

Problems

Problem 3.1 Show that the integers modulo a prime number p form a finite field.

Problem 3.2 Show that the set of all real numbers of the form $a + b\sqrt{2}$, where a and b are rational numbers, is a field. If a and b are restricted to the integers show that this set is a ring, but is not a field.

3.2 Vector spaces

A **vector space** $(V, \mathbb{K})$ consists of an additive abelian group V whose elements $u, v, \ldots$ are called **vectors** together with a field $\mathbb{K}$ whose elements are termed **scalars**. The law of

3.2 Vector spaces

composition $u + v$ defining the abelian group is called **vector addition**. There is also an operation $\mathbb{K} \times V \to V$ called **scalar multiplication**, which assigns a vector $au \in V$ to any pair $a \in \mathbb{K}$, $u \in V$. The identity element 0 for vector addition, satisfying $0 + u = u$ for all vectors u, is termed the **zero vector**, and the inverse of any vector u is denoted $-u$. In principle there can be a minor confusion in the use of the same symbol $+$ for vector addition and scalar addition, and the same symbol 0 both for the zero vector and the zero scalar. It should, however, always be clear from the context which is being used. A similar remark applies to scalar multiplication au and field multiplication of scalars ab. The full list of axioms to be satisfied by a vector space is:

(VS1) For all $u, v, w \in V$ and $a, b, c \in \mathbb{K}$,

$$u + (v + w) = (u + v) + w \qquad a + (b + c) = (a + b) + c \qquad a(bc) = (ab)c$$
$$u + v = v + u \qquad a + b = b + a \qquad ab = ba$$
$$u + 0 = 0 + u = u \qquad a + 0 = 0 + a = a \qquad a1 = 1a = a$$
$$u + (-u) = 0; \qquad a + (-a) = 0; \qquad a(b + c) = ab + ac.$$

(VS2) $a(u + v) = au + av$.
(VS3) $(a + b)u = au + bu$.
(VS4) $a(bv) = (ab)v$.
(VS5) $1v = v$.

A vector space $(V, \mathbb{K})$ is often referred to as a **vector space V over a field $\mathbb{K}$** or simply a vector space V when the field of scalars is implied by some introductory phrase such as 'let V be a real vector space', or 'V is a complex vector space'.

Since $v = (1 + 0)v = v + 0v$ it follows that $0v = 0$ for any vector $v \in V$. Furthermore, $(-1)v$ is the additive inverse of v since, by (VS3), $(-1)v + v = (-1 + 1)v = 0v = 0$. It is also common to write $u - v$ in place of $u + (-v)$, so that $u - u = 0$. Vectors are often given distinctive notations such as $\mathbf{u}, \mathbf{v}, \ldots$ or $\vec{u}, \vec{v}, \ldots$, etc. to distinguish them from scalars, but we will only adopt such notations in specific instances.

Example 3.5 The set $\mathbb{K}^n$ of all n-tuples $\mathbf{x} = (x_1, x_2, \ldots, x_n)$ where $x_i \in \mathbb{K}$ is a vector space, with vector addition and scalar multiplication defined by

$$\mathbf{x} + \mathbf{y} = (x_1 + y_1, x_2 + y_2, \ldots, x_n + y_n),$$
$$a\mathbf{x} = (ax_1, ax_2, \ldots, ax_n).$$

Specific instances are $\mathbb{R}^n$ or $\mathbb{C}^n$. Sometimes the vectors of $\mathbb{K}^n$ will be represented by $n \times 1$ column matrices and there are some advantages in denoting the components by superscripts,

$$\mathbf{x} = \begin{pmatrix} x^1 \\ x^2 \\ \vdots \\ x^n \end{pmatrix}.$$

Vector spaces

Scalar multiplication and addition of vectors is then

$$a\mathbf{x} = \begin{pmatrix} ax^1 \\ ax^2 \\ \vdots \\ ax^n \end{pmatrix}, \quad \mathbf{x} + \mathbf{y} = \begin{pmatrix} x^1 + y^1 \\ x^2 + y^2 \\ \vdots \\ x^n + y^n \end{pmatrix}.$$

Exercise: Verify that all axioms of a vector space are satisfied by $\mathbb{K}^n$.

Example 3.6 Let $\mathbb{K}^\infty$ denote the set of all sequences $u = (u_1, u_2, u_3, \ldots)$ where $u_i \in \mathbb{K}$. This is a vector space if vector addition and scalar multiplication are defined as in Example 3.5:

$$u + v = (u_1 + v_1,\ u_2 + v_2,\ u_3 + v_3, \ldots),$$
$$a u = (au_1,\ au_2,\ au_3, \ldots).$$

Example 3.7 The set of all $m \times n$ matrices over the field $\mathbb{K}$, denoted $M^{(m,n)}(\mathbb{K})$, is a vector space. In this case vectors are denoted by $\mathsf{A} = [a_{ij}]$ where $i = 1, \ldots, m$, $j = 1, \ldots, n$ and $a_{ij} \in \mathbb{K}$. Addition and scalar multiplication are defined by:

$$\mathsf{A} + \mathsf{B} = [a_{ij} + b_{ij}], \qquad c\mathsf{A} = [c\, a_{ij}].$$

Although it may seem a little strange to think of a matrix as a 'vector', this example is essentially no different from Example 3.5, except that the sequence of numbers from the field $\mathbb{K}$ is arranged in a rectangular array rather than a row or column.

Example 3.8 Real-valued functions on $\mathbb{R}^n$, denoted $\mathcal{F}(\mathbb{R}^n)$, form a vector space over $\mathbb{R}$. As described in Section 1.4, the vectors in this case can be thought of as functions of n arguments,

$$f(\mathbf{x}) = f(x_1, x_2, \ldots, x_n),$$

and vector addition $f + g$ and scalar multiplication af are defined in the obvious way,

$$(f + g)(\mathbf{x}) = f(\mathbf{x}) + g(\mathbf{x}), \qquad (af)(\mathbf{x}) = af(\mathbf{x}).$$

The verification of the axioms of a vector space is a straightforward exercise.

More generally, if S is an arbitrary set, then the set $\mathcal{F}(S, \mathbb{K})$ of all $\mathbb{K}$-valued functions on S forms a vector space over $\mathbb{K}$. For example, the set of complex-valued functions on $\mathbb{R}^n$, denoted $\mathcal{F}(\mathbb{R}^n, \mathbb{C})$, is a complex vector space. We usually denote $\mathcal{F}(S, \mathbb{R})$ simply by $\mathcal{F}(S)$, taking the real numbers as the default field. If S is a finite set $S = \{1, 2, \ldots, n\}$, then $\mathcal{F}(S, \mathbb{K})$ is equivalent to the vector space $\mathbb{K}^n$, setting $u_i = u(i)$ for any $u \in \mathcal{F}(S, \mathbb{K})$.

When the vectors can be uniquely specified by a finite number of scalars from the field $\mathbb{K}$, as in Examples 3.5 and 3.7, the vector space is said to be *finite dimensional*. The number of independent components needed to specify an arbitrary vector is called the *dimension* of the space; e.g., $\mathbb{K}^n$ has dimension n, while $M^{m,n}(\mathbb{K})$ is of dimension mn. On the other hand, in Examples 3.6 and 3.8 it is clearly impossible to specify the vectors by a finite number of scalars and these vector spaces are said to be *infinite dimensional*. A rigorous definition of these terms will be given in Section 3.5.

Example 3.9 A set M is called a **module** over a *ring* R if it satisfies all the axioms (VS1)–(VS5) with R replacing the field $\mathbb{K}$. Axiom (VS5) is only included if the ring has an identity. This concept is particularly useful when R is a ring of real or complex-valued functions on a set S such as the rings $\mathcal{F}(S)$ or $\mathcal{F}(S, \mathbb{C})$ in Example 3.8.

A typical example of a module is the following. Let $\mathcal{C}(\mathbb{R}^n)$ be the ring of *continuous* real-valued functions on $\mathbb{R}^n$, sometimes called *scalar fields*, and let $\mathcal{V}^n$ be the set of all n-tuples of real-valued continuous functions on $\mathbb{R}^n$. A typical element of $\mathcal{V}^n$, called a *vector field* on $\mathbb{R}^n$, can be written

$$\mathbf{v}(\mathbf{x}) = (v_1(\mathbf{x}), v_2(\mathbf{x}), \ldots, v_n(\mathbf{x}))$$

where each $v_i(\mathbf{x})$ is a continuous real-valued function on $\mathbb{R}^n$. Vector fields can be added in the usual way and multiplied by scalar fields,

$$\mathbf{u} + \mathbf{v} = (u_1(\mathbf{x}) + v_1(\mathbf{x}), \ldots, u_n(\mathbf{x}) + v_n(\mathbf{x})),$$
$$f(\mathbf{x})\mathbf{v}(\mathbf{x}) = (f(\mathbf{x})v_1(\mathbf{x}), f(\mathbf{x})v_2(\mathbf{x}), \ldots, f(\mathbf{x})v_n(\mathbf{x})).$$

The axioms (VS1)–(VS5) are easily verified, showing that $\mathcal{V}^n$ is a module over $\mathcal{C}(\mathbb{R}^n)$. This *module* is finite dimensional in the sense that only a finite number of component scalar fields are needed to specify any vector field. Of course $\mathcal{V}^n$ also has the structure of a vector space over the field $\mathbb{R}$, similar to the vector space $\mathcal{F}(\mathbb{R}^n)$ in Example 3.8, but as a vector space it is *infinite dimensional*.

3.3 Vector space homomorphisms

If V and W are two vector spaces over the same field $\mathbb{K}$, a map $T : V \to W$ is called **linear**, or a **vector space homomorphism** from V into W, if

$$T(au + bv) = aTu + bTv \qquad (3.1)$$

for all $a, b \in \mathbb{K}$ and all $u, v \in V$. The notation Tu on the right-hand side is commonly used in place of $T(u)$. Vector space homomorphisms play a similar role to group homomorphisms in that they preserve the basic operations of vector addition and scalar multiplication that define a vector space. They are the morphisms of the *category of vector spaces*.

Since $T(u + 0) = Tu = Tu + T0$, it follows that the zero vector of V goes to the zero vector of W under a linear map, $T0 = 0$. Note, however, that the zero vectors on the two sides of this equation lie in different spaces and are, strictly speaking, different vectors.

A linear map $T : V \to W$ that is one-to-one and onto is called a **vector space isomorphism**. In this case the inverse map $T^{-1} : W \to V$ must also be linear, for if $u, v \in W$ let

$u' = T^{-1}u$, $v' = T^{-1}v$ then

$$\begin{aligned}T^{-1}(au+bv) &= T^{-1}(aTu'+bTv') \\ &= T^{-1}(T(au'+bv')) \\ &= \mathrm{id}_V(au'+bv') \\ &= au'+bv' \\ &= aT^{-1}u+bT^{-1}v.\end{aligned}$$

Two vector spaces V and W are called **isomorphic**, written $V \cong W$, if there exists a vector space isomorphism $T : V \to W$. Two isomorphic vector spaces are essentially identical in all their properties.

Example 3.10 Consider the set $\mathcal{P}_n(x)$ of all real-valued polynomials of degree $\leq n$,

$$f(x) = a_0 + a_1 x + a_2 x^2 + \cdots + a_n x^n.$$

Polynomials of degree $\leq n$ can be added and multiplied by scalars in the obvious way,

$$\begin{aligned}f(x)+g(x) &= (a_0+b_0)+(a_1+b_1)x+(a_2+b_2)x^2+\cdots+(a_n+b_n)x^n, \\ cf(x) &= ca_0+ca_1x+ca_2x^2+\cdots+ca_nx^n,\end{aligned}$$

making $\mathcal{P}_n(x)$ into a vector space. The map $S : \mathcal{P}_n(x) \to \mathbb{R}^{n+1}$ defined by

$$S(a_0+a_1x+a_2x^2+\cdots+a_nx^n) = (a_0, a_1, a_2, \ldots, a_n)$$

is one-to-one and onto and clearly preserves basic vector space operations,

$$S(f(x)+g(x)) = S(f(x))+S(g(x)), \qquad S(af(x)) = aS(f(x)).$$

Hence S is a vector space isomorphism, and $\mathcal{P}_n(x) \cong \mathbb{R}^{n+1}$.

The set $\hat{\mathbb{R}}^\infty$ of all sequences of real numbers $(a_0, a_1, \ldots)$ having only a finite number of non-zero members $a_i \neq 0$ is a vector space, using the same rules of vector addition and scalar multiplication given for $\mathbb{R}^\infty$ in Example 3.6. The elements of $\hat{\mathbb{R}}^\infty$ are real sequences of the form $(a_0, a_1, \ldots, a_m, 0, 0, 0, \ldots)$. Let $\mathcal{P}(x)$ be the set of all real polynomials, $\mathcal{P}(x) = \mathcal{P}_0(x) \cup \mathcal{P}_1(x) \cup \mathcal{P}_2(x) \cup \ldots$ This is clearly a vector space with respect to the standard rules of addition of polynomials and scalar multiplication. The map $S : \hat{\mathbb{R}}^\infty \to \mathcal{P}$ defined by

$$S : (a_0, a_1, \ldots, a_m, 0, 0, \ldots) \mapsto a_0 + a_1 x + \cdots + a_m x^m$$

is an isomorphism.

It is simple to verify that the inclusion maps defined in Section 1.4,

$$i_1 : \hat{\mathbb{R}}^\infty \to \mathbb{R}^\infty \quad \text{and} \quad i_2 : \mathcal{P}_n(x) \to \mathcal{P}(x)$$

are vector space homomorphisms.

Let $L(V, W)$ denote the set of all linear maps from V to W. If T, S are linear maps from V to W, addition $T+S$ and scalar multiplication aT are defined by

$$(T+S)(u) = Tu + Su, \qquad (aT)u = aTu.$$

3.3 Vector space homomorphisms

The set $L(V, W)$ is a vector space with respect to these operations. Other common notations for this space are $\text{Hom}(V, W)$ and $\text{Lin}(V, W)$.

Exercise: Verify that $L(V, W)$ satisfies all the axioms of a vector space.

If $T \in L(U, V)$ and $S \in L(V, W)$, define their **product** to be the composition map $ST = S \circ T : U \to W$,

$$(ST)u = S(Tu).$$

This map is clearly linear since

$$ST(au + bv) = S(aTu + bTv) = aSTu + bSTv.$$

If S and T are invertible linear maps then so is their product ST, and $(ST)^{-1} : W \to U$ satisfies

$$(ST)^{-1} = T^{-1}S^{-1}, \qquad (3.2)$$

since

$$T^{-1}S^{-1}ST = T^{-1}\text{id}_V T = T^{-1}T = \text{id}_U.$$

Linear maps $S : V \to V$ are called **linear operators** on V. They form the vector space $L(V, V)$. If S is an invertible linear operator on V it is called a **linear transformation** on V. It may be thought of as a vector space isomorphism of V onto itself, or an *automorphism* of V. The linear transformations of V form a group with respect to the product law of composition, called the **general linear group** on V and denoted $GL(V)$. The group properties are easily proved:

Closure: if S and T are linear transformations of V then so is ST, since (a) it is a linear map, and (b) it is invertible by Eq. (3.2).
Associativity: this is true of all maps (see Section 1.4).
Unit: the identity map id_V is linear and invertible.
Inverse: as shown above, the inverse T^{-1} of any vector space isomorphism T is linear.

Note, however, that $GL(V)$ is *not* a vector space, since the zero operator that sends every vector in V to the zero vector 0 is not invertible and therefore does not belong to $GL(V)$.

Problems

Problem 3.3 Show that the infinite dimensional vector space $\mathbb{R}^\infty$ is isomorphic with a proper subspace of itself.

Problem 3.4 On the vector space $\mathcal{P}(x)$ of polynomials with real coefficients over a variable x, let x be the operation of multiplying by the polynomial x, and let D be the operation of differentiation,

$$x : f(x) \mapsto xf(x), \qquad D : f(x) \mapsto \frac{df(x)}{dx}.$$

Show that both of these are linear operators over $\mathcal{P}(x)$ and that $Dx - xD = I$, where I is the identity operator.

3.4 Vector subspaces and quotient spaces

A **(vector) subspace** W of a vector space V is a subset that is a vector space in its own right, with respect to the operations of vector addition and scalar multiplication defined on V. There is a simple criterion for determining whether a subset is a subspace:

A subset W is a subspace of V if and only if $u + av \in W$ for all $a \in \mathbb{K}$ and all $u, v \in W$.

For, setting $a = 1$ shows that W is closed under vector addition, while $u = 0$ implies that it is closed with respect to scalar multiplication. Closure with respect to these two operations is sufficient to demonstrate that W is a vector subspace: the zero vector $0 \in W$ since $0 = 0v \in W$ for any $v \in W$; the inverse vector $-u = (-1)u \in W$ for every $u \in W$, and the remaining vector space axioms (VS1)–(VS5) are all satisfied by W since they are inherited from V.

Example 3.11 Let $U = \{(u_1, u_2, \ldots, u_m, 0, \ldots, 0)\} \subseteq \mathbb{R}^n$ be the subset of n-vectors whose last $n - m$ components all vanish. U is a vector subspace of $\mathbb{R}^n$, since

$$(u_1, \ldots, u_m, 0, \ldots, 0) + a(v_1, \ldots, v_m, 0, \ldots, 0)$$
$$= (u_1 + av_1, \ldots, u_m + av_m, 0, \ldots, 0) \in U.$$

This subspace is isomorphic to $\mathbb{R}^m$, through the isomorphism

$$T : (u_1, \ldots, u_m, 0, \ldots, 0) \mapsto (u_1, \ldots, u_m).$$

Exercise: Show that $\mathbb{R}^n$ is isomorphic to a subspace of $\mathbb{R}^\infty$ for every $n > 0$.

Example 3.12 Let V be a vector space over a field $\mathbb{K}$, and $u \in V$ any vector. The set $U = \{cu \mid c \in \mathbb{K}\}$ is a subspace of V, since for any pair of scalars $c, c' \in \mathbb{K}$,

$$cu + a(c'u) = cu + (ac')u = (c + ac')u \in U.$$

Exercise: Show that the set $\{(x, y, z) \mid x + 2y + 3z = 0\}$ forms a subspace of $\mathbb{R}^3$, while the subset $\{(x, y, z) \mid x + 2y + 3z = 1\}$ does not.

Example 3.13 The set of all continuous real-valued functions on $\mathbb{R}^n$, denoted $\mathcal{C}(\mathbb{R}^n)$, is a subspace of $\mathcal{F}(\mathbb{R}^n)$ defined in Example 3.8, for if f and g are any pair of continuous functions on $\mathbb{R}^n$ then so is any linear combination $f + ag$ where $a \in \mathbb{R}$.

Exercise: Show that the vector space of all real polynomials $\mathcal{P}(x)$, defined in Example 3.10, is a vector subspace of $\mathcal{C}(\mathbb{R})$.

Given two subspaces U and W of a vector space V, their set-theoretical intersection $U \cap W$ forms a vector subspace of V, for if $u, w \in U \cap W$ then any linear combination $u + aw$ belongs to each subspace U and W separately. This argument can easily be extended to show that the intersection $\bigcap_{i \in I} U_i$ of any family of subspaces is a subspace of V.

3.4 Vector subspaces and quotient spaces

Complementary subspaces and quotient spaces

While the intersection of any pair of subspaces U and W is a vector subspace of V, this is not true of their set-theoretical union $U \cup V$ – consider, for example, the union of the two subspaces $\{(c, 0) \mid c \in \mathbb{R}\}$ and $\{(0, c) \mid c \in \mathbb{R}\}$ of $\mathbb{R}^2$. Instead, we can define the **sum** $U + W$ of any pair of subspaces to be the 'smallest' vector space that contains $U \cup W$,

$$U + W = \{u + w \mid u \in U, \, w \in W\}.$$

This is a vector subspace, for if $u = u_1 + w_1$ and $v = u_2 + w_2$ belong to $U + W$, then

$$u + av = (u_1 + w_1) + a(u_2 + w_2) = (u_1 + au_2) + (w_1 + aw_2) \in U + W.$$

Two subspaces U and W of V are said to be **complementary** if every vector $v \in V$ has a *unique* decomposition $v = u + w$ where $u \in U$ and $w \in W$. V is then said to be the **direct sum** of the subspaces U and W, written $V = U \oplus W$.

Theorem 3.1 *U and W are complementary subspaces of V if and only if (i) $V = U + W$ and (ii) $U \cap W = \{0\}$.*

Proof: If U and W are complementary subspaces then (i) is obvious, and if there exists a non-zero vector $u \in U \cap V$ then the zero vector would have alternative decompositions $0 = 0 + 0$ and $0 = u + (-u)$. Conversely, if (i) and (ii) hold then the decomposition $v = u + w$ is unique, for if $v = u' + w'$ then $u - u' = w - w' \in U \cap W$. Hence $u - u' = w - w' = 0$, so $u = u'$ and $w = w'$. ∎

Example 3.14 Let $\mathbb{R}_1$ be the subspace of $\mathbb{R}^n$ consisting of vectors of the form $\{(x_1, 0, 0, \ldots, 0) \mid x_1 \in \mathbb{R}\}$, and S_1 the subspace

$$S_1 = \{(0, x_2, x_3, \ldots, x_n) \mid (x_i \in \mathbb{R})\}.$$

Then $\mathbb{R}^n = \mathbb{R}_1 \oplus S_1$. Continuing in a similar way S_1 may be written as a direct sum of $\mathbb{R}_2 = \{(0, x_2, 0, \ldots, 0)\}$ and a subspace S_2. We eventually arrive at the direct sum decomposition

$$\mathbb{R}^n = \mathbb{R}_1 \oplus \mathbb{R}_2 \oplus \cdots \oplus \mathbb{R}_n \cong \mathbb{R} \oplus \mathbb{R} \oplus \cdots \oplus \mathbb{R}.$$

If U and W are arbitrary vector spaces it is possible to define their direct sum in a constructive way, sometimes called their **external direct sum**, by setting

$$U \oplus W = U \times W = \{(u, w) \mid u \in U, w \in W\}$$

with vector addition and scalar multiplication defined by

$$(u, w) + (u', w') = (u + u', w + w'), \qquad a(u, w) = (au, aw).$$

The map $\varphi : U \to \hat{U} = \{(u, 0) \mid u \in U\} \subset U \oplus W$ defined by $\varphi(u) = (u, 0)$ is clearly an isomorphism. Hence we may identify U with the subspace $\hat{U}$, and similarly W is identifiable with $\hat{W} = \{(0, w) \mid w \in W\}$. With these identifications the constructive notion of direct sum is equivalent to the 'internally defined' version, since $U \oplus W = \hat{U} \oplus \hat{W}$.

The real number system $\mathbb{R}$ can be regarded as a real vector space in which scalar multiplication is simply multiplication of real numbers – vectors and scalars are indistinguishable

Vector spaces

in this instance. Since the subspaces $\mathbb{R}_i$ defined in Example 3.14 are clearly isomorphic to $\mathbb{R}$ for each $i = 1, \ldots, n$, the decomposition given in that example can be written

$$\mathbb{R}^n \cong \underbrace{\mathbb{R} \oplus \mathbb{R} \oplus \cdots \oplus \mathbb{R}}_{n}.$$

For any given vector subspace there always exist complementary subspaces. We give the proof here as an illustration of the use of Zorn's lemma, Theorem 1.6, but it is somewhat technical and the reader will lose little continuity by moving on if they feel so inclined.

Theorem 3.2 *Given a subspace $W \subseteq V$ there always exists a complementary subspace $U \subseteq V$ such that $V = U \oplus W$.*

Proof: Given a vector subspace W of V, let $\mathcal{U}$ be the collection of all vector subspaces $U \subseteq V$ such that $U \cap W = \{0\}$. The set $\mathcal{U}$ can be partially ordered by set inclusion as in Example 1.5. Furthermore, if $\{U_i \,|\, i \in I\}$ is any totally ordered subset of $\mathcal{U}$ such that for every pair $i, j \in I$ we have either $U_i \subseteq U_j$ or $U_j \subseteq U_i$, then their union is bounded above by

$$\tilde{U} = \bigcup_{i \in I} U_i.$$

The set $\tilde{U}$ is a vector subspace of V, for if $u \in \tilde{U}$ and $v \in \tilde{U}$ then there exists a member U_i of the totally ordered family such that both vectors must belong to the same member U_i – if $u \in U_j$ and $v \in U_k$ then set $i = k$ if $U_j \subseteq U_k$, else set $i = j$. Hence $u + av \in U_i \subseteq \tilde{U}$ for all $a \in \mathbb{K}$. By Zorn's lemma we conclude that there exists a maximal subspace $U \in \mathcal{U}$.

It remains to show that U is complementary to W. Suppose not; then there exists a vector $v' \in V$ that cannot be expressed in the form $v' = u + w$ where $u \in U$, $w \in W$. Let U' be the vector subspace defined by

$$U' = \{av' + u \,|\, u \in U\} = \{av'\} \oplus U.$$

It belongs to the family $\mathcal{U}$, for if $U' \cap W \neq \{0\}$ then there would exist a non-zero vector $w' = av' + u$ belonging to W. This implies $v' = a^{-1}(w' - u)$, in contradiction to the requirement that v' cannot be expressed as a sum of vectors from U and W. Hence we have strict inclusion $U \subset U'$, contradicting the maximality of U. Thus U is a subspace complementary to W, as required. ∎

This proof has a distinctly non-constructive feel to it, which is typical of proofs invoking Zorn's lemma. A more direct way to arrive at a vector space complementary to a given subspace W is to define an equivalence relation $\equiv_W$ on V by

$$u \equiv_W v \quad \text{iff} \quad u - v \in W.$$

Checking the equivalence properties is easy:

Reflexive: $u - u = 0 \in W$ for all $u \in V$,
Symmetric: $u - v \in W \implies v - u = -(u - v) \in W$,
Transitive: $u - v \in W$ and $v - w \in W \implies u - w = (u - v) + (v - w) \in W$.

3.4 Vector subspaces and quotient spaces

The equivalence class to which u belongs is written $u + W$, where

$$u + W = \{u + w \mid w \in W\},$$

and is called a **coset** of W. This definition is essentially identical to that given in Section 2.5 for the case of an abelian group. It is possible to form the sum of cosets and multiply them by scalars, by setting

$$(u + W) + (v + W) = (u + v) + W, \qquad a(u + W) = (au) + W.$$

For consistency, it is necessary to show that these definitions are independent of the choice of coset representative. For example, if $u \equiv_W u'$ and $v \equiv_W v'$ then $(u' + v') \equiv_W (u + v)$, for

$$(u' + v') - (u + v) = (u' - u) + (v' - v) \in W.$$

Hence

$$(u' + W) + (v' + W) = (u' + v') + W = (u + v) + W = (u + W) + (v + W).$$

Similarly $au' \equiv_W au$ since $au' - au = a(u' - u) \in W$ and

$$a(u' + W) = (au') + W = (au) + W = a(u + W).$$

The task of showing that the set of cosets is a vector space with respect to these operations is tedious but undemanding. For example, the distributive law (VS2) follows from

$$a((u + W) + (v + W)) = a((u + v) + W)$$
$$= a(u + v) + W$$
$$= (au + av) + W$$
$$= ((au) + W) + ((av) + W)$$
$$= a(u + W) + a(v + W).$$

The rest of the axioms follow in like manner, and are left as exercises. The vector space of cosets of W is called the **quotient space** of V by W, denoted V/W.

To picture a quotient space let U be any subspace of V that is complementary to W. Every element of V/W can be written *uniquely* as a coset $u + W$ where $u \in U$. For, if $v + W$ is any coset, let $v = u + w$ be the unique decomposition of v into vectors from U and W respectively, and it follows that $v + W = u + W$ since $v \equiv_W u$. The map $T : U \to V/W$ defined by $T(u) = u + W$ describes an isomorphism between U and V/W. For, if $u + W = u' + W$ where $u, u' \in U$ then $u - u' \in W \cap U$, whence $u = u'$ since U and W are complementary subspaces.

Exercise: Complete the details to show that the map $T : U \to V/W$ is linear, one-to-one and onto, so that $U \cong V/W$.

This argument also shows that all complementary spaces to a given subspace W are isomorphic to each other. The quotient space V/W is a method for constructing the 'canonical complement' to W.

Example 3.15 While V/W is in a sense complementary to W it is not a subspace of V and, indeed, there is no *natural* way of identifying it with any subspace complementary to W. For example, let $W = \{(x, y, 0) \mid x, y \in \mathbb{R}\}$ be the subspace $z = 0$ of $\mathbb{R}^3$. Its cosets are planes $z = a$, parallel to the x–y plane, and it is these planes that constitute the 'vectors' of V/W. The subspace $U = \{(0, 0, z) \mid z \in \mathbb{R}\}$ is clearly complementary to W and is isomorphic to V/W using the map

$$(0, 0, a) \mapsto (0, 0, a) + W = \{(x, y, a) \mid x, y \in \mathbb{R}\}.$$

However, there is no natural way of identifying V/W with a complementary subspace such as U. For example, the space $U' = \{(0, 2z, z) \mid z \in \mathbb{R}\}$ is also complementary to W since $U' \cap W = \{0\}$ and every vector $(a, b, c) \in \mathbb{R}^3$ has the decomposition

$$(a, b, c) = (a, b - 2c, 0) + (0, 2c, c), \qquad (a, b - 2c, 0) \in W, \quad (0, 2c, c) \in U'.$$

Again, $U' \cong V/W$, under the map

$$(0, 2c, c) \mapsto (0, 2c, c) + W = (0, 0, c) + W.$$

Note how the 'W-component' of (a, b, c) depends on the choice of complementary subspace; $(a, b, 0)$ with respect to U, and $(a, b - 2c, 0)$ with respect to U'.

Images and kernels of linear maps

The **image** of a linear map $T : V \to W$ is defined to be the set

$$\operatorname{im} T = T(V) = \{w \mid w = Tv\} \subseteq W.$$

The set $\operatorname{im} T$ is a subspace of W, for if $w, w' \in \operatorname{im} T$ then

$$w + aw' = Tv + aTv' = T(v + av') \in \operatorname{im} T.$$

The **kernel** of the map T is defined as the set

$$\ker T = T^{-1}(0) = \{v \in V \mid Tv = 0\} \subseteq V.$$

This is also a subspace of V, for if $v, v' \in \ker T$ then $T(v + av') = Tv + aTv' = 0 + 0 = 0$. The two spaces are related by the identity

$$\operatorname{im} T \cong V/\ker T. \tag{3.3}$$

Proof: Define the map $\tilde{T} : V/\ker T \to \operatorname{im} T$ by

$$\tilde{T}(v + \ker T) = Tv.$$

This map is well-defined since it is independent of the choice of coset representative v,

$$v + \ker T = v' + \ker T \implies v - v' \in \ker T$$
$$\implies T(v - v') = 0$$
$$\implies Tv = Tv'.$$

3.4 Vector subspaces and quotient spaces

and is clearly linear. It is onto and one-to-one, for every element of im T is of the form $Tv = \tilde{T}(v + \ker T)$ and

$$\tilde{T}(v + \ker T) = \tilde{T}(v' + \ker T) \implies Tv = Tv'$$
$$\implies v - v' \in \ker T$$
$$\implies v + \ker T = v' + \ker T.$$

Hence $\tilde{T}$ is a vector space isomorphism, which proves Eq. (3.3). ∎

Example 3.16 Let $V = \mathbb{R}^3$ and $W = \mathbb{R}^2$, and define the map $T : V \to W$ by

$$T(x, y, z) = (x + y + z, 2x + 2y + 2z).$$

The subspace im T of $\mathbb{R}^2$ consists of the set of all vectors of the form $(a, 2a)$, where $a \in \mathbb{R}$, while ker T is the subset of all vectors $(x, y, z) \in V$ such that $x + y + z = 0$ – check that these do form a subspace of V. If $v = (x, y, z)$ and $a = x + y + z$, then $v - ae \in \ker T$ where $e = (1, 0, 0) \in V$, since

$$T(v - ae) = T(x, y, z) - T(x + y + z, 0, 0) = (0, 0).$$

Furthermore a is the unique value having this property, for if $a' \neq a$ then $v - a'e \notin \ker T$. Hence every coset of ker T has a unique representative of the form ae and may be written uniquely in the form $ae + \ker T$. The isomorphism $\tilde{T}$ defined in the above proof is given by

$$\tilde{T}(ae + \ker T) = (a, 2a) = a(1, 2).$$

Problems

Problem 3.5 If L, M and N are vector subspaces of V show that

$$L \cap (M + (L \cap N)) = L \cap M + L \cap N$$

but it is not true in general that

$$L \cap (M + N) = L \cap M + L \cap N.$$

Problem 3.6 Let $V = U \oplus W$, and let $v = u + w$ be the unique decomposition of a vector v into a sum of vectors from $u \in U$ and $w \in W$. Define the *projection operators* $P_U : V \to U$ and $P_W : V \to W$ by

$$P_U(v) = u, \qquad P_W(v) = w.$$

Show that

(a) $P_U^2 = P_U$ and $P_W^2 = P_W$.
(b) Show that if $P : V \to V$ is an operator satisfying $P^2 = P$, said to be an *idempotent* operator, then there exists a subspace U such that $P = P_U$. [*Hint*: Set $U = \{u \mid Pu = u\}$ and $W = \{w \mid Pw = 0\}$ and show that these are complementary subspaces such that $P = P_U$ and $P_W = \text{id}_V - P$.]

Vector spaces

3.5 Bases of a vector space

Subspace spanned by a set

If A is any subset of a vector space V define the **subspace spanned or generated by** A, denoted $L(A)$, as the set of all *finite* linear combinations of elements of A,

$$L(A) = \left\{ \sum_{i=1}^{n} a^i v_i \,|\, a^i \in \mathbb{K},\ v_i \in A,\ n = 1, 2, \ldots \right\}.$$

The word 'finite' is emphasized here because no meaning can be attached to infinite sums until we have available the concept of 'limit' (see Chapters 10 and 13). We may think of $L(A)$ as the intersection of all subspaces of V that contain A – essentially, it is the 'smallest' vector subspace containing A. At first sight the notation whereby the indices on the coefficients a^i of the linear combinations have been set in the superscript position may seem a little peculiar, but we will eventually see that judicious and systematic placements of indices can make many expressions much easier to manipulate.

Exercise: If M and N are subspaces of V show that their sum $M + N$ is identical with the span of their union, $M + N = L(M \cup N)$.

The vector space V is said to be **finite dimensional** [7] if it can be spanned by a finite set, $V = L(A)$, where $A = \{v_1, v_2, \ldots, v_n\}$. Otherwise we say V is **infinite dimensional**. When V is finite dimensional its **dimension**, $\dim V$, is defined to be the smallest number n such that V is spanned by a set consisting of just n vectors.

Example 3.17 $\mathbb{R}^n$ is finite dimensional, since it can be generated by the set of 'unit vectors',

$$A = \{e_1 = (1, 0, \ldots, 0),\quad e_2 = (0, 1, \ldots, 0),\quad \ldots,\quad e_n = (0, 0, \ldots, 1)\}.$$

Since any vector u can be written

$$u = (u_1, u_2, \ldots, u_n) = u_1 e_1 + u_2 e_2 + \cdots + u_n e_n,$$

these vectors span $\mathbb{R}^n$, and $\dim \mathbb{R}^n \leq n$. We will see directly that $\dim \mathbb{R}^n = n$, as to be expected.

Example 3.18 $\mathbb{R}^\infty$ is clearly infinite dimensional. It is not even possible to span this space with the set of vectors $A = \{e_1, e_2, \ldots\}$, where

$$e_1 = (1, 0, \ldots),\quad e_2 = (0, 1, \ldots),\quad \ldots.$$

The reason is that any finite linear combination of these vectors will only give rise to vectors having at most a finite number of non-zero components. The set of all those vectors that are finite linear combinations of vectors from A does in fact form an infinite dimensional subspace of $\mathbb{R}^\infty$, but it is certainly not the whole space. The space spanned by A is precisely the subspace $\hat{\mathbb{R}}^\infty$ defined in Example 3.10.

Exercise: If V is a vector space and $u \in V$ is any non-zero vector show that $\dim V = 1$ if and only if every vector $v \in V$ is proportional to u; i.e., $v = au$ for some $a \in \mathbb{K}$.

3.5 Bases of a vector space

Exercise: Show that the set of functions on the real line, $\mathcal{F}(\mathbb{R})$, is an infinite dimensional vector space.

Basis of a vector space

A set of vectors A is said to be **linearly independent**, often written 'l.i.', if every finite subset of vectors $\{v_1, v_2, \ldots, v_k\} \subseteq A$ has the property that

$$\sum_{i=1}^{k} a^i v_i = 0 \implies a^j = 0 \text{ for all } j = 1, \ldots, k.$$

In other words, the zero vector 0 cannot be written as a non-trivial linear combination of these vectors. The zero vector can never be a member of a l.i. set since $a0 = 0$ for any $a \in \mathbb{K}$. If A is a finite set of vectors, $A = \{v_1, v_2, \ldots, v_n\}$, it is sufficient to set $k = n$ in the above definition. A subset E of a vector space V is called a **basis** if it is linearly independent and spans the whole of V. A set of vectors is said to be **linearly dependent** if it is not l.i.

Example 3.19 The set of vectors $\{e_1 = (1, 0, \ldots, 0), \ldots, e_n = (0, 0, \ldots 0, 1)\}$ span $\mathbb{K}^n$, since every vector $v = (v_1, v_2, \ldots, v_n)$ can be expressed as a linear combination

$$v = v_1 e_1 + v_2 e_2 + \cdots + v_n e_n.$$

They are linearly independent, for if $v = 0$ then we must have $v_1 = v_2 = \cdots v_n = 0$. Hence $e_1, \ldots, e_n$ is a basis of $\mathbb{K}^n$.

Exercise: Show that the vectors $f_1 = (1, 0, 0)$, $f_2 = (1, 1, -1)$ and $f_3 = (1, 1, 1)$ are l.i. and form a basis of $\mathbb{R}^3$.

It is perhaps surprising to learn that even infinite dimensional vector space such as $\mathbb{R}^\infty$ always has a basis. Just try and construct a basis! The set $A = \{e_1 = (1, 0, 0, \ldots), e_2 = (0, 1, 0, \ldots), \ldots\}$ clearly won't do, since any vector having an infinite number of non-zero components cannot be a *finite* linear combination of these vectors. We omit the proof as it is heavily dependent on Zorn's lemma and such bases are only of limited use. For the rest of this section we only consider bases in finite dimensional spaces.

Theorem 3.3 *Let V be a finite dimensional vector space of dimension n. A subset $E = \{e_1, e_2, \ldots, e_n\}$ spans V if and only if it is linearly independent.*

Proof: *Only if*: Assume $V = L(E)$, so that every vector $v \in V$ is a linear combination

$$v = \sum_{i=1}^{n} v^i e_i.$$

The set E is then linearly independent, for suppose there exists a vanishing linear combination

$$\sum_{i=1}^{n} a^i e_i = 0 \quad (a^i \in \mathbb{K})$$

73

Vector spaces

where, say, $a^1 \neq 0$. Replacing e_1 by $e_1 = b^2 e_2 + \cdots + b^n e_n$ where $b^i = -a^i/a^1$, we find

$$v = \sum_{j=2}^{n} \bar{v}^j e_j \quad \text{where} \quad \bar{v}^j = v^j + b^j v^1.$$

Thus $E' = E - \{e_1\}$ spans V, contradicting the initial hypothesis that V cannot be spanned by a set of fewer than n vectors on account of dim $V = n$.

If: Assume E is linearly independent. Our aim is to show that it spans all of V. Since dim $V = n$ there must exist a set $F = \{f_1, f_2, \ldots, f_n\}$ of exactly n vectors spanning V. By the above argument this set is l.i. Expand e_1 in terms of the vectors from F,

$$e_1 = a^1 f_1 + a^2 f_2 + \cdots + a^n f_n, \tag{3.4}$$

where, by a permutation of the vectors of the basis F, we may assume that $a^1 \neq 0$. The set $F' = \{e_1, f_2, \ldots, f_n\}$ is a basis for V:

(a) F' is linearly independent, for if there were a vanishing linear combination

$$c e_1 + c^2 f_2 + \cdots + c^n f_n = 0,$$

then substituting (3.4) gives

$$c a^1 f_1 + (c a^2 + c^2) f_2 + \cdots + (c a^n + c^n) f_n = 0.$$

By linear independence of $\{f_1, \ldots, f_n\}$ and $a^1 \neq 0$ it follows that $c = 0$, and subsequently that $c^2 = \cdots = c^n = 0$.

(b) The set F' spans the vector space V since by Eq. (3.4)

$$f_1 = \frac{1}{a^1}(e_1 - a^2 f_2 - \cdots - a^n f_n),$$

and every $v \in V$ must be a linear combination of $\{e_1, f_2, \ldots, f_n\}$ since it is spanned by $\{f_1, \ldots, f_n\}$.

Continuing, e_2 must be a unique linear combination

$$e_2 = b^1 e_1 + b^2 f_2, + \cdots + b^n f_n.$$

Since, by hypothesis, e_1 and e_2 are linearly independent, at least one of the coefficients $b^2, \ldots, b^n$ must be non-zero, say $b^2 \neq 0$. Repeating the above argument we see that the set $F'' = \{e_1, e_2, f_3, \ldots, f_n\}$ is a basis for V. Continue the process n times to prove that $E = F^{(n)} = \{e_1, e_2, \ldots, e_n\}$ is a basis for V. ∎

Corollary 3.4 *If $E = \{e_1, e_2, \ldots, e_n\}$ is a basis of the vector space V then* dim $V = n$.

Proof: Suppose that dim $V = m < n$. The set of vectors $E' = \{e_1, e_2, \ldots, e_m\}$ is l.i., since it is a subset of the l.i. set E. Hence, by Theorem 3.3 it spans V, since it consists of exactly $m = $ dim V vectors. But this is impossible since, for example, the vector e_n cannot be a linear combination of the vectors in E'. Hence we must have dim $V \geq n$. However, by the definition of dimension it is impossible to have dim $V > n$; hence dim $V = n$. ∎

Exercise: Show that if $A = \{v_1, v_2, \ldots, v_m\}$ is an l.i. set of vectors then $m \leq n = $ dim V.

3.5 Bases of a vector space

Theorem 3.5 *Let V be a finite dimensional vector space, $n = \dim V$. If $\{e_1, \ldots, e_n\}$ is a basis of V then each vector $v \in V$ has a unique decomposition*

$$v = \sum_{i=1}^{n} v^i e_i, \quad v^i \in \mathbb{K}. \tag{3.5}$$

*The n scalars $v^i \in \mathbb{K}$ are called the **components** of the vector v with respect to this basis.*

Proof: Since the e_i span V, every vector v has a decomposition of the form (3.5). If there were a second such decomposition,

$$v = \sum_{i=1}^{n} v^i e_i = \sum_{i=1}^{n} w^i e_i$$

then

$$\sum_{i=1}^{n} (v^i - w^i) e_i = 0.$$

Since the e_i are linearly independent, each coefficient of this sum must vanish, $v^i - w^i = 0$. Hence $v^i = w^i$, and the decomposition is unique. ∎

Theorem 3.6 *If V and W are finite dimensional then they are isomorphic if and only if they have the same dimension.*

Proof: Suppose V and W have the same dimension n. Let $\{e_i\}$ be a basis of V and $\{f_i\}$ a basis of W, where $i = 1, 2, \ldots, n$. Set $T : V \to W$ to be the linear map defined by $Te_1 = f_1, Te_2 = f_2, \ldots, Te_n = f_n$. This map extends uniquely to all vectors in V by linearity,

$$T\left(\sum_{i=1}^{n} v^i e_i\right) = \sum_{i=1}^{n} v^i Te_i = \sum_{i=1}^{n} v^i f_i$$

and is clearly one-to-one and onto. Thus T is an isomorphism between V and W.

Conversely suppose V and W are isomorphic vector spaces, and let $T : V \to W$ be a linear map having inverse $T^{-1} : W \to V$. If $\{e_1, e_2, \ldots, e_n\}$ is a basis of V, we show that $\{f_i = Te_i\}$ is a basis of W:

(a) The vectors $\{f_i\}$ are linearly independent, for suppose there exist scalars $a^i \in \mathbb{K}$ such that

$$\sum_{i=1}^{n} a^i f_i = 0.$$

Then,

$$0 = T^{-1}\left(\sum_{i=1}^{n} a^i f_i\right) = \sum_{i=1}^{n} a^i T^{-1} f_i = \sum_{i=1}^{n} a^i e_i$$

and from the linear independence of $\{e_i\}$ it follows that $a^1 = a^2 = \cdots = a^n = 0$.

(b) To show that the vectors $\{f_i\}$ span W let w be any vector in W and set $v = T^{-1}w \in V$.

Vector spaces

Since $\{e_i\}$ spans V there exist scalars v^i such that

$$v = \sum_{i=1}^{n} v^i e_i.$$

Applying the map T to this equation results in

$$w = \sum_{i=1}^{n} v^i T e_i = \sum_{i=1}^{n} v^i f_i,$$

which shows that the set $\{f_1, \ldots, f_n\}$ spans W.

By Corollary 3.4 it follows that $\dim W = \dim V = n$ since both vector spaces have a basis consisting of n vectors. ∎

Example 3.20 By Corollary 3.4 the space $\mathbb{K}^n$ is n-dimensional since, as shown in Example 3.19, the set $\{e_i = (0, 0, \ldots, 0, 1, 0, \ldots, 0)\}$ is a basis. Using Theorem 3.6 every n-dimensional vector space V over the field $\mathbb{K}$ is isomorphic to $\mathbb{K}^n$, which may be thought of as the archetypical n-dimensional vector space over $\mathbb{K}$. Every basis $\{f_1, f_2, \ldots, f_n\}$ of V establishes an isomorphism $T : V \to \mathbb{K}^n$ defined by

$$Tv = (v^1, v^2, \ldots, v^n) \in \mathbb{K}^n \quad \text{where} \quad v = \sum_{i=1}^{n} v^i f_i.$$

Example 3.20 may lead the reader to wonder why we bother at all with the abstract vector space machinery of Section 3.2, when all properties of a finite dimensional vector space V could be referred to the space $\mathbb{K}^n$ by simply picking a basis. This would, however, have some unfortunate consequences. Firstly, there are infinitely many bases of the vector space V, each of which gives rise to a different isomorphism between V and $\mathbb{K}^n$. There is nothing *natural* in the correspondence between the two spaces, since there is no general way of singling out a preferred basis for the vector space V. Furthermore, any vector space concept should ideally be given a basis-independent definition, else we are always faced with the task of showing that it is independent of the choice of basis. For these reasons we will persevere with the 'invariant' approach to vector space theory.

Matrix of a linear operator

Let $T : V \to V$ be a linear operator on a finite dimensional vector space V. Given a basis $\{e_1, e_2, \ldots, e_n\}$ of V define the **components** T^a_k of the linear operator T with respect to this basis by setting

$$T e_j = \sum_{i=1}^{n} T^i{}_j e_i. \tag{3.6}$$

By Theorem 3.5 the components $T^i{}_j$ are uniquely defined by these equations, and the square $n \times n$ matrix $\mathsf{T} = [T^i{}_j]$ is called the **matrix of** T with respect to the basis $\{e_i\}$. It is usual to take the superscript i as the 'first' index, labelling rows, while the subscript j labels the columns, and for this reason it is generally advisable to leave some horizontal spacing between these two indices. In Section 2.3 the components of a matrix were denoted by

3.5 Bases of a vector space

subscripted symbols such as $A = [a_{ij}]$, but in general vector spaces it is a good idea to display the components of a matrix representing a linear operator T in this 'mixed script' notation.

If $v = \sum_{k=1}^{n} v^k e_k$ is an arbitrary vector of V then its image vector $\tilde{v} = Tv = \sum_{j=1}^{m} w^j e_j$ is given by

$$\tilde{v} = Tv = T\left(\sum_{j=1}^{n} v^j e_j\right) = \sum_{j=1}^{n} v^j T e_j = \sum_{j=1}^{n}\sum_{i=1}^{n} v^j T^i{}_j e_i,$$

and the components of $\tilde{v}$ are given by

$$\tilde{v}^i = (Tv)^i = \sum_{j=1}^{n} T^i{}_j v^j. \tag{3.7}$$

If we write the components of v and $\tilde{v}$ as column vectors or $n \times 1$ matrices, $\mathbf{v}$ and $\tilde{\mathbf{v}}$,

$$\mathbf{v} = \begin{pmatrix} v^1 \\ v^2 \\ \vdots \\ v^n \end{pmatrix}, \quad \tilde{\mathbf{v}} = \begin{pmatrix} \tilde{v}^1 \\ \tilde{v}^2 \\ \vdots \\ \tilde{v}^n \end{pmatrix},$$

then Eq. (3.7) is the componentwise representation of the matrix equation

$$\tilde{\mathbf{v}} = \mathsf{T}\mathbf{v}. \tag{3.8}$$

The matrix of the composition of two operators $ST \equiv S \circ T$ is given by

$$ST(e_i) = \sum_{j=1}^{n} S(T^j{}_i e_j)$$
$$= \sum_{j=1}^{n} T^j{}_i S e_j$$
$$= \sum_{j=1}^{n}\sum_{k=1}^{n} T^j{}_i S^k{}_j e_k$$
$$= \sum_{k=1}^{n} (ST)^k{}_i e_k$$

where

$$(ST)^k{}_i = \sum_{j=1}^{n} S^k{}_j T^j{}_i. \tag{3.9}$$

This can be recognized as the componentwise formula for the matrix product ST.

Example 3.21 Care should be taken when reading off the components of the matrix T from (3.6) as it is very easy to come up mistakenly with the 'transpose' array. For example, if a transformation T of a three-dimensional vector space is defined by its effect on a basis

Vector spaces

e_1, e_2, e_3,

$$Te_1 = e_1 - e_2 + e_3$$
$$Te_2 = e_1 - e_3$$
$$Te_3 = e_2 + 2e_3,$$

then its matrix with respect to this basis is

$$T = \begin{pmatrix} 1 & 1 & 0 \\ -1 & 0 & 1 \\ 1 & -1 & 2 \end{pmatrix}.$$

The result of applying T to a vector $u = xe_1 + ye_2 + ze_3$ is

$$Tu = xTe_1 + yTe_2 + zTe_3 = (x+y)e_1 + (-x+z)e_2 + (x-y+2z)e_3,$$

which can also be obtained by multiplying the matrix T and the column vector $u = (x, y, z)^T$,

$$Tu = \begin{pmatrix} 1 & 1 & 0 \\ -1 & 0 & 1 \\ 1 & -1 & 2 \end{pmatrix} \begin{pmatrix} x \\ y \\ z \end{pmatrix} = \begin{pmatrix} x+y \\ -x+z \\ x-y+2z \end{pmatrix}.$$

If S is a transformation given by

$$Se_1 = e_1 + 2e_3$$
$$Se_2 = e_2$$
$$Se_3 = e_1 - e_2$$

whose matrix with respect to this basis is

$$S = \begin{pmatrix} 1 & 0 & 1 \\ 0 & 1 & -1 \\ 2 & 0 & 0 \end{pmatrix},$$

the product of these two transformations is found from

$$STe_1 = Se_1 - Se_2 + Se_3 = 2e_1 - 2e_2 + 2e_3$$
$$STe_2 = Se_1 - Se_3 \qquad\qquad = e_2 + 2e_3$$
$$STe_3 = Se_2 + 2Se_3 \qquad\;\; = 2e_1 - e_2.$$

Thus the matrix of ST is the matrix product of S and T,

$$ST = \begin{pmatrix} 2 & 0 & 2 \\ -2 & 1 & -1 \\ 2 & 2 & 0 \end{pmatrix} = \begin{pmatrix} 1 & 0 & 1 \\ 0 & 1 & -1 \\ 2 & 0 & 0 \end{pmatrix} \begin{pmatrix} 1 & 1 & 0 \\ -1 & 0 & 1 \\ 1 & -1 & 2 \end{pmatrix}.$$

Exercise: In Example 3.21 compute TS by calculating $T(Se_i)$ and also by evaluating the matrix product of T and S.

3.5 Bases of a vector space

Exercise: If V is a finite dimensional vector space, dim $V = n$, over the field $\mathbb{K} = \mathbb{R}$ or $\mathbb{C}$, show that the group $GL(V)$ of linear transformations of V is isomorphic to the matrix group of invertible $n \times n$ matrices, $GL(n, \mathbb{K})$.

Basis extension theorem

While specific bases should not be used in general definitions of vector space concepts if at all possible, there are specific instances when the singling out of a basis can prove of great benefit. The following theorem is often useful, in that it allows us to extend any l.i. set to a basis. In particular, it implies that if $v \in V$ is any non-zero vector, one can always find a basis such that $e_1 = v$.

Theorem 3.7 *Let $A = \{v_1, v_2, \ldots, v_m\}$ be any l.i. subset of V, where $m \le n = \dim V$. Then there exists a basis $E = \{e_1, e_2, \ldots, e_n\}$ of V such that $e_1 = v_1$, $e_2 = v_2, \ldots, e_m = v_m$.*

Proof: If $m = n$ then by Theorem 3.3 the set E is a basis of V and there is nothing to show. Assuming $m < n$, we set $e_1 = v_1$, $e_2 = v_2, \ldots, e_m = v_m$. By Corollary 3.4 the set A cannot span V since it consists of fewer than n elements, and there must exist a vector $e_{m+1} \in V$ that is not a linear combination of $e_1, \ldots, e_m$. The set $A' = \{e_1, e_2, \ldots, e_{m+1}\}$ is l.i., for if

$$a^1 e_1 + \cdots + a^m e_m + a^{m+1} e_{m+1} = 0,$$

then we must have $a^{m+1} = 0$, else e_{m+1} would be a linear combination of $e_1, \ldots, e_m$. The linear independence of $e_1, \ldots, e_m$ then implies that $a^1 = a^2 = \cdots = a^m = 0$. If $m + 1 < n$ continue adding vectors that are linearly independent of those going before, until we arrive at a set $E = A^{(n-m)}$, which is l.i. and has n elements. This set must be a basis and the process can be continued no further. ∎

The following examples illustrate how useful this theorem can be in applications.

Example 3.22 Let W be a k-dimensional vector subspace of a vector space V of dimension n. We will demonstrate that the dimension of the factor space V/W, known as the **codimension** of W, is $n - k$. By Theorem 3.7 it is possible to find a basis $\{e_1, e_2, \ldots, e_n\}$ of V such that the first k vectors $e_1, \ldots, e_k$ are a basis of W. Then $\{e_{k+1} + W, e_{k+2} + W, \ldots, e_n + W\}$ forms a basis for V/W since every coset $v + W$ can be written

$$v + W = v^{k+1}(e_{k+1} + W) + v^{k+2}(e_{k+2} + W) + \cdots + v^n(e_n + W)$$

where $v = \sum_{i=1}^{n} v^i e_i$ is the unique expansion given by Theorem 3.5. These cosets therefore span V/W. They are also l.i., for if

$$0 + W = a^{k+1}(e_{k+1} + W) + a^{k+2}(e_{k+2} + W) + \cdots + a^n(e_n + W)$$

then $a^{k+1} e_{k+1} + a^{k+2} e_{k+2} + \cdots + a^n e_n \in W$, which implies that there exist $b^1, \ldots, b^k$ such that

$$a^{k+1} e_{k+1} + a^{k+2} e_{k+2} + \cdots + a^n e_n = b^1 e_1 + \cdots + b^k e_k.$$

By the linear independence of $e_1, \ldots, e_n$ we have that $a^{k+1} = a^{k+2} = \cdots = a^n = 0$. The desired result now follows,

$$\operatorname{codim} W \equiv \dim(V/W) = n - k = \dim V - \dim W.$$

Example 3.23 Let $A : V \to V$ be a linear operator on a finite dimensional vector space V. Define its **rank** $\rho(A)$ to be the dimension of its image im A, and its **nullity** $\nu(A)$ to be the dimension of its kernel ker A,

$$\rho(A) = \dim \operatorname{im} A, \qquad \nu(A) = \dim \ker A.$$

By Theorem 3.7 there exists a basis $\{e_1, e_2, \ldots, e_n\}$ of V such that the first ν vectors $e_1, \ldots, e_\nu$ form a basis of ker A such that $Ae_1 = Ae_2 = \cdots = Ae_\nu = 0$. For any vector $u = \sum_{i=1}^n u^i e_i$

$$Au = \sum_{i=\nu+1}^n u^i Ae_i,$$

and im $A = L(\{Ae_{\nu+1}, \ldots, Ae_n\})$. Furthermore the vectors $Ae_{\nu+1}, \ldots, Ae_n$ are l.i., for if there were a non-trivial linear combination

$$\sum_{i=\nu+1}^n b^i Ae_i = A\left(\sum_{i=\nu+1}^n b^i e_i\right) = 0$$

then $\sum_{i=\nu+1}^n b^i e_i \in \ker A$, which is only possible if all $b^i = 0$. Hence $\dim \operatorname{im} A = n - \dim \ker A$, so that

$$\rho(A) = n - \nu(A) \quad \text{where} \quad n = \dim V. \tag{3.10}$$

Problems

Problem 3.7 Show that the vectors $(1, x)$ and $(1, y)$ in $\mathbb{R}^2$ are linearly dependent iff $x = y$. In $\mathbb{R}^3$, show that the vectors $(1, x, x^2)$, $(1, y, y^2)$ and $(1, z, z^2)$ are linearly dependent iff $x = y$ or $y = z$ or $x = z$.

Generalize these statements to $(n+1)$ dimensions.

Problem 3.8 Let V and W be any vector spaces, which are possibly infinite dimensional, and $T : V \to W$ a linear map. Show that if M is a l.i. subset of W, then $T^{-1}(M) = \{v \in V \mid Tv \in M\}$ is a linearly independent subset of V.

Problem 3.9 Let V and W be finite dimensional vector spaces of dimensions n and m respectively, and $T : V \to W$ a linear map. Given a basis $\{e_1, e_2, \ldots, e_n\}$ of V and a basis $\{f_1, f_2, \ldots, f_m\}$ of W, show that the equations

$$Te_k = \sum_{a=1}^m T_k^a f_a \quad (k = 1, 2, \ldots, n)$$

serve to uniquely define the $m \times n$ matrix of components $\mathsf{T} = [T_k^a]$ of the linear map T with respect to these bases.

3.6 Summation convention and transformation of bases

If $v = \sum_{k=1}^{n} v^k e_k$ is an arbitrary vector of V show that the components of its image vector $w = Tv$ are given by

$$w^a = (Tv)^a = \sum_{k=1}^{n} T^a_k v^k.$$

Write this as a matrix equation.

Problem 3.10 Let V be a four-dimensional vector space and $T : V \to V$ a linear operator whose effect on a basis $e_1, \ldots, e_4$ is

$$Te_1 = 2e_1 - e_4$$
$$Te_2 = -2e_1 + e_4$$
$$Te_3 = -2e_1 + e_4$$
$$Te_4 = e_1.$$

Find a basis for ker T and im T and calculate the rank and nullity of T.

3.6 Summation convention and transformation of bases

Summation convention

In the above formulae summation over an index such as i or j invariably occurs on a pair of equal indices that are oppositely placed in the superscript and subscript position. Of course it is not inconceivable to have a summation between indices on the same level but, as we shall see, it is unlikely to happen in a natural way. In fact, this phenomenon occurs with such regularity that it is possible to drop the summation sign $\sum_{i=1}^{n}$ whenever the same index i appears in opposing positions without running into any serious misunderstandings, a convention first proposed by Albert Einstein (1879–1955) in the theory of general relativity, where multiple summation signs of this type arise repeatedly in the use of tensor calculus (see Chapter 18). The principal rule of Einstein's **summation convention** is:

If, in any expression, a superscript is equal to a subscript then it will be assumed that these indices are summed over from 1 to n where n is the dimension of the space.

Repeated indices are called **dummy** or **bound**, while those appearing singly are called **free**. Free indices are assumed to take all values over their range, and we omit statements such as $(i, j = 1, 2, \ldots, n)$. For example, for any vector u and basis $\{e_i\}$ it is acceptable to write

$$u = u^i e_i \equiv \sum_{i=1}^{n} u^i e_i.$$

The index i is a dummy index, and can be replaced by any other letter having the same range,

$$u = u^i e_i = u^j e_j = u^k e_k.$$

Vector spaces

For example, writing out Eqs. (3.6), (3.7) and (3.9) in this convention,

$$T e_j = T^i{}_j e_i, \tag{3.11}$$

$$\tilde{v}^i = (Tv)^i = T^i{}_j v^j, \tag{3.12}$$

$$(ST)^k{}_i = S^k{}_j T^j{}_i. \tag{3.13}$$

In more complicated expressions such as

$$T^{ijk} S_{hij} \equiv \sum_{i=1}^{n} \sum_{j=1}^{n} T^{ijk} S_{hij} \quad (h, k = 1, \ldots, n)$$

i and j are dummy indices and h and k are free. It is possible to replace the dummy indices

$$T^{ilk} S_{hil} \quad \text{or} \quad T^{mik} S_{hmi}, \text{ etc.}$$

without in any way changing the meaning of the expression. In such replacements any letter of the alphabet other than one already used as a free index in that expression can be used, but you should always stay within a specified alphabet such as Roman, Greek, upper case Roman, etc., and sometimes even within a particular range of letters.

Indices should not be repeated on the same level, and in particular no index should ever appear more than twice in any expression. This would occur in $V^j T_{ij}$ if the dummy index j were replaced by the already occurring free index i to give the non-permissible $V^i T_{ii}$. Although expressions such as $V_j T_{ij}$ should not occur, there can be exceptions; for example, in cartesian tensors all indices occur in the subscript position and the summation convention is often modified to apply to expressions such as $V_j T_{ij} \equiv \sum_{j=1}^{n} V_j T_{ij}$.

In an equation relating indexed expressions, a given free index should appear in the same position, either as a superscript or subscript, on each expression of the equation. For example, the following are examples of equations that are *not* permissible unless there are mitigating explanations:

$$T^i = S_i, \qquad T^j + U_j F^{jk} = S^j, \qquad T^{kk}{}_k = S^k.$$

A free index in an equation can be changed to any symbol not already used as a dummy index in any part of the equation. However, the change must be made simultaneously in all expressions appearing in the equation. For example, the equation

$$Y_j = T^k_j X_k$$

can be replaced by

$$Y_i = T^j_i X_j$$

without changing its meaning, as both equations are a shorthand for the n equations

$$Y_1 = T^1_1 X_1 + T^2_1 X_2 + \cdots + T^n_1 X_n$$
$$\cdots = \cdots$$
$$Y_n = T^1_n X_1 + T^2_n X_2 + \cdots + T^n_n X_n.$$

3.6 Summation convention and transformation of bases

Among the most useful identities in the summation convention are those concerning the **Kronecker delta** δ^i_j defined by

$$\delta^i_j = \begin{cases} 1 & \text{if } i = j, \\ 0 & \text{if } i \neq j. \end{cases} \qquad (3.14)$$

These are the components of the unit matrix, $I = [\delta^i_j]$. This is the matrix of the identity operator id_V with respect to any basis $\{e_1, \ldots, e_n\}$. The Kronecker delta often acts as an 'index replacement operator'; for example,

$$T^{ij}_k \delta^m_i = T^{mj}_k,$$
$$T^{ij}_k \delta^k_l = T^{ij}_l,$$
$$T^{ij}_k \delta^k_j = T^{ij}_j = T^{ik}_k.$$

To understand these rules, consider the first equation. On the left-hand side the index i is a dummy index, signifying summation from $i = 1$ to $i = n$. Whenever $i \neq m$ in this sum we have no contribution since $\delta^m_i = 0$, while the contribution from $i = m$ results in the right-hand side. The remaining equations are proved similarly.

Care should be taken with the expression δ^i_i. If we momentarily suspend the summation convention, then obviously $\delta^i_i = 1$, but with the summation convention in operation the i is a dummy index, so that

$$\delta^i_i = \delta^1_1 + \delta^2_2 + \cdots + \delta^n_n = 1 + 1 + \cdots + 1.$$

Hence

$$\delta^i_i = n = \dim V. \qquad (3.15)$$

In future, the summation convention will always be assumed to apply unless a rider like 'summation convention suspended' is imposed for some reason.

Basis transformations

Consider a change of basis

$$E = \{e_1, e_2, \ldots, e_n\} \longrightarrow E' = \{e'_1, e'_2, \ldots, e'_n\}.$$

By Theorem 3.5 each of the original basis vectors e_i has a unique linear expansion in terms of the new basis,

$$e_i = A^j_{\ i} e'_j, \qquad (3.16)$$

where $A^j_{\ i}$ represents the jth component of the vector e_i with respect to the basis E'. Of course, the summation convention has now been adopted.

What happens to the components of a typical vector v under such a change of basis? Substituting Eq. (3.16) into the component expansion of v results in

$$v = v^i e_i = v^i A^j_{\ i} e'_j = v'^j e'_j,$$

where

$$v'^j = A^j{}_i v^i. \tag{3.17}$$

This law of transformation of components of a vector v is sometimes called the **contravariant transformation law of components**, a curious and somewhat old-fashioned terminology that possibly defies common sense. Equation (3.17) should be thought of as a 'passive' transformation, since only the *components* of the vector change, not the physical vector itself. On the other hand, a linear transformation $S : V \to V$ of the vector space can be thought of as moving actual vectors around, and for this reason is referred to as an 'active' transformation.

Nevertheless, it is still possible to think of (3.17) as a matrix equation if we represent the components v^i and v'^j of the vector v as *column vectors*

$$\mathbf{v} = \begin{pmatrix} v^1 \\ v^2 \\ \vdots \\ v^n \end{pmatrix}, \quad \mathbf{v}' = \begin{pmatrix} v'^1 \\ v'^2 \\ \vdots \\ v'^n \end{pmatrix},$$

and the transformation coefficients $A^j{}_i$ as an $n \times n$ matrix

$$\mathbf{A} = \begin{pmatrix} A^1{}_1 & A^1{}_2 & \cdots & A^1{}_n \\ A^2{}_1 & A^2{}_2 & \cdots & A^2{}_n \\ \cdots & \cdots & \cdots & \cdots \\ A^n{}_1 & A^n{}_2 & \cdots & A^n{}_n \end{pmatrix}.$$

Equation (3.17) can then be written as a matrix equation

$$\mathbf{v}' = \mathbf{A}\mathbf{v}. \tag{3.18}$$

Note, however, that $\mathbf{A} = [A^j{}_i]$ is a matrix of coefficients representing the old basis $\{e_i\}$ in terms of the new basis $\{e'_j\}$. It is not the matrix of components of a linear operator.

Example 3.24 Let V be a three-dimensional vector space with basis $\{e_1, e_2, e_3\}$. Vectors belonging to V can be set in correspondence with the 3×1 column vectors by

$$v = v^1 e_1 + v^2 e_2 + v^3 e_3 \longleftrightarrow \mathbf{v} = \begin{pmatrix} v^1 \\ v^2 \\ v^3 \end{pmatrix}.$$

Let $\{e'_i\}$ be a new basis defined by

$$e'_1 = e_1, \qquad e'_2 = e_1 + e_2 - e_3, \qquad e'_3 = e_1 + e_2 + e_3.$$

Solving for e_i in terms of the e'_j gives

$$e_1 = e'_1$$
$$e_2 = -e'_1 + \tfrac{1}{2}(e'_2 + e'_3)$$
$$e_3 = \tfrac{1}{2}(-e'_2 + e'_3),$$

3.6 Summation convention and transformation of bases

and the components of the matrix $\mathsf{A} = [A^j{}_i]$ can be read off using Eq. (3.16),

$$\mathsf{A} = \begin{pmatrix} 1 & -1 & 0 \\ 0 & \frac{1}{2} & -\frac{1}{2} \\ 0 & \frac{1}{2} & \frac{1}{2} \end{pmatrix}.$$

A general vector v is written in the e'_i basis as

$$\begin{aligned} v &= v^1 e_1 + v^2 e_2 + v^3 e_3 \\ &= v^1 e'_1 + v^2 \left(-e'_1 + \tfrac{1}{2}(e'_2 + e'_3)\right) + v^3 \tfrac{1}{2}(-e'_2 + e'_3) \\ &= (v^1 - v^2) e'_1 + \tfrac{1}{2}(v^2 - v^3) e'_2 + \tfrac{1}{2}(v^2 + v^3) e'_3 \\ &= v'^1 e'_1 + v'^2 e'_2 + v'^3 e'_3, \end{aligned}$$

where

$$\mathbf{v}' \equiv \begin{pmatrix} v'^1 \\ v'^2 \\ v'^3 \end{pmatrix} = \begin{pmatrix} v^1 - v^2 \\ \tfrac{1}{2}(v^2 - v^3) \\ \tfrac{1}{2}(v^2 + v^3) \end{pmatrix} = \mathsf{A}\mathbf{v}.$$

We will denote the inverse matrix to $\mathsf{A} = [A^i{}_j]$ by $\mathsf{A}' = [A'^i{}_k] = \mathsf{A}^{-1}$. Using the summation convention, the inverse matrix relations

$$\mathsf{A}'\mathsf{A} = \mathsf{I}, \qquad \mathsf{A}\mathsf{A}' = \mathsf{I}$$

may be written componentwise as

$$A'^k{}_j A^j{}_i = \delta^k_i, \qquad A^i{}_k A'^k{}_j = \delta^i_j. \tag{3.19}$$

From (3.16)

$$A'^i{}_k e_i = A'^i{}_k A^j{}_i e'_j = \delta^j_k e'_j = e'_k,$$

which can be rewritten as

$$e'_j = A'^k{}_j e_k. \tag{3.20}$$

Exercise: From Eq. (3.19) or Eq. (3.20) derive the inverse transformation law of vector components

$$v^i = A'^i{}_j v'^j. \tag{3.21}$$

We are now in a position to derive the transformation law of components of a linear operator $T : V \to V$. The matrix components of T with respect to the new basis, denoted $\mathsf{T}' = [T'^j{}_i]$, are given by

$$T e'_i = T'^j{}_i e'_j,$$

Vector spaces

and using Eqs. (3.16) and (3.20) we have

$$Te'_i = T(A'^k{}_i e_k)$$
$$= A'^k{}_i T e_k$$
$$= A'^k{}_i T^m{}_k e_m$$
$$= A'^k{}_i T^m{}_k A^j{}_m e'_j.$$

Hence

$$T'^j{}_i = A^j{}_m T^m{}_k A'^k{}_i, \qquad (3.22)$$

or in matrix notation, since $A' = A^{-1}$,

$$T' = ATA' = ATA^{-1}. \qquad (3.23)$$

Equation (3.23) is the *passive view* – it represents the change in *components* of an operator under a change of basis. With a different interpretation Eq. (3.23) could however be viewed as an operator equation. If we treat the basis $\{e_1, \ldots, e_n\}$ as fixed and regard A as being the matrix representing an operator whose effect on vector components is given by

$$v'^i = A^i{}_j v^j \iff \mathbf{v}' = A\mathbf{v},$$

then Eq. (3.23) represents a *change of operator*, called a **similarity transformation**. If $\mathbf{x}' = A\mathbf{x}$, $\mathbf{y}' = A\mathbf{y}$, then

$$\mathbf{y}' = AT\mathbf{x} = ATA^{-1}A\mathbf{x} = T'\mathbf{x}',$$

and $T' = ATA^{-1}$ is the operator that relates the transforms, under A, of any pair of vectors $\mathbf{x}$ and $\mathbf{y}$ that were originally related through the operator T. This is called the *active view* of Eq. (3.23). The two views are often confused in physics, mainly because operators are commonly identified with their matrices. The following example should help to clarify any lingering confusions.

Example 3.25 Consider a clockwise rotation of axes in $\mathbb{R}^2$ through an angle θ,

$$\begin{aligned} e'_1 &= \cos\theta\, e_1 - \sin\theta\, e_2 \\ e'_2 &= \sin\theta\, e_1 + \cos\theta\, e_2 \end{aligned} \iff \begin{aligned} e_1 &= \cos\theta\, e'_1 + \sin\theta\, e'_2 \\ e_2 &= -\sin\theta\, e'_1 + \cos\theta\, e'_2. \end{aligned}$$

The matrix of this basis transformation is

$$A = [A^i{}_j] = \begin{pmatrix} \cos\theta & -\sin\theta \\ \sin\theta & \cos\theta \end{pmatrix}$$

and the components of any position vector

$$\mathbf{x} = \begin{pmatrix} x \\ y \end{pmatrix}$$

3.6 Summation convention and transformation of bases

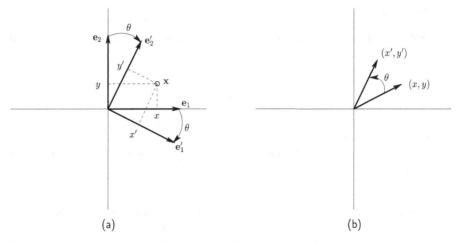

Figure 3.1 Active and passive views of a transformation

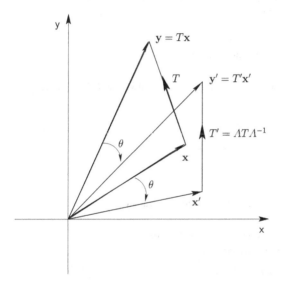

Figure 3.2 Active view of a similarity transformation

change by

$$\mathbf{x}' = \mathbf{A}\mathbf{x} \iff \begin{array}{l} x' = \cos\theta\, x - \sin\theta\, y \\ y' = \sin\theta\, x + \cos\theta\, y. \end{array}$$

This is the passive view. On the other hand, if we regard A as the matrix of components of an operator with respect to fixed axes e_1, e_2, then it represents a physical rotation of the space by an angle θ in a *counterclockwise* direction, opposite to the rotation of the axes in the passive view. Figure 3.1 demonstrates the apparent equivalence of these two views, while Fig. 3.2 illustrates the active view of a similarity transformation $T' = ATA^{-1}$ on a linear operator $T : \mathbb{R}^2 \to \mathbb{R}^2$.

Vector spaces

Problems

Problem 3.11 Let $\{e_1, e_2, e_3\}$ be a basis of a three-dimensional vector space V. Show that the vectors $\{e'_1, e'_2, e'_3\}$ defined by

$$e'_1 = e_1 + e_3$$
$$e'_2 = 2e_1 + e_2$$
$$e'_3 = 3e_2 + e_3$$

also form a basis of V.

What are the elements of the matrix $\mathsf{A} = [A^j{}_i]$ in Eq. (3.16)? Calculate the components of the vector

$$v = e_1 - e_2 + e_3$$

with respect to the basis $\{e'_1, e'_2, e'_3\}$, and verify the column vector transformation $\mathbf{v'} = \mathsf{A}\mathbf{v}$.

Problem 3.12 Let $T : V \to W$ be a linear map between vector spaces V and W. If $\{e_i \,|\, i = 1, \ldots, n\}$ is a basis of V and $\{f_a \,|\, a = 1, \ldots, m\}$ a basis of W, how does the matrix T, defined in Problem 3.9, transform under a transformation of bases

$$e_i = A^j{}_i e'_j, \quad f_a = B^b{}_a f'_b \,?$$

Express your answer both in component and in matrix notation.

Problem 3.13 Let e_1, e_2, e_3 be a basis for a three-dimensional vector space and e'_1, e'_2, e'_3 a second basis given by

$$e'_1 = e_3,$$
$$e'_2 = e_2 + 2e_3,$$
$$e'_3 = e_1 + 2e_2 + 3e_3.$$

(a) Express the e_i in terms of the e'_j, and write out the transformation matrices $\mathsf{A} = [A^i{}_j]$ and $\mathsf{A}' = \mathsf{A}^{-1} = [A'^i{}_j]$.
(b) If $u = e_1 + e_2 + e_3$, compute its components in the e'_i basis.
(c) Let T be the linear transformation defined by

$$Te_1 = e_2 + e_3, \quad Te_2 = e_3 + e_1, \quad Te_3 = e_1 + e_2.$$

What is the matrix of components of T with respect to the basis e_i?
(d) By evaluating Te'_1, etc. in terms of the e'_j, write out the matrix of components of T with respect to the e'_j basis and verify the similarity transformation $\mathsf{T}' = \mathsf{A}\mathsf{T}\mathsf{A}^{-1}$.

3.7 Dual spaces

Linear functionals

A **linear functional** φ on a vector space V over a field $\mathbb{K}$ is a linear map $\varphi : V \to \mathbb{K}$, where the field of scalars $\mathbb{K}$ is regarded as a one-dimensional vector space, spanned by the element 1,

$$\varphi(au + bv) = a\varphi(u) + b\varphi(v). \qquad (3.24)$$

3.7 Dual spaces

The use of the phrase *linear functional* in place of 'linear function' is largely adopted with infinite dimensional spaces in mind. For example, if V is the space of continuous functions on the interval $[0, 1]$ and $K(x)$ is an integrable function on $[0, 1]$, let $\varphi_K : V \to \mathbb{K}$ be defined by

$$\varphi_K(f) = \int_0^1 K(y) f(y) \, dy.$$

As the linear map φ_K is a function whose argument is another function, the terminology 'linear functional' seems more appropriate. In this case it is common to write the action of φ_K on the function f as $\varphi_K[f]$ in place of $\varphi_K(f)$.

Theorem 3.8 *If $\varphi : V \to \mathbb{K}$ is any linear functional then its kernel $\ker \varphi$ has codimension 1. Conversely, any subspace $W \subset V$ of codimension 1 defines a linear functional φ on V uniquely up to a scalar factor, such that $W = \ker \varphi$.*

Proof: The first part of the theorem follows from Example 3.22 and Eq. (3.3),

$$\text{codim}(\ker \varphi) = \dim(V/(\ker \varphi)) = \dim(\text{im}\,\varphi) = \dim \mathbb{K} = 1.$$

To prove the converse, let u be any vector not belonging to W – if no such vector exists then $W = V$ and the codimension is 0. The set of cosets $au + W = a(u + W)$ where $a \in \mathbb{K}$ form a one-dimensional vector space that must be identical with all of V/W. Every vector $v \in V$ therefore has a unique decomposition $v = au + w$ where $w \in W$, since

$$v = a'u + w' \implies (a - a')u = w - w' \implies a = a' \text{ and } w = w'.$$

A linear functional φ having kernel W has $\varphi(W) = 0$ and $\varphi(u) = c \neq 0$, for if $c = 0$ then the kernel of φ is all of V. Furthermore, given any non-zero scalar $c \in \mathbb{K}$, these two requirements define a linear functional on V as its value on any vector $v \in V$ is uniquely determined by

$$\varphi(v) = \varphi(au + w) = ac.$$

If φ' is any other such linear functional, having $c' = \varphi'(u) \neq 0$, then $\varphi' = (c'/c)\varphi$ since $\varphi'(v) = ac' = (c'/c)\varphi(v)$. ∎

This proof even applies, as it stands, to infinite dimensional spaces, although in that case it is usual to impose the added stipulation that linear functionals be continuous (see Section 10.9 and Chapter 13).

Exercise: If ω and ρ are linear functionals on V, show that

$$\ker \omega = \ker \rho \iff \omega = a\rho \quad \text{for some } a \in \mathbb{K}.$$

Example 3.26 Let $V = \mathbb{K}^n$, where n is any integer ≥ 2 or possibly $n = \infty$. For convenience, we will take V to be the space of row vectors of length n here. Let W be the subspace of vectors

$$W = \{(x_1, x_2, \dots) \in \mathbb{K}^n \mid x_1 + x_2 = 0\}.$$

Vector spaces

This is a subspace of codimension 1, for if we set $u = (1, 0, 0, \ldots)$, then any vector v can be written

$$v = (v_1, v_2, v_3, \ldots) = au + w$$

where $a = v_1 + v_2$ and $w = (-v_2, v_2, v_3, \ldots)$. This decomposition is unique, for if $au + w = a'u + w'$ then $(a - a')u = w' - w \in W$. Hence $a = a'$ and $w = w'$, since $u \notin W$. Every coset $v + W$ can therefore be uniquely expressed as $a(u + W)$, and W has codimension 1.

Let $\varphi : V \to \mathbb{K}$ be the linear functional such that $\varphi(W) = 0$ and $\varphi(u) = 1$. Then $\varphi(v) = \varphi(au + w) = a\varphi(u) + \varphi(w) = a$, so that

$$\varphi\big((x_1, x_2, x_3, \ldots)\big) = x_1 + x_2.$$

The kernel of φ is evidently W, and every other linear functional φ' having kernel W is of the form $\varphi\big((x_1, x_2, x_3, \ldots)\big) = c(x_1 + x_2)$.

The dual space of a vector space

As for general linear maps, it is possible to add linear functionals and multiply them by scalars,

$$(\varphi + \omega)(u) = \varphi(u) + \omega(u), \qquad (a\omega)(u) = a\omega(u).$$

With respect to these operations the set of linear functionals on V forms a vector space over $\mathbb{K}$ called the **dual space** of V, usually denoted V^*. In keeping with earlier conventions, other possible notations for this space are $L(V, \mathbb{K})$ or $\mathrm{Hom}(V, \mathbb{K})$. Frequently, linear functionals on V will be called **covectors**, and in later chapters we will have reason to refer to them as **1-forms**.

Let V be a finite dimensional vector space, $\dim V = n$, and $\{e_1, e_2, \ldots, e_n\}$ any basis for V. A linear functional ω on V is uniquely defined by assigning its values on the basis vectors,

$$w_1 = \omega(e_1), \quad w_2 = \omega(e_2), \quad \ldots, \quad w_n = \omega(e_n),$$

since the value on any vector $u = u^i e_i \equiv \sum_{i=1}^n u^i e_i$ can be determined using linearity (3.24),

$$\omega(u) = \omega(u^i e_i) = u^i \omega(e_i) = w_i u^i \equiv \sum_{i=1}^n w_i u^i. \tag{3.25}$$

Define n linear functionals $\varepsilon^1, \varepsilon^2, \ldots, \varepsilon^n$ by

$$\varepsilon^i(e_j) = \delta^i{}_j \tag{3.26}$$

where $\delta^i{}_j$ is the Kronecker delta defined in Eq. (3.14). Note that these equations uniquely define each linear functional ε^i, since their values are assigned on each basis vector e_j in turn.

Theorem 3.9 *The n-linear functionals $\{\varepsilon^1, \varepsilon^2, \ldots, \varepsilon^n\}$ form a basis of V^*, called the **dual basis** to $\{e_1, \ldots, e_n\}$. Hence $\dim V^* = n = \dim V$.*

3.7 Dual spaces

Proof: Firstly, suppose there are scalars $a_i \in \mathbb{K}$ such that

$$a_i \varepsilon^i = 0.$$

Applying the linear functional on the left-hand side of this equation to an arbitrary basis vector e_j,

$$0 = a_i \varepsilon^i(e_j) = a_i \delta^i_j = a_j,$$

shows that the linear functionals $\{\varepsilon^1, \ldots, \varepsilon^n\}$ are linearly independent. Furthermore these linear functions span V^* since every linear functional ω on V can be written

$$\omega = w_i \varepsilon^i \quad \text{where} \quad w_i = \omega(e_i). \tag{3.27}$$

This follows from

$$\begin{aligned}
w_i \varepsilon^i(u) &= w_i \varepsilon^i(u^j e_j) \\
&= w_i u^j \varepsilon^i(e_j) \\
&= w_i u^j \delta^i_j \\
&= w_i u^i \\
&= \omega(u) \quad \text{by Eq. (3.25)}.
\end{aligned}$$

Thus ω and $\omega' = w_i \varepsilon^i$ have the same effect on every vector $u = u^i e_i$; they are therefore identical linear functionals. The proposition that $\dim V^* = n$ follows from Corollary 3.4. ∎

Exercise: Show that the expansion (3.27) is unique; i.e., if $\omega = w'_i \varepsilon^i$ then $w'_i = w_i$ for each $i = 1, \ldots, n$.

Given a basis $E = \{e_i\}$, we will frequently refer to the n numbers $w_i = \omega(e_i)$ as the **components of the linear functional** ω in this basis. Alternatively, we can think of them as the components of ω with respect to the dual basis in V^*. The formula (3.25) has a somewhat deceptive 'dot product' feel about it. In Chapter 5 a dot product will be correctly defined as a product between vectors from the *same* vector space, while (3.25) is a product between vectors from different spaces V and V^*. It is, in fact, better to think of the components u^i of a vector u from V as forming a column vector, while the components w_j of a linear functional ω form a row vector. The above product then makes sense as a matrix product between a $1 \times n$ row matrix and an $n \times 1$ column matrix. While a vector is often thought of geometrically as a directed line segment, often represented by an arrow, this is not a good way to think of a covector. Perhaps the best way to visualize a linear functional is as a set of parallel planes of vectors determined by $\omega(v) = \text{const.}$ (see Fig. 3.3).

Dual of the dual

We may enquire whether this dualizing process can be continued to generate further vector spaces such as the dual of the dual space V^{**}, etc. For finite dimensional spaces the process essentially stops at the first dual, for there is a completely natural way in which V can be

Vector spaces

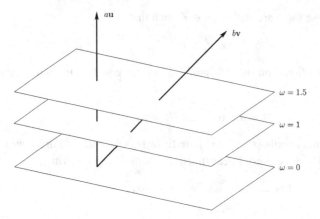

Figure 3.3 Geometrical picture of a linear functional

identified with V^{**}. To understand how V itself can be regarded as the dual space of V^*, define a linear map $\bar{v} : V^* \to \mathbb{K}$ corresponding to any vector $v \in V$ by

$$\bar{v}(\omega) = \omega(v) \quad \text{for all } \omega \in V^*. \tag{3.28}$$

The map $\bar{v}$ is a linear functional on V^*, since

$$\bar{v}(a\omega + b\rho) = (a\omega + b\rho)(v)$$
$$= a\omega(v) + b\rho(v)$$
$$= a\bar{v}(\omega) + b\bar{v}(\rho).$$

The map $\beta : V \to V^{**}$ defined by $\beta(v) = \bar{v}$ is linear, since

$$\beta(au + bv)(\varphi) = \overline{au + bv}(\varphi) \quad (\varphi \in V^*)$$
$$= \varphi(au + bv)$$
$$= a\varphi(u) + b\varphi(v)$$
$$= a\bar{u}(\varphi) + b\bar{v}(\varphi)$$
$$= a\beta(u)(\varphi) + b\beta(v)(\varphi).$$

As this holds for arbitrary covectors φ we have $\beta(au + bv) = a\beta(u) + b\beta(v)$. Furthermore if $e_1, e_2, \ldots, e_n$ is a basis of V with dual basis $\varepsilon^1, \varepsilon^2, \ldots, \varepsilon^n$, then

$$\bar{e}_i(\varepsilon^j) = \varepsilon^j(e_i) = \delta^j_i,$$

and it follows from Theorem 3.9 that $\{\beta(e_i) = \bar{e}_i\}$ is the basis of V^{**} dual to the basis $\{\varepsilon^j\}$ of V^*. The map $\beta : V \to V^{**}$ is therefore onto since every $f \in V^{**}$ can be written in the form

$$f = u^i \beta(e_i) = \beta(u) = \bar{u} \quad \text{where} \quad u = u^i e_i.$$

3.7 Dual spaces

Since $\{\beta(e_i)\}$ is a basis, it follows from Theorem 3.5 that the components u^i, and therefore the vector u, are uniquely determined by f. The map β is thus a vector space isomorphism, as it is both onto and one-to-one.

We have shown that $V \cong V^{**}$. In itself this is to be expected since these spaces have the same dimension, but the significant thing to note is that since the defining Eq. (3.28) makes no mention of any particular choice of basis, the correspondence between V and V^{**} is totally *natural*. There is therefore no ambiguity in *identifying* $\bar{v}$ with v, and rewriting (3.28) as

$$v(\omega) = \omega(v).$$

This reciprocity between the two spaces V and V^* lends itself to the following alternative notations, which will be used interchangeably throughout this book:

$$\langle \omega, v \rangle \equiv \langle v, \omega \rangle \equiv v(\omega) \equiv \omega(v). \tag{3.29}$$

However, it should be pointed out that the identification of V and V^{**} will only work for *finite* dimensional vector spaces. In infinite dimensional spaces, every vector may be regarded as a linear functional on V^* in a natural way, but the converse is not true – there exist linear functionals on V^* that do not correspond to vectors from V.

Transformation law of covector components

By Theorem 3.9 the spaces V and V^* are in one-to-one correspondence since they have the same dimension, but unlike that described above between V and V^{**} this correspondence is not *natural*. For example, if $v = v^i e_i$ is a vector in V let υ be the linear functional whose components in the dual basis are exactly the same as the components of the original vector, $\upsilon = v^i \varepsilon^i$. While the map $v \mapsto \upsilon$ is a vector space isomorphism, the same rule applied with respect to a different basis $\{e'_i\}$ will generally lead to a different correspondence between vectors and covectors. Thus, given an arbitrary vector $v \in V$ there is no *basis-independent* way of pointing to a covector partner in V^*. Essentially this arises from the fact that the law of transformation of components of a linear functional is different from the transformation law of components for a vector.

We have seen in Section 3.6 that the transformation of a basis can be written by Eqs. (3.16) and (3.20),

$$e_i = A^j{}_i e'_j, \qquad e'_j = A'^k{}_j e_k \tag{3.30}$$

where $[A^i{}_k]$ and $[A'^i{}_k]$ are related through the inverse matrix equations (3.19). Let $\{\varepsilon^i\}$ and $\{\varepsilon'^i\}$ be the dual bases corresponding to the bases $\{e_j\}$ and $\{e'_j\}$ respectively of V,

$$\varepsilon^i(e_j) = \delta^i{}_j, \qquad \varepsilon'^i(e'_j) = \delta^i{}_j. \tag{3.31}$$

Vector spaces

Set
$$\varepsilon^i = B^i{}_j \varepsilon'^j,$$

and substituting this and Eq. (3.30) into the first identity of (3.31) gives, after replacing the index j by k,

$$\begin{aligned}\delta^i_k &= B^i{}_j \varepsilon'^j(e_k) \\ &= B^i{}_j \varepsilon'^j(A^l{}_k e'_l) \\ &= B^i{}_j A^l{}_k \delta^j_l \\ &= B^i{}_j A^j{}_k.\end{aligned}$$

Hence $B^i{}_j = A'^i{}_j$ and the transformation of the dual basis is

$$\varepsilon^i = A'^i{}_j \varepsilon'^j. \tag{3.32}$$

Exercise: Show the transform inverse to (3.32)

$$\varepsilon'^j = A^j{}_k \varepsilon^k. \tag{3.33}$$

If $\omega = w_i \varepsilon^i$ is a linear functional having components w_i with respect to the first basis, then

$$\omega = w_i \varepsilon^i = w_i A'^i{}_j \varepsilon'^j = w'_i \varepsilon'^i$$

where

$$w'_i = A'^j{}_i w_j. \tag{3.34}$$

This is known as the **covariant vector transformation law of components**. Its inverse is

$$w_j = A^k{}_j w'_k. \tag{3.35}$$

These equations are to be compared with the contravariant vector transformation law of components of a vector $v = v^i e_i$, given by Eqs. (3.17) and (3.21),

$$v'^j = A^j{}_i v^i, \qquad v^i = A'^i{}_j v^j. \tag{3.36}$$

Exercise: Verify directly from (3.34) and (3.36) that Eq. (3.25) is basis-independent,

$$\omega(u) = w_i u^i = w'_j u'^j.$$

Exercise: Show that if the components of ω are displayed as a $1 \times n$ row matrix $\mathbf{w}^T = (w_1, w_2, \ldots, w_n)$ then the transformation law (3.34) can be written as a matrix equation

$$\mathbf{w}'^T = \mathbf{w}^T \mathsf{A}.$$

3.7 Dual spaces

Problems

Problem 3.14 Find the dual basis to the basis of $\mathbb{R}^3$ having column vector representation

$$e_1 = \begin{pmatrix} 1 \\ 1 \\ 1 \end{pmatrix}, \quad e_2 = \begin{pmatrix} 1 \\ 0 \\ -1 \end{pmatrix}, \quad e_3 = \begin{pmatrix} 0 \\ -1 \\ 1 \end{pmatrix}.$$

Problem 3.15 Let $\mathcal{P}(x)$ be the vector space of real polynomials $f(x) = a_0 + a_1 x + \cdots + a_n x^n$. If $(b^0, b^1, b^2, \ldots)$ is any sequence of real numbers, show that the map $\beta : \mathcal{P}(x) \to \mathbb{R}$ given by

$$\beta(f(x)) = \sum_{i=0}^{n} b^i a_i$$

is a linear functional on $\mathcal{P}(x)$.

Show that *every* linear functional β on $\mathcal{P}(x)$ can be obtained in this way from such a sequence and hence that $(\hat{\mathbb{R}}^\infty)^* = \mathbb{R}^\infty$.

Problem 3.16 Define the **annihilator** $S^\perp$ of a subset $S \subseteq V$ as the set of all linear functionals that vanish on S,

$$S^\perp = \{\omega \in V^* \mid \omega(u) = 0 \quad \forall u \in S\}.$$

(a) Show that for any subset S, $S^\perp$ is a vector subspace of V^*.
(b) If $T \subseteq S$, show that $S^\perp \subseteq T^\perp$.
(c) If U is a vector subspace of V, show that $(V/U)^* \cong U^\perp$. [*Hint*: For each ω in $U^\perp$ define the element $\bar\omega \in (V/U)^*$ by $\bar\omega(v + U) = \omega(v)$.]
(d) Show that $U^* \cong V^*/U^\perp$.
(e) If V is finite dimensional with $\dim V = n$ and W is any subspace of V with $\dim W = m$, show that $\dim W^\perp = n - m$. [*Hint*: Use a basis adapted to the subspace W by Theorem 3.7 and consider its dual basis in V^*.]
(f) Adopting the natural identification of V and V^{**}, show that $(W^\perp)^\perp = W$.

Problem 3.17 Let u be a vector in the vector space V of dimension n.

(a) If ω is a linear functional on V such that $a = \omega(u) \neq 0$, show that a basis $e_1, \ldots, e_n$ can be chosen such that

$$u = e_1 \quad \text{and} \quad \omega = a\varepsilon^1$$

where $\{\varepsilon^1, \ldots, \varepsilon^n\}$ is the dual basis. [*Hint*: Apply Theorem 3.7 to the vector u and try a further basis transformation of the form $e'_1 = e_1,\ e'_2 = e_2 + a_2 e_1, \ldots,\ e'_n = e_n + a_n e_1$.]
(b) If $a = 0$, show that the basis may be chosen such that

$$u = e_1 \quad \text{and} \quad \omega = \varepsilon^2.$$

Problem 3.18 For the three-dimensional basis transformation of Problem 3.13 evaluate the ε'^j dual to e'_i in terms of the dual basis ε^j. What are the components of the linear functional $\omega = \varepsilon^1 + \varepsilon^2 + \varepsilon^3$ with respect to the new dual basis?

Vector spaces

Problem 3.19 If $A : V \to V$ is a linear operator, define its **transpose** to be the linear map $A' : V^* \to V^*$ such that

$$A'\omega(u) = \omega(Au), \quad \forall u \in V, \, \omega \in V^*.$$

Show that this relation uniquely defines the linear operator A' and that

$$O' = O, \quad (\mathrm{id}_V)' = \mathrm{id}_{V^*}, \quad (aB + bA)' = aB' + bA', \quad \forall a, b \in \mathbb{K}.$$

(a) Show that $(BA)' = A'B'$.
(b) If A is an invertible operator then show that $(A')^{-1} = (A^{-1})'$.
(c) If V is finite dimensional show that $A'' = A$, if we make the natural identification of V^{**} and V.
(d) Show that the matrix of components of the transpose map A' with respect to the dual basis is the transpose of the matrix of A, $\mathsf{A}' = \mathsf{A}^T$.
(e) Using Problem 3.16 show that $\ker A' = (\mathrm{im}\, A)^\perp$.
(f) Use (3.10) to show that the rank of A' equals the rank of A.

Problem 3.20 The *row rank* of a matrix is defined as the maximum number of linearly independent rows, while its *column rank* is the maximum number of linearly independent columns.

(a) Show that the rank of a linear operator A on a finite dimensional vector space V is equal to the column rank of its matrix A with respect to any basis of V.
(b) Use parts (d) and (f) of Problem 3.19 to show that the row rank of a square matrix is equal to its column rank.

Problem 3.21 Let S be a linear operator on a vector space V.

(a) Show that the rank of S is one, $\rho(S) = 1$, if and only if there exists a non-zero vector u and a non-zero linear functional α such that

$$S(v) = u\alpha(v).$$

(b) With respect to any basis $\{e_i\}$ of V and its dual basis $\{\varepsilon^j\}$, show that

$$S^i{}_j = u^i a_j \quad \text{where} \quad u = u^i e_i, \quad \alpha = a_j \varepsilon^j.$$

(c) Show that every linear operator A of rank r can be written as a sum of r linear operators of rank one.
(d) Show that the last statement is equivalent to the assertion that for every matrix A of rank r there exist column vectors $\mathbf{u}_i$ and $\mathbf{a}_i$ $(i = 1, \ldots, r)$ such that

$$\mathsf{S} = \sum_{i=1}^{r} \mathbf{u}_i \mathbf{a}_i{}^T.$$

References

[1] G. Birkhoff and S. MacLane. *A Survey of Modern Algebra*. New York, MacMillan, 1953.
[2] N. B. Haaser and J. A. Sullivan. *Real Analysis*. New York, Van Nostrand Reinhold Company, 1971.
[3] S. Hassani. *Foundations of Mathematical Physics*. Boston, Allyn and Bacon, 1991.

References

[4] R. Geroch. *Mathematical Physics*. Chicago, The University of Chicago Press, 1985.

[5] S. Lang. *Algebra*. Reading, Mass., Addison-Wesley, 1965.

[6] L. H. Loomis and S. Sternberg. *Advanced Calculus*. Reading, Mass., Addison-Wesley, 1968.

[7] P. R. Halmos. *Finite-dimensional Vector Spaces*. New York, D. Van Nostrand Company, 1958.

4 Linear operators and matrices

Given a basis $\{e_1, e_2, \ldots, e_n\}$ of a finite dimensional vector space V, we recall from Section 3.5 that the **matrix of components** $\mathsf{T} = [T^i{}_j]$ of a linear operator $T : V \to V$ with respect to this basis is defined by Eq. (3.6) as:

$$T e_j = T^i{}_j e_i, \tag{4.1}$$

and under a transformation of basis,

$$e_i = A^j{}_i e'_j, \qquad e'_i = A'^k{}_i e_k, \tag{4.2}$$

where

$$A'^k{}_j A^j{}_i = \delta^k_i, \qquad A^i{}_k A'^k{}_j = \delta^i{}_j, \tag{4.3}$$

the components of any linear operator T transform by

$$T'^j{}_i = A^j{}_m T^m{}_k A'^k{}_i. \tag{4.4}$$

The matrices $\mathsf{A} = [A^j{}_i]$ and $\mathsf{A}' = [A'^k{}_i]$ are inverse to each other, $\mathsf{A}' = \mathsf{A}^{-1}$, and (4.4) can be written in matrix notation as a *similarity transformation*

$$\mathsf{T}' = \mathsf{A}\mathsf{T}\mathsf{A}^{-1}. \tag{4.5}$$

The main task of this chapter will be to find a basis that provides a standard representation of any given linear operator, called the *Jordan canonical form*. This representation is uniquely determined by the operator and encapsulates all its essential properties. The proof given in Section 4.2 is rather technical and may be skipped on first reading. It would, however, be worthwhile to understand its appearance, summarized at the end of that section, as it has frequent applications in mathematical physics. Good references for linear operators and matrices in general are [1–3], while a detailed discussion of the Jordan canonical form can be found in [4].

It is important to realize that we are dealing with linear operators on *free vector spaces*. This concept will be defined rigorously in Chapter 6, but essentially it means that the vector spaces have no further structure imposed on them. A number of concepts such as 'symmetric', 'hermitian' and 'unitary', which often appear in matrix theory, have no place in free vector spaces. For example, the requirement that T be a symmetric matrix would read $T^i{}_j = T^j{}_i$ in components, an awkward-looking relation that violates the rules given in Section 3.6. In Chapter 5 we will find a proper context for notions such as 'symmetric transformations' and 'hermitian transformations'.

4.1 Eigenspaces and characteristic equations

Invariant subspaces

A subspace U of V is said to be **invariant** under a linear operator $S : V \to V$ if

$$SU = \{Su \mid u \in U\} \subseteq U.$$

In this case, the action of S restricted to the subspace U, $S|_U$, gives rise to a linear operator on U.

Example 4.1 Let V be a three-dimensional vector space with basis $\{e_1, e_2, e_3\}$, and S the operator defined by

$$Se_1 = e_2 + e_3,$$
$$Se_2 = e_1 + e_3,$$
$$Se_3 = e_1 + e_2.$$

Let U be the subspace of all vectors of the form $(a+b)e_1 + be_2 + (-a+b)e_3$, where a and b are arbitrary scalars. This subspace is spanned by $f_1 = e_1 - e_3$ and $f_2 = e_1 + e_2 + e_3$ and is invariant under S, since

$$Sf_1 = -f_1, \; Sf_2 = 2f_2 \implies S(af_1 + bf_2) = -af_1 + 2bf_2 \in U.$$

Exercise: Show that if both U and W are invariant subspaces of V under an operator S then so is their intersection $U \cap W$ and their sum $U + W = \{v = u + w \mid u \in U, w \in W\}$.

Suppose $\dim U = m < n = \dim V$ and let $\{e_1, \ldots, e_m\}$ be a basis of U. By Theorem 3.7 this basis can be extended to a basis $\{e_1, \ldots, e_n\}$ spanning all of V. The invariance of U under S implies that the first m basis vectors are transformed among themselves,

$$Se_a = \sum_{b=1}^{m} S^b_a e_b \quad (a \leq m).$$

In such a basis, the components S^k_i of the operator S vanish for $i \leq m, k > m$, and the $n \times n$ matrix $\mathsf{S} = [S^k_i]$ has the *upper block diagonal* form

$$\mathsf{S} = \begin{pmatrix} \mathsf{S}_1 & \mathsf{S}_3 \\ \mathsf{O} & \mathsf{S}_2 \end{pmatrix}.$$

The submatrix S_1 is the $m \times m$ matrix of components of $S|_U$ expressed in the basis $\{e_1, \ldots, e_m\}$, while S_3 and S_2 are submatrices of orders $m \times p$ and $p \times p$, respectively, where $p = n - m$, and O is the zero $p \times m$ matrix.

If $V = U \oplus W$ is a decomposition with *both* U and W invariant under S, then choose a basis $\{e_1, \ldots, e_m, e_{m+1}, \ldots, e_n\}$ of V such that the first m vectors span U while the last $p = n - m$ vectors span W. Then $S^k_i = 0$ whenever $i > m$ and $k \leq m$, and the matrix of the operator S has *block diagonal* form

$$\mathsf{S} = \begin{pmatrix} \mathsf{S}_1 & \mathsf{O} \\ \mathsf{O} & \mathsf{S}_2 \end{pmatrix}.$$

Linear operators and matrices

Example 4.2 In Example 4.1 set $f_3 = e_3$. The vectors f_1, f_2, f_3 form a basis adapted to the invariant subspace spanned by f_1 and f_2,

$$Sf_1 = -f_1, \quad Sf_2 = 2f_2, \quad Sf_3 = f_2 - f_3,$$

and the matrix of S has the upper block diagonal form

$$S = \begin{pmatrix} -1 & 0 & 0 \\ 0 & 2 & 1 \\ 0 & 0 & -1 \end{pmatrix}.$$

On the other hand, the one-dimensional subspace W spanned by $f_3' = e_3 - \frac{1}{2}e_1 - \frac{1}{2}e_2$ is invariant since $Sf_3' = -f_3'$, and in the basis $\{f_1, f_2, f_3'\}$ adapted to the invariant decomposition $V = U \oplus W$ the matrix S takes on block diagonal form

$$S = \begin{pmatrix} -1 & 0 & 0 \\ 0 & 2 & 1 \\ 0 & 0 & -1 \end{pmatrix}.$$

Eigenvectors and eigenvalues

Given an operator $S: V \to V$, a scalar $\lambda \in \mathbb{K}$ is said to be an **eigenvalue** of S if there exists a non-zero vector v such that

$$Sv = \lambda v \quad (v \neq 0) \tag{4.6}$$

and v is called an **eigenvector** of S corresponding to the eigenvalue λ. Eigenvectors are those non-zero vectors that are 'stretched' by an amount λ on application of the operator S. It is important to stipulate $v \neq 0$ since the equation (4.6) *always* holds for the zero vector, $S0 = 0 = \lambda 0$.

For any scalar $\lambda \in \mathbb{K}$, let

$$V_\lambda = \{u \mid Su = \lambda u\}. \tag{4.7}$$

The set V_λ is a vector subspace, for

$$Su = \lambda u \text{ and } Sv = \lambda v \implies S(u + av) = Su + aSv$$
$$\implies S(u + av) = \lambda u + a\lambda v = \lambda(u + av) \quad \text{for all } a \in \mathbb{K}.$$

For every λ, the subspace V_λ is invariant under S,

$$u \in V_\lambda \implies Su = \lambda u \implies S(Su) = \lambda Su \implies Su \in V_\lambda.$$

V_λ consists of the set of all eigenvectors having eigenvalue λ, supplemented with the zero vector $\{0\}$. If λ is not an eigenvalue of S, then $V_\lambda = \{0\}$.

4.1 Eigenspaces and characteristic equations

If $\{e_1, e_2, \ldots, e_n\}$ is any basis of the vector space V, and $v = v^i e_i$ any vector of V, let **v** be the column vector of components

$$\mathbf{v} = \begin{pmatrix} v^1 \\ v^2 \\ \vdots \\ v^n \end{pmatrix}.$$

By Eq. (3.8) the matrix equivalent of (4.6) is

$$\mathbf{Sv} = \lambda \mathbf{v}, \tag{4.8}$$

where **S** is the matrix of components of S. Under a change of basis (4.2) we have from Eqs. (3.18) and (4.5),

$$\mathbf{v'} = \mathbf{Av}, \qquad \mathbf{S'} = \mathbf{ASA}^{-1}.$$

Hence, if **v** satisfies (4.8) then **v'** is an eigenvector of **S'** with the same eigenvalue λ,

$$\mathbf{S'v'} = \mathbf{ASA}^{-1}\mathbf{Av} = \mathbf{ASv} = \lambda \mathbf{Av} = \lambda \mathbf{v'}.$$

This result is not unexpected, since Eq. (4.8) and its primed version are simply representations with respect to different bases of the same basis-independent equation (4.6).

Define the *n*th power S^n of an operator inductively, by setting $S^0 = \text{id}_V$ and

$$S^n = S \circ S^{n-1} = SS^{n-1}.$$

Thus $S^1 = S$ and $S^2 = SS$, etc. If $p(x) = a_0 + a_1 x + a_2 x^2 + \cdots + a_n x^n$ is any polynomial with coefficients $a_i \in \mathbb{K}$, the operator polynomial $p(S)$ is defined in the obvious way,

$$p(S) = a_0 + a_1 S + a_2 S^2 + \cdots + a_n S^n.$$

If λ is an eigenvalue of S and v a corresponding eigenvector, then v is an eigenvector of any power S^n corresponding to eigenvalue λ^n. For $n = 0$,

$$S^0 v = \text{id}_V v = \lambda^0 v \quad \text{since} \quad \lambda^0 = 1,$$

and the proof follows by induction: assume $S^{n-1} v = \lambda^{n-1} v$ then by linearity

$$S^n v = S S^{n-1} v = S \lambda^{n-1} v = \lambda^{n-1} S v = \lambda^{n-1} \lambda v = \lambda^n v.$$

For a polynomial $p(x)$, it follows immediately that v is an eigenvector of the operator $p(S)$ with eigenvalue $p(\lambda)$,

$$p(S)v = p(\lambda)v. \tag{4.9}$$

Characteristic equation

The matrix equation (4.8) can be written in the form

$$(\mathbf{S} - \lambda \mathbf{I})\mathbf{v} = \mathbf{0}. \tag{4.10}$$

Linear operators and matrices

A necessary and sufficient condition for this equation to have a non-trivial solution $v \neq 0$ is

$$f(\lambda) = \det(S - \lambda I) = \begin{vmatrix} S_1^1 - \lambda & S_2^1 & \cdots & S_n^1 \\ S_1^2 & S_2^2 - \lambda & \cdots & S_n^2 \\ \vdots & \vdots & & \vdots \\ S_1^n & S_2^n & \cdots & S_n^n - \lambda \end{vmatrix} = 0, \qquad (4.11)$$

called the **characteristic equation** of S. The function $f(\lambda)$ is a polynomial of degree n in λ,

$$f(\lambda) = (-1)^n (\lambda^n - S_k^k \lambda^{n-1} + \cdots + (-1)^n \det S), \qquad (4.12)$$

known as the **characteristic polynomial** of S.

If the field of scalars is the complex numbers, $\mathbb{K} = \mathbb{C}$, then the fundamental theorem of algebra implies that there exist complex numbers $\lambda_1, \lambda_2, \ldots, \lambda_n$ such that

$$f(\lambda) = (-1)^n (\lambda - \lambda_1)(\lambda - \lambda_2) \ldots (\lambda - \lambda_n).$$

As some of these roots of the characteristic equation may appear repeatedly, we can write the characteristic polynomial in the form

$$f(z) = (-1)^n (z - \lambda_1)^{p_1} (z - \lambda_2)^{p_2} \ldots (z - \lambda_m)^{p_m}$$
$$\text{where} \quad p_1 + p_2 + \cdots + p_m = n. \qquad (4.13)$$

Since for each $\lambda = \lambda_i$ there exists a non-zero complex vector solution v to the linear set of equations given by (4.10), the eigenvalues of S must all come from the set of roots $\{\lambda_1, \ldots, \lambda_m\}$. The positive integer p_i is known as the **multiplicity** of the eigenvalue λ_i.

Example 4.3 When the field of scalars is the real numbers $\mathbb{R}$ there will not in general be real eigenvectors corresponding to complex roots of the characteristic equation. For example, let A be the operator on $\mathbb{R}^2$ defined by the following action on the standard basis vectors $e_1 = (1, 0)$ and $e_2 = (0, 1)$,

$$Ae_1 = e_2, \quad Ae_2 = -e_1 \implies A = \begin{pmatrix} 0 & -1 \\ 1 & 0 \end{pmatrix}.$$

The characteristic polynomial is

$$f(z) = \begin{vmatrix} -z & -1 \\ 1 & -z \end{vmatrix} = z^2 + 1,$$

whose roots are $z = \pm i$. The operator A thus has no real eigenvalues and eigenvectors. However, if we regard the field of scalars as being $\mathbb{C}$ and treat A as operating on $\mathbb{C}^2$, then it has complex eigenvectors

$$u = e_1 - ie_2, \quad Au = iu,$$
$$v = e_1 + ie_2, \quad Av = -iv.$$

4.1 Eigenspaces and characteristic equations

It is worth noting that, since $A^2e_1 = Ae_2 = -e_1$ and $A^2e_2 = -Ae_1 = -e_2$, the operator A satisfies its own characteristic equation

$$A^2 + \mathrm{id}_{\mathbb{R}^2} = 0 \implies \mathsf{A}^2 + \mathsf{I} = 0.$$

This is a simple example of the important *Cayley–Hamilton theorem* – see Theorem 4.3 below.

Example 4.4 Let V be a three-dimensional complex vector space with basis e_1, e_2, e_3, and $S : V \to V$ the operator whose matrix with respect to this basis is

$$\mathsf{S} = \begin{pmatrix} 1 & 1 & 0 \\ 0 & 1 & 0 \\ 0 & 0 & 2 \end{pmatrix}.$$

The characteristic polynomial is

$$f(z) = \begin{vmatrix} 1-z & 1 & 0 \\ 0 & 1-z & 0 \\ 0 & 0 & 2-z \end{vmatrix} = -(z-1)^2(z-2).$$

Hence the eigenvalues are 1 and 2, and it is trivial to check that the eigenvector corresponding to 2 is e_3. Let $u = xe_1 + ye_2 + ze_3$ be an eigenvector with eigenvalue 1,

$$\mathsf{S}\mathbf{u} = \mathbf{u} \quad \text{where} \quad \mathbf{u} = \begin{pmatrix} x \\ y \\ z \end{pmatrix},$$

then

$$x + y = x, \quad y = y, \quad 2z = z.$$

Hence $y = z = 0$ and $u = xe_1$. Thus, even though the eigenvalue $\lambda = 1$ has multiplicity 2, all corresponding eigenvectors are multiples of e_1.

Note that while e_2 is not an eigenvector, it is annihilated by $(S - \mathrm{id}_V)^2$, for

$$Se_2 = e_1 + e_2 \implies (S - \mathrm{id}_V)e_2 = e_1$$
$$\implies (S - \mathrm{id}_V)^2 e_2 = (S - \mathrm{id}_V)e_1 = 0.$$

Operators of the form $S - \lambda_i \mathrm{id}_V$ and their powers $(S - \lambda \mathrm{id}_V)^m$, where λ_i are eigenvalues of S, will make regular appearances in what follows. These operators evidently commute with each other and there is no ambiguity in writing them as $(S - \lambda_i)^m$.

Theorem 4.1 *Any set of eigenvectors corresponding to distinct eigenvalues of an operator S is linearly independent.*

Proof: Let $\{f_1, f_2, \ldots, f_k\}$ be a set of eigenvectors of S corresponding to eigenvalues $\lambda_1, \lambda_2, \ldots, \lambda_k$, no pair of which are equal,

$$Sf_i = \lambda_i f_i \quad (i = 1, \ldots, k),$$

and let $c_1, c_2, \ldots, c_k$ be scalars such that

$$c_1 f_1 + c_2 f_2 + \cdots + c_k f_k = 0.$$

Linear operators and matrices

If we apply the polynomial $P_1(S) = (S - \lambda_2)(S - \lambda_3)\ldots(S - \lambda_k)$ to this equation, then all terms except the first are annihilated, leaving

$$c_1 P_1(\lambda_1) f_1 = 0.$$

Hence

$$c_1(\lambda_1 - \lambda_2)(\lambda_1 - \lambda_3)\ldots(\lambda_1 - \lambda_k) f_1 = 0,$$

and since $f_1 \neq 0$ and all the factors $(\lambda_1 - \lambda_i) \neq 0$ for $i = 2, \ldots, k$, it follows that $c_1 = 0$. Similarly, $c_2 = \cdots = c_k = 0$, proving linear independence of $f_1, \ldots, f_k$. ∎

If the operator $S : V \to V$ has n distinct eigenvalues $\lambda_1, \ldots, \lambda_n$ where $n = \dim V$, then Theorem 4.1 shows the eigenvectors $f_1, f_2, \ldots, f_n$ are l.i. and form a basis of V. With respect to this basis the matrix of S is diagonal and its eigenvalues lie along the diagonal,

$$S = \begin{pmatrix} \lambda_1 & 0 & \cdots & 0 \\ 0 & \lambda_2 & \cdots & 0 \\ \vdots & & \ddots & \vdots \\ 0 & 0 & \cdots & \lambda_n \end{pmatrix}.$$

Conversely, any operator whose matrix is diagonalizable has a basis of eigenvectors (the eigenvalues need not be distinct for the converse). The more difficult task lies in the classification of those cases such as Example 4.4, where an eigenvalue λ has multiplicity $p > 1$ but there are less than p independent eigenvectors corresponding to it.

Minimal annihilating polynomial

The space of linear operators $L(V, V)$ is a vector space of dimension n^2 since it can be put into one-to-one correspondence with the space of $n \times n$ matrices. Hence the first n^2 powers $I \equiv \mathrm{id}_V = S^0$, $S = S^1$, $S^2, \ldots, S^{n^2}$ of any linear operator S on V cannot be linearly independent since there are $n^2 + 1$ operators in all. Thus S must satisfy a polynomial equation,

$$P(S) = c_0 I + c_1 S + c_2 S^2 + \cdots + c_{n^2} S^{n^2} = 0,$$

not all of whose coefficients $c_0, c_1, \ldots, c_n$ vanish.

Exercise: Show that the matrix equivalent of any such polynomial equation is basis-independent by showing that any similarity transform $S' = ASA^{-1}$ of S satisfies the same polynomial equation, $P(S') = 0$.

Let

$$\Delta(S) = S^k + c_1 S^{k-1} + \cdots + c_k I = 0$$

be the polynomial equation with leading coefficient 1 of *lowest* degree $k \leq n^2$, satisfied by S. The polynomial $\Delta(S)$ is unique, for if

$$\Delta'(S) = S^k + c'_1 S^{k-1} + \cdots + c'_k I = 0$$

4.1 Eigenspaces and characteristic equations

is another such polynomial equation, then on subtracting these two equations we have

$$(\Delta - \Delta')(S) = (c_1 - c'_1)S^{k-1} + (c_2 - c'_2)S^{k-2} + \cdots + (c_k - c'_k) = 0,$$

which is a polynomial equation of degree $< k$ satisfied by S. Hence $c_1 = c'_1, c_2 = c'_2, \ldots, c_k = c'_k$. The unique polynomial $\Delta(z) = z^k + c_1 z^{k-1} + \cdots + c_k$ is called the **minimal annihilating polynomial of S**.

Theorem 4.2 *A scalar λ is an eigenvalue of an operator S over a vector space V if and only if it is a root of the minimal annihilating polynomial $\Delta(z)$.*

Proof: If λ is an eigenvalue of S, let $u \neq 0$ be any corresponding eigenvector, $Su = \lambda u$. Since $0 = \Delta(S)u = \Delta(\lambda)u$ it follows that $\Delta(\lambda) = 0$.

Conversely, if λ is a root of $\Delta(z) = 0$ then there exists a polynomial $\Delta'(z)$ such that

$$\Delta(z) = (z - \lambda)\Delta'(z),$$

and since $\Delta'(z)$ has lower degree than $\Delta(z)$ it cannot annihilate S,

$$\Delta'(S) \neq 0.$$

Therefore, there exists a vector $u \in V$ such that $\Delta'(S)u = v \neq 0$, and

$$0 = \Delta(S)u = (S - \lambda)\Delta'(S)u = (S - \lambda)v.$$

Hence $Sv = \lambda v$, and λ is an eigenvalue of S with eigenvector v. ∎

It follows from this theorem that the minimal annihilating polynomial of an operator S on a complex vector space can be written in the form

$$\Delta(z) = (z - \lambda_1)^{k_1}(z - \lambda_2)^{k_2} \ldots (z - \lambda_m)^{k_m} \quad (k_1 + k_2 + \cdots + k_m = k) \tag{4.14}$$

where $\lambda_1, \lambda_2, \ldots, \lambda_m$ run through all the distinct eigenvalues of S. The various factors $(z - \lambda_i)^{k_i}$ are called the **elementary divisors** of S. The following theorem shows that the characteristic polynomial is always divisible by the minimal annihilating polynomial; that is, for each $i = 1, 2, \ldots, m$ the coefficient $k_i \leq p_i$ where p_i is the multiplicity of the ith eigenvalue.

Theorem 4.3 (Cayley–Hamilton) *Every linear operator S over a finite dimensional vector space V satisfies its own characteristic equation*

$$f(S) = (S - \lambda_1)^{p_1}(S - \lambda_2)^{p_2} \ldots (S - \lambda_m)^{p_m} = 0.$$

Equivalently, every $n \times n$ matrix S satisfies its own characteristic equation

$$f(\mathsf{S}) = (\mathsf{S} - \lambda_1 \mathsf{I})^{p_1}(\mathsf{S} - \lambda_2 \mathsf{I})^{p_2} \ldots (\mathsf{S} - \lambda_m \mathsf{I})^{p_m} = 0.$$

Proof: Let $e_1, e_2, \ldots, e_n$ be any basis of V, and let $\mathsf{S} = [S^k{}_j]$ be the matrix of components of S with respect to this basis,

$$Se_j = S^k{}_j e_k.$$

This equation can be written as

$$(S^k{}_j \mathrm{id}_V - \delta^k{}_j S)e_k = 0,$$

Linear operators and matrices

or alternatively as

$$T^k_j(S)e_k = 0, \qquad (4.15)$$

where

$$T^k_j(z) = S^k_j - \delta^k_j z.$$

Set $R(z) = [R^k_j(z)]$ to be the matrix of cofactors of $T(z) = [T^k_j(z)]$, such that

$$R^j_i(z)T^k_j(z) = \delta^k_i \det T(z) = \delta^k_i f(z).$$

The components $R^k_j(z)$ are polynomials of degree $\leq (n-1)$ in z, and multiplying both sides of Eq. (4.15) by $R^j_i(S)$ gives

$$R^j_i(S)T^k_j(S)e_k = \delta^k_i f(S)e_k = f(S)e_i = 0.$$

Since the e_i span V we have the desired result, $f(S) = 0$. The matrix version is simply the component version of this equation. ∎

Example 4.5 Let A be the matrix operator on the space of complex 4×1 column vectors given by

$$A = \begin{pmatrix} i & \alpha & 0 & 0 \\ 0 & i & 0 & 0 \\ 0 & 0 & i & 0 \\ 0 & 0 & 0 & -1 \end{pmatrix}.$$

Successive powers of A are

$$A^2 = \begin{pmatrix} -1 & 2i\alpha & 0 & 0 \\ 0 & -1 & 0 & 0 \\ 0 & 0 & -1 & 0 \\ 0 & 0 & 0 & 1 \end{pmatrix}, \quad A^3 = \begin{pmatrix} -i & -3\alpha & 0 & 0 \\ 0 & -i & 0 & 0 \\ 0 & 0 & -i & 0 \\ 0 & 0 & 0 & -1 \end{pmatrix}$$

and it is straightforward to verify that the matrices I, A, A^2 are linearly independent, while

$$A^3 = (-1 + 2i)A^2 + (1 + 2i)A + I.$$

Hence the minimal annihilating polynomial of A is

$$\Delta(z) = z^3 + (1 - 2i)z^2 - (1 + 2i)z - 1 = (z+1)(z-i)^2.$$

The elementary divisors of A are thus $z + 1$ and $(z - i)^2$ and the eigenvalues are -1 and i. Computation of the characteristic polynomial reveals that

$$f(z) = \det(A - zI) = (z+1)(z-i)^3,$$

which is divisible by $\Delta(z)$ in agreement with Theorem 4.3.

4.2 Jordan canonical form

Problem

Problem 4.1 The **trace** of an $n \times n$ matrix $\mathsf{T} = [T^i_j]$ is defined as the sum of its diagonal elements,

$$\operatorname{tr} \mathsf{T} = T^i_i = T^1_1 + T^2_2 + \cdots + T^n_n.$$

Show that

(a) $\operatorname{tr}(ST) = \operatorname{tr}(TS)$.
(b) $\operatorname{tr}(ATA^{-1}) = \operatorname{tr} T$.
(c) If $T : V \to V$ is any operator define its trace to be the trace of its matrix with respect to a basis $\{e_i\}$. Show that this definition is independent of the choice of basis, so that there is no ambiguity in writing $\operatorname{tr} T$.
(d) If $f(z) = a_0 + a_1 z + a_2 z^2 + \cdots + (-1)^n z^n$ is the characteristic polynomial of the operator T, show that $\operatorname{tr} T = (-1)^{n-1} a_{n-1}$.
(e) If T has eigenvalues $\lambda_1, \ldots, \lambda_m$ with multiplicities $p_1, \ldots, p_m$, show that

$$\operatorname{tr} T = \sum_{i=1}^m p_i \lambda_i.$$

4.2 Jordan canonical form

Block diagonal form

Let S be a linear operator over a complex vector space V, with characteristic polynomial and minimal annihilating polynomial given by (4.13) and (4.14), respectively. The restriction to complex vector spaces ensures that all roots of the polynomials $f(z)$ and $\Delta(z)$ are eigenvalues. A canonical form can also be derived for operators on real vector spaces, but it relies heavily on the complex version.

If $(z - \lambda_i)^{k_i}$ is the ith elementary divisor of S, define the subspace

$$V_i = \{u \mid (S - \lambda_i)^{k_i} u = 0\}.$$

This subspace is invariant under S, for

$$\begin{aligned} u \in V_i &\Longrightarrow (S - \lambda_i)^{k_i} u = 0 \\ &\Longrightarrow (S - \lambda_i)^{k_i} Su = S(S - \lambda_i)^{k_i} u = 0 \\ &\Longrightarrow Su \in V_i. \end{aligned}$$

Our first task will be to show that V is a direct sum of these invariant subspaces,

$$V = V_1 \oplus V_2 \oplus \ldots$$

Lemma 4.4 *If $(z - \lambda_i)^{k_i}$ is an elementary divisor of the operator S and $(S - \lambda_i)^{k_i + r} u = 0$ for some $r > 0$, then $(S - \lambda_i)^{k_i} u = 0$.*

Proof: There is clearly no loss of generality in setting $i = 1$ in the proof of this result. The proof proceeds by induction on r.

Case $r = 1$: Let u be any vector such that $(S - \lambda_1)^{k_1 + 1} u = 0$ and set

$$v = (S - \lambda_1)^{k_1} u.$$

Since $(S - \lambda_1)v = 0$, this vector satisfies the eigenvector equation, $Sv = \lambda_1 v$. If $\Delta(S)$ is the minimal annihilating polynomial of S, then

$$0 = \Delta(S)u = (S - \lambda_2)^{k_2} \ldots (S - \lambda_m)^{k_m} v$$
$$= (\lambda_1 - \lambda_2)^{k_2} \ldots (\lambda_1 - \lambda_m)^{k_m} v.$$

As all $\lambda_i \neq \lambda_j$ for $i \neq j$ it follows that $v = 0$, which proves the case $r = 1$.

Case $r > 1$: Suppose the lemma has been proved for $r - 1$. Then

$$(S - \lambda_1)^{k_1+r} u = 0 \implies (S - \lambda_1)^{k_1+r-1}(S - \lambda_1)u = 0$$
$$\implies (S - \lambda_1)^{k_1}(S - \lambda_1)u = 0 \quad \text{by induction hypothesis}$$
$$\implies (S - \lambda_1)^{k_1} u = 0 \quad \text{by the case } r = 1,$$

which concludes the proof of the lemma. ∎

If $\dim(V_1) = p$ let $h_1, \ldots, h_p$ be a basis of V_1 and extend to a basis of V using Theorem 3.7:

$$h_1, h_2, \ldots, h_p, h_{p+1}, \ldots, h_n. \tag{4.16}$$

Of course if $(z - \lambda_1)^{k_1}$ is the only elementary divisor of S then $p = n$, since $V_1 = V$ as every vector is annihilated by $(S - \lambda_1)^{k_1}$. If, however, $p < n$ we will show that the vectors

$$h_1, h_2, \ldots, h_p, \hat{h}_1, \hat{h}_2, \ldots, \hat{h}_{n-p} \tag{4.17}$$

form a basis of V, where

$$\hat{h}_a = (S - \lambda_1)^{k_1} h_{p+a} \quad (a = 1, \ldots, n-p). \tag{4.18}$$

Since the vectors listed in (4.17) are n in number it is only necessary to show that they are linearly independent. Suppose that for some constants $c^1, \ldots, c^p, \hat{c}^1, \ldots, \hat{c}^{n-p}$

$$\sum_{i=1}^{p} c^i h_i + \sum_{a=1}^{n-p} \hat{c}^a \hat{h}_a = 0. \tag{4.19}$$

Apply $(S - \lambda_1)^{k_1}$ to this equation. The first sum on the left is annihilated since it belongs to V_1, resulting in

$$(S - \lambda_1)^{k_1} \sum_{a=1}^{n-p} \hat{c}^a \hat{h}_a = (S - \lambda_1)^{2k_1} \sum_{a=1}^{n-p} \hat{c}^a h_{p+a} = 0,$$

and since $k_1 > 0$ we conclude from Lemma 4.4 that

$$(S - \lambda_1)^{k_1} \sum_{a=1}^{n-p} \hat{c}^a h_{p+a} = 0.$$

Hence $\sum_{a=1}^{n-p} \hat{c}^a h_{p+a} \in V_1$, and there exist constants $d^1, d^2, \ldots, d^p$ such that

$$\sum_{a=1}^{n-p} \hat{c}^a h_{p+a} = \sum_{i=1}^{p} d^i h_i.$$

4.2 Jordan canonical form

As the set $\{h_1, \ldots, h_n\}$ is by definition a basis of V, these constants must vanish: $\hat{c}^a = d^i = 0$ for all $a = 1, \ldots, n-p$ and $i = 1, \ldots, p$. Substituting into (4.19), it follows from the linear independence of $h_1, \ldots, h_p$ that $c^1, \ldots, c^p$ all vanish as well. This proves the linear independence of the vectors in (4.17).

Let $W_1 = L(\hat{h}_1, \hat{h}_2, \ldots, \hat{h}_{n-p})$. By Eq. (4.18) every vector $x \in W_1$ is of the form

$$x = (S - \lambda_1)^{k_1} y$$

since this is true of each of the vectors spanning W_1. Conversely, suppose $x = (S - \lambda_1)^{k_1} y$ and let $\{y^1, \ldots, y^n\}$ be the components of y with respect to the original basis (4.16); then

$$x = (S - \lambda_1)^{k_1} \left(\sum_i y^i h_i + \sum_a y^{p+a} h_{p+a} \right) = \sum_a y^{p+a} \hat{h}_a \in W_1.$$

Hence W_1 consists precisely of all vectors of the form $x = (S - \lambda_1)^{k_1} y$ where y is an arbitrary vector of V. Furthermore W_1 is an invariant subspace of V, for if $x \in W_1$ then

$$x = (S - \lambda_1)^{k_1} y \implies Sx = (S - \lambda_1)^{k_1} Sy \implies Sx \in W_1.$$

Hence W_1 and V_1 are complementary invariant subspaces, $V = V_1 \oplus W_1$, and the matrix of S with respect to the basis (4.17) has block diagonal form

$$S = \begin{pmatrix} S_1 & 0 \\ 0 & T_1 \end{pmatrix}$$

where S_1 is the matrix of $S_1 = S|_{V_1}$ and T_1 is the matrix of $T_1 = S|_{W_1}$.

Now on the subspace V_1 we have, by definition,

$$(S_1 - \lambda_1)^{k_1} = (S - \lambda_1)^{k_1} \Big|_{V_1} = 0.$$

Hence λ_1 is the only eigenvalue of S_1, for if $u \in V_1$ is an eigenvector of S_1 corresponding to an eigenvalue σ,

$$S_1 u = \sigma u,$$

then

$$(S_1 - \lambda_1)^{k_1} u = (\sigma - \lambda_1)^{k_1} u = 0,$$

from which it follows that $\sigma = \lambda_1$ since $u \neq 0$. The characteristic equation of S_1 is therefore

$$\det(S_1 - zI) = (-1)^p (\lambda_1 - z)^p. \tag{4.20}$$

Furthermore, the operator

$$(T_1 - \lambda_1)^{k_1} = (S - \lambda_1)^{k_1} \Big|_{W_1} : W_1 \to W_1$$

is invertible. For, let x be an arbitrary vector in W_1 and set $x = (S - \lambda_1)^{k_1} y$ where $y \in V$. Let $y = y_1 + y_2$ be the unique decomposition such that $y_1 \in V_1$ and $y_2 \in W_1$; then

$$x = (S - \lambda_1)^{k_1} y_2 = (T_1 - \lambda_1)^{k_1} y_2,$$

Linear operators and matrices

and since any surjective (onto) linear operator on a finite dimensional vector space is bijective (one-to-one), the map $(T_1 - \lambda_1)^{k_1}$ must be invertible on W_1. Hence

$$\det(\mathsf{T}_1 - \lambda_1 \mathsf{I}) \neq 0 \tag{4.21}$$

and λ_1 cannot be an eigenvalue of T_1.

The characteristic equation of S is

$$\det(\mathsf{S} - z\mathsf{I}) = \det(\mathsf{S}_1 - z\mathsf{I}) \det(\mathsf{T}_1 - z\mathsf{I})$$

and from (4.20) and (4.21) the only way the right-hand side can equal the expression in Eq. (4.13) is if

$$p_1 = p \quad \text{and} \quad \det(\mathsf{T}_1 - z\mathsf{I}) = (-1)^{n-p}(z - \lambda_2)^{p_2} \ldots (z - \lambda_m)^{p_m}.$$

Hence, the dimension p_i of each space V_i is equal to the multiplicity of the eigenvalue λ_i and from the Cayley–Hamilton Theorem 4.3 it follows that $p_i \geq k_i$.

Repeating this process on T_1, and proceeding inductively, it follows that

$$V = V_1 \oplus V_2 \oplus \cdots \oplus V_m,$$

and setting

$$h_{11}, h_{12}, \ldots, h_{1p_1}, h_{21}, \ldots, h_{2p_2}, \ldots, h_{m1}, \ldots, h_{mp_m}$$

to be a basis adapted to this decomposition, the matrix of S has block diagonal form

$$\mathsf{S} = \begin{pmatrix} \mathsf{S}_1 & 0 & \cdots & 0 \\ 0 & \mathsf{S}_2 & \cdots & 0 \\ \cdots & \cdots & \cdots & \cdots \\ 0 & 0 & \cdots & \mathsf{S}_m \end{pmatrix}. \tag{4.22}$$

The restricted operators $S_i = S|_{V_i}$ each have a minimal polynomial equation of the form

$$(S_i - \lambda_i)^{k_i} = 0,$$

so that

$$S_i = \lambda_i \mathrm{id}_i + N_i \quad \text{where} \quad N_i^{k_i} = 0. \tag{4.23}$$

Nilpotent operators

Any operator N satisfying an equation of the form $N^k = 0$ is called a **nilpotent operator**. The matrix N of any nilpotent operator satisfies $\mathsf{N}^k = 0$, and is called a **nilpotent matrix**. From Eq. (4.23), each $p_i \times p_i$ matrix S_i in the decomposition (4.22) is a multiple of the unit matrix plus a nilpotent matrix N_i,

$$\mathsf{S}_i = \lambda_i \mathsf{I} + \mathsf{N}_i \quad \text{where} \quad (\mathsf{N}_i)^{k_i} = 0.$$

We next find a basis that expresses the matrix of any nilpotent operator in a standard (canonical) form.

4.2 Jordan canonical form

Let U be a finite dimensional space, not necessarily complex, $\dim U = p$, and N a nilpotent operator on U. Set k to be the smallest positive integer such that $N^k = 0$. Evidently $k = 1$ if $N = 0$, while if $N \neq 0$ then $k > 1$ and $N^{k-1} \neq 0$. Define the subspaces X_i of U by

$$X_i = \{x \in U \mid N^i x = 0\} \quad (i = 0, 1, \ldots, k).$$

These subspaces form an increasing sequence,

$$\{0\} = X_0 \subset X_1 \subset X_2 \subset \cdots \subset X_k = U$$

and all are invariant under N, for if $u \in X_i$ then Nu also belongs to X_i since $N^i Nu = NN^i u = N0 = 0$. The set inclusions are strict inclusion in every case, for suppose that $X_i = X_{i+1}$ for some $i \leq k-1$. Then for any vector $x \in X_{i+2}$ we have

$$N^{i+2}x = N^{i+1}Nx = 0 \implies N^i Nx = 0 \implies x \in X_{i+1}.$$

Hence $X_{i+1} = X_{i+2}$, and continuing inductively we find that $X_i = \cdots = X_{k-1} = X_k = U$. This leads to the conclusion that $N^{k-1}x = 0$ for all $x \in U$, which contradicts the assumption that $N^{k-1} \neq 0$. Hence none of the subspaces X_i can be equal to each other.

We call a set of vectors $v_1, v_2, \ldots, v_s$ belonging to X_i linearly independent with respect to X_{i-1} if

$$a^1 v_1 + \cdots + a^s v_s \in X_{i-1} \implies a^1 = \cdots = a^s = 0.$$

Lemma 4.5 *Set $r_i = \dim X_i - \dim X_{i-1} > 0$ so that $p = r_1 + r_2 + \cdots + r_k$. Then r_i is the maximum number of vectors in X_i that can form a set that is linearly independent with respect to X_{i-1}.*

Proof: Let $\dim X_{i-1} = q$, $\dim X_i = q' > q$ and let $u_1, \ldots, u_q$ be a basis of X_{i-1}. Suppose $\{v_1, \ldots, v_r\}$ is a maximal set of vectors l.i. with respect to X_{i-1}; that is, a set that cannot be extended to a larger such set. Such a maximal set must exist since any set of vectors that is l.i. with respect to X_{i-1} is linearly independent and therefore cannot exceed q in number. We show that $S = \{u_1, \ldots, u_q, v_1, \ldots, v_r\}$ is a basis of X_i:
(a) S is a l.i. set since

$$\sum_{i=1}^{q} a^i u_i + \sum_{a=1}^{r} b^a v_a = 0$$

implies firstly that all b^a vanish by the requirement that the vectors v_a are l.i. with respect to X_{i-1}, and secondly all the $a^i = 0$ because the u_i are l.i.
(b) S spans X_i else there would exist a vector x that cannot be expressed as a linear combination of vectors of S, and $S \cup \{x\}$ would be linearly independent. In that case, the set of vectors $\{v_1, \ldots, v_r, x\}$ would be l.i. with respect to X_{i-1}, for if $\sum_{a=1}^{r} b^a v_a + bx \in X_{i-1}$ then from the linear independence of $S \cup \{x\}$ all $b^a = 0$ and $b = 0$. This contradicts the maximality of $v_1, \ldots, v_r$, and S must span the whole of X_i.
This proves the lemma. ∎

Linear operators and matrices

Let $\{h_1, \ldots, h_{r_k}\}$ be a maximal set of vectors in X_k that is l.i. with respect to X_{k-1}. From Lemma 4.5 we have $r_k = \dim X_k - \dim X_{k-1}$. The vectors

$$h'_1 = Nh_1, \ h'_2 = Nh_2, \ldots, \ h'_{r_k} = Nh_{r_k}$$

all belong to X_{k-1} and are l.i. with respect to X_{k-2}, for if

$$a^1 h'_1 + a^2 h'_2 + \cdots + a^{r_k} h'_{r_k} \in X_{k-2}$$

then

$$N^{k-1}(a^1 h_1 + a^2 h_2 + \cdots + a^{r_k} h_{r_k}) = 0,$$

from which it follows that

$$a^1 h_1 + a^2 h_2 + \cdots + a^{r_k} h_{r_k} \in X_{k-1}.$$

Since $\{h_1, \ldots, h_{r_k}\}$ are l.i. with respect to X_{k-1} we must have

$$a^1 = a^2 = \cdots = a^{r_k} = 0.$$

Hence $r_{k-1} \geq r_k$. Applying the same argument to all other X_i gives

$$r_k \leq r_{k-1} \leq \cdots \leq r_2 \leq r_1. \tag{4.24}$$

Now complete the set $\{h'_1, \ldots, h'_{r_k}\}$ to a maximal system of vectors in X_{k-1} that is l.i. with respect to X_{k-2},

$$h'_1, \ldots, h'_{r_k}, h'_{r_k+1}, \ldots, h'_{r_{k-1}}.$$

Similarly, define the vectors $h''_i = Nh'_i$ ($i = 1, \ldots, r_{k-1}$) and extend to a maximal system $\{h''_1, h''_2, \ldots, h''_{r_{k-2}}\}$ in X_{k-2}. Continuing in this way, form a series of $r_1 + r_2 + \cdots + r_k = p = \dim U$ vectors that are linearly independent and form a basis of U and may be displayed in the following scheme:

$$\begin{array}{ccccccc}
h_1 & \ldots & h_{r_k} \\
h'_1 & \ldots & h'_{r_k} & \ldots & h'_{r_{k-1}} \\
h''_1 & \ldots & h''_{r_k} & \ldots & h''_{r_{k-1}} & \ldots & h''_{r_{k-2}} \\
\ldots & \ldots & \ldots & \ldots & \ldots & \ldots & \ldots \\
h_1^{(k-1)} & \ldots & h_{r_k}^{(k-1)} & \ldots & h_{r_{k-1}}^{(k-1)} & \ldots & \ldots & \ldots & h_{r_1}^{(k-1)}.
\end{array}$$

Let U_a be the subspace generated by the ath column where $a = 1, \ldots, r_1$. These subspaces are all invariant under N, since $Nh_a^{(j)} = h_a^{(j+1)}$, and the bottom elements $h_a^{(k-1)} \in X_1$ are annihilated by N,

$$Nh_a^{(k-1)} \in X_0 = \{0\}.$$

Since the vectors $h_a^{(i)}$ are linearly independent and form a basis for U, the subspaces U_a are non-intersecting, and

$$U = U_1 \oplus U_2 \oplus \cdots \oplus U_{r_1},$$

4.2 Jordan canonical form

where the dimension $d(a) = \dim U_a$ of the ath subspace is given by height of the ath column. In particular, $d(1) = k$ and

$$\sum_{a=1}^{r_1} d(a) = p.$$

If a basis is chosen in U_a by proceeding up the ath column starting from the vector in the bottom row,

$$f_{a1} = h_a^{(k-1)}, \; f_{a2} = h_a^{(k-2)}, \ldots, f_{ad(a)} = h_a^{(k-d(a))},$$

then the matrix of $N_a = N\big|_{U_a}$ has all components zero except for 1's in the superdiagonal,

$$N_a = \begin{pmatrix} 0 & 1 & 0 & \cdots & \\ 0 & 0 & 1 & \cdots & \\ & & \ddots & & \\ 0 & 0 & \cdots & & 1 \\ 0 & 0 & \cdots & & 0 \end{pmatrix}. \tag{4.25}$$

Exercise: Check this matrix representation by remembering that the components of the matrix of an operator M with respect to a basis $\{u_i\}$ are given by

$$Mu_i = M^j{}_i u_j.$$

Now set $M = N_a$, $u_1 = f_{a1}$, $u_2 = f_{a2}, \ldots, u_{d(a)} = f_{ad(a)}$ and note that $Mu_1 = 0$, $Mu_2 = u_1$, etc.

Selecting a basis for U that runs through the subspaces $U_1, \ldots, U_{r_1}$ in order,

$$e_1 = f_{11}, \; e_2 = f_{12}, \ldots, e_k = f_{1k}, \; e_{k+1} = f_{21}, \ldots, e_p = f_{r_1 d(r_1)},$$

the matrix of N appears in block diagonal form

$$N = \begin{pmatrix} N_1 & & & \\ & N_2 & & \\ & & \ddots & \\ & & & N_{r_1} \end{pmatrix} \tag{4.26}$$

where each submatrix N_a has the form (4.25).

Jordan canonical form

Let V be a complex vector space and $S : V \to V$ a linear operator on V. To summarize the above conclusions: there exists a basis of V such that the operator S has matrix S in block diagonal form (4.22), and each S_i has the form $S_i = \lambda_i I + N_i$, where N_i is a nilpotent matrix. The basis can then be further specialized such that each nilpotent matrix is in turn decomposed into a block diagonal form (4.26) such that the submatrices along the diagonal all have the form (4.25). This is called the **Jordan canonical form** of the matrix S.

Linear operators and matrices

In other words, if S is an arbitrary $n \times n$ complex matrix, then there exists a non-singular complex matrix A such that ASA^{-1} is in Jordan form. The essential features of the matrix S can be summarized by the following **Segré characteristics**:

Eigenvalues	λ_1	...	λ_m
Multiplicities	p_1	...	p_m
	$(d_{11}\ldots d_{1r_1})$	...	$(d_{m1}\ldots d_{mr_m})$

where r_i is the number of eigenvectors corresponding to the eigenvalue λ_i and

$$\sum_{a=1}^{r_i} d_{ia} = p_i, \qquad \sum_{i=1}^{m} p_i = n = \dim V.$$

The Segré characteristics are determined entirely by properties of the operator S such as its eigenvalues and its elementary divisors. It is important, however, to realize that the Jordan canonical form only applies in the context of a complex vector space since it depends critically on the fundamental theorem of algebra. For a real matrix there is no guarantee that a *real* similarity transformation will convert it to the Jordan form.

Example 4.6 Let S be a transformation on a four-dimensional vector space having matrix with respect to a basis $\{e_1, e_2, e_3, e_4\}$ whose components are

$$S = \begin{pmatrix} 1 & 1 & -1 & 0 \\ 0 & 1 & 0 & -1 \\ 1 & 0 & 1 & 1 \\ 0 & 1 & 0 & 1 \end{pmatrix}.$$

The characteristic equation can be written in the form

$$\det(S - \lambda I) = ((\lambda - 1)^2 + 1)^2 = 0,$$

which has two roots $\lambda = 1 \pm i$, both of which are repeated roots. Each root corresponds to just a single eigenvector, written in column vector form as

$$\mathbf{f}_1 = \begin{pmatrix} 1 \\ 0 \\ -i \\ 0 \end{pmatrix} \quad \text{and} \quad \mathbf{f}_3 = \begin{pmatrix} 1 \\ 0 \\ i \\ 0 \end{pmatrix},$$

satisfying

$$S\mathbf{f}_1 = (1+i)\mathbf{f}_1 \quad \text{and} \quad S\mathbf{f}_3 = (1-i)\mathbf{f}_3.$$

Let $\mathbf{f}_2$ and $\mathbf{f}_4$ be the vectors

$$\mathbf{f}_2 = \begin{pmatrix} 0 \\ 1 \\ 0 \\ -i \end{pmatrix} \quad \text{and} \quad \mathbf{f}_4 = \begin{pmatrix} 0 \\ 1 \\ 0 \\ i \end{pmatrix},$$

4.2 Jordan canonical form

and we find that

$$Sf_2 = f_1 + (1+i)f_2 \quad \text{and} \quad Sf_4 = f_3 + (1-i)f_4.$$

Expressing these column vectors in terms of the original basis

$$f_1 = e_1 - ie_3, \quad f_2 = e_2 - ie_4, \quad f_3 = e_1 + ie_3, \quad f_4 = e_2 + ie_4$$

provides a new basis with respect to which the matrix of operator S has block diagonal Jordan form

$$S' = \begin{pmatrix} 1+i & 1 & 0 & 0 \\ 0 & 1+i & 0 & 0 \\ 0 & 0 & 1-i & 1 \\ 0 & 0 & 0 & 1-i \end{pmatrix}.$$

The matrix A needed to accomplish this form by the similarity transformation $S' = ASA^{-1}$ is found by solving for the e_i in terms of the f_j,

$$e_1 = \tfrac{1}{2}(f_1 + f_3), \quad e_2 = \tfrac{1}{2}(f_2 + f_4), \quad e_3 = \tfrac{1}{2}i(f_1 - f_3), \quad e_4 = \tfrac{1}{2}i(f_2 - f_4),$$

which can be written

$$e_j = A^i{}_j f_i \quad \text{where} \quad A = [A^i{}_j] = \tfrac{1}{2} \begin{pmatrix} 1 & 0 & i & 0 \\ 0 & 1 & 0 & i \\ 1 & 0 & -i & 0 \\ 0 & 1 & 0 & -i \end{pmatrix}.$$

The matrix S' is summarized by the Segré characteristics:

$1+i$	$1-i$
2	2
(2)	(2)

Exercise: Verify that $S' = ASA^{-1}$ in Example 4.6.

Problems

Problem 4.2 On a vector space V let S and T be two commuting operators, $ST = TS$.

(a) Show that if v is an eigenvector of T then so is Sv.
(b) Show that a basis for V can be found such that the matrices of both S and T with respect to this basis are in upper triangular form.

Problem 4.3 For the operator $T : V \to V$ on a four-dimensional vector space given in Problem 3.10, show that no basis exists such that the matrix of T is diagonal. Find a basis in which the matrix

Linear operators and matrices

of T has the Jordan form

$$\begin{pmatrix} 0 & 0 & 0 & 0 \\ 0 & 0 & 0 & 0 \\ 0 & 0 & \lambda & 1 \\ 0 & 0 & 0 & \lambda \end{pmatrix}$$

for some λ, and calculate the value of λ.

Problem 4.4 Let S be the matrix

$$S = \begin{pmatrix} i-1 & 1 & 0 & 0 \\ -1 & 1+i & 0 & 0 \\ -1-2i & 2i & -i & 1 \\ 2i-1 & 1 & 0 & -i \end{pmatrix}.$$

Find the minimal annihilating polynomial and the characteristic polynomial of this matrix, its eigenvalues and eigenvectors, and find a basis that reduces it to its Jordan canonical form.

4.3 Linear ordinary differential equations

While no techniques exist for solving general differential equations, systems of linear ordinary differential equations with constant coefficients are completely solvable with the help of the Jordan form. Such systems can be written in the form

$$\dot{\mathbf{x}} \equiv \frac{d\mathbf{x}}{dt} = \mathbf{A}\mathbf{x}, \tag{4.27}$$

where $\mathbf{x}(t)$ is an $n \times 1$ column vector and $\mathbf{A}$ an $n \times n$ matrix of real constants. Initially it is best to consider this as an equation in complex variables $\mathbf{x}$, even though we may only be seeking real solutions. Greater details of the following discussion, as well as applications to non-linear differential equations, can be found in [4, 5].

Try for a solution of (4.27) in exponential form

$$\mathbf{x}(t) = e^{\mathbf{A}t}\mathbf{x}_0, \tag{4.28}$$

where $\mathbf{x}_0$ is an arbitrary constant vector, and the exponential of a matrix is defined by the convergent series

$$e^{\mathbf{S}} = \mathbf{I} + \mathbf{S} + \frac{\mathbf{S}^2}{2!} + \frac{\mathbf{S}^3}{3!} + \ldots \tag{4.29}$$

If $\mathbf{S}$ and $\mathbf{T}$ are two commuting matrices, $\mathbf{ST} = \mathbf{TS}$, it then follows just as for real or complex scalars that $e^{\mathbf{S}\mathbf{T}} = e^{\mathbf{S}}e^{\mathbf{T}}$.

The initial value at $t = 0$ of the solution given by Eq. (4.28) is clearly $\mathbf{x}(0) = \mathbf{x}_0$. If $\mathbf{P}$ is any invertible $n \times n$ matrix, then $\mathbf{y} = \mathbf{P}\mathbf{x}$ satisfies the differential equation

$$\dot{\mathbf{y}} = \mathbf{A}'\mathbf{y} \quad \text{where} \quad \mathbf{A}' = \mathbf{P}\mathbf{A}\mathbf{P}^{-1},$$

and gives rise to the solution

$$\mathbf{y} = e^{\mathbf{A}'t}\mathbf{y}_0 \quad \text{where} \quad \mathbf{y}_0 = \mathbf{P}\mathbf{x}_0.$$

4.3 Linear ordinary differential equations

If P is chosen such that A' has the Jordan form

$$A' = \begin{pmatrix} \lambda_1 I + N_{11} & & \\ & \ddots & \\ & & \lambda_m I + N_{mr_m} \end{pmatrix}$$

where the N_{ij} are nilpotent matrices then, since $\lambda_i I$ commutes with N_{ij} for every i, j, the exponential term has the form

$$e^{A't} = \begin{pmatrix} e^{\lambda_1 t} e^{N_{11} t} & & \\ & \ddots & \\ & & e^{\lambda_m t} e^{N_{mr_m} t} \end{pmatrix}.$$

If N is a $k \times k$ Jordan matrix having 1's along the superdiagonal as in (4.25), then N^2 has 1's in the next diagonal out and each successive power of N pushes this diagonal of 1's one place further until N^k vanishes altogether,

$$N^2 = \begin{pmatrix} 0 & 0 & 1 & 0 & \ldots \\ 0 & 0 & 0 & 1 & \ldots \\ & & & \ddots & \\ 0 & 0 & 0 & \ldots & 0 \end{pmatrix}, \ldots, N^{k-1} = \begin{pmatrix} 0 & 0 & 0 & \ldots & 1 \\ 0 & 0 & 0 & \ldots & 0 \\ & & & \ddots & \\ & & & \ldots & 0 \end{pmatrix}, N^k = O.$$

Hence

$$e^{Nt} = \begin{pmatrix} 1 & t & \frac{t^2}{2} & \ldots & \frac{t^{k-1}}{(k-1)!} \\ 0 & 1 & t & \ldots & \\ & & & \ddots & \\ 0 & 0 & & \ldots & 1 \end{pmatrix},$$

and the solution (4.28) can be expressed as a linear supposition of solutions of the form

$$\mathbf{x}_r(t) = \mathbf{w}_r(t) e^{\lambda_i t} \tag{4.30}$$

where

$$\mathbf{w}_r(t) = \frac{t^{r-1}}{(r-1)!} \mathbf{h}_1 + \frac{t^{r-2}}{(r-2)!} \mathbf{h}_2 + \cdots + t \mathbf{h}_{r-1} + \mathbf{h}_r. \tag{4.31}$$

If A is a *real* matrix then the matrices P and A' are in general complex, but given real initial values $\mathbf{x}_0$ the solution having these values at $t = 0$ is

$$\mathbf{x} = \mathsf{P}^{-1} \mathbf{y} = \mathsf{P}^{-1} e^{A't} \mathsf{P} \mathbf{x}_0,$$

which must necessarily be real by the existence and uniqueness theorem of ordinary differential equations. Alternatively, for A real, both the real and imaginary parts of any complex solution $\mathbf{x}(t)$ are solutions of the linear differential equation (4.27), which may be separated by the identity

$$e^{\lambda t} = \cos \lambda t + i \sin \lambda t.$$

Linear operators and matrices

Two-dimensional autonomous systems

Consider the special case of a planar (two-dimensional) system (4.27) having constant coefficients, known as an **autonomous system**,

$$\dot{\mathbf{x}} = \mathbf{A}\mathbf{x} \quad \text{where} \quad \mathbf{A} = \begin{pmatrix} a_{11} & a_{12} \\ a_{21} & a_{22} \end{pmatrix}, \quad \mathbf{x} = \begin{pmatrix} x_1 \\ x_2 \end{pmatrix}.$$

Both the matrix $\mathbf{A}$ and vector $\mathbf{x}$ are assumed to be real. A **critical point** $\mathbf{x}_0$ refers to any constant solution $\mathbf{x} = \mathbf{x}_0$ of (4.27). The analysis of autonomous systems breaks up into a veritable zoo of cases and subcases. We consider the case where the matrix $\mathbf{A}$ is non-singular, for which the only critical point is $\mathbf{x}_0 = \mathbf{0}$. Both eigenvalues λ_1 and λ_2 are $\neq 0$, and the following possibilities arise.

(1) $\lambda_1 \neq \lambda_2$ and both eigenvalues are real. In this case the eigenvectors $\mathbf{h}_1$ and $\mathbf{h}_2$ form a basis of $\mathbb{R}^2$ and the general solution is

$$\mathbf{x} = c_1\, e^{\lambda_1 t}\mathbf{h}_1 + c_2\, e^{\lambda_2 t}\mathbf{h}_2.$$

(1a) If $\lambda_2 < \lambda_1 < 0$ the critical point is called a **stable node**.
(1b) If $\lambda_2 > \lambda_1 > 0$ the critical point is called an **unstable node**.
(1c) If $\lambda_2 < 0 < \lambda_1$ the critical point is called a **saddle point**.

These three cases are shown in Fig. 4.1, after the basis of the vector space axes has been transformed to lie along the vectors $\mathbf{h}_i$.

(2) $\lambda_1 = \lambda$, $\lambda_2 = \bar{\lambda}$ where λ is complex. The eigenvectors are then complex conjugate to each other since $\mathbf{A}$ is a real matrix,

$$\mathbf{A}\mathbf{h} = \lambda \mathbf{h} \quad \Rightarrow \quad \mathbf{A}\bar{\mathbf{h}} = \bar{\lambda}\bar{\mathbf{h}},$$

and the arbitrary real solution is

$$\mathbf{x} = c\, e^{\lambda t}\mathbf{h} + \bar{c}\, e^{\bar{\lambda} t}\bar{\mathbf{h}}.$$

If we set

$$\mathbf{h} = \tfrac{1}{2}(\mathbf{h}_1 - i\mathbf{h}_2), \quad \lambda = \mu + i\nu, \quad c = R e^{i\alpha}$$

where $\mathbf{h}_1, \mathbf{h}_2, \mu, \nu, R > 0$ and α are all real quantities, then the solution $\mathbf{x}$ has the form

$$\mathbf{x} = R\, e^{\mu t}\big(\cos(\nu t + \alpha)\mathbf{h}_1 + \sin(\nu t + \alpha)\mathbf{h}_2\big).$$

(2a) $\mu < 0$: This is a logarithmic spiral approaching the critical point $\mathbf{x} = \mathbf{0}$ as $t \to \infty$, and is called a **stable focus**.
(2b) $\mu > 0$: Again the solution is a logarithmic spiral but arising from the critical point $\mathbf{x} = \mathbf{0}$ as $t \to -\infty$, called an **unstable focus**.
(2c) $\mu = 0$: With respect to the basis $\mathbf{h}_1, \mathbf{h}_2$, the solution is a set of circles about the origin. When the original basis $\mathbf{e}_1 = \begin{pmatrix} 1 \\ 0 \end{pmatrix}$, $\mathbf{e}_2 = \begin{pmatrix} 0 \\ 1 \end{pmatrix}$ is used, the solutions are a set of ellipses and the critical point is called a **vortex point**.

These solutions are depicted in Fig. 4.2.

4.3 Linear ordinary differential equations

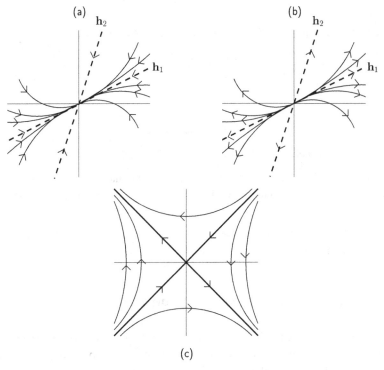

Figure 4.1 (a) Stable node, (b) unstable node, (c) saddle point

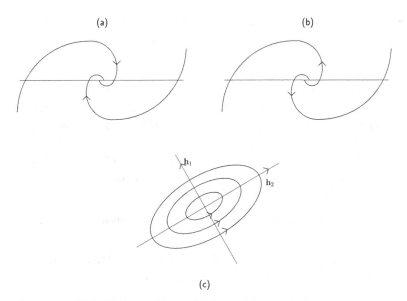

Figure 4.2 (a) Stable focus, (b) unstable focus, (c) vortex point

Linear operators and matrices

Problems

Problem 4.5 Verify that (4.30), (4.31) is a solution of $\dot{\mathbf{x}}_r = \mathbf{A}\mathbf{x}_r(t)$ provided

$$\mathbf{A}\mathbf{h}_1 = \lambda_i \mathbf{h}_1$$
$$\mathbf{A}\mathbf{h}_2 = \lambda_i \mathbf{h}_2 + \mathbf{h}_1$$
$$\vdots$$
$$\mathbf{A}\mathbf{h}_r = \lambda_i \mathbf{h}_r + \mathbf{h}_{r-1}$$

where λ_i is an eigenvalue of $\mathbf{A}$.

Problem 4.6 Discuss the remaining cases for two-dimensional autonomous systems: (a) $\lambda_1 = \lambda_2 = \lambda \neq 0$ and (i) two distinct eigenvectors $\mathbf{h}_1$ and $\mathbf{h}_2$, (ii) only one eigenvector $\mathbf{h}_1$; (b) $\mathbf{A}$ a singular matrix. Sketch the solutions in all instances.

Problem 4.7 Classify all three-dimensional autonomous systems of linear differential equations having constant coefficients.

4.4 Introduction to group representation theory

Groups appear most frequently in physics through their actions on vector spaces, known as *representations*. More specifically, a **representation** of any group G on a vector space V is a homomorphism T of G into the group of linear automorphisms of V,

$$T : G \to GL(V).$$

For every group element g we then have a corresponding linear transformation $T(g) : V \to V$ such that

$$T(g)T(h)v = T(gh)v \quad \text{for all } g, h \in G, \ v \in V.$$

Essentially, a representation of an abstract group is a way of providing a concrete model of the elements of the group as linear transformations of a vector space. The representation is said to be **faithful** if it is one-to-one; that is, if ker $T = \{e\}$. While in principal V could be either a real or complex vector space, we will mostly consider representations on complex vector spaces. If V is finite-dimensional its dimension n is called the **degree** of the representation. We will restrict attention almost entirely to representations of finite degree. Group representation theory is developed in much greater detail in [6–8].

Exercise: Show that any representation T induces a faithful representation of the factor group $G/\ker T$ on V.

Two representations $T_1 : G \to V_1$ and $T_2 : G \to V_2$ are said to be **equivalent**, written $T_1 \sim T_2$, if there exists a vector space isomorphism $A : V_1 \to V_2$ such that

$$T_2(g)A = AT_1(g), \quad \text{for all } g \in G. \tag{4.32}$$

If $V_1 = V_2 = V$ then $T_2(g) = AT_1(g)A^{-1}$. For finite dimensional representations the matrices representing T_2 are then derived from those representing T_1 by a similarity

4.4 Introduction to group representation theory

transformation. In this case the two representations can be thought of as essentially identical, since they are related simply by a change of basis.

Any operator A, even if it is singular, which satisfies Eq. (4.32) is called an **intertwining operator** for the two representations. This condition is frequently depicted by a commutative diagram

$$\begin{array}{ccc} V_1 & \xrightarrow{A} & V_2 \\ T_1(g) \downarrow & & \downarrow T_2(g) \\ V_1 & \xrightarrow{A} & V_2 \end{array}$$

Irreducible representations

A subspace W of V is said to be **invariant** under the action G, or **G-invariant**, if it is invariant under each linear transformation $T(g)$,

$$T(g)W \subseteq W \quad \text{for all } g \in G.$$

For every $g \in G$ the map $T(g)$ is surjective, $T(g)W = W$, since $w = T(g)(T(g^{-1}w))$ for every vector $w \in W$. Hence the restriction of $T(g)$ to W is an automorphism of W and provides another representation of G, called a **subrepresentation** of G, denoted $T_W : G \to GL(W)$. The whole space V and the trivial subspace $\{0\}$ are clearly G-invariant for any representation T on V. If these are the only invariant subspaces the representation is said to be **irreducible**.

If W is an invariant subspace of a representation T on V then a representation is induced on the quotient space V/W, defined by

$$T_{V/W}(g)(v + W) = T(g)v + W \quad \text{for all } g \in G, v \in V.$$

Exercise: Verify that this definition is independent of the choice of representative from the coset $v + W$, and that it is indeed a representation.

Let V be finite dimensional, $\dim V = n$, and $W' \cong V/W$ be a complementary subspace to W, such that $V = W \oplus W'$. From Theorem 3.7 and Example 3.22 there exists a basis whose first $r = \dim W$ vectors span W while the remaining $n - r$ span W'. The matrices of the representing transformations with respect to such a basis will have the form

$$\mathsf{T}(g) = \begin{pmatrix} \mathsf{T}_W(g) & \mathsf{S}(g) \\ \mathsf{O} & \mathsf{T}_{W'}(g) \end{pmatrix}.$$

The submatrices $\mathsf{T}_{W'}(g)$ form a representation on the subspace W' that is equivalent to the quotient space representation, but W' is not in general G-invariant because of the existence of the off-block diagonal matrices G. If $\mathsf{S}(g) \neq \mathsf{O}$ then it is essentially impossible to recover the original representation purely from the subrepresentations on W and W'. Matters are much improved, however, if the complementary subspace W' is G-invariant as well as W. In this case the representing matrices have the block diagonal form in a basis adapted to W

Linear operators and matrices

and W',

$$T(g) = \begin{pmatrix} T_W(g) & 0 \\ 0 & T_{W'}(g) \end{pmatrix}$$

and the representation is said to be **completely reducible**.

Example 4.7 Let the map $T : \mathbb{R} \to GL(\mathbb{R}^2)$ be defined by

$$T : a \mapsto T(a) = \begin{pmatrix} 1 & a \\ 0 & 1 \end{pmatrix}.$$

This is a representation since $T(a)T(b) = T(a+b)$. The subspace of vectors of the form $\begin{pmatrix} x \\ 0 \end{pmatrix}$ is invariant, but there is no complementary invariant subspace – for example, vectors of the form $\begin{pmatrix} 0 \\ y \end{pmatrix}$ are not invariant under the matrices $T(a)$. Equivalently, it follows from the Jordan canonical form that no matrix A exists such that $AT(1)A^{-1}$ is diagonal. The representation T is thus an example of a representation that is reducible but not completely reducible.

Example 4.8 The symmetric group of permutations on three objects, denoted S_3, has a representation T on a three-dimensional vector space V spanned by vectors e_1, e_2 and e_3, defined by

$$T(\pi)e_i = e_{\pi(i)}.$$

In this basis the matrix of the transformation $T(\pi)$ is $\mathsf{T} = [T_i^j(\pi)]$, where

$$T(\pi)e_i = T_i^j(\pi)e_j.$$

Using cyclic notation for permutations the elements of S_3 are $e = \text{id}$, $\pi_1 = (1\ 2\ 3)$, $\pi_2 = (1\ 3\ 2)$, $\pi_3 = (1\ 2)$, $\pi_4 = (2\ 3)$, $\pi_5 = (1\ 3)$. Then $T(e)e_i = e_i$, so that $T(e)$ is the identity matrix I, while $T(\pi_1)e_1 = e_2$, $T(\pi_1)e_2 = e_3$, $T(\pi_1)e_3 = e_1$, etc. The matrix representations of all permutations of S_3 are

$$T(e) = \begin{pmatrix} 1 & 0 & 0 \\ 0 & 1 & 0 \\ 0 & 0 & 1 \end{pmatrix}, \quad T(\pi_1) = \begin{pmatrix} 0 & 0 & 1 \\ 1 & 0 & 0 \\ 0 & 1 & 0 \end{pmatrix}, \quad T(\pi_2) = \begin{pmatrix} 0 & 1 & 0 \\ 0 & 0 & 1 \\ 1 & 0 & 0 \end{pmatrix},$$

$$T(\pi_3) = \begin{pmatrix} 0 & 1 & 0 \\ 1 & 0 & 0 \\ 0 & 0 & 1 \end{pmatrix}, \quad T(\pi_4) = \begin{pmatrix} 1 & 0 & 0 \\ 0 & 0 & 1 \\ 0 & 1 & 0 \end{pmatrix}, \quad T(\pi_5) = \begin{pmatrix} 0 & 0 & 1 \\ 0 & 1 & 0 \\ 1 & 0 & 0 \end{pmatrix}.$$

Let $v = v^i e_i$ be any vector, then $T(\pi)v = T_i^j(\pi)v^i e_j$ and the action of the matrix $T(\pi)$ is left multiplication on the column vector

$$\mathbf{v} = \begin{pmatrix} v^1 \\ v^2 \\ v^3 \end{pmatrix}.$$

4.4 Introduction to group representation theory

We now find the invariant subspaces of this representation. In the first place any one-dimensional invariant subspace must be spanned by a vector v that is an eigenvector of each operator $T(\pi_i)$. In matrices,

$$T(\pi_1)\mathbf{v} = \alpha\mathbf{v} \implies v^1 = \alpha v^2, \quad v^2 = \alpha v^3, \quad v^3 = \alpha v^1,$$

whence

$$v^1 = \alpha v^2 = \alpha^2 v^3 = \alpha^3 v^1.$$

Similarly $v^2 = \alpha^3 v^2$ and $v^3 = \alpha^3 v^3$, and since $\mathbf{v} \neq \mathbf{0}$ we must have that $\alpha^3 = 1$. Since $\alpha \neq 0$ it follows that all three components v^1, v^2 and v^3 are non-vanishing. A similar argument gives

$$T(\pi_3)v = \beta v \implies v^1 = \beta v^2, \quad v^2 = \beta v^1, \quad v^3 = \beta v^3,$$

from which $\beta^2 = 1$, and $\alpha\beta = 1$ since $v^1 = \alpha v^2 = \alpha\beta v^1$. The only pair of complex numbers α and β satisfying these relations is $\alpha = \beta = 1$. Hence $v^1 = v^2 = v^3$ and the only one-dimensional invariant subspace is that spanned by $v = e_1 + e_2 + e_3$.

We shall now show that this representation is completely reducible by choosing the basis

$$f_1 = e_1 + e_2 + e_3, \qquad f_2 = e_1 - e_2, \qquad f_3 = e_1 + e_2 - 2e_3.$$

The inverse transformation is

$$e_1 = \frac{1}{3}f_1 + \frac{1}{2}f_2 + \frac{1}{6}f_3, \qquad e_2 = \frac{1}{3}f_1 - \frac{1}{2}f_2 + \frac{1}{6}f_3, \qquad e_3 = \frac{1}{3}f_1 - \frac{1}{3}f_3,$$

and the matrices representing the elements of S_3 are found by calculating the effect of the various transformations on the basis elements f_i. For example,

$$T(e)f_1 = f_1, \qquad T(e)f_2 = f_2, \qquad T(e)f_3 = f_3,$$
$$T(\pi_1)f_1 = T(\pi_1)(e_1 + e_2 + e_3)$$
$$= e_2 + e_3 + e_1 = f_1,$$
$$T(\pi_1)f_2 = T(\pi_1)(e_1 - e_2)$$
$$= e_2 - e_3 = \frac{1}{2}f_2 + \frac{1}{2}f_3,$$
$$T(\pi_1)f_3 = T(\pi_1)(e_1 + e_2 - 2e_3)$$
$$= e_2 + e_3 - 2e_1 = -\frac{3}{2}f_2 - \frac{1}{2}f_3, \text{ etc.}$$

Continuing in this way for all $T(\pi_i)$ we arrive at the following matrices:

$$T(e) = \begin{pmatrix} 1 & 0 & 0 \\ 0 & 1 & 0 \\ 0 & 0 & 1 \end{pmatrix}, \quad T(\pi_1) = \begin{pmatrix} 1 & 0 & 0 \\ 0 & -\frac{1}{2} & -\frac{3}{2} \\ 0 & \frac{1}{2} & -\frac{1}{2} \end{pmatrix}, \quad T(\pi_2) = \begin{pmatrix} 1 & 0 & 0 \\ 0 & -\frac{1}{2} & \frac{3}{2} \\ 0 & -\frac{1}{2} & -\frac{1}{2} \end{pmatrix},$$

$$T(\pi_3) = \begin{pmatrix} 1 & 0 & 0 \\ 0 & -1 & 0 \\ 0 & 0 & 1 \end{pmatrix}, \quad T(\pi_4) = \begin{pmatrix} 1 & 0 & 0 \\ 0 & \frac{1}{2} & \frac{3}{2} \\ 0 & \frac{1}{2} & -\frac{1}{2} \end{pmatrix}, \quad T(\pi_5) = \begin{pmatrix} 1 & 0 & 0 \\ 0 & \frac{1}{2} & -\frac{3}{2} \\ 0 & -\frac{1}{2} & -\frac{1}{2} \end{pmatrix}.$$

The two-dimensional subspace spanned by f_2 and f_3 is thus invariant under the action of S_3 and the representation T is completely reducible.

Exercise: Show that the representation T restricted to the subspace spanned by f_2 and f_3 is irreducible, by showing that there is no invariant one-dimensional subspace spanned by f_2 and f_3.

Schur's lemma

The following key result and its corollary are useful in the classification of irreducible representations of groups.

Theorem 4.6 (Schur's lemma) *Let $T_1 : G \to GL(V_1)$ and $T_2 : G \to GL(V_2)$ be two irreducible representations of a group G, and $A : V_1 \to V_2$ an intertwining operator such that*

$$T_2(g)A = AT_1(g) \quad \text{for all} \quad g \in G.$$

Then either $A = 0$ or A is an isomorphism, in which case the two representations are equivalent, $T_1 \sim T_2$.

Proof: Let $v \in \ker A \subseteq V_1$. Then

$$AT_1(g)v = T_2(g)Av = 0$$

so that $T_1(g)v \in \ker A$. Hence $\ker A$ is an invariant subspace of the representation T_1. As T_1 is an irreducible representation we have that either $\ker A = V_1$, in which case $A = 0$, or $\ker A = \{0\}$. In the latter case A is one-to-one. To show it is an isomorphism it is only necessary to show that it is onto. This follows from the fact that $\operatorname{im} A \subset V_2$ is an invariant subspace of the representation T_2,

$$T_2(g)(\operatorname{im} A) = T_2(g)A(V_1) = AT_1(g)(V_1) \subseteq A(V_1) = \operatorname{im} A.$$

Since T_2 is an irreducible representation we have either $\operatorname{im} A = \{0\}$ or $\operatorname{im} A = V_2$. In the first case $A = 0$, while in the second A is onto. Schur's lemma is proved. ∎

Corollary 4.7 *Let $T : G \to GL(V)$ be a representation of a finite group G on a complex vector space V and $A : V \to V$ an operator that commutes with all $T(g)$; that is, $AT(g) = T(g)A$ for all $g \in G$. Then $A = \alpha \operatorname{id}_V$ for some complex scalar α.*

Proof: Set $V_1 = V_2 = V$ and $T_1 = T_2 = T$ in Schur's lemma. Since $AT(g) = T(g)A$ we have

$$(A - \alpha \operatorname{id}_V)T(g) = T(g)(A - \alpha \operatorname{id}_V)$$

since id_V commutes with all linear operators on V. By Theorem 4.6 either $A - \alpha \operatorname{id}_V$ is invertible or it is zero. Let α be an eigenvalue of A – for operators on a complex vector space this is always possible. The operator $A - \alpha \operatorname{id}_V$ is not invertible, for if it is applied to a corresponding eigenvector the result is the zero vector. Hence $A - \alpha \operatorname{id}_V = 0$, which is the desired result. ∎

It should be observed that the proof of this corollary only holds for complex representations since real matrices do not necessarily have any real eigenvalues.

Example 4.9 If G is a finite abelian group then all its irreducible representations are one-dimensional. This follows from Corollary 4.7, for if $T : G \to GL(V)$ is any representation of G then any $T(h)$ ($h \in G$) commutes with all $T(g)$ and is therefore a multiple of the identity,

$$T(h) = \alpha(h)\text{id}_V.$$

Hence any vector $v \in V$ is an eigenvector of $T(h)$ for all $h \in G$ and spans an invariant one-dimensional subspace of V. Thus, if $\dim V > 1$ the representation T cannot be irreducible.

References

[1] P. R. Halmos. *Finite-dimensional Vector Spaces*. New York, D. Van Nostrand Company, 1958.
[2] S. Hassani. *Foundations of Mathematical Physics*. Boston, Allyn and Bacon, 1991.
[3] F. P. Hildebrand. *Methods of Applied Mathematics*. Englewood Cliffs, N. J., Prentice-Hall, 1965.
[4] L. S. Pontryagin. *Ordinary Differential Equations*. New York, Addison-Wesley, 1962.
[5] D. A. Sánchez. *Ordinary Differential Equations and Stability Theory: An Introduction*. San Francisco, W. H. Freeman and Co., 1968.
[6] S. Lang. *Algebra*. Reading, Mass., Addison-Wesley, 1965.
[7] M. Hammermesh. *Group Theory and its Applications to Physical Problems*. Reading, Mass., Addison-Wesley, 1962.
[8] S. Sternberg. *Group Theory and Physics*. Cambridge, Cambridge University Press, 1994.

5 Inner product spaces

In matrix theory it is common to say that a matrix is *symmetric* if it is equal to its transpose, $S^T = S$. This concept does not however transfer meaningfully to the matrix of a linear operator on a vector space unless some extra structure is imposed on that space. For example, let $S : V \to V$ be an operator whose matrix $S = [S^i{}_j]$ is symmetric with respect to a specific basis. Under a change of basis $e_i = A^j{}_i e'_j$ the transformed matrix is $S' = ASA^{-1}$, while for the transpose matrix

$$S'^T = (ASA^{-1})^T = (A^{-1})^T SA^T.$$

Hence $S'^T \neq S'$ in general. We should hardly be surprised by this conclusion for, as commented at the beginning of Chapter 4, the component equation $S^i{}_j = S^j{}_i$ violates the index conventions of Section 3.6.

Exercise: Show that S' is symmetric if and only if S commutes with $A^T A$,

$$SA^T A = A^T AS.$$

Thus the concept of a 'symmetric operator' is not invariant under general basis transformations, but it is invariant with respect to orthogonal basis transformations, $A^T = A^{-1}$.

If V is a complex vector space it is similarly meaningless to talk of an operator $H : V \to V$ as being 'hermitian' if its matrix H with respect to some basis $\{e_i\}$ is *hermitian*, $H = H^\dagger$.

Exercise: Show that the hermitian property is not in general basis invariant, but is preserved under *unitary* transformations, $e_i = U^j{}_i e'_j$ where $U^{-1} = U^\dagger$.

In this chapter we shall see that symmetric and hermitian matrices play a different role in vector space theory, in that they represent *inner products* instead of operators [1–3]. Matrices representing inner products are best written with both indices on the subscript level, $G = G^T = [g_{ij}]$ and $H = H^\dagger = [h_{ij}]$. The requirements of symmetry $g_{ji} = g_{ij}$ and hermiticity $h_{ji} = \overline{h_{ij}}$ are not then at odds with the index conventions.

5.1 Real inner product spaces

Let V be a *real* finite dimensional vector space with dim $V = n$. A **real inner product**, often referred to simply as an **inner product** when there is no danger of confusion, on the

5.1 Real inner product spaces

vector space V is a map $V \times V \to \mathbb{R}$ that assigns a real number $u \cdot v \in \mathbb{R}$ to every pair of vectors $u, v \in V$ satisfying the following three conditions:

(RIP1) The map is symmetric in both arguments, $u \cdot v = v \cdot u$.
(RIP2) The distributive law holds, $u \cdot (av + bw) = au \cdot v + bu \cdot w$.
(RIP3) If $u \cdot v = 0$ for all $v \in V$ then $u = 0$.

A real vector space V together with an inner product defined on it is called a **real inner product space**. The inner product is also distributive on the first argument for, by conditions (RIP1) and (RIP2),

$$(au + bv) \cdot w = w \cdot (au + bv) = aw \cdot u + bw \cdot v = au \cdot w + bv \cdot w.$$

We often refer to this linearity in both arguments by saying that the inner product is **bilinear**.

As a consequence of property (RIP3) the inner product is said to be **non-singular** and is often referred to as **pseudo-Euclidean**. Sometimes (RIP3) is replaced by the stronger condition

(RIP3') $u \cdot u > 0$ for all vectors $u \neq 0$.

In this case the inner product is said to be **positive definite** or **Euclidean**, and a vector space V with such an inner product defined on it is called a **Euclidean vector space**. Condition (RIP3') implies condition (RIP3), for if there exists a non-zero vector u such that $u \cdot v = 0$ for all $v \in V$ then $u \cdot u = 0$ (on setting $v = u$), which violates (RIP3'). Positive definiteness is therefore a stronger requirement than non-singularity.

Example 5.1 The space of ordinary 3-vectors **a**, **b**, etc. is a Euclidean vector space, often denoted $\mathbb{E}^3$, with respect to the usual scalar product

$$\mathbf{a} \cdot \mathbf{b} = a_1 b_1 + a_2 b_2 + a_3 b_3 = |\mathbf{a}| \, |\mathbf{b}| \cos \theta$$

where $|\mathbf{a}|$ is the length or magnitude of the vector **a** and θ is the angle between **a** and **b**. Conditions (RIP1) and (RIP2) are simple to verify, while (RIP3') follows from

$$\mathbf{a} \cdot \mathbf{a} = \mathbf{a}^2 = (a_1)^2 + (a_2)^2 + (a_3)^2 > 0 \quad \text{if} \quad \mathbf{a} \neq 0.$$

This generalizes to a positive definite inner product on $\mathbb{R}^n$,

$$\mathbf{a} \cdot \mathbf{b} = a_1 b_1 + a_2 b_2 + \cdots + a_n b_n = \sum_{i=1}^{n} a_i b_i,$$

the resulting Euclidean vector space denoted by $\mathbb{E}^n$.

The **magnitude** of a vector w is defined as $w \cdot w$. Note that in a pseudo-Euclidean space the magnitude of a non-vanishing vector may be negative or zero, but in a Euclidean space it is always a positive quantity. The *length* of a vector in a Euclidean space is defined to be the square root of the magnitude.

Two vectors u and v are said to be **orthogonal** if $u \cdot v = 0$. By requirement (RIP3) there is no non-zero vector u that is orthogonal to every vector in V. A pseudo-Euclidean inner product may allow for the existence of self-orthogonal or **null vectors** $u \neq 0$ having zero magnitude $u \cdot u = 0$, but this possibility is clearly ruled out in a Euclidean vector space.

Inner product spaces

In Chapter 9 we shall see that Einstein's special theory of relativity postulates a pseudo-Euclidean structure for space-time known as *Minkowski space*, in which null vectors play a significant role.

Components of a real inner product

Given a basis $\{e_1, \ldots, e_n\}$ of an inner product space V, set

$$g_{ij} = e_i \cdot e_j = g_{ji}, \qquad (5.1)$$

called the **components of the inner product with respect to the basis** $\{e_i\}$. The inner product is completely specified by the components of the symmetric matrix, for if $u = u^i e_i$, $v = v^j e_j$ are any pair of vectors then, on using (RIP1) and (RIP2), we have

$$u \cdot v = g_{ij} u^i v^j. \qquad (5.2)$$

If we write the components of the inner product as a symmetric matrix

$$\mathbf{G} = [g_{ij}] = [e_i \cdot e_j] = [e_j \cdot e_i] = [g_{ji}] = \mathbf{G}^T,$$

and display the components of the vectors u and v in column form as $\mathbf{u} = [u^i]$ and $\mathbf{v} = [v^j]$, then the inner product can be written in matrix notation,

$$u \cdot v = \mathbf{u}^T \mathbf{G} \mathbf{v}.$$

Theorem 5.1 *The matrix* $\mathbf{G}$ *is non-singular if and only if condition (RIP3) holds.*

Proof: To prove the *if* part, assume that $\mathbf{G}$ is singular, $\det[g_{ij}] = 0$. Then there exists a non-trivial solution u^j to the linear system of equations

$$g_{ij} u^j \equiv \sum_{j=1}^{n} g_{ij} u^j = 0.$$

The vector $u = u^j e_j$ is non-zero and orthogonal to all $v = v^i e_i$,

$$u \cdot v = g(u, v) = g_{ij} u^i v^j = 0,$$

in contradiction to (RIP3).

Conversely, assume the matrix $\mathbf{G}$ is non-singular and that there exists a vector u violating (RIP3); $u \neq 0$ and $u \cdot v = 0$ for all $v \in V$. Then, by Eq. (5.2), we have

$$u_j v^j = 0 \quad \text{where} \quad u_j = g_{ij} u^i = g_{ji} u^i$$

for arbitrary values of v^j. Hence $u_j = 0$ for $j = 1, \ldots, n$. However, this implies a non-trivial solution to the set of linear equations $g_{ji} u^i = 0$, which is contrary to the non-singularity assumption, $\det[g_{ij}] \neq 0$. ∎

Orthonormal bases

Under a change of basis

$$e_i = A^j{}_i e'_j, \qquad e'_j = A'^k{}_j e_k, \qquad (5.3)$$

5.1 Real inner product spaces

the components g_{ij} transform by

$$g_{ij} = e_i \cdot e_j = (A^k{}_i e'_k) \cdot (A^l{}_j e'_l) \qquad (5.4)$$
$$= A^k{}_i g'_{kl} A^l{}_j,$$

where $g'_{kl} = e'_k \cdot e'_l$. In matrix notation this equation reads

$$\mathsf{G} = \mathsf{A}^T \mathsf{G}' \mathsf{A}. \qquad (5.5)$$

Using $\mathsf{A}' = [A'^j{}_k] = \mathsf{A}^{-1}$, the transformed matrix G' can be written

$$\mathsf{G}' = \mathsf{A}'^T \mathsf{G} \mathsf{A}'. \qquad (5.6)$$

An **orthonormal basis** $\{e_1, e_2, \ldots, e_n\}$, for brevity written 'o.n. basis', consists of vectors all of magnitude ± 1 and orthogonal to each other in pairs,

$$g_{ij} = e_i \cdot e_j = \eta_i \delta_{ij} \quad \text{where} \quad \eta_i = \pm 1, \qquad (5.7)$$

where the summation convention is temporarily suspended. We occasionally do this when a relation is referred to a specific class of bases.

Theorem 5.2 *In any finite dimensional real inner product space $(V, \cdot)$, with $\dim V = n$, there exists an orthonormal basis $\{e_1, e_2, \ldots, e_n\}$ satisfying Eq. (5.7).*

Proof: The method is by a procedure called **Gram–Schmidt orthonormalization**, an algorithmic process for constructing an o.n. basis starting from any arbitrary basis $\{u_1, u_2, \ldots, u_n\}$. For Euclidean inner products the procedure is relatively straightforward, but the possibility of vectors having zero magnitudes in general pseudo-Euclidean spaces makes for added complications.

Begin by choosing a vector u such that $u \cdot u \neq 0$. This is always possible because if $u \cdot u = 0$ for all $u \in V$, then for any pair of vectors u, v

$$0 = (u + v) \cdot (u + v) = u \cdot u + 2u \cdot v + v \cdot v = 2u \cdot v,$$

which contradicts the non-singularity condition (RIP3). For the first step of the Gram–Schmidt procedure we normalize this vector,

$$e_1 = \frac{u}{\sqrt{|u \cdot u|}} \quad \text{and} \quad \eta_1 = e_1 \cdot e_1 = \pm 1.$$

In the Euclidean case any non-zero vector u will do for this first step, and $e_1 \cdot e_1 = 1$.

Let V_1 be the subspace of V consisting of vectors orthogonal to e_1,

$$V_1 = \{w \in V \mid w \cdot e_1 = 0\}.$$

This is a vector subspace, for if w and w' are orthogonal to e_1 then so is any linear combination of the form $w + aw'$,

$$(w + aw') \cdot e_1 = w \cdot e_1 + aw' \cdot e_1 = 0.$$

For any $v \in V$, the vector $v' = v - ae_1 \in V_1$ where $a = \eta_1(v \cdot e_1)$, since $v' \cdot e_1 = v \cdot e_1 - (\eta_1)^2 v \cdot e_1 = 0$. Furthermore, the decomposition $v = a\,e_1 + v'$ into a component parallel

Inner product spaces

to e_1 and a vector orthogonal to e_1 is unique, for if $v = a' e_1 + v''$ where $v'' \in V_1$ then

$$(a' - a)e_1 = v'' - v'.$$

Taking the inner product of both sides with e_1 gives firstly $a' = a$, and consequently $v'' = v'$.

The inner product restricted to V_1, as a map $V_1 \times V_1 \to \mathbb{R}$, is an inner product on the vector subspace V_1. Conditions (RIP1) and (RIP2) are trivially satisfied if the vectors u, v and w are restricted to vectors belonging to V_1. To show (RIP3), that this inner product is non-singular, let $v' \in V_1$ be a vector such that $v' \cdot w' = 0$ for all $w' \in V_1$. Then v' is orthogonal to every vector in $w \in V$ for, by the decomposition

$$w = \eta_1(w \cdot e_1)e_1 + w',$$

we have $v' \cdot w = 0$. By condition (RIP3) for the inner product on V this implies $v' = 0$, as required.

Repeating the above argument, there exists a vector $u' \in V_1$ such that $u' \cdot u' \neq 0$. Set

$$e_2 = \frac{u'}{\sqrt{|u' \cdot u'|}}$$

and $\eta_2 = e_2 \cdot e_2 = \pm 1$. Clearly $e_2 \cdot e_1 = 0$ since $e_2 \in V_1$. Defining the subspace V_2 of vectors orthogonal to e_1 and e_2, the above argument can be used again to show that the restriction of the inner product to V_2 satisfies (RIP1)–(RIP3). Continue this procedure until n orthonormal vectors $\{e_1, e_2, \ldots, e_n\}$ have been produced. These vectors must be linearly independent, for if there were a vanishing linear combination $a^i e_i = 0$, then performing the inner product of this equation with any e_j gives $a^j = 0$. By Theorem 3.3 these vectors form a basis of V. At this stage of the orthonormalization process $V_n = \{0\}$, as there can be no vector that is orthogonal to every $e_1, \ldots, e_n$, and the procedure comes to an end. ∎

The following theorem shows that for a fixed inner product space, apart from the order in which they appear, the coefficients η_i are the same in all orthonormal frames.

Theorem 5.3 (Sylvester) *The number of $+$ and $-$ signs among the η_i is independent of the choice of orthonormal basis.*

Proof: Let $\{e_i\}$ and $\{f_j\}$ be two orthonormal bases such that

$$e_1 \cdot e_1 = \cdots = e_r \cdot e_r = +1, \qquad e_{r+1} \cdot e_{r+1} = \cdots = e_n \cdot e_n = -1,$$
$$f_1 \cdot f_1 = \cdots = f_s \cdot f_s = +1, \qquad f_{s+1} \cdot f_{s+1} = \cdots = f_n \cdot f_n = -1.$$

If $s > r$ then the vectors $f_1, \ldots, f_s$ and $e_{r+1}, \ldots, e_n$ are a set of $s + n - r > n = \dim V$ vectors and there must be a non-trivial linear relation between them,

$$a^1 f_1 + \cdots + a^s f_s + b^1 e_{r+1} + \cdots + b^{n-r} e_n = 0.$$

The a^i cannot all vanish since the e_i form an l.i. set. Similarly, not all the b^j will vanish. Setting

$$u = a^1 f_1 + \cdots + a^s f_s = -b^1 e_{r+1} - \cdots - b^{n-r} e_n \neq 0$$

5.1 Real inner product spaces

we have the contradiction

$$u \cdot u = \sum_{i=1}^{s}(a^i)^2 > 0 \quad \text{and} \quad u \cdot u = -\sum_{j=1}^{n-r}(b^j)^2 < 0.$$

Hence $r = s$ and the two bases must have exactly the same number of $+$ and $-$ signs. ∎

If r is the number of $+$ signs and s the number of $-$ signs then their difference $r - s$ is called the **index** of the inner product. Sylvester's theorem shows that it is an invariant of the inner product space, independent of the choice of o.n. basis. For a Euclidean inner product, $r - s = n$, although the word 'Euclidean' is also applied to the negative definite case, $r - s = -n$. If $r - s = \pm(n - 2)$, the inner product is called **Minkowskian**.

Example 5.2 In a Euclidean space the Gram–Schmidt procedure is carried out as follows:

$$f_1 = u_1 \qquad e_1 = \frac{f_1}{\sqrt{f_1 \cdot f_1}} \qquad \eta_1 = e_1 \cdot e_1 = 1,$$

$$f_2 = u_2 - (e_1 \cdot u_2)e_1 \qquad e_2 = \frac{f_2}{\sqrt{f_2 \cdot f_2}} \qquad \eta_2 = e_2 \cdot e_2 = 1,$$

$$f_3 = u_3 - (e_1 \cdot u_3)e_1 - (e_2 \cdot u_3)e_2 \qquad e_3 = \frac{f_3}{\sqrt{f_3 \cdot f_3}} \qquad \eta_3 = e_3 \cdot e_3 = 1, \text{ etc.}$$

Since each vector has positive magnitude, all denominators $\sqrt{f_i \cdot f_i} > 0$, and each step is well-defined. Each vector e_i is a unit vector and is orthogonal to each previous e_j ($j < i$).

Example 5.3 Consider an inner product on a three-dimensional space having components in a basis u_1, u_2, u_3

$$\mathbf{G} = [g_{ij}] = [u_i \cdot u_j] = \begin{pmatrix} 0 & 1 & 1 \\ 1 & 0 & 1 \\ 1 & 1 & 0 \end{pmatrix}.$$

The procedure given in Example 5.2 obviously fails as each basis vector is a null vector, $u_1 \cdot u_1 = u_2 \cdot u_2 = u_3 \cdot u_3 = 0$, and cannot be normalized to a unit vector.

Firstly, we find a vector u such that $u \cdot u \ne 0$. Any vector of the form $u = u_1 + au_2$ with $a \ne 0$ will do, since

$$u \cdot u = u_1 \cdot u_1 + 2au_1 \cdot u_2 + u_2 \cdot u_2 = 2a.$$

Setting $a = 1$ gives $u = u_1 + u_2$ and $u \cdot u = 2$. The first step in the orthonormalization process is then

$$e_1 = \frac{1}{\sqrt{2}}(u_1 + u_2), \qquad \eta_1 = e_1 \cdot e_1 = 1.$$

There is of course a significant element of arbitrariness in this as the choice of u is by no means unique; for example, choosing $a = \frac{1}{2}$ leads to $e_1 = u_1 + \frac{1}{2}u_2$.

The subspace V_1 of vectors orthogonal to e_1 consists of vectors of the form $v = au_1 + bu_2 + cu_3$ such that

$$v \cdot e_1 \propto v \cdot u = (au_1 + bu_2 + cu_3) \cdot (u_1 + u_2) = a + b + 2c = 0.$$

Inner product spaces

Setting, for example, $c = 0$ and $a = -b = 1$ results in $v = u_1 - u_2$. The magnitude of v is $v \cdot v = -2$ and normalizing gives

$$e_2 = \frac{1}{\sqrt{2}}(u_1 - u_2), \qquad \eta_2 = e_2 \cdot e_2 = -1, \qquad e_2 \cdot e_1 = 0.$$

Finally, we need a vector $w = au_1 + bu_2 + cu_3$ that is orthogonal to both e_1 and e_2. These two requirements imply that $a = b = -c$, and setting $c = 1$ results in $w = u_1 + u_2 - u_3$. Normalizing w results in

$$w \cdot w = (u_1 + u_2 - u_3) \cdot (u_1 + u_2 - u_3) = -2$$

$$\implies e_3 = \frac{1}{\sqrt{2}}(u_1 + u_2 - u_3), \quad \eta_3 = e_3 \cdot e_3 = -1.$$

The components of the inner product in this o.n. basis are therefore

$$G' = [g'_{ij}] = [e_i \cdot e_j] = \begin{pmatrix} 1 & 0 & 0 \\ 0 & -1 & 0 \\ 0 & 0 & -1 \end{pmatrix}.$$

The index of the inner product is $1 - 2 = -1$.

Any pair of orthonormal bases $\{e_i\}$ and $\{e'_i\}$ are connected by a basis transformation

$$e_i = L^j{}_i e'_j,$$

such that

$$g_{ij} = e_i \cdot e_j = e'_i \cdot e'_j = g'_{ij} = \text{diag}(\eta_1, \ldots, \eta_n).$$

From Eq. (5.4) we have

$$g_{ij} = g_{kl} L^k{}_i L^l{}_j, \tag{5.8}$$

or its matrix equivalent

$$G = L^T G L. \tag{5.9}$$

For a Euclidean metric $G = I$, and L is an orthogonal transformation, while for a Minkoswkian metric with $n = 4$ the transformations are *Lorentz* transformations discussed in Section 2.7. As was shown in Chapter 2, these transformations form the groups $O(n)$ and $O(3, 1)$ respectively. The general pseudo-orthogonal inner product results in a group $O(p, q)$ of **pseudo-orthogonal transformations of type** (p, q).

Problems

Problem 5.1 Let $(V, \cdot)$ be a real Euclidean inner product space and denote the length of a vector $x \in V$ by $|x| = \sqrt{x \cdot x}$. Show that two vectors u and v are orthogonal iff $|u + v|^2 = |u|^2 + |v|^2$.

5.2 Complex inner product spaces

Problem 5.2 Let

$$G = [g_{ij}] = [u_i \cdot u_j] = \begin{pmatrix} 0 & 1 & 0 \\ 1 & 0 & -1 \\ 0 & -1 & 1 \end{pmatrix}$$

be the components of a real inner product with respect to a basis u_1, u_2, u_3. Use Gram–Schmidt orthogonalization to find an orthonormal basis e_1, e_2, e_3, expressed in terms of the vectors u_i, and find the index of this inner product.

Problem 5.3 Let G be the symmetric matrix of components of a real inner product with respect to a basis u_1, u_2, u_3,

$$G = [g_{ij}] = [u_i \cdot u_j] = \begin{pmatrix} 1 & 0 & 1 \\ 0 & -2 & 1 \\ 1 & 1 & 0 \end{pmatrix}.$$

Using Gram–Schmidt orthogonalization, find an orthonormal basis e_1, e_2, e_3 expressed in terms of the vectors u_i.

Problem 5.4 Define the concept of a 'symmetric operator' $S: V \to V$ as one that satisfies

$$(Su) \cdot v = u \cdot (Sv) \qquad \text{for all } u, v \in V.$$

Show that this results in the component equation

$$S^k_i g_{kj} = g_{ik} S^k_j,$$

equivalent to the matrix equation

$$S^T G = GS.$$

Show that for an orthonormal basis in a Euclidean space this results in the usual notion of symmetry, but fails for pseudo-Euclidean spaces.

Problem 5.5 Let V be a Minkowskian vector space of dimension n with index $n - 2$ and let $k \ne 0$ be a null vector ($k \cdot k = 0$) in V.

(a) Show that there is an orthonormal basis $e_1, \ldots, e_n$ such that

$$k = e_1 - e_n.$$

(b) Show that if u is a 'timelike' vector, defined as a vector with negative magnitude $u \cdot u < 0$, then u is not orthogonal to k.
(c) Show that if v is a null vector such that $v \cdot k = 0$, then $v \propto k$.
(d) If $n \ge 4$ which of these statements generalize to a space of index $n - 4$?

5.2 Complex inner product spaces

We now consider a complex vector space V, which in the first instance may be infinite dimensional. Vectors will continue to be denoted by lower case Roman letters such as u and v, but complex scalars will be denoted by Greek letters such as $\alpha, \beta, \ldots$ from the early part of the alphabet. The word **inner product**, or **scalar product**, on a complex vector space

Inner product spaces

V will be reserved for a map $V \times V \to \mathbb{C}$ that assigns to every pair of vectors $u, v \in V$ a complex scalar $\langle u \mid v \rangle$ satisfying

(IP1) $\langle u \mid v \rangle = \overline{\langle v \mid u \rangle}$.
(IP2) $\langle u \mid \alpha v + \beta w \rangle = \alpha \langle u \mid v \rangle + \beta \langle u \mid w \rangle$ for all complex numbers α, β.
(IP3) $\langle u \mid u \rangle \geq 0$ and $\langle u \mid u \rangle = 0$ iff $u = 0$.

The condition (IP1) implies $\langle u \mid u \rangle$ is always real, a necessary condition for (IP3) to make any sense. From (IP1) and (IP2)

$$\langle \alpha v + \beta w \mid u \rangle = \overline{\langle u \mid \alpha v + \beta w \rangle}$$
$$= \overline{\alpha \langle u \mid v \rangle + \beta \langle u \mid w \rangle}$$
$$= \overline{\alpha} \overline{\langle u \mid v \rangle} + \overline{\beta} \, \overline{\langle w \mid v \rangle},$$

so that

$$\langle \alpha v + \beta w \mid u \rangle = \overline{\alpha} \langle v \mid u \rangle + \overline{\beta} \langle w \mid u \rangle. \tag{5.10}$$

This property is often described by saying that the inner product is **antilinear** with respect to the first argument.

A complex vector space with an inner product will simply be called an **inner product space**. If V is finite dimensional it is often called a **finite dimensional Hilbert space**, but for infinite dimensional spaces the term *Hilbert space* only applies if the space is *complete* (see Chapter 13).

Mathematicians more commonly adopt a notation (u, v) in place of our angular bracket notation, and demand linearity in the *first* argument, with antilinearity in the second. Our conventions follow that which is most popular with physicists and takes its origins in Dirac's 'bra' and 'ket' terminology for quantum mechanics (see Chapter 14).

Example 5.4 On $\mathbb{C}^n$ set

$$\langle (\alpha_1, \ldots, \alpha_n) \mid (\beta_1, \ldots, \beta_n) \rangle = \sum_{i=1}^{n} \overline{\alpha_i} \beta_i.$$

Conditions (IP1)–(IP3) are easily verified. We shall see directly that this is the archetypal finite dimensional inner product space. Every finite dimensional inner product space has a basis such that the inner product takes this form.

Example 5.5 A complex-valued function $\varphi : [0, 1] \to \mathbb{C}$ is said to be *continuous* if both the real and imaginary parts of the function $\varphi(x) = f(x) + ig(x)$ are continuous. Let $\mathcal{C}[0, 1]$ be the set of continuous complex-valued functions on the real line interval $[0, 1]$, and define an inner product

$$\langle \varphi \mid \psi \rangle = \int_0^1 \overline{\varphi(x)} \psi(x) \, dx.$$

Conditions (IP1) and (IP2) are simple to prove, but in order to show (IP3) it is necessary to show that

$$\int_0^1 |f(x)|^2 + |g(x)|^2 \, dx = 0 \implies f(x) = g(x) = 0, \quad \forall x \in [0, 1].$$

5.2 Complex inner product spaces

If $f(a) \neq 0$ for some $0 \leq a \leq 1$ then, by continuity, there exists an interval $[a - \epsilon, a]$ or an interval $[a, a + \epsilon]$ on which $|f(x)| > \frac{1}{2}|f(a)|$. Then

$$\int_0^1 |\varphi(x)|^2 \, dx > \frac{1}{2}\epsilon |f(a)|^2 + \int_0^1 |g(x)|^2 \, dx > 0.$$

Hence $f(x) = 0$ for all $x \in [0, 1]$. The proof that $g(x) = 0$ is essentially identical.

Example 5.6 A complex-valued function on the real line, $\varphi : \mathbb{R} \to \mathbb{C}$, is said to be square integrable if $|\varphi|^2$ is an integrable function on any closed interval of $\mathbb{R}$ and $\int_{-\infty}^{\infty} |\varphi(x)|^2 dx < \infty$. The set $L^2(\mathbb{R})$ of square integrable complex-valued functions on the real line is a complex vector space, for if α is a complex constant and φ and ψ are any pair of square integrable functions, then

$$\int_{-\infty}^{\infty} |\varphi(x) + \alpha \psi(x)|^2 \, dx \leq \int_{-\infty}^{\infty} |\varphi(x)|^2 \, dx + |\alpha|^2 \int_{-\infty}^{\infty} |\psi(x)|^2 \, dx < \infty.$$

On $L^2(\mathbb{R})$ define the inner product

$$\langle \varphi | \psi \rangle = \int_{-\infty}^{\infty} \overline{\varphi(x)} \psi(x) \, dx.$$

This is well-defined for any pair of square integrable functions φ and ψ for, after some algebraic manipulation, we find that

$$\overline{\varphi}\psi = \tfrac{1}{2}\left(|\varphi + \psi|^2 - i|\varphi + i\psi|^2 - (1 - i)(|\varphi|^2 + |\psi|^2)\right).$$

Hence the integral of the left-hand side is equal to a sum of integrals on the right-hand side, each of which has been shown to exist.

The properties (IP1) and (IP2) are trivial to show but the proof of (IP3) along the lines given in Example 5.5 will not suffice here since we do not stipulate continuity for the functions in $L^2(\mathbb{R})$. For example, the function $f(x)$ defined by $f(x) = 0$ for all $x \neq 0$ and $f(0) = 1$ is a positive non-zero function whose integral vanishes. The remedy is to 'identify' any two real functions f and g having the property that $\int_{-\infty}^{\infty} |f(x) - g(x)|^2 dx = 0$. Such a pair of functions will be said to be equal **almost everywhere**, and $L^2(\mathbb{R})$ must be interpreted as consisting of equivalence classes of complex-valued functions whose real and imaginary parts are equal almost everywhere. A more complete discussion will be given in Chapter 13, Example 13.4.

Exercise: Show that the relation $f \equiv g$ iff $\int_{-\infty}^{\infty} |f(x) - g(x)|^2 dx = 0$ is an equivalence relation on $L^2(\mathbb{R})$.

Norm of a vector

The **norm** of a vector u in an inner product space, denoted $\|u\|$, is defined to be the non-negative real number

$$\|u\| = \sqrt{\langle u | u \rangle} \geq 0. \tag{5.11}$$

Inner product spaces

From (IP2) and Eq. (5.10) it follows immediately that

$$\|\alpha u\| = |\alpha|\, \|u\|. \tag{5.12}$$

Theorem 5.4 (Cauchy–Schwarz inequality) *For any pair of vectors u, v in an inner product space*

$$|\langle u\,|\,v\rangle| \le \|u\|\,\|v\|. \tag{5.13}$$

Proof: By (IP3), (IP2) and Eq. (5.10) we have for all $\lambda \in \mathbb{C}$

$$0 \le \langle u + \lambda v\,|\,u + \lambda v\rangle$$
$$= \langle u\,|\,u\rangle + \lambda\langle u\,|\,v\rangle + \bar{\lambda}\langle v\,|\,u\rangle + \lambda\bar{\lambda}\langle v\,|\,v\rangle.$$

Substituting the particular value

$$\lambda = -\frac{\langle v\,|\,u\rangle}{\langle v\,|\,v\rangle}$$

gives the inequality

$$0 \le \langle u\,|\,u\rangle - \frac{\langle v\,|\,u\rangle\langle u\,|\,v\rangle}{\langle v\,|\,v\rangle} - \frac{\overline{\langle v\,|\,u\rangle}\langle v\,|\,u\rangle}{\langle v\,|\,v\rangle} + \frac{|\langle v\,|\,u\rangle|^2}{\langle v\,|\,v\rangle}$$
$$= \langle u\,|\,u\rangle - \frac{|\langle v\,|\,u\rangle|^2}{\langle v\,|\,v\rangle}.$$

Hence, from (IP1),

$$|\langle u\,|\,v\rangle|^2 = |\langle v\,|\,u\rangle|^2 \ge \langle u\,|\,u\rangle\langle v\,|\,v\rangle$$

and the desired result follows from (5.11) on taking the square roots of both sides of this inequality. ∎

Corollary 5.5 *Equality in (5.13) can only result if u and v are proportional to each other,*

$$|\langle u\,|\,v\rangle| = \|u\|\,\|v\| \iff u = \alpha v, \quad \text{for some } \alpha \in \mathbb{C}.$$

Proof: If $u = \alpha v$ then from (5.11) and (5.12) we have

$$|\langle u\,|\,v\rangle| = |\langle \alpha v\,|\,v\rangle| = |\alpha|\langle v\,|\,v\rangle = \|\alpha v\|\,\|v\| = \|u\|\,\|v\|.$$

Conversely, if $|\langle u\,|\,v\rangle| = \|u\|\,\|v\|$ then,

$$|\langle u\,|\,v\rangle|^2 = \|u\|^2\,\|v\|^2$$

and reversing the steps in the proof of Lemma 5.4 with inequalities replaced by equalities gives

$$\langle u + \lambda v\,|\,u + \lambda v\rangle = 0 \quad \text{where} \quad \lambda = -\frac{\langle v\,|\,u\rangle}{\langle v\,|\,v\rangle}.$$

By (IP3) we conclude that $u = -\lambda v$ and the proposition follows with $\alpha = -\lambda$. ∎

5.2 Complex inner product spaces

Theorem 5.6 (Triangle inequality) *For any pair of vectors u and v in an inner product space,*

$$\|u + v\| \leq \|u\| + \|v\|. \tag{5.14}$$

Proof:
$$\begin{aligned}(\|u + v\|)^2 &= \langle u + v \,|\, u + v\rangle \\ &= \langle u \,|\, u\rangle + \langle v \,|\, v\rangle + \langle v \,|\, u\rangle + \langle u \,|\, v\rangle \\ &= \|u\|^2 + \|v\|^2 + 2\mathrm{Re}(\langle u \,|\, v\rangle) \\ &\leq \|u\|^2 + \|v\|^2 + 2|\langle u \,|\, v\rangle| \\ &\leq \|u\|^2 + \|v\|^2 + 2\|u\|\|v\| \quad \text{by Eq. (5.13)} \\ &= (\|u\| + \|v\|)^2\,.\end{aligned}$$

The triangle inequality (5.14) follows on taking square roots. ∎

Orthonormal bases

Let V be a finite dimensional inner product space with basis $e_1, e_2, \ldots, e_n$. Define the **components** of the inner product with respect to this basis to be

$$h_{ij} = \langle e_i \,|\, e_j\rangle = \overline{\langle e_j \,|\, e_i\rangle} = \overline{h_{ji}}. \tag{5.15}$$

The matrix of components $\mathsf{H} = [h_{ij}]$ is clearly hermitian,

$$\mathsf{H} = \mathsf{H}^\dagger \quad \text{where} \quad \mathsf{H}^\dagger = \overline{\mathsf{H}}^T.$$

Under a change of basis (5.3) we have

$$\langle e_i \,|\, e_j\rangle = \overline{A^k{}_i} \langle e'_k \,|\, e'_m\rangle A^m{}_j$$

and the components of the inner product transform as

$$h_{ij} = \overline{A^k{}_i} h'_{km} A^m{}_j. \tag{5.16}$$

An identical argument can be used to express the primed components in terms of unprimed components,

$$h'_{ij} = \langle e'_i \,|\, e'_j\rangle = \overline{A'^k{}_i} h_{km} A'^m{}_j. \tag{5.17}$$

These equations have matrix equivalents,

$$\mathsf{H} = \mathsf{A}^\dagger \mathsf{H}' \mathsf{A}, \qquad \mathsf{H}' = \mathsf{A}'^\dagger \mathsf{H} \mathsf{A}', \tag{5.18}$$

where $\mathsf{A}' = \mathsf{A}^{-1}$.

Exercise: Show that the hermitian nature of the matrix H is unchanged by a transformation (5.18).

Two vectors u and v are said to be **orthogonal** if $\langle u \,|\, v\rangle = 0$. A basis $e_1, e_2, \ldots, e_n$ is called an **orthonormal basis** if the vectors all have unit norm and are orthogonal to each

Inner product spaces

other,

$$\langle e_i | e_j \rangle = \delta_{ij} = \begin{cases} 1 & \text{if } i = j, \\ 0 & \text{if } i \ne j. \end{cases}$$

Equivalently, a basis is orthonormal if the matrix of components of the inner product with respect to the basis is the unit matrix, $\mathsf{H} = \mathsf{I}$.

Starting with an arbitrary basis $\{u_1, u_2, \ldots, u_n\}$, it is always possible to construct an orthonormal basis by a process known as **Schmidt orthonormalization**, which closely mirrors the Gram–Schmidt process for Euclidean inner products, outlined in Example 5.2. Sequentially, the steps are:

1. Set $f_1 = u_1$, then $e_1 = \dfrac{f_1}{\|f_1\|}$.
2. Set $f_2 = u_2 - \langle e_1 | u_2 \rangle e_1$, which is orthogonal to e_1 since

$$\langle e_1 | f_2 \rangle = \langle e_1 | u_2 \rangle - \langle e_1 | u_2 \rangle \|e_1\| = 0.$$

Normalize f_2 by setting $e_2 = f_2 / \|f_2\|$.

3. Set $f_3 = u_3 - \langle e_1 | u_3 \rangle e_1 - \langle e_2 | u_3 \rangle e_2$, which is orthogonal to both e_1 and e_2. Normalize to give $e_3 = \dfrac{f_3}{\|f_3\|}$.

4. Continue in this way until

$$f_n = u_n - \sum_{i=1}^{n-1} \langle e_i | u_n \rangle e_i \quad \text{and} \quad e_n = \dfrac{f_n}{\|f_n\|}.$$

Since each vector e_i is a unit vector and is orthogonal to all the e_j for $j < i$ defined by previous steps, they form an o.n. set. It is easily seen that any vector $v = v^i u_i$ of V is a linear combination of the e_i since each u_j is a linear combination of $e_1, \ldots, e_j$. Hence the vectors $\{e_i\}$ form a basis by Theorem 3.3 since they span V and are n in number.

With respect to an orthonormal basis the inner product of any pair of vectors $u = u^i e_i$ and $v = v^j e_j$ is given by

$$\langle u | v \rangle = \langle u^i e_i | v^j e_j \rangle = \overline{u^i} v^j \langle e_i | e_j \rangle = \overline{u^i} v^j \delta_{ij}.$$

Hence

$$\langle u | v \rangle \sum_{i=1}^{n} \overline{u^i} v^i = \overline{u^1} v^1 + \overline{u^2} v^2 + \cdots + \overline{u^n} v^n,$$

which is equivalent to the standard inner product defined on $\mathbb{C}^n$ in Example 5.4.

Example 5.7 Let an inner product have the following components in a basis u_1, u_2, u_3:

$h_{11} = \langle u_1 | u_1 \rangle = 1$ $\quad h_{12} = \langle u_1 | u_2 \rangle = 0$ $\quad h_{13} = \langle u_1 | u_3 \rangle = \frac{1}{2}(1+i)$

$h_{21} = \langle u_2 | u_1 \rangle = 0$ $\quad h_{22} = \langle u_2 | u_2 \rangle = 2$ $\quad h_{23} = \langle u_2 | u_3 \rangle = 0$

$h_{31} = \langle u_3 | u_1 \rangle = \frac{1}{2}(1-i)$ $\quad h_{32} = \langle u_3 | u_2 \rangle = 0$ $\quad h_{33} = \langle u_3 | u_3 \rangle = 1.$

Before proceeding it is important to realize that this inner product does in fact satisfy the positive definite condition (IP3). This would not be true, for example, if we had

5.2 Complex inner product spaces

given $h_{13} = \overline{h_{31}} = 1+i$, for then the vector $v = u_1 - \dfrac{1-i}{\sqrt{2}} u_3$ would have negative norm $\langle v | v \rangle = 2(1 - \sqrt{2}) < 0$.

In the above inner product, begin by setting $e_1 = f_1 = u_1$. The next vector is

$$f_2 = u_2 - \langle e_1 | u_2 \rangle e_1 = u_2, \qquad e_2 = \dfrac{u_2}{\|u_2\|} = \dfrac{1}{\sqrt{2}} u_2.$$

The last step is to set

$$f_3 = u_3 - \langle e_1 | u_3 \rangle e_1 - \langle e_2 | u_3 \rangle e_2 = u_3 - \tfrac{1}{2}(1+i) u_1$$

which has norm squared

$$(\|f_3\|)^2 = \langle u_3 | u_3 \rangle - \tfrac{1}{2}(1-i)\langle u_1 | u_3 \rangle - \tfrac{1}{2}(1+i)\langle u_3 | u_1 \rangle + \tfrac{1}{4}(1-i)(1+i)\langle u_1 | u_1 \rangle$$
$$= 1 - \tfrac{1}{4}(1-i)(1+i) - \tfrac{1}{4}(1+i)(1-i) + \tfrac{1}{2} = \tfrac{1}{2}.$$

Hence $e_3 = f_3/\|f_3\| = \sqrt{2}(u_3 - (1+i)u_1)$ completes the orthonormal basis.

The Schmidt orthonormalization procedure actually provides a good method for proving positive definiteness, since the process breaks down at some stage, producing a vector with non-positive norm if the inner product does not satisfy (IP3).

Exercise: Try to perform the Schmidt orthonormalization on the above inner product suggested with the change $h_{13} = \overline{h_{31}} = 1+i$, and watch it break down!

Unitary transformations

A linear operator $U : V \to V$ on an inner product space is said to be **unitary** if it preserves inner products,

$$\langle Uu | Uv \rangle = \langle u | v \rangle, \quad \forall u, v \in V. \tag{5.19}$$

Unitary operators clearly preserve the norm of any vector v,

$$\|Uv\| = \sqrt{\langle Uv | Uv \rangle} = \sqrt{\langle v | v \rangle} = \|v\|.$$

In fact it can be shown that a linear operator U is unitary if and only if it is norm preserving (see Problem 5.7).

A unitary operator U transforms any orthonormal basis $\{e_i\}$ into another o.n. basis $e'_i = Ue_i$, since

$$\langle e'_i | e'_j \rangle = \langle Ue_i | Ue_j \rangle = \langle e_i | e_j \rangle = \delta_{ij}. \tag{5.20}$$

The set $\{e'_1, \ldots, e'_n\}$ is linearly independent, and is thus a basis, for if $\alpha^i e'_i = 0$ then $\langle e'_j | \alpha^i e'_i \rangle = \alpha^j = 0$. The map U is onto since every vector $u = u'^i e'_i = U(u'^i e_i)$, and one-to-one since $Uv = 0 \Rightarrow v^i e'_i = 0 \Rightarrow v = v^i e_i = 0$. Hence every unitary operator U is invertible.

Inner product spaces

With respect to an orthonormal basis $\{e_i\}$ the components of the linear transformation U, defined by $Ue_i = U^k_i e_k$, form a unitary matrix $\mathsf{U} = [U^k_i]$:

$$\delta_{ij} = \langle Ue_i | Ue_j \rangle = \overline{U^k_i} U^m_j \langle e_k | e_m \rangle$$
$$= \overline{U^k_i} U^m_j \delta_{km} = \sum_{k=1}^{n} \overline{U^k_i} U^k_j,$$

or, in terms of matrices,

$$\mathsf{I} = \mathsf{U}^\dagger \mathsf{U}.$$

If $\{e_i\}$ and $\{e'_j\}$ are any pair of orthonormal bases, then the linear operator U defined by $e'_i = Ue_i$ is unitary since for any pair of vectors $u = u^i e_i$ and $v = v^j e_j$

$$\langle Uu | Uv \rangle = \overline{u^i} v^j \langle e'_i | e'_j \rangle$$
$$= \overline{u^i} v^j \delta_{ij}$$
$$= \overline{u^i} v^j \langle e_i | e_j \rangle = \langle u | v \rangle.$$

Thus all orthonormal bases are uniquely related by unitary transformations.

In the language of Section 3.6 this is the *active view*, wherein vectors are 'physically' moved about in the inner product space by the unitary transformation. In the related *passive view*, the change of basis is given by (5.3) – it is the *components* of vectors that are transformed, not the vectors themselves. If both bases are orthonormal the components of an inner product, given by Eq. (5.16), are $h_{ij} = h'_{ij} = \delta_{ij}$, and setting $A^k_i = U^k_i$ in Eq. (5.18) implies the matrix $\mathsf{U} = [U^k_i]$ is unitary,

$$\mathsf{I} = \mathsf{H} = \mathsf{U}^\dagger \mathsf{H}' \mathsf{U} = \mathsf{U}^\dagger \mathsf{I} \mathsf{U} = \mathsf{U}^\dagger \mathsf{U}.$$

Thus, from both the active and passive viewpoint, orthonormal bases are related by unitary matrices.

Problems

Problem 5.6 Show that the norm defined by an inner product satisfies the **parallelogram law**

$$\|u + v\|^2 + \|u - v\|^2 = 2\|u\|^2 + 2\|v\|^2.$$

Problem 5.7 On an inner product space show that

$$4\langle u | v \rangle = \|u + v\|^2 - \|u - v\|^2 - i\|u + iv\|^2 + i\|u - iv\|^2.$$

Hence show that a linear transformation $U : V \to V$ is unitary iff it is norm preserving,

$$\langle Uu | Uv \rangle = \langle u | v \rangle, \quad \forall u, v \in V \iff \|Uv\| = \|v\|, \quad \forall v \in V.$$

Problem 5.8 Show that a pair of vectors u and v in a complex inner product space are orthogonal iff

$$\|\alpha u + \beta v\|^2 = \|\alpha u\|^2 + \|\beta v\|^2, \quad \forall \alpha, \beta \in \mathbb{C}.$$

Find a non-orthogonal pair of vectors u and v in a complex inner product space such that $\|u + v\|^2 = \|u\|^2 + \|v\|^2$.

5.3 Representations of finite groups

Problem 5.9 Show that the formula

$$\langle A|B\rangle = \text{tr}(B\,A^\dagger)$$

defines an inner product on the vector space of $m \times n$ complex matrices $M(m, n)$.

(a) Calculate $\|I_n\|$ where I_n is the $n \times n$ identity matrix.
(b) What characterizes matrices orthogonal to I_n?
(c) Show that all unitary $n \times n$ matrices U have the same norm with respect to this inner product.

Problem 5.10 Let S and T be complex inner product spaces and let $U : S \to T$ be a linear map such that $\|Ux\| = \|x\|$. Prove that

$$\langle Ux | Uy \rangle = \langle x | y \rangle \quad \text{for all } x, y \in S.$$

Problem 5.11 Let V be a complex vector space with an 'indefinite inner product', defined as an inner product that satisfies (IP1), (IP2) but with (IP3) replaced by the non-singularity condition

(IP3') $\langle u | v \rangle = 0$ for all $v \in V$ implies that $u = 0$.

(a) Show that similar results to Theorems 5.2 and 5.3 can be proved for such an indefinite inner product.
(b) If there are p +1's along the diagonal and q −1's, find the defining relations for the group of transformations $U(p, q)$ between orthonormal basis.

Problem 5.12 If V is an inner product space, an operator $K : V \to V$ is called **self-adjoint** if

$$\langle u | Kv \rangle = \langle Ku | v \rangle$$

for any pair of vectors $u, v \in V$. Let $\{e_i\}$ be an arbitrary basis, having $\langle e_i | e_j \rangle = h_{ij}$, and set $Ke_k = K^j_{\,k} e_j$. Show that if $\mathsf{H} = [h_{ij}]$ and $\mathsf{K} = [K^k_{\,j}]$ then

$$\mathsf{HK} = \mathsf{K}^\dagger \mathsf{H} = (\mathsf{HK})^\dagger.$$

If $\{e_i\}$ is an orthonormal basis, show that K is a hermitian matrix.

5.3 Representations of finite groups

If G is a finite group, it turns out that every finite dimensional representation is equivalent to a representation by *unitary* transformations on an inner product space – known as a **unitary representation**. For, let T be a representation on any finite dimensional vector space V, and let $\{e_i\}$ be any basis of V. Define an inner product $(u|v)$ on V by setting $\{e_i\}$ to be an orthonormal set,

$$(u|v) = \sum_{i=1}^{n} \overline{u^i} v^i \quad \text{where} \quad u = u^i e_i,\ v = v^j e_j \quad (u^i, v^j \in \mathbb{C}). \tag{5.21}$$

Of course there is no reason why the linear transformations $T(g)$ should be unitary with respect to this inner product, but they *will* be unitary with respect to the inner product $\langle u | v \rangle$

Inner product spaces

formed by 'averaging over the group',

$$\langle u | v \rangle = \frac{1}{|G|} \sum_{a \in G} (T(a)u | T(a)v), \tag{5.22}$$

where $|G|$ is the *order* of the group G (the number of elements in G). This follows from

$$\langle T(g)u | T(g)v \rangle = \frac{1}{|G|} \sum_{a \in G} (T(a)T(g)u | T(a)T(g)v)$$

$$= \frac{1}{|G|} \sum_{a \in G} (T(ag)u | T(ag)v)$$

$$= \frac{1}{|G|} \sum_{b \in G} (T(b)u | T(b)v)$$

$$= \langle u | v \rangle$$

since, as a ranges over the group G, so does $b = ag$ for any fixed $g \in G$.

Theorem 5.7 *Any finite dimensional representation of a finite group G is completely reducible into a direct sum of irreducible representations.*

Proof: Using the above device we may assume that the representation is unitary on a finite dimensional Hilbert space V with inner product $\langle \cdot | \cdot \rangle$. If W is a vector subspace of V, define its **orthogonal complement** $W^\perp$ to be the set of vectors orthogonal to W,

$$W^\perp = \{u \mid \langle u | w \rangle = 0, \ \forall w \in W\}.$$

$W^\perp$ is clearly a vector subspace, for if α is an arbitrary complex number then

$$u, v \in W^\perp \implies \langle u + \alpha v | w \rangle = \langle u | w \rangle + \alpha \langle v | w \rangle = 0 \implies u + \alpha v \in W^\perp.$$

By selecting an orthonormal basis such that the first dim W vectors belong to W, it follows that the remaining vectors of the basis span $W^\perp$. Hence W and $W^\perp$ are orthogonal and complementary subspaces, $H = W \oplus W^\perp$. If W is a G-invariant subspace, then $W^\perp$ is also G-invariant, For, if $u \in W^\perp$ then for any $w \in W$,

$$0 = \langle T(g)u | w \rangle = \langle T(g)u | T(g)T(g)^{-1}w \rangle$$

$$= \langle u | T(g^{-1})w \rangle \quad \text{since } T(g) \text{ is unitary}$$

$$= 0$$

since $T(g^{-1})w \in W$ by the G-invariance of W. Hence $T(g)u \in W^\perp$.

Now pick W to be the G-invariant subspace of V of smallest dimension, not counting the trivial subspace $\{0\}$. The representation induced on W must be irreducible since it can have no proper G-invariant subspaces, as they would need to have smaller dimension. If $W = V$ then the representation T is irreducible. If $W \neq V$ its orthogonal complement $W^\perp$ is either irreducible, in which case the proof is finished, or it has a non-trivial invariant subspace W'. Again pick the invariant subspace of smallest dimension and continue in this fashion until V is a direct sum of irreducible subspaces,

$$V = W \oplus W' \oplus W'' \oplus \cdots$$

The representation T decomposes into subrepresentations $T\big|_{W^{(i)}}$. ∎

5.3 Representations of finite groups

Orthogonality relations

The components of the matrices of irreducible group representatives satisfy a number of important orthogonality relationships, which are the cornerstone of the classification procedure of group representations. We will give just a few of these relations; others can be found in [4, 5].

Let T_1 and T_2 be irreducible representations of a finite group G on complex vector spaces V_1 and V_2 respectively. If $\{e_i \mid i = 1, \ldots, n_1 = \dim V_1\}$ and $\{f_a \mid a = 1, \ldots, n_2 = \dim V_2\}$ are bases of these two vector spaces, we will write the representative matrices as $\mathsf{T}_1(g) = [T_{(1)}{}^j{}_i]$ and $\mathsf{T}_2(g) = [T_{(2)}{}^b{}_a]$ where

$$T_1(g)e_i = T_{(1)}{}^j{}_i e_j \quad \text{and} \quad T_2(g)f_a = T_{(2)}{}^b{}_a f_b.$$

If $A : V_1 \to V_2$ is any linear map, define its 'group average' $\tilde{A} : V_1 \to V_2$ to be the linear map

$$\tilde{A} = \frac{1}{|G|} \sum_{g \in G} T_2(g) A T_1(g^{-1}).$$

Then if h is any element of the group G,

$$T_2(h)\,\tilde{A} T_1(h^{-1}) = \frac{1}{|G|} \sum_{g \in G} T_2(hg) A T_1((hg)^{-1})$$

$$= \frac{1}{|G|} \sum_{g' \in G} T_2(g') A T_1((g')^{-1})$$

$$= \tilde{A}.$$

Hence $\tilde{A}$ is an intertwining operator,

$$T_2(h)\tilde{A} = \tilde{A} T_1(h) \quad \text{for all } h \in G,$$

and by Schur's lemma, Theorem 4.6, if $T_1 \nsim T_2$ then $\tilde{A} = 0$. On the other hand, from the corollary to Schur's lemma, 4.7, if $V_1 = V_2 = V$ and $T_1 = T_2$ then $\tilde{A} = c\,\mathrm{id}_V$. The matrix version of this equation with respect to any basis of V is $\tilde{\mathsf{A}} = c\,\mathsf{I}$, and taking the trace gives

$$c = \frac{1}{n} \operatorname{tr} \tilde{\mathsf{A}}.$$

However

$$\operatorname{tr} \tilde{\mathsf{A}} = \frac{1}{|G|} \operatorname{tr} \sum_{g \in G} \mathsf{T}(g) \mathsf{A} \mathsf{T}^{-1}(g)$$

$$= \frac{1}{|G|} \sum_{g \in G} \operatorname{tr}(\mathsf{T}^{-1}(g) \mathsf{T}(g) \mathsf{A})$$

$$= \frac{1}{|G|} \sum_{g \in G} \operatorname{tr} \mathsf{A} = \operatorname{tr} \mathsf{A},$$

whence

$$c = \frac{1}{n} \operatorname{tr} \mathsf{A}.$$

Inner product spaces

If $T_1 \sim T_2$, expressing A and $\tilde{A}$ in terms of the bases e_i and f_a,
$$Ae_i = A^a{}_i f_a \quad \text{and} \quad \tilde{A}e_i = \tilde{A}^a{}_i f_a,$$
the above consequence of Schur's lemma can be written
$$\tilde{A}^a{}_i = \frac{1}{|G|} \sum_{g \in G} T_{(2)}{}^a{}_b(g) A^b{}_j T_{(1)}{}^j{}_i(g^{-1}) = 0.$$

As A is an arbitrary operator the matrix elements $A^b{}_j$ are arbitrary complex numbers, so that
$$\frac{1}{|G|} \sum_{g \in G} T_{(2)}{}^a{}_b(g) T_{(1)}{}^j{}_i(g^{-1}) = 0. \tag{5.23}$$

If $T_1 = T_2 = T$ and $n = \dim V$ is the degree of the representation we have
$$\tilde{A}^j{}_i = \frac{1}{|G|} \sum_{g \in G} T^j{}_k(g) A^k{}_l T^l{}_i(g^{-1}) = \frac{1}{n} A^k{}_k \delta^j{}_i.$$

As $A^k{}_l$ are arbitrary,
$$\frac{1}{|G|} \sum_{g \in G} T^j{}_k(g) T^l{}_i(g^{-1}) = \frac{1}{n} \delta^j{}_i \delta^l{}_k. \tag{5.24}$$

If $\langle \cdot | \cdot \rangle$ is the invariant inner product defined by a representation T on a vector space V by (5.22), and $\{e_i\}$ is any basis such that
$$\langle e_i | e_j \rangle = \delta_{ji},$$
then the unitary condition $\langle T(g)u | T(g)v \rangle = \langle u | v \rangle$ implies
$$\sum_k \overline{T_{ki}(g)} T_{kj}(g) = \delta_{ij},$$
where indices on T are all lowered. In matrices
$$\mathsf{T}^\dagger(g) \mathsf{T}(g) = \mathsf{I},$$
whence
$$\mathsf{T}(g^{-1}) = (\mathsf{T}(g))^{-1} = \mathsf{T}^\dagger(g),$$
or equivalently
$$T_{ji}(g^{-1}) = \overline{T_{ij}(g)}. \tag{5.25}$$

Substituting this relation for T_1 in place of T into (5.23), with all indices now lowered, gives
$$\frac{1}{|G|} \sum_{g \in G} \overline{T_{(1)ij}(g)} T_{(2)ab}(g) = 0. \tag{5.26}$$

Similarly if $T_1 = T_2 = T$, Eqs. (5.25) and (5.24) give
$$\frac{1}{|G|} \sum_{g \in G} \overline{T_{ij}(g)} T_{kl}(g) = \frac{1}{n} \delta_{ik} \delta_{jl}. \tag{5.27}$$

5.3 Representations of finite groups

The left-hand sides of Eqs. (5.26) and (5.27) have the appearance of an inner product, and this is in fact so. Let $\mathcal{F}(G)$ be the space of all complex-valued functions on G,

$$\mathcal{F}(G) = \{\phi \,|\, \phi : G \to \mathbb{C}\}$$

with inner product

$$(\phi, \psi) = \frac{1}{|G|} \sum_{a \in G} \overline{\phi(a)} \psi(a). \tag{5.28}$$

It is easy to verify that the requirements (IP1)–(IP3) hold for this inner product, namely

$$(\phi, \psi) = \overline{(\psi, \phi)},$$
$$(\phi, \alpha\psi) = \alpha(\phi, \psi),$$
$$(\phi, \phi) \geq 0 \quad \text{and} \quad (\phi, \phi) = 0 \quad \text{iff} \quad \phi = 0.$$

The matrix components T_{ji} of any representation with respect to an o.n. basis form a set of n^2 complex-valued functions on G, and Eqs. (5.26) and (5.27) read

$$(T_{(1)ij}, T_{(2)ab}) = 0 \quad \text{if} \quad T_1 \not\sim T_2, \tag{5.29}$$

and

$$(T_{ij}, T_{kl}) = \frac{1}{n} \delta_{ik} \delta_{jl}. \tag{5.30}$$

Example 5.8 Consider the group S_3 with notation as in Example 4.8. The invariant inner product (5.22) on the space spanned by e_1, e_2 and e_3 is given by

$$\langle u \,|\, v \rangle = \tfrac{1}{6} \sum_\pi \sum_{i=1}^3 \overline{(T(\pi)u)^i} (T(\pi)v)^i$$
$$= \tfrac{1}{6} \left(\overline{u^1} v^1 + \overline{u^2} v^2 + \overline{u^3} v^3 + \overline{u^3} v^3 + \overline{u^1} v^1 + \overline{u^2} v^2 + \dots \right)$$
$$= \overline{u^1} v^1 + \overline{u^2} v^2 + \overline{u^3} v^3.$$

Hence $\{e_i\}$ forms an orthonormal basis for this inner product,

$$h_{ij} = \langle e_i \,|\, e_j \rangle = \delta_{ij}.$$

It is only because S_3 runs through *all* permutations that the averaging process gives the same result as the inner product defined by (5.21). A similar conclusion would hold for the action of S_n on an n-dimensional space spanned by $\{e_1, \dots, e_n\}$, but these vectors would *not* in general be orthonormal with respect to the inner product (5.22) defined by an arbitrary subgroup of S_n.

As seen in Example 4.8, the vector $f_1 = e_1 + e_2 + e_3$ spans an invariant subspace with respect to this representation of S_3. As in Example 4.8, the vectors f_1, $f_2 = e_1 - e_2$ and $f_3 = e_1 + e_2 - 2e_3$ are mutually orthogonal,

$$\langle f_1 \,|\, f_2 \rangle = \langle f_1 \,|\, f_3 \rangle = \langle f_2 \,|\, f_3 \rangle = 0.$$

Hence the subspace spanned by f_2 and f_3 is orthogonal to f_1, and from the proof of Theorem 5.7, it is also invariant. Form an o.n. set by normalizing their lengths to

Inner product spaces

unity,

$$f_1' = \frac{f_1}{\sqrt{3}}, \quad f_2' = \frac{f_2}{\sqrt{2}}, \quad f_3' = \frac{f_3}{\sqrt{6}}.$$

The representation T_1 on the one-dimensional subspace spanned by f_1' is clearly the trivial one, whereby every group element is mapped to the number 1,

$$T_1(\pi) = 1 \quad \text{for all } \pi \in S_3.$$

The matrices of the representation T_2 on the invariant subspace spanned by $h_1 = f_2'$ and $h_2 = f_3'$ are easily found from the 2×2 parts of the matrices given in Example 4.8 by transforming to the renormalized basis,

$$T_2(e) = \begin{pmatrix} 1 & 0 \\ 0 & 1 \end{pmatrix}, \quad T_2(\pi_1) = \begin{pmatrix} -\frac{1}{2} & -\frac{\sqrt{3}}{2} \\ \frac{\sqrt{3}}{2} & -\frac{1}{2} \end{pmatrix}, \quad T_2(\pi_2) = \begin{pmatrix} -\frac{1}{2} & \frac{\sqrt{3}}{2} \\ -\frac{\sqrt{3}}{2} & -\frac{1}{2} \end{pmatrix},$$

$$T_2(\pi_3) = \begin{pmatrix} -1 & 0 \\ 0 & 1 \end{pmatrix}, \quad T_2(\pi_4) = \begin{pmatrix} \frac{1}{2} & \frac{\sqrt{3}}{2} \\ \frac{\sqrt{3}}{2} & -\frac{1}{2} \end{pmatrix}, \quad T_2(\pi_5) = \begin{pmatrix} \frac{1}{2} & -\frac{\sqrt{3}}{2} \\ -\frac{\sqrt{3}}{2} & -\frac{1}{2} \end{pmatrix}.$$

It is straightforward to verify (5.29):

$$\left(T_{(1)11}, T_{(2)11}\right) = \frac{1}{6}\left(1 - \frac{1}{2} - \frac{1}{2} - 1 + \frac{1}{2} + \frac{1}{2}\right) = 0,$$

$$\left(T_{(1)11}, T_{(2)12}\right) = \frac{1}{6}\left(0 - \frac{\sqrt{3}}{2} + \frac{\sqrt{3}}{2} + 0 + \frac{\sqrt{3}}{2} - \frac{\sqrt{3}}{2}\right) = 0, \text{ etc.}$$

From the exercise following Example 4.8 the representation T_2 is an irreducible representation with $n = 2$ and the relations (5.30) are verified as follows:

$$\left(T_{(2)11}, T_{(2)11}\right) = \frac{1}{6}\left(1^2 + \left(-\frac{1}{2}\right)^2 + \left(-\frac{1}{2}\right)^2 + (-1)^2 + \frac{1^2}{2} + \frac{1^2}{2}\right) = \frac{3}{6} = \frac{1}{2}\delta_{11}\delta_{11} = \frac{1}{2}$$

$$\left(T_{(2)12}, T_{(2)12}\right) = \frac{1}{6}\left(\left(-\frac{\sqrt{3}}{2}\right)^2 + \left(\frac{\sqrt{3}}{2}\right)^2 + \left(\frac{\sqrt{3}}{2}\right)^2 + \left(-\frac{\sqrt{3}}{2}\right)^2\right) = \frac{1}{2} = \frac{1}{2}\delta_{11}\delta_{22}$$

$$\left(T_{(2)11}, T_{(2)12}\right) = \frac{1}{6}\left(1 \cdot 0 - \frac{1}{2} \cdot \left(-\frac{\sqrt{3}}{2}\right) - \frac{1}{2} \cdot \frac{\sqrt{3}}{2} - 1 \cdot 0 + \frac{1}{2} \cdot \frac{\sqrt{3}}{2} + \frac{1}{2} \cdot \left(-\frac{\sqrt{3}}{2}\right)\right)$$

$$= 0 = \frac{1}{2}\delta_{11}\delta_{12}, \text{ etc.}$$

Theorem 5.8 *There are a finite number N of inequivalent irreducible representations of a finite group, and*

$$\sum_{\mu=1}^{N}(n_\mu)^2 \leq |G|, \tag{5.31}$$

where $n_\mu = \dim V_\mu$ are the degrees of the inequivalent representations.

Proof: Let $T_1 : G \to GL(V_1)$, $T_2 : G \to GL(V_2)$, ... be inequivalent irreducible representations of G. If the basis on each vector space V_μ is chosen to be orthonormal with respect to the inner product $\langle \cdot | \cdot \rangle$ for each $\mu = 1, 2, \ldots$, then (5.29) and (5.30) may be

5.3 Representations of finite groups

summarized as the single equation

$$\left(T_{(\mu)ij},\ T_{(\nu)ab}\right) = \frac{1}{n_\mu}\delta_{\mu\nu}\delta^{ia}\delta_{jb}.$$

Hence for each μ the $T_{(\mu)ij}$ consist of $(n_\mu)^2$ mutually orthogonal functions in $\mathcal{F}(G)$ that are orthogonal to all $T_{(\nu)ab}$ for $\nu \neq \mu$. There cannot therefore be more than $\dim \mathcal{F}(G) = |G|$ of these, giving the desired inequality (5.31). Clearly there are at most a finite number N of such representations. ∎

It may in fact be shown that the inequality (5.31) can be replaced by equality, a very useful identity in the enumeration of irreducible representations of a finite group. Details of the proof as well as further orthogonality relations and applications of group representation theory to physics may be found in [4, 5].

Problems

Problem 5.13 For a function $\phi : G \to \mathbb{C}$, if we set $g\phi$ to be the function $(g\phi)(a) = \phi(g^{-1}a)$ show that $(gg')\phi = g(g'\phi)$. Show that the inner product (5.28) is G-invariant, $(g\phi, g\psi) = (\phi, \psi)$ for all $g \in G$.

Problem 5.14 Let the **character** of a representation T of a group G on a vector space V be the function $\chi : G \to \mathbb{C}$ defined by

$$\chi(g) = \operatorname{tr} T(g) = T^i_i(g).$$

(a) Show that the character is independent of the choice of basis and is a member of $\mathcal{F}(G)$, and that characters of equivalent representations are identical. Show that $\chi(e) = \dim V$.

(b) Any complex-valued function on G that is constant on conjugacy classes (see Section 2.4) is called a **central function**. Show that characters are central functions.

(c) Show that with respect to the inner product (5.28), characters of any pair of inequivalent irreducible representations $T_1 \not\sim T_2$ are orthogonal to each other, $(\chi_1, \chi_2) = 0$, while the character of any irreducible representation T has unit norm $(\chi, \chi) = 1$.

(d) From Theorem 5.8 and Theorem 5.7 every unitary representation T can be decomposed into a direct sum of inequivalent irreducible unitary representations $T_\mu : G \to GL(V_\mu)$,

$$T \sim m_1 T_1 \oplus m_2 T_2 \oplus \cdots \oplus m_N T_N \quad (m_\mu \geq 0).$$

Show that the **multiplicities** m_μ of the representations T_μ are given by

$$m_\mu = (\chi, \chi_\mu) = \frac{1}{|G|}\sum_{g \in G} \overline{\chi(g)}\chi_\mu(g)$$

and T is irreducible if and only if its character has unit magnitude, $(\chi, \chi) = 1$. Show that T and T' have no irreducible representations in common in their decompositions if and only if their characters are orthogonal.

References

[1] P. R. Halmos. *Finite-dimensional Vector Spaces*. New York, D. Van Nostrand Company, 1958.
[2] P. R. Halmos. *Introduction to Hilbert Space*. New York, Chelsea Publishing Company, 1951.
[3] L. H. Loomis and S. Sternberg. *Advanced Calculus*. Reading, Mass., Addison-Wesley, 1968.
[4] M. Hammermesh. *Group Theory and its Applications to Physical Problems*. Reading, Mass., Addison-Wesley, 1962.
[5] S. Sternberg. *Group Theory and Physics*. Cambridge, Cambridge University Press, 1994.

6 Algebras

In this chapter we allow for yet another law of composition to be imposed on vector spaces, whereby the product of any two vectors results in another vector from the same vector space. Structures of this kind are generically called *algebras* and arise naturally in a variety of contexts [1, 2].

6.1 Algebras and ideals

An **algebra** consists of a vector space $\mathcal{A}$ over a field $\mathbb{K}$ together with a *law of composition* or **product** of vectors, $\mathcal{A} \times \mathcal{A} \to \mathcal{A}$, denoted

$$(A, B) \mapsto AB \in \mathcal{A} \quad (A, B \in \mathcal{A}),$$

which satisfies a pair of distributive laws:

$$A(aB + bC) = aAB + bAC, \quad (aA + bB)C = aAC + bBC \quad (6.1)$$

for all scalars $a, b \in \mathbb{K}$ and vectors A, B and C. In the right-hand sides of Eq. (6.1) quantities such as aAB are short for $a(AB)$; this is permissible on setting $b = 0$, which gives the identities $aAB = (aA)B = A(aB)$. In Section 3.2 it was shown that $0A = A0 = O$ for all $A \in \mathcal{A}$, taking careful note of the difference between the zero vector O and the zero scalar 0. Hence

$$OA = (0A)A = 0(AA) = O, \quad AO = A(0A) = 0AA = O.$$

We have used capital letters A, B, etc. to denote vectors because algebras most frequently arise in spaces of linear operators over a vector space V. There is, however, nothing in principle to prevent the more usual notation $u, v, \ldots$ for vectors and to write uv for their product. The vector product has been denoted by a simple juxtaposition of vectors, but other notations such as $A \times B$, $A \otimes B$, $A \wedge B$ and $[A, B]$ may arise, depending upon the context. The algebra is said to be **associative** if $A(BC) = (AB)C$ for all $A, B, C \in \mathcal{A}$. It is called **commutative** if $AB = BA$ for all $A, B \in \mathcal{A}$.

Example 6.1 On the vector space of ordinary three-dimensional vectors $\mathbb{R}^3$ define the usual vector product $\mathbf{u} \times \mathbf{v}$ by

$$(\mathbf{u} \times \mathbf{v})_i = \sum_{j=1}^{3} \sum_{k=1}^{3} \epsilon_{ijk} u_j v_k$$

Algebras

where

$$\epsilon_{ijk} = \begin{cases} 0 & \text{if any pair of indices } i, j, k \text{ are equal,} \\ 1 & \text{if } ijk \text{ is an even permutation of 123,} \\ -1 & \text{if } ijk \text{ is an odd permutation of 123.} \end{cases}$$

The vector space $\mathbb{R}^3$ with this law of composition is a non-commutative, non-associative algebra. The product is non-commutative since

$$\mathbf{u} \times \mathbf{v} = -\mathbf{v} \times \mathbf{u},$$

and it is non-associative as

$$(\mathbf{u} \times \mathbf{v}) \times \mathbf{w} - \mathbf{u} \times (\mathbf{v} \times \mathbf{w}) = (\mathbf{u} \cdot \mathbf{v})\mathbf{w} - (\mathbf{v} \cdot \mathbf{w})\mathbf{u}$$

does not vanish in general.

Example 6.2 The vector space $L(V, V)$ of linear operators on a vector space V forms an associative algebra where the product AB is defined in the usual way,

$$(AB)u = A(Bu).$$

The distributive laws (6.1) follow trivially and the associative law $A(BC) = (AB)C$ holds for all linear transformations. It is, however, non-commutative as $AB \neq BA$ in general.

Similarly the set of all $n \times n$ real matrices $\mathcal{M}_n$ forms an algebra with respect to matrix multiplication, since it may be thought of as being identical with $L(\mathbb{R}^n, \mathbb{R}^n)$ where $\mathbb{R}^n$ is the vector space of $n \times 1$ column vectors. If the field of scalars is the complex numbers, we use $\mathcal{M}_n(\mathbb{C})$ to denote the algebra of $n \times n$ complex matrices.

If $\mathcal{A}$ is a finite dimensional algebra and $E_1, E_2, \ldots, E_n$ any basis, then let C_{ij}^k be a set of scalars defined by

$$E_i E_j = C_{ij}^k E_k. \tag{6.2}$$

The scalars $C_{ij}^k \in \mathbb{K}$, uniquely defined as the components of the vector $E_i E_j$ with respect to the given basis, are called the **structure constants** of the algebra with respect to the basis $\{E_i\}$. This is a common way of defining an algebra for, once the structure constants are specified with respect to any basis, we can generate the product of any pair of vectors $A = a^i E_i$ and $B = b^j E_j$ by the distributive law (6.1),

$$AB = (a^i E_i)(b^j E_j) = a^i b^j E_i E_j = \left(a^i b^j C_{ij}^k\right) E_k.$$

Exercise: Show that an algebra is commutative iff the structure constants are symmetric in the subscripts, $C_{ij}^k = C_{ji}^k$.

Let $\mathcal{A}$ and $\mathcal{B}$ be any pair of algebras. A linear map $\varphi : \mathcal{A} \to \mathcal{B}$ is called an **algebra homomorphism** if it preserves products, $\varphi(AB) = \varphi(A)\varphi(B)$.

Exercise: Show that for any pair of scalars a, b and vectors A, B, C

$$\varphi\big(A(aB + bC)\big) = a\varphi(A)\varphi(B) + b\varphi(A)\varphi(C).$$

6.1 Algebras and ideals

A **subalgebra** $\mathcal{B}$ of $\mathcal{A}$ is a vector subspace that is closed under the law of composition,

$$A \in \mathcal{B}, \ B \in \mathcal{B} \Longrightarrow AB \in \mathcal{B}.$$

Exercise: Show that if $\varphi : \mathcal{A} \to \mathcal{B}$ is an algebra homomorphism then the image set $\varphi(\mathcal{A}) \subseteq \mathcal{B}$ is a subalgebra of $\mathcal{B}$.

A homomorphism φ is called an **algebra isomorphism** if it is one-to-one and onto; the two algebras $\mathcal{A}$ and $\mathcal{B}$ are then said to be **isomorphic**.

Example 6.3 On the vector space $\mathbb{R}^\infty$ define a law of multiplication

$$(a_0, a_1, a_2, \ldots)(b_0, b_1, b_2, \ldots) = (c_0, c_1, c_2, \ldots)$$

where

$$c_p = a_0 b_p + a_1 b_{p-1} + \cdots + a_p b_0.$$

Setting $A = (a_0, a_1, a_2, \ldots)$, $B = (b_0, b_1, b_2, \ldots)$, it is straightforward to verify Eq. (6.1) and the commutative law $AB = BA$. Hence with this product law, $\mathbb{R}^\infty$ is a commutative algebra. Furthermore this algebra is associative,

$$(A(BC))_p = \sum_{i=0}^{p} a_i (bc)_{p-i}$$

$$= \sum_{i=0}^{p} \sum_{j=0}^{p-i} a_i b_j c_{p-i-j}$$

$$= \sum_{i+j+k=p} a_i b_j c_k$$

$$= \sum_{j=0}^{p} \sum_{i=0}^{p-j} a_j b_{p-i-j} c_i$$

$$= ((AB)C)_p.$$

The infinite dimensional vector space of all real polynomials $\mathcal{P}$ is also a commutative and associative algebra, whereby the product of a polynomial $f(x) = a_0 + a_1 x + a_2 x^2 + \cdots + a_n x^n$ of degree n and $g(x) = b_0 + b_1 x + b_2 x^2 + \cdots + b_m x^m$ of degree m results in a polynomial $f(x)g(x)$ of degree $m+n$ in the usual way. On explicitly carrying out the multiplication of two such polynomials it follows that the map $\varphi : \mathcal{P} \to \mathbb{R}^\infty$ defined by

$$\varphi(f(x)) = (a_0, a_1, a_2, \ldots, a_n, 0, 0, \ldots)$$

is an algebra homomorphism. In Example 3.10 it was shown that the map φ (denoted S in that example) establishes a vector space isomorphism between $\mathcal{P}$ and the vector space $\hat{\mathbb{R}}^\infty$ of sequences having only finitely many non-zero terms. If $A \in \hat{\mathbb{R}}^\infty$ let us call its *length* the largest natural number p such that $a_p \ne 0$. From the law of composition it follows that if A has length p and B has length q then AB is a vector of length $\le p + q$. The space $\hat{\mathbb{R}}^\infty$ is a subalgebra of $\mathbb{R}^\infty$, and is isomorphic to the algebra $\mathcal{P}$.

Algebras

Ideals and factor algebras

A vector subspace $\mathcal{B}$ of $\mathcal{A}$ is a subalgebra if it is closed with respect to products, a property that may be written $\mathcal{BB} \subseteq \mathcal{B}$. A vector subspace $\mathcal{L}$ of $\mathcal{A}$ is called a **left ideal** if

$$L \in \mathcal{L}, \ A \in \mathcal{A} \implies AL \in \mathcal{L},$$

or, in the above notation, $\mathcal{AL} \subseteq \mathcal{L}$. Similarly a **right ideal** $\mathcal{R}$ is a subspace such that

$$\mathcal{RA} \subseteq \mathcal{R}.$$

A **two-sided ideal** or simply an **ideal** is a subspace $\mathcal{I}$ that is both a left and right-sided ideal. An ideal is always a subalgebra, but the converse is not true.

Ideals play a role in algebras parallel to that played by normal subgroups in group theory (see Section 2.5). To appreciate this correspondence let $\varphi : \mathcal{A} \to \mathcal{B}$ be an algebra homomorphism between any two algebras. As in Section 3.4, define the **kernel** $\ker \varphi$ of the linear map φ to be the vector subspace of $\mathcal{A}$ consisting of those vectors that are mapped to the zero element O' of $\mathcal{B}$, namely $\ker \varphi = \varphi^{-1}(O')$.

Theorem 6.1 *The kernel of an algebra homomorphism $\varphi : \mathcal{A} \to \mathcal{B}$ is an ideal of $\mathcal{A}$. Conversely, if $\mathcal{I}$ is an ideal of $\mathcal{A}$ then there is a natural algebra structure defined on the factor space $\mathcal{A}/\mathcal{I}$ such that the map $\varphi : \mathcal{A} \to \mathcal{A}/\mathcal{I}$ whereby $A \mapsto [A] \equiv A + \mathcal{I}$ is a homomorphism with kernel $\mathcal{I}$.*

Proof: The vector subspace $\ker \varphi$ is a left ideal of $\mathcal{A}$, for if $B \in \ker \varphi$ and $A \in \mathcal{A}$ then $AB \in \ker \varphi$, for

$$\varphi(AB) = \varphi(A)\varphi(B) = \varphi(A)O' = O'.$$

Similarly $\ker \varphi$ is a right ideal.

If $\mathcal{I}$ is an ideal of $\mathcal{A}$, denote the typical elements of $\mathcal{A}/\mathcal{I}$ by the coset $[A] = A + \mathcal{I}$ and define an algebra structure on $\mathcal{A}/\mathcal{I}$ by setting $[A][B] = [AB]$. This product rule is 'natural' in the sense that it is independent of the choice of representative from $[A]$ and $[B]$, for if $[A'] = [A]$ and $[B'] = [B]$ then $A' \in A + \mathcal{I}$ and $B' \in B + \mathcal{I}$. Using the fact that $\mathcal{I}$ is both a left and right ideal, we have

$$A'B' \in (A + \mathcal{I})(B + \mathcal{I}) = AB + A\mathcal{I} + \mathcal{I}B + \mathcal{I}\mathcal{I} = AB + \mathcal{I}.$$

Hence $[A'][B'] = [A'B'] = [AB] = [A][B]$. The map $\varphi : \mathcal{A} \to \mathcal{A}/\mathcal{I}$ defined by $\varphi(A) = [A]$ is clearly a homomorphism, and its kernel is $\varphi^{-1}([O]) = \mathcal{I}$. ∎

6.2 Complex numbers and complex structures

The complex numbers $\mathbb{C}$ form a two-dimensional commutative and associative algebra over the real numbers, with a basis $\{1, i\}$ having the defining relations

$$1^2 = 11 = 1, \quad i1 = 1i = i, \quad i^2 = ii = -1.$$

6.2 Complex numbers and complex structures

Setting $E_1 = 1$, $E_2 = i$ the structure constants are

$$C^1_{11} = 1 \quad C^1_{12} = C^1_{21} = 0 \quad C^1_{22} = -1$$
$$C^2_{11} = 0 \quad C^2_{12} = C^2_{21} = 1 \quad C^2_{22} = 0.$$

It is common to write the typical element $xE_1 + yE_2 = x\mathbf{1} + xi$ simply as $x + iy$ and Eq. (6.1) gives the standard rule for complex multiplication,

$$(u + iv)(x + iy) = ux - vy + i(uy + vx).$$

Exercise: Verify that this algebra is commutative and associative.

Every non-zero complex number $\alpha = x + iy$ has an **inverse** α^{-1} with the property $\alpha \alpha^{-1} = \alpha^{-1} \alpha = 1$. Explicitly,

$$\alpha^{-1} = \frac{\bar{\alpha}}{|\alpha|^2},$$

where

$$\bar{\alpha} = x - iy$$

and

$$|\alpha| = \sqrt{\alpha \bar{\alpha}} = \sqrt{x^2 + y^2}$$

are the **complex conjugate** and **modulus** of α, respectively.

Any algebra in which all non-zero vectors have an inverse is called a **division algebra**, since for any pair of elements A, B ($B \ne O$) it is possible to define $A/B = AB^{-1}$. The complex numbers are the only associative, commutative division algebra of dimension > 1 over the real numbers $\mathbb{R}$.

Exercise: Show that an associative, commutative division algebra is a field.

Example 6.4   There is a different, but occasionally useful representation of the complex numbers as matrices. Let I and J be the matrices

$$\mathsf{I} = \begin{pmatrix} 1 & 0 \\ 0 & 1 \end{pmatrix}, \quad \mathsf{J} = \begin{pmatrix} 0 & 1 \\ -1 & 0 \end{pmatrix}.$$

It is a trivial matter to verify that

$$\mathsf{J}\mathsf{I} = \mathsf{I}\mathsf{J} = \mathsf{J}, \quad \mathsf{I}^2 = \mathsf{I}, \quad \mathsf{J}^2 = -\mathsf{I}, \tag{6.3}$$

and the subalgebra of $\mathcal{M}_2$ generated by these two matrices is isomorphic to the algebra of complex numbers. The isomorphism can be displayed as

$$x + iy \longleftrightarrow x\mathsf{I} + y\mathsf{J} = \begin{pmatrix} x & y \\ -y & x \end{pmatrix}.$$

Exercise: Check that the above map is an isomorphism by verifying that

$$(u + iv)(x + iy) \longleftrightarrow (u\mathsf{I} + v\mathsf{J})(x\mathsf{I} + v\mathsf{J}).$$

Complexification of a real vector space

Define the **complexification** V^C of a real vector space V as the set of all ordered pairs $w = (u, v) \in V \times V$ with vector addition and scalar product by complex numbers defined as

$$(u, v) + (u', v') = (u + u', v + v'),$$
$$(a + ib)(u, v) = (au - bv, bu + av),$$

for all $u, u', v, v' \in V$ and $a, b \in \mathbb{R}$. This process of transforming any real vector space into a complex space is totally natural, independent of choice of basis.

Exercise: Verify that the axioms (VS1)–(VS6) in Section 3.2 are satisfied for V^C with the complex numbers $\mathbb{C}$ as the field of scalars. Most axioms are trivial, but (VS4) requires proof:

$$(c + id)((a + ib)(u, v)) = ((c + id)(a + ib))(u, v).$$

Essentially what we have done here is to 'expand' the original vector space by permitting multiplication with complex scalars. There is no ambiguity in adopting the notation $w = u + iv$ for $w = (u, v)$, since

$$(a + ib)(u + iv) \equiv (a + ib)(u, v) = (au - bv, bu + av) \equiv au - bv + i(bu + av).$$

If V is finite dimensional and $n = \dim V$ then V^C is also finite dimensional and has the same dimension as V. For, let $\{e_i \,|\, i = 1, \ldots, n\}$ be any basis of V. These vectors clearly span V^C, for if $u = u^j e_j$ and $v = v^j e_j$ are any pair of vectors in V then

$$u + iv = (u^j + iv^j)e_j.$$

Furthermore, the vectors $\{e_j\}$ are linearly independent over the field of complex numbers, for if $(u^j + iv^j)e_j = 0$ then $(u^j e_j, v^j e_j) = (0, 0)$. Hence $u^j e_j = 0$ and $v^j e_j = 0$, so that $u^j = v^j = 0$ for all $j = 1, \ldots, n$. Thus $\{e_1, e_2, \ldots, e_n\}$ also forms a basis for V^C.

In the complexification V^C of a real space V we can define *complex conjugation* by

$$\overline{w} = \overline{u + iv} = u - iv.$$

In an arbitrary complex vector space, however, there is no natural, basis-independent, way of defining complex conjugation of vectors. For example, if we set the complex conjugate of a vector $u = u^j e_j$ to be $\overline{u} = \overline{u^j} e_j$, this definition will give a different answer in the basis $\{ie_j\}$ since

$$u = (-iu^j)(ie_j) \implies \overline{u} = (\overline{iu^j})(ie_j) = -\overline{u^j} e_j.$$

Thus the concept of complex conjugation of vectors requires prior knowledge of the 'real part' of a complex vector space. The complexification of a real space has precisely the required extra structure needed to define complex conjugation of vectors, but there is no natural way of reversing the complexification process to produce a real vector space of the same dimension from any given complex vector space.

6.2 Complex numbers and complex structures

Complex structure on a vector space

One way of creating a real vector space V^R from a complex vector space V is to forget altogether about the possibility of multiplying vectors by complex numbers and only allow scalar multiplication with real numbers. In this process a pair of vectors u and iu must be regarded as linearly independent vectors in V^R for any non-zero vector $u \in V$. Thus if V is finite dimensional and dim $V = n$, then V^R is $2n$-dimensional, for if $\{e_1, \ldots, e_n\}$ is a basis of V then

$$e_1, e_2, \ldots, e_n, ie_1, ie_2, \ldots, ie_n$$

is readily shown to be a l.i. set of vectors spanning V^R.

To reverse this 'realification' of a complex vector space, observe firstly that the operator $J : V^R \to V^R$ defined by $Jv = iv$ satisfies the relation $J^2 = -\mathrm{id}_{V^R}$. We now show that given any operator on a real vector space having this property, it is possible to define a passage to a complex vector space. This process is not to be confused with the complexification of a vector space, but there is a connection with it (see Problem 6.2).

If V is a real vector space, any operator $J : V \to V$ such that $J^2 = -\mathrm{id}_V$ is called a **complex structure** on V. A complex structure J can be used to convert V into a complex vector space V_J by defining addition of vectors $u + v$ just as in the real space V, and scalar multiplication of vectors by complex numbers through

$$(a + ib)v = av + bJv.$$

It remains to prove that V_J is a complex vector space; for example, to show axiom (VS4) of Section 3.2

$$\begin{aligned}(a+ib)((c+id)v) &= a(cv + dJv) + bJ(cv + dJv) \\ &= (ac - bd)v + (ad + bc)Jv \\ &= (ac - bd + i(ad + bc))v \\ &= ((a+ib)(c+id))v.\end{aligned}$$

Most other axioms are trivial.

A complex structure is always an invertible operator since

$$JJ^3 = J^4 = (-\mathrm{id}_V)^2 = \mathrm{id}_V \implies J^{-1} = J^3.$$

Furthermore if dim $V = n$ and $\{e_1, \ldots, e_n\}$ is any basis of V then the matrix $\mathsf{J} = [J_i^j]$ defined by $Je_i = J_i^j e_j$ satisfies

$$\mathsf{J}^2 = -\mathsf{I}.$$

Taking determinants gives

$$(\det \mathsf{J})^2 = \det(-\mathsf{I}) = (-1)^n,$$

which is only possible for a real matrix J if n is an even number, $n = 2m$. Thus a real vector space can only have a complex structure if it is even dimensional.

As a set, the original real vector space V is identical to the complex space V_J, but scalar multiplication is restricted to the reals. It is in fact the real space constructed from V_J by the above realification process,

$$V = (V_J)^R.$$

Hence the dimension of the complex vector space V_J is half that of the real space from which it comes, $\dim V_J = m = \frac{1}{2} \dim V$.

Problems

Problem 6.1 The following is an alternative method of defining the algebra of complex numbers. Let $\mathcal{P}$ be the associative algebra consisting of real polynomials on the variable x, defined in Example 6.3. Set $\mathcal{C}$ to be the ideal of $\mathcal{P}$ generated by $x^2 + 1$; i.e., the set of all polynomials of the form $f(x)(x^2 + 1)g(x)$. Show that the linear map $\phi : \mathbb{C} \to \mathcal{P}/\mathcal{C}$ defined by

$$\phi(i) = [x] = x + \mathcal{C}, \qquad \phi(1) = [1] = 1 + \mathcal{C}$$

is an algebra isomorphism.

Which complex number is identified with the polynomial class $[1 + x + 3x^2 + 5x^3] \in \mathcal{P}/\mathcal{C}$?

Problem 6.2 Let J be a complex structure on a real vector space V, and set

$$V(J) = \{v = u - iJu \mid u \in V\} \subseteq V^C, \qquad \bar{V}(J) = \{v = u + iJu \mid u \in V\}.$$

(a) Show that $V(J)$ and $\bar{V}(J)$ are complex vector subspaces of V^C.
(b) Show that $v \in V(J) \Rightarrow Jv = iv$ and $v \in \bar{V}(J) \Rightarrow Jv = -iv$.
(c) Prove that the complexification of V is the direct sum of $V(J)$ and $\bar{V}(J)$,

$$V^C = V(J) \oplus \bar{V}(J).$$

Problem 6.3 If V is a real vector space and U and $\bar{U}$ are complex conjugate subspaces of V^C such that $V^C = U \oplus \bar{U}$, show that there exists a complex structure J for V such that $U = V(J)$ and $\bar{U} = \bar{V}(J)$, where $V(J)$ and $\bar{V}(J)$ are defined in the previous problem.

Problem 6.4 Let J be a complex structure on a real vector space V of dimension $n = 2m$. Let $u_1, u_2, \ldots, u_m$ be a basis of the subspace $V(J)$ defined in Problem 6.2, and set

$$u_a = e_a - ie_{m+a} \quad \text{where} \quad e_a, e_{m+a} \in V \quad (a = 1, \ldots, m).$$

Show that the matrix $J_0 = [J_i^j]$ of the complex structure, defined by $Je_i = J_i^j e_j$ where $i = 1, 2, \ldots, n = 2m$, has the form

$$J_0 = \begin{pmatrix} 0 & I \\ -I & 0 \end{pmatrix}.$$

Show that the matrix of any complex structure with respect to an arbitrary basis has the form

$$J = AJ_0 A^{-1}.$$

6.3 Quaternions and Clifford algebras

Quaternions

In 1842 Hamilton showed that the next natural generalization to the complex numbers must occur in four dimensions. Let $\mathcal{Q}$ be the associative algebra over $\mathbb{R}$ generated by four elements $\{1, i, j, k\}$ satisfying

$$i^2 = j^2 = k^2 = -1,$$
$$ij = k, \quad jk = i, \quad ki = j,$$
$$1^2 = 1, \quad 1i = i, \quad 1j = j, \quad 1k = k. \tag{6.4}$$

The element 1 may be regarded as being identical with the real number 1. From these relations and the associative law it follows that

$$ji = -k, \quad kj = -i, \quad ik = -j.$$

To prove the first identity use the defining relation $jk = i$ and the associative law,

$$ji = j(jk) = (jj)k = j^2 k = (-1)k = -k.$$

The other identities follow in a similar way. The elements of this algebra are called **quaternions**; they form a non-commutative algebra since $ij - ji = 2k \neq 0$.

Exercise: Write out the structure constants of the quaternion algebra for the basis $E_1 = 1$, $E_2 = i$, $E_3 = j$, $E_4 = k$.

Every quaternion can be written as

$$Q = q_0 1 + q_1 i + q_2 j + q_3 k = q_0 + \mathbf{q}$$

where q_0 is known as its **scalar part** and $\mathbf{q} = q_1 i + q_2 j + q_3 k$ is its **vector part**. Define the **conjugate quaternion** $\overline{Q}$ by

$$\overline{Q} = q_0 1 - q_1 i - q_2 j - q_3 k = q_0 - \mathbf{q}.$$

Pure quaternions are those of the form $Q = q_1 i + q_2 j + q_3 k = \mathbf{q}$, for which the scalar part vanishes. If $\mathbf{p}$ and $\mathbf{q}$ are pure quaternions then

$$\mathbf{pq} = -\mathbf{p} \cdot \mathbf{q} + \mathbf{p} \times \mathbf{q}, \tag{6.5}$$

a formula in which both the scalar product and cross product of ordinary 3-vectors make an appearance.

Exercise: Prove Eq. (6.5).

For full quaternions

$$PQ = (p_0 + \mathbf{p})(q_0 + \mathbf{q})$$
$$= p_0 q_0 - \mathbf{p} \cdot \mathbf{q} + p_0 \mathbf{q} + q_0 \mathbf{p} + \mathbf{p} \times \mathbf{q}. \tag{6.6}$$

Algebras

Curiously, the scalar part of PQ is the four-dimensional Minkowskian scalar product of special relativity

$$\tfrac{1}{2}(PQ + \overline{PQ}) = p_0 q_0 - p_1 q_1 - p_2 q_2 - p_3 q_3 = p_0 q_0 - \mathbf{p} \cdot \mathbf{q}.$$

To show that quaternions form a division algebra, define the **magnitude** $|Q|$ of a quaternion Q by

$$|Q|^2 = Q\overline{Q} = q_0^2 + q_1^2 + q_2^2 + q_3^2.$$

The right-hand side is clearly a non-negative quantity that vanishes if and only if $Q = 0$.

Exercise: Show that $|\overline{Q}| = |Q|$.

The **inverse** of any non-zero quaternion Q is

$$Q^{-1} = \frac{\overline{Q}}{|Q|^2},$$

since

$$Q^{-1}Q = QQ^{-1} = \frac{\overline{Q}Q}{|Q|^2} = 1.$$

Hence, as claimed, quaternions form a division algebra.

Clifford algebras

Let V be a real vector space with inner product $u \cdot v$, and $e_1, e_2, \ldots, e_n$ an orthonormal basis,

$$g_{ij} = e_i \cdot e_j = \begin{cases} \pm 1 & \text{if } i = j, \\ 0 & \text{if } i \neq j. \end{cases}$$

The **Clifford algebra** associated with this inner product space, denoted C_g, is defined as the associative algebra generated by $1, e_1, e_2, \ldots, e_n$ with the product rules

$$e_i e_j + e_j e_i = 2g_{ij} 1, \qquad 1 e_i = e_i 1 = e_i. \tag{6.7}$$

The case $n = 1$ and $g_{11} = -1$ gives rise to the complex numbers on setting $i = e_1$. The algebra of quaternions arises on setting $n = 2$ and $g_{ij} = -\delta_{ij}$, and making the identifications

$$i \equiv e_1, \qquad j \equiv e_2, \qquad k \equiv e_1 e_2 = -e_2 e_1.$$

Evidently $k = ij = -ji$, while other quaternionic identities in Eq. (6.4) are straightforward to show. For example,

$$ki = e_1 e_2 e_1 = -e_1 e_1 e_2 = e_2 = j, \text{ etc.}$$

Thus Clifford algebras are a natural generalization of complex numbers and quaternions. They are not, however, division algebras – the only possible higher dimensional division algebra turns out to be non-associative and is known as an **octonian**.

6.3 Quaternions and Clifford algebras

The Clifford algebra C_g is spanned by successive products of higher orders $e_i e_j$, $e_i e_j e_k$, etc. However, since any pair $e_i e_j = -e_j e_i$ for $i \neq j$, it is possible to keep commuting neighbouring elements of any product $e_{i_1} e_{i_2} \ldots e_{i_r}$ until they are arranged in increasing order $i_1 \leq i_2 \leq \cdots \leq i_r$, with at most a change of sign occurring in the final expression. Furthermore, whenever an equal pair appear next to each other, $e_i e_i$, they can be replaced by $g_{ii} = \pm 1$, so there is no loss of generality in assuming $i_1 < i_2 < \cdots < i_r$. The whole algebra is therefore spanned by

$$1, \{e_i \mid i = 1, \ldots, n\}, \{e_i e_j \mid i < j\}, \{e_i e_j e_k \mid i < j < k\}, \ldots$$
$$\{e_{i_1} e_{i_2} \ldots e_{i_r} \mid i_1 < i_2 < \cdots < i_r\}, \ldots, e_1 e_2 \ldots e_n.$$

Each basis element can be labelled e_A where A is any subset of the integers $\{1, 2, \ldots, n\}$, the empty set corresponding to the unit scalar, $e_\emptyset \equiv 1$. From Example 1.1 we have the dimension of C_g is 2^n. The definition of Clifford algebras given here depends on the choice of basis for V. It is possible to give a basis-independent definition but this involves the concept of a free algebra (see Problem 7.5 of the next chapter).

The most important application of Clifford algebras in physics is the relativistic theory of the spin $\frac{1}{2}$ particles. In 1928, Paul Dirac (1902–1984) sought a linear first-order equation for the electron,

$$\gamma^\mu \partial_\mu \psi = -m_e \psi \quad \text{where} \quad \mu = 1, 2, 3, 4, \ \partial_\mu \equiv \frac{\partial}{\partial x^\mu}.$$

In order that this equation imply the relativistic *Klein–Gordon equation*,

$$g^{\mu\nu} \partial_\mu \partial_\nu \psi = -m_e^2 \psi$$

where

$$[g^{\mu\nu}] = [g_{\mu\nu}] = \begin{pmatrix} 1 & 0 & 0 & 0 \\ 0 & 1 & 0 & 0 \\ 0 & 0 & 1 & 0 \\ 0 & 0 & 0 & -1 \end{pmatrix},$$

it is required that the coefficients γ^μ satisfy

$$\gamma^\mu \gamma^\nu + \gamma^\nu \gamma^\mu = 2 g^{\mu\nu}.$$

The elements γ_μ defined by 'lowering the index', $\gamma_\mu = g_{\mu\rho} \gamma^\rho$, must satisfy

$$\gamma_\mu \gamma_\nu + \gamma_\nu \gamma_\mu = 2 g_{\mu\nu}$$

and can be used to generate a Clifford algebra with $n = 4$. Such a Clifford algebra has $2^4 = 16$ dimensions. If one attempts to represent this algebra by a set of matrices, the lowest possible order turns out to be 4×4 matrices. The vectorial quantities ψ on which these matrices act are known as *spinors*; they have at least four components, a fact related to the concept of relativistic spin. The greatest test for Dirac's theory was the prediction of antiparticles known as *positrons*, shown to exist experimentally by Anderson in 1932.

Algebras

Problems

Problem 6.5 Show the 'anticommutation law' of conjugation,

$$\overline{PQ} = \overline{Q}\,\overline{P}.$$

Hence prove

$$|PQ| = |P||Q|.$$

Problem 6.6 Show that the set of 2×2 matrices of the form

$$\begin{pmatrix} z & w \\ -\bar{w} & \bar{z} \end{pmatrix},$$

where z and w are complex numbers, forms an algebra of dimension 4 over the real numbers.

(a) Show that this algebra is isomorphic to the algebra of quaternions by using the bijection

$$Q = a + bi + cj + dk \longleftrightarrow \begin{pmatrix} a + ib & c + id \\ -c + id & a - ib \end{pmatrix}.$$

(b) Using this matrix representation prove the identities given in Problem 6.5.

Problem 6.7 Find a quaternion Q such that

$$Q^{-1}iQ = j, \qquad Q^{-1}jQ = k.$$

[*Hint*: Write the first equation as $iQ = Qj$.] For this Q calculate $Q^{-1}kQ$.

Problem 6.8 Let e_A and e_B where $A, B \subseteq \{1, 2, \ldots, n\}$ be two basis elements of the Clifford algebra associated with the Euclidean inner product space having $g_{ij} = \delta_{ij}$. Show that $e_A e_B = \pm e_C$ where $C = A \cup B - A \cap B$. Show that a plus sign appears in this rule if the number of pairs

$$\{(i_r, j_s) \mid i_r \in A, \ j_s \in B, \ i_r > j_s\}$$

is even, while a minus sign occurs if this number of pairs is odd.

6.4 Grassmann algebras

Multivectors

Hermann Grassmann took a completely different direction to generalize Hamilton's quaternion algebra (1844), one in which there is no need for an inner product. Grassmann's idea was to regard entire subspaces of a vector space as single algebraic objects and to define a method of multiplying them together. The resulting algebra has far-reaching applications, particularly in differential geometry and the theory of integration on manifolds (see Chapters 16 and 17). In this chapter we present Grassmann algebras in a rather intuitive way, leaving a more formal presentation to Chapter 8.

Let V be any real vector space. For any pair of vectors $u, v \in V$ define an abstract quantity $u \wedge v$, subject to the following identifications for all vectors $u, v, w \in V$ and scalars

6.4 Grassmann algebras

$a, b \in \mathbb{R}$:

$$(au + bv) \wedge w = au \wedge w + bv \wedge w, \tag{6.8}$$

$$u \wedge v = -v \wedge u. \tag{6.9}$$

Any quantity $u \wedge v$ will be known as a **simple 2-vector** or **bivector**. Taking into account the identities (6.8) and (6.9), we denote by $\Lambda^2(V)$ the vector space generated by the set of all simple 2-vectors. Without fear of confusion, we denote vector addition in $\Lambda^2(V)$ by the same symbol $+$ as for the vector space V. Every element A of $\Lambda^2(V)$, generically known as a **2-vector**, can be written as a sum of bivectors,

$$A = \sum_{i=1}^{r} a_i u_i \wedge v_i = a_1 u_1 \wedge v_1 + a_2 u_2 \wedge v_2 + \cdots + a_r u_r \wedge v_r. \tag{6.10}$$

From (6.8) and (6.9) the wedge operation is obviously linear in the second argument,

$$u \wedge (av + bw) = au \wedge v + bu \wedge w. \tag{6.11}$$

Also, for any vector u,

$$u \wedge u = -u \wedge u \implies u \wedge u = 0.$$

As an intuitive aid it is useful to think of a simple 2-vector $u \wedge v$ as representing the subspace or 'area element' spanned by u and v. If u and v are proportional to each other, $v = au$, the area element collapses to a line and vanishes, in agreement with $u \wedge v = au \wedge u = 0$. No obvious geometrical picture presents itself for non-simple 2-vectors, and a sum of simple bivectors such as that in Eq. (6.10) must be thought of in a purely formal way.

If V has dimension n and $\{e_1, \ldots, e_n\}$ is a basis of V then for any pair of vectors $u = u^i e_i$, $v = v^j e_j$

$$u \wedge v = u^i v^j e_i \wedge e_j = -v^j u^i e_j \wedge e_i.$$

Setting

$$e_{ij} = e_i \wedge e_j = -e_{ji} \quad (i, j = 1, 2, \ldots, n)$$

it follows from Eqs. (6.10) and (6.8) that all elements of $\Lambda^2(V)$ can be written as a linear combination of the bivectors e_{ij}

$$A = A^{ij} e_{ij}.$$

Furthermore, since $e_{ij} = -e_{ji}$, each term in this sum can be converted to a sum of terms

$$A = \sum_{i<j} (A^{ij} - A^{ji}) e_{ij},$$

and the space of 2-vectors, $\Lambda^2(V)$, is spanned by the set

$$E_2 = \{e_{ij} \mid 1 \le i < j \le n\}.$$

As E_2 consists of $\binom{n}{2}$ elements it follows that

$$\dim(\Lambda^2(V)) \le \binom{n}{2} = \frac{n(n-1)}{2}.$$

Algebras

In the tensorial approach of Chapter 8 it will emerge that the set E_2 is linearly independent, whence

$$\dim(\Lambda^2(V)) = \binom{n}{2}.$$

The space of r-**vectors**, $\Lambda^r(V)$, is defined in an analogous way. For any set of r vectors $u_1, u_2, \ldots, u_r$, let the **simple r-vector** spanned by these vectors be defined as the abstract object $u_1 \wedge u_2 \wedge \cdots \wedge u_r$, and define a general r-**vector** to be a formal linear sum of simple r-vectors,

$$A = \sum_{J=1}^{N} a_J u_{J1} \wedge u_{J2} \wedge \cdots \wedge u_{Jr} \quad \text{where} \quad a_J \in \mathbb{R}, u_{Ji} \in V.$$

In forming such sums we impose linearity in the first argument,

$$(au_1 + bu'_1) \wedge u_2 \wedge \cdots \wedge u_r = au_1 \wedge u_2 \wedge \cdots \wedge u_r + bu'_1 \wedge u_2 \wedge \cdots \wedge u_r, \quad (6.12)$$

and skew symmetry in any pair of vectors,

$$u_1 \wedge \cdots \wedge u_i \wedge \cdots \wedge u_j \wedge \cdots \wedge u_r = -u_1 \wedge \cdots \wedge u_j \wedge \cdots \wedge u_i \wedge \cdots \wedge u_r. \quad (6.13)$$

As for 2-vectors, linearity holds on each argument separately

$$u_1 \wedge \cdots \wedge (au_i + bu'_i) \wedge \cdots \wedge u_r = au_1 \wedge \cdots \wedge u_i \wedge \cdots \wedge u_r + bu_1 \wedge \cdots \wedge u'_i \wedge \cdots \wedge u_r,$$

and for a general permutation π of $1, 2, \ldots, r$

$$u_1 \wedge u_2 \wedge \cdots \wedge u_r = (-1)^\pi u_{\pi(1)} \wedge u_{\pi(2)} \wedge \cdots \wedge u_{\pi(r)}. \quad (6.14)$$

If any two vectors among $u_1, \ldots, u_r$ are equal then $u_1 \wedge u_2 \wedge \cdots \wedge u_r$ vanishes,

$$u_1 \wedge \cdots \wedge u_i \wedge \cdots \wedge u_i \wedge \cdots \wedge u_r = -u_1 \wedge \cdots \wedge u_i \wedge \cdots \wedge u_i \wedge \cdots \wedge u_r = 0. \quad (6.15)$$

Again, it is possible to think of a simple r-vector as having the geometrical interpretation of an r-dimensional subspace or volume element spanned by the vectors $u_1, \ldots, u_r$. The general r-vector is a formal sum of such volume elements.

If V has dimension n and $\{e_1, \ldots, e_n\}$ is a basis of V, it is convenient to define the r-vectors

$$e_{i_1 i_2 \ldots i_r} = e_{i_1} \wedge e_{i_2} \cdots \wedge e_{i_r}.$$

For any permutation π of $1, 2, \ldots, r$ Eq. (6.14) implies that

$$e_{i_1 i_2 \ldots i_r} = (-1)^\pi e_{i_{\pi(1)} i_{\pi(2)} \ldots i_{\pi(r)}},$$

while if any pair of indices are equal, say $i_a = i_b$ for some $1 \le a < b \le r$, then $e_{i_1 i_2 \ldots i_r} = 0$. For example

$$e_{123} = -e_{321} = e_{231}, \text{ etc.} \qquad e_{112} = e_{233} = 0, \text{ etc.}$$

By a permutation of vectors e_j the r-vector $e_{i_1 i_2 \ldots i_r}$ may be brought to a form in which $i_1 < i_2 < \cdots < i_r$, to within a possible change of sign. Since the simple r-vector spanned

6.4 Grassmann algebras

by vectors $u_1 = u_1^i e_i, \ldots, u_r = u_r^i e_i$ is given by

$$u_1 \wedge u_2 \wedge \cdots \wedge u_r = u_1^{i_1} u_2^{i_2} \ldots u_r^{i_r} e_{i_1 i_2 \ldots i_r},$$

the vector space $\Lambda^r(V)$ is spanned by the set

$$E_r = \{e_{i_1 i_2 \ldots i_r} \mid 1 \le i_1 < i_2 < \cdots < i_r \le n\}.$$

As for the case $r = 2$, every r-vector A can be written, using the summation convention,

$$A = A^{i_1 i_2 \ldots i_r} e_{i_1 i_2 \ldots i_r}, \tag{6.16}$$

which can be recast in the form

$$A = \sum_{i_1 < i_2 < \cdots < i_r} \tilde{A}^{i_1 i_2 \ldots i_r} e_{i_1 i_2 \ldots i_r} \tag{6.17}$$

where

$$\tilde{A}^{i_1 i_2 \ldots i_r} = \sum_\sigma (-1)^\sigma A^{i_{\sigma(1)} i_{\sigma(2)} \ldots i_{\sigma(r)}}.$$

When written in the second form, the components $\tilde{A}^{i_1 i_2 \ldots i_r}$ are totally skew symmetric,

$$\tilde{A}^{i_1 i_2 \ldots i_r} = (-1)^\pi \tilde{A}^{i_{\pi(1)} i_{\pi(2)} \ldots i_{\pi(r)}}$$

for any permutation π of $(1, 2, \ldots, r)$. As there are no further algebraic relationships present with which to simplify the r-vectors in E_r, we may again assume that the $e_{i_1 i_2 \ldots i_r}$ are linearly independent. The dimension of $\Lambda^r(V)$ is then the number of ways in which r values can be selected from the n index values $\{1, 2, \ldots, n\}$, i.e.

$$\dim \Lambda^r(V) = \binom{n}{r} = \frac{n!}{r!(n-r)!}.$$

For $r > n$ the dimension is zero, $\dim \Lambda^r(V) = 0$, since each basis r-vector $e_{i_1 i_2 \ldots i_r} = e_{i_1} \wedge e_{i_2} \wedge \cdots \wedge e_{i_r}$ must vanish by (6.15) since some pair of indices must be equal.

Exterior product

Setting the original vector space V to be $\Lambda^1(V)$ and denoting the field of scalars by $\Lambda^0(V) \equiv \mathbb{R}$, we define the vector space

$$\Lambda(V) = \Lambda^0(V) \oplus \Lambda^1(V) \oplus \Lambda^2(V) \oplus \cdots \oplus \Lambda^n(V).$$

The elements of $\Lambda(V)$, called **multivectors**, can be uniquely written in the form

$$A = A_0 + A_1 + A_2 + \cdots + A_n \quad \text{where} \quad A_r \in \Lambda^r(V).$$

The dimension of $\Lambda(V)$ is found by the binomial theorem,

$$\dim(\Lambda(V)) = \sum_{r=0}^n \binom{n}{r} = (1+1)^n = 2^n. \tag{6.18}$$

Define a law of composition $A \wedge B$ for any pair of multivectors A and B, called **exterior product**, satisfying the following rules:

Algebras

(EP1) If $A = a \in \mathbb{R} = \Lambda^0(V)$ and $B \in \Lambda^r(V)$ the exterior product is defined as scalar multiplication, $a \wedge B = B \wedge a = aB$.

(EP2) If $A = u_1 \wedge u_2 \wedge \cdots \wedge u_r$ is a simple r-vector and $B = v_1 \wedge v_2 \wedge \cdots \wedge v_s$ a simple s-vector then their exterior product is defined as the simple $(r+s)$-vector

$$A \wedge B = u_1 \wedge \cdots \wedge u_r \wedge v_1 \wedge \cdots \wedge v_s.$$

(EP3) The exterior product $A \wedge B$ is linear in both arguments,

$$(aA + bB) \wedge C = aA \wedge C + bB \wedge C,$$
$$A \wedge (aB + bC) = aA \wedge B + bA \wedge C.$$

Property (EP3) makes $\Lambda(V)$ into an algebra with respect to exterior product, called the **Grassmann algebra** or **exterior algebra** over V.

By (EP2), the product of a basis r-vector and basis s-vector is

$$e_{i_1 \ldots i_r} \wedge e_{j_1 \ldots j_s} = e_{i_1 \ldots i_r j_1 \ldots j_s}, \tag{6.19}$$

and since

$$e_{i_1 i_2 \ldots i_r} \wedge \left(e_{j_1 j_2 \ldots j_s} \wedge e_{k_1 k_2 \ldots k_t} \right) = \left(e_{i_1 i_2 \ldots i_r} \wedge e_{j_1 j_2 \ldots j_s} \right) \wedge e_{k_1 k_2 \ldots k_t}$$
$$= e_{i_1 i_2 \ldots i_r j_1 j_2 \ldots j_s k_1 k_2 \ldots k_t}$$

the associative law follows for all multivectors by the linearity condition (EP3),

$$A \wedge (B \wedge C) = (A \wedge B) \wedge C \quad \text{for all } A, B, C \in \Lambda(V).$$

Thus $\Lambda(V)$ is an associative algebra. The property that the product of an r-vector and an s-vector always results in an $(r+s)$-vector is characteristic of what is commonly called a **graded algebra**.

Example 6.5 General products of multivectors are straightforward to calculate from the exterior products of basis elements. Some simple examples are

$$e_i \wedge e_j = e_{ij} = -e_{ji} = -e_j \wedge e_i,$$
$$e_i \wedge e_i = 0 \quad \text{for all } i = 1, 2, \ldots, n,$$
$$e_1 \wedge e_{23} = e_{123} = -e_{132} = -e_{213} = -e_2 \wedge e_{13},$$
$$e_{14} \wedge e_{23} = e_{1423} = -e_{1324} = -e_{13} \wedge e_{24} = e_{1234} = e_{12} \wedge e_{34},$$
$$e_{24} \wedge e_{14} = e_{2414} = 0,$$
$$(ae_1 + be_{23}) \wedge (a' + b'e_{34}) = aa'e_1 + a'be_{23} + ab'e_{134}.$$

Properties of exterior product

If A is an r-vector and B an s-vector, they satisfy the 'anticommutation rule'

$$A \wedge B = (-1)^{rs} B \wedge A. \tag{6.20}$$

6.4 Grassmann algebras

Since every r-vector is by definition a linear combination of simple r-vectors it is only necessary to prove Eq. (6.20) for simple r-vectors and s-vectors

$$A = x_1 \wedge x_2 \wedge \cdots \wedge x_r, \quad B = y_1 \wedge y_2 \wedge \cdots \wedge y_s.$$

Successively perform r interchanges of positions of each vector y_i to bring it in front of x_1, and we have

$$\begin{aligned} A \wedge B &= x_1 \wedge x_2 \wedge \cdots \wedge x_r \wedge y_1 \wedge y_2 \wedge \cdots \wedge y_s \\ &= (-1)^r y_1 \wedge x_1 \wedge x_2 \wedge \cdots \wedge x_r \wedge y_2 \wedge \cdots \wedge y_s \\ &= (-1)^{2r} y_1 \wedge y_2 \wedge x_1 \wedge x_2 \wedge \cdots \wedge x_r \wedge y_3 \wedge \cdots \wedge y_s \\ &= \cdots \\ &= (-1)^{sr} y_1 \wedge y_2 \wedge \cdots \wedge y_s \wedge x_1 \wedge x_2 \wedge \cdots \wedge x_r \\ &= (-1)^{rs} B \wedge A, \end{aligned}$$

as required. Hence an r-vector and an s-vector anticommute, $A \wedge B = -B \wedge A$, if both r and s are odd. They commute if either one of them has even degree.

The following theorem gives a particularly quick and useful method for deciding whether or not a given set of vectors is linearly independent.

Theorem 6.2 *Vectors $u_1, u_2, \ldots, u_r$ are linearly dependent if and only if their wedge product vanishes,*

$$u_1 \wedge u_2 \wedge \cdots \wedge u_r = 0.$$

Proof: 1. If the vectors are linearly dependent then without loss of generality we may assume that u_1 is a linear combination of the others,

$$u_1 = a^2 u_2 + a^3 u_3 + \cdots + a^r u_r.$$

Hence

$$\begin{aligned} u_1 \wedge u_2 \wedge \cdots \wedge u_r &= \sum_{i=2}^{r} a^i u_i \wedge u_2 \wedge \cdots \wedge u_r \\ &= \sum_{i=2}^{r} \pm a^i u_2 \wedge \cdots \wedge u_i \wedge u_i \wedge \cdots \wedge u_r \\ &= 0. \end{aligned}$$

This proves the *only if* part of the theorem.

2. Conversely, suppose $u_1, \ldots, u_r$ ($r \le n$) are linearly independent. By Theorem 3.7 there exists a basis $\{e_j\}$ of V such that

$$e_1 = u_1, \quad e_2 = u_2, \quad \ldots, \quad e_r = u_r.$$

Since $e_1 \wedge e_2 \wedge \cdots \wedge e_r$ is a basis vector of $\Lambda^r(V)$ it cannot vanish. ∎

Example 6.6 If e_1, e_2 and e_3 are three basis vectors of a vector space V then the vectors $e_1 + e_3$, $e_2 + e_3$, $e_1 + e_2$ are linearly independent, for

$$\begin{aligned} (e_1 + e_3) \wedge (e_2 + e_3) \wedge (e_1 + e_2) &= (e_{12} + e_{13} + e_{32}) \wedge (e_1 + e_2) \\ &= e_{132} + e_{321} \\ &= 2e_{132} \ne 0. \end{aligned}$$

Algebras

On the other hand, the vectors $e_1 - e_3$, $e_2 - e_3$, $e_1 - e_2$ are linearly dependent since

$$(e_1 - e_3) \wedge (e_2 - e_3) \wedge (e_1 - e_2) = (e_{12} - e_{13} - e_{32}) \wedge (e_1 - e_2)$$
$$= e_{132} - e_{321}$$
$$= 0.$$

We return to the subject of exterior algebra in Chapter 8.

Problems

Problem 6.9 Let $\{e_1, e_2, \ldots, e_n\}$ be a basis of a vector space V of dimension $n \geq 5$. By calculating their wedge product, decide whether the following vectors are linearly dependent or independent:

$$e_1 + e_2 + e_3, \quad e_2 + e_3 + e_4, \quad e_3 + e_4 + e_5, \quad e_1 + e_3 + e_5.$$

Can you find a linear relation among them?

Problem 6.10 Let W be a vector space of dimension 4 and $\{e_1, e_2, e_3, e_4\}$ a basis. Let A be the 2-vector on W,

$$A = e_2 \wedge e_1 + a e_1 \wedge e_3 + e_2 \wedge e_3 + c e_1 \wedge e_4 + b e_2 \wedge e_4.$$

Write out explicitly the equations $A \wedge u = 0$ where $u = u^1 e_1 + u^2 e_2 + u^3 e_3 + u^4 e_4$ and show that they have a non-trivial solution if and only if $c = ab$. In this case find two vectors u and v such that $A = u \wedge v$.

Problem 6.11 Let U be a subspace of V spanned by linearly independent vectors $\{u_1, u_2, \ldots, u_p\}$.

(a) Show that the p-vector $E_U = u_1 \wedge u_2 \wedge \cdots \wedge u_p$ is defined uniquely up to a factor by the subspace U in the sense that if $\{u'_1, u'_2, \ldots, u'_p\}$ is any other linearly independent set spanning U then the p-vector $E'_U \equiv u'_1 \wedge \cdots \wedge u'_p$ is proportional to E_U; i.e., $E'_U = c E_U$ for some scalar c.
(b) Let W be a q-dimensional subspace of V, with corresponding q-vector E_W. Show that $U \subseteq W$ if and only if there exists a $(q - p)$-vector F such that $E_W = E_U \wedge F$.
(c) Show that if $p > 0$ and $q > 0$ then $U \cap W = \{0\}$ if and only if $E_U \wedge E_W \neq 0$.

6.5 Lie algebras and Lie groups

An important class of *non-associative algebras* is due to the Norwegian mathematician Sophus Lie (1842–1899). Lie's work on transformations of surfaces (known as *contact transformations*) gave rise to a class of continuous groups now known as *Lie groups*. These encompass essentially all the important groups that appear in mathematical physics, such as the orthogonal, unitary and symplectic groups. This subject is primarily a branch of differential geometry and a detailed discussion will appear in Chapter 19. Lie's principal discovery was that Lie groups were related to a class of non-associative algebras that in turn are considerably easier to classify. These algebras have come to be known as *Lie algebras*. A more complete discussion of Lie algebra theory and, in particular, the Cartan–Dynkin classification for *semisimple* Lie algebras can be found in [3–5].

6.5 Lie algebras and Lie groups

Lie algebras

A **Lie algebra** $\mathcal{L}$ is a real or complex vector space with a law of composition or **bracket product** $[X, Y]$ satisfying

(LA1) $[X, Y] = -[Y, X]$ (antisymmetry).
(LA2) $[X, aY + bZ] = a[X, Y] + b[X, Z]$ (distributive law).
(LA3) $[X, [Y, Z]] + [Y, [Z, X]] + [Z, [X, Y]] = 0$ (**Jacobi identity**).

By (LA1) the bracket product is also distributive on the first argument,

$$[aX + bY, Z] = -[Z, aX + bY] = -a[Z, X] - b[Z, Y] = a[X, Z] + b[Y, Z].$$

Lie algebras are therefore algebras in the general sense, since Eq. (6.1) holds for the bracket product. The Jacobi identity replaces the associative law.

Example 6.7 Any associative algebra, such as the algebra $\mathcal{M}_n$ of $n \times n$ matrices discussed in Example 6.2, can be converted to a Lie algebra by defining the bracket product to be the **commutator** of two elements

$$[X, Y] = XY - YX.$$

Conditions (LA1) and (LA2) are trivial to verify, while the Jacobi identity (LA3) is straightforward:

$$\begin{aligned}
&[X, [Y, Z]] + [Y, [Z, X]] + [Z, [X, Y]] \\
&= X(YZ - ZY) - (YZ - ZY)X + Y(ZX - XZ) \\
&\quad - (ZX - XZ)Y + Z(XY - YX) - (XY - YX)Z \\
&= XYZ - XZY - YZX + ZYX + YZX - YXZ \\
&\quad - ZXY + XZY + ZXY - ZYX - XYZ + YXZ \\
&= 0.
\end{aligned}$$

The connection between brackets and commutators motivates the terminology that if a Lie algebra $\mathcal{L}$ has all bracket products vanishing, $[X, Y] = 0$ for all $X, Y \in \mathcal{L}$, it is said to be **abelian**.

Given a basis $X_1, \ldots, X_n$ of $\mathcal{L}$, let $C^k_{ij} = -C^k_{ji}$ be the structure constants with respect to this basis,

$$[X_i, X_j] = C^k_{ij} X_k. \tag{6.21}$$

Given the structure constants, it is possible to calculate the bracket product of any pair of vectors $A = a^i X_i$ and $B = b^j X_j$:

$$[A, B] = a^i b^j [X_i, X_j] = a^i b^j C^k_{ij} X_k. \tag{6.22}$$

A Lie algebra is therefore abelian if and only if all its structure constants vanish. It is important to note that structure constants depend on the choice of basis and are generally different in another basis $X'_i = A'^{j}_{i} X_j$.

Algebras

Example 6.8 Consider the set T_2 of real upper triangular 2×2 matrices, having the form

$$A = \begin{pmatrix} a & b \\ 0 & c \end{pmatrix}.$$

Since

$$\left[\begin{pmatrix} a & b \\ 0 & c \end{pmatrix}, \begin{pmatrix} d & e \\ 0 & f \end{pmatrix} \right] = \begin{pmatrix} 0 & ae + bf - bd - ce \\ 0 & 0 \end{pmatrix}, \quad (6.23)$$

these matrices form a Lie algebra with respect to the commutator product. The following three matrices form a basis of this Lie algebra:

$$X_1 = \begin{pmatrix} 0 & 1 \\ 0 & 0 \end{pmatrix}, \quad X_2 = \begin{pmatrix} 1 & 0 \\ 0 & 0 \end{pmatrix}, \quad X_3 = \begin{pmatrix} 0 & 0 \\ 0 & 1 \end{pmatrix},$$

having the commutator relations

$$[X_1, X_2] = -X_1, \quad [X_1, X_3] = X_1, \quad [X_2, X_3] = O.$$

The corresponding structure constants are

$$C^1_{12} = -C^1_{21} = -1, \quad C^1_{13} = -C^1_{31} = 1, \quad \text{all other } C^i_{jk} = 0.$$

Consider a change of basis to

$$X'_1 = X_2 + X_3 = \begin{pmatrix} 1 & 0 \\ 0 & 1 \end{pmatrix},$$

$$X'_2 = X_1 + X_3 = \begin{pmatrix} 0 & 1 \\ 0 & 1 \end{pmatrix},$$

$$X'_3 = X_1 + X_2 = \begin{pmatrix} 1 & 1 \\ 0 & 0 \end{pmatrix}.$$

The commutation relations are

$$[X'_1, X'_2] = O, \quad [X'_1, X'_3] = O, \quad [X'_2, X'_3] = -2X_1 = X'_1 - X'_2 - X'_3,$$

with corresponding structure constants

$$C'^1_{23} = -C'^2_{23} = -C'^3_{23} = 1, \quad \text{all other } C'^i_{jk} = 0.$$

As before, an **ideal** $\mathcal{I}$ of a Lie algebra $\mathcal{L}$ is a subset such that $[X, Y] \in \mathcal{I}$ for all $X \in \mathcal{L}$ and all $Y \in \mathcal{I}$, a condition written more briefly as $[\mathcal{L}, \mathcal{I}] \subseteq \mathcal{I}$. From (LA1) it is clear that any right or left ideal must be two-sided. If $\mathcal{I}$ is an ideal of $\mathcal{L}$ it is possible to form a **factor Lie algebra** on the space of cosets $X + \mathcal{I}$, with bracket product defined by

$$[X + \mathcal{I}, Y + \mathcal{I}] = [X, Y] + \mathcal{I}.$$

This product is independent of the choice of representative from the cosets $X + \mathcal{I}$ and $Y + \mathcal{I}$, since

$$[X + \mathcal{I}, Y + \mathcal{I}] = [X, Y] + [X, \mathcal{I}] + [\mathcal{I}, Y] + [\mathcal{I}, \mathcal{I}] \subseteq [X, Y] + \mathcal{I}.$$

6.5 Lie algebras and Lie groups

Example 6.9 In Example 6.8 let $\mathcal{B}$ be the subset of matrices of the form

$$\begin{pmatrix} 0 & x \\ 0 & 0 \end{pmatrix}.$$

From the product rule (6.23) it follows that $\mathcal{B}$ forms an ideal of $\mathcal{T}_2$. Every coset $\mathsf{X} + \mathcal{B}$ clearly has a diagonal representative

$$\mathsf{X} = \begin{pmatrix} a & 0 \\ 0 & b \end{pmatrix},$$

and since diagonal matrices always commute with each other, the factor algebra is abelian,

$$[\mathsf{X} + \mathcal{B}, \mathsf{Y} + \mathcal{B}] = \mathsf{O} + \mathcal{B}.$$

The linear map $\varphi : \mathcal{T}_2 \to \mathcal{T}_2/\mathcal{B}$ defined by

$$\varphi : \begin{pmatrix} a & x \\ 0 & b \end{pmatrix} \longmapsto \begin{pmatrix} a & 0 \\ 0 & b \end{pmatrix} + \mathcal{B}$$

is a Lie algebra homomorphism, since by Eq. (6.23) it follows that $[\mathsf{X}, \mathsf{Y}] \in \mathcal{B}$ for any pair of upper triangular matrices X and Y, and

$$\varphi([\mathsf{X}, \mathsf{Y}]) = \mathsf{O} + \mathcal{B} = [\varphi(\mathsf{X}), \varphi(\mathsf{Y})].$$

The kernel of the homomorphism φ consists of those matrices having zero diagonal elements,

$$\ker \varphi = \mathcal{B}.$$

This example is an illustration of Theorem 6.1.

Matrix Lie groups

In Chapter 19 we will give a rigorous definition of a *Lie group*, but for the present purpose we may think of a Lie group as a group whose elements depend continuously on n real parameters $\lambda_1, \lambda_2, \ldots, \lambda_n$. For simplicity we will assume the group to be a matrix group, whose elements can typically be written as

$$\mathsf{A} = \Gamma(\lambda_1, \lambda_2, \ldots, \lambda_n).$$

The identity is taken to be the element corresponding to the origin $\lambda_1 = \lambda_2 = \cdots = \lambda_n = 0$:

$$\mathsf{I} = \Gamma(0, 0, \ldots, 0).$$

Example 6.10 The general member of the rotation group is an orthogonal 3×3 matrix A that can be written in terms of three angles ψ, θ, ϕ,

$$\mathsf{A} = \begin{pmatrix} \cos\phi\cos\psi & \cos\phi\sin\psi & -\sin\phi \\ \sin\theta\sin\phi\cos\psi - \cos\theta\sin\psi & \sin\theta\sin\phi\sin\psi + \cos\theta\cos\psi & \sin\theta\cos\phi \\ \cos\theta\sin\phi\cos\psi + \sin\theta\sin\psi & \cos\theta\sin\phi\sin\psi - \sin\theta\cos\psi & \cos\theta\cos\phi \end{pmatrix}.$$

Algebras

As required, the identity element I corresponds to $\theta = \phi = \psi = 0$. These angles are similar but not identical to the standard Euler angles of classical mechanics, which have an unfortunate degeneracy at the identity element. Group elements near the identity have the form $A = I + \epsilon X$ ($\epsilon \ll 1$), where

$$I = AA^T = (I + \epsilon X)(I + \epsilon X^T) = I + \epsilon(X + X^T) + O(\epsilon^2).$$

If we only keep terms to first order in this equation then X must be antisymmetric; $X = -X^T$.

Although the product of two antisymmetric matrices is not in general antisymmetric, the set of $n \times n$ antisymmetric matrices is closed with respect to commutator products and forms a Lie algebra:

$$[X, Y]^T = (XY - YX)^T = Y^T X^T - X^T Y^T = (-Y)(-X) + X(-Y) = YX - XY = -[X, Y].$$

The Lie algebra of 3×3 antisymmetric matrices may be thought of as representing 'infinitesimal rotations', or orthogonal matrices 'near the identity'. Every 3×3 antisymmetric matrix X can be written in the form

$$X = \begin{pmatrix} 0 & x^3 & -x^2 \\ -x^3 & 0 & x^1 \\ x^2 & -x^1 & 0 \end{pmatrix} = \sum_{i=1}^{3} x^i X_i$$

where

$$X_1 = \begin{pmatrix} 0 & 0 & 0 \\ 0 & 0 & 1 \\ 0 & -1 & 0 \end{pmatrix}, \quad X_2 = \begin{pmatrix} 0 & 0 & -1 \\ 0 & 0 & 0 \\ 1 & 0 & 0 \end{pmatrix}, \quad X_3 = \begin{pmatrix} 0 & 1 & 0 \\ -1 & 0 & 0 \\ 0 & 0 & 0 \end{pmatrix}. \tag{6.24}$$

The basis elements X_i are called *infinitesimal generators* of the group and satisfy the following commutation relations:

$$[X_1, X_2] = -X_3, \quad [X_2, X_3] = -X_1, \quad [X_3, X_1] = -X_2. \tag{6.25}$$

This example is typical of the procedure for creating a Lie algebra from the group elements 'near the identity'. More generally, if G is a matrix Lie group whose elements depend on n continuous parameters

$$A = \Gamma(\lambda_1, \lambda_2, \ldots, \lambda_n) \quad \text{with} \quad I = \Gamma(0, 0, \ldots, 0),$$

define the **infinitesimal generators** by

$$X_i = \left.\frac{\partial \Gamma}{\partial \lambda_i}\right|_{\lambda=0} \quad (\lambda \equiv (\lambda_1, \lambda_2, \ldots, \lambda_n)), \tag{6.26}$$

so that elements near the identity can be written

$$A = I + \sum_{i=1}^{n} \epsilon a^i X_i + O(\epsilon^2).$$

The group structure of G implies the commutators of the X_i are always linear combinations of the X_i, satisfying Eq. (6.21) for some structure constants $C^k_{ji} = -C^k_{ij}$. The proof will be given in Chapter 19.

6.5 Lie algebras and Lie groups

One-parameter subgroups

A **one-parameter subgroup** of a Lie group G is the image $\varphi(\mathbb{R})$ of a homomorphism $\varphi : \mathbb{R} \to G$ of the additive group of real numbers into G. Writing the elements of a one-parameter subgroup of a matrix Lie group simply as $\mathsf{A}(t) = \Gamma(a^1(t), a^2(t), \ldots, a^n(t))$, the homomorphism property requires that

$$\mathsf{A}(t+s) = \mathsf{A}(t)\mathsf{A}(s). \tag{6.27}$$

It can be shown that through every element g in a neighbourhood of the identity of a Lie group there exists a one-parameter subgroup φ such that $g = \varphi(1)$.

Applying the operation $\left.\dfrac{d}{ds}\right|_{s=0}$ to Eq. (6.27) results in

$$\left.\frac{d}{ds}\mathsf{A}(t+s)\right|_{s=0} = \left.\frac{d}{dt}\mathsf{A}(t+s)\right|_{s=0} = \mathsf{A}(t)\left.\frac{d\mathsf{A}(s)}{ds}\right|_{s=0}.$$

Hence

$$\frac{d}{dt}\mathsf{A}(t) = \mathsf{A}(t)\mathsf{X} \tag{6.28}$$

where

$$\mathsf{X} = \left.\frac{d A(s)}{ds}\right|_{s=0} = \sum_i \left.\frac{\partial \Gamma(\lambda)}{\partial \lambda^i}\right|_{\lambda=0} \left.\frac{d a^i(s)}{ds}\right|_{s=0}$$

$$= \sum_i \left.\frac{d a^i}{ds}\right|_{s=0} X_i.$$

The unique solution of the differential equation (6.28) that satisfies the boundary condition $\mathsf{A}(0) = \mathsf{I}$ is $\mathsf{A}(t) = e^{t\mathsf{X}}$, where the exponential of the matrix $t\mathsf{X}$ is defined by the power series

$$e^{t\mathsf{X}} = \mathsf{I} + t\mathsf{X} + \frac{1}{2!}t^2\mathsf{X}^2 + \frac{1}{3!}t^3\mathsf{X}^3 + \cdots$$

The group property $e^{(t+s)\mathsf{X}} = e^{t\mathsf{X}}e^{s\mathsf{X}}$ follows from the fact that if A and B are any pair of commuting matrices $\mathsf{A}\mathsf{B} = \mathsf{B}\mathsf{A}$ then $e^\mathsf{A} e^\mathsf{B} = e^{\mathsf{A}+\mathsf{B}}$.

In a *neighbourhood of the identity* consisting of group elements all connected to the identity by one-parameter subgroups, it follows that any group element A_1 can be written as the exponential of a Lie algebra element

$$\mathsf{A}_1 = \mathsf{A}(1) = e^\mathsf{X} = \mathsf{I} + \mathsf{X} + \frac{1}{2!}\mathsf{X}^2 + \frac{1}{3!}\mathsf{X}^3 + \cdots \quad \text{where} \quad \mathsf{X} \in \mathcal{G}.$$

Given a Lie algebra, say by specifying its structure constants, it is possible to reverse this process and construct the connected neighbourhood of the identity of a unique Lie group. Since the structure constants are a finite set of numbers, as opposed to the complicated set of *functions* needed to specify the group products, it is generally much easier to classify Lie groups by their Lie algebras than by their group products.

Algebras

Example 6.11 In Example 6.10 the one-parameter group e^{tX_1} generated by the infinitesimal generator X_1 is found by calculating the first few powers

$$X_1 = \begin{pmatrix} 0 & 0 & 0 \\ 0 & 0 & 1 \\ 0 & -1 & 0 \end{pmatrix}, \quad X_1^2 = \begin{pmatrix} 0 & 0 & 0 \\ 0 & -1 & 0 \\ 0 & 0 & -1 \end{pmatrix}, \quad X_1^3 = \begin{pmatrix} 0 & 0 & 0 \\ 0 & 0 & -1 \\ 0 & 1 & 0 \end{pmatrix} = -X_1,$$

as all higher powers follow a simple rule

$$X_1^4 = -X_1^2, \quad X_1^5 = X_1, \quad X_1^6 = X_1^2, \text{ etc.}$$

From the exponential expansion

$$e^{tX_1} = I + tX_1 + \frac{1}{2!}t^2 X_1^2 + \frac{1}{3!}t^3 X_1^3 + \cdots$$

it is possible to calculate all components

$$\left(e^{tX_1}\right)_{11} = 1$$

$$\left(e^{tX_1}\right)_{22} = 1 - \frac{t^2}{2!} + \frac{t^4}{4!} - \cdots = \cos t$$

$$\left(e^{tX_1}\right)_{23} = 0 + t - \frac{t^3}{3!} + \frac{t^5}{5!} - \cdots = \sin t, \text{ etc.}$$

Hence

$$e^{tX_1} = \begin{pmatrix} 1 & 0 & 0 \\ 0 & \cos t & \sin t \\ 0 & -\sin t & \cos t \end{pmatrix},$$

which represents a rotation by the angle t about the x-axis. It is straightforward to verify the one-parameter group law

$$e^{tX_1} e^{sX_1} = e^{(t+s)X_1}.$$

Exercise: Show that e^{tX_2} and e^{tX_3} represent rotations by angle t about the y-axis and z-axis respectively.

Complex Lie algebras

While most of the above discussion assumes real Lie algebras, it can apply equally to complex Lie algebras. As seen in Section 6.2, it is always possible to regard a complex vector space $\mathcal{G}$ as being a real space $\mathcal{G}^R$ of twice the number of dimensions, by simply restricting the field of scalars to the real numbers. In this way any complex Lie algebra of dimension n can also be considered as being a real Lie algebra of dimension $2n$. It is important to be aware of whether it is the real or complex version of a given Lie algebra that is in question.

Example 6.12 In Example 2.15 of Chapter 2 it was seen that the 2×2 unitary matrices form a group $SU(2)$. For unitary matrices near the identity, $U = I + \epsilon A$,

$$I = UU^\dagger = I + \epsilon(A + A^\dagger) + O(\epsilon^2).$$

6.5 Lie algebras and Lie groups

Hence A must be *anti-hermitian*,

$$A + A^\dagger = O.$$

Special unitary matrices are required to have the further restriction that their determinant is 1,

$$\det U = \begin{vmatrix} 1+\epsilon a_{11} & \epsilon a_{12} \\ \epsilon a_{12} & 1+\epsilon a_{22} \end{vmatrix} = 1+\epsilon(a_{11}+a_{22})+O(\epsilon^2),$$

and the matrix A must be trace-free as well as being anti-hermitian,

$$A = \begin{pmatrix} ic & b+ia \\ -b+ia & -ic \end{pmatrix} \quad (a,b,c \in \mathbb{R}).$$

Such matrices form a *real* Lie algebra, as they constitute a real vector space and are closed with respect to commutator product,

$$[A, A'] = AA' - A'A = \begin{pmatrix} 2i(ba'-ab') & 2(ac'-ca')+2i(cb'-bc') \\ -2(ac'-ca')+2i(cb'-bc') & -2i(ba'-ab') \end{pmatrix}.$$

Any trace-free anti-hermitian matrix may be cast in the form

$$A = ia\sigma_1 + ib\sigma_2 + ic\sigma_3$$

where σ_i are the **Pauli matrices**,

$$\sigma_1 = \begin{pmatrix} 0 & 1 \\ 1 & 0 \end{pmatrix}, \quad \sigma_2 = \begin{pmatrix} 0 & -i \\ i & 0 \end{pmatrix}, \quad \sigma_3 = \begin{pmatrix} 1 & 0 \\ 0 & -1 \end{pmatrix} \tag{6.29}$$

whose commutation relations are easily calculated,

$$[\sigma_1, \sigma_2] = 2i\sigma_3, \quad [\sigma_2, \sigma_3] = 2i\sigma_1, \quad [\sigma_3, \sigma_1] = 2i\sigma_2. \tag{6.30}$$

Although this Lie algebra consists of complex matrices, note that it is *not* a complex Lie algebra since multiplying an anti-hermitian matrix by a complex number does not in general result in an anti-hermitian matrix. However multiplying by real scalars does retain the anti-hermitian property. A basis for this Lie algebra is

$$X_1 = \tfrac{1}{2}i\sigma_1, \quad X_2 = \tfrac{1}{2}i\sigma_2, \quad X_3 = \tfrac{1}{2}i\sigma_3,$$

and the general Lie algebra element A has the form

$$A = 2aX_1 + 2bX_2 + 2cX_3 \quad (a,b,c \in \mathbb{R}).$$

By (6.30) the commutation relations between the X_k are

$$[X_1, X_2] = -X_3, \quad [X_2, X_3] = -X_1, \quad [X_3, X_1] = -X_2,$$

which shows that this Lie algebra is in fact isomorphic to the Lie algebra of the group of 3×3 orthogonal matrices given in Example 6.10. Denoting these real Lie algebras by $\mathcal{SU}(2)$ and $\mathcal{SO}(3)$ respectively, we have

$$\mathcal{SU}(2) \cong \mathcal{SO}(3).$$

Algebras

However, the underlying groups are not isomorphic in this case, although there does exist a homomorphism $\varphi : SU(2) \to SO(3)$ whose kernel consists of just the two elements $\pm I$. This is the so-called **spinor representation** of the rotation group. Strictly speaking it is not a representation of the rotation group – rather, it asserts that there is a representation of $SU(2)$ as the rotation group in $\mathbb{R}^3$.

Example 6.13 A genuinely complex Lie algebra is $\mathcal{SL}(2, \mathbb{C})$, the Lie algebra of the group of 2×2 complex unimodular matrices. As in the preceding example the condition of unimodularity, or having determinant 1, implies that the infinitesimal generators are trace-free,

$$\det(I + \epsilon A) = 1 \implies \operatorname{tr} A = a_{11} + a_{22} = 0.$$

The set of complex trace-free matrices form a complex Lie algebra since (a) it forms a complex vector space, and (b) it is closed under commutator products by Eq. (2.15),

$$\operatorname{tr}[A, B] = \operatorname{tr}(AB) - \operatorname{tr}(BA) = 0.$$

This complex Lie algebra is spanned by

$$Y_1 = \tfrac{1}{2} i\sigma_1 = \tfrac{1}{2}\begin{pmatrix} 0 & i \\ i & 0 \end{pmatrix}, \quad Y_2 = \tfrac{1}{2} i\sigma_2 = \tfrac{1}{2}\begin{pmatrix} 0 & 1 \\ -1 & 0 \end{pmatrix}, \quad Y_3 = \tfrac{1}{2} i\sigma_3 = \tfrac{1}{2}\begin{pmatrix} i & 0 \\ 0 & -i \end{pmatrix},$$

for if $A = [A_{ij}]$ is trace-free then

$$A = \alpha Y_1 + \beta Y_2 + \gamma Y_3$$

where

$$\alpha = -i(A_{12} + A_{21}), \quad \beta = A_{12} - A_{21}, \quad \gamma = -2i A_{11}.$$

The Lie algebra $\mathcal{SL}(2, \mathbb{C})$ is isomorphic as a complex Lie algebra to the Lie algebra $\mathcal{SO}(3, \mathbb{C})$ of infinitesimal complex orthogonal transformations. The latter Lie algebra is spanned, as a complex vector space, by the same matrices X_i defined in Eq. (6.26) to form a basis of the real Lie algebra $\mathcal{SO}(3)$. Since, by (6.30), the commutation relations of the Y_i are

$$[Y_1, Y_2] = -Y_3, \quad [Y_2, Y_3] = -Y_1, \quad [Y_3, Y_1] = -Y_2,$$

comparison with Eq. (6.25) shows that the linear map $\varphi : \mathcal{SL}(2, \mathbb{C}) \to \mathcal{SO}(3, \mathbb{C})$ defined by $\varphi(Y_i) = X_i$ is a Lie algebra isomorphism.

However, as a real Lie algebra the story is quite different since the matrices Y_1, Y_2 and Y_3 defined above are not sufficient to span $\mathcal{SL}(2, \mathbb{C})^R$. If we supplement them with the matrices

$$Z_1 = \tfrac{1}{2}\sigma_1 = \tfrac{1}{2}\begin{pmatrix} 0 & 1 \\ 1 & 0 \end{pmatrix}, \quad Z_2 = \tfrac{1}{2}\sigma_2 = \tfrac{1}{2}\begin{pmatrix} 0 & -i \\ i & 0 \end{pmatrix}, \quad Z_3 = \tfrac{1}{2}\sigma_3 = \tfrac{1}{2}\begin{pmatrix} 1 & 0 \\ 0 & 1 \end{pmatrix},$$

then every member of $\mathcal{SO}(3, \mathbb{C})$ can be written uniquely in the form

$$A = a_1 Y_1 + a_2 Y_2 + a_3 Y_3 + b_1 Z_1 + b_2 Z_2 + b_3 Z_3 \quad (a_i, b_i \in \mathbb{R})$$

6.5 Lie algebras and Lie groups

where
$$b_3 = A_{11} + \overline{A_{11}},$$
$$a_3 = -i(A_{11} - \overline{A_{11}}),$$
$$b_1 = A_{12} + A_{21} + \overline{A_{12} + A_{21}}, \text{ etc.}$$

Hence the Y_i and Z_i span $\mathcal{SL}(2, \mathbb{C})$ as a real vector space, which is a real Lie algebra determined by the commutation relations

$$[Y_1, Y_2] = -Y_3 \quad [Y_2, Y_3] = -Y_1 \quad [Y_3, Y_1] = -Y_2, \quad (6.31)$$
$$[Z_1, Z_2] = Y_3 \quad [Z_2, Z_3] = Y_1 \quad [Z_3, Z_1] = Y_2, \quad (6.32)$$
$$[Y_1, Z_2] = -Z_3 \quad [Y_2, Z_3] = -Z_1 \quad [Y_3, Z_1] = -Z_2, \quad (6.33)$$
$$[Y_1, Z_3] = Z_2 \quad [Y_2, Z_1] = Z_3 \quad [Y_3, Z_2] = Z_1, \quad (6.34)$$
$$[Y_1, Z_1] = 0 \quad [Y_2, Z_2] = 0 \quad [Y_3, Z_3] = 0. \quad (6.35)$$

Example 6.14 Lorentz transformations are defined in Section 2.7 by

$$\mathbf{x}' = L\mathbf{x}, \qquad G = L^T G L$$

where

$$G = \begin{pmatrix} 1 & 0 & 0 & 0 \\ 0 & 1 & 0 & 0 \\ 0 & 0 & 1 & 0 \\ 0 & 0 & 0 & -1 \end{pmatrix}.$$

Hence infinitesimal Lorentz transformations $L = I + \epsilon A$ satisfy the equation

$$A^T G + G A = O,$$

which reads in components

$$A_{ij} + A_{ji} = 0, \qquad A_{4i} - A_{i4} = 0, \qquad A_{44} = 0$$

where indices i, j range from 1 to 3. It follows that the Lie algebra of the Lorentz group is spanned by six matrices

$$Y_1 = \begin{pmatrix} 0 & 0 & 0 & 0 \\ 0 & 0 & 1 & 0 \\ 0 & -1 & 0 & 0 \\ 0 & 0 & 0 & 0 \end{pmatrix}, \quad Y_2 = \begin{pmatrix} 0 & 0 & -1 & 0 \\ 0 & 0 & 0 & 0 \\ 1 & 0 & 0 & 0 \\ 0 & 0 & 0 & 0 \end{pmatrix}, \quad Y_3 = \begin{pmatrix} 0 & 1 & 0 & 0 \\ 1 & 0 & 0 & 0 \\ 0 & 0 & 0 & 0 \\ 0 & 0 & 0 & 0 \end{pmatrix},$$

$$Z_1 = \begin{pmatrix} 0 & 0 & 0 & 1 \\ 0 & 0 & 0 & 0 \\ 0 & 0 & 0 & 0 \\ 1 & 0 & 0 & 0 \end{pmatrix}, \quad Z_2 = \begin{pmatrix} 0 & 0 & 0 & 0 \\ 0 & 0 & 0 & 1 \\ 0 & 0 & 0 & 0 \\ 0 & 1 & 0 & 0 \end{pmatrix}, \quad Z_3 = \begin{pmatrix} 0 & 0 & 0 & 0 \\ 0 & 0 & 0 & 0 \\ 0 & 0 & 0 & 1 \\ 0 & 0 & 1 & 0 \end{pmatrix}.$$

These turn out to have exactly the same commutation relations (6.31)–(6.35) as the generators of $\mathcal{SL}(2, \mathbb{C})$ in the previous example. Hence the *real* Lie algebra $\mathcal{SL}(2, \mathbb{C})$ is isomorphic to the Lie algebra of the Lorentz group $SO(3, 1)$. Since the complex Lie algebras $\mathcal{SL}(2, \mathbb{C})$

Algebras

and $SO(3, \mathbb{C})$ were shown to be isomorphic in Example 6.13, their real versions must also be isomorphic. We thus have the interesting sequence of isomorphisms of real Lie algebras,

$$SO(3,1) \cong SL(2,\mathbb{C}) \cong SO(3,\mathbb{C}).$$

Problems

Problem 6.12 As in Example 6.12, $n \times n$ unitary matrices satisfy $UU^\dagger = I$ and those near the identity have the form

$$U = I + \epsilon A \quad (\epsilon \ll 1)$$

where A is anti-hermitian, $A = -A^\dagger$.

(a) Show that the set of anti-hermitian matrices form a Lie algebra with respect to the commutator $[A, B] = AB - BA$ as bracket product.

(b) The four *Pauli matrices* σ_μ ($\mu = 0, 1, 2, 3$) are defined by

$$\sigma_0 = \begin{pmatrix} 1 & 0 \\ 0 & 1 \end{pmatrix}, \quad \sigma_1 = \begin{pmatrix} 0 & 1 \\ 1 & 0 \end{pmatrix}, \quad \sigma_2 = \begin{pmatrix} 0 & -i \\ i & 0 \end{pmatrix}, \quad \sigma_3 = \begin{pmatrix} 1 & 0 \\ 0 & -1 \end{pmatrix}.$$

Show that $Y_\mu = \frac{1}{2} i \sigma_\mu$ form a basis of the Lie algebra of $U(2)$ and calculate the structure constants.

(c) Show that the one-parameter subgroup generated by Y_1 consists of matrices of the form

$$e^{tY_1} = \begin{pmatrix} \cos \frac{1}{2}t & i \sin \frac{1}{2}t \\ i \sin \frac{1}{2}t & \cos \frac{1}{2}t \end{pmatrix}.$$

Calculate the one-parameter subgroups generated by Y_2, Y_3 and Y_0.

Problem 6.13 Let u be an $n \times 1$ column vector. A non-singular matrix A is said to *stretch* u if it is an eigenvector of A,

$$Au = \lambda u.$$

Show that the set of all non-singular matrices that stretch u forms a group with respect to matrix multiplication, called the *stretch group of* u.

(a) Show that the 2×2 matrices of the form

$$\begin{pmatrix} a & a+c \\ b+c & b \end{pmatrix} \quad (c \neq 0, a+b+c \neq 0)$$

form the stretch group of the 2×1 column vector $u = \begin{pmatrix} 1 \\ -1 \end{pmatrix}$.

(b) Show that the Lie algebra of this group is spanned by the matrices

$$X_1 = \begin{pmatrix} 1 & 1 \\ 0 & 0 \end{pmatrix}, \quad X_2 = \begin{pmatrix} 0 & 0 \\ 1 & 1 \end{pmatrix}, \quad X_3 = \begin{pmatrix} 0 & 1 \\ 1 & 0 \end{pmatrix}.$$

Calculate the structure constants for this basis.

(c) Write down the matrices that form the one-parameter subgroups e^{tX_1} and e^{tX_3}.

Problem 6.14 Show that 2×2 trace-free matrices, having tr $A = A_{11} + A_{22} = 0$, form a Lie algebra with respect to bracket product $[A, B] = AB - BA$.

(a) Show that the following matrices form a basis of this Lie algebra:

$$X_1 = \begin{pmatrix} 1 & 0 \\ 0 & -1 \end{pmatrix}, \quad X_2 = \begin{pmatrix} 0 & 1 \\ 0 & 0 \end{pmatrix}, \quad X_3 = \begin{pmatrix} 0 & 0 \\ 1 & 0 \end{pmatrix}$$

and compute the structure constants for this basis.

(b) Compute the one-parameter subgroups e^{tX_1}, e^{tX_2} and e^{tX_3}.

Problem 6.15 Let $\mathcal{L}$ be the Lie algebra spanned by the three matrices

$$X_1 = \begin{pmatrix} 0 & 1 & 0 \\ 0 & 0 & 0 \\ 0 & 0 & 0 \end{pmatrix}, \quad X_2 = \begin{pmatrix} 0 & 0 & 0 \\ 0 & 0 & 1 \\ 0 & 0 & 0 \end{pmatrix}, \quad X_3 = \begin{pmatrix} 0 & 0 & 1 \\ 0 & 0 & 0 \\ 0 & 0 & 0 \end{pmatrix}.$$

Write out the structure constants for this basis, with respect to the usual matrix commutator bracket product.

Write out the three one-parameter subgroups e^{tX_i} generated by these basis elements, and verify in each case that they do in fact form a one-parameter group of matrices.

References

[1] R. Geroch. *Mathematical Physics*. Chicago, The University of Chicago Press, 1985.
[2] S. Lang. *Algebra*. Reading, Mass., Addison-Wesley, 1965.
[3] S. Helgason. *Differential Geometry and Symmetric Spaces*. New York, Academic Press, 1962.
[4] N. Jacobson. *Lie Algebras*. New York, Interscience, 1962.
[5] H. Samelson. *Notes on Lie Algebras*. New York, D. Van Nostrand Reinhold Company, 1969.

7 Tensors

In Chapter 3 we saw that any vector space V gives rise to other vector spaces such as the dual space $V^* = L(V, \mathbb{R})$ and the space $L(V, V)$ of all linear operators on V. In this chapter we will consider a more general class of spaces constructed from a vector space V, known as *tensor spaces*, of which these are particular cases. In keeping with modern mathematical practice, tensors and their basic operations will be defined invariantly, but we will also relate it to the 'old-fashioned' multicomponented formulation that is often better suited to applications in physics [1].

There are two significantly different approaches to tensor theory. Firstly, the method of Section 7.1 defines the tensor product of two vector spaces as a factor space of a *free vector space* [2]. While somewhat abstract in character, this is an essentially constructive procedure. In particular, it can be used to gain a deeper understanding of associative algebras, and supplements the material of Chapter 6. Furthermore, it applies to infinite dimensional vector spaces. The second method defines tensors as *multilinear maps* [3–5]. Readers may find this second approach the easier to understand, and there will be no significant loss in comprehension if they move immediately to Section 7.2. For finite dimensional vector spaces the two methods are equivalent [6].

7.1 Free vector spaces and tensor spaces

Free vector spaces

If S is an arbitrary set, the concept of a *free vector space on S* over a field $\mathbb{K}$ can be thought of intuitively as the set of all 'formal finite sums'

$$a^1 s_1 + a^2 s_2 + \cdots + a^n s_n \quad \text{where} \quad n = 0, 1, 2, \ldots; \ a^i \in \mathbb{K}, \ s_i \in S.$$

The word 'formal' means that if S is an algebraic structure that already has a concept of addition or scalar multiplication defined on it, then the scalar product and summation in the *formal* sum bears no relation to these.

More rigorously, the **free vector space** $F(S)$ on a set S is defined as the set of all functions $f : S \to \mathbb{K}$ that vanish at all but a finite number of elements of S. Clearly $F(S)$ is a vector space with the usual definitions,

$$(f+g)(s) = f(s) + g(s), \quad (af)(s) = af(s).$$

7.1 Free vector spaces and tensor spaces

It is spanned by the characteristic functions $\chi_t \equiv \chi_{\{t\}}$ (see Example 1.7)

$$\chi_t(s) = \begin{cases} 1 & \text{if } s = t, \\ 0 & \text{if } s \neq t, \end{cases}$$

since any function having non-zero values at just a finite number of places $s_1, s_2, \ldots, s_n$ can be written uniquely as

$$f = f(s_1)\chi_{s_1} + f(s_2)\chi_{s_2} + \cdots + f(s_n)\chi_{s_n}.$$

Evidently the elements of $F(S)$ are in one-to-one correspondence with the 'formal finite sums' alluded to above.

Example 7.1 The vector space $\hat{\mathbb{R}}^\infty$ defined in Example 3.10 is isomorphic with the free vector space on any countably infinite set $S = \{s_1, s_2, s_3, \ldots\}$, since the map $\sigma : \hat{\mathbb{R}}^\infty \to F(S)$ defined by

$$\sigma(a^1, a^2, \ldots, a^n, 0, 0, \ldots) = \sum_{i=1}^n a^i \chi_{s_i}$$

is linear, one-to-one and onto.

The tensor product $V \otimes W$

Let V and W be two vector spaces over a field $\mathbb{K}$. Imagine forming a product $v \otimes w$ between elements of these two vector spaces, called their 'tensor product', subject to the rules

$$(av + bv') \otimes w = av \otimes w + bv' \otimes w, \qquad v \otimes (aw + bw') = av \otimes w + bv \otimes w'.$$

The main difficulty with this simple idea is that we have no idea of what space $v \otimes w$ belongs to. The concept of a free vector space can be used to give a proper definition for this product.

Let $F(V \times W)$ be the free vector space over $V \times W$. This vector space is, in a sense, much too 'large' since pairs such as $a(v, w)$ and (av, w), or $(v + v', w)$ and $(v, w) + (v', w)$, are totally unrelated in $F(V \times W)$. To reduce the vector space to sensible proportions, we define U to be the vector subspace of $F(V \times W)$ generated by all elements of the form

$$(av + bv', w) - a(v, w) - b(v', w) \quad \text{and} \quad (v, aw + bw') - a(v, w) - b(v, w')$$

where, for notational simplicity, we make no distinction between a pair (v, w) and its characteristic function $\chi_{(v,w)} \in F(V \times W)$. The subspace U contains essentially all vector combinations that are to be identified with the zero element. The **tensor product** of V and W is defined to be the factor space

$$V \otimes W = F(V \times W)/U.$$

The **tensor product** $v \otimes w$ **of a pair of vectors** $v \in V$ and $w \in W$ is defined as the equivalence class or coset in $V \otimes W$ to which (v, w) belongs,

$$v \otimes w = [(v, w)] = (v, w) + U.$$

Tensors

This product is *bilinear*,

$$(av + bv') \otimes w = av \otimes w + bv' \otimes w,$$
$$v \otimes (aw + bw') = av \otimes w + bv \otimes w'.$$

To show the first identity,

$$\begin{aligned}(av + bv') \otimes w &= (av + bv', w) + U \\ &= (av + bv', w) - ((av + bv', w) - a(v, w) - b(v', w)) + U \\ &= a(v, w) + b(v', w) + U \\ &= av \otimes w + bv' \otimes w,\end{aligned}$$

and the second identity is similar.

If V and W are both finite dimensional let $\{e_i \mid i = 1, \ldots, n\}$ be a basis for V, and $\{f_a \mid a = 1, \ldots, m\}$ a basis of W. Every tensor product $v \otimes w$ can, through bilinearity, be written

$$v \otimes w = (v^i e_i) \otimes (w^a f_a) = v^i w^a (e_i \otimes f_a). \tag{7.1}$$

We will use the term **tensor** to describe the general element of $V \otimes W$. Since every tensor A is a finite sum of elements of the form $v \otimes w$ it can, on substituting (7.1), be expressed in the form

$$A = A^{ia} e_i \otimes f_a. \tag{7.2}$$

Hence the tensor product space $V \otimes W$ is spanned by the nm tensors $\{e_i \otimes f_a \mid i = 1, \ldots, n, \ a = 1, \ldots, m\}$.

Furthermore, these tensors form a basis of $V \otimes W$ since they are linearly independent. To prove this statement, let (ρ, φ) be any ordered pair of linear functionals $\rho \in V^*, \varphi \in W^*$. Such a pair defines a linear functional on $F(V \times W)$ by setting

$$(\rho, \varphi)(v, w) = \rho(v)\varphi(w),$$

and extending to all of $F(V \times W)$ by linearity,

$$(\rho, \varphi)\left[\sum_r a^r (v_r, w_r)\right] = \sum_r a^r \rho(v_r)\varphi(w_r).$$

This linear functional vanishes on the subspace U and therefore 'passes to' the tensor product space $V \otimes W$ by setting

$$(\rho, \varphi)\left[\sum_r a^r v_r \otimes w_r\right] = \sum_r a^r (\rho, \varphi)((v_r, w_r)).$$

Let ε^k, φ^b be the dual bases in V^* and W^* respectively; we then have

$$\begin{aligned}A^{ia} e_i \otimes f_a = 0 &\implies (\varepsilon^j, \varphi^b)(A^{ia} e_i \otimes f_a) = 0 \\ &\implies A^{ia} \delta^j_i \delta^b_a = 0 \\ &\implies A^{jb} = 0.\end{aligned}$$

7.1 Free vector spaces and tensor spaces

Hence the tensors $e_i \otimes f_a$ are l.i. and form a basis of $V \otimes W$. The dimension of $V \otimes W$ is given by

$$\dim(V \otimes W) = \dim V \dim W.$$

Setting $V = W$, the elements of $V \otimes V$ are called **contravariant tensors of degree 2** on V. If $\{e_i\}$ is a basis of V every contravariant tensor of degree 2 has a unique expansion

$$T = T^{ij} e_i \otimes e_j,$$

and the real numbers T^{ij} are called the *components* of the tensor with respect to this basis. Similarly each element S of $V^* \otimes V^*$ is called a **covariant tensor of degree 2** on V and has a unique expansion with respect to the dual basis ε^i,

$$S = S_{ij} \varepsilon^i \otimes \varepsilon^j.$$

Dual representation of tensor product

Given a pair of vector spaces V_1 and V_2 over a field $\mathbb{K}$, a map $T : V_1 \times V_2 \to \mathbb{K}$ is said to be **bilinear** if

$$T(av_1 + bv'_1, v_2) = aT(v_1, v_2) + bT(v'_1, v_2),$$
$$T(v_1, av_2 + bv'_2,) = aT(v_1, v_2) + bT(v_1, v'_2),$$

for all $v_1, v'_1 \in V_1$, $v_2, v'_2 \in V_2$, and $a, b \in \mathbb{K}$. Bilinear maps can be added and multiplied by scalars in a manner similar to that for linear functionals given in Section 3.7, and form a vector space that will be denoted $(V_1, V_2)^*$.

Every pair of vectors (v, w) where $v \in V, w \in W$ defines a bilinear map $V^* \times W^* \to \mathbb{K}$, by setting

$$(v, w) : (\rho, \varphi) \mapsto \rho(v)\varphi(w).$$

We can extend this correspondence to all of $F(V \times W)$ in the obvious way by setting

$$\sum_r a^r (v_r, w_r) : (\rho, \varphi) \mapsto \sum_r a^r \rho(v_r)\varphi(w_r),$$

and since the action of any generators of the subspace U, such as $(av + bv', w) - a(v, w) - b(v', w)$, clearly vanishes on $V^* \times W^*$, the correspondence passes in a unique way to the tensor product space $V \otimes W = F(V, W)/U$. That is, every tensor $A = \sum_r a^r v_r \otimes w_r$ defines a bilinear map $V^* \times W^* \to \mathbb{K}$, by setting

$$A(\rho, \varphi) = \sum_r a^r \rho(v_r)\varphi(w_r).$$

This linear mapping from $V \otimes W$ into the space of bilinear maps on $V^* \times W^*$ is also one-to-one in the case of finite dimensional vector spaces. For, suppose $A(\rho, \varphi) = B(\rho, \varphi)$ for all $\rho \in V^*, \varphi \in W^*$. Let e_i, f_a be bases of V and W respectively, and ε^j, φ^a be the dual bases. Writing $A = A^{ia} e_i \otimes f_a = 0$, $B = B^{ia} e_i \otimes f_a = 0$, we have

$$A^{ia} \rho(e_i)\varphi(f_a) = B^{ia} \rho(e_i)\varphi(f_a)$$

Tensors

for all linear functionals $\rho \in V^*$, $\varphi \in W^*$. If we set $\rho = \varepsilon^j$, $\varphi = \varphi^b$ then

$$A^{ia}\delta_i^j \delta_a^b = B^{ia}\delta_i^j \delta_a^b,$$

resulting in $A^{jb} = B^{jb}$. Hence $A = B$, and for finite dimensional vector spaces V and W we have shown that the linear correspondence between $V \otimes W$ and $(V^*, W^*)^*$ is one-to-one,

$$V \otimes W \cong (V^*, W^*)^*. \tag{7.3}$$

This isomorphism does not hold for infinite dimensional spaces.

A tedious but straightforward argument results in the associative law for tensor products of three vectors

$$u \otimes (v \otimes w) = (u \otimes v) \otimes w.$$

Hence the tensor product of three or more vector spaces is defined in a unique way,

$$U \otimes V \otimes W = U \otimes (V \otimes W) = (U \otimes V) \otimes W.$$

For finite dimensional spaces it may be shown to be isomorphic with the space of maps $A : U^* \times V^* \times W^* \to \mathbb{K}$ that are linear in each argument separately.

Free associative algebras

Let $\mathcal{F}(V)$ be the infinite direct sum of vector spaces

$$\mathcal{F}(V) = V^{(0)} \oplus V^{(1)} \oplus V^{(2)} \oplus V^{(3)} \oplus \dots$$

where $\mathbb{K} = V^{(0)}$, $V = V^{(1)}$ and

$$V^{(r)} = \underbrace{V \otimes V \otimes \dots \otimes V}_{r}.$$

The typical member of this infinite direct sum can be written as a *finite* formal sum of tensors from the tensor spaces $V^{(r)}$,

$$a + u + A_2 + \dots + A_r.$$

To define a product rule on $\mathcal{F}(V)$ set

$$\underbrace{u_1 \otimes u_2 \otimes \dots \otimes u_r}_{\in V^{(r)}} \underbrace{v_1 \otimes v_2 \otimes \dots \otimes v_s}_{\in V^{(s)}} = \underbrace{u_1 \otimes \dots \otimes u_r \otimes v_1 \otimes \dots \otimes v_s}_{\in V^{(r+s)}}$$

and extend to all of $\mathcal{F}(V)$ by linearity. The distributive law (6.1) is automatically satisfied, making $\mathcal{F}(V)$ with this product structure into an associative algebra. The algebra $\mathcal{F}(V)$ has in essence no 'extra rules' imposed on it other than simple juxtaposition of elements from V and multiplication by scalars. It is therefore called the **free associative algebra** over V. All associative algebras can be constructed as a factor algebra of the free associative algebra over a vector space. The following example illustrates this point.

Example 7.2 If V is the one-dimensional free vector space over the reals on the singleton set $S = \{x\}$ then the free associative algebra over $\mathcal{F}(V)$ is in one-to-one correspondence

7.1 Free vector spaces and tensor spaces

with the algebra of real polynomials $\mathcal{P}$, Example 6.3, by setting

$$a_0 + a_1 x + a_2 x^2 + \cdots + a_n x^n \equiv a_0 + a_1 x + a_2 x \otimes x + \cdots + a_n \underbrace{x \otimes x \otimes \cdots \otimes x}_{n}.$$

This correspondence is an algebra isomorphism since the product defined on $\mathcal{F}(V)$ by the above procedure will be identical with multiplication of polynomials. For example,

$$(ax + bx \otimes x \otimes x)(c + dx \otimes x) = acx + (ad+bc)x \otimes x \otimes x + bdx \otimes x \otimes x \otimes x \otimes x$$
$$\equiv acx + (ad+bc)x^3 + bdx^5$$
$$= (ax + bx^3)(c + dx^2), \text{ etc.}$$

Set $\mathcal{C}$ to be the ideal of $\mathcal{P}$ generated by $x^2 + 1$, consisting of all polynomials of the form $f(x)(x^2 + 1)g(x)$. By identifying i with the polynomial class $[x] = x + \mathcal{C}$ and real numbers with the class of constant polynomials $a \to [a]$, the algebra of complex numbers is isomorphic with the factor algebra $\mathcal{P}/\mathcal{C}$, for

$$i^2 \equiv [x]^2 = [x^2] = [x^2 - (x^2+1)] = [-1] \equiv -1.$$

Grassmann algebra as a factor algebra of free algebras

The definition of Grassmann algebra given in Section 6.4 is unsatisfactory in two key aspects. Firstly, in the definition of exterior product it is by no means obvious that the rules (EP1)–(EP3) produce a well-defined and unique product on $\Lambda(V)$. Secondly, the matter of linear independence of the basis vectors $e_{i_1 i_2 \ldots i_r}$ $(i_1 < i_2 < \cdots < i_r)$ had to be postulated separately in Section 6.4. The following discussion provides a more rigorous foundation for Grassmann algebras, and should clarify these issues.

Let $\mathcal{F}(V)$ be the free associative algebra over a real vector space V, and let $\mathcal{S}$ be the ideal generated by all elements of $\mathcal{F}(V)$ of the form $u \otimes T \otimes v + v \otimes T \otimes u$ where $u, v \in V$ and $T \in \mathcal{F}(V)$. The general element of $\mathcal{S}$ is

$$S \otimes u \otimes T \otimes v \otimes U + S \otimes v \otimes T \otimes u \otimes U$$

where $u, v \in V$ and $S, T, U \in \mathcal{F}(V)$. The ideal $\mathcal{S}$ essentially identifies those elements of $\mathcal{F}(V)$ that will vanish when the tensor product $\otimes$ is replaced by the wedge product $\wedge$.

Exercise: Show that the ideal $\mathcal{S}$ is generated by all elements of the form $w \otimes T \otimes w$ where $w \in V$ and $T \in \mathcal{F}(V)$. [*Hint*: Set $w = u + v$.]

Define the *Grassmann algebra* $\Lambda(V)$ to be the factor algebra

$$\Lambda(V) = \mathcal{F}(V)/\mathcal{S}, \qquad (7.4)$$

and denote the induced associative product by $\wedge$,

$$[A] \wedge [B] = [A \otimes B] \qquad (7.5)$$

where $[A] \equiv A + \mathcal{S}$, $[B] \equiv B + \mathcal{S}$. As in Section 6.4, the elements $[A]$ of the factor algebra are called **multivectors**. There is no ambiguity in dropping the square brackets,

Tensors

$A \equiv [A]$ and writing $A \wedge B$ for $[A] \wedge [B]$. The algebra $\Lambda(V)$ is the direct sum of subspaces corresponding to tensors of degree r,

$$\Lambda(V) = \Lambda^0(V) \oplus \Lambda^1(V) \oplus \Lambda^2(V) \oplus \ldots,$$

where

$$\Lambda^r(V) = [V^{(r)}] = \{A + \mathcal{S} \mid A \in V^{(r)}\}$$

whose elements are called r-**vectors**. If A is an r-vector and B an s-vector then $A \wedge B$ is an $(r+s)$-vector.

Since, by definition, $u \otimes v + v \otimes u$ is a member of $\mathcal{S}$, we have

$$u \wedge v = [u \otimes v] = [-v \otimes u] = -v \wedge u,$$

for all $u, v \in V$. Hence $u \wedge u = 0$ for all $u \in V$.

Exercise: Prove that if A, B and C are any multivectors then for all $u, v \in V$

$$A \wedge u \wedge B \wedge v \wedge C + A \wedge v \wedge B \wedge u \wedge C = 0,$$

and $A \wedge u \wedge B \wedge u \wedge C = 0$.

Exercise: From the corresponding rules of tensor product show that exterior product is associative and distributive.

From the associative law

$$(u_1 \wedge u_2 \wedge \cdots \wedge u_r) \wedge (v_1 \wedge v_2 \wedge \cdots \wedge v_s) = u_1 \wedge \cdots \wedge u_r \wedge v_1 \wedge \cdots \wedge v_s,$$

in agreement with (EP2) of Section 6.4. This provides a basis-independent definition for exterior product on any finite dimensional vector space V, having the desired properties (EP1)–(EP3). Since every r-vector is the sum of simple r-vectors, the space of r-vectors $\Lambda^r(V)$ is spanned by

$$E_r = \{e_{i_1 i_2 \ldots i_r} \mid i_1 < i_2 < \cdots < i_r\},$$

where

$$e_{i_1 i_2 \ldots i_r} = e_{i_1} \wedge e_{i_2} \cdots \wedge e_{i_r},$$

as shown in Section 6.4. It is left as an exercise to show that the set E_r does indeed form a basis of the space of r-vectors (see Problem 7.6). Hence, as anticipated in Section 6.4, the dimension of the space of r-vectors is

$$\dim \Lambda^r(V) = \binom{n}{r} = \frac{n!}{r!(n-r)!},$$

and the dimension of the Grassmann algebra $\Lambda(V)$ is 2^n.

7.1 Free vector spaces and tensor spaces

Problems

Problem 7.1 Show that the direct sum $V \oplus W$ of two vector spaces can be defined from the free vector space as $F(V \times W)/U$ where U is a subspace generated by all linear combinations of the form

$$(av + bv', aw + bw') - a(v, w) - b(v', w').$$

Problem 7.2 Prove the so-called *universal property* of free vector spaces. Let $\varphi : S \to F(S)$ be the map that assigns to any element $s \in S$ its characteristic function $\chi_s \in F(S)$. If V is any vector space and $\alpha : S \to V$ any map from S to V, then there exists a unique linear map $T : F(S) \to V$ such that $\alpha = T \circ \varphi$, as depicted by the *commutative diagram*

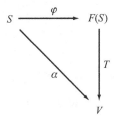

Show that this process is reversible and may be used to define the free vector space on S as being the unique vector space $F(S)$ for which the above commutative diagram holds.

Problem 7.3 Let $\mathcal{F}(V)$ be the free associative algebra over a vector space V.

(a) Show that there exists a linear map $I : V \to \mathcal{F}(V)$ such that if $\mathcal{A}$ is any associative algebra over the same field $\mathbb{K}$ and $S : V \to \mathcal{A}$ a linear map, then there exists a unique algebra homomorphism $\alpha : \mathcal{F}(V) \to \mathcal{A}$ such that $S = \alpha \circ I$.
(b) Depict this property by a commutative diagram.
(c) Show the converse: any algebra $\mathcal{F}$ for which there is a map $I : V \to \mathcal{F}$ such that the commutative diagram holds for an arbitrary linear map S is isomorphic with the free associative algebra over V.

Problem 7.4 Give a definition of quaternions as a factor algebra of the free algebra on a three-dimensional vector space.

Problem 7.5 The Clifford algebra $\mathcal{C}_g$ associated with an inner product space V with scalar product $g(u, v) \equiv u \cdot v$ can be defined in the following way. Let $\mathcal{F}(V)$ be the free associative algebra on V and $\mathcal{C}$ the two-sided ideal generated by all elements of the form

$$A \otimes (u \otimes v + v \otimes u - 2g(u, v)1) \otimes B \quad (A, B \in \mathcal{F}(V)).$$

The Clifford algebra in question is now defined as the factor space $\mathcal{F}(V)/\mathcal{C}$. Verify that this algebra is isomorphic with the Clifford algebra as defined in Section 6.3, and could serve as a basis-independent definition for the Clifford algebra associated with a real inner product space.

Problem 7.6 Show that E_2 is a basis of $\Lambda^2(V)$. In outline: define the maps $\varepsilon^{kl} : V^2 \to \mathbb{R}$ by

$$\varepsilon^{kl}(u, v) = \varepsilon^k(u)\varepsilon^l(v) - \varepsilon^l(u)\varepsilon^k(v) = u^k v^l - u^l v^k.$$

Extend by linearity to the tensor space $V^{(2,0)}$ and show there is a natural passage to the factor space, $\hat{\varepsilon}^{kl} : \Lambda^2(V) = V^{(2,0)}/S^2 \to \mathbb{R}$. If a linear combination from E_2 were to vanish,

$$\sum_{i<j} A^{ij} e_i \wedge e_j = 0,$$

apply the map $\hat{\varepsilon}^{kl}$ to this equation, to show that all coefficients A^{ij} must vanish separately.

Tensors

Indicate how the argument may be extended to show that if $r \leq n = \dim V$ then E_r is a basis of $\Lambda^r(V)$.

7.2 Multilinear maps and tensors

The dual representation of tensor product allows for an alternative definition of tensor spaces and products. Key to this approach is the observation in Section 3.7, that every finite dimensional vector space V has a natural isomorphism with V^{**} whereby a vector v acts as a linear functional on V^* through the identification

$$v(\omega) = \omega(v) = \langle v, \omega \rangle = \langle \omega, v \rangle.$$

Multilinear maps and tensor spaces of type (r, s)

Let $V_1, V_2, \ldots, V_N$ be vector spaces over the field $\mathbb{R}$. A map

$$T : V_1 \times V_2 \times \cdots \times V_N \to \mathbb{R}$$

is said to be **multilinear** if it is linear in each argument separately,

$$T(v_1, \ldots, v_{i-1}, av_i + bv'_i, \ldots, v_N)$$
$$= aT(v_1, \ldots, v_{i-1}, v_i, \ldots, v_N) + bT(v_1, \ldots, v_{i-1}, v'_i, \ldots, v_N).$$

Multilinear maps can be added and multiplied by scalars in the usual fashion,

$$(aT + bS)(v_1, v_2, \ldots, v_s) = aT(v_1, v_2, \ldots, v_s) + bS(v_1, v_2, \ldots, v_s),$$

and form a vector space, denoted

$$V_1^* \otimes V_2^* \otimes \cdots \otimes V_N^*,$$

called the **tensor product of the dual spaces** $V_1^*, V_2^*, \ldots, V_N^*$. When $N = 1$, the word 'multilinear' is simply replaced with the word 'linear' and the notation is consistent with the concept of the dual space V^* defined in Section 3.7 as the set of linear functionals on V. If we identify every vector space V_i with its double dual V_i^{**}, the **tensor product of the vector spaces** $V_1, V_2, \ldots, V_N$, denoted $V_1 \otimes V_2 \otimes \cdots \otimes V_N$, is then the set of multilinear maps from $V_1^* \times V_2^* \times \cdots \times V_N^*$ to $\mathbb{R}$.

Let V be a vector space of dimension n over the field $\mathbb{R}$. Setting

$$V_1 = V_2 = \cdots = V_r = V^*, \qquad V_{r+1} = V_{r+2} = \cdots = V_{r+s} = V, \quad \text{where} \quad N = r + s,$$

we refer to any multilinear map

$$T : \underbrace{V^* \times V^* \times \cdots \times V^*}_{r} \times \underbrace{V \times V \times \cdots \times V}_{s} \to \mathbb{R}$$

as a **tensor of type** (r, s) on V. The integer $r \geq 0$ is called the **contravariant degree** and $s \geq 0$ the **covariant degree** of T. The vector space of tensors of type (r, s) is

7.2 Multilinear maps and tensors

denoted
$$V^{(r,s)} = \underbrace{V \otimes V \otimes \cdots \otimes V}_{r} \otimes \underbrace{V^* \otimes V^* \otimes \cdots \otimes V^*}_{s}.$$

This definition is essentially equivalent to the dual representation of the definition in Section 7.1. Both definitions are totally 'natural' in that they do not require a choice of basis on the vector space V.

It is standard to set $V^{(0,0)} = \mathbb{R}$; that is, tensors of type $(0, 0)$ will be identified as scalars. Tensors of type $(0, 1)$ are linear functionals (covectors)
$$V^{(0,1)} \equiv V^*,$$
while tensors of type $(1, 0)$ can be regarded as ordinary vectors
$$V^{(1,0)} \equiv V^{**} \equiv V.$$

Covariant tensors of degree 2

A tensor of type $(0, 2)$ is a bilinear map $T : V \times V \to \mathbb{R}$. In keeping with the terminology of Section 7.1, such a tensor may be referred to as a **covariant tensor of degree 2** on V. Linearity in each argument reads
$$T(av + bw, u) = aT(v, u) + bT(w, u) \quad \text{and} \quad T(v, au + bw) = aT(v, u) + bT(v, w).$$

If $\omega, \rho \in V^*$ are linear functionals over V, let their **tensor product** $\omega \otimes \rho$ be the covariant tensor of degree 2 defined by
$$\omega \otimes \rho \, (u, v) = \omega(u) \, \rho(v).$$

Linearity in the first argument follows from
$$\begin{aligned}
\omega \otimes \rho \, (au + bv, w) &= \omega(au + bv)\rho(w) \\
&= (a\omega(u) + b\omega(v))\rho(w) \\
&= a\omega(u)\rho(w) + b\omega(v)\rho(w) \\
&= a\omega \otimes \rho \, (u, w) + b\omega \otimes \rho \, (v, w).
\end{aligned}$$

A similar argument proves linearity in the second argument v.

Example 7.3 Tensor product is not a commutative operation since in general $\omega \otimes \rho \neq \rho \otimes \omega$. For example, let e_1, e_2 be a basis of a two-dimensional vector space V, and let $\varepsilon^1, \varepsilon^2$ be the dual basis of V^*. If
$$\omega = 3\varepsilon^1 + 2\varepsilon^2, \qquad \rho = \varepsilon^1 - \varepsilon^2$$
then
$$\begin{aligned}
\omega \otimes \rho \, (u, v) &= \omega(u^1 e_1 + u^2 e_2) \rho(v^1 e_1 + v^2 e_2) \\
&= (3u^1 + 2u^2)(v^1 - v^2) \\
&= 3u^1 v^1 - 3u^1 v^2 + 2u^2 v^1 - 2u^2 v^2
\end{aligned}$$

and
$$\rho \otimes \omega(u, v) = (u^1 - u^2)(3v^1 + 3v^2)$$
$$= 3u^1v^1 + 2u^1v^2 - 3u^2v^1 - 2u^2v^2$$
$$\neq \omega \otimes \rho(u, v).$$

More generally, let $e_1, \ldots, e_n$ be a basis of the vector space V and $\varepsilon^1, \ldots, \varepsilon^n$ its dual basis, defined by

$$\varepsilon^i(e_j) = \langle \varepsilon^i, e_j \rangle = \delta^i_j \quad (i, j = 1, \ldots, n). \tag{7.6}$$

Theorem 7.1 *The tensor products*

$$\varepsilon^i \otimes \varepsilon^j \quad (i, j = 1, \ldots, n)$$

form a basis of the vector space $V^{(0,2)}$, which therefore has dimension n^2.

Proof: The tensors $\varepsilon^i \otimes \varepsilon^j$ $(i, j = 1, \ldots, n)$ are linearly independent, for if

$$a_{ij} \varepsilon^i \otimes \varepsilon^j \equiv \sum_{i=1}^{n} \sum_{j=1}^{n} a_{ij} \varepsilon^i \otimes \varepsilon^j = 0,$$

then for each $1 \le k, l \le n$,

$$0 = a_{ij} \varepsilon^i \otimes \varepsilon^j (e_k, e_l) = a_{ij} \delta^i_k \delta^j_l = a_{kl}.$$

Furthermore, if T is any covariant tensor of degree 2 then

$$T = T_{ij} \varepsilon^i \otimes \varepsilon^j \quad \text{where} \quad T_{ij} = T(e_i, e_j) \tag{7.7}$$

since for any pair of vectors $u = u^i e_i$, $v = v^j e_j$ in V,

$$(T - T_{ij} \varepsilon^i \otimes \varepsilon^j)(u, v) = T(u, v) - T_{ij} \varepsilon^i(u) \varepsilon^j(v)$$
$$= T(u^i e_i, v^j e_j) - T_{ij} u^i v^j$$
$$= u^i v^j T_{e_i, e_j} - T_{ij} u^i v^j$$
$$= u^i v^j T_{ij} - T_{ij} u^i v^j = 0.$$

Hence the n^2 tensors $\varepsilon^i \otimes \varepsilon^j$ are linearly independent and span $V^{(0,2)}$. They therefore form a basis of $V^{(0,2)}$. ∎

The coefficients T_{ij} in the expansion $T = T_{ij} \varepsilon^i \otimes \varepsilon^j$ are uniquely given by the expression on the right in Eq. (7.7), for if $T = T'_{ij} \varepsilon^i \otimes \varepsilon^j$ then by linear independence of the $\varepsilon^i \otimes \varepsilon^j$

$$(T'_{ij} - T_{ij}) \varepsilon^i \otimes \varepsilon^j = 0 \implies T'_{ij} - T_{ij} = 0 \implies T'_{ij} = T_{ij}.$$

They are called the **components of T with respect to the basis $\{e_i\}$**. For any vectors $u = u^i e_i$, $v = v^j e_j$

$$T(u, v) = T_{ij} u^i v^j. \tag{7.8}$$

Example 7.4 For the linear functional ω and ρ given in Example 7.3, we can write

$$\omega \otimes \rho = (3\varepsilon^1 + 2\varepsilon^2) \otimes (\varepsilon^1 - \varepsilon^2)$$
$$= 3\varepsilon^1 \otimes \varepsilon^1 - 3\varepsilon^1 \otimes \varepsilon^2 + 2\varepsilon^2 \otimes \varepsilon^1 - 2\varepsilon^2 \otimes \varepsilon^2$$

7.2 Multilinear maps and tensors

and similarly

$$\rho \otimes \omega = 3\varepsilon^1 \otimes \varepsilon^1 + 2\varepsilon^1 \otimes \varepsilon^2 - 3\varepsilon^2 \otimes \varepsilon^1 - 2\varepsilon^2 \otimes \varepsilon^2.$$

Hence the components of the tensor products $\omega \otimes \rho$ and $\rho \otimes \omega$ with respect to the basis tensors $\varepsilon^i \otimes \varepsilon^j$ may be displayed as arrays,

$$[(\omega \otimes \rho)_{ij}] = \begin{pmatrix} 3 & -3 \\ 2 & -2 \end{pmatrix}, \quad [(\rho \otimes \omega)_{ij}] = \begin{pmatrix} 3 & 2 \\ 3 & -2 \end{pmatrix}.$$

Exercise: Using the components $(\omega \otimes \rho)_{12}$, etc. in the preceding example verify the formula (7.8) for $\omega \otimes \rho\,(u, v)$. Do the same for $\rho \otimes \omega\,(u, v)$.

In general, if $\omega = w_i \varepsilon^i$ and $\rho = r_j \varepsilon^j$ then

$$\omega \otimes \rho = (w_i \varepsilon^i) \otimes (r_j \varepsilon^j) = w_i r_j \varepsilon^i \otimes \varepsilon^j,$$

and the components of $\omega \otimes \rho$ are

$$(\omega \otimes \rho)_{ij} = w_i r_j. \tag{7.9}$$

This $n \times n$ array of components is formed by taking all possible component-by-component products of the two linear functionals.

Exercise: Prove Eq. (7.9) by evaluating $(\omega \otimes \rho)_{ij} = (\omega \otimes \rho)(e_i, e_j)$.

Example 7.5 Let $(V, \cdot)$ be a real inner product space, as in Section 5.1. The map $g : V \times V \to \mathbb{R}$ defined by

$$g(u, v) = u \cdot v$$

is obviously bilinear, and is a covariant tensor of degree 2 called the **metric tensor** of the inner product. The components $g_{ij} = e_i \cdot e_j = g(e_i, e_j)$ of the inner product are the components of the metric tensor with respect to the basis $\{e_i\}$,

$$g = g_{ij}\, \varepsilon^i \otimes \varepsilon^j,$$

while the inner product of two vectors is

$$u \cdot v = g(u, v) = g_{ij} u^i v^j.$$

The metric tensor is **symmetric**,

$$g(u, v) = g(v, u).$$

Exercise: Show that a tensor T is symmetric if and only if its components form a symmetric array,

$$T_{ij} = T_{ji} \quad \text{for all } i, j = 1, \ldots, n.$$

Example 7.6 Let T be a covariant tensor of degree 2. Define the map $\bar{T} : V \to V^*$ by

$$\langle \bar{T}v, u \rangle = T(u, v) \quad \text{for all } u, v \in V. \tag{7.10}$$

Tensors

Here $\bar{T}(v)$ has been denoted more simply by $\bar{T}v$. The map $\bar{T}$ is clearly linear, $\bar{T}(au + bv) = a\bar{T}u + b\bar{T}v$, since

$$\langle \bar{T}(au + bv), w \rangle = T(w, au + bv) = aT(w, u) + bT(w, v) = \langle a\bar{T}u + b\bar{T}v, w \rangle$$

holds for all $w \in V$. Conversely, given a linear map $\bar{T} : V \to V^*$, Eq. (7.10) defines a tensor T since $T(u, v)$ so defined is linear both in u and v. Thus every covariant tensor of degree 2 can be identified with an element of $L(V, V^*)$.

Exercise: In components, show that $(\bar{T}v)_i = T_{ij}v^j$.

Contravariant tensors of degree 2

A **contravariant tensor of degree 2** on V, or tensor of type $(2, 0)$, is a bilinear real-valued map S over V^*:

$$S : V^* \times V^* \to \mathbb{R}.$$

Then, for all $a, b \in \mathbb{R}$ and all $\omega, \rho, \theta \in V^*$

$$S(a\omega + b\rho, \theta) = aS(\omega, \theta) + bS(\rho, \theta) \quad \text{and} \quad S(\omega, a\rho + b\theta) = aS(\omega, \rho) + bS(\omega, \theta).$$

If u and v are any two vectors in V, then their **tensor product** $u \otimes v$ is the contravariant tensor of degree 2 defined by

$$u \otimes v\,(\omega, \rho) = u(\omega)\,v(\rho) = \omega(u)\,\rho(v). \tag{7.11}$$

If $e_1, \ldots, e_n$ is a basis of the vector space V with dual basis $\{\varepsilon^j\}$ then, just as for $V^{(0,2)}$, the tensors $e_i \otimes e_j$ form a basis of the space $V^{(2,0)}$, and every contravariant tensor of degree 2 has a unique expansion

$$S = S^{ij} e_i \otimes e_j \quad \text{where} \quad S^{ij} = S(\varepsilon^i, \varepsilon^j).$$

The scalars S^{ij} are called the **components** of the tensor S with respect to the basis $\{e_i\}$.

Exercise: Provide detailed proofs of these statements.

Exercise: Show that the components of the tensor product of two vectors is given by

$$(u \otimes v)^{ij} = u^i v^j.$$

Example 7.7 It is possible to identify contravariant tensors of degree 2 with linear maps from V^* to V. If S is a tensor of type $(0, 2)$ define a map $\bar{S} : V^* \to V$ by

$$(\bar{S}\rho)(\omega) \equiv \omega(\bar{S}\rho) = S(\omega, \rho) \quad \text{for all } \omega, \rho \in V^*.$$

The proof that this correspondence is one-to-one is similar to that given in Example 7.6.

Example 7.8 Let $(V, \cdot)$ be a real inner product space with metric tensor g as defined in Example 7.5. Let $\bar{g} : V \to V^*$ be the map defined by g using Example 7.6,

$$\langle \bar{g}v, u \rangle = g(u, v) = u \cdot v \quad \text{for all } u, v \in V. \tag{7.12}$$

7.2 Multilinear maps and tensors

From the non-singularity condition (SP3) of Section 5.1 the kernel of this map is $\{0\}$, from which it follows that it is one-to-one. Furthermore, because the dimensions of V and V^* are identical, $\bar{g}$ is onto and therefore invertible. As shown in Example 7.7 its inverse $\bar{g}^{-1} : V^* \to V$ defines a tensor g^{-1} of type $(2, 0)$ by

$$g^{-1}(\omega, \rho) = \langle \omega, \bar{g}^{-1}\rho \rangle.$$

From the symmetry of the metric tensor g and the identities

$$\bar{g}\bar{g}^{-1} = \mathrm{id}_{V^*}, \qquad \bar{g}^{-1}\bar{g} = \mathrm{id}_V$$

it follows that g^{-1} is also a symmetric tensor, $g^{-1}(\omega, \rho) = g^{-1}(\rho, \omega)$:

$$\begin{aligned}
g^{-1}(\omega, \rho) &= \langle \omega, \bar{g}^{-1}\rho \rangle \\
&= \langle \bar{g}\bar{g}^{-1}\omega, \bar{g}^{-1}\rho \rangle \\
&= g(\bar{g}^{-1}\rho, \bar{g}^{-1}\omega) \\
&= g(\bar{g}^{-1}\omega, \bar{g}^{-1}\rho) \\
&= \langle \bar{g}\bar{g}^{-1}\rho, \bar{g}^{-1}\omega \rangle \\
&= \langle \rho, \bar{g}^{-1}\omega \rangle \\
&= g^{-1}(\rho, \omega).
\end{aligned}$$

It is usual to denote the components of the inverse metric tensor with respect to any basis $\{e_i\}$ by the symbol g^{ij}, so that

$$g^{-1} = g^{ij} e_i \otimes e_j \quad \text{where} \quad g^{ij} = g^{-1}(\varepsilon_i, \varepsilon_j) = g^{ji}. \tag{7.13}$$

From Example 7.6 we have

$$\langle \bar{g}e_i, e_j \rangle = g(e_j, e_i) = g_{ji},$$

whence

$$\bar{g}(e_i) = g_{ji}\varepsilon^j.$$

Similarly, from Example 7.7

$$\langle \bar{g}^{-1}\varepsilon^j, \varepsilon^k \rangle = g^{-1}(\varepsilon^k, \varepsilon^j) = g^{kj} \implies \bar{g}^{-1}(\varepsilon^j) = g^{kj} e_k.$$

Hence

$$e_i = \bar{g}^{-1} \circ \bar{g}(e_i) = \bar{g}^{-1}(g_{ji}\varepsilon^j) = g_{ji}g^{kj} e_k.$$

Since $\{e_i\}$ is a basis of V, or from Eq. (7.6), we conclude that

$$g^{kj} g_{ji} = \delta^k_i, \tag{7.14}$$

and the matrices $[g^{ij}]$ and $[g_{ij}]$ are inverse to each other.

Tensors

Mixed tensors

A tensor R of covariant degree 1 and contravariant degree 1 is a bilinear map $R : V^* \times V \to \mathbb{R}$:

$$R(a\omega + b\rho, u) = aR(\omega, u) + bR(\rho, u)$$
$$R(\omega, au + bv) = aR(\omega, u) + bR(\omega, v),$$

sometimes referred to as a **mixed tensor**. Such tensors are of type $(1, 1)$, belonging to the vector space $V^{(1,1)}$.

For a vector $v \in V$ and covector $\omega \in V^*$, define their tensor product $v \otimes \omega$ by

$$v \otimes \omega (\rho, u) = v(\rho)\omega(u).$$

As in the preceding examples it is straightforward to show that $e_i \otimes \varepsilon^j$ form a basis of $V^{(1,1)}$, and every mixed tensor R has a unique decomposition

$$R = R^i{}_j e_i \otimes \varepsilon^j \quad \text{where} \quad R^i{}_j = R(\varepsilon^i, e_j).$$

Example 7.9 Every tensor R of type $(1, 1)$ defines a map $\bar{R} : V \to V$ by

$$\langle \bar{R}u, \omega \rangle \equiv \omega(\bar{R}u) = R(\omega, u).$$

The proof that there is a one-to-one correspondence between such maps and tensors of type $(1, 1)$ is similar to that given in Examples 7.6 and 7.7. Operators on V and tensors of type $(1, 1)$ can be thought of as essentially identical, $V^{(1,1)} \cong L(V, V)$.

If $\{e_i\}$ is a basis of V then setting

$$\bar{R}e_i = \bar{R}^j{}_i e_j$$

we have

$$\langle \bar{R}e_j, \varepsilon^i \rangle = R(\varepsilon^i, e_j) = R^i{}_j$$

and

$$\langle \bar{R}e_j, \varepsilon^i \rangle = \langle \bar{R}^k{}_j e_k, \varepsilon^i \rangle = \bar{R}^k{}_j \delta^i{}_k = \bar{R}^j{}_i.$$

Hence $R^k{}_i = \bar{R}^k{}_i$ and it follows that

$$\bar{R}e_i = R^j{}_i e_j.$$

On comparison with Eq. (3.6) it follows that the components of a mixed tensor R are the same as the matrix components of the associated operator on V.

Exercise: Show that a tensor R of type $(1, 1)$ defines a map $\tilde{R} : V^* \to V^*$.

Example 7.10 Define the map $\delta : V^* \times V \to \mathbb{R}$ by

$$\delta(\omega, v) = \omega(v) = v(\omega) = \langle \omega, v \rangle.$$

7.3 Basis representation of tensors

This map is clearly linear in both arguments and therefore constitutes a tensor of type $(1, 1)$. If $\{e_i\}$ is any basis of V with dual basis $\{\varepsilon^j\}$, then it is possible to set

$$\delta = e_i \otimes \varepsilon^i \equiv e_1 \otimes \varepsilon^1 + e_2 \otimes \varepsilon^2 + \cdots + e_n \otimes \varepsilon^n$$

since

$$e_i \otimes \varepsilon^i(\omega, v) = e_i(\omega) \varepsilon^i(v) = w_i v^i = \omega(v).$$

An alternative expression for δ is

$$\delta = e_i \otimes \varepsilon^i = \delta^i{}_j e_i \otimes \varepsilon^j$$

from which the components of the mixed tensor δ are precisely the Kronecker delta $\delta^i{}_j$. As no specific choice of basis has been made in this discussion the components of the tensor δ are 'invariant', in the sense that they do not change under basis transformations.

Exercise: Show that the map $\bar{\delta} : V \to V$ that corresponds to the tensor δ according to Example 7.9 is the identity map

$$\bar{\delta} = \mathrm{id}_V.$$

Problems

Problem 7.7 Let $\bar{T}$ be the linear map defined by a covariant tensor T of degree 2 as in Example 7.6. If $\{e_i\}$ is a basis of V and $\{\varepsilon^j\}$ the dual basis, define the matrix of components of $\bar{T}$ with respect to these bases as $[\bar{T}_{ji}]$ where

$$\bar{T}(e_i) = \bar{T}_{ji} \varepsilon^j.$$

Show that the components of the tensor T in this basis are identical with the components as a map, $T_{ij} = \bar{T}_{ij}$.

Similarly if S is a contravariant tensor of degree 2 and $\bar{S}$ the linear map defined in Example 7.7, show that the components $\bar{S}^{ij}$ are identical with the tensor components S^{ij}.

Problem 7.8 Show that every tensor R of type $(1, 1)$ defines a map $\tilde{R} : V^* \to V^*$ by

$$\langle \tilde{R}\omega, u \rangle = R(\omega, u)$$

and show that for a natural definition of components of this map, $\tilde{R}^k{}_i = R^k{}_i$.

Problem 7.9 Show that the definition of tensor product of two vectors $u \times v$ given in Eq. (7.11) agrees with that given in Section 7.1 after relating the two concepts of tensor by isomorphism.

7.3 Basis representation of tensors

We now construct a basis of $V^{(r,s)}$ from any given basis $\{e_i\}$ of V and its dual basis $\{\varepsilon^j\}$, and display tensors of type (r, s) with respect to this basis. While the expressions that arise often turn out to have a rather complicated appearance as multicomponented objects, this

Tensors

is simply a matter of becoming accustomed to the notation. It is still the represention of tensors most frequently used by physicists.

Tensor product

If T is a tensor of type (r, s) and S is a tensor of type (p, q) then define $T \otimes S$, called their **tensor product**, to be the tensor of type $(r + p, s + q)$ defined by

$$(T \otimes S)(\omega^1, \ldots, \omega^r, \rho^1, \ldots, \rho^p, u_1, \ldots, u_s, v_1, \ldots, v_q)$$
$$= T(\omega^1, \ldots, \omega^r, u_1, \ldots, u_s) S(\rho^1, \ldots, \rho^p, v_1, \ldots, v_q). \quad (7.15)$$

This product generalizes the definition of tensor products of vectors and covectors in the previous secion. It is readily shown to be associative

$$T \otimes (S \otimes R) = (T \otimes S) \otimes R,$$

so there is no ambiguity in writing expressions such as $T \otimes S \otimes R$.

If $e_1, \ldots, e_n$ is a basis for V and $\varepsilon^1, \ldots, \varepsilon^n$ the dual basis of V^* then the tensors

$$e_{i_1} \otimes \cdots \otimes e_{i_r} \otimes \varepsilon^{j_1} \otimes \cdots \otimes \varepsilon^{j_s} \quad (i_1, i_2, \ldots, j_1, \ldots, j_s = 1, 2, \ldots, n)$$

form a basis of $V^{(r,s)}$, since every tensor T of type (r, s) has a unique expansion

$$T = T^{i_1 \ldots i_r}_{j_1 \ldots j_s} e_{i_1} \otimes \cdots \otimes e_{i_r} \otimes \varepsilon^{j_1} \otimes \cdots \otimes \varepsilon^{j_s} \quad (7.16)$$

where

$$T^{i_1 \ldots i_r}_{j_1 \ldots j_s} = T(\varepsilon^{i_1}, \ldots, \varepsilon^{i_r}, e_{j_1}, \ldots, e_{j_s}) \quad (7.17)$$

are called the **components** of the tensor T with respect to the basis $\{e_i\}$ of V.

Exercise: Prove these statements in full detail. Despite the apparent complexity of indices the proof is essentially identical to that given for the case of $V^{(0,2)}$ in Theorem 7.1.

The components of a linear combination of two tensors of the same type are given by

$$(T + aS)^{ij\ldots}_{kl\ldots} = (T + aS)(\varepsilon^i, \varepsilon^j, \ldots, e_k, e_l, \ldots)$$
$$= T(\varepsilon^i, \varepsilon^j, \ldots, e_k, e_l, \ldots) + aS(\varepsilon^i, \varepsilon^j, \ldots, e_k, e_l, \ldots)$$
$$= T^{ij\ldots}_{kl\ldots} + aS^{ij\ldots}_{kl\ldots}.$$

The components of the tensor product of two tensors T and S are given by

$$(T \otimes S)^{ij\ldots pq\ldots}_{kl\ldots mn\ldots} = T^{ij\ldots}_{kl\ldots} S^{pq\ldots}_{mn\ldots}$$

The proof follows from Eq. (7.15) on setting $\omega^1 = \varepsilon^i, \omega^2 = \varepsilon^j, \ldots, \rho^1 = \varepsilon^k, \ldots, u_1 = e_p, u_2 = e_q$, etc.

Exercise: Show that in components, a multilinear map T has the expression

$$T(\omega, \rho, \ldots, u, v, \ldots) = T^{ij\ldots}_{kl\ldots} w_i r_j \ldots u^k v^l \ldots \quad (7.18)$$

where $\omega = w_i \varepsilon^i, \rho = r_j \varepsilon^j, u = u^k e_k$, etc.

7.3 Basis representation of tensors

Change of basis

Let $\{e_i\}$ and $\{e'_j\}$ be two bases of V related by

$$e_i = A'^j{}_i e'_j, \qquad e'_j = A'^k{}_j e_k \qquad (7.19)$$

where the matrices $[A'^i{}_j]$ and $[A''^i{}_j]$ are inverse to each other,

$$A'^k{}_j A'^j{}_i = A^k{}_j A''^j{}_i = \delta^k_i. \qquad (7.20)$$

As shown in Chapter 3 the dual basis transforms by Eq. (3.32),

$$\varepsilon'^j = A^j{}_k \varepsilon^k \qquad (7.21)$$

and under the transformation laws components of vectors $v = v^i e_i$ and covectors $\omega = w_j \varepsilon^j$ are

$$v'^j = A^j{}_i v^i, \qquad w'_i = A''^j{}_i w_j. \qquad (7.22)$$

The terminology 'contravariant' and 'covariant' transformation laws used in Chapter 3 is motivated by the fact that vectors and covectors are tensors of contravariant degree 1 and covariant degree 1 respectively.

If $T = T^{ij} e_i \otimes e_j$ is a tensor of type (2, 0) then

$$\begin{aligned} T &= T^{ij} e_i \otimes e_j \\ &= T^{ij} A^k{}_i e'_k \otimes A^l{}_j e'_l \\ &= T'^{kl} e'_k \otimes e'_l \end{aligned}$$

where

$$T'^{kl} = T^{ij} A^k{}_i A^l{}_j. \qquad (7.23)$$

Exercise: Alternatively, show this result from Eq. (7.21) and

$$T'^{kl} = T(\varepsilon'^k, \varepsilon'^l).$$

Similarly the components of a covariant tensor $T = T_{ij} \varepsilon^i \otimes \varepsilon^j$ of degree 2 transform as

$$T'_{kl} = T_{ij} A''^i{}_k A''^j{}_l. \qquad (7.24)$$

Exercise: Show (7.24) (i) by transformation of ε^i using Eq. (7.21), and (ii) from $T'_{ij} = T(e_i, e_j)$ using Eq. (7.19).

In the same way, the components of a mixed tensor $T = T^i{}_j e_i \otimes \varepsilon^j$ can be shown to have the transformation law

$$T'^k{}_l = T^i{}_j A^k{}_i A''^j{}_l. \qquad (7.25)$$

Exercise: Show Eq. (7.25).

Tensors

Before giving the transformation law of components of a general tensor it is useful to establish a convention known as the **kernel index notation**. In this notation we denote the indices on the transformed bases and dual bases by a primed index, $\{e'_{i'} \mid i' = 1, \ldots, n\}$ and $\{\varepsilon'^{j'} \mid j' = 1, \ldots, n\}$. The primes on the 'kernel' letters e and ε are essentially superfluous and little meaning is lost in dropping them, simply writing $e_{i'}$ and $\varepsilon^{j'}$ for the transformed bases. The convention may go even further and require that the primed indices range over an indexed set of natural numbers $i' = 1', \ldots, n'$. These practices may seem a little bizarre and possibly confusing. Accordingly, we will only follow a 'half-blown' kernel index notation, with the key requirement that primed indices be used on transformed quantities. The main advantage of the kernel index notation is that it makes the transformation laws of tensors easier to commit to memory.

Instead of Eq. (7.19) we now write the basis transformations as

$$e_i = A'^{i'}_i e'_{i'}, \qquad e'_{j'} = A'^j_{j'} e_j \qquad (7.26)$$

where the matrix array $\mathbf{A} = [A'^{j'}_i]$ is always written with the primed index in the superscript position, while its inverse $\mathbf{A}^{-1} = [A'^k_{j'}]$ has the primed index as a subscript. The relations (7.20) between these are now written

$$A'^k_{j'} A'^{j'}_i = \delta^k_i, \qquad A'^{j'}_k A'^k_{i'} = \delta^{j'}_{i'}, \qquad (7.27)$$

which take the place of (7.20).

The dual basis satisfies

$$\varepsilon'^{i'}(e'_{j'}) = \delta^{i'}_{j'}$$

and is related to the original basis by Eq. (7.21), which reads

$$\varepsilon^i = A'^i_{j'} \varepsilon'^{j'}, \qquad \varepsilon'^{i'} = A'^{i'}_j \varepsilon^j. \qquad (7.28)$$

The transformation laws of vectors and covectors (7.22) are replaced by

$$v'^{i'} = A'^{i'}_j v^j, \qquad w'_{j'} = A'^i_{j'} w_i. \qquad (7.29)$$

Exercise: If $e_1 = e'_1 + e'_2$, $e_2 = e'_2$ are a basis transformation on a two-dimensional vector space V, write out the matrices $[A'^{i'}_j]$ and $[A'^i_{j'}]$ and the transformation equation for the components of a contravariant vector v^i and a covariant vector w_j.

The tensor transformation laws (7.23), (7.24) and (7.25) can be replaced by

$$T'^{i'j'} = A'^{i'}_i A'^{j'}_j T^{ij},$$

$$T'_{i'j'} = A'^i_{i'} A'^j_{j'} T_{ij},$$

$$T'^{i'}_{k'} = A'^{i'}_i A'^k_{k'} T^i_k.$$

When transformation laws are displayed in this notation the placement of the indices immediately determines whether $A'^{i'}_i$ or $A'^i_{j'}$ is to be used, as only one of them will give rise to a formula obeying the conventions of summation convention and kernel index notation.

7.3 Basis representation of tensors

Exercise: Show that the components of the tensor δ are the same in all bases by

(a) showing $e_i \otimes \varepsilon^i = e'_{i'} \otimes \varepsilon'^{i'}$, and
(b) using the transformation law Eq. (7.25).

Now let T be a general tensor of type (r, s),

$$T = T^{i_1...i_r}{}_{j_1...j_s} e_{i_1} \otimes \cdots \otimes e_{i_r} \otimes \varepsilon^{j_1} \otimes \cdots \otimes \varepsilon^{j_s}$$

where

$$T^{i_1...i_r}{}_{j_1...j_s} = T(\varepsilon^{i_1}, \ldots, \varepsilon^{i_r}, e_{j_1}, \ldots e_{j_s}).$$

The separation in spacing between contravariant indices and covariant indices is not strictly necessary but has been done partly for visual display and also to anticipate a further operation called 'raising and lowering indices', which is available in inner product spaces. The transformation of the components of T is given by

$$T'^{i'_1,\ldots,i'_r}{}_{j'_1,\ldots,j'_s} = T(\varepsilon'^{i'_1}, \ldots, \varepsilon'^{i'_r}, e'_{j'_1}, \ldots, e'_{j'_s})$$

$$= T(A^{i'_1}_{i_1}\varepsilon^{i_1}, \ldots, A^{i'_r}_{i_r}\varepsilon^{i_r}, A'^{j_1}_{j'_1}e_{j_1}, \ldots, A'^{j_s}_{j'_s}e_{j_s})$$

$$= A^{i'_1}_{i_1} \ldots A^{i'_r}_{i_r} A'^{j_1}_{j'_1} \ldots A'^{j_s}_{j'_s} T^{i_1...i_r}{}_{j_1...j_s}. \tag{7.30}$$

The general tensor transformation law of components merely replicates the contravariant and covarient transformation law given in (7.29) for each contravariant and covariant index separately. The final formula (7.30) compactly expresses a multiple summation that can represent an enormous number of terms, even in quite simple cases. For example in four dimensions a tensor of type $(3, 2)$ has $4^{3+2} = 1024$ components. Its transformation law therefore consists of 1024 separate formulae, each of which has in it a sum of 1024 terms that themselves are products of six indexed entities. Including all indices and primes on indices, the total number of symbols used would be that occurring in about 20 typical books.

Problems

Problem 7.10 Let e_1, e_2 and e_3 be a basis of a vector space V and $e'_{i'}$ a second basis given by

$$e'_1 = e_1 - e_2,$$
$$e'_2 = e_3,$$
$$e'_3 = e_1 + e_2.$$

(a) Display the transformation matrix $A' = [A^{i'}_i]$.
(b) Express the original basis e_i in terms of the $e'_{i'}$ and write out the transformation matrix $A = [A^{j'}_j]$.
(c) Write the old dual basis ε^i in terms of the new dual basis $\varepsilon'^{i'}$ and conversely.
(d) What are the components of the tensors $T = e_1 \otimes e_2 + e_2 \otimes e_1 + e_3 \otimes e_3$ and $S = e_1 \otimes \varepsilon^1 + 3e_1 \otimes \varepsilon^3 - 2e_2 \otimes \varepsilon^3 - e_3 \otimes \varepsilon^1 + 4e_3 \otimes \varepsilon^2$ in terms of the basis e_i and its dual basis?
(e) What are the components of these tensors in terms of the basis $e'_{i'}$ and its dual basis?

Tensors

Problem 7.11 Let V be a vector space of dimension 3, with basis e_1, e_2, e_3. Let T be the contravariant tensor of rank 2 whose components in this basis are $T^{ij} = \delta^{ij}$, and let S be the covariant tensor of rank 2 whose components are given by $S_{ij} = \delta_{ij}$ in this basis. In a new basis e'_1, e'_2, e'_3 defined by

$$e'_1 = e_1 + e_3$$
$$e'_2 = 2e_1 + e_2$$
$$e'_3 = 3e_2 + e_3$$

calculate the components $T^{i'j'}$ and $S'_{i'j'}$.

Problem 7.12 Let $T : V \to V$ be a linear operator on a vector space V. Show that its components T^i_j given by Eq. (3.6) are those of the tensor $\hat{T}$ defined by

$$\hat{T}(\omega, v) = \langle \omega, Tv \rangle.$$

Prove that they are also the components with respect to the dual basis of a linear operator $T^* : V^* \to V^*$ defined by

$$\langle T^*\omega, v \rangle = \langle \omega, Tv \rangle.$$

Show that tensors of type (r, s) are in one-to-one correspondence with linear maps from $V^{(s,0)}$ to $V^{(r,0)}$, or equivalently from $V^{(0,r)}$ to $V^{(0,s)}$.

Problem 7.13 Let $T : V \to V$ be a linear operator on a vector space V. Show that its components T^i_j defined through Eq. (3.6) transform as those of a tensor of rank $(1,1)$ under an arbitrary basis transformation.

Problem 7.14 Show directly from Eq. (7.14) and the transformation law of components g_{ij}

$$g'_{j'k'} = g_{jk} A'^{j}_{j'} A'^{k}_{k'},$$

that the components of an inverse metric tensor g^{ij} transform as a contravariant tensor of degree 2,

$$g'^{i'k'} = A^{i'}_l g^{lk} A^{k'}_k.$$

7.4 Operations on tensors

Contraction

The process of tensor product (7.15) creates tensors of higher degree from those of lower degrees,

$$\otimes : V^{(r,s)} \times V^{(p,q)} \to V^{(r+p,s+q)}.$$

We now describe an operation that lowers the degree of tensor. Firstly, consider a mixed tensor $T = T^i_j e_i \otimes \varepsilon^j$ of type $(1, 1)$. Its **contraction** is defined to be a scalar denoted $C^1_1 T$, given by

$$C^1_1 T = T(\varepsilon^i, e_i) = T(\varepsilon^1, e_1) + \cdots + T(\varepsilon^n, e_n).$$

7.4 Operations on tensors

Although a basis of V and its dual basis have been used in this definition, it is independent of the choice of basis, for if $e'_{i'} = A'^i{}_{i'} e_i$ is any other basis then

$$\begin{aligned} T(\varepsilon'^{i'}, e'_{i'}) &= T(A^{i'}{}_i \varepsilon^i, A'^k{}_{i'} e_k) \\ &= A^{i'}{}_i A'^k{}_{i'} T(\varepsilon^i, e_k) \\ &= \delta^k_i T(\varepsilon^i, e_k) \quad \text{using Eq. (7.27)} \\ &= T(\varepsilon^i, e_i). \end{aligned}$$

In components, contraction is written

$$C^1_1 T = T^i{}_i = T^1_1 + T^2_2 + \cdots T^n_n.$$

This is a basis-independent expression since

$$T'^{i'}{}_{i'} = T^i{}_j A^{i'}{}_i A'^j{}_{i'} = T^i{}_j \delta^j_i = T^i{}_i.$$

Exercise: If $T = u \otimes \omega$, show that its contraction is $C^1_1 T = \omega(u)$.

More generally, for a tensor T of type (r, s) with both $r > 0$ and $s > 0$ one can define its (p, q)-**contraction** $(1 \leq p \leq r, 1 \leq q \leq s)$ to be the tensor $C^p_q T$ of type $(r-1, s-1)$ defined by

$$(C^p_q T)(\omega^1, \ldots, \omega^{r-1}, v_1, \ldots, v_{s-1})$$
$$= \sum_{k=1}^n T(\omega^1, \ldots, \omega^{p-1}, \varepsilon^k, \omega^{p+1}, \ldots, \omega^{r-1},$$
$$v_1, \ldots, v_{q-1}, e_k, v_{q+1}, \ldots, v_{s-1}). \tag{7.31}$$

Exercise: Show that the definition of $C^p_q T$ is independent of choice of basis. The proof is essentially identical to that for the case $r = s = 1$.

On substituting $\omega^1 = \varepsilon^{i_1}, \ldots, v_1 = e_{j_1}, \ldots$, etc., we arrive at an expression for the (p, q)-contraction in terms of components,

$$(C^p_q T)^{i_1 \ldots i_{r-1}}{}_{j_1 \ldots j_{s-1}} = T^{i_1 \ldots i_{p-1} k i_{p+1} \ldots i_{r-1}}{}_{j_1 \ldots j_{q-1} k j_{q+1} \ldots j_{s-1}}. \tag{7.32}$$

Example 7.11 Let T be a tensor of type $(2, 3)$ having components $T^{ij}{}_{klm}$. Set $A = C^1_2 T$, $B = C^2_3 T$ and $D = C^1_1 T$. In terms of the components of T,

$$A^j{}_{km} = T^{ij}{}_{kim},$$
$$B^i{}_{kl} = T^{ij}{}_{klj},$$
$$D^j{}_{lm} = T^{ij}{}_{ilm}.$$

Typical contraction properties of the special mixed tensor δ defined in Example 7.10 are illustrated in the following formulae:

$$\delta^i{}_j T^j{}_{kl} = T^i{}_{kl},$$
$$\delta^i{}_j S^{lm}{}_{ik} = S^{lm}{}_{jk},$$
$$\delta^i{}_i = \underbrace{1 + 1 + \cdots + 1}_n = n = \dim V.$$

Exercise: Write these equations in C^p_q form.

199

Tensors

Raising and lowering indices

Let V be a real inner product space with metric tensor $g = g_{ij}\varepsilon^i \otimes \varepsilon^j$ such that

$$u \cdot v = g_{ij}u^i v^j = C_1^1 C_2^2 g \otimes u \otimes v.$$

By Theorem 5.1 the components g_{ij} are a *non-singular* matrix, so that $\det[g_{ij}] \neq 0$. As shown in Example 7.8 there is a tensor g^{-1} whose components, written $g^{ij} = g^{ji}$, form the inverse matrix $\mathbf{G}^{-1}$. Given a vector $u = u^i e_i$, the components of the covector $C_2^1(g \otimes u)$ can be written

$$u_i = g_{ij}u^j,$$

a process that is called **lowering the index**. Conversely, given a covector $\omega = w_i \varepsilon^i$, the vector $C_1^2 g^{-1} \otimes \omega$ can be written in components

$$w^i = g^{ij}w_j,$$

and is called **raising the index**. Lowering and raising indices in succession, in either order, has no effect as

$$u^i = \delta^i{}_j u^j = g^{ik}g_{kj}u^j = g^{ik}u_k.$$

This is important, for without this property, the convention of retaining the same kernel letter u in a raising or lowering operation would be quite untenable.

Exercise: Show that lowering the index on a vector u is equivalent to applying the map $\bar{g}$ in Example 7.6 to u, while raising the index of a covector ω is equivalent to the map $\bar{g}^{-1}$ of Example 7.8.

The tensors g and g^{-1} can be used to raise and lower indices of tensors in general, for example

$$T_i{}^j = g_{ik}T^{kj} = g^{jk}T_{ik} = g_{ik}g^{jl}T^k{}_l, \text{ etc.}$$

It is strongly advised to space out the upper and lower indices of mixed tensors for this process, else it will not be clear which 'slot' an index should be raised or lowered into. For example

$$S_i{}^j{}_p{}^m = S^{kjq}{}_l g_{ik}g_{qp}g^{ml}.$$

If no metric tensor is specified there is no distinction in the relative ordering of covariant and contravariant indices and they can simply be placed one above the other or, as often done above, the contravariant indices may be placed first followed by the covariant indices. Given the capability to raise and lower indices, however, it is important to space all indices correctly. Indeed, by lowering all superscripts every tensor can be displayed in a purely covariant form. Alternatively, it can be displayed in a purely contravariant form by raising every subscript. However, unless the indices are correctly spaced we would not know where the different indices in either of these forms came from in the original 'unlowered' tensor.

Example 7.12 It is important to note that while $\delta^i{}_j$ are components of a mixed tensor the symbol δ_{ij} does *not* represent components of a tensor of covariant degree 2. We therefore try to avoid using this symbol in general tensor analysis. However, by Theorem 5.2, for a

7.4 Operations on tensors

Euclidean inner product space with positive definite metric tensor g it is always possible to find an orthonormal basis $\{e_1, e_2, \ldots, e_n\}$ such that

$$e_i \cdot e_j = g_{ij} = \delta_{ij}.$$

In this case a special restricted tensor theory called **cartesian tensors** is frequently employed in which only orthonormal bases are permitted and basis transformations are restricted to orthogonal transformations. In this theory δ_{ij} can be treated as a tensor. The inverse metric tensor then also has the same components $g^{ij} = \delta_{ij}$ and the lowered version of any component index is identical with its raised version,

$$T_{\ldots i \ldots} = g_{ij} T_{\ldots \ldots}^{\ j} = \delta_{ij} T_{\ldots \ldots}^{\ j} = T^{\ldots i \ldots}.$$

Thus every cartesian tensor may be written with all its indices in the lower position, $T_{ijk\ldots}$, since raising an index has no effect on the values of the components of the tensor.

In cartesian tensors it is common to adopt the summation convention for repeated indices even when they are both subscripts. For example in the standard vector theory of three-dimensional Euclidean space commonly used in mechanics and electromagnetism, one adopts conventions such as

$$\mathbf{a} \cdot \mathbf{b} = a_i b_i \equiv a_1 b_1 + a_2 b_2 + a_3 b_3,$$

and

$$\mathbf{a} \times \mathbf{b} = \epsilon_{ijk} a_j b_k \equiv \sum_{i=1}^{3} \sum_{j=1}^{3} \epsilon_{ijk} a_j b_k,$$

where the alternating symbol ϵ_{ijk} is defined by

$$\epsilon_{ijk} = \begin{cases} 1 & \text{if } ijk \text{ is an even permutation of } 123, \\ -1 & \text{if it is an odd permutation of } 123, \\ 0 & \text{if any pair of } ijk \text{ are equal}. \end{cases}$$

It will be shown in Chapter 8 that with respect to proper orthogonal transformations ϵ_{ijk} is a cartesian tensor of type $(0, 3)$.

Symmetries

A tensor S of type $(0, 2)$ is called **symmetric** if $S(u, v) = S(v, u)$ for all vectors u, v in V, while a tensor A of type $(0, 2)$ is called **antisymmetric** if $A(u, v) = -A(v, u)$.

Exercise: In terms of components show that S is a symmetric tensor iff $S_{ij} = S_{ji}$ and A is antisymmetric iff $A_{ij} = -A_{ji}$.

Any tensor T of type $(0, 2)$ can be decomposed into a symmetric and antisymmetric part, $T = \mathcal{S}(T) + \mathcal{A}(T)$, where

$$\mathcal{S}(T)(u, v) = \tfrac{1}{2}(T(u, v) + T(v, u)),$$
$$\mathcal{A}(T)(u, v) = \tfrac{1}{2}(T(u, v) - T(v, u)).$$

Tensors

It is immediate that these tensors are symmetric and antisymmetric respectively. Setting $u = e_i$, $v = e_j$, this decomposition becomes

$$T_{ij} = T(e_i, e_j) = T_{(ij)} + T_{[ij]},$$

where

$$T_{(ij)} = S(T)_{ij} = \tfrac{1}{2}(T_{ij} + T_{ji})$$

and

$$T_{[ij]} = A(T)_{ij} = \tfrac{1}{2}(T_{ij} - T_{ji}).$$

A similar discussion applies to tensors of type (2, 0), having components T^{ij}, but one cannot talk of symmetries of a mixed tensor.

Exercise: Show that $T^i_j = T^j_i$ is not a tensor equation, since it is not invariant under basis transformations.

If A is an antisymmetric tensor of type (0, 2) and S a symmetric tensor of type (2, 0) then their total contraction vanishes,

$$C^1_1 C^2_2 A \otimes S \equiv A_{ij} S^{ij} = 0 \tag{7.33}$$

since

$$A_{ij} S^{ij} = -A_{ji} S^{ji} = -A_{ij} S^{ij}.$$

Problems

Problem 7.15 Let g_{ij} be the components of an inner product with respect to a basis u_1, u_2, u_3

$$g_{ij} = u_i \cdot u_j = \begin{pmatrix} 1 & 0 & 1 \\ 0 & 1 & 0 \\ 1 & 0 & 0 \end{pmatrix}.$$

(a) Find an orthonormal basis of the form $e_1 = u_1$, $e_2 = u_2$, $e_3 = au_1 + bu_2 + cu_3$ such that $a > 0$, and find the index of this inner product.
(b) If $v = u_1 + \tfrac{1}{2} u_3$ find its lowered components v_i.
(c) Express v in terms of the orthonormal basis found above, and write out its lowered components with respect to that basis.

Problem 7.16 Let g be a metric tensor on a vector space V and define T to be the tensor

$$T = ag^{-1} \otimes g + \delta \otimes u \otimes \omega$$

where u is a non-zero vector of V and ω is a covector.

(a) Write out the components $T^{ij}{}_{kl}$ of the tensor T.
(b) Evaluate the components of the following four contractions:

$$A = C^1_1 T, \qquad B = C^1_2 T, \qquad C = C^2_1 T, \qquad D = C^2_2 T$$

and show that $B = C$.
(c) Show that $D = 0$ iff $\omega(u) = -a$. Hence show that if $T^{ij}_{kl} u^l u_j = 0$, then $D = 0$.
(d) Show that if $n = \dim V > 1$ then $T^{ij}_{kl} u^l u_j = 0$ if and only if $a = \omega(u) = 0$ or $u_i u^i = 0$.

202

Problem 7.17 On a vector space V of dimension n let T be a tensor of rank $(1, 1)$, S a symmetric tensor of rank $(0, 2)$ and δ the usual 'invariant tensor' of rank $(1, 1)$. Write out the components $R^{ij}{}_{klmr}$ of the tensor

$$R = T \otimes S \otimes \delta + S \otimes \delta \otimes T + \delta \otimes T \otimes S.$$

Perform the contraction of this tensor over i and k, using any available contraction properties of $\delta^i{}_j$. Perform a further contraction over the indices j and r.

Problem 7.18 Show that covariant symmetric tensors of rank 2, satisfying $T_{ij} = T_{ji}$, over a vector space V of dimension n form a vector space of dimension $n(n + 1)/2$.

(a) A tensor S of type $(0, r)$ is called *totally symmetric* if $S_{i_1 i_2 \ldots i_r}$ is left unaltered by any interchange of indices. What is the dimension of the vector space spanned by the totally symmetric tensors on V?
(b) Find the dimension of the vector space of covariant tensors of rank 3 having the cyclic symmetry

$$T(u, v, w) + T(v, w, u) + T(w, u, v) = 0.$$

References

[1] J. L. Synge and A. Schild. *Tensor Calculus*. Toronto, University of Toronto Press, 1959.
[2] R. Geroch. *Mathematical Physics*. Chicago, The University of Chicago Press, 1985.
[3] S. Hassani. *Foundations of Mathematical Physics*. Boston, Allyn and Bacon, 1991.
[4] E. Nelson. *Tensor Analysis*. Princeton, Princeton University Press, 1967.
[5] L. H. Loomis and S. Sternberg. *Advanced Calculus*. Reading, Mass., Addison-Wesley, 1968.
[6] S. Sternberg. *Lectures on Differential Geometry*. Englewood Cliffs, N.J., Prentice-Hall, 1964.

8 Exterior algebra

In Section 6.4 we gave an intuitive introduction to the concept of Grassmann algebra $\Lambda(V)$ as an associative algebra of dimension 2^n constructed from a vector space V of dimension n. Certain difficulties, particularly those relating to the definition of exterior product, were cleared up by the more formal approach to the subject in Section 7.1. In this chapter we propose a definition of Grassmann algebra entirely of tensors [1–5]. This presentation has a more 'concrete' constructive character, and to distinguish it from the previous treatments we will use the term *exterior algebra* over V to describe it from here on.

8.1 r-Vectors and r-forms

A tensor of type $(r, 0)$ is said to be **antisymmetric** if, as a multilinear function, it changes sign whenever any pair of its arguments are interchanged,

$$A(\alpha^1, \ldots, \alpha^i, \ldots, \alpha^j, \ldots, \alpha^r) = -A(\alpha^1, \ldots, \alpha^j, \ldots, \alpha^i, \ldots, \alpha^r). \quad (8.1)$$

Equivalently, if π is any permutation of $1, \ldots, r$ then

$$A(\alpha^{\pi(1)}, \alpha^{\pi(2)}, \ldots, \alpha^{\pi(r)}) = (-1)^\pi A(\alpha^1, \alpha^2, \ldots, \alpha^r).$$

To express these conditions in components, let $\{e_i\}$ be any basis of V and $\{\varepsilon^j\}$ its dual basis. Setting $\alpha^1 = \varepsilon^{i_1}$, $\alpha^2 = \varepsilon^{i_2}, \ldots$ in (8.1), a tensor A is antisymmetric if it changes sign whenever any pair of component indices is interchanged,

$$A^{i_1 \ldots j \ldots k \ldots i_r} = -A^{i_1 \ldots k \ldots j \ldots i_r}.$$

For any permutation π of $1, \ldots, r$ we have

$$A^{i_{\pi(1)} \ldots i_{\pi(r)}} = (-1)^\pi A^{i_1 \ldots i_r}.$$

Antisymmetric tensors of type $(r, 0)$ are also called r-**vectors**, forming a vector space denoted $\Lambda^r(V)$. Ordinary vectors of V are 1-vectors and scalars will be called 0-vectors.

A similar treatment applies to antisymmetric tensors of type $(0, r)$, called r-**forms**. These are usually denoted by Greek letters α, β, etc., and satisfy

$$\alpha(v_{\pi(1)}, \ldots, v_{\pi(r)}) = (-1)^\pi \alpha(v_1, \ldots, v_r),$$

8.1 r-Vectors and r-forms

or in terms of components

$$\alpha_{i_{\pi(1)}\ldots i_{\pi(r)}} = (-1)^\pi \alpha_{i_1\ldots i_r}.$$

Linear functionals, or covectors, are called 1-forms and scalars are 0-forms. The vector space of r-forms is denoted $\Lambda^{*r}(V) \equiv \Lambda^r(V^*)$.

As shown in Eq. (7.33), the total contraction of a 2-form α and a symmetric tensor S of type (2, 0) vanishes,

$$\alpha_{ij} S^{ij} = 0.$$

The same holds true of more general contractions such as that between an r-form α and a tensor S of type $(s, 0)$ that is symmetric in any pair of indices; for example, if $S^{ikl} = S^{lki}$ then

$$\alpha_{ijkl} S^{ikl} = 0.$$

The antisymmetrization operator $\mathcal{A}$

Let T be any totally contravariant tensor of degree r; that is, of type $(r, 0)$. Its **antisymmetric part** is defined to be the tensor $\mathcal{A}T$ given by

$$\mathcal{A}T(\omega^1, \omega^2, \ldots, \omega^r) = \frac{1}{r!} \sum_\sigma (-1)^\sigma T(\omega^{\sigma(1)}, \omega^{\sigma(2)}, \ldots, \omega^{\sigma(r)}), \quad (8.2)$$

where the summation on the right-hand side runs through all permutations σ of $1, 2, \ldots, r$. If π is any permutation of $1, 2, \ldots, r$ then

$$\mathcal{A}T(\alpha^{\pi(1)}, \alpha^{\pi(2)}, \ldots, \alpha^{\pi(r)}) = \frac{1}{r!} \sum_\sigma (-1)^\sigma T(\alpha^{\pi\sigma(1)}, \alpha^{\pi\sigma(2)}, \ldots, \alpha^{\pi\sigma(r)})$$

$$= \frac{1}{r!} \sum_{\sigma'} (-1)^\pi (-1)^{\sigma'} T(\alpha^{\sigma'(1)}, \alpha^{\sigma'(2)}, \ldots, \alpha^{\sigma'(r)})$$

$$= (-1)^\pi \mathcal{A}T(\alpha^1, \alpha^2, \ldots, \alpha^r),$$

since $\sigma' = \pi\sigma$ runs through all permutations of $1, 2, \ldots, r$ and $(-1)^{\sigma'} = (-1)^\pi (-1)^\sigma$. Hence $\mathcal{A}T$ is an antisymmetric tensor.

The **antisymmetrization operator** $\mathcal{A}: V^{(r,0)} \to \Lambda^r(V) \subseteq V^{(r,0)}$ is clearly a linear operator on $V^{(r,0)}$,

$$\mathcal{A}(aT + bS) = a\mathcal{A}(T) + b\mathcal{A}(S),$$

and since the antisymmetric part of an r-vector A is always A itself, it is **idempotent**

$$\mathcal{A}^2 = \mathcal{A}.$$

Thus $\mathcal{A}$ is a *projection operator* (see Problem 3.6). This property generalizes to the following useful theorem:

Theorem 8.1 *If T is a tensor of type $(r, 0)$ and S a tensor of type $(s, 0)$, then*

$$\mathcal{A}(\mathcal{A}T \otimes S) = \mathcal{A}(T \otimes S), \quad \mathcal{A}(T \otimes \mathcal{A}S) = \mathcal{A}(T \otimes S).$$

Exterior algebra

Proof: We will prove the first equation, the second being essentially identical. Let $\omega^1, \omega^2, \ldots, \omega^{r+s}$ be any $r+s$ covectors. Then

$$AT \otimes S(\omega^1, \omega^2, \ldots, \omega^{r+s}) = \frac{1}{r!} \sum_\sigma (-1)^\sigma T(\omega^{\sigma(1)}, \ldots, \omega^{\sigma(r)}) S(\omega^{r+1}, \ldots, \omega^{r+s}).$$

Treating each permutation σ in this sum as a permutation σ' of $1, 2, \ldots, r+s$ that leaves the last s numbers unchanged, this equation can be written

$$AT \otimes S(\omega^1, \omega^2, \ldots, \omega^{r+s}) = \frac{1}{r!} \sum_{\sigma'} (-1)^{\sigma'} T(\omega^{\sigma'(1)}, \ldots, \omega^{\sigma'(r)}) S(\omega^{\sigma'(r+1)}, \ldots, \omega^{\sigma'(r+s)}).$$

Now for each permutation σ', as ρ ranges over all permutations of $1, 2, \ldots, r+s$, the product $\pi = \rho\sigma'$ also ranges over all such permutations, and $(-1)^\pi = (-1)^\rho (-1)^{\sigma'}$. Hence

$$\mathcal{A}(AT \otimes S)(\omega^1, \omega^2, \ldots, \omega^{r+s})$$
$$= \frac{1}{(r+s)!} \sum_\rho (-1)^\rho \frac{1}{r!} \sum_{\sigma'} (-1)^{\sigma'} T(\omega^{\rho\sigma'(1)}, \ldots, \omega^{\rho\sigma'(r)}) S(\omega^{\rho\sigma'(r+1)}, \ldots, \omega^{\rho\sigma'(r+s)})$$
$$= \frac{1}{r!} \sum_{\sigma'} \frac{1}{(r+s)!} \sum_\pi (-1)^\pi T(\omega^{\pi(1)}, \ldots, \omega^{\pi(r)}) S(\omega^{\pi(r+1)}, \ldots, \omega^{\pi(r+s)}),$$

since there are $r!$ permutations of type σ', each making an identical contribution. Hence

$$\mathcal{A}(AT \otimes S)(\omega^1, \omega^2, \ldots, \omega^{r+s})$$
$$= \frac{1}{(r+s)!} \sum_\pi (-1)^\pi T(\omega^{\pi(1)}, \ldots, \omega^{\pi(r)}) S(\omega^{\pi(r+1)}, \ldots, \omega^{\pi(r+s)})$$
$$= \mathcal{A}(T \otimes S)(\omega^1, \omega^2, \ldots, \omega^{r+s}),$$

as required. ∎

The same symbol $\mathcal{A}$ can also be used to represent the projection operator $\mathcal{A}: V^{(0,r)} \to \Lambda^{*r}$ defined by

$$\mathcal{A}T(u_1, u_2, \ldots, u_r) = \frac{1}{r!} \sum_\sigma (-1)^\sigma T(u_{\sigma(1)}, u_{\sigma(2)}, \ldots, u_{\sigma(r)}).$$

Theorem 8.1 has a natural counterpart for tensors T of type $(0, r)$ and S of type $(0, s)$.

8.2 Basis representation of r-vectors

Let $\{e_i\}$ be any basis of V with dual basis $\{\varepsilon^j\}$, then setting $\omega^1 = \varepsilon^{i_1}, \omega^1 = \varepsilon^{i_1}, \ldots, \omega^r = \varepsilon^{i_r}$ in Eq. (8.2) results in an equation for the components of any tensor T of type $(r, 0)$

$$(\mathcal{A}T)^{i_1 i_2 \ldots i_r} = T^{[i_1 i_2 \ldots i_r]} \equiv \frac{1}{r!} \sum_\sigma (-1)^\sigma T^{i_{\sigma(1)} i_{\sigma(2)} \ldots i_{\sigma(r)}}.$$

8.2 Basis representation of r-vectors

From the properties of the antisymmetrization operator, the square bracketing of any set of indices satisfies

$$T^{[i_{\pi(1)}i_{\pi(2)}\ldots i_r]} = (-1)^\pi T^{[i_{\pi(1)}i_{\pi(2)}\ldots i_{\pi(r)}]}, \quad \text{for any permutation } \pi$$
$$T^{[[i_1 i_2 \ldots i_r]]} = T^{[i_1 i_2 \ldots i_r]}.$$

If A is an r-vector then $\mathcal{A}A = A$, or in components,

$$A^{i_1 i_2 \ldots i_r} = A^{[i_1 i_2 \ldots i_r]}.$$

Similar statements apply to tensors of covariant type, for example

$$T_{[ij]} = \tfrac{1}{2}(T_{ij} - T_{ji}),$$
$$T_{[ijk]} = \tfrac{1}{6}(T_{ijk} + T_{jki} + T_{kij} - T_{ikj} - T_{jik} - T_{kji}).$$

Theorem 8.1 can be expressed in components as

$$T^{[[i_1 i_2 \ldots i_r]} S^{j_{r+1} j_{r+2} \ldots j_{r+s}]} = T^{[i_1 i_2 \ldots i_r} S^{j_{r+1} j_{r+2} \ldots j_{r+s}]},$$

or, with a slight generalization, square brackets occurring anywhere within square brackets may always be eliminated,

$$T^{[i_1 \ldots [i_k \ldots i_{k+l}] \ldots i_r]} = T^{[i_1 \ldots i_k \ldots i_{k+l} \ldots i_r]}.$$

By the antisymmetry of its components every r-vector A can be written

$$\begin{aligned} A &= A^{i_1 i_2 \ldots i_r} e_{i_1} \otimes e_{i_2} \otimes \cdots \otimes e_{i_r} \\ &= A^{i_1 i_2 \ldots i_r} e_{i_1 i_2 \ldots i_r} \end{aligned} \tag{8.3}$$

where

$$e_{i_1 i_2 \ldots i_r} = \frac{1}{r!} \sum_\sigma (-1)^\sigma e_{i_{\sigma(1)}} \otimes e_{i_{\sigma(2)}} \otimes \cdots \otimes e_{i_{\sigma(r)}}. \tag{8.4}$$

For example

$$e_{12} = \tfrac{1}{2}(e_1 \otimes e_2 - e_2 \otimes e_1),$$
$$\begin{aligned} e_{123} = \tfrac{1}{6}\big(&e_1 \otimes e_2 \otimes e_3 - e_1 \otimes e_3 \otimes e_2 + e_2 \otimes e_3 \otimes e_1 \\ &- e_2 \otimes e_1 \otimes e_3 + e_3 \otimes e_1 \otimes e_2 - e_3 \otimes e_2 \otimes e_1\big), \text{ etc.} \end{aligned}$$

The r-vectors $e_{i_1 \ldots i_r}$ have the property

$$e_{i_1 \ldots i_r} = \begin{cases} 0 & \text{if any pair of indices are equal,} \\ (-1)^\pi e_{i_{\pi(1)} \ldots i_{\pi(r)}} & \text{for any permutation } \pi \text{ of } 1, 2, \ldots, r. \end{cases} \tag{8.5}$$

Hence the expansion (8.3) can be reduced to one in which every term has $i_1 < i_2 < \cdots < i_r$,

$$A = r! \sum \cdots \sum_{i_1 < i_2 < \cdots < i_r} A^{i_1 i_2 \ldots i_r} e_{i_1 i_2 \ldots i_r}. \tag{8.6}$$

Hence $\Lambda^r(V)$ is spanned by the set

$$E_r = \{e_{i_1 i_2 \ldots i_r} \mid i_1 < i_2 < \cdots < i_r\}.$$

207

Exterior algebra

Furthermore this set is linearly independent, for if there were a linear relation

$$0 = \underbrace{\sum \cdots \sum}_{i_1 < i_2 < \cdots < i_r} B^{i_1 i_2 \ldots i_r} e_{i_1 i_2 \ldots i_r},$$

application of this multilinear function to arguments $\varepsilon^{j_1}, \varepsilon^{j_2}, \ldots, \varepsilon^{j_r}$ with $j_1 < j_2 \cdots < j_r$ gives

$$0 = \sum \cdots \sum_{i_1 < i_2 < \cdots < i_r} B^{i_1 i_2 \ldots i_r} \delta_{i_1}^{j_1} \delta_{i_2}^{j_2} \ldots \delta_{i_r}^{j_r} = B^{j_1 j_2 \ldots j_r}.$$

Hence E_r forms a basis of $\Lambda^r(V)$.

The dimension of the vector space $\Lambda^r(V)$ is the number of subsets $\{i_1 < i_2 < \cdots < i_r\}$ occurring in the first n integers $\{1, 2, \ldots, n\}$,

$$\dim \Lambda^r(V) = \binom{n}{r} = \frac{n!}{r!(n-r)!}.$$

In particular $\dim \Lambda^n(V) = 1$, while $\dim \Lambda^{(n+k)}(V) = 0$ for all $k > 0$. The latter follows from the fact that if $r > n$ then all r-vectors $e_{i_1 i_2 \ldots i_r}$ vanish, since some pair of indices must be equal.

An analogous argument shows that the set of r-forms

$$\mathcal{E}_r = \{\varepsilon^{i_1 \ldots i_r} \mid i_i < \cdots < i_r\}, \tag{8.7}$$

where

$$\varepsilon^{i_1 \ldots i_r} = \frac{1}{r!} \sum_\pi (-1)^\pi \varepsilon^{i_{\pi(1)}} \otimes \varepsilon^{i_{\pi(2)}} \otimes \cdots \otimes \varepsilon^{i_{\pi(r)}}, \tag{8.8}$$

is a basis of $\Lambda^{*r}(V)$ and the dimension of the space of r-forms is also $\binom{n}{r}$.

8.3 Exterior product

The vector space $\Lambda(V)$ is defined to be the direct sum

$$\Lambda(V) = \Lambda^0(V) \oplus \Lambda^1(V) \oplus \Lambda^2(V) \oplus \cdots \oplus \Lambda^n(V).$$

Elements of $\Lambda(V)$ are called **multivectors**, written

$$A = A_0 + A_1 + \cdots + A_n \quad \text{where} \quad A_r \in \Lambda^r(V).$$

As shown in Section 6.4,

$$\dim(\Lambda(V)) = \sum_{r=0}^n \binom{n}{r} = 2^n.$$

For any r-vector A and s-vector B we define their **exterior product** or **wedge product** $A \wedge B$ to be the $(r+s)$-vector

$$A \wedge B = \mathcal{A}(A \otimes B), \tag{8.9}$$

8.3 Exterior product

and extend to all of $\Lambda(V)$ by linearity,

$$(aA + bB) \wedge C = aA \wedge C + bB \wedge C, \qquad A \wedge (aB + bC) = aA \wedge B + bA \wedge C.$$

The wedge product of a 0-vector, or scalar, a with an r-vector A is simply scalar multiplication, since

$$a \wedge A = \mathcal{A}(a \otimes A) = \mathcal{A}(aA) = a\mathcal{A}A = aA.$$

For general multivectors $(\sum_a A_a)$ and $(\sum_b B_b)$ we have

$$\left(\sum_a A_a\right) \wedge \left(\sum_b B_b\right) = \sum_a \sum_b A_a \wedge B_b.$$

The associative law holds by Theorem 8.1 and the associative law for tensor products,

$$\begin{aligned} A \wedge (B \wedge C) &= \mathcal{A}(A \otimes \mathcal{A}(B \otimes C)) \\ &= \mathcal{A}(A \otimes (B \otimes C)) \\ &= \mathcal{A}((A \otimes B) \otimes C) \\ &= \mathcal{A}(\mathcal{A}(A \otimes B) \otimes C) \\ &= (A \wedge B) \wedge C. \end{aligned}$$

The space $\Lambda(V)$ with wedge product $\wedge$ is therefore an associative algebra, called the **exterior algebra** over V. There is no ambiguity in writing expressions such as $A \wedge B \wedge C$. Since the exterior product has the property

$$\wedge : \Lambda^r(V) \times \Lambda^s(V) \to \Lambda^{r+s}(V),$$

it is called a *graded product* and the exterior algebra $\Lambda(V)$ is called a *graded algebra*.

Example 8.1 If u and v are vectors then their exterior product has the property

$$\begin{aligned} (u \wedge v)(\omega, \rho) &= \mathcal{A}(u \otimes v)(\omega, \rho) \\ &= \mathcal{A}(u \otimes v)(\omega, \rho) \\ &= \tfrac{1}{2}(u(\omega)v(\rho) - u(\rho)v(\omega)), \end{aligned}$$

whence

$$u \wedge v = \tfrac{1}{2}(u \otimes v - v \otimes u) = -v \wedge u. \tag{8.10}$$

Obviously $u \wedge u = 0$. Setting $\omega = \varepsilon^i$ and $\rho = \varepsilon^j$ in the derivation of (8.10) gives

$$(u \wedge v)^{ij} = \tfrac{1}{2}(u^i v^j - u^j v^i).$$

In many textbooks exterior product $A \wedge B$ is defined as $\dfrac{(r+s)!}{r!s!}\mathcal{A}(A \otimes B)$, in which case the factor $\tfrac{1}{2}$ does not appear in these formulae.

The anticommutation property (8.10) is easily generalized to show that for any permutation π of $1, 2, \ldots, r$,

$$u_{\pi(1)} \wedge u_{\pi(2)} \wedge \cdots \wedge u_{\pi(r)} = (-1)^\pi u_1 \wedge u_2 \wedge \cdots \wedge u_r. \tag{8.11}$$

Exterior algebra

The basis r-vectors $e_{i_1\ldots i_r}$ defined in Eq. (8.4) can clearly be written

$$e_{i_1 i_2\ldots i_r} = e_{i_1} \wedge e_{i_2} \wedge \cdots \wedge e_{i_r}, \tag{8.12}$$

and the permutation property (8.5) is equivalent to Eq. (8.11).

For any pair of basis elements $e_{i_1\ldots i_r} \in \Lambda^r(V)$ and $e_{j_1\ldots j_s} \in \Lambda^s(V)$ it follows immediately from Eq. (8.12) that

$$e_{i_1\ldots i_r} \wedge e_{j_1\ldots j_s} = e_{i_1\ldots i_r j_1\ldots j_s}. \tag{8.13}$$

These expressions permit us to give a unique expression for the exterior products of arbitrary multivectors, for if A is an r-vector and B an s-vector,

$$A = r! \underbrace{\sum \cdots \sum}_{i_1 < i_2 < \cdots < i_r} A^{i_1\ldots i_r} e_{i_1\ldots i_r}, \qquad B = s! \underbrace{\sum \cdots \sum}_{j_1 < j_2 < \cdots < j_s} B^{j_1\ldots j_s} e_{j_1\ldots j_s},$$

then

$$A \wedge B = r! s! \underbrace{\sum \cdots \sum}_{i_1 < \cdots < i_r} \underbrace{\sum \cdots \sum}_{j_1 < \cdots < j_s} A^{i_1\ldots i_r} B^{j_1\ldots j_s} e_{i_1\ldots i_r j_1\ldots j_s}. \tag{8.14}$$

Alternatively, the formula for wedge product can be written

$$\begin{aligned} A \wedge B &= A^{i_1 i_2\ldots i_r} e_{i_1 i_2\ldots i_r} \wedge B^{j_1 j_2\ldots j_s} e_{j_1 j_2\ldots j_s} \\ &= A^{i_1\ldots i_r} B^{j_1\ldots j_s} e_{i_1\ldots i_r j_1\ldots j_s} \\ &= A^{[i_1\ldots i_r} B^{j_1\ldots j_s]} e_{i_1} \otimes \cdots \otimes e_{i_r} \otimes e_{j_1} \otimes \cdots \otimes e_{j_s} \end{aligned}$$

on using Eq. (8.4). The tensor components of $A \wedge B$ are thus

$$(A \wedge B)^{i_1\ldots i_r j_1\ldots j_s} = A^{[i_1\ldots i_r} B^{j_1\ldots j_s]}. \tag{8.15}$$

Example 8.2 If u and v are 1-vectors, then

$$(u \wedge v)^{ij} = u^{[i} v^{j]} = \tfrac{1}{2}(u^i v^j - u^j v^i)$$

as in Example 8.1. For exterior product of a 1-vector u and a 2-vector A we find, using the skew symmetry $A^{jk} = -A^{kj}$,

$$\begin{aligned} (u \wedge A)^{ijk} &= u^{[i} A^{jk]} \\ &= \tfrac{1}{6}\left(u^i A^{jk} - u^i A^{kj} + u^j A^{ki} - u^j A^{ik} + u^k A^{ij} - u^k A^{ji}\right) \\ &= \tfrac{1}{3}\left(u^i A^{jk} + u^j A^{ki} + u^k A^{ij}\right). \end{aligned}$$

The wedge product of three vectors is

$$\begin{aligned} u \wedge v \wedge w = \tfrac{1}{6}(& u \otimes v \otimes w + v \otimes w \otimes u + w \otimes u \otimes v \\ & - u \otimes w \otimes v - w \otimes v \otimes u - v \otimes u \otimes w). \end{aligned}$$

In components,

$$\begin{aligned} (u \wedge v \wedge w)^{ijk} &= u^{[i} v^j w^{k]} \\ &= \tfrac{1}{6}\left(u^i v^j w^k - u^i v^k w^j + u^j v^k w^i - u^j v^i w^k + u^k v^i w^j - u^k v^j w^i\right). \end{aligned}$$

8.3 Exterior product

Continuing in this way, the wedge product of any r vectors $u_1, u_2, \ldots, u_r$ results in the r-vector

$$u_1 \wedge u_2 \wedge \cdots \wedge u_r = \frac{1}{r!} \sum_\pi (-1)^\pi u_{\pi(1)} \otimes u_{\pi(2)} \otimes \cdots \otimes u_{\pi(r)}$$

which has components

$$(u_1 \wedge u_2 \wedge \cdots \wedge u_r)^{i_1 i_2 \ldots i_r} = u_1^{[i_1} u_2^{i_2} \ldots u_r^{i_r]}.$$

The anticommutation rule for vectors, $u \wedge v = -v \wedge u$, generalizes for an r-vector A and s-vector B to

$$A \wedge B = (-1)^{rs} B \wedge A. \tag{8.16}$$

This result has been proved in Section 6.4, Eq. (6.20). It follows from

$$A \wedge B = A^{i_1 \ldots i_r} B^{j_1 \ldots j_s} e_{i_1 \ldots i_r j_1 \ldots j_s}$$
$$= (-1)^{rs} B^{j_1 \ldots j_s} A^{i_1 \ldots i_r} e_{j_1 \ldots j_s i_1 \ldots i_r}$$

since rs interchanges are needed to bring the indices $j_1, \ldots, j_r$ in front of the indices $i_1, \ldots, i_s$.

If r is even then A commutes with all multivectors, while a pair of odd degree multivectors always anticommute. For example if u is a 1-vector and A a 2-vector, then $u \wedge A = A \wedge u$, since

$$(u \wedge A)^{ijk} = u^{[i} A^{jk]} = A^{[jk} u^{i]} = A^{[ij} u^{k]} = (A \wedge u)^{ijk}.$$

The space of *multiforms* is defined in a totally analogous manner,

$$\Lambda^*(V) = \Lambda(V^*) = \Lambda_0^*(V) \oplus \Lambda_1^*(V) \oplus \Lambda_2^*(V) \oplus \cdots \oplus \Lambda_n^*(V),$$

with an exterior product

$$\alpha \wedge \beta = \mathcal{A}(\alpha \otimes \beta) \tag{8.17}$$

having identical properties to the wedge product on multivectors,

$$\alpha \wedge (\beta \wedge \gamma) = (\alpha \wedge \beta) \wedge \gamma \quad \text{and} \quad \alpha \wedge \beta = (-1)^{rs} \beta \wedge \alpha.$$

The basis r-forms defined in Eq. (8.7) can be written as

$$\varepsilon^{i_1 \ldots i_r} = \varepsilon^{i_1} \wedge \cdots \wedge \varepsilon^{i_r},$$

and the component expression for exterior product of a pair of forms is

$$(\alpha \wedge \beta)_{i_1 \ldots i_r j_1 \ldots j_s} = \alpha_{[i_1 \ldots i_r} \beta_{j_1 \ldots j_s]}.$$

Simple p-vectors and subspaces

A **simple** p-vector is one that can be written as a wedge product of 1-vectors,

$$A = v_1 \wedge v_2 \wedge \cdots \wedge v_p, \quad v_i \in \Lambda^1(V) = V.$$

Exterior algebra

Similarly a **simple** p-form α is one that is decomposable into a wedge product of 1-forms,
$$\alpha = \alpha^1 \wedge \alpha^2 \wedge \cdots \wedge \alpha^p, \quad \alpha^i \in \Lambda^{*1}(V) = V^*.$$

Let W be a p-dimensional subspace of V. For any basis $e_1, e_2, \ldots, e_p$ of W, define the p-vector $E_W = e_1 \wedge e_2 \wedge \cdots \wedge e_p$. If $e'_1, e'_2, \ldots, e'_p$ is a second basis then for some coefficients $B^i_{i'}$

$$e'_{i'} = \sum_{i=1}^{p} B^i_{i'} e_i,$$

and the p-vector corresponding to this basis is

$$\begin{aligned}
E'_W &= e'_1 \wedge \cdots \wedge e'_p \\
&= \sum_{i_1} \cdots \sum_{i_p} B^{i_1}_1 B^{i_2}_2 \ldots B^{i_p}_p e_{i_1} \wedge e_{i_2} \wedge e_{i_p} \\
&= \sum_{\pi} (-1)^{\pi} B^{i_\pi(1)}_1 B^{i_\pi(2)}_2 \ldots B^{i_\pi(p)}_p e_1 \wedge e_2 \wedge \cdots \wedge e_p \\
&= \det[B^i_{j'}] E_W.
\end{aligned}$$

Hence the subspace W corresponds uniquely, up to a multiplying factor, to a simple p-vector E_W.

Theorem 8.2 *A vector u belongs to W if and only if $u \wedge E_W = 0$.*

Proof: This statement is an immediate corollary of Theorem 6.2. ∎

Problems

Problem 8.1 Express components of the exterior product of two 2-vectors $A = A^{ij} e_{ij}$ and $B = B^{kl} e_{kl}$ as a sum of six terms,

$$(A \wedge B)^{ijkl} = \tfrac{1}{6}(A^{ij} B^{kl} + A^{ik} B^{lj} + \ldots).$$

How many terms would be needed for a product of a 2-vector and a 4-vector? Show that in general the components of the exterior product of an r-vector and an s-vector can be expressed as a sum of $\dfrac{(r+s)!}{r! s!}$ terms.

Problem 8.2 Let V be a four-dimensional vector space with basis $\{e_1, e_2, e_3, e_4\}$, and A a 2-vector on V.

(a) Show that a vector u satisfies the equation
$$A \wedge u = 0$$
if and only if there exists a vector v such that
$$A = u \wedge v.$$
[*Hint*: Pick a basis such that $e_1 = u$.]

(b) If
$$A = e_2 \wedge e_1 + ae_1 \wedge e_3 + e_2 \wedge e_3 + ce_1 \wedge e_4 + be_2 \wedge e_4$$

write out explicitly the equations $A \wedge u = 0$ where $u = u^1 e_1 + u^2 e_2 + u^3 e_3 + u^4 e_4$ and show that they have a solution if and only if $c = ab$. In this case find two vectors u and v such that $A = u \wedge v$.

(c) In general show that the 4-vector $A \wedge A = 8\alpha e_1 \wedge e_2 \wedge e_3 \wedge e_4$ where

$$\alpha = A^{12} A^{34} + A^{23} A^{14} + A^{31} A^{24},$$

and

$$\det[A^{ij}] = \alpha^2.$$

(d) Show that A is the wedge product of two vectors $A = u \wedge v$ if and only if $A \wedge A = 0$.

Problem 8.3 Prove **Cartan's lemma**, that if $u_1, u_2, \ldots, u_p$ are linearly independent vectors and $v_1, \ldots, v_p$ are vectors such that

$$u_1 \wedge v_1 + u_2 \wedge v_2 + \cdots + u_p \wedge v_p = 0,$$

then there exists a symmetric set of coefficients $A_{ij} = A_{ji}$ such that

$$v_i = \sum_{j=1}^{p} A_{ij} u_j.$$

[*Hint*: Extend the u_i to a basis for the whole vector space V.]

Problem 8.4 If V is an n-dimensional vector space and A a 2-vector, show that there exists a basis $e_1, e_2, \ldots, e_n$ of V such that

$$A = e_1 \wedge e_2 + e_3 \wedge e_4 + \ldots e_{2r-1} \wedge e_{2r},$$

for some number $2r$, called the *rank* of A.

(a) Show that the rank only depends on the 2-vector A, not on the choice of basis, by showing that $A^r \ne 0$ and $A^{r+1} = 0$ where

$$A^p = \underbrace{A \wedge A \wedge \cdots \wedge A}_{p}.$$

(b) If $f_1, f_2, \ldots, f_n$ is any basis of V and $A = A^{ij} f_i \otimes f_j$ where $A^{ij} = -A^{ji}$, show that the rank of the matrix of components $\mathsf{A} = [A^{ij}]$ coincides with the rank as defined above.

Problem 8.5 Let V be an n-dimensional space and A an arbitrary $(n-1)$-vector.
(a) Show that the subspace V_A of vectors u such that $u \wedge A = 0$ has dimension $n-1$.
(b) Show that every $(n-1)$-vector A is decomposable, $A = v_1 \wedge v_2 \wedge \cdots \wedge v_{n-1}$ for some vectors $v_1, \ldots, v_{n-1} \in V$. [*Hint*: Take a basis for $e_1, \ldots, e_n$ of V such that the first $n-1$ vectors span the subspace V_A, which is always possible by Theorem 3.7, and expand A in terms of this basis.]

8.4 Interior product

Let u be a vector in V and α an r-form. We define the **interior product** $i_u \alpha$ to be an $(r-1)$-form defined by

$$(i_u \alpha)(u_2, \ldots, u_r) = r\alpha(u, u_2, \ldots, u_r). \tag{8.18}$$

Exterior algebra

The interior product of a vector with a scalar is assumed to vanish, $i_u a = 0$ for all $a \in \Lambda^{*0}(V) \equiv \mathbb{R}$. The component expression with respect to any basis $\{e_i\}$ of the interior product of a vector with an r-form is given by

$$(i_u \alpha)_{i_2 \ldots i_r} = (i_u \alpha)(e_{i_2}, \ldots, e_{i_r})$$
$$= r\alpha(u^i e_i, e_{i_2}, \ldots, e_{i_r})$$
$$= r u^i \alpha_{i i_2 \ldots i_r}.$$

Hence

$$i_u \alpha = r C_1^1 (u \otimes \alpha),$$

where C_1^1 is the $(1, 1)$ contraction operator.

Performing the interior product with two vectors in succession on any r-form α has the property

$$i_u(i_v \alpha) = -i_v(i_u \alpha), \tag{8.19}$$

for

$$(i_u(i_v \alpha))(u_3, \ldots, u_r) = (r-1)(i_v \alpha)(u, u_3, \ldots, u_r)$$
$$= (r-1) r \alpha(v, u, u_3, \ldots, u_r)$$
$$= -(r-1) r \alpha(u, v, u_3, \ldots, u_r)$$
$$= -(i_v(i_u \alpha))(u_3, \ldots, u_r).$$

It follows immediately that $(i_u)^2 \equiv i_u i_u = 0$.

Another important identity, for an arbitrary r-form α and s-form β, is the **antiderivation law**

$$i_u(\alpha \wedge \beta) = (i_u \alpha) \wedge \beta + (-1)^r \alpha \wedge (i_u \beta). \tag{8.20}$$

Proof: Let $u_1, u_2, \ldots, u_{r+s}$ be arbitrary vectors. By Eqs. (8.18) and (8.17)

$$(i_{u_1}(\alpha \wedge \beta))(u_2, \ldots, u_{r+s}) = (r+s) \mathcal{A}(\alpha \otimes \beta)(u_1, u_2, \ldots, u_{r+s})$$
$$= \frac{1}{(r+s-1)!} \sum_\sigma (-1)^\sigma \alpha(u_{\sigma(1)}, \ldots, u_{\sigma(r)}) \beta(u_{\sigma(r+1)}, \ldots, u_{\sigma(r+s)}).$$

For each $1 \le a \le r+s$ let γ_a be the cyclic permutation $(1\,2\,\ldots\,a)$. If σ is any permutation such that $\sigma(a) = 1$ then $\sigma = \sigma' \gamma_a$ where $\sigma'(1) = \sigma(a) = 1$. The signs of the permutations σ and σ' are related by $(-1)^\sigma = (-1)^{\sigma'} (-1)^{a+1}$, and the sum of permutations in the above equation may be written as

$$\frac{1}{(r+s-1)!} \left(\sum_{a=1}^{r} \sum_{\sigma'} (-1)^{\sigma'} (-1)^{a+1} \alpha\big(u_{\sigma'(2)}, \ldots, u_{\sigma'(a)}, u_1 \ldots, u_{\sigma'(r)}\big) \right.$$
$$\times \beta\big(u_{\sigma'(r+1)}, \ldots, u_{\sigma'(r+s)}\big) + \sum_{b=1}^{s} \sum_{\sigma'} (-1)^{\sigma'} (-1)^{r+b+1} \alpha\big(u_{\sigma'(2)}, \ldots, u_{\sigma'(r+1)}\big)$$
$$\left. \times \beta\big(u_{\sigma'(r+2)}, \ldots, u_{\sigma'(r+b)}, u_1, \ldots, u_{\sigma'(r+s)}\big) \right).$$

By cyclic permutations u_1 can be brought to the first argument of α and β respectively, introducing factors $(-1)^{a+1}$ and $(-1)^{b+1}$ in the two sums, to give

$$\frac{1}{(r+s-1)!}\left(r\sum_{\sigma'}(-1)^{\sigma'}\alpha\big(u_1,u_{\sigma'(2)},\ldots,u_{\sigma'(a-1)},u_{\sigma'(a+1)},\ldots,u_{\sigma'(r)}\big)\right.$$
$$\times \beta\big(u_{\sigma'(r+1)},\ldots,u_{\sigma'(r+s)}\big)+s(-1)^r\sum_{\sigma'}(-1)^{\sigma'}\alpha\big(u_{\sigma'(1)},\ldots,u_{\sigma'(r)}\big)$$
$$\left.\times \beta\big(u_1,u_{\sigma'(r+1)},\ldots,u_{\sigma'(r+b-1)},u_{\sigma'(r+b+1)},\ldots,u_{\sigma'(r+s)}\big)\right),$$

where σ' ranges over all permutations of $(2,3,\ldots,r+s)$. Thus

$$i_{u_1}(\alpha\wedge\beta)(u_2,\ldots,u_{r+s}) = \mathcal{A}\big((i_{u_1}\alpha)\otimes\beta + (-1)^r\alpha\otimes(i_{u_1}\beta)\big)(u_2,\ldots,u_{r+s})$$
$$= (i_{u_1}\alpha)\wedge\beta + (-1)^r\alpha\wedge(i_{u_1}\beta)(u_2,\ldots,u_{r+s}).$$

Equation (8.20) follows on setting $u = u_1$. ∎

8.5 Oriented vector spaces

n-Vectors and n-forms

Let V be a vector space of dimension n with basis $\{e_1,\ldots,e_n\}$ and dual basis $\{\varepsilon^1,\ldots,\varepsilon^n\}$. Since the spaces $\Lambda^n(V)$ and $\Lambda^{*n}(V)$ are both one-dimensional, the n-vector

$$E = e_{12\ldots n} = e_1\wedge e_2\wedge\cdots\wedge e_n$$

forms a basis of $\Lambda^n(V)$, while the n-form

$$\Omega = \varepsilon^{12\ldots n} = \varepsilon^1\wedge\cdots\wedge\varepsilon^n$$

is a basis of $\Lambda^{*n}(V)$. These will sometimes be referred to as **volume elements** associated with this basis.

Exercise: Show that every non-zero n-vector is the volume element associated with some basis of V.

Given a basis $\{e_1,\ldots,e_n\}$, every n-vector A has a unique expansion

$$A = aE = a\,e_1\wedge e_2\wedge\cdots\wedge e_n$$
$$= \frac{a}{n!}\sum_\sigma (-1)^\sigma e_{\sigma(1)}\otimes e_{\sigma(2)}\otimes\cdots\otimes e_{\sigma(n)}$$
$$= \frac{a}{n!}\epsilon^{i_1 i_2\ldots i_n} e_{i_1}\otimes e_{i_2}\otimes\cdots\otimes e_{i_n}$$

where the ϵ-symbols, or **Levi–Civita symbols**, $\epsilon^{i_1 i_2\ldots i_n}$ and $\epsilon_{i_1 i_2\ldots i_n}$ are defined by

$$\epsilon^{i_1 i_2\ldots i_n} = \epsilon_{i_1 i_2\ldots i_n} = \begin{cases} 1 & \text{if } i_1\ldots i_n \text{ is an even permutation of } 1,2,\ldots,n, \\ -1 & \text{if } i_1\ldots i_n \text{ is an odd permutation of } 1,2,\ldots,n, \\ 0 & \text{if any pair of indices are equal.} \end{cases} \quad (8.21)$$

The ϵ-symbols are clearly antisymmetric in any pair of indices.

Exterior algebra

Exercise: Show that any n-form β has a unique expansion

$$\beta = b\Omega = \frac{b}{n!} \epsilon_{i_1...i_n} \varepsilon^{i_1} \otimes \cdots \otimes \varepsilon^{i_n}.$$

Every n-vector and n-form has tensor components proportional to the ϵ-symbols,

$$A^{i_1...i_n} = \frac{a}{n!} \epsilon^{i_1...i_n},$$

$$\beta_{i_1...i_n} = \frac{b}{n!} \epsilon_{i_1...i_n},$$

and setting $a = b = 1$ we have

$$E^{i_1...i_n} = \frac{1}{n!} \epsilon^{i_1...i_n}, \qquad \Omega_{i_1...i_n} = \frac{1}{n!} \epsilon_{i_1...i_n}. \tag{8.22}$$

Transformation laws of n-vectors and n-forms

The transformation matrix $\mathsf{A} = [A^{i'}{}_i]$ appearing in Eq. (7.26) satisfies

$$A^{i'_1}{}_{i_1} A^{i'_2}{}_{i_2} \ldots A^{i'_n}{}_{i_n} \epsilon^{i_1 i_2 ... i_n} = \det[A^{i'}{}_i] \, \epsilon^{i'_1 i'_2 ... i'_n}. \tag{8.23}$$

Proof: If $i'_1 = 1, i'_2 = 2, \ldots, i'_n = n$ then the left-hand side is the usual expansion of the determinant of the matrix A as a sum of products of its elements taken from different rows and columns with appropriate $\pm$ signs, while the right-hand side is $\det \mathsf{A} \epsilon^{12...n} = \det \mathsf{A}$. From the antisymmetry of the epsilon symbol in any pair of indices i and j we have

$$\ldots A^{i'}{}_i \ldots A^{j'}{}_j \ldots \epsilon^{...i...j...} = - \ldots A^{i'}{}_j \ldots A^{j'}{}_i \ldots \epsilon^{...i...j...} = - \ldots A^{j'}{}_i \ldots A^{i'}{}_j \ldots \epsilon^{...i...j...},$$

and the whole expression is antismmetric in any pair of indices i', j'. In particular, it vanishes if $i' = j'$. Hence if π is any permutation of $(1, 2, \ldots, n)$ such that $i'_1 = \pi(1), i'_2 = \pi(2), \ldots, i'_n = \pi(n)$ then both sides of Eq. (8.23) are multiplied by the sign of the permutation $(-1)^\pi$, while if any pair of indices $i'_1 \ldots i'_n$ are equal, both sides of the equation vanish. ■

If $A = aE$ is any n-vector, then from the law of transformation of tensor components, Eq. (7.30),

$$A^{i'_1...i'_n} = A^{i_1...i_n} A^{i'_1}{}_{i_1} \ldots A^{i'_n}{}_{i_n}$$

$$= \frac{a}{n!} \epsilon^{i_1...i_n} A^{i'_1}{}_{i_1} \ldots A^{i'_n}{}_{i_n}$$

$$= \det[A^{i'}{}_i] \frac{a}{n!} \epsilon^{i'_1...i'_n}.$$

Setting $A = a'E'$, the factor a is seen to transform as

$$a' = a \det[A^{i'}{}_i] = a \det \mathsf{A}. \tag{8.24}$$

If $a = 1$ we arrive at the transformation law of volume elements,

$$E = \det \mathsf{A} \, E', \qquad E' = \det \mathsf{A}^{-1} \, E. \tag{8.25}$$

8.5 Oriented vector spaces

A similar formula to Eq. (8.23) holds for the inverse matrix $A' = [A'^i{}_{j'}]$,

$$A'^{i_1}_{i'_1} A'^{i_2}_{i'_2} \ldots A'^{i_n}_{i'_n} \epsilon_{i_1 i_2 \ldots i_n} = \det[A'^i{}_{j'}] \, \epsilon_{i'_1 i'_2 \ldots i'_n}, \tag{8.26}$$

and the transformation law for an n-form $\beta = b\Omega = \frac{b}{n!}\epsilon^{12\ldots n} = b'\Omega'$ is

$$b' = b \det[A'^i{}_{j'}] = b \det \mathsf{A}^{-1} \quad \text{that is} \quad b = b' \det \mathsf{A}, \tag{8.27}$$

$$\Omega = \det \mathsf{A}^{-1} \, \Omega', \quad \Omega' = \det \mathsf{A} \, \Omega. \tag{8.28}$$

Exercise: Prove Eqs. (8.26)–(8.28).

Note, from Eqs. (8.23) and (8.26), that the ϵ-symbols do not transform as components of tensors under general basis transformations. They do however transform as tensors with respect to the restricted set of unimodular transformations, having $\det \mathsf{A} = \det[A^{j'}_i] = 1$. In particular, for cartesian tensors they transform as tensors provided only proper orthogonal transformations, rotations, are permitted. The term 'tensor density' is sometimes used to refer to entities that include determinant factors like those in (8.23) and (8.26), while scalar quantities that transform like a or b in (8.24) and (8.27) are referred to as 'densities'.

Oriented vector spaces

Two bases $\{e_i\}$ and $\{e'_{i'}\}$ are said to have the **same orientation** if the transformation matrix $\mathsf{A} = [A^{i'}_j]$ in Eq. (7.26) has positive determinant, $\det \mathsf{A} > 0$; otherwise they are said to be **oppositely oriented**. Writing $\{e_i\} o \{e'_{i'}\}$ iff $\{e_i\}$ and $\{e'_{i'}\}$ have the same orientation, it is straightforward to show that o is an equivalence relation and divides the set of all bases on V into two equivalence classes, called **orientations**. A vector space V together with a choice of orientation is called an **oriented vector space**. Any basis belonging to the selected orientation will be said to be **positively oriented**, while oppositely oriented bases will be called **negatively oriented**.

Example 8.3 Euclidean three-dimensional space $\mathbb{E}^3$ together with choice of a right-handed orthonormal basis $\{\mathbf{i}, \mathbf{j}, \mathbf{k}\}$ is an oriented vector space. The orientation consists of the set of all bases related to $\{\mathbf{i}, \mathbf{j}, \mathbf{k}\}$ through a positive determinant transformation. A left-handed set of axes has opposite orientation since the basis transformation will involve a reflection, having negative determinant.

Example 8.4 Let V be an n-dimensional vector space. Denote the set of all volume elements on V by $\dot{\Lambda}^n(V)$ and $\dot{\Lambda}^{*n}(V)$. Two non-zero n-vectors A and B can be said to **have the same orientation** if $A = cB$ with $c > 0$. This clearly provides an equivalence relation on $\dot{\Lambda}^n(V)$, dividing it into two non-intersecting equivalence classes. A selection of one of these two classes is an alternative way of specifying an orientation on a vector space V, for we may stipulate that a basis $\{e_i\}$ has positive orientation if $A = aE$ with $a > 0$ for all volume elements A in the chosen class. By Eqs. (8.24) and (8.25), this is equivalent to dividing the set of bases on V into two classes.

217

Exterior algebra

Now let V be an oriented n-dimensional real inner product space having index $t = r - s$ where $n = r + s$. If $\{e_1, e_2, \ldots, e_n\}$ is any positively oriented orthonormal frame such that

$$e_i \cdot e_j = \eta_{ij} = \begin{cases} \eta_i = \pm 1 & \text{if } i = j, \\ 0 & \text{if } i \neq j, \end{cases}$$

then r is the number of $+1$'s and s the number of -1's among the η_i. As pseudo-orthogonal transformations all have determinant ± 1, those relating positively oriented orthonormal frames must have det $= 1$. Hence, by Eq. (8.25), the volume element $E = e_1 \wedge e_2 \wedge \cdots \wedge e_n$ is independent of the choice of positively oriented orthonormal basis and is entirely determined by the inner product and the orientation on V.

By Eq. (8.22), the components of the volume element E with respect to any positively oriented orthonormal basis are

$$E^{i_1 i_2 \ldots i_n} = \frac{1}{n!} \epsilon^{i_1 i_2 \ldots i_n}.$$

With respect to an arbitrary positively oriented basis e'_i, not necessarily orthonormal, the components of E are, by Eqs. (7.30) and (8.23),

$$E'^{i'_1 i'_2 \ldots i'_n} = \det[A_i^{i'}] \frac{1}{n!} \epsilon^{i'_1 i'_2 \ldots i'_n}. \tag{8.29}$$

Take note that these are the components of the volume element E determined by the original orthonormal basis expressed with respect to the new basis, *not* the components of the volume element $e'_1 \wedge e'_2 \wedge \cdots \wedge e'_n$ determined by the new basis. It is possible to arrive at a formula for the components on the right-hand side of Eq. (8.29) that is independent of the transformation matrix $\mathbf{A}$. Consider the transformation of components of the metric tensor, defined by $u \cdot v = g_{ij} u^i v^j$,

$$g'_{i'j'} = g_{ij} A'^i_{i'} A'^j_{j'},$$

which can be written in matrix form

$$\mathbf{G}' = (\mathbf{A}^{-1})^T \mathbf{G} \mathbf{A}^{-1} \quad \text{where} \quad \mathbf{G} = [g_{ij}], \; \mathbf{G}' = [g'_{i'j'}].$$

On taking determinants

$$g' = g(\det \mathbf{A})^{-2} \quad \text{where} \quad g = \det \mathbf{G} = \pm 1, \; g' = \det \mathbf{G}',$$

and subsituting in (8.29) we have

$$E'^{i'_1 i'_2 \ldots i'_n} = \frac{1}{n! \sqrt{|g'|}} \epsilon^{i'_1 i'_2 \ldots i'_n}.$$

Eliminating the primes, it follows that the components of the volume element E defined by the inner product can be written in an arbitrary positively oriented basis as

$$E^{i_1 i_2 \ldots i_n} = \frac{1}{n! \sqrt{|g|}} \epsilon^{i_1 i_2 \ldots i_n}. \tag{8.30}$$

8.5 Oriented vector spaces

On lowering the indices of E we have

$$E_{i_1 i_2 \ldots i_n} = g_{i_1 j_1} g_{i_2 j_2} \cdots g_{i_n j_n} E^{j_1 j_2 \ldots j_n}$$

$$= \frac{1}{n! \sqrt{|g|}} g_{i_1 j_1} g_{i_2 j_2} \cdots g_{i_n j_n} \epsilon^{j_1 j_2 \ldots j_n}$$

$$= \frac{1}{n! \sqrt{|g|}} g \epsilon_{i_1 i_2 \ldots i_n}.$$

Since the sign of the determinant g is equal to $(-1)^s$ we have, in any positively oriented basis,

$$E_{i_1 i_2 \ldots i_n} = (-1)^s \frac{\sqrt{|g|}}{n!} \epsilon_{i_1 i_2 \ldots i_n}. \tag{8.31}$$

Exercise: Show that the components of the n-form $\Omega = \varepsilon^{12\ldots n}$ defined by a positively oriented o.n. basis are

$$\Omega_{i_1 i_2 \ldots i_n} = \frac{\sqrt{|g|}}{n!} \epsilon_{i_1 i_2 \ldots i_n} = (-1)^s E_{i_1 i_2 \ldots i_n}. \tag{8.32}$$

ϵ-Symbol identities

The ϵ-symbols satisfy a number of fundamental identities, the most general of which is

$$\epsilon_{i_1 \ldots i_n} \epsilon^{j_1 \ldots j_n} = \delta^{j_1 \ldots j_n}_{i_1 \ldots i_n}, \tag{8.33}$$

where the generalized δ-symbol is defined by

$$\delta^{j_1 \ldots j_r}_{i_1 \ldots i_r} = \begin{cases} 1 & \text{if } j_1 \ldots j_r \text{ is an even permutation of } i_1 \ldots i_r, \\ -1 & \text{if } j_1 \ldots j_r \text{ is an odd permutation of } i_1 \ldots i_r, \\ 0 & \text{otherwise.} \end{cases} \tag{8.34}$$

Total contraction of (8.33) over all indices gives

$$\epsilon_{i_1 \ldots i_n} \epsilon^{i_1 \ldots i_n} = n!. \tag{8.35}$$

Contracting (8.33) over the first $n-1$ indices gives

$$\epsilon_{i_1 \ldots i_{n-1} j} \epsilon^{i_1 \ldots i_{n-1} k} = (n-1)! \delta^k_j, \tag{8.36}$$

for if $k \neq j$ each term in the summation $\delta^{i_1 \ldots i_{n-1} k}_{i_1 \ldots i_{n-1} j}$ vanishes since in every summand either one pair of superscripts or one pair of subscripts must be equal, while if $k = j$ the expression is a sum of $(n-1)!$ terms each of value $+1$.

The most general contraction identity arising from (8.34) is

$$\epsilon_{i_1 \ldots i_{n-r} j_1 \ldots j_r} \epsilon^{i_1 \ldots i_{n-r} k_1 \ldots k_r} = (n-r)! \delta^{k_1 \ldots k_r}_{j_1 \ldots j_r}, \tag{8.37}$$

where the δ-symbol on the right-hand side can be expressed in terms of Kronecker

Exterior algebra

deltas,

$$\delta^{k_1 \ldots k_r}_{j_1 \ldots j_r} = \sum_\sigma (-1)^\sigma \delta^{k_1}_{j_{\sigma(1)}} \delta^{k_2}_{j_{\sigma(2)}} \ldots \delta^{k_r}_{j_{\sigma(r)}}$$
$$= \delta^{k_1}_{j_1} \delta^{k_2}_{j_2} \ldots \delta^{k_r}_{j_r} - \delta^{k_1}_{j_2} \delta^{k_2}_{j_1} \ldots \delta^{k_r}_{j_r} + \ldots, \qquad (8.38)$$

a sum of $r!$ terms in which the j_i indices run over every permutation of $j_1, j_2, \ldots, j_r$.

Example 8.5 In three dimensions we have

$$\epsilon_{ijk}\epsilon^{lmn} = \delta_i^l \delta_j^m \delta_k^n - \delta_i^l \delta_j^n \delta_k^m + \delta_i^m \delta_j^n \delta_k^l - \delta_i^m \delta_j^l \delta_k^n + \delta_i^n \delta_j^l \delta_k^m - \delta_i^n \delta_j^m \delta_k^l,$$
$$\epsilon_{ijk}\epsilon^{imn} = \delta_j^m \delta_k^n - \delta_k^m \delta_j^n,$$
$$\epsilon_{ijk}\epsilon^{ijn} = 2\delta_k^n,$$
$$\delta_{ijk}\epsilon^{ijk} = 6.$$

The last three identities are particularly useful in cartesian tensors where the summation convention is used on repeated subscripts, giving

$$\epsilon_{ijk}\epsilon_{imn} = \delta_{mj}\delta_{nk} - \delta_{mk}\delta_{nj}, \qquad \epsilon_{ijk}\epsilon_{ijn} = 2\delta_{nk}, \qquad \delta_{ijk}\epsilon_{ijk} = 6.$$

For example, the *vector product* $\mathbf{u} \times \mathbf{v}$ of two vectors $\mathbf{u} = u_i \mathbf{e}_i$ and $\mathbf{v} = v_i \mathbf{e}_i$ is defined as the vector whose components are given by

$$(\mathbf{u} \times \mathbf{v})_k = \epsilon_{ijk}(u \wedge v)^{ij} = \epsilon_{kij} u_i v_j.$$

The vector identity

$$\mathbf{u} \times (\mathbf{v} \times \mathbf{w}) = (\mathbf{u} \cdot \mathbf{w})\mathbf{v} - (\mathbf{u} \cdot \mathbf{v})\mathbf{w}$$

follows from

$$\big(\mathbf{u} \times (\mathbf{v} \times \mathbf{w})\big)_i = \epsilon_{ijk} u_j \epsilon_{klm} v_l w_m$$
$$= (\delta_{il}\delta_{jm} - \delta_{im}\delta_{jl}) u_j v_l w_m = (u_j w_j) v_i - (u_j v_j) w_i = \big((\mathbf{u} \cdot \mathbf{w})\mathbf{v} - (\mathbf{u} \cdot \mathbf{v})\mathbf{w}\big)_i.$$

Exercise: Show the identity $(\mathbf{u} \times \mathbf{v})^2 = \mathbf{u}^2 \mathbf{v}^2 - (\mathbf{u} \cdot \mathbf{v})^2$.

8.6 The Hodge dual

Inner product of p-vectors

The coupling between linear functionals (1-forms) and vectors, denoted

$$\langle u, \omega \rangle \equiv \langle \omega, u \rangle = \omega(u) = u^i w_i,$$

can be extended to define a product between p-vectors A and p-forms β,

$$\langle A, \beta \rangle = p! C_1^1 C_2^2 \ldots C_p^p A \otimes \beta = p! A^{i_1 i_2 \ldots i_p} \beta_{i_1 i_2 \ldots i_p}. \qquad (8.39)$$

8.6 The Hodge dual

For each fixed p-form β the map $A \mapsto \langle A, \beta \rangle$ clearly defines a linear functional on the vector space $\Lambda^p(V)$.

Exercise: Show that

$$\langle u_1 \wedge u_2 \wedge \cdots \wedge u_p, \beta \rangle = p! \beta(u_1, u_2, \ldots, u_p). \tag{8.40}$$

Theorem 8.3 *If A is a p-vector, β a $(p+1)$-form and u an arbitrary vector, then*

$$\langle A, i_u \beta \rangle = \langle u \wedge A, \beta \rangle. \tag{8.41}$$

Proof: For a simple p-vector, $A = u_1 \wedge u_2 \wedge \cdots \wedge u_p$, using Eqs. (8.40) and (8.18),

$$\langle u_1 \wedge u_2 \wedge \cdots \wedge u_p, i_u \beta \rangle = p!(i_u \beta)(u_1, u_2, \ldots, u_p)$$
$$= (p+1)! \beta(u, u_1, \ldots, u_p)$$
$$= \langle u \wedge u_1 \wedge \cdots \wedge u_p, \beta \rangle.$$

Since every p-vector A is a sum of simple p-vectors, this generalizes to arbitrary p-vectors by linearity. ∎

If V is an inner product space it is possible to define the **inner product** of two p-vectors A and B to be

$$(A, B) = \langle A, \beta \rangle \tag{8.42}$$

where β is the tensor formed from B by lowering indices,

$$\beta_{i_1 i_2 \ldots i_p} = B_{i_1 i_2 \ldots i_p} = g_{i_1 j_1} g_{i_2 j_2} \cdots g_{i_p j_p} B^{j_1 j_2 \ldots j_p}.$$

Lemma 8.4 *Let X and Y be simple p-vectors, $X = x_1 \wedge x_2 \wedge \cdots \wedge x_p$ and $Y = y_1 \wedge y_2 \wedge \cdots \wedge y_p$, then*

$$(X, Y) = \det[x_i \cdot y_j]. \tag{8.43}$$

Proof: With respect to a basis $\{e_i\}$

$$(X, Y) = p! x_1^{[i_1} x_2^{i_2} \ldots x_p^{i_p]} g_{i_1 j_1} g_{i_2 j_2} \cdots g_{i_p j_p} y_1^{[j_1} y_2^{j_2} \ldots y_p^{j_p]}$$
$$= p! x_1^{[i_1} x_2^{i_2} \ldots x_p^{i_p]} y_{1 i_1} y_{2 i_2} \cdots y_{p i_p}$$
$$= \sum_\sigma (-1)^\sigma x_1^{i_{\sigma(1)}} x_2^{i_{\sigma(2)}} \ldots x_p^{i_{\sigma(p)}} y_{1 i_1} y_{2 i_2} \cdots y_{p i_p}$$
$$= \det[\langle x_i, y_j \rangle] \quad \text{where } y_j = y_{jk} \varepsilon^k$$
$$= \det[g_{kl} x_i^k y_j^l]$$
$$= \det[x_i \cdot y_j]. \quad \blacksquare$$

Theorem 8.5 *The map $A, B \mapsto (A, B)$ makes $\Lambda^p(V)$ into a real inner product space.*

Proof: From Eqs. (8.42) and (8.39)

$$(A, B) = p! A^{i_1 i_2 \ldots i_p} g_{i_1 j_1} g_{i_2 j_2} \cdots g_{i_p j_p} B^{j_1 j_2 \ldots j_p} = (B, A)$$

so that (A, B) is a symmetric bilinear function of A and B. It remains to show that $(.,.)$ is non-singular, satisfying (SP3) of Section 5.1.

Exterior algebra

Let $e_1, e_2, \ldots, e_n$ be an orthonormal basis,

$$e_i \cdot e_j = \eta_i \delta_{ij}, \quad \eta_i = \pm 1,$$

and for any arbitrary increasing sequences of indices $\mathbf{h} = h_1 < h_2 < \cdots < h_p$ set $e_\mathbf{h}$ to be the basis p-vector

$$e_\mathbf{h} = e_{h_1} \wedge e_{h_2} \wedge \cdots \wedge e_{h_p}.$$

If $\mathbf{h} = h_1 < h_2 < \cdots < h_p$ and $\mathbf{k} = k_1 < k_2 < \cdots < k_p$ are any pair of increasing sequences of indices and $h_i \notin \{k_1, k_2, \ldots, k_p\}$ for some i, then, by Lemma 8.4

$$(e_\mathbf{h}, e_\mathbf{k}) = \det[e_{h_i} \cdot e_{k_j}] = 0$$

since the ith row of the determinant vanishes completely. On the other hand, if $\mathbf{h} = \mathbf{k}$ we have

$$(e_\mathbf{h}, e_\mathbf{h}) = \eta_{h_1} \eta_{h_2} \cdots \eta_{h_p}.$$

In summary,

$$(e_\mathbf{h}, e_\mathbf{k}) = \det[e_{h_i} \cdot e_{k_j}] = \begin{cases} 0 & \text{if } \mathbf{h} \neq \mathbf{k}, \\ \eta_{h_1} \eta_{h_2} \cdots \eta_{h_p} & \text{if } \mathbf{h} = \mathbf{k}, \end{cases} \quad (8.44)$$

so $E_p = \{e_\mathbf{h} \mid h_1 < h_2 < \cdots < h_p\}$ forms an orthonormal basis with respect to the inner product on $\Lambda^p(V)$. The matrix of the inner product is non-singular with respect to this basis since it is diagonal with ± 1's along the diagonal. ∎

Exercise: Show that $(E, E) = (-1)^s$ where $E = e_{12\ldots n}$ and s is the number of $-$ signs in g_{ij}.

An inner product can of course be defined on $\Lambda^{*p}(V)$ in exactly the same way as for $\Lambda^p(V)$,

$$(\alpha, \beta) = \langle A, \beta \rangle = p! A^{i_1 i_2 \ldots i_p} \beta_{i_1 i_2 \ldots i_p} \quad (8.45)$$

where A is the tensor formed from α by raising indices,

$$A^{i_1 i_2 \ldots i_p} = \alpha^{i_1 i_2 \ldots i_p} = g^{i_1 j_1} g^{i_2 j_2} \cdots g^{i_p j_p} \alpha_{j_1 j_2 \ldots j_p}.$$

Exercise: Show that Theorem 8.3 can be expressed in the alternative form: for any p-form α, $(p+1)$-form β and vector u

$$(\alpha, i_u \beta) = (\bar{g}u \wedge \alpha, \beta) \quad (8.46)$$

where $\bar{g}u$ is the 1-form defined in Example 7.8 by lowering the index of u^i.

The Hodge star operator

Let V be an oriented inner product space and $\{e_1, e_2, \ldots, e_n\}$ be a positively oriented orthonormal basis, $e_i \cdot e_j \delta_{ij} \eta_i$, with associated volume element $E = e_{12\ldots n} = e_1 \wedge e_2 \wedge$

8.6 The Hodge dual

$\cdots \wedge e_n$. For any $A \in \Lambda^p(V)$ the map $f_A : \Lambda^{n-p}(V) \to \mathbb{R}$ defined by $A \wedge B = f_A(B)E$ is linear since

$$f_A(B + aC)E = A \wedge (B + aC) = A \wedge B + a A \wedge C = \bigl(f_A(B) + a f_A(C)\bigr)E.$$

Thus f_A is a linear functional on $\Lambda^{n-p}(V)$, and as the inner product $(.,.)$ on $\Lambda^{n-p}(V)$ is non-singular there exists a unique $(n-p)$-vector $*A$ such that $f_A(B) = (*A, B)$. The $(n-p)$-vector $*A$ is uniquely determined by this equation, and is frequently referred to as the **(Hodge) dual** of A,

$$A \wedge B = (*A, B)E \qquad \text{for all } B \in \Lambda^{(n-p)}(V). \tag{8.47}$$

The one-to-one map $* : \Lambda^p(V) \to \Lambda^{n-p}(V)$ is called the **Hodge star operator**; it assigns an $(n-p)$-vector to each p-vector, and vice versa. This reciprocity is only possible because the dimensions of these two vector spaces are identical,

$$\dim \Lambda^p(V) = \dim \Lambda^{n-p}(V) = \binom{n}{p} = \binom{n}{n-p} = \frac{n!}{p!(n-p)!}.$$

Since E is independent of the choice of positively oriented orthonormal basis, the Hodge dual is a basis-independent concept.

To calculate the components of the dual with respect to a positively oriented orthonormal basis $e_1, \ldots, e_n$, set $\mathbf{i} = i_1 < i_2 < \cdots < i_p$ and $\mathbf{j} = j_1 < j_2 < \cdots < j_{n-p}$. Since

$$e_\mathbf{i} \wedge e_\mathbf{j} = \epsilon_{i_1 i_2 \ldots i_p j_1 \ldots j_{n-p}} E,$$

we have from Eq. (8.47) that

$$(*e_\mathbf{i}, e_\mathbf{j}) = \epsilon_{i_1 i_2 \ldots i_p j_1 \ldots j_{n-p}},$$

and Eq. (8.44) gives

$$*e_\mathbf{i} = \epsilon_{i_1 i_2 \ldots i_p j_1 \ldots j_{n-p}} (e_\mathbf{j}, e_\mathbf{j}) e_\mathbf{j} \tag{8.48}$$

where $\mathbf{j} = j_1 < j_2 < \cdots < j_{n-p}$ is the complementary set of indices to $i_1 < \cdots < i_p$. By a similar argument

$$*e_\mathbf{j} = \epsilon_{j_1 j_2 \ldots j_{n-p} i_1 \ldots i_p}(e_\mathbf{i}, e_\mathbf{i}) e_\mathbf{i},$$

and, temporarily suspending the summation convention,

$$**e_\mathbf{i} = \epsilon_{i_1 i_2 \ldots i_p j_1 \ldots j_{n-p}} \epsilon_{j_1 j_2 \ldots j_{n-p} i_1 \ldots i_p} (e_\mathbf{i}, e_\mathbf{i})(e_\mathbf{j}, e_\mathbf{j}) e_\mathbf{i}$$
$$= (-1)^{p(n-p)} \eta_{i_1} \eta_{i_2} \cdots \eta_{i_p} \eta_{j_1} \eta_{j_2} \cdots \eta_{j_{n-p}} e_\mathbf{i}$$
$$= (-1)^{p(n-p)+s} e_\mathbf{i}$$

where s is the number of -1's in g_{ij}. The coefficient s can also be written as $s = \tfrac{1}{2}(n-t)$ where t is the index of the metric. As any p-vector A is a linear combination of the $e_\mathbf{i}$, we have the identity

$$** A = (-1)^{p(n-p)+s} A. \tag{8.49}$$

Exterior algebra

Theorem 8.6 *For any p-vectors A and B we also have the following identities*:

$$A \wedge *B = B \wedge *A = (-1)^s (A, B) E, \qquad (8.50)$$

and

$$(*A, *B) = (-1)^s (A, B). \qquad (8.51)$$

Proof: The first part of (8.50) follows from

$$A \wedge *B = (*A, *B)E = (*B, *A)E = B \wedge *A.$$

The second part of (8.50) follows on using Eqs. (8.16) and (8.49),

$$A \wedge *B = (-1)^{p(n-p)} * B \wedge A$$
$$= (-1)^{p(n-p)} * B \wedge A$$
$$= (-1)^{p(n-p)} (**B, A)E$$
$$= (-1)^s (A, B)E.$$

Using Eqs. (8.47) and (8.50) we have

$$(*A, *B)E = A \wedge *B = (-1)^s (A, B)E$$

and (8.51) follows at once since $E \neq 0$. ■

The component prescription for the Hodge dual in an o.n. basis is straightforward, and is left as an exercise

$$*A^{j_1 \ldots j_{n-p}} = \frac{(-1)^s}{(n-p)!} \epsilon^{i_1 \ldots i_p j_1 \ldots j_{n-p}} A_{i_1 \ldots i_p}. \qquad (8.52)$$

Writing this equation as the component form of the tensor equation,

$$*A^{j_1 \ldots j_{n-p}} = \frac{(-1)^s n!}{(n-p)!} E^{i_1 \ldots i_p j_1 \ldots j_{n-p}} A_{i_1 \ldots i_p},$$

and using Eq. (8.30), we can express the Hodge dual in an arbitrary basis:

$$*A^{j_1 \ldots j_{n-p}} = \frac{(-1)^s}{(n-p)!\sqrt{|g|}} \epsilon^{i_1 \ldots i_p j_1 \ldots j_{n-p}} A_{i_1 \ldots i_p}. \qquad (8.53)$$

Exercise: Show that on lowering indices, (8.53) can be written

$$*A_{j_1 \ldots j_{n-p}} = \frac{\sqrt{|g|}}{(n-p)!} \epsilon_{i_1 \ldots i_p j_1 \ldots j_{n-p}} A^{i_1 \ldots i_p}. \qquad (8.54)$$

Example 8.6 Treating 1 as the basis 0-vector, we obtain from (8.48) that

$$*1 = \epsilon_{12 \ldots n}(E, E)E = (-1)^s E.$$

Conversely the dual of the volume element E is 1,

$$*E = \epsilon_{12 \ldots n} 1 = 1.$$

These two formulae agree with the double star formula (8.49) on setting $p = n$ or $p = 0$.

8.6 The Hodge dual

Example 8.7 In three-dimensional cartesian tensors, $s = 0$ and all indices are in the subscript position:

$$(*1)_{ijk} = \epsilon_{ijk},$$

$$A = A_i e_i \implies (*A)_{ij} = \tfrac{1}{2!}\epsilon_{kij} A_k = \tfrac{1}{2}\epsilon_{ijk} A_k,$$

$$A = A_{ij} e_i \otimes e_j \implies (*A)_i = \epsilon_{jki} A_{jk} = \epsilon_{ijk} A_{jk},$$

$$*E = \tfrac{1}{3!}\epsilon_{ijk}\epsilon_{ijk} = 1.$$

The concept of *vector product* of any two vectors $\mathbf{u} \times \mathbf{v}$ is the dual of the wedge product, for

$$\mathbf{u} \equiv u = u_i e_i, \; \mathbf{v} \equiv v = v_j e_j \implies (u \wedge v)_{ij} = \tfrac{1}{2}(u_i v_j - u_j v_i),$$

whence

$$*(u \wedge v)_i = \epsilon_{ijk}(u \wedge v)_{jk}$$
$$= \tfrac{1}{2}\epsilon_{ijk}(u_j v_k - u_k v_j)$$
$$= \epsilon_{ijk} u_j v_k$$
$$= (\mathbf{u} \times \mathbf{v})_i.$$

Example 8.8 In four-dimensional Minkowski space, with $s = 1$, the formulae for duals of various p-vectors are

$$(*1)^{ijkl} = -\tfrac{1}{4!}\epsilon^{ijkl} - -\tfrac{1}{24}\epsilon^{ijkl},$$

$$A = A^i e_i \implies (*A)^{ijk} = \tfrac{-1}{3!}\epsilon^{lijk} A_l = \tfrac{1}{6}\epsilon^{ijkl} A_l,$$

$$B = B^{ijk} e_i \otimes e_j \otimes e_k \implies (*B)^i = -\epsilon^{jkli} B_{jkl} = \epsilon^{ijkl} B_{jkl},$$

$$F = F^{ij} e_i \otimes e_j \implies (*F)^{ij} = \tfrac{-1}{2!}\epsilon^{klij} F_{kl} = -\tfrac{1}{2}\epsilon^{ijkl} F_{kl},$$

$$*E = -\epsilon^{ijkl}\tfrac{-1}{4!}\epsilon_{ijkl} = \tfrac{4!}{4!} = 1.$$

Note that if the components F^{ij} are written out in the following 'electromagnetic form', the significance of which will become clear in Chapter 9,

$$[F^{ij}] = \begin{pmatrix} 0 & B_3 & -B_2 & -E_1 \\ -B_3 & 0 & B_1 & -E_2 \\ B_2 & -B_1 & 0 & -E_3 \\ E_1 & E_2 & E_3 & 0 \end{pmatrix} \implies [F_{ij}] = \begin{pmatrix} 0 & B_3 & -B_2 & E_1 \\ -B_3 & 0 & B_1 & E_2 \\ B_2 & -B_1 & 0 & E_3 \\ -E_1 & -E_2 & -E_3 & 0 \end{pmatrix}$$

then the dual tensor essentially permutes electric and magnetic fields, up to a sign change,

$$*F^{ij} = \begin{pmatrix} 0 & -E_3 & E_2 & -B_1 \\ E_3 & 0 & -E_1 & -B_2 \\ -E_2 & E_1 & 0 & -B_3 \\ B_1 & B_2 & B_3 & 0 \end{pmatrix}.$$

Exterior algebra

The index lowering map $\bar{g} : V \to V^*$ can be uniquely extended to a linear map from p-vectors to p-forms,

$$\bar{g} : \Lambda^p(V) \to \Lambda^{*p}(V)$$

by requiring that

$$\bar{g}(u \wedge v \wedge \cdots \wedge w) = \bar{g}(u) \wedge \bar{g}(v) \wedge \cdots \wedge \bar{g}(w).$$

In components it has the expected effect

$$(\bar{g}A)_{i_1 i_2 \ldots i_p} = g_{i_1 j_1} g_{i_2 j_2} \cdots g_{i_p j_p} A^{j_1 j_2 \ldots j_p}$$

and from Eqs. (8.42) and (8.45) it follows that

$$(\bar{g}(A), \bar{g}(B)) = (A, B). \tag{8.55}$$

If the Hodge star operator is defined on forms by requiring that it commutes with the lowering operator $\bar{g}$,

$$*\bar{g}(A) = \bar{g}(*A) \tag{8.56}$$

then we find the defining relation for the Hodge star of a p-form α in a form analogous to Eq. (8.47),

$$\alpha \wedge \beta = (-1)^s (*\alpha, \beta)\Omega. \tag{8.57}$$

The factor $(-1)^s$ that enters this equation is essentially due to the fact that the 'lowered' basis vectors $\bar{g}(e_i)$ have a different orientation to the dual basis ε^i if s is an odd number, while they have the same orientation when s is even.

Problems

Problem 8.6 Show that the quantity (A, β) defined in Eq. (8.39) vanishes for all p-vectors A if and only if $\beta = 0$. Hence show that the correspondence between linear functionals on $\Lambda(V)$ and p-forms is one-to-one,

$$(\Lambda^p(V))^* \cong \Lambda^{*p}(V).$$

Problem 8.7 Show that the interior product between basis vectors e_i and $\varepsilon^{i_1 i_2 \ldots i_r}$ is given by

$$i_{e_i} \varepsilon^{i_1 \ldots i_r} = \begin{cases} 0 & \text{if } i \notin \{i_1, \ldots, i_r\}, \\ (-1)^{a-1} \varepsilon^{i_1 \ldots i_{a-1} i_{a+1} \ldots i_r} & \text{if } i = i_a. \end{cases}$$

Problem 8.8 Prove Eq. (8.52).

Problem 8.9 Every p-form α can be regarded as a linear functional on $\Lambda^p(V)$ through the action $\alpha(A) = \langle A, \alpha \rangle$. Show that the basis ε^i is dual to the basis e_j where $\mathbf{i} = i_1 < i_2 < \cdots < i_p$, $\mathbf{j} = j_1 < j_2 < \cdots < j_p$,

$$\langle e_\mathbf{j}, \varepsilon^\mathbf{i} \rangle = \delta^\mathbf{i}_\mathbf{j} \equiv \delta^{i_1}_{j_1} \delta^{i_2}_{j_2} \cdots \delta^{i_p}_{j_p}.$$

Verify that

$$\sum_{i_1 < i_2 < \cdots < i_p} \cdots \sum e_{i_1 i_2 \ldots i_p} \varepsilon^{i_1 i_2 \ldots i_p} = \dim \Lambda^p(V).$$

Problem 8.10 Show that if u, v and w are vectors in an n-dimensional real inner product space then

(a) $(u \wedge v, u \wedge v) = (u \cdot u)(v \cdot v) - (u \cdot v)^2$.
(b) $u \wedge *(v \wedge w) = (u \cdot w) * v - (u \cdot v) * w$.
(c) Which identities do these equations reduce to in three-dimensional cartesian vectors?

Problem 8.11 Let g_{ij} be the Minkowski metric on a four-dimensional space, having index 2 (so that there are three + signs and one − sign).

(a) By calculating the inner products $(e_{i_1 i_2}, e_{j_1, j_2})$, using (8.44) show that there are three +1's −1's in these inner products, and the index of the inner product defined on the six-dimensional space of bivectors $\Lambda^2(V)$ is therefore 0.
(b) What is the index of the inner product on $\Lambda^2(V)$ if V is n-dimensional and g_{ij} has index t? [*Ans.*: $\frac{1}{2}(t^2 - n)$.]

Problem 8.12 Show that in an arbitrary basis the component representation of the dual of a p-form α is

$$(*\alpha)_{j_1 j_2 \ldots j_{n-p}} = \frac{\sqrt{|g|}}{(n-p)!} \epsilon_{i_1 \ldots i_p j_1 j_2 \ldots j_{n-p}} \alpha^{i_1 \ldots i_p}. \tag{8.58}$$

Problem 8.13 If u is any vector, and α any p-form show that

$$i_u * \alpha = *(\alpha \wedge \bar{g}(u)).$$

References

[1] R. W. R. Darling. *Differential Forms and Connections*. New York, Cambridge University Press, 1994.
[2] H. Flanders. *Differential Forms*. New York, Dover Publications, 1989.
[3] S. Hassani. *Foundations of Mathematical Physics*. Boston, Allyn and Bacon, 1991.
[4] L. H. Loomis and S. Sternberg. *Advanced Calculus*. Reading, Mass., Addison-Wesley, 1968.
[5] E. Nelson. *Tensor Analysis*. Princeton, Princeton University Press, 1967.

9 Special relativity

In Example 7.12 we saw that a Euclidean inner product space with positive definite metric tensor g gives rise to a restricted tensor theory called **cartesian tensors**, wherein all bases $\{e_i\}$ are required to be orthonormal and basis transformations $e_i = A^{i'}_i e_{i'}$ are restricted to orthogonal transformations. Cartesian tensors may be written with all their indices in the lower position, $T_{ijk...}$ and it is common to adopt the summation convention for repeated indices even though both are subscripts.

In a general pseudo-Euclidean inner product space we may also restrict ourselves to orthonormal bases wherein

$$g_{ij} = \begin{cases} \pm 1 & \text{if } i = j, \\ 0 & \text{if } i \neq j, \end{cases}$$

so that only pseudo-orthogonal transformation matrices $A = [A^{i'}_i]$ are allowed. The resulting tensor theory is referred to as a *restricted tensor theory*. For example, in a four-dimensional Minkowskian vector space the metric tensor in an orthonormal basis is

$$g_{ij} = \begin{pmatrix} 1 & 0 & 0 & 0 \\ 0 & 1 & 0 & 0 \\ 0 & 0 & 1 & 0 \\ 0 & 0 & 0 & -1 \end{pmatrix},$$

and the associated restricted tensors are commonly called **4-tensors**. In 4-tensor theory there is a simple connection between covariant and contravariant indices, for example

$$U^1 = U_1, \quad U^2 = U_2, \quad U^3 = U_3, \quad U^4 = -U_4,$$

but the distinction between the two types of indices must still be maintained. In this chapter we give some applications of 4-tensor theory in Einstein's special theory of relativity [1–3].

9.1 Minkowski space-time

In Newtonian mechanics an **inertial frame** is a one-to-one correspondence between physical events and points of $\mathbb{R}^4$, each event being assigned coordinates (x, y, z, t) such that the motion of any free particle is represented by a *rectilinear* path $\mathbf{r} = \mathbf{u}t + \mathbf{r}_0$. This is Newton's first law of motion. Coordinate transformations $(x, y, z, t) \to (x', y', z', t')$

9.1 Minkowski space-time

between inertial frames are called **Galilean transformations**, shown in Example 2.29 to have the form

$$t' = t + a, \qquad \mathbf{r}' = \mathsf{A}\mathbf{r} - \mathbf{v}t + \mathbf{b} \tag{9.1}$$

where a is a real constant, $\mathbf{v}$ and $\mathbf{b}$ are constant vectors, and A is a 3×3 orthogonal matrix, $\mathsf{A}^T \mathsf{A} = \mathsf{I}$.

If there is no rotation, $\mathsf{A} = \mathsf{I}$ in (9.1), then a rectilinear motion $\mathbf{r} = \mathbf{u}t + \mathbf{r}_0$ is transformed to

$$\mathbf{r}' = \mathbf{u}'t' + \mathbf{r}'_0$$

where $\mathbf{r}'_0 = \mathbf{r}_0 + \mathbf{b} - a(\mathbf{u} - \mathbf{v})$ and

$$\mathbf{u}' = \mathbf{u} - \mathbf{v}.$$

This is known as the law of **transformation of velocities** and its inverse form,

$$\mathbf{u} = \mathbf{u}' + \mathbf{v}$$

is called the Newtonian law of **addition of velocities**.

In 1888 the famous Michelson–Morley experiment, using light beams oppositely directed at different points of the Earth's orbit, failed to detect any motion of the Earth relative to an 'aether' postulated to be an absolute rest frame for the propagation of electromagnetic waves. The apparent interpretation that the speed of light be constant under transformations between inertial frames in relative motion is clearly at odds with Newton's law of addition of velocities. Eventually the resolution of this problem came in the form of **Einstein's principle of relativity** (1905). This is essentially an extension of Galileo's and Newton's ideas on invariance of mechanics, made to include electromagnetic fields (of which light is a particular manifestation). The geometrical interpretation due to Hermann Minkowski (1908) is the version we will discuss in this chapter.

Poincaré and Lorentz transformations

In classical mechanics we assume that events (x, y, z, t) form a Galilean space-time, as described in Example 2.29. In relativity the structure is somewhat different. Instead of separate spatial and temporal intervals there is a single interval defined between pairs of events, written

$$\Delta s^2 = \Delta x^2 + \Delta y^2 + \Delta z^2 - c^2 \Delta t^2$$

where c is the velocity of light ($c \approx 3 \times 10^8$ m s^{-1}). This singles out events connected by a light signal as satisfying $\Delta s^2 = 0$. Setting $(x^1 = x, x^2 = y, x^3 = z, x^4 = ct)$, the interval reads

$$\Delta s^2 = g_{\mu\nu} \Delta x^\mu \Delta x^\nu, \tag{9.2}$$

Special relativity

where

$$G = [g_{\mu\nu}] = \begin{pmatrix} 1 & 0 & 0 & 0 \\ 0 & 1 & 0 & 0 \\ 0 & 0 & 1 & 0 \\ 0 & 0 & 0 & -1 \end{pmatrix}.$$

Throughout this chapter, Greek indices μ, ν, etc. will range from 1 to 4 while indices $i, j, k \ldots$ range from 1 to 3. The set $\mathbb{R}^4$ with this interval structure is called **Minkowski space-time**, or simply **Minkowski space**. The geometrical version of the principle of relativity says that the set of **events** forms a Minkowski space-time. The definition of Minkoswki space as given here is not altogether satisfactory. We will give a more precise definition directly, in terms of an *affine space*.

The restricted class of coordinate systems for which the space-time interval has the form (9.2) will be called **inertial frames**. We will make the assumption, as in Newtonian mechanics, that free particles have rectilinear paths with respect to inertial frames in Minkowski space-time. As shown in Example 2.30, transformations preserving (9.2) are of the form

$$x'^{\mu'} = L^{\mu'}_{\nu} x^{\nu} + a^{\mu'} \tag{9.3}$$

where the coefficients $L^{\mu'}_{\nu}$ satisfy

$$g_{\rho\sigma} = g_{\mu'\nu'} L^{\mu'}_{\rho} L^{\nu'}_{\sigma}. \tag{9.4}$$

Equation (9.3) is known as a **Poincaré transformation**, while the linear transformations $x'^{\mu'} = L^{\mu'}_{\rho} x^{\rho}$ that arise on setting $a^{\mu'} = 0$ are called **Lorentz transformations**.

We define the **light cone** C_p at an event $p = (x, y, z, t)$ to be the set of points connected to p by *light signals*,

$$C_p = \{p' = (x', y', z', ct') \mid \Delta x^2 + \Delta y^2 + \Delta z^2 - c^2 \Delta t^2 = 0\}$$

where $\Delta x = x' - x$, $\Delta y = y' - y$, etc. Events p' on C_p can be thought of either as a receiver of light signals from p, or as a transmitter of signals that arrive at p. Poincaré transformations clearly preserve the light cone C_p at any event p.

As for Eq. (5.9), the matrix version of (9.4) is (see also Example 2.30)

$$G = L^T G L, \tag{9.5}$$

where $G = [g_{\mu\nu}]$ and $L = [L^{\mu'}_{\nu}]$. Taking determinants, we have $\det L = \pm 1$. It is further possible to subdivide Lorentz transformations into those having $L^{4'}_{4} \geq 1$ and those having $L^{4'}_{4} \leq -1$ (see Problem 9.2). Those Lorentz tansformations for which both $\det L = +1$ and $L^{4'}_{4} \geq 1$ are called **proper Lorentz transformations**. They are analogous to rotations about the origin in Euclidean space. All other Lorentz transformations are called **improper**.

Affine geometry

There is an important distinction to be made between Minkowski space and a Minkowskian vector space as defined in Section 5.1. Most significantly, Minkowski space is not a vector space since events do not combine linearly in any natural sense. For example, consider

9.1 Minkowski space-time

two events q and p, having coordinates q^μ and p^μ with respect to some inertial frame. If the linear combination $q + bp$ is defined in the obvious way as being the event having coordinates $(q + bp)^\mu = q^\mu + bp^\mu$, then under a Poincaré transformation (9.3)

$$q'^{\mu'} + bp'^{\mu'} = L^{\mu'}_\nu(q^\nu + bp^\nu) + (1+b)a^{\mu'} \neq L^{\mu'}_\nu(q + bp)^\nu + a^{\mu'}.$$

In particular, the origin $q^\mu = 0$ of Minkowski space has no invariant meaning since it is transformed to a non-zero point under a general Poincaré transformation. The difference of any pair of points, $q^\mu - p^\mu$, does however always undergo a linear transformation

$$q'^{\mu'} - p'^{\mu'} = L^{\mu'}_\nu(q^\nu - p^\nu)$$

and can be made to form a genuine vector space. Loosely speaking, a structure in which *differences* of points are defined and form a vector space is termed an *affine space*.

More precisely, we define an **affine space** to be a pair (M, V) consisting of a set M and a vector space V, such that V acts freely and transitively on M as an abelian group of transformations. The operation of V on M is written $+ : M \times V \to M$, and is required to satisfy

$$p + (u + v) = (p + u) + v, \qquad p + 0 = p$$

for all $p \in M$. There is then no ambiguity in writing expressions such as $p + u + v$. Recall from Section 2.6 that a *free* action means that if $p + u = p$ ($p \in M$) then $u = 0$, while the action is *transitive* if for any pair of points $p, q \in M$ there exists a vector $u \in V$ such that $q = p + u$. The vector u in this equation is necessarily unique, for if $q = p + u = p + u'$ then $p + u - u' = p$, and since the action is free it follows that $u = u'$.

Let p_0 be a fixed point of M. For any point $p \in M$ let $x(p) \in V$ be the unique vector such that $p = p_0 + x(p)$. This establishes a one-to-one correspondence between the underlying set M of an affine space and the vector space V acting on it. If e_i is any basis for V then the real functions $p \mapsto x^i(p)$ where $x(p) = x^i(p)e_i$ are said to be **coordinates on** M determined by the basis e_i and the **origin** p_0.

As anticipated above, in an affine space it is always possible to define the difference of any pair of points $q - p$. Given a fixed point $p_0 \in M$ let $x(p)$ and $x(q)$ be the unique vectors in V such that $p = p_0 + x(p)$ and $q = p_0 + x(q)$, and define the **difference** of two points of M to be the vector $q - p = x(q) - x(p) \in V$. This definition is independent of the choice of fixed point p_0, for if p'_0 is a second fixed point such that $p_0 = p'_0 + v$ then

$$p = p'_0 + v + x(p) = p'_0 + x'(p),$$
$$q = p'_0 + v + x(q) = p'_0 + x'(q),$$

and

$$x'(q) - x'(p) = v + x(q) - v - x(p) = x(q) - x(p) = q - p.$$

Minkowski space and 4-tensors

Minkowski space can now be defined as an affine space (M, V) where V is a four-dimensional Minkowskian vector space having metric tensor g, acting freely and transitively

Special relativity

on the set M. If $\{e_1, e_2, e_3, e_4\}$ is an orthonormal basis of V such that

$$g_{\mu\nu} = g(e_\mu, e_\nu) = \begin{cases} 1 & \text{if } \mu = \nu < 4 \\ -1 & \text{if } \mu = \nu = 4 \\ 0 & \text{if } \mu \neq \nu \end{cases}$$

we say an **inertial frame** is a choice of fixed point $p_0 \in M$, called the **origin**, together with the coordinates $x^\mu(p)$ on M defined by

$$p = p_0 + x^\mu(p)e_\mu \quad (p \in M).$$

The **interval** between any two events q and p in M is defined by

$$\Delta s^2 = g(p - q, p - q).$$

This is independent of the choice of fixed point p_0 or orthonormal frame e_μ, since it depends only on the vector difference between p and q and the metric tensor g. In an inertial frame the interval may be expressed in terms of coordinates

$$\Delta s^2 = g_{\mu\nu}\Delta x^\mu \Delta x^\nu \quad \text{where} \quad \Delta x^\mu = x^\mu(q) - x^\mu(p) = x^\mu(q - p). \tag{9.6}$$

Under a Lorentz transformation $e_\nu = L^{\mu'}_\nu e'_{\mu'}$ and a change of origin $p_0 = p'_0 + a^{\mu'} e'_{\mu'}$ we have for an arbitrary point p

$$p = p_0 + x^\nu(p)e_\nu$$
$$= p'_0 + a^{\mu'} e'_{\mu'} + x^\nu(q) L^{\mu'}_\nu e'_{\mu'}$$
$$= p'_0 + x'^{\mu'}(p)e'_{\mu'}$$

where $x'^{\mu'}(p)$ is given by the Poincaré transformation

$$x'^{\mu'}(p) = L^{\mu'}_\nu x^\nu(p) + a^{\mu'}. \tag{9.7}$$

It is a simple matter to verify that the coordinate expression (9.6) for Δs^2 is invariant with respect to Poincaré transformations (9.7).

Elements $v = v^\mu e_\mu$ of V will be termed **4-vectors**. With respect to a Poincaré transformation (9.7) the components v^μ transform as

$$v'^{\mu'} = L^{\mu'}_\nu v^\nu,$$

where $L^{\mu'}_\nu$ satisfy (9.4). The inverse transformations are

$$v^\nu = L'^\nu_{\rho'} v'^{\rho'} \quad \text{where} \quad L'^\nu_{\rho'} L^{\rho'}_\mu = \delta^\nu_\mu.$$

Elements of $V(r, s)$, defined in Chapter 7, are termed **4-tensors of type** (r, s). Since we restrict attention to orthonormal bases of V, the components $T^{\sigma\tau\ldots}{}_{\mu\nu\ldots}$ of a 4-tensor are only required to transform as a tensor with respect to the Lorentz transformations,

$$T'^{\sigma'\rho'\ldots}{}_{\mu'\nu'\ldots} = T^{\sigma\rho\ldots}{}_{\mu\nu\ldots} L'^{\sigma'}_\sigma L'^{\rho'}_\rho \ldots L^\mu_{\mu'} L^\nu_{\nu'} \ldots$$

4-tensors of type $(0, 1)$ are called **4-covectors**, and 4-tensors of type $(0, 0)$ will be termed **4-scalars** or simply **scalars**. The important thing about 4-tensors, as for general tensors,

9.1 Minkowski space-time

is that if a 4-tensor equation can be shown to hold in one particular frame it holds in all frames. This is an immediate consequence of the homogeneous transformation law of components.

By Eq. (9.4) g is a covariant 4-tensor of rank 2 since its components $g_{\mu\nu}$ transform as

$$g'_{\mu'\nu'} = g_{\mu\nu} L'^{\mu}{}_{\mu'} L'^{\nu}{}_{\nu'}$$

where $g'_{\mu'\nu'} = g_{\mu'\nu'}$. The inverse metric $g^{\mu\nu}$, defined by

$$g^{\mu\rho} g_{\rho\nu} = \delta^{\mu}_{\nu},$$

has identical components to $g_{\mu\nu}$ and is a contravariant tensor of rank 2,

$$g'^{\mu'\nu'} = g^{\mu\nu} L^{\mu'}{}_{\mu} L^{\nu'}{}_{\nu}, \qquad g^{\mu\nu} = g'^{\mu'\nu'} L'^{\mu}{}_{\mu'} L'^{\nu}{}_{\nu'}.$$

We will use $g^{\mu\nu}$ and $g_{\mu\nu}$ to raise and lower indices of 4-tensors; for example,

$$U^{\mu} = g^{\mu\nu} U_{\nu}, \qquad W_{\mu} = g_{\mu\nu} W^{\nu}, \qquad T_{\mu\nu}{}^{\rho} = g_{\mu\alpha} g^{\rho\beta} T^{\alpha}{}_{\nu\beta}.$$

Given two 4-vectors $A = A^{\mu} e_{\mu}$, $B = B^{\nu} e_{\nu}$, define their **inner product** to be the scalar

$$g(A, B) = A^{\mu} B_{\mu} = g_{\mu\nu} A^{\mu} B^{\nu} = A_{\mu} B^{\mu} = g^{\mu\nu} A_{\mu} B_{\nu}.$$

We say the vectors are **orthogonal** if $A^{\mu} B_{\mu} = 0$. The **magnitude** of a 4-vector A^{μ} is defined to be $g(A, A) = A^{\mu} A_{\mu}$. A non-zero 4-vector A^{μ} is called

spacelike if $g(A, A) = A^{\mu} A_{\mu} > 0$,
timelike if $g(A, A) = A^{\mu} A_{\mu} < 0$,
null if $g(A, A) = A^{\mu} A_{\mu} = 0$.

The set of all null 4-vectors is called the **null cone**. This is a subset of the vector space V of 4-vectors. The concept of a **light cone at** $p \in M$, defined in Section 2.30, is the set of points of M that are connected to p by a null vector, $C_p = \{q \mid g(q - p, q - p) = 0\} \subset M$. Figure 9.1 shows how the null cone separates 4-vectors into the various classes. Timelike or null vectors falling within or on the upper half of the null cone are called **future-pointing**, while those in the lower half are **past-pointing**.

Spacelike vectors, however, lie outside the null cone and form a continuously connected region of V, making it impossible to define invariantly the concept of a future-pointing or past-pointing spacelike vector – see Problem 9.2.

Problems

Problem 9.1 Show that

$$[L^{\mu'}{}_{\nu}] = \begin{pmatrix} 1 & 0 & -\alpha & \alpha \\ 0 & 1 & -\beta & \beta \\ \alpha & \beta & 1-\gamma & \gamma \\ \alpha & \beta & -\gamma & 1+\gamma \end{pmatrix} \qquad \text{where} \quad \gamma = \tfrac{1}{2}(\alpha^2 + \beta^2)$$

Special relativity

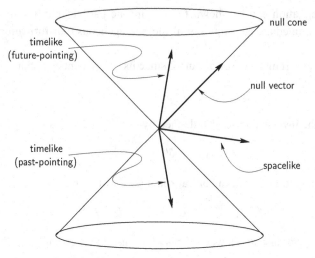

Figure 9.1 The null cone in Minkowski space

is a Lorentz transformation for all values of α and β. Find those 4-vectors V^μ whose components are unchanged by all Lorentz transformations of this form.

Problem 9.2 Show that for *any* Lorentz transformation $L^{\mu'}_\nu$ one must have either

$$L^4_{\;4} \geq 1 \quad \text{or} \quad L^4_{\;4} \leq -1.$$

(a) Show that those transformations having $L^4_{\;4} \geq 1$ have the property that they preserve the concept of 'before' and 'after' for timelike separated events by demonstrating that they preserve the sign of Δx^4.
(b) What is the effect of a Lorentz transformation having $L^4_{\;4} \leq -1$?
(c) Is there any meaning, independent of the inertial frame, to the concepts of 'before' and 'after' for spacelike separated events?

Problem 9.3 Show that (i) if T^α is a timelike 4-vector it is always possible to find a Lorentz transformation such that $T'^{\alpha'}$ will have components $(0, 0, 0, a)$ and (ii) if N^α is a null vector then it is always possible to find a Lorentz transformation such that $N'^{\alpha'}$ has components $(0, 0, 1, 1)$.
   Let U^α and V^α be 4-vectors. Show the following:

(a) If $U^\alpha V_\alpha = 0$ and U^α is timelike, then V^α is spacelike.
(b) If $U^\alpha V_\alpha = 0$ and U^α and V^α are both null vectors, then they are proportional to each other.
(c) If U^α and V^α are both timelike future-pointing then $U_\alpha V^\alpha < 0$ and $U^\alpha + V^\alpha$ is timelike.
(d) Find other statements similar to the previous assertions when U^α and V^α are taken to be various combinations of null, future-pointing null, timelike future-pointing, spacelike, etc.

Problem 9.4 If the 4-component of a 4-vector equation $A^4 = B^4$ is shown to hold in all inertial frames, show that all components are equal in all frames, $A^\mu = B^\mu$.

9.2 Relativistic kinematics

Special Lorentz transformations

From time to time we will call upon specific types of Lorentz transformations. The following two examples present the most commonly used types.

Example 9.1 Time-preserving Lorentz transformations have $t' = t$, or equivalently $x'^4 = x^4$. Such transformations have $L_4^4 = 1$, $L_i^4 = 0$ for $i = 1, 2, 3$, and substituting in Eq. (9.4) with $\rho = \sigma = 4$ gives

$$\sum_{i'=1}^{3}(L_4^{i'})^2 - (L_4^4)^2 = -1 \implies \sum_{i'=1}^{3}(L_4^{i'})^2 = 0.$$

This can only hold if $L_4^{i'} = 0$ for $i' = 1, 2, 3$. Hence

$$\mathsf{L} = [L_\nu^{\mu'}] = \begin{pmatrix} & & & 0 \\ & [a_{ij}] & & 0 \\ & & & 0 \\ 0 & 0 & 0 & 1 \end{pmatrix} \quad (i, j, \cdots = 1, 2, 3) \tag{9.8}$$

where $\mathsf{A} = [a_{ij}]$ is an orthogonal 3×3 matrix, $\mathsf{A}^T \mathsf{A} = \mathsf{I}$, which follows on substituting L in Eq. (9.5). If $\det \mathsf{L} = \det \mathsf{A} = +1$ these transformations are spatial rotations, while if $\det \mathsf{A} = -1$ they are space reflections.

Example 9.2 Lorentz transformations that leave the y and z coordinates unchanged are of the form

$$\mathsf{L} = [L_\nu^{\mu'}] = \begin{pmatrix} L_1^1 & 0 & 0 & L_4^1 \\ 0 & 1 & 0 & 0 \\ 0 & 0 & 1 & 0 \\ L_1^4 & 0 & 0 & L_4^4 \end{pmatrix}.$$

Substituting in Eq. (9.4) gives

$$L_1^1 L_1^1 - L_1^4 L_1^4 = g_{11} = 1, \tag{9.9}$$
$$L_1^1 L_4^1 - L_1^4 L_4^4 = g_{14} = g_{41} = 0, \tag{9.10}$$
$$L_4^1 L_4^1 - L_4^4 L_4^4 = g_{44} = -1. \tag{9.11}$$

From (9.11), we have $(L_4^4)^2 = 1 + \sum_{i=1}^{3}(L_4^i)^2 \geq 1$, and assuming $L_4^4 \geq 1$ it is possible to set $L_4^4 = \cosh \alpha$ for some real number α. Then $L_4^1 = \pm\sqrt{\cosh^2 \alpha - 1} = \sinh \alpha$ on choosing α with the appropriate sign. Similarly, (9.9) implies that $L_1^1 = \cosh \beta$, $L_1^4 = \sinh \beta$ and (9.10) gives

$$0 = \sinh \alpha \cosh \beta - \cosh \alpha \sinh \beta = \sinh(\alpha - \beta) \implies \alpha = \beta.$$

Let v be the unique real number defined by

$$\tanh \alpha = -\frac{v}{c},$$

Special relativity

then trigonometric identities give that $|v| < c$ and

$$\cosh\alpha = \gamma, \qquad \sinh\alpha = -\gamma\frac{v}{c}$$

where

$$\gamma = \frac{1}{\sqrt{1 - v^2/c^2}}. \tag{9.12}$$

The resulting Lorentz transformations have the form

$$\mathsf{L} = [L^{\mu'}_{\nu}] = \begin{pmatrix} \gamma & 0 & 0 & -\gamma\frac{v}{c} \\ 0 & 1 & 0 & 0 \\ 0 & 0 & 1 & 0 \\ -\gamma\frac{v}{c} & 0 & 0 & \gamma \end{pmatrix}, \tag{9.13}$$

and are known as **boosts** with velocity v in the x-direction. Written out explicitly in x, y, z, t coordinates they read

$$x' = \gamma(x - vt), \qquad y' = y, \qquad z' = z, \qquad t' = \gamma\left(t - \frac{v}{c^2}x\right). \tag{9.14}$$

The inverse transformation is obtained on replacing v by $-v$

$$x = \gamma(x' + vt'), \qquad y = y', \qquad z = z', \qquad t = \gamma\left(t' + \frac{v}{c^2}x'\right). \tag{9.15}$$

The parameter v plays the role of a relative velocity between the two frames since the spatial origin ($x' = 0, y' = 0, z' = 0$) in the primed frame satisfies the equation $x = vt$ in the unprimed frame. As the relative velocity v must always be less than c we have the first indication that according to relativity theory, the velocity of light c is a limiting velocity for material particles.

Exercise: Verify that performing two Lorentz transformations with velocities v_1 and v_2 in the x-directions in succession is equivalent to a single Lorentz transformation with velocity

$$v = \frac{v_1 + v_2}{1 + v_1 v_2/c^2}.$$

Relativity of time, length and velocity

Two events $p = (x_1, y_1, z_1, ct_1)$ and $q = (x_2, y_2, z_2, ct_2)$ are called *simultaneous* with respect to an inertial frame K if $\Delta t = t_2 - t_1 = 0$. Consider a second frame K' related to K by a boost, (9.14). These equations are linear and therefore apply to coordinate differences,

$$\Delta x' = \gamma(\Delta x - v\Delta t), \qquad \Delta t' = \gamma\left(\Delta t - \frac{v}{c^2}\Delta x\right).$$

Hence,

$$\Delta t = 0 \implies \Delta t' = -\gamma\frac{v}{c^2}\Delta x \ne 0 \text{ if } x_1 \ne x_2,$$

demonstrating the effect known as **relativity of simultaneity**: simultaneity of spatially separated points is not an absolute concept.

9.2 Relativistic kinematics

Consider now a clock at rest in K' marking off successive 'ticks' at events (x', y', z', ct'_1) and (x', y', z', ct'_2). The time difference according to K is given by (9.15),

$$\Delta t = \gamma\left(\Delta t' + \frac{v}{c^2}\Delta x'\right) = \gamma \Delta t' \quad \text{if} \quad \Delta x' = 0.$$

That is,

$$\Delta t = \frac{\Delta t'}{\sqrt{1 - \frac{v^2}{c^2}}} \geq \Delta t', \tag{9.16}$$

an effect known as **time dilatation** – a moving clock appears to slow down. Equivalently, a stationary clock in K appears to run slow according to the moving observer K'.

Now consider a rod of length $\ell = \Delta x$ at rest in K. Again, using the inverse boost transformation (9.15) we have

$$\ell = \Delta x = \gamma(\Delta x' + v\Delta t') = \gamma \Delta x' \quad \text{if} \quad \Delta t' = 0.$$

The rod's length with respect to K' is determined by considering simultaneous moments $t'_1 = t'_2$ at the end points,

$$\ell' = \Delta x' = \frac{\ell}{\gamma} = \sqrt{1 - \frac{v^2}{c^2}}\,\ell \leq \ell. \tag{9.17}$$

The common interpretation of this result is that the length of a rod is contracted when viewed by a moving observer, an effect known as the **Lorentz–Fitzgerald contraction**. By reversing the roles of K and K' it is similarly found that a moving rod is contracted in the direction of its motion. The key to this effect is that, by the relativity of simultaneity, pairs of events on the histories of the ends of the rod that are simultaneous with respect to K' differ from simultaneous pairs in the frame K. Since there is no contraction perpendicular to the motion, a moving volume V will undergo a contraction

$$V' = \sqrt{1 - \frac{v^2}{c^2}}\,V. \tag{9.18}$$

This is the most useful application of the Lorentz–Fitzgerald contraction.

Exercise: Give the reverse arguments to the above; that a clock at rest runs slow relative to a moving observer, and that a moving rod appears contracted.

Let a particle have velocity $\mathbf{u} = (u_x, u_y, u_z)$ with respect to K, and $\mathbf{u}' = (u'_x, u'_y, u'_z)$ with respect to K'. Setting

$$u_x = \frac{dx}{dt}, \quad u'_x = \frac{dx'}{dt'}, \quad u_y = \frac{dy}{dt}, \quad \text{etc.}$$

and using the Lorentz transformations (9.14), we have

$$u'_x = \frac{u_x - v}{1 - u_x v/c^2}, \quad u'_y = \frac{u_y}{\gamma(1 - u_x v/c^2)}, \quad u'_z = \frac{u_z}{\gamma(1 - u_x v/c^2)}. \tag{9.19}$$

Comparing with the Newtonian discussion at the beginning of this chapter it is natural to call this the **relativistic law of transformation of velocities**. Similarly on using the

Special relativity

inverse Lorentz transformations (9.15), we arrive at the **relativistic law of addition of velocities**:

$$u_x = \frac{u'_x + v}{1 + u'_x v/c^2}, \quad u_y = \frac{u'_y}{\gamma(1 + u'_x v/c^2)}, \quad u_z = \frac{u'_z}{\gamma(1 + u'_x v/c^2)}. \quad (9.20)$$

The same result can be obtained from (9.19) by replacing v by $-v$ and interchanging primed and unprimed velocities.

For a particle moving in the x–y plane set $u_x = u \cos\theta$, $u_y = u \sin\theta$, $u_z = 0$ and $u'_x = u' \cos\theta'$, $u'_y = u' \sin\theta'$, $u'_z = 0$. If $u' = c$ it follows from Eq. (9.20) that $u = c$, and the velocity of light is independent of the motion of the observer as required by Einstein's principle of relativity. The second equation of (9.20) gives a relation between the θ and θ', the angles the light beam subtends with the x- and x'-directions respectively:

$$\sin\theta = \frac{\sin\theta'}{1 + (v/c)\cos\theta'}\sqrt{1 - \frac{v^2}{c^2}}. \quad (9.21)$$

This formula is known as the **relativistic aberration of light**. If $\dfrac{v}{c} \ll 1$ then

$$\delta\theta = \theta - \theta' \approx -\frac{v}{c}\sin\theta',$$

a Newtonian formula for aberration of light, which follows simply from the triangle addition law of velocities and was used by the astronomer Bradley nearly 300 years ago to estimate the velocity of light.

Problems

Problem 9.5 From the law of transformation of velocities, Eq. (9.19), show that the velocity of light in an arbitrary direction is invariant under boosts.

Problem 9.6 If two intersecting light beams appear to be making a non-zero angle ϕ in one frame K, show that there always exists a frame K' whose motion relative to K is in the plane of the beams such that the beams appear to be directed in opposite directions.

Problem 9.7 A source of light emits photons uniformly in all directions in its own rest frame.

(a) If the source moves with velocity v with respect to an inertial frame K, show the 'headlight effect': half the photons seem to be emitted in a forward cone whose semi-angle is given by $\cos\theta = v/c$.
(b) In films of the *Star Wars* genre, star fields are usually seen to be swept backwards around a rocket as it accelerates towards the speed of light. What would such a rocketeer really see as his velocity $v \to c$?

Problem 9.8 If two separate events occur at the same time in some inertial frame S, prove that there is no limit on the time separations assigned to these events in other frames, but that their space separation varies from infinity to a minimum that is measured in S. With what speed must an observer travel in order that two simultaneous events at opposite ends of a 10-metre room appear to differ in time by 100 years?

9.3 Particle dynamics

Problem 9.9 A supernova is seen to explode on Andromeda galaxy, while it is on the western horizon. Observers A and B are walking past each other, A at 5 km/h towards the east, B at 5 km/h towards the west. Given that Andromeda is about a million light years away, calculate the difference in time attributed to the supernova event by A and B. Who says it happened earlier?

Problem 9.10 Twin A on the Earth and twin B who is in a rocketship moving away from him at a speed of $\frac{1}{2}c$ separate from each other at midday on their common birthday. They decide to each blow out candles exactly four years from B's departure.

(a) What moment in B's time corresponds to the event P that consists of A blowing his candle out? And what moment in A's time corresponds to the event Q that consists of B blowing her candle out?
(b) According to A which happened earlier, P or Q? And according to B?
(c) How long will A have to wait before he *sees* his twin blowing her candle out?

9.3 Particle dynamics

World-lines and proper time

Let $I = [a, b]$ be any closed interval of the real line $\mathbb{R}$. A continuous map $\sigma : I \to M$ is called a **parametrized curve** in Minkowski space (M, V). In an inertial frame generated by a basis e_μ of V such a curve may be written as four real functions $x^\mu \circ \sigma : I \to \mathbb{R}$. We frequently write these functions as $x^\mu(\lambda)$ ($a \le \lambda \le b$) in place of $x^\mu(\sigma(\lambda))$, and generally assume them to be differentiable.

If the parametrized curve σ passes through the event p having coordinates p^μ, so that $p^\mu = x^\mu(\lambda_0) \equiv x^\mu(\sigma(\lambda_0))$ for some $\lambda_0 \in I$, define the **tangent 4-vector** to the curve at p to be the 4-vector U given by

$$U = U^\mu e_\mu \in V \quad \text{where} \quad U^\mu = \left.\frac{dx^\mu}{d\lambda}\right|_{\lambda = \lambda_0}.$$

This definition is independent of the choice of orthonormal basis e_μ on V, for if $e'_{\mu'}$ is a second o.n. basis related by a Lorentz transformation $e_\nu = L^{\mu'}_\nu e'_{\mu'}$, then

$$U' = U'^{\mu'} e'_{\mu'} = \left.\frac{dx'^{\mu'}(\lambda)}{d\lambda}\right|_{\lambda=\lambda_0} L^\mu_{\mu'} e_\mu = \left.\frac{dx^\mu}{d\lambda}\right|_{\lambda=\lambda_0} e_\mu = U.$$

The parametrized curve σ is called **timelike, spacelike** or **null** at p if its tangent 4-vector at p is timelike, spacelike or null, respectively. The path of a material particle will be assumed to be timelike at all events through which it passes, and is frequently referred to as the particle's **world-line** (see Fig. 9.2). This assumption amounts to the requirement that the particle's velocity is always less than c, for

$$0 > g(U(\lambda), U(\lambda)) = g_{\mu\nu} \frac{dx^\mu(\lambda)}{d\lambda} \frac{dx^\nu(\lambda)}{d\lambda} = \left(\frac{dt}{d\lambda}\right)^2 \left(\sum_{i=1}^3 \left(\frac{dx^i(t)}{dt}\right)^2 - c^2 \right),$$

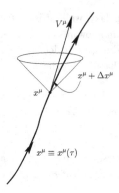

Figure 9.2   World-line of a material particle

on setting $t = x^4/c = t(\lambda)$. Hence

$$v^2 = \sum_{i=1}^{3}\left(\frac{dx^i}{dt}\right)^2 < c^2.$$

For two neighbouring events on the world-line, $x^\mu(\lambda)$ and $x^\mu(\lambda + \Delta\lambda)$, set

$$\Delta\tau^2 = -\frac{1}{c^2}\Delta s^2 = -\frac{1}{c^2}g_{\mu\nu}\Delta x^\mu \Delta x^\nu > 0,$$

where

$$\Delta x^\mu = x^\mu(\lambda + \Delta\lambda) - x^\mu(\lambda).$$

In the limit $\Delta\lambda \to 0$

$$\Delta\tau^2 \to -\frac{1}{c^2}g_{\mu\nu}\frac{dx^\mu(\lambda)}{d\lambda}\frac{dx^\nu(\lambda)}{d\lambda}(\Delta\lambda)^2 = -\frac{1}{c^2}(v^2 - c^2)\left(\frac{dt}{d\lambda}\right)^2\Delta\lambda^2.$$

Hence

$$\Delta\tau \to \sqrt{1 - \frac{v^2}{c^2}}\Delta t = \frac{1}{\gamma}\Delta t. \tag{9.22}$$

Since the velocity of the particle is everywhere less than c, the relativistic law of transformation of velocities (9.19) can be used to find a combination of rotation and boost, (9.8) and (9.14), which transforms the particle's velocity to zero at any given point $p = \sigma(\lambda_0)$ on the particle's path. Any such inertial frame in which the particle is momentarily at rest is known as an **instantaneous rest frame** or i.r.f. at p. The i.r.f will of course vary from point to point on a world-line, unless the velocity is constant along it. Since $\mathbf{v} = \mathbf{0}$ in an i.r.f. we have from (9.22) that $\Delta\tau/\Delta t \to 1$ as $\Delta t \to 0$. Thus $\Delta\tau$ measures the time interval registered on an inertial clock instantaneously comoving with the particle. It is generally interpreted as the time measured on a clock *carried* by the particle from x^μ to $x^\mu + \Delta x^\mu$.

9.3 Particle dynamics

The factor $1/\gamma$ in Eq. (9.22) represents the time dilatation effect of Eq. (9.16) on such a clock due to its motion relative to the external inertial frame. The total time measured on a clock carried by the particle from event p to event q is given by

$$\tau_{pq} = \int_p^q d\tau = \int_{t_p}^{t_q} \frac{dt}{\gamma},$$

and is called the **proper time** from p to q. If we fix the event p and let q vary along the curve then proper time can be used as a parameter along the curve,

$$\tau = \int_{t_p}^t \frac{dt}{\gamma} = \tau(t). \tag{9.23}$$

The tangent 4-vector $V = V^\mu e_\mu$ calculated with respect to this special parameter is called the **4-velocity** of the particle,

$$V^\mu = \frac{dx^\mu}{d\tau} = \gamma \frac{dx^\mu}{dt} = \gamma(\mathbf{v}, c). \tag{9.24}$$

Unlike coordinate time t, proper time τ is a true scalar parameter independent of inertial frame; hence the components of 4-velocity V^μ transform as a contravariant 4-vector

$$V'^{\mu'} = L^{\mu'}_{\nu} V^\nu.$$

From Eq. (9.24) the magnitude of the 4-velocity always has constant magnitude $-c^2$,

$$g(V, V) = V^\mu V_\mu = (v^2 - c^2)\gamma^2 = -c^2. \tag{9.25}$$

The **4-acceleration** of a particle is defined to be the contravariant 4-vector $A = A^\mu e_\mu$ with components

$$A^\mu = \frac{dV^\mu}{d\tau} = \frac{d^2 x^\mu(\tau)}{d\tau^2}. \tag{9.26}$$

Expressing these components in terms of the coordinate time parameter t gives

$$A^\mu = \gamma \left(\frac{d\gamma}{dt} \mathbf{v} + \gamma \frac{d\mathbf{v}}{dt}, c \frac{d\gamma}{dt} \right). \tag{9.27}$$

The 4-vectors A and V are orthogonal to each other since

$$\frac{d}{d\tau}(g(V, V)) = \frac{d}{d\tau}(-c^2) = 0,$$

and expanding the left-hand side gives $g(A, V) + g(V, A) = 2g(A, V)$, so that

$$g(A, V) = A^\mu V_\mu = 0. \tag{9.28}$$

Exercise: Show that in an i.r.f. the components of 4-velocity and 4-acceleration are given by

$$V^\mu = (\mathbf{0}, c), \qquad A^\mu = (\mathbf{a}, 0) \quad \text{where} \quad \mathbf{a} = \frac{d\mathbf{v}}{dt},$$

and verify that the 4-vectors A and V are orthogonal to each other.

Special relativity

Relativistic particle dynamics

We assume each particle has a constant scalar m attached to it, called its **rest mass**. This may be thought of as the Newtonian mass in an instantaneous rest frame of the particle, satisfying Newton's second law $\mathbf{F} = m\mathbf{a}$ for any imposed force $\mathbf{F}$ in that frame. The **4-momentum** of the particle is defined to be the 4-vector having components $P^\mu = mV^\mu$ where $V = V^\mu e_\mu$ is the 4-velocity of the particle,

$$P^\mu = \left(\mathbf{p}, \frac{E}{c}\right) \tag{9.29}$$

where

$$\mathbf{p} = m\gamma \mathbf{v} = \frac{m\mathbf{v}}{\sqrt{1 - v^2/c^2}} = \text{momentum}, \tag{9.30}$$

$$E = m\gamma c^2 = \frac{mc^2}{\sqrt{1 - v^2/c^2}} = \text{energy}. \tag{9.31}$$

For $v \ll c$ the momentum reduces to the Newtonian formula $\mathbf{p} = m\mathbf{v}$ and the energy can be written as $E \approx mc^2 + \frac{1}{2}mv^2 + \ldots$ The energy contribution $E = mc^2$, which arises even when the particle is at rest, is called the particle's **rest-energy**.

Exercise: Show the following identities:

$$g(P, P) = P^\mu P_\mu = -m^2 c^2, \quad E = \sqrt{p^2 c^2 + m^2 c^4}, \quad \mathbf{p} = \frac{E\mathbf{v}}{c^2}. \tag{9.32}$$

The relations (9.32) make sense even in the limit $v \to c$ provided the particle has zero rest mass, $m = 0$. Such particles will be termed **photons**, and satisfy the relations

$$E = pc, \quad \mathbf{p} = \frac{E}{c}\mathbf{n} \quad \text{where} \quad \mathbf{n} \cdot \mathbf{n} = 1. \tag{9.33}$$

Here $\mathbf{n}$ is called the **direction of propagation** of the photon. The 4-momentum of a photon has the form

$$P^\mu = \left(\mathbf{p}, \frac{E}{c}\right) = \frac{E}{c}(\mathbf{n}, 1),$$

and is clearly a null vector, $P^\mu P_\mu = 0$.

In analogy with Newton's law $\mathbf{F} = m\mathbf{a}$, it is sometimes useful to define a **4-force** $F = F^\mu e_\mu$ having components

$$F^\mu = \frac{dP^\mu}{d\tau} = mA^\mu. \tag{9.34}$$

By Eq. (9.28) the 4-force is always orthogonal to the 4-velocity. Defining **3-force** $\mathbf{f}$ in the usual way by

$$\mathbf{f} = \frac{d\mathbf{p}}{dt}$$

9.3 Particle dynamics

and using $\frac{d}{d\tau} = \gamma \frac{d}{dt}$ we obtain

$$F^\mu = \gamma \left(\mathbf{f}, \frac{1}{c}\frac{dE}{dt} \right). \tag{9.35}$$

Problems

Problem 9.11 Using the fact that the 4-velocity $V^\mu = \gamma(u)(u_x, u_y, u_z, c)$ transforms as a 4-vector, show from the transformation equation for V'^4 that the transformation of u under boosts is

$$\frac{\gamma(u')}{\gamma(u)} = \gamma(v)\left(1 - \frac{vu_x}{c^2}\right).$$

From the remaining transformation equations for $V'^{i'}$ derive the law of transformation of velocities (9.19).

Problem 9.12 Let K' be a frame with velocity v relative to K in the x-direction.

(a) Show that for a particle having velocity u', acceleration a' in the x'-direction relative to K', its acceleration in K is

$$a = \frac{a'}{[\gamma(1 + vu'/c^2)]^3}.$$

(b) A rocketeer leaves Earth at $t = 0$ with constant acceleration g at every moment relative to his instantaneous rest frame. Show that his motion relative to the Earth is given by

$$x = \frac{c^2}{g}\left(\sqrt{1 + \frac{g^2}{c^2}t^2} - 1\right).$$

(c) In terms of his own proper time τ show that

$$x = \frac{c^2}{g}\left(\cosh\frac{g}{c}\tau - 1\right).$$

(d) If he proceeds for 10 years of his life, decelerates with $g = 9.80$ m s^{-2} for another 10 years to come to rest, and returns in the same way, taking 40 years in all, how much will people on Earth have aged on his return? How far, in light years, will he have gone from Earth?

Problem 9.13 A particle is in hyperbolic motion along a world-line whose equation is given by

$$x^2 - c^2t^2 = a^2, \qquad y = z = 0.$$

Show that

$$\gamma = \frac{\sqrt{a^2 + c^2 t^2}}{a}$$

and that the proper time starting from $t = 0$ along the path is given by

$$\tau = \frac{a}{c}\cosh^{-1}\frac{ct}{a}.$$

Evaluate the particle's 4-velocity V^μ and 4-acceleration A^μ. Show that A^μ has constant magnitude.

Special relativity

Problem 9.14 For a system of particles it is generally assumed that the **conservation of total 4-momentum** holds in any localized interaction,

$$\sum_a P^\mu_{(a)} = \sum_b Q^\mu_{(b)}.$$

Use Problem 9.4 to show that the law of conservation of 4-momentum holds for a given system provided the law of energy conservation holds in all inertial frames. Also show that the law of conservation of momentum in all frames is sufficient to guarantee conservation of 4-momentum.

Problem 9.15 A particle has momentum **p**, energy E in a frame K.

(a) If K' is an inertial frame having velocity **v** relative to K, use the transformation law of the momentum 4-vector $P^\mu = \left(\mathbf{p}, \dfrac{E}{c}\right)$ to show that

$$E' = \gamma(E - \mathbf{v}\cdot\mathbf{p}), \qquad \mathbf{p}'_\perp = \mathbf{p}_\perp \quad\text{and}\quad \mathbf{p}'_\| = \gamma\left(\mathbf{p}_\| - \dfrac{E}{c^2}\mathbf{v}\right),$$

where $\mathbf{p}_\perp$ and $\mathbf{p}_\|$ are the components of $\mathbf{p}$ respectively perpendicular and parallel to **v**.

(b) If the particle is a photon, use these transformations to derive the aberration formula

$$\cos\theta' = \dfrac{\cos\theta - v/c}{1 - \cos\theta\,(v/c)}$$

where θ is the angle between **p** and **v**.

Problem 9.16 Use $F^\mu V_\mu = 0$ to show that

$$\mathbf{f}\cdot\mathbf{v} = \dfrac{dE}{dt}.$$

Also show this directly from the definitions (9.30) and (9.31) of **p**, E.

9.4 Electrodynamics

4-Tensor fields

A **4-tensor field** of type (r, s) consists of a map $T : M \to V^{(r,s)}$. We can think of this as a 4-tensor assigned at each point of space-time. The components of a 4-tensor field are functions of space-time coordinates

$$T^{\mu\nu\ldots}{}_{\rho\sigma\ldots} = T^{\mu\nu\ldots}{}_{\rho\sigma\ldots}(x^\alpha).$$

Define the **gradient** of a 4-tensor field T to be the 4-tensor field of type $(r, s+1)$, having components

$$T^{\mu\nu\ldots}{}_{\rho\sigma\ldots,\tau} = \dfrac{\partial}{\partial x^\tau} T^{\mu\nu\ldots}{}_{\rho\sigma\ldots}.$$

9.4 Electrodynamics

This is a 4-tensor field since a Poincaré transformation (9.7) induces the transformation

$$T^{\mu'\cdots}{}_{\rho'\ldots,\tau'} = \frac{\partial}{\partial x'^{\tau'}}\left(T^{\alpha\cdots}{}_{\beta\ldots}L^{\mu'}_{\alpha}\ldots L'^{\beta}{}_{\rho'}\ldots\right)$$

$$= \frac{\partial x^{\gamma}}{\partial x'^{\tau'}}\frac{\partial}{\partial x^{\gamma}}\left(T^{\alpha\cdots}{}_{\beta\ldots}L^{\mu'}_{\alpha}\ldots L'^{\beta}{}_{\rho'}\ldots\right)$$

$$= T^{\alpha\cdots}{}_{\beta\ldots,\gamma}L'^{\gamma}{}_{\tau'}L^{\mu'}_{\alpha}\ldots L'^{\beta}{}_{\rho'}\ldots$$

For example, if $f : M \to \mathbb{R}$ is a scalar field, its gradient is a 4-covector field,

$$f_{,\mu} = \frac{\partial f(x^{\alpha})}{\partial x^{\mu}}.$$

Example 9.3 A 4-vector field, $J = J^{\mu}(x^{\alpha})e_{\mu}$, is said to be **divergence-free** if

$$J^{\mu}{}_{,\mu} = 0.$$

Setting $j_i = J^i$ ($i = 1, 2, 3$) and $\rho = \frac{1}{c}J^4$, the divergence-free condition reads

$$\frac{\partial \rho}{\partial t} + \nabla \cdot \mathbf{j} = 0, \tag{9.36}$$

known both in hydrodynamics and electromagnetism as the **equation of continuity**. Interpreting ρ as the *charge density* or charge per unit volume, $\mathbf{j}$ is the *current density*. The charge per unit time crossing unit area normal to the unit vector $\mathbf{n}$ is given by $\mathbf{j} \cdot \mathbf{n}$. Equation (9.36) implies *conservation of charge* – the rate of increase of charge in a volume $\mathcal{V}$ equals the flux of charge entering through the boundary surface $\mathcal{S}$:

$$\frac{dq}{dt} = \int_{\mathcal{V}} \frac{\partial \rho}{\partial t}\,dV = -\int_{\mathcal{V}} \nabla \cdot \mathbf{j}\,dV = -\int_{\mathcal{S}} \mathbf{j} \cdot d\mathbf{S}.$$

Electromagnetism

As in Example 9.3, let there be a continuous distribution of electric charge present in Minkowski space-time, having **charge density** $\rho(\mathbf{r}, t)$ and **charge flux density** or **current density** $\mathbf{j} = \rho \mathbf{v}$, where $\mathbf{v}(\mathbf{r}, t)$ is the velocity field of the fluid. The total charge of a system is a scalar quantity – else an unionized gase would not generally be electrically neutral. Charge density in a local instantaneous rest frame of the fluid at any event p is denoted $\rho_0(p)$ and is known as **proper charge density**. It may be assumed to be a scalar quantity, since it is defined in a specific inertial frame at p. On the other hand, the charge density ρ is given by

$$\rho = \lim_{\Delta V \to 0} \frac{\Delta q}{\Delta V}$$

where, by the length–volume contraction effect (9.18),

$$\Delta V = \frac{1}{\gamma}\Delta V_0.$$

Special relativity

Since charge is a scalar quantity, $\Delta q = \Delta q_0$, charge density and proper charge density are related by

$$\rho = \lim_{\Delta V_0 \to 0} \frac{\Delta q_0}{V_0/\gamma} = \gamma \rho_0.$$

If the charged fluid has a 4-velocity field $V = V^\mu(x^\alpha)e_\mu$, define the **4-current** J to be the 4-vector field having components

$$J^\mu = \rho_0 V^\mu.$$

From Eq. (9.24) together with the above we have

$$J^\mu = (\mathbf{j}, \rho c),$$

and by Example 9.3, conservation of charge is equivalent to requiring the 4-current be divergence-free,

$$J^\mu_{,\mu} = 0 \iff \nabla \cdot \mathbf{j} + \frac{\partial \rho}{\partial t} = 0.$$

In **electrodynamics** we are given a 4-current field $J = J^\mu e_\mu$ representing the charge density and current of the electric charges present, also known as the **source field**, and an antisymmetric 4-tensor field $F = F_{\mu\nu}(x^\alpha)\varepsilon^\mu \otimes \varepsilon^\nu$ such that $F_{\mu\nu} = -F_{\nu\mu}$, known as the **electromagnetc field**, satisfying the **Maxwell equations**:

$$F_{\mu\nu,\rho} + F_{\nu\rho,\mu} + F_{\rho\mu,\nu} = 0, \tag{9.37}$$

$$F^{\mu\nu}{}_{,\nu} = \frac{4\pi}{c} J^\mu, \tag{9.38}$$

where $F^{\mu\nu} = g^{\mu\alpha} g^{\nu\beta} F_{\alpha\beta}$. Units adopted here are the Gaussian units, which are convenient for the formal presentation of the subject.

The first set (9.37) is known as the **source-free Maxwell equations**, while the second set (9.38) relates electromagnetic field and sources. It is common to give explicit symbols for the components of the electromagnetic field tensor

$$F_{\mu\nu} = \begin{pmatrix} 0 & B_3 & -B_2 & E_1 \\ -B_3 & 0 & B_1 & E_2 \\ B_2 & -B_1 & 0 & E_3 \\ -E_1 & -E_2 & -E_3 & 0 \end{pmatrix}, \quad \text{i.e. set } F_{12} = B_3, \text{ etc.} \tag{9.39}$$

The 3-vector fields $\mathbf{E} = (E_1, E_2, E_3)$ and $\mathbf{B} = (B_1, B_2, B_3)$ are called the **electric** and **magnetic** fields, respectively. The source-free Maxwell equations (9.37) give non-trivial equations only when all three indices μ, ν and ρ are unequal, giving four independent equations

$$(\mu, \nu, \rho) = (1, 2, 3) \implies \nabla \cdot \mathbf{B} = 0, \tag{9.40}$$

$$(\mu, \nu, \rho) = (2, 3, 4), \text{ etc.} \implies \nabla \times \mathbf{E} + \frac{1}{c}\frac{\partial \mathbf{B}}{\partial t} = 0. \tag{9.41}$$

9.4 Electrodynamics

The second set of Maxwell equations (9.38) imply charge conservation for, on commuting partial derivatives and using the antisymmetry of $F^{\mu\nu}$, we have

$$J^{\mu}{}_{,\mu} = \frac{c}{4\pi} F^{\mu\nu}{}_{,\nu\mu} = \frac{c}{8\pi}(F^{\mu\nu} - F^{\nu\mu})_{,\mu\nu} = 0.$$

Using $F_{i4} = -F_{4i} = E_i$ and $F_{ij} = \epsilon_{ijk} B_k$, Eqs. (9.38) reduce to the vector form of Maxwell equations

$$\nabla \cdot \mathbf{E} = 4\pi\rho, \qquad (9.42)$$

$$-\frac{1}{c}\frac{\partial \mathbf{E}}{\partial t} + \nabla \times \mathbf{B} = \frac{4\pi}{c}\mathbf{j}. \qquad (9.43)$$

Exercise: Show Eqs. (9.42) and (9.43).

There are essentially two independent **invariants** that can be constructed from an electromagnetic field,

$$F_{\mu\nu}F^{\mu\nu} \quad \text{and} \quad *F_{\mu\nu}F^{\mu\nu}$$

where the **dual electromagnetic tensor** $*F_{\mu\nu}$ is given in Example 8.8. Substituting electric and magnetic field components we find

$$F_{\mu\nu}F^{\mu\nu} = 2(\mathbf{B}^2 - \mathbf{E}^2) \quad \text{and} \quad *F_{\mu\nu}F^{\mu\nu} = -4\mathbf{E} \cdot \mathbf{B}. \qquad (9.44)$$

Exercise: Show that the source-free Maxwell equations (9.37) can be written in the dual form

$$*F^{\mu\nu}{}_{,\nu} = 0.$$

The equation of motion of a charged particle, charge q, is given by the **Lorentz force equation**

$$\frac{d}{d\tau}P_{\mu} = \frac{q}{c}F_{\mu\nu}V^{\nu} = F_{\mu} \qquad (9.45)$$

where the 4-momentum P_{μ} has components $(\mathbf{p}, -\mathcal{E}/c)$. Energy is written here as $\mathcal{E}$ so that no confusion with the magnitude of electric field can arise. Using Eq. (9.34) for components of the 4-force F_{μ} we find that

$$\mathbf{f} = \frac{d\mathbf{p}}{dt} = q\left(\mathbf{E} + \frac{1}{c}\mathbf{v} \times \mathbf{B}\right) \qquad (9.46)$$

and taking $\cdot \mathbf{v}$ of this equation gives rise to the energy equation (see Problem 9.16)

$$\frac{d\mathcal{E}}{dt} = \mathbf{f} \cdot \mathbf{v} = q\mathbf{E} \cdot \mathbf{v}.$$

Potentials and gauge transformations

The source-free equations (9.37) are true if and only if in a neighbourhood of any event there exists a 4-covector field $A_{\mu}(x^{\alpha})$, called the **4-potential**, such that

$$F_{\mu\nu} = A_{\nu,\mu} - A_{\mu,\nu}. \qquad (9.47)$$

Special relativity

The *if* part of this statement is simple, for (9.47) implies, on commuting partial derivatives,

$$F_{\mu\nu,\rho} + F_{\nu\rho,\mu} + F_{\rho\mu,\nu} = A_{\nu,\mu\rho} - A_{\mu,\nu\rho} + A_{\rho,\nu\mu} - A_{\nu,\rho\mu} + A_{\mu,\rho\nu} - A_{\rho,\mu\nu} = 0.$$

The converse will be postponed till Chapter 17, Theorem 17.5.

Exercise: Setting $A_\mu = (A_1, A_2, A_3, -\phi) = (\mathbf{A}, -\phi)$, show that Eq. (9.47) reads

$$\mathbf{B} = \nabla \times \mathbf{A}, \qquad \mathbf{E} = -\nabla\phi - \frac{1}{c}\frac{\partial \mathbf{A}}{\partial t}. \tag{9.48}$$

$\mathbf{A}$ is known as the *vector potential*, and ϕ as the **scalar potential**.

If the 4-vector potential of an electromagnetic field is altered by addition of the gradient of a scalar field ψ

$$\tilde{A}_\mu = A_\mu + \psi_{,\mu} \tag{9.49}$$

then the electromagnetic tensor $F_{\mu\nu}$ remains unchanged

$$\tilde{F}_{\mu\nu} = \tilde{A}_{\nu,\mu} - \tilde{A}_{\mu,\nu} = A_{\nu,\mu} + \psi_{,\nu\mu} - A_{\mu,\nu} - \psi_{,\mu\nu} = F_{\mu\nu}.$$

A transformation (9.49), which has no effect on the electromagnetic field, is called a **gauge transformation**.

Exercise: Write the gauge transformation (9.49) in terms of the vector and scalar potential,

$$\tilde{\mathbf{A}} = \mathbf{A} + \nabla\psi, \qquad \tilde{\phi} = \phi - \frac{1}{c}\frac{\partial \psi}{\partial t},$$

and check that $\mathbf{E}$ and $\mathbf{B}$ given by Eq. (9.48) are left unchanged by these transformations.

Under a gauge transformation, the divergence of A^μ transforms as

$$\tilde{A}^\mu{}_{,\mu} = A^\mu{}_{,\mu} + \Box\psi$$

where

$$\Box\psi = \psi^{,\mu}{}_{,\mu} = g^{\mu\nu}\psi_{,\mu\nu} = \nabla^2\psi - \frac{1}{c^2}\frac{\partial^2\psi}{\partial t^2}.$$

The operator $\Box$ is called the **wave operator** or **d'Alembertian**. If we choose ψ to be any solution of the inhomogeneous wave equation

$$\Box\psi = -A^\mu{}_{,\mu} \tag{9.50}$$

then $\tilde{A}^\mu{}_{,\mu} = 0$. Ignoring the tilde over A, any choice of 4-potential A^μ that satisfies

$$A^\mu{}_{,\mu} = \nabla \cdot \mathbf{A} + \frac{1}{c}\frac{\partial \phi}{\partial t} = 0 \tag{9.51}$$

is called a **Lorentz gauge**. Since solutions of the inhomogeneous wave equation (9.50) are always locally available, we may always adopt a Lorentz gauge if we wish. It should, however, be pointed out that the 4-potential A_μ is not uniquely determined by the Lorentz gauge condition (9.51), for it is still possible to add a further gradient $\bar{\psi}_{,\mu}$ provided $\bar{\psi}$ is a solution of the wave equation, $\Box\bar{\psi} = 0$. This is said to be the available **gauge freedom** in the Lorentz gauge.

9.4 Electrodynamics

In terms of a 4-potential, the source-free part of the Maxwell equations (9.37) is automatically satisfied, while the source-related part (9.38) reads

$$F^{\mu\nu}{}_{,\nu} = A^{\nu,\mu}{}_{,\nu} - A^{\mu,\nu}{}_{,\nu} = \frac{4\pi}{c} J^{\mu}.$$

If A^{μ} is in a Lorentz gauge (9.51), then the first term in the central expression vanishes and the Maxwell equations reduce to inhomogeneous wave equations,

$$\Box A^{\mu} = -\frac{4\pi}{c} J^{\mu}, \qquad A^{\mu}{}_{,\mu} = 0, \tag{9.52}$$

or in terms of vector and scalar potentials

$$\Box \mathbf{A} = -\frac{4\pi}{c}\mathbf{j}, \quad \Box \phi = -4\pi\rho, \quad \nabla \cdot \mathbf{A} + \frac{1}{c}\frac{\partial \phi}{\partial t} = 0. \tag{9.53}$$

In the case of a vacuum, $\rho = 0$ and $\mathbf{j} = 0$, the Maxwell equations read

$$\Box \mathbf{A} = 0, \qquad \Box \phi = 0.$$

Problems

Problem 9.17 Show that with respect to a rotation (9.8) the electric and magnetic fields **E** and **B** transform as 3-vectors,

$$E'_i = a_{ij} E_j, \qquad B'_i = a_{ij} B_j.$$

Problem 9.18 Under a boost (9.13) show that the 4-tensor transformation law for $F_{\mu\nu}$ or $F^{\mu\nu}$ gives rise to

$$E'_1 = F'_{14} = E_1, \qquad E'_2 = \gamma\left(E_2 - \frac{v}{c}B_3\right), \qquad E'_3 = \gamma\left(E_2 + \frac{v}{c}B_2\right),$$

$$B'_1 = F'_{23} = B_1, \qquad B'_2 = \gamma\left(B_2 + \frac{v}{c}E_3\right), \qquad B'_3 = \gamma\left(B_2 - \frac{v}{c}E_2\right).$$

Decomposing **E** and **B** into components parallel and perpendicular to $\mathbf{v} = (v, 0, 0)$, show that these transformations can be expressed in vector form:

$$\mathbf{E}'_{\parallel} = \mathbf{E}_{\parallel}, \qquad \mathbf{E}'_{\perp} = \gamma\left(\mathbf{E}_{\perp} + \frac{1}{c}\mathbf{v} \times \mathbf{B}\right),$$

$$\mathbf{B}'_{\parallel} = \mathbf{B}_{\parallel}, \qquad \mathbf{B}'_{\perp} = \gamma\left(\mathbf{B}_{\perp} - \frac{1}{c}\mathbf{v} \times \mathbf{E}\right).$$

Problem 9.19 It is possible to use transformation of **E** and **B** under boosts to find the field of a uniformly moving charge. Consider a charge q travelling with velocity **v**, which without loss of generality may be taken to be in the x-direction. Let $\mathbf{R} = (x - vt, y, z)$ be the vector connecting charge to field point $\mathbf{r} = (x, y, z)$. In the rest frame of the charge, denoted by primes, suppose the field is the coulomb field

$$\mathbf{E}' = \frac{q\mathbf{r}'}{r'^3}, \qquad \mathbf{B}' = 0$$

where

$$\mathbf{r}' = (x', y', z') = \left(\frac{x - vt}{\sqrt{1 - v^2/c^2}}, y, z\right).$$

Special relativity

Apply the transformation law for **E** and **B** derived in Problem 9.18 to show that

$$E = \frac{qR(1 - v^2/c^2)}{R^3(1 - (v^2/c^2)\sin^2\theta)^{3/2}} \quad \text{and} \quad B = \frac{1}{c}v \times E,$$

where θ is the angle between **R** and **v**. At a given distance R where is most of the electromagnetic field concentrated for highly relativistic velocities $v \approx c$?

Problem 9.20 A particle of rest mass m, charge q is in motion in a uniform constant magnetic field $\mathbf{B} = (0, 0, B)$. Show from the Lorentz force equation that the energy $\mathcal{E}$ of the particle is constant, and its motion is a helix about a line parallel to **B**, with angular frequency

$$\omega = \frac{qcB}{\mathcal{E}}.$$

Problem 9.21 Let **E** and **B** be perpendicular constant electric and magnetic fields, $\mathbf{E} \cdot \mathbf{B} = 0$.

(a) If $B^2 > E^2$ show that a transformation to a frame K' having velocity $\mathbf{v} = k\mathbf{E} \times \mathbf{B}$ can be found such that $\mathbf{E}'$ vanishes.
(b) What is the magnitude of $\mathbf{B}'$ after this transformation?
(c) If $E^2 > B^2$ find a transformation that makes $\mathbf{B}'$ vanish.
(d) What happens if $E^2 = B^2$?
(e) A particle of charge q is in motion in a crossed constant electric and magnetic field $\mathbf{E} \cdot \mathbf{B} = 0$, $B^2 > E^2$. From the solution of Problem 9.20 for a particle in a constant magnetic field, describe its motion.

Problem 9.22 An electromagnetic field $F_{\mu\nu}$ is said to be of 'electric type' at an event p if there exists a unit timelike 4-vector U_μ at p, $U_\alpha U^\alpha = -1$, and a spacelike 4-vector field E_μ orthogonal to U^μ such that

$$F_{\mu\nu} = U_\mu E_\nu - U_\nu E_\mu, \qquad E_\alpha U^\alpha = 0.$$

(a) Show that any purely electric field, i.e. one having $\mathbf{B} = \mathbf{0}$, is of electric type.
(b) If $F_{\mu\nu}$ is of electric type at p, show that there is a velocity **v** such that

$$\mathbf{B} = \frac{\mathbf{v}}{c} \times \mathbf{E} \quad (|\mathbf{v}| < c).$$

Using Problem 9.18 show that there is a Lorentz transformation that transforms the electromagnetic field to one that is purely electric at p.
(c) If $F_{\mu\nu}$ is of electric type everywhere with U^μ a constant vector field, and satisfies the Maxwell equations *in vacuo*, $J^\mu = 0$, show that the vector field E^μ is divergence-free, $E^\nu{}_{,\nu} = 0$.

Problem 9.23 Use the gauge freedom $\Box\psi = 0$ in the Lorentz gauge to show that it is possible to set $\phi = 0$ and $\nabla \cdot \mathbf{A} = 0$. This is called a *radiation gauge*.

(a) What gauge freedoms are still available to maintain the radiation gauge?
(b) Suppose **A** is independent of coordinates x and y in the radiation gauge. Show that the Maxwell equations have solutions of the form

$$\mathbf{E} = (E_1(u), E_2(u), 0), \qquad \mathbf{B} = (-E_2(u), E_1(u), 0)$$

where $u = ct - z$ and $E_i(u)$ are arbitrary differentiable functions.
(c) Show that these solutions may be interpreted as right-travelling electromagnetic waves.

9.5 Conservation laws and energy–stress tensors

Conservation of charge

Consider a general four-dimensional region Ω of space-time with boundary 3-surface $\partial\Omega$. The **four-dimensional Gauss theorem** (see Chapter 17) asserts that for any vector field A^α

$$\iiiint_\Omega A^\alpha{}_{,\alpha}\, dx^1\, dx^2\, dx^3\, dx^4 = \iiint_{\partial\Omega} A^\alpha dS_\alpha. \tag{9.54}$$

If $\partial\Omega$ has the parametric form $x^\alpha = x^\alpha(\lambda_1, \lambda_2, \lambda_3)$, the vector 3-volume element dS_α is defined by

$$dS_\alpha = \epsilon_{\alpha\beta\gamma\delta} \frac{\partial x^\beta}{\partial \lambda_1}\frac{\partial x^\gamma}{\partial \lambda_2}\frac{\partial x^\delta}{\partial \lambda_3}\, d\lambda_1\, d\lambda_2\, d\lambda_3,$$

with the four-dimensional epsilon symbol $\epsilon_{\alpha\beta\gamma\delta}$ defined by Eq. (8.21). Since the epsilon symbol transforms as a tensor with respect to basis transformations having determinant 1, it is a 4-tensor if we restrict ourselves to proper Lorentz transformations, and it follows that dS_α is a 4-vector. Furthermore, dS_α is orthogonal to the 3-surface $\partial\Omega$, for any 4-vector X^α tangent to the 3-surface has a linear decomposition

$$X^\alpha = \sum_{i=1}^{3} c_i \frac{\partial x^\delta}{\partial \lambda_i},$$

and by the total antisymmetry of $\epsilon_{\alpha\beta\gamma\delta}$ it follows that

$$dS_\alpha X^\alpha = \sum_{i=1}^{3} c_i \epsilon_{\alpha\beta\gamma\delta} \frac{\partial x^\alpha}{\partial \lambda_i}\frac{\partial x^\beta}{\partial \lambda_1}\frac{\partial x^\gamma}{\partial \lambda_2}\frac{\partial x^\delta}{\partial \lambda_3} = 0.$$

The four-dimensional Gauss theorem is a natural generalization of the well-known three-dimensional result. In Chapter 17, it will become clear that this theorem is independent of the choice of parametrization λ_i on $\partial\Omega$.

A 3-surface S is called **spacelike** if its orthogonal 3-volume element dS_α is a timelike 4-covector. The reason for this terminology is that a 4-vector orthogonal to three linearly independent spacelike 4-vectors must be timelike. The archetypal spacelike 3-surface is given by the equation $t = \text{const.}$ in a given inertial frame. Parametrically the surface may be given by $x = \lambda_1, y = \lambda_2, z = \lambda_3$ and its 3-volume element is

$$dS_\alpha = \epsilon_{\alpha 123}\, dx\, dy\, dz = (0, 0, 0, -dx\, dy\, dz).$$

Given a current 4-vector $J^\alpha = (\mathbf{j}, c\rho)$, satisfying the divergence-free condition $J^\alpha{}_{,\alpha} = 0$, it is natural to define the 'total charge' over an arbitrary spacelike 3-surface S to be

$$Q = -\frac{1}{c} \iiint_S J^\alpha dS_\alpha, \tag{9.55}$$

as this gives the expected $Q = \iiint \rho\, dx\, dy\, dz$ when S is a surface of type $t = \text{const.}$

Special relativity

Let Ω be a 4-volume enclosed by two spacelike surfaces S and S' having infinite extent. Using the four-dimensional Gauss theorem and the divergence-free condition $J^\alpha{}_{,\alpha} = 0$ we obtain the law of **conservation of charge**,

$$Q' - Q = \frac{1}{c}\left(\iiint_S J^\alpha \, dS_\alpha - \iiint_{S'} J^\alpha \, dS_\alpha\right) = \frac{1}{c}\iiiint_\Omega J^\alpha{}_{,\alpha} \, dx^1 \, dx^2 \, dx^3 \, dx^4 = 0$$

where the usual physical assumption is made that the 4-current J^α vanishes at spatial infinity $|\mathbf{r}| \to \infty$. This implies that there are no contributions from the timelike 'sides at infinity' to the 3-surface integral over ∂S. Note that in Minkowski space, dS^α is required to be 'inwards-pointing' on the spacelike parts of the boundary, S and S', as opposed to the more usual outward pointing requirement in three-dimensional Euclidean space.

As seen in Example 9.3 and Section 9.4 there is a converse to this result: given a conserved quantity Q, generically called 'charge', then $J^\alpha = (\mathbf{j}, c\rho)$, where ρ is the charge density and $\mathbf{j}$ the charge flux density, form the components of a divergence-free 4-vector field, $J^\alpha{}_{,\alpha} = 0$.

Energy–stress tensors

Assume now that the total 4-momentum P^μ of a system is conserved. Treating its components as four separate conserved 'charges', we are led to propose the existence of a quantity $T^{\mu\nu}$ such that

$$T^{\mu\nu}{}_{,\nu} = 0 \tag{9.56}$$

and the total 4-momentum associated with any spacelike surface S is given by

$$P^\mu = -\frac{1}{c}\iiint_S T^{\mu\nu} \, dS_\nu. \tag{9.57}$$

In order to ensure that Eq. (9.56) be a tensorial equation it is natural to postulate that $T^{\mu\nu}$ is a 4-tensor field, called the **energy–stress tensor** of the system. This will also guarantee that the quantity P^μ defined by (9.57) is a 4-vector. For a surface $t = \text{const.}$ we have

$$P^\mu = \left(\mathbf{p}, \frac{E}{c}\right) = \frac{1}{c}\int_{t=\text{const.}} T^{\mu 4} \, d^3 x$$

and the physical interpretation of the components of the energy–stress tensor $T^{\mu\nu}$ are

T^{44} = energy density,
$T^{4i} = \dfrac{1}{c} \times$ energy flux density,
$T^{i4} = c \times$ momentum density,
$T^{ij} = j$th component of flux of ith component of momentum = stress tensor.

It is usual to require that $T^{\mu\nu}$ are components of a symmetric tensor, $T^{\mu\nu} = T^{\nu\mu}$. The argument for this centres around the concept of **angular 4-momentum**, which for a continuous distribution of matter is defined to be

$$M^{\mu\nu} = \iiint_S x^\mu \, dP^\nu - x^\nu \, dP^\mu \equiv -\frac{1}{c}\iiint_S (x^\mu T^{\nu\rho} - x^\nu T^{\mu\rho}) \, dS_\rho = -M^{\nu\mu}.$$

9.5 Conservation laws and energy–stress tensors

Conservation of angular 4-momentum $M^{\mu\nu}$ is equivalent to

$$0 = (x^\mu T^{\nu\rho} - x^\nu T^{\mu\rho})_{,\rho} = \delta^\mu_\rho T^{\nu\rho} - \delta^\nu_\rho T^{\mu\rho} = T^{\nu\mu} - T^{\mu\nu}.$$

Example 9.4 Consider a fluid having 4-velocity $V^\mu = \gamma(\mathbf{v}, c)$ where $\mathbf{v} = \mathbf{v}(\mathbf{r}, t)$. Let the local rest mass density (as measured in the i.r.f.) be $\rho(\mathbf{r}, t)$. In the i.r.f. at any point of the fluid the energy density is given by ρc^2, and since there is no energy flux in the i.r.f. we may set $T^{4i} = 0$. By the symmetry of $T^{\mu\nu}$ there will also be no momentum density T^{i4} and the energy–stress tensor has the form

$$T^{\mu\nu} = \begin{pmatrix} & & & 0 \\ & T_{ij} & & 0 \\ & & & 0 \\ 0 & 0 & 0 & \rho c^2 \end{pmatrix} = \begin{pmatrix} P_1 & & & 0 \\ & P_2 & & 0 \\ & & P_3 & 0 \\ 0 & 0 & 0 & \rho c^2 \end{pmatrix},$$

where the diagonalization of the 3×3 matrix $[T_{ij}]$ can be achieved by a rotation of axes. The P_i are called the **principal pressures** at that point. If they are all equal, $P_1 = P_2 = P_3 = P$, then the fluid is said to be a **perfect fluid** and P is simply called the **pressure**. In that case

$$T^{\mu\nu} = \left(\rho + \frac{1}{c^2}P\right) V^\mu V^\nu + P g^{\mu\nu}, \tag{9.58}$$

as may be checked by verifying that this equation holds in the i.r.f. at any point, in which frame $V^\mu = (0, 0, 0, c)$. Since (9.58) is a 4-tensor equation it must hold in all inertial frames.

Exercise: Verify that the conservation laws $T^{\mu\nu}{}_{,\nu} = 0$ reduce for $v \ll c$ to the equation of continuity and Euler's equation

$$\frac{\partial \rho}{\partial t} + \nabla \cdot (\rho \mathbf{v}) = 0,$$

$$\rho \left(\frac{\partial \mathbf{v}}{\partial t} + (\mathbf{v} \cdot \nabla)\mathbf{v} \right) = -\nabla P.$$

Example 9.5 The energy–stress tensor of the electromagnetic field is given by

$$T^{\mu\nu} = \frac{1}{4\pi} \left(F^\mu{}_\rho F^{\nu\rho} - \tfrac{1}{4} g^{\mu\nu} F_{\rho\sigma} F^{\rho\sigma} \right) = T^{\nu\mu}. \tag{9.59}$$

The energy density of the electromagnetic field is thus

$$\epsilon = T^{44} = \frac{1}{16\pi} \left(4 F^4{}_i F^{4i} - g^{44} F_{\rho\sigma} F^{\rho\sigma} \right) = \frac{1}{16\pi} \left(4 \mathbf{E}^2 + 2(\mathbf{B}^2 - \mathbf{E}^2) \right) = \frac{\mathbf{E}^2 + \mathbf{B}^2}{8\pi}$$

and the energy flux density has components

$$c T^{4i} = \frac{c}{4\pi} F^4{}_j F^{ij} = \frac{c}{4\pi} E_j \epsilon_{ijk} B_k = \frac{c}{4\pi} (\mathbf{E} \times \mathbf{B})_i.$$

The vector $\mathbf{S} = \dfrac{c}{4\pi}(\mathbf{E} \times \mathbf{B})$ is known as the **Poynting vector**. The spatial components T_{ij}

Special relativity

are known as the *Maxwell stress tensor*

$$T^{ij} = T_{ij} = \frac{1}{16\pi}\left(4(F_{ik}F_j{}^k + F_{i4}F_j{}^4) - \delta_{ij}2(\mathbf{B}^2 - \mathbf{E}^2)\right)$$

$$= \frac{1}{4\pi}\left(-E_i E_j - B_i B_j + \tfrac{1}{2}\delta_{ij}(\mathbf{E}^2 + \mathbf{B}^2)\right).$$

The total 4-momentum of an electromagnetic field over a spacelike surface S is calculated from Eq. (9.57).

Exercise: Show that the average pressure $P = \tfrac{1}{3}\sum_i T_{ii}$ of an electromagnetic field is equal to $\tfrac{1}{3} \times$ energy density. Show that this also follows from the fact that $T^{\mu\nu}$ is trace-free, $T^\mu{}_\mu = 0$.

For further developments in relativistic classical field theory the reader is referred to [4, 5].

Problems

Problem 9.24 Show that as a consequence of the Maxwell equations,

$$T^\beta{}_{\alpha,\beta} = -\frac{1}{c}F_{\alpha\gamma}J^\gamma$$

where $T^\beta{}_\alpha$ is the electromagnetic energy–stress tensor (9.59), and when no charges and currents are present it satisfies Eq. (9.56). Show that the $\alpha = 4$ component of this equation has the form

$$\frac{\partial \epsilon}{\partial t} + \nabla \cdot \mathbf{S} = -\mathbf{j} \cdot \mathbf{E}$$

where ϵ = energy density and $\mathbf{S}$ = Poynting vector. Interpret this equation physically.

Problem 9.25 For a plane wave, Problem 9.23, show that

$$T_{\alpha\beta} = \epsilon\, n_\alpha n_\beta$$

where $\epsilon = E^2/4\pi$ and $n^\alpha = (\mathbf{n}, 1)$ is the null vector pointing in the direction of propagation of the wave. What pressure does the wave exert on a wall placed perpendicular to the path of the wave?

References

[1] W. Kopczyński and A. Trautman. *Spacetime and Gravitation*. Chichester, John Wiley & Sons, 1992.
[2] W. Rindler. *Introduction to Special Relativity*. Oxford, Oxford University Press, 1991.
[3] R. K. Sachs and H. Wu. *General Relativity for Mathematicians*. New York, Springer-Verlag, 1977.
[4] L. D. Landau and E. M. Lifshitz. *The Classical Theory of Fields*. Reading, Mass., Addison-Wesley, 1971.
[5] W. Thirring. *A Course in Mathematical Physics, Vol. 2: Classical Field Theory*. New York, Springer-Verlag, 1979.

10 Topology

Up till now we have focused almost entirely on the role of algebraic structures in mathematical physics. Occasionally, as in the previous chapter, it has been necessary to use some differential calculus, but this has not been done in any systematic way. Concepts such as *continuity* and *differentiability*, central to the area of mathematics known as *analysis*, are essentially geometrical in nature and require the use of *topology* for their rigorous definition. In broad terms, a *topology* is a structure imposed on a set to allow for the definition of *convergence* and *limits* of sequences or subsets. A space with a topology defined on it will be called a *topological space*, and a *continuous map* between topological spaces is one that essentially preserves limit points of subsets. The most general approach to this subject turns out to be through the concept of *open sets*.

Consider a two-dimensional surface S embedded in Euclidean three-dimensional space $\mathbb{E}^3$. In this case we have an intuitive understanding of a 'continuous deformation' of the surface as being a transformation of the surface that does not involve any tearing or pasting. Topology deals basically with those properties that are invariant under continuous deformations of the surface. Metric properties are not essential to the concept of continuity, and since operations such as 'stretching' are permissible, topology is sometimes called 'rubber sheet geometry'. In this chapter we will also define the concept of a *metric space*. Such a space always has a naturally defined topology associated with it, but the converse is not true in general – it is quite possible to define topology on a space without having a concept of distance defined on the space.

10.1 Euclidean topology

The archetypal model for a topological space is the real line and the Euclidean plane $\mathbb{R}^2$. On the real line $\mathbb{R}$, an **open interval** is any set $(a, b) = \{x \in \mathbb{R} \mid a < x < b\}$. A set $U \subseteq \mathbb{R}$ is called a **neighbourhood** of $x \in \mathbb{R}$ if there exists $\epsilon > 0$ such that the open interval $(x - \epsilon, x + \epsilon)$ is a subset of U. We say a sequence of real numbers $\{x_n\}$ **converges to** $x \in \mathbb{R}$, written $x_n \to x$, if for every $\epsilon > 0$ there exists an integer $N > 0$ such that $|x - x_n| < \epsilon$ for all $n > N$; that is, for sufficiently large n the sequence x_n enters and stays in every neighbourhood U of x. The point x is then said to be the **limit** of the sequence $\{x_n\}$.

Exercise: Show that the limit of a sequence is unique: if $x_n \to x$ and $x_n \to x'$ then $x = x'$.

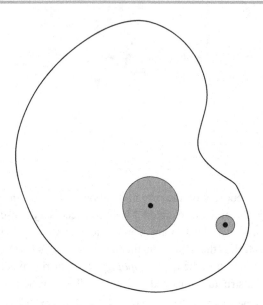

Figure 10.1 Points in an open set can be 'thickened' to an open ball within the set

Similar definitions apply to the Euclidean plane $\mathbb{R}^2$, where we set $|\mathbf{y} - \mathbf{x}| = \sqrt{(y_1 - x_1)^2 + (y_2 - x_2)^2}$. In this case, open intervals are replaced by **open balls**

$$B_r(\mathbf{x}) = \{\mathbf{y} \in \mathbb{R}^2 \mid |\mathbf{y} - \mathbf{x}| < r\}$$

and a set $U \subseteq \mathbb{R}^2$ is said to be a **neighbourhood** of $\mathbf{x} \in \mathbb{R}^2$ if there exists a real number $\epsilon > 0$ such that the open ball $B_\epsilon(\mathbf{x}) \subseteq U$. A sequence of points $\{\mathbf{x}_n\}$ **converges to** $\mathbf{x} \in \mathbb{R}^2$, or $\mathbf{x}$ is the **limit** of the sequence $\{\mathbf{x}_n\}$, if for every $\epsilon > 0$ there exists an integer $N > 0$ such that

$$\mathbf{x}_n \in B_\epsilon(\mathbf{x}) \quad \text{for all} \quad n > N.$$

Again we write $\mathbf{x}_n \to \mathbf{x}$, and the definition is equivalent to the statement that for every neighbourhood U of $\mathbf{x}$ there exists $N > 0$ such that $\mathbf{x}_n \in U$ for all $n > N$.

An **open set** U in $\mathbb{R}$ or $\mathbb{R}^2$ is a set that is a neighbourhood of every point in it. Intuitively, U is open in $\mathbb{R}$ (resp. $\mathbb{R}^2$) if every point in U can be 'thickened out' to an open interval (resp. open ball) within U (see Fig. 10.1). For example, the unit ball $B_1(\mathbf{O}) = \{\mathbf{y} \mid |\mathbf{y}|^2 < 1\}$ is an open set since, for every point $\mathbf{x} \in B_1(\mathbf{O})$ the open ball $B_\epsilon(\mathbf{x}) \subseteq B_1(\mathbf{O})$ where $\epsilon = 1 - |\mathbf{x}| > 0$.

On the real line it may be shown that the most general open set consists of a union of non-intersecting open intervals,

$$\ldots, (a_{-1}, a_0), \ (a_1, a_2), \ (a_3, a_4), \ (a_5, a_6), \ldots$$

where $\ldots a_{-1} < a_0 \leq a_1 < a_2 \leq a_3 < a_4 \leq a_5 < a_6 \leq \ldots$. In $\mathbb{R}^2$ open sets cannot be so simply categorized, for while every open set is a union of open balls, the union need not be disjoint.

10.2 General topological spaces

In standard analyis, a function $f : \mathbb{R} \to \mathbb{R}$ is said to be *continuous at x* if for every $\epsilon > 0$ there exists $\delta > 0$ such that

$$|y - x| < \delta \implies |f(y) - f(x)| < \epsilon.$$

Hence, for every $\epsilon > 0$, the inverse image set $f^{-1}(f(x) - \epsilon, f(x) + \epsilon)$ is a neighbourhood of x, since it includes an open interval $(x - \delta, x + \delta)$ centred on x. As every neighbourhood of $f(x)$ contains an interval of the form $(f(x) - \epsilon, f(x) + \epsilon)$ the function f is continuous at x if and only if the inverse image of every neighbourhood of $f(x)$ is a neighbourhood of x. A function $f : \mathbb{R} \to \mathbb{R}$ is said to be **continuous on** $\mathbb{R}$ if it is continuous at every point $x \in \mathbb{R}$.

Theorem 10.1 *A function $f : \mathbb{R} \to \mathbb{R}$ is continuous on $\mathbb{R}$ if and only if the inverse image $V = f^{-1}(U)$ of every open set $U \subseteq \mathbb{R}$ is an open subset of $\mathbb{R}$.*

Proof: Let f be continuous on $\mathbb{R}$. Since an open set U is a neighbourhood of every point $y \in U$, its inverse image $V = f^{-1}(U)$ must be a neighbourhood of every point $x \in V$. Hence V is an open set.

Conversely let $f : \mathbb{R} \to \mathbb{R}$ be any function having the property that $V = f^{-1}(U)$ is open for every open set $U \subseteq \mathbb{R}$. Then for any $x \in \mathbb{R}$ and every $\epsilon > 0$ the inverse image under f of the open interval $(f(x) - \epsilon, f(x) + \epsilon)$ is an open set including x. It therefore contains an open interval of the form $(x - \delta, x + \delta)$, so that f is continuous at x. Since x is an arbitrary point, the function f is continuous on $\mathbb{R}$. ∎

In general topology this will be used as the defining characteristic of a continuous map. In $\mathbb{R}^2$ the treatment is almost identical. A function $f : \mathbb{R}^2 \to \mathbb{R}^2$ is said to be *continuous at* **x** if for every $\epsilon > 0$ there exists a real number $\delta > 0$ such that

$$|\mathbf{y} - \mathbf{x}| < \delta \implies |f(\mathbf{y}) - f(\mathbf{x})| < \epsilon.$$

An essentially identical proof to that given in Theorem 10.1 shows that a function f is *continuous on* $\mathbb{R}^2$ if and only if the inverse image $f^{-1}(U)$ of every open set $U \subseteq \mathbb{R}^2$ is an open subset of $\mathbb{R}^2$. The same applies to real-valued functions $f : \mathbb{R}^2 \to \mathbb{R}$. Thus continuity of functions can be described entirely by their inverse action on open sets. For this reason, open sets are regarded as the key ingredients of a topological space. Experience from Euclidean spaces and surfaces embedded in them has taught mathematicians that the most important properties of open sets can be summarized in a few simple rules, which are set out in the next section (see also [1–8]).

10.2 General topological spaces

Given a set X, a **topology** on X consists of a family of subsets $\mathcal{O}$, called **open sets**, which satisfy the following conditions:

(Top1) The empty set $\emptyset$ is open and the entire space X is open, $\{\emptyset, X\} \subset \mathcal{O}$.
(Top2) If U and V are open sets then so is their intersection $U \cap V$,

$$U \in \mathcal{O} \text{ and } V \in \mathcal{O} \implies U \cap V \in \mathcal{O}.$$

(Top3) If $\{V_i \mid i \in I\}$ is any family of open sets then their union $\bigcup_{i \in I} V_i$ is open.

257

Successive application of (Top2) implies that the intersection of any finite number of open sets is open, but $\mathcal{O}$ is not in general closed with respect to infinite intersections of open sets. On the other hand, $\mathcal{O}$ is closed with respect to arbitrary unions of open sets. The pair $(X, \mathcal{O})$, where $\mathcal{O}$ is a topology on X, is called a **topological space**. We often refer simply to a topological space X when the topology $\mathcal{O}$ is understood. The elements of the underlying space X are normally referred to as **points**.

Example 10.1 Define $\mathcal{O}$ to be the collection of subsets U of the real line $\mathbb{R}$ having the property that for every $x \in U$ there exists an open interval $(x - \epsilon, x + \epsilon) \subseteq U$ for some $\epsilon > 0$. These sets agree with the definition of open sets given in Section 10.1. The empty set is assumed to belong to $\mathcal{O}$ by default, while the whole line $\mathbb{R}$ is evidently open since every point lies in an open interval. Thus (Top1) holds for the family $\mathcal{O}$. To prove (Top2) let U and V be open sets such that $U \cap V \neq \emptyset$, the case where $U \cap V = \emptyset$ being trivial. For any $x \in U \cap V$ there exist positive numbers ϵ_1 and ϵ_2 such that

$$(x - \epsilon_1, x + \epsilon_1) \subseteq U \quad \text{and} \quad (x - \epsilon_2, x + \epsilon_2) \subseteq V.$$

If $\epsilon = \min(\epsilon_1, \epsilon_2)$ then $(x - \epsilon, x + \epsilon) \subseteq U \cap V$, hence $U \cap V$ is an open set.

For (Top3), let U be the union of an arbitrary collection of open sets $\{U_i \mid i \in I\}$. If $x \in U$, then $x \in U_j$ for some $j \in I$ and there exists $\epsilon > 0$ such that $(x - \epsilon, x + \epsilon) \subseteq U_j \subseteq U$. Hence U is open and the family $\mathcal{O}$ forms a topology for $\mathbb{R}$. It is often referred to as the **standard topology** on $\mathbb{R}$. Any open interval (a, b) where $a < b$ is an open set, for if $x \in (a, b)$ then $(x - \epsilon, x + \epsilon) \subset (a, b)$ for $\epsilon = \frac{1}{2}\min(x - a, b - x)$. A similar argument shows that the intervals may also be of semi-infinite extent, such as $(-\infty, a)$ or (b, ∞). Notice that infinite intersections of open sets do not generally result in an open set. For example, an isolated point $\{a\}$ is not an open set since it contains no finite open interval, yet it is the intersection of an infinite sequence of open intervals such as

$$(a - 1, a + 1), \ (a - \tfrac{1}{2}, a + \tfrac{1}{2}), \ (a - \tfrac{1}{3}, a + \tfrac{1}{3}), \ (a - \tfrac{1}{4}, a + \tfrac{1}{4}), \ldots$$

Similar arguments can be used to show that the open sets defined on $\mathbb{R}^2$ in Section 10.1 form a topology. Similarly, in $\mathbb{R}^n$ we define a topology where a set U is said to be open if for every point $\mathbf{x} \in U$ there exists an open ball

$$B_r(\mathbf{x}) = \{\mathbf{y} \in \mathbb{R}^2 \mid |\mathbf{y} - \mathbf{x}| < r\} \subset U,$$

where

$$|\mathbf{y} - \mathbf{x}| = \sqrt{(y_1 - x_1)^2 + (y_2 - x_2)^2 + \cdots + (y_n - x_n)^2}.$$

This topology will again be termed the **standard topology** on $\mathbb{R}^n$.

Example 10.2 Consider the family $\mathcal{O}'$ of all open intervals on $\mathbb{R}$ of the form $(-a, b)$ where $a, b > 0$, together with the empty set. All these intervals contain the origin 0. It is not hard to show that (Top1)–(Top3) hold for this family and that $(X, \mathcal{O}')$ is a topological space. This space is not very 'nice' in some of its properties. For example no two points $x, y \in \mathbb{R}$ lie in non-intersecting neighbourhoods. In a sense all points of the line are 'arbitrarily close' to each other in this topology.

10.2 General topological spaces

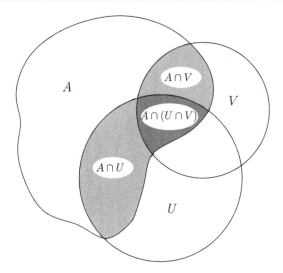

Figure 10.2 Relative topology induced on a subset of a topological space

A subset V is called **closed** if its complement $X - V$ is open. The empty set and the whole space are clearly closed sets, since they are both open sets and are the complements of each other. The intersection of an arbitrary family of closed sets is closed, as it is the complement of a union of open sets. However, only finite unions of closed sets are closed in general.

Example 10.3 Every *closed interval* $[a, b] = \{x \mid a \le x \le b\}$ where $-\infty < a \le b < \infty$ is a closed set, as it is the complement of the open set $(-\infty, a) \cup (b, \infty)$. Every singleton set consisting of an isolated point $\{a\} \equiv [a, a]$ is closed. Closed intervals $[a, b]$ are not open sets since the end points a or b do not belong to any open interval included in $[a, b]$.

If A is any subset of X the **relative topology** on A, or **topology induced** on A, is the topology whose open sets are

$$\mathcal{O}_A = \{A \cap U \mid U \in \mathcal{O}\}.$$

Thus a set is open in the relative topology on A iff it is the intersection of A and an open set U in X (see Fig. 10.2). That these sets form a topology on A follows from the following three facts:

1. $\emptyset \cap A = \emptyset$, $X \cap A = A$.
2. $(U \cap A) \cap (V \cap A) = (U \cap V) \cap A$.
3. $\bigcup_{i \in I}(U_i \cap A) = \left(\bigcup_{i \in I} U_i\right) \cap A$.

A subset A of X together with the relative topology $\mathcal{O}_A$ induced on it is called a **subspace** of $(X, \mathcal{O})$.

Example 10.4 The relative topology on the half-open interval $A = [0, 1) \subset \mathbb{R}$, induced on A by the standard topology on $\mathbb{R}$, is the union of half-open intervals of the form $[0, a)$

259

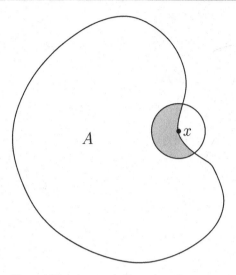

Figure 10.3 Accumulation point of a set

where $0 < a < 1$, and all intervals of the form (a, b) where $0 < a < b \le 1$. Evidently some of the open sets in this topology are not open in $\mathbb{R}$.

Exercise: Show that if $A \subseteq X$ is an open set then all open sets in the relative topology on A are open in X.

Exercise: If A is a closed set, show that every closed set in the induced topology on A is closed in X.

A point x is said to be an **accumulation point** of a set A if every open neighbourhood U of x contains points of A other than x itself, as shown in Fig. 10.3. What this means is that x may or may not lie in A, but points of A 'cluster' arbitrarily close to it (sometimes it is also called a *cluster point* of A). A related concept is commonly applied to sequences of points $x_n \in X$. We say that the sequence $x_n \in X$ **converges to** $x \in X$ or that x is a **limit point** of $\{x_n\}$, denoted $x_n \to x$, if for every open neighbourhood U of x there is an integer N such that $x_n \in U$ for all $n \ge N$. This differs from an accumulation point in that we could have $x_n = x$ for all $n > n_0$ for some n_0.

The **closure** of any set A, denoted $\overline{A}$, is the union of the set A and all its accumulation points. The **interior** of A is the union of all open sets $U \subseteq A$, denoted A^o. The difference of these two sets, $b(A) = \overline{A} - A^o$, is called the **boundary** of A.

Theorem 10.2 *The closure of any set A is a closed set. The interior A^o is the largest open set included in A. The boundary $b(A)$ is a closed set.*

Proof: Let x be any point not in $\overline{A}$. Since x is not in A and is not an accumulation point of A, it has an open neighbourhood U_x not intersecting A. Furthermore, U_x cannot contain any other accumulation point of A else it would be an open neighbourhood of that point not intersecting A. Hence the complement $X - \overline{A}$ of the closure of A is the union of the open sets U_x. It is therefore itself an open set and its complement $\overline{A}$ is a closed set.

10.2 General topological spaces

Since the interior A^o is a union of open sets, it is an open set by (Top3). If U is any open set such that $U \subseteq A$ then, by definition, $U \subseteq A^o$. Thus A^o is the largest open subset of A. Its complement is closed and the boundary $b(A) = \overline{A} \cap (X - A^o)$ is necessarily a closed set. ∎

Exercise: Show that a set A is closed if and only if it contains its boundary, $A \supseteq b(A)$.

Exercise: A set A is open if and only if $A \cap b(A) = \emptyset$.

Exercise: Show that all accumulation points of A lie in the boundary $b(A)$.

Exercise: Show that a point x lies in the boundary of A iff every neighbourhood of x contains points both in A and not in A.

Example 10.5 The closure of the open ball $B_a(\mathbf{x}) \subset \mathbb{R}^n$ (see Example 10.1) is the *closed ball*

$$\overline{B_a}(\mathbf{x}) = \{\mathbf{y} \mid |\mathbf{y} - \mathbf{x}| \leq a\}.$$

Since every open ball is an open set, it is its own interior, $B_a^o(\mathbf{x}) = B_a(\mathbf{x})$ and its boundary is the $(n-1)$-sphere of radius a, centre $\mathbf{x}$,

$$b(B_a(\mathbf{x})) = S_a^{n-1}(\mathbf{x}) = \{y \mid |\mathbf{y} - \mathbf{x}| = a\}.$$

Example 10.6 A set whose closure is the entire space X is said to be **dense** in X. For example, since every real number has rational numbers arbitrarily close to it, the rational numbers $\mathbb{Q}$ are a countable set that is dense in the set of real numbers. In higher dimensions the situation is similar. The set of points with rational coordinates $\mathbb{Q}^n$ is a countable set that is dense in $\mathbb{R}^n$.

Exercise: Show that $\mathbb{Q}$ is neither an open or closed set in $\mathbb{R}$.

Exercise: Show that $\mathbb{Q}^o = \emptyset$ and $b(\mathbb{Q}) = \mathbb{R}$.

It is sometimes possible to compare different topologies $\mathcal{O}_1$ and $\mathcal{O}_2$ on a set X. We say $\mathcal{O}_1$ is **finer** or **stronger** than $\mathcal{O}_2$ if $\mathcal{O}_1 \supseteq \mathcal{O}_2$. Essentially, $\mathcal{O}_1$ has more open sets than $\mathcal{O}_2$. In this case we also say that $\mathcal{O}_2$ is **coarser** or **weaker** than $\mathcal{O}_1$.

Example 10.7 All topologies on a set X lie somewhere between two extremes, the discrete and indiscrete topologies. The **indiscrete** or **trivial** topology consists simply of the empty set and the whole space itself, $\mathcal{O}_1 = \{\emptyset, X\}$. It is the coarsest possible topology on X – if $\mathcal{O}$ is any other topology then $\mathcal{O}_1 \subseteq \mathcal{O}$ by (Top1). The **discrete topology** consists of all subsets of X, $\mathcal{O}_2 = 2^X$. This topology is the finest possible topology on X, since it includes all other topologies $\mathcal{O}_2 \supseteq \mathcal{O}$. For both topologies (Top1)–(Top3) are trivial to verify.

Given a set X, and an arbitrary collection of subsets $\mathcal{U}$, we can ask for the weakest topology $\mathcal{O}(\mathcal{U})$ containing $\mathcal{U}$. This topology is the intersection of all topologies that contain

$\mathcal{U}$ and is called the **topology generated** by $\mathcal{U}$. It is analogous to the concept of the vector subspace $L(M)$ generated by an arbitrary subset M of a vector space V (see Section 3.5).

A contructive way of defining $\mathcal{O}(\mathcal{U})$ is the following. Firstly, adjoin the empty set $\emptyset$ and the entire space X to $\mathcal{U}$ if they are not already in it. Next, extend $\mathcal{U}$ to a family $\hat{\mathcal{U}}$ consisting of all finite intersections $U_1 \cap U_2 \cap \cdots \cap U_n$ of sets $U_i \in \mathcal{U} \cup \{\emptyset, X\}$. Finally, the set $\mathcal{O}(\mathcal{U})$ consisting of arbitrary unions of sets from $\hat{\mathcal{U}}$ forms a topology. To prove (Top2),

$$\bigcup_{i \in I}\left(\bigcap_{a=1}^{n_i} U_{ia}\right) \cap \bigcup_{j \in J}\left(\bigcap_{b=1}^{n_j} V_{jb}\right) = \bigcup_{i \in I}\bigcup_{j \in J}\left(U_{i1} \cap \cdots \cap U_{in_i} \cap V_{j1} \cap \cdots \cap V_{jn_j}\right).$$

Property (Top3) follows immediately from the contruction.

Example 10.8 On the real line $\mathbb{R}$, the family $\mathcal{U}$ of all open intervals generates the standard topology since every open set is a union of open sets of the form $(x - \epsilon, x + \epsilon)$. Similarly, the standard topology on $\mathbb{R}^2$ is generated by the set of open balls

$$\mathcal{U} = \{B_a(\mathbf{r}) \mid > 0, \ \mathbf{r} = (x, y) \in \mathbb{R}^2\}.$$

To prove this statement we must show that every set that is an intersection of two open balls $B_a(\mathbf{r})$ and $B_b(\mathbf{r}')$ is a union of open balls from $\mathcal{U}$. If $\mathbf{x} \in B_a(\mathbf{r})$, let $\epsilon < a$ be such that $B_\epsilon(\mathbf{x}) \subset B_a(\mathbf{r})$. Similarly if $\mathbf{x} \in B_b(\mathbf{r}')$, let $\epsilon' < a$ be such that $B_{\epsilon'}(\mathbf{x}) \subset B_b(\mathbf{r}')$. Hence, if $\mathbf{x} \in B_a(\mathbf{r}) \cap B_b(\mathbf{r}')$ then $B_{\epsilon''}(\mathbf{x}) \subset B_a(\mathbf{r}) \cap B_b(\mathbf{r}')$ where $\epsilon'' = \min(\epsilon, \epsilon')$. The proof easily generalizes to intersections of any finite number of open balls. Hence the standard topology of $\mathbb{R}^2$ is generated by the set of all open balls. The extension to $\mathbb{R}^n$ is straightforward.

Exercise: Show that the discrete topology on X is generated by the family of all singleton sets $\{x\}$ where $x \in X$.

A set A is said to be a **neighbourhood** of $x \in X$ if there exists an open set U such that $x \in U \subset A$. If A itself is open it is called an **open neighbourhood** of x. A topological space X is said to be **first countable** if every point $x \in X$ has a countable collection $U_1(x), U_2(x), \ldots$ of open neighbourhoods of x such that every open neighbourhood U of x includes one of these neighbourhoods $U \supset U_n(x)$. A stronger condition is the following: a topological space $(X, \mathcal{O})$ is said to be **second countable** or **separable** if there exists a countable set $U_1, U_2, U_3, \ldots$ that generates the topology of X.

Example 10.9 The standard topology of the Euclidean plane $\mathbb{R}^2$ is separable, since it is generated by the set of all rational open balls,

$$\mathcal{B}_{\text{rat}} = \{B_a(\mathbf{r}) \mid a > 0 \in \mathbb{Q}, \mathbf{r} = (x, y) \text{ s.t. } x, y \in \mathbb{Q}\}.$$

The set $\mathcal{B}_{\text{rat}}$ is countable as it can be put in one-to-one correspondence with a subset of $\mathbb{Q}^3$. Since the rational numbers are dense in the real numbers, every point $\mathbf{x}$ of an open set U lies in a rational open ball. Thus every open set is a union of rational open balls. By a similar argument to that used in Example 10.8 it is straightforward to prove that the intersection of two sets from $\mathcal{B}_{\text{rat}}$ is a union of rational open balls. Hence $\mathbb{R}^2$ is separable. Similarly, all spaces $\mathbb{R}^n$ where $n \geq 1$ are separable.

10.2 General topological spaces

Let X and Y be two topological spaces. Theorem 10.1 motivates the following definition: a function $f : X \to Y$ is said to be **continuous** if the inverse image $f^{-1}(U)$ of every open set U in Y is open in X. If f is one-to-one and its inverse $f^{-1} : Y \to X$ is continuous, the function is called a **homeomorphism** and the topological spaces X and Y are said to be **homeomorphic** or **topologically equivalent**, written $X \cong Y$. The main task of *topology* is to find **topological invariants** – properties that are preserved under homeomorphisms. They may be real numbers, algebraic structures such as groups or vector spaces constructed from the topological space, or specific properties such as *compactness* and *connectedness*. The ultimate goal is to find a set of topological invariants that characterize a topological space. In the language of category theory, Section 1.7, continuous functions are the morphisms of the category whose objects are topological spaces, and homeomorphisms are the isomorphism of this category.

Example 10.10 Let $f : X \to Y$ be a continuous function between topological spaces. If the topology on X is discrete then every function f is continuous, for no matter what the topology on Y, every inverse image set $f^{-1}(U)$ is open in X. Similarly if the topology on Y is indiscrete than the function f is always continuous since the only inverse images in X of open sets are $f^{-1}(\emptyset) = \emptyset$ and $f^{-1}(Y) = X$, which are always open sets by (Top1).

Problems

Problem 10.1 Give an example in $\mathbb{R}^2$ of each of the following:

(a) A family of open sets whose intersection is a closed set that is not open.
(b) A family of closed sets whose union is an open set that is not closed.
(c) A set that is neither open nor closed.
(d) A countable dense set.
(e) A sequence of continuous functions $f_n : \mathbb{R}^2 \to \mathbb{R}$ whose limit is a discontinuous function.

Problem 10.2 If $\mathcal{U}$ generates the topology on X show that $\{A \cap U \mid U \in \mathcal{U}\}$ generates the relative topology on A.

Problem 10.3 Let X be a topological space and $A \subset B \subset X$. If B is given the relative topology, show that the relative topology induced on A by B is identical to the relative topology induced on it by X.

Problem 10.4 Show that for any subsets U, V of a topological space $\overline{U \cup V} = \overline{U} \cup \overline{V}$. Is it true that $\overline{U \cap V} = \overline{U} \cap \overline{V}$? What corresponding statements hold for the interior and boundaries of unions and intersections of sets?

Problem 10.5 If A is a dense set in a topological space X and $U \subseteq X$ is open, show that $U \subseteq \overline{A \cap U}$.

Problem 10.6 Show that a map $f : X \to Y$ between two topological spaces X and Y is continuous if and only if $f(\overline{U}) \subseteq \overline{f(U)}$ for all sets $U \subseteq X$. Show that f is a homeomorphism only if $f(\overline{U}) = \overline{f(U)}$ for all sets $U \subseteq X$.

Problem 10.7 Show the following:

(a) In the trivial topology, every sequence x_n converges to every point of the space $x \in X$.

Topology

(b) In R^2 the family of open sets consisting of all open balls centred on the origin $B_r(0)$ is a topology. Any sequence $\mathbf{x}_n \to \mathbf{x}$ converges to all points on the circle of radius $|\mathbf{x}|$ centred on the origin.

(c) If C is a closed set of a topological space X it contains all limit points of sequences $x_n \in C$.

(d) Let $f : X \to Y$ be a continuous function between topological spaces X and Y. If $x_n \to x$ is any convergent sequence in X then $f(x_n) \to f(x)$ in Y.

Problem 10.8 If W, X and Y are topological spaces and the functions $f : W \to X, g : X \to Y$ are both continuous, show that the function $h = g \circ f : W \to Y$ is continuous.

10.3 Metric spaces

To generalize the idea of 'distance' as it appears in $\mathbb{R}$ and $\mathbb{R}^2$, we define a **metric space** [9] to be a set M with a **distance function** or **metric** $d : M \times M \to \mathbb{R}$ such that

(Met1) $d(x, y) \geq 0$ for all $x, y \in M$.
(Met2) $d(x, y) = 0$ if and only if $x = y$.
(Met3) $d(x, y) = d(y, x)$.
(Met4) $d(x, y) + d(y, z) \geq d(x, z)$.

Condition (Met4) is called the **triangle inequality** – the length of any side of a triangle xyz is less than the sum of the other two sides. For every x in a metric space (M, d) and positive real number $a > 0$ we define the **open ball** $B_a(x) = \{y \mid d(x, y) < a\}$.

In n-dimensional Euclidean space $\mathbb{R}^n$ the distance function is given by

$$d(\mathbf{x}, \mathbf{y}) = |\mathbf{x} - \mathbf{y}| = \sqrt{(x_1 - y_1)^2 + (x_2 - y_2)^2 + \cdots + (x_n - y_n)^2},$$

but the following could also serve as acceptable metrics:

$$d_1(\mathbf{x}, \mathbf{y}) = |x_1 - y_1| + |x_2 - y_2| + \cdots + |x_n - y_n|,$$
$$d_2(\mathbf{x}, \mathbf{y}) = \max(|x_1 - y_1|, |x_2 - y_2|, \ldots, |x_n - y_n|).$$

Exercise: Show that $d(\mathbf{x}, \mathbf{y})$, $d_1(\mathbf{x}, \mathbf{y})$ and $d_2(\mathbf{x}, \mathbf{y})$ satisfy the metric axioms (Met1)–(Met4).

Exercise: In $\mathbb{R}^2$ sketch the open balls $B_1((0, 0))$ for the metrics d, d_1 and d_2.

If (M, d) is a metric space, then a subset $U \subset M$ is said to be open if and only if for every $x \in U$ there exists an open ball $B_\epsilon(x) \subseteq U$. Just as for $\mathbb{R}^2$, this defines a natural topology on M, called the **metric topology**. This topology is generated by the set of all open balls $B_a(x) \subset M$. The proof closely follows the argument in Example 10.8.

In a metric space (M, d), a sequence x_n converges to a point x if and only if $d(x_n, x) \to 0$ as $n \to \infty$. Equivalently, $x_n \to x$ if and only if for every $\epsilon > 0$ the sequence eventually enters and stays in the open ball $B_\epsilon(x)$. In a metric space the limit point x of a sequence x_n is unique, for if $x_n \to x$ and $x_n \to y$ then $d(x, y) \leq d(x, x_n) + d(x_n, y)$ by the triangle inequality. By choosing n large enough we have $d(x, y) < \epsilon$ for any $\epsilon > 0$. Hence $d(x, y) = 0$, and $x = y$ by (Met2). For this reason, the concept of convergent sequences is more useful in metric spaces than in general topological spaces (see Problem 10.7).

10.4 Induced topologies

In a metric space (M, d) let x_n be a sequence that converges to some point $x \in M$. Then for every $\epsilon > 0$ there exists a positive integer N such that $d(x_n, x_m) < \epsilon$ for all $n, m > N$. For, let N be an integer such that $d(x_k, x) < \frac{1}{2}\epsilon$ for all $k > N$, then

$$d(x_n, x_m) \leq d(x_n, x) + d(x, x_m) < \epsilon \quad \text{for all } n, m > N.$$

A sequence having this property, $d(x_n, x_m) \to 0$ as $n, m \to \infty$, is termed a **Cauchy sequence**.

Example 10.11 Not every Cauchy sequence need converge to a point of M. For example, in the open interval $(0, 1)$ with the usual metric topology, the sequence $x_n = 2^{-n}$ is a Cauchy sequence yet it does not converge to any point in the open interval. A metric space (M, d) is said to be **complete** if every Cauchy sequence $x_1, x_2, \ldots$ converges to a point $x \in M$. Completeness is not a topological property. For example the real line $\mathbb{R}$ is a complete metric space, and the Cauchy sequence 2^{-n} has the limit 0 in $\mathbb{R}$. The topological spaces $\mathbb{R}$ and $(0, 1)$ are homeomorphic, using the map $\varphi : x \mapsto \tan \frac{1}{2}\pi(2x - 1)$. However one space is complete while the other is not with respect to the metrics generating their topologies.

Problems

Problem 10.9 Show that every metric space is first countable. Hence show that every subset of a metric space can be written as the intersection of a countable collection of open sets.

Problem 10.10 If $\mathcal{U}_1$ and $\mathcal{U}_2$ are two families of subsets of a set X, show that the topologies generated by these families are homeomorphic if every member of $\mathcal{U}_2$ is a union of sets from $\mathcal{U}_1$ and vice versa. Use this property to show that the metric topologies on $\mathbb{R}^n$ defined by the metrics d, d_1 and d_2 are all homeomorphic.

Problem 10.11 A topological space X is called **normal** if for every pair of disjoint closed subsets A and B there exist disjoint open sets U and V such that $A \subset U$ and $B \subset V$. Show that every metric space is normal.

10.4 Induced topologies

Induced topologies and topological products

Given a topological space $(X, \mathcal{O})$ and a map $f : Y \to X$ from an arbitrary set Y into X, we can ask for the weakest topology on Y for which this map is continuous – it is useless to ask for the finest such topology since, as shown in Example 10.10, the discrete topology on Y always achieves this end. This is known as the topology **induced** on Y by the map f. Let $\mathcal{O}_f$ be the family of all inverse images of open sets of X,

$$\mathcal{O}_f = \{f^{-1}(U) \,|\, U \in \mathcal{O}\}.$$

Since f is required to be continuous, all members of this collection must be open in the induced topology. Furthermore, $\mathcal{O}_f$ is a topology on Y since (i) property (Top1) is trivial, as $\emptyset = f^{-1}(\emptyset)$ and $Y = f^{-1}(X)$; (ii) the axioms (Top2) and (Top3) follow from the set-theoretical identities

$$f^{-1}(U \cap V) = f^{-1}(U) \cap f^{-1}(V) \quad \text{and} \quad \bigcup_{i \in I} f^{-1}(U_i) = f^{-1}\left(\bigcup_{i \in I} U_i\right).$$

Hence $\mathcal{O}_f$ is a topology on Y and is included in any other topology such that the map f is continuous. It must be the topology induced on Y by the map f since it is the coarsest possible such topology.

Example 10.12 Let $(X, \mathcal{O})$ be any topological space and A any subset of X. In the topology induced on A by the natural inclusion map $i_A : A \to X$ defined by $i_A(x) = x$ for all $x \in A$, a subset B of A is open iff it is the intersection of A with an open set of X; that is, $B = A \cap U$ where U is open in X. This is precisely the relative topology on A defined in Section 10.2. The relative topology is thus the coarsest topology on A for which the inclusion map is continuous.

More generally, for a collection of maps $\{f_i : Y \to X_i \mid i \in I\}$ where X_i are topological spaces, the weakest topology on Y such that all these maps are continuous is said to be the topology **induced** by these maps. To create this topology it is necessary to consider the set of all inverse images of open sets $\mathcal{U} = \{f_i^{-1}(U_i) \mid U_i \in \mathcal{O}_i\}$. This collection of sets is not itself a topology in general, the topology generated by these sets *will* be the coarsest topology on Y such that each function f_i is continuous.

Given two topological spaces $(X, \mathcal{O}_X)$ and $(Y, \mathcal{O}_Y)$, let $\text{pr}_1 : X \times Y \to X$ and $\text{pr}_2 : X \times Y \to Y$ be the natural projection maps defined by

$$\text{pr}_1(x, y) = x \quad \text{and} \quad \text{pr}_2(x, y) = y.$$

The **product topology** on the set $X \times Y$ is defined as the topology induced by these two maps. The space $X \times Y$ together with the product topology is called the **topological product** of X and Y. It is the coarsest topology such that the projection maps are continuous. The inverse image under pr_1 of an open set U in X is a 'vertical strip' $U \times Y$, while the inverse image of an open set V in Y under pr_2 is a 'horizontal strip' $X \times V$. The intersection of any pair of these strips is a set of the form $U \times V$ where U and V are open sets from X and Y respectively (see Fig. 10.4). Since the topology generated by the vertical and horizontal strips consists of all possible unions of such intersections, it follows that in the product topology a subset $A \subset X \times Y$ is open if for every point $(x, y) \in A$ there exist open sets $U \in \mathcal{O}_X$ and $V \in \mathcal{O}_Y$ such that $(x, y) \in U \times V \subseteq A$.

Given an arbitrary collection of sets $\{X_i \mid i \in I\}$, their **cartesian product** $P = \prod_{i \in I} X_i$ is defined as the set of maps $f : I \to \bigcup_i X_i$ such that $f(i) \in X_i$ for each $i \in I$. For a finite number of sets, taking $I = \{1, 2, \ldots, n\}$, this concept is identical with the set of n-tuples from $X_1 \times X_2 \times \cdots \times X_n$. The product topology on P is the topology induced by the projection maps $\text{pr}_i : P \to X_i$ defined by $\text{pr}_i(f) = f(i)$. This topology is coarser than the topology generated by all sets of the form $\prod_{i \in I} U_i$ where U_i is an open subset of X_i.

Example 10.13 Let S^1 be the unit circle in $\mathbb{R}^2$ defined by $x^2 + y^2 = 1$, with the relative topology. The product space $S^1 \times S^1$ is homeomorphic to the torus T^2 or 'donut', with topology induced from its embedding as a subset of $\mathbb{R}^3$. This can be seen by embedding S^1 in the $z = 0$ plane of $\mathbb{R}^3$, and attaching a vertical unit circle facing outwards from each

10.4 Induced topologies

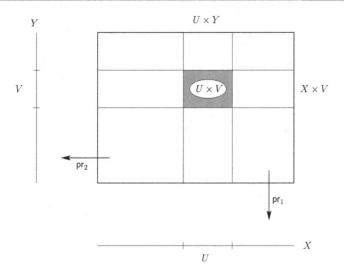

Figure 10.4   Product of two topological spaces

point on S^1. As the vertical circles 'sweep around' the horizontal cirle the resulting circle is clearly a torus.

The following is an occasionally useful theorem.

Theorem 10.3   *If X and Y are topological spaces then for each point $x \in X$ the **injection map** $\iota_x : Y \to X \times Y$ defined by $\iota_x(y) = (x, y)$ is continuous. Similarly the map $\iota'_y : X \to X \times Y$ defined by $\iota'_y(x) = (x, y)$ is continuous.*

Proof: Let U and V be open subsets of X and Y respectively. Then

$$(\iota_x)^{-1}(U \times V) = \begin{cases} V & \text{if } x \in U, \\ \emptyset & \text{if } x \notin U. \end{cases}$$

Since every open subset of $U \times V$ in the product topology is a union of sets of type $U \times V$, it follows that the inverse image under ι_x of every open set in $X \times Y$ is an open subset of Y. Hence the map ι_x is continuous. Similarly for the map ι'_y. ∎

Topology by identification

We may also reverse the above situation. Let $(X, \mathcal{O})$ be a topological space, and $f : X \to Y$ a map from X onto an arbitrary set Y. In this case the topology on Y **induced** by f is defined to be the *finest* topology such that f is continuous. This topology consists of all subsets $U \subseteq Y$ such that $f^{-1}(U)$ is open in X; that is, $\mathcal{O}_Y = \{f^{-1}(U) \mid U \in \mathcal{O}\}$.

Exercise: Show that $\mathcal{O}_Y$ is a topology on Y.

Exercise: Show that $\mathcal{O}_Y$ is the strongest topology such that f is continuous.

Topology

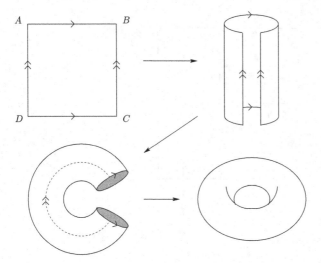

Figure 10.5 Construction of a torus by identification of opposite sides of a square

A common instance of this type of induced topology occurs when there is an equivalence relation E defined on a topological space X. Let $[x] = \{y \mid yEx\}$ be the equivalence class containing the point $x \in X$. In the factor space $X/E = \{[x] \mid x \in X\}$ define the topology obtained **by identification** from X to be the topology induced by the natural map $i_E : X \to X/E$ associating each point $x \in X$ with the equivalence class to which it belongs, $i_E(x) = [x]$. In this topology a subset A is open iff its inverse image $i_E^{-1}(A) = \{x \in X \mid [x] \in A\} = \bigcup_{[x] \in A}[x]$ is an open subset of X. That is, a subset of equivalence classes $A = \{[x]\}$ is open in the identification topology on X/E iff the union of the sets $[x]$ that belong to A is an open subset of X.

Exercise: Verify directly from the last statement that the axioms (Top1)–(Top3) are satisfied by this topology on X/E.

Example 10.14 As in Example 1.4, we say two points (x, y) and (x', y') in the plane $\mathbb{R}^2$ are equivalent if their coordinates differ by integral amounts,

$$(x, y) \equiv (x'y') \quad \text{iff} \quad x - x' = n, \ y - y' = m \quad (n, m \in \mathbb{Z}).$$

The topology on the space $\mathbb{R}^2/\equiv$ obtained by identification can be pictured as the unit square with opposite sides identified (see Fig. 10.5). To understand that this is a representation of the torus T^2, consider a square rubber sheet. Identifying sides AD and BC is equivalent to joining these two sides together to form a cylinder. The identification of AB and CD is now equivalent to identifying the circular edges at the top and bottom of the cylinder. In three-dimensions this involves bending the cylinder until top and bottom join up to form the inner tube of a tyre – remember, distances or metric properties need not be preserved for a topological transformation. The n-torus T^n can similarly be defined as the topological

10.5 Hausdorff spaces

space obtained by identification from the corresponding equivalence relation on R^n, whereby points are equivalent if their coordinates differ by integers.

Example 10.15 Let $\dot{\mathbb{R}}^3 = \mathbb{R}^3 - \{0\}$ be the set of non-zero 3-triples of real numbers given the relative topology in $\mathbb{R}^3$. Define an equivalence relation on $\dot{\mathbb{R}}^3$ whereby $(x, y, z) \equiv (x', y', z')$ iff there exists a real number $\lambda \neq 0$ such that $x = \lambda x'$, $y = \lambda y'$ and $z = \lambda z'$. The factor space $P^2 = \dot{\mathbb{R}}^3/\equiv$ is known as the **real projective plane**.

Each equivalence class $[(x, y, z)]$ is a straight line through the origin that meets the unit 2-sphere S^2 in two diametrically opposite points. Define an equivalence relation on S^2 by identifying diametrically opposite points, $(x, y, z) \sim (-x, -y, -z)$ where $x^2 + y^2 + z^2 = 1$. The topology on P^2 obtained by identification from $\dot{\mathbb{R}}^3$ is thus identical with that of the 2-sphere S^2 with diametrically opposite points identified.

Generalizing, we define **real projective n-space** P^n to be $\dot{\mathbb{R}}^{n+1}/\equiv$ where $(x_1, x_2, \ldots, x_{n+1}) \equiv (x'_1, x'_2, \ldots, x'_{n+1})$ if and only if there exists $\lambda \neq 0$ such that $x_1 = \lambda x'_1$, $x_2 = \lambda x'_2, \ldots, x_{n+1} = \lambda x'_{n+1}$. This space can be thought of as the set of all straight lines through the origin in $\mathbb{R}^{n+1}$. The topology of P^n is homeomorphic with that of the n-sphere S^n with opposite points identified.

Problems

Problem 10.12 If $f : X \to Y$ is a continuous map between topological spaces, we define its **graph** to be the set $G = \{(x, f(x)) \mid x \in X\} \subseteq X \times Y$. Show that if G is given the relative topology induced by the topological product $X \times Y$ then it is homeomorphic to the topological space X.

Problem 10.13 Let X and Y be topological spaces and $f : X \times Y \to X$ a continuous map. For each fixed $a \in X$ show that the map $f_a : Y \to X$ defined by $f_a(y) = f(a, y)$ is continuous.

10.5 Hausdorff spaces

In some topologies, for example the indiscrete topology, there are so few open sets that different points cannot be separated by non-intersecting neighbourhoods. To remedy this situation, conditions known as *separation axioms* are sometimes imposed on topological spaces. One of the most common of these is the **Hausdorff condition**: for every pair of points $x, y \in X$ there exist open neighbourhoods U of x and V of y such that $U \cap V = \emptyset$. A topological space satisfying this property is known as a **Hausdorff space**. In an intuitive sense, no pair of distinct points of a Hausdorff space are 'arbitrarily close' to each other.

A typical 'nice' property of Hausdorff spaces is the fact that the limit of any convergent sequence $x_n \to x$, defined in Problem 10.7, is unique. Suppose, for example, that $x_n \to x$ and $x_n \to x'$ in a Hausdorff space X. If $x \neq x'$ let U and U' be disjoint open neighbourhoods such that $x \in U$ and $x' \in U'$, and N an integer such that $x_n \in U$ for all $n > N$. Since $x_n \notin U'$ for all $n > N$ the sequence x_n cannot converge to x'. Hence $x = x'$.

In a Hausdorff space every singleton set $\{x\}$ is a closed set, for let $Y = X - \{x\}$ be its complement. Every point $y \in Y$ has an open neighbourhood U_y that does not intersect some open neighbourhood of x. In particular $x \notin U_y$. By (Top3) the union of all these open neighbourhoods, $Y = \bigcup_{y \in Y} U_y = X - \{x\}$, is open. Hence $\{x\} = X - Y$ is closed since it is the complement of an open set.

Exercise: Show that on a finite set X, the only Hausdorff topology is the discrete topology. For this reason, finite topologies are of limited interest.

Theorem 10.4 *Every metric space (X, d) is a Hausdorff space.*

Proof: Let $x, y \in X$ be any pair of unequal points and let $\epsilon = \frac{1}{4} d(x, y)$. The open balls $U = B_\epsilon(x)$ and $V = B_\epsilon(y)$ are open neighbourhoods of x and y respectively. Their intersection is empty, for if $z \in U \cap V$ then $d(x, z) < \epsilon$ and $d(y, z) < \epsilon$, which contradicts the triangle inequality (Met4),

$$d(x, y) \leq d(x, z) + d(z, y) \leq 2\epsilon < \tfrac{1}{2} d(x, y).$$

∎

An immediate consequence of this theorem is that the standard topology on $\mathbb{R}^n$ is Hausdorff for all $n > 0$.

Theorem 10.5 *If X and Y are topological spaces and $f : X \to Y$ is a one-to-one continuous mapping, then X is Hausdorff if Y is Hausdorff.*

Proof: Let x and x' be any pair of distinct points in X and set $y = f(x)$, $y' = f(x')$. Since f is one-to-one these are distinct points of Y. If Y is Hausdorff there exist non-intersecting open neighbourhoods U_y and $U_{y'}$ in Y of y and y' respectively. The inverse images of these sets under f are open neighbourhoods of x and x' respectively that are non-intersecting, since $f^{-1}(U_y) \cap f^{-1}(U_{y'}) = f^{-1}(U_y \cap U_{y'}) = f^{-1}(\emptyset) = \emptyset$. ∎

This shows that the Hausdorff condition is a genuine topological property, invariant under topological transformations, for if $f : X \to Y$ is a homeomorphism then $f^{-1} : Y \to X$ is continuous and one-to-one.

Corollary 10.6 *Any subspace of a Hausdorff space is Hausdorff in the relative topology.*

Proof: Let A be any subset of a topological space X. In the relative topology the inclusion map $i_A : A \to X$ is continuous. Since it is one-to-one, Theorem 10.5 implies that A is Hausdorff. ∎

Theorem 10.7 *If X and Y are Hausdorff topological spaces then their topological product $X \times Y$ is Hausdorff.*

Proof: Let (x, y) and (x', y') be any distinct pair of points in $X \times Y$, so that either $x \neq x'$ or $y \neq y'$. Suppose that $x \neq x'$. There then exist open sets U and U' in X such that $x \in U$, $x' \in U'$ and $U \cap U' = \emptyset$. The sets $U \times Y$ and $U' \times Y$ are disjoint open neighbourhoods of (x, y) and (x', y') respectively. Similarly, if $y \neq y'$ a pair of disjoint neighbourhoods of the form $X \times V$ and $X \times V'$ can be found that separate the two points. ∎

Problems

Problem 10.14 If Y is a Hausdorff topological space show that every continuous map $f : X \to Y$ from a topological space X with indiscrete topology into Y is a *constant map*; that is, a map of the form $f(x) = y_0$ where y_0 is a fixed element of Y.

Problem 10.15 Show that if $f : X \to Y$ and $g : X \to Y$ are continuous maps from a topological space X into a Hausdorff space Y then the set of points A on which these maps agree, $A = \{x \in X \,|\, f(x) = g(x)\}$, is closed. If A is a dense subset of X show that $f = g$.

10.6 Compact spaces

A collection of sets $\mathcal{U} = \{U_i \,|\, i \in I\}$ is said to be a **covering** of a subset A of a topological space X if every point $x \in A$ belongs to some member of the collection. If every member $\mathcal{U}$ is an open set it is called an **open covering**. A subset of the covering, $\mathcal{U}' \subseteq \mathcal{U}$, which covers A is referred to as a **subcovering**. If $\mathcal{U}'$ consists of finitely many sets $\{U_1, U_2, \ldots, U_n\}$ it is called a **finite subcovering**.

A topological space $(X, \mathcal{O})$ is said to be **compact** if every open covering of X contains a finite subcovering. The motivation for this definition lies in the following theorem, the proof of which can be found in standard books on analysis [10–12].

Theorem 10.8 (Heine–Borel) *A subset A of $\mathbb{R}^n$ is closed and bounded (included in a central ball, $A \subset B_a(0)$ for some $a > 0$) if and only if every open covering $\mathcal{U}$ of A has a finite subcovering.*

Theorem 10.9 *Every closed subspace A of a compact space X is compact in the relative topology.*

Proof: Let $\mathcal{U}$ be any covering of A by sets that are open in the relative topology. Each member of this covering must be of the form $U \cap A$, where U is open in Y. The sets $\{U\}$ together with the open set $X - A$ form an open covering of X that, by compactness of X, must have a finite subcovering $\{U_1, U_2, \ldots, U_n, X - A\}$. The sets $\{U_1 \cap A, \ldots, U_n \cap A\}$ are thus a finite subfamily of the original open covering $\mathcal{U}$ of A. Hence A is compact in the relative topology. ∎

Theorem 10.10 *If $f : X \to Y$ is a continuous map from a compact topological space X into a topological space Y, then the image set $f(X) \subseteq Y$ is compact in the relative topology.*

Proof: Let $\mathcal{U}$ be any covering of $f(X)$ consisting entirely of open sets in the relative topology. Each member of this covering is of the form $U \cap f(X)$, where U is open in Y. Since f is continuous, the sets $f^{-1}(U)$ form an open covering of X. By compactness of X, a finite subfamily $\{f^{-1}(U_i) \,|\, i = 1, \ldots, n\}$ serves to cover X, and the corresponding sets $U_i \cap f(X)$ evidently form a finite subcovering of $f(X)$. ∎

Compactness is therefore a topological property, invariant under homeomorphisms.

Example 10.16 If E is an equivalence relation on a compact topological space X, the map $i_E : X \to X/E$ is continuous in the topology on X/E obtained by identification from X. By

Theorem 10.10 the topological space X/E is compact. For example, the torus T^2 formed by identifying opposite sides of the closed and compact unit square in $\mathbb{R}^2$ is a compact space.

Theorem 10.11 *The topological product $X \times Y$ is compact if and only if both X and Y are compact.*

Proof: If $X \times Y$ is compact then X and Y are compact by Theorem 10.10 since both the projection maps $\mathrm{pr}_1 : X \times Y \to X$ and $\mathrm{pr}_2 : X \times Y \to Y$ are continuous in the product topology.

Conversely, suppose X and Y are compact. Let $\mathcal{W} = \{W_i \mid i \in I\}$ be an open covering of $X \times Y$. Since each set W_i is a union of such sets of the form $U \times V$ where U and V are open sets of X and Y respectively, the family of all such sets $U_j \times V_j$ ($j \in J$) that are subsets of W_i for some $i \in I$ is an open cover of $X \times Y$. Given any point $y \in Y$, the set of all U_j such that $y \in V_j$ is an open cover of X, and since X is compact there exists a finite subcover $\{U_{j_1}, U_{j_2}, \ldots, U_{j_n}\}$. The set $A_y = V_{j_1} \cap V_{j_2} \cap \cdots \cap V_{j_n}$ is an open set in Y by condition (Top2), and $y \in A_y$ since $y \in V_{j_k}$ for each $k = 1, \ldots, n$. Thus the family of sets $\{A_y \mid y \in Y\}$ forms an open cover of Y. As Y is compact, there is a finite subcovering $A_{y_1}, A_{y_2}, \ldots, A_{y_m}$. The totality of all the sets $U_{j_k} \times V_{j_k}$ associated with these sets A_{y_a} forms a finite open covering of $X \times Y$. For each such set select a corresponding member W_i of the original covering $\mathcal{W}$ of which it is a subset. The result is a finite subcovering of $X \times Y$, proving that $X \times Y$ is compact. ∎

Somewhat surprisingly, this statement extends to arbitrary infinite products (*Tychonoff's theorem*). The interested reader is referred to [8] or [2] for a proof of this more difficult result.

Theorem 10.12 *Every infinite subset of a compact topological space has an accumulation point.*

Proof: Suppose X is a compact topological space and $A \subset X$ has no accumulation point. The aim is to show that A is a finite set. Since every point in $x \in A - X$ has an open neighbourhood U_x such that $U_x \cap A = \emptyset$ it follows that $A \subseteq X$ is closed since its complement $A - X = \bigcup_{x \in X - A} U_x$ is open. Hence, by Theorem 10.9, A is compact. Since each point $a \in A$ is not an accumulation point, there exists an open neighbourhood U_a of a such that $U_a \cap A = \{a\}$. Hence each singleton $\{a\}$ is an open set in the relative topology induced on A, and the relative topology on A is therefore the discrete topology. The singleton sets $\{a \mid a \in A\}$ therefore form an open covering of A, and since A is compact there must be a finite subcovering $\{a_1\}, \{a_2\}, \ldots, \{a_n\}$. Thus $A = \{a_1, a_2, \ldots, a_n\}$ is a finite set. ∎

Theorem 10.13 *Every compact subspace of a Hausdorff space is closed.*

Proof: Let X be a Hausdorff space and A a compact subspace in the relative topology. If $a \in A$ and $x \in X - A$ then there exist disjoint open sets U_a and V_a such that $a \in U_a$ and $x \in V_a$. The family of open sets $U_a \cap A$ is an open covering of A in the relative topology. Since A is compact there is a finite subcovering $\{U_{a_1} \cap A, \ldots, U_{a_n} \cap A\}$. The intersection of the corresponding neighbourhoods $W = V_{a_1} \cap \cdots \cap V_{a_n}$ is an open set that contains x. As all its points lie outside every $U_{a_i} \cap A$ we have $W \cap A = \emptyset$. Thus every point $x \in X - A$ has an open neighbourhood with no points in A. Hence A includes all its accumulation points and must be a closed set. ∎

In a metric space (M, d) we will say a subset A is **bounded** if $\sup\{d(x, y) \mid x, y \in A\} < \infty$.

Theorem 10.14 *Every compact subspace of a metric space is closed and bounded.*

Proof: Let A be a compact subspace of a metric space (M, d). Since M is a Hausdorff space by Theorem 10.4 it follows by the previous theorem that A is closed. Let $\mathcal{U} = \{B_1(a) \cap A \mid a \in A\}$ be the open covering of A consisting of intersections of A with unit open balls centred on points of A. Since A is compact, a finite number of these open balls $\{B_1(a_1), B_1(a_2), \ldots, B_1(a_n)\}$ can be selected to cover A. Let the greatest distance between any pair of these points be $D = \max d(a_i, a_j)$. For any pair of points $a, b \in A$, if $a \in B_1(a_k)$ and $b \in B_1(a_l)$ then by the triangle inequality

$$d(a, b) \leq d(a, a_i) + d(a_i, a_j) + \cdots + d(a_j, b) \leq D + 2.$$

Thus A is a bounded set. ∎

Problems

Problem 10.16 Show that every compact Hausdorff space is normal (see Problem 10.11).

Problem 10.17 Show that every one-to-one continuous map $f : X \to Y$ from a compact space X onto a Hausdorff space Y is a homeomorphism.

10.7 Connected spaces

Intuitively, we can think of a topological space X as being 'disconnected' if it can be decomposed into two disjoint subsets $X = A \cup B$ without these sets having any boundary points in common. Since the boundary of a set is at the same time the boundary of the complement of that set, the only way such a decomposition can occur is if there exists a set A other than the empty set or the whole space X that has no boundary points at all. Since $b(A) = \overline{A} - A^o$, the only way $b(A) = \emptyset$ can occur is if $\overline{A} = A^o$. As $\overline{A} \subseteq A \subseteq A^o$, the set A must equal both its closure and interior; in particular, it would need to be both open and closed at the same time. This motivates the following definition: a topological space X is said to be **connected** if the only subsets that are both open and closed are the empty set and the space X itself. A space is said to be **disconnected** if it is not connected. In other words, X is disconnected if $X = A \cup B$ where A and B are disjoint sets that are both open and closed. A subset $A \subset X$ is said to be **connected** if it is connected in the relative topology.

Example 10.17 The indiscrete topology on any set X is connected, since the only open sets are $\emptyset$ or X. The discrete topology on any set X having more than one point is disconnected since every non-empty subset is both open and closed.

Example 10.18 The real numbers $\mathbb{R}$ are connected in the standard topology. To show this, let $A \subset \mathbb{R}$ be both open and closed. If $x \in A$ set y to be the least upper bound of those real numbers such that $[x, y) \subset A$. If $y < \infty$ then y is an accumulation point of A, and therefore $y \in A$ since A is closed. However, since A is an open set there exists an interval $(y - a, y + a) \subset A$. Thus $[x, y + a) \subset A$, contradicting the stipulation that y is the least upper bound. Hence $y = \infty$. Similarly $(-\infty, x] \subset A$ and the only possibility is that $A = \emptyset$ or $A = \mathbb{R}$.

Theorem 10.15 *The closure of a connected set is connected.*

Proof: Let $A \subset X$ be a connected set. Suppose U is a subset of the closure $\overline{A}$ of A, which is both open and closed in $\overline{A}$, and let $V = \overline{A} - U$ be the complement of U in $\overline{A}$. Since A is connected and the sets $U \cap A$ and $V \cap A$ are both open and closed in A, one of them must be the empty set, while the other is the whole set A. If, for example, $V \cap A = \emptyset$ then $U \cap A = A$, so that $A \subset U \subseteq \overline{A}$. Since U is closed in $\overline{A}$ we must have that $U = \overline{A}$ and $V = \overline{A} - U = \emptyset$. If $U \cap A = \emptyset$ then $U = \emptyset$ by an identical argument. Hence $\overline{A}$ is connected. ∎

The following theorem is used in many arguments to do with connectedness of topological spaces or their subspaces. Intuitively, it says that connectedness is retained if any number of connected sets are 'attached' to a given connected set.

Theorem 10.16 *Let A_0 be any connected subset of a topological space X and $\{A_i \mid i \in I\}$ any family of connected subsets of X such that $A_0 \cap A_i \ne \emptyset$ for each member of the family. Then the set $A = A_0 \cup \left(\bigcup_{i \in I} A_i\right)$ is a connected subset of X.*

Proof: Suppose $A = U \cup V$ where U and V are disjoint open sets in the relative topology on A. For all $i \in I$ the sets $U \cap A_i$ and $V \cap A_i$ are disjoint open sets of A_i whose union is A_i. Since A_i is connected, either $U \cap A_i = \emptyset$ or $V \cap A_i = \emptyset$. This also holds for A_0: either $U \cap A_0 = \emptyset$ or $V \cap A_0 = \emptyset$, say the latter. Then $U \cap A_0 = A_0$, so that $A_0 \subseteq U$. Since $A_0 \cap A_i \ne \emptyset$ we have $U \cap A_i \ne \emptyset$ for all $i \in I$. Hence $V \cap A_i = \emptyset$ and $U \cap A_i = A_i$; that is, $A_i \subseteq U$ for all $i \in I$. Hence $U = A$ and $V = \emptyset$, showing that A is a connected subset of X. ∎

A theorem similar to Theorem 10.10 is available for connectedness: the image of a connected space under a continuous map is connected. This also shows that connectedness is a topological property, invariant under homeomorphisms.

Theorem 10.17 *If $f : X \to Y$ is a continuous map from a connected topological space X into a topological space Y, its image set $f(X)$ is a connected subset of Y.*

Proof: Let B be any non-empty subset of $f(X)$ that is both open and closed in the relative topology. This means there exists an open set $U \subseteq Y$ and a closed set $C \subseteq Y$ such that $B = U \cap f(X) = C \cap f(X)$. Since f is a continuous map, the inverse image set $f^{-1}(B) = f^{-1}(U) = f^{-1}(C)$ is both open and closed in X. As X is connected it follows that $B = f(X)$; hence $f(X)$ is connected. ∎

A useful application of these theorems is to show the topological product of two connected spaces is connected.

10.7 Connected spaces

Theorem 10.18 *The topological product $X \times Y$ of two topological spaces is connected if and only if both X and Y are connected spaces.*

Proof: By Theorem 10.3, the maps $\iota_x : Y \to X \times Y$ and $\iota'_y : X \to X \times Y$ defined by $\iota_x(y) = \iota'_y(x) = (x, y)$ are both continuous. Suppose that both X and Y are connected topological spaces. Select a fixed point $y_0 \in Y$. By Theorem 10.17 the set of points $X \times y_0 = \{(x, y_0) \mid x \in X\} = \iota'_{y_0}(X)$ is a connected subset of $X \times Y$. Similarly, the sets $\{x \times Y = \iota_x(Y) \mid x \in X\}$ are connected subsets of $X \times Y$, each of which intersects $X \times y_0$ in the point (x, y_0). The union of these sets is clearly $X \times Y$, which by Theorem 10.16 must be connected.

Conversely, suppose $X \times Y$ is connected. The spaces X and Y are both connected, by Theorem 10.17, since they are the images of the continuous projection maps $\text{pr}_1 : X \times Y \to X$ and $\text{pr}_2 : X \times Y \to Y$, respectively. ∎

Example 10.19 The spaces $\mathbb{R}^n$ are connected by Example 10.18 and Theorem 10.18. To show that the 2-sphere S^2 is connected consider the 'punctured' spheres $S' = S^2 - \{N = (0, 0, 1)\}$ and $S'' = S^2 - \{S = (0, 0, -1)\}$ by removing the north and south poles, respectively. The set S' is connected since it is homeomorphic to the plane $\mathbb{R}^2$ under stereographic projection (Fig. 10.6),

$$x' = \frac{x}{1-z}, \quad y' = \frac{y}{1-z} \quad \text{where} \quad z = \pm\sqrt{1 - x^2 - y^2}, \tag{10.1}$$

which has continuous inverse

$$x = \frac{2x'}{r'^2 + 1}, \quad y = \frac{2y'}{r'^2 + 1}, \quad z = \frac{r'^2 - 1}{r'^2 + 1} \quad (r'^2 = x'^2 + y'^2). \tag{10.2}$$

Similarly S'' is connected since it is homeomorphic to $\mathbb{R}^2$. As $S' \cap S'' \neq \emptyset$ and $S^2 = S' \cup S''$ it follows from Theorem 10.16 that S^2 is a connected subset of $\mathbb{R}^3$. A similar argument can be used to show that the n-sphere S^n is a connected topological space for all $n \geq 1$.

A **connected component** C of a topological space X is a maximal connected set; that is, C is a connected subset of X such that if $C' \supseteq C$ is any connected superset of C then $C' = C$. A **connected component of a subset** $A \subset X$ is a connected component with respect to the relative topology on A. By Theorem 10.15 it is immediate that any connected component is a closed set, since it implies that $C = \overline{C}$. A topological space X is connected if and only if the whole space X is its only connected component. In the discrete topology the connected components consist of all singleton sets $\{x\}$.

Exercise: Show that any two distinct components A and B are *separated*, in the sense that $\overline{A} \cap B = A \cap \overline{B} = \emptyset$.

Theorem 10.19 *Each connected subset A of a topological space lies in a unique connected component.*

Proof: Let C be the union of all connected subsets of X that contain the set A. Since these sets all intersect the connected subset A it follows from Theorem 10.16 that C is a connected set. It is clearly maximal, for if there exists a connected set C_1 such that $C \subseteq C_1$, then C_1 is in the family of sets of which C is the union, so that $C \supset C_1$. Hence $C = C_1$.

Topology

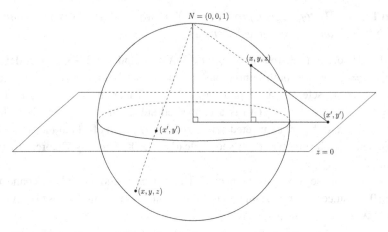

Figure 10.6 Stereographic projection from the north pole of a sphere

To prove uniqueness, suppose C' were another connected component such that $C' \supset A$. By Theorem 10.16, $C \cup C'$ is a connected set and by maximality of C and C' we have $C \cup C' = C = C'$. ∎

Problems

Problem 10.18 Show that a topological space X is connected if and only if every continuous map $f : X \to Y$ of X into a discrete topological space Y consisting of at least two points is a constant map (see Problem 10.14).

Problem 10.19 From Theorem 10.16 show that the unit circle S^1 is connected, and that the punctured n-space $\dot{\mathbb{R}}^n = \mathbb{R}^n - \{0\}$ is connected for all $n > 1$. Why is this not true for $n = 1$?

Problem 10.20 Show that the real projective space defined in Example 10.15 is connected, Hausdorff and compact.

Problem 10.21 Show that the rational numbers $\mathbb{Q}$ are a disconnected subset of the real numbers. Are the irrational points a disconnected subset of $\mathbb{R}$? Show that the connected components of the rational numbers $\mathbb{Q}$ consist of singleton sets $\{x\}$.

10.8 Topological groups

There are a number of useful ways in which topological and algebraic structure can be combined. The principal requirement connecting the two types of structure is that the functions representing the algebraic laws of composition be continuous with respect to the topology imposed on the underlying set. In this section we combine group theory with topology.

A **topological group** is a set G that is both a group and a Hausdorff topological space such that the map $\psi : G \times G \to G$ defined by $\psi(g, h) = gh^{-1}$ is continuous. The topological group G is called **discrete** if the underlying topology is discrete.

10.8 Topological groups

The maps $\phi : G \to G$ and $\tau : G \to G$ defined by $\phi(g, h) = gh$ and $\tau(g) = g^{-1}$ are both continuous. For, by Theorem 10.3 the injection map $i : G \to G \times G$ defined by $i(h) = \iota_e(h) = (e, h)$ is continuous. The map τ is therefore continuous since it is a composition of continuous maps, $\tau : \psi \circ i$. Since $\phi(g, h) = \psi(g, \tau(h))$ it follows immediately that ϕ is also a continuous map.

Exercise: Show that τ is a homeomorphism of G.

Exercise: If ϕ and τ are continuous maps, show that ψ is continuous.

Example 10.20 The additive group $\mathbb{R}^n$, where the 'product' is vector addition

$$\phi(\mathbf{x}, \mathbf{y}) = (x^1, \ldots, x^n) + (y^1, \ldots, y^n) = (x^1 + y^1, \ldots, x^n + y^n)$$

and the inverse map is

$$\tau(\mathbf{x}) = -\mathbf{x} = (-x^1, \ldots, -x^n),$$

is an abelian topological group with respect to the Euclidean topology on $\mathbb{R}^n$. The n-torus $T^n = \mathbb{R}^n/\mathbb{Z}^n$ is also an abelian topological group, where group composition is addition modulo 1.

Example 10.21 The set $M_n(\mathbb{R})$ of $n \times n$ real matrices has a topology homeomorphic to the Euclidean topology on $\mathbb{R}^{n^2}$. The determinant map $\det : M_n(\mathbb{R}) \to \mathbb{R}$ is clearly continuous since $\det A$ is a polynomial function of the components of A. Hence the general linear group $GL(n, \mathbb{R})$ is an open subset of $M_n(\mathbb{R})$ since it is the inverse image of the open set $\dot{\mathbb{R}} = \mathbb{R} - \{0\}$ under the determinant map. If $GL(n, \mathbb{R})$ is given the induced relative topology in $M_n(\mathbb{R})$ then the map ψ reads in components,

$$(\psi(\mathsf{A}, \mathsf{B}))_{ij} = \sum_{k=1}^{n} A_{ik} \left(B^{-1}\right)_{kj}.$$

These are continuous functions since $\left(B^{-1}\right)_{ij}$ are rational polynomial functions of the components B_{ij} with non-vanishing denominator $\det \mathsf{B}$.

A subgroup H of G together with its relative topology is called a **topological subgroup** of G. To show that any subgroup H becomes a topological subgroup with respect to the relative topology, let $U' = H \cap U$ where U is an arbitrary open subset of G. By continuity of the map ϕ, for any pair of points $g, h \in H$ such that $gh \in U' \subset U$ there exist open sets A and B of G such that $A \times B \subset \phi^{-1}(U)$. It follows that $\phi(A' \times B') \subset H \cap U$, where $A' = A \cap H$, $B' = B \cap H$, and the continuity of $\phi\big|_H$ is immediate. Similarly the inverse map τ is continuous when restricted to H. If H is a closed set in G, it is called a **closed subgroup** of G.

For each $g \in G$ let the **left translation** $L_g : G \to G$ be the map

$$L_g(h) \equiv L_g h = gh,$$

as defined in Example 2.25. The map L_g is continuous since it is the composition of two continuous maps, $L_g = \phi \circ \iota_g$, where $\iota_g : G \to G \times G$ is the injection map $\iota_g(h) = (g, h)$

(see Theorem 10.3). It is clearly one-to-one, for $gh = gh' \Longrightarrow h = g^{-1}gh' = h'$, and its inverse is the continuous map $L_{g^{-1}}$. Hence L_g is a homeomorphism. Similarly, every **right translation** $R_g : G \to G$ defined by $R_g h = hg$ is a homeomorphism of G, as is the **inner automorphism** $C_g : G \to G$ defined by $C_g h = ghg^{-1} = L_h \circ R_{h^{-1}}(g)$.

Connected component of the identity

If G is a topological group we will denote by G_0 the connected component containing the identity element e, simply referred to as the **component of the identity**.

Theorem 10.20 *Let G be a topological group, and G_0 the component of the identity. Then G_0 is a closed normal subgroup of G.*

Proof: By Theorem 10.17 the set $G_0 g^{-1}$ is connected, since it is a continuous image under right translation by g^{-1} of a connected set. If $g \in G_0$ then $e = gg^{-1} \in G_0 g^{-1}$. Hence $G_0 g^{-1}$ is a closed connected subset containing the identity e, and must therefore be a subset of G_0. We have therefore $G_0 G_0^{-1} \subseteq G_0$, showing that G_0 is a subgroup of G. Since it is a connected component of G it is a closed set. Thus, G_0 is a closed subgroup of G.

For any $g \in G$, the set $gG_0 g^{-1}$ is connected as it is the image of G_0 under the inner automorphism $h \mapsto C_g(h)$. Since this set contains the identity e, we have $gG_0 g^{-1} \subseteq G_0$, and G_0 is a normal subgroup. ∎

A topological space X is said to be **locally connected** if every neighbourhood of every point of X contains a connected open neighbourhood. A topological group G is locally connected if it is locally connected at the identity e, for if V is a connected open neighbourhood of e then $gV = L_g V$ is a connected open neighbourhood of any selected point $g \in G$. If K is any subset of a group G, we call the smallest subgroup of G that contains K the **subgroup generated by** K. It is the intersection of all subgroups of G that contain K.

Theorem 10.21 *In any locally connected group G the component of the identity G_0 is generated by any connected neighbourhood of the identity e.*

Proof: Let V be any connected neighbourhood of e, and H the subgroup generated by V. For any $g \in H$, the left coset $gV = L_g V \subset H$ is a neighbourhood of g since L_g is a homeomorphism. Hence H is an open subset of G. On the other hand, if H is an open subgroup of G it is also closed since it is the complement in G of the union of all cosets of H that differ from H itself. Thus H is both open and closed. It is therefore the connected component of the identity, G_0. ∎

Let H be a closed subgroup of a topological group G, we can give the factor space G/H the natural topology induced by the canonical projection map $\pi : g \mapsto gH$. This is the finest topology on G/H such that π is a continuous map. In this topology a collection of cosets $U \subseteq G/H$ is open if and only if their union is an open subset of G. Clearly π is an **open map** with respect to this topology, meaning that $\pi(V)$ is open for all open sets $V \subseteq G$.

10.9 Topological vector spaces

Theorem 10.22 *If G is a topological group and H a closed connected subgroup such that the factor space G/H is connected, then G is connected.*

Proof: Suppose G is not connected. There then exist open sets U and V such that $G = U \cup V$, with $U \cap V = \emptyset$. Since π is an open map the sets $\pi(U)$ and $\pi(V)$ are open in G/H and $G/H = \pi(U) \cup \pi(V)$. But G/H is connected, $\pi(U) \cap \pi(V) \neq \emptyset$, so there exists a coset $gH \in \pi(U) \cap \pi(V)$. As a subset of G this coset clearly meets both U and V, and $gH = (gH \cap U) \cup (gH \cap V)$, contradicting the fact that gH is connected (since it is the image under the continuous map L_g of a connected set H). Hence G is connected. ∎

Example 10.22 The general linear group $GL(n, \mathbb{R})$ is not connected since the determinant map $\det : GL(n, \mathbb{R}) \to \mathbb{R}$ has image $\dot{\mathbb{R}} = \mathbb{R} - \{0\}$, which is a disconnected set. The component G_0 of the identity I is the set of $n \times n$ matrices with determinant > 0, and the group of components is discrete

$$GL(n, \mathbb{R})/G_0 \cong \{1, -1\} = Z_2.$$

Note, however, that the complex general linear group $GL(n, \mathbb{C})$ is connected, as may be surmised from the fact that the Jordan canonical form of any non-singular complex matrix can be continuously deformed to the identity matrix I.

The special orthogonal groups $SO(n)$ are all connected. This can be shown by induction on the dimension n. Evidently $SO(1) = \{1\}$ is connected. Assume that $SO(n)$ is connected. It will be shown in Chapter 19, Example 19.10, that $SO(n+1)/SO(n)$ is homeomorphic to the n-sphere S^n. As this is a connected set (see Example 10.19) it follows from Theorem 10.22 that $SO(n+1)$ is connected. By induction, $SO(n)$ is a connected group for all $n = 1, 2, \ldots$ However the orthogonal groups $O(n)$ are not connected, the component of the identity being $SO(n)$ while the remaining orthogonal matrices have determinant -1.

Similarly, $SU(1) = \{1\}$ and $SU(n+1)/SU(n) \cong S^{2n-1}$, from which it follows that all special unitary groups $SU(n)$ are connected. By Theorem 10.22 the unitary groups $U(n)$ are also all connected, since $U(n)/SU(n) = S^1$ is connected.

Problem

Problem 10.22 If G_0 is the component of the identity of a locally connected topological group G, the factor group G/G_0 is called the **group of components** of G. Show that the group of components is a discrete topological group with respect to the topology induced by the natural projection map $\pi : g \mapsto gG_0$.

10.9 Topological vector spaces

A **topological vector space** is a vector space V that has a Hausdorff topology defined on it, such that the operations of vector addition and scalar multiplication are continuous functions on their respective domains with respect to this topology,

$$\psi : V \times V \to V \quad \text{defined by} \quad \psi(u, v) = u + v,$$
$$\tau : \mathbb{K} \times V \to V \quad \text{defined by} \quad \tau(\lambda, v) = \lambda v.$$

We will always assume that the field of scalars is either the real or complex numbers, $\mathbb{K} = \mathbb{R}$ or $\mathbb{K} = \mathbb{C}$; in the latter case the topology is the standard topology in $\mathbb{R}^2$.

Recall from Section 10.3 that a sequence of vectors $v_n \in V$ is called **convergent** if there exists a vector $v \in V$, called its **limit**, such that for every open neighbourhood U of v there is an integer N such that $v_n \in U$ for all $n \geq N$. We also say the sequence **converges to** v, denoted

$$v_n \to v \quad \text{or} \quad \lim_{n \to \infty} v_n = v.$$

The following properties of convergent sequences are easily proved:

$$v_n \to v \quad \text{and} \quad v_n \to v' \implies v = v', \tag{10.3}$$

$$v_n = v \text{ for all } n \implies v_n \to v, \tag{10.4}$$

$$\text{if } \{v'_n\} \text{ is a subsequence of } v_n \to v \text{ then } v'_n \to v, \tag{10.5}$$

$$u_n \to u, \quad v_n \to v \implies u_n + \lambda v_n \to u + \lambda v, \tag{10.6}$$

where $\lambda \in \mathbb{K}$ is any scalar. Also, if λ_n is a convergent sequence of scalars in $\mathbb{K}$ then

$$\lambda_n \to \lambda \implies \lambda_n u \to \lambda u. \tag{10.7}$$

Example 10.23 The vector spaces $\mathbb{R}^n$ are topological vector spaces with respect to the Euclidean topology. It is worth giving the full proof of this statement, as it sets the pattern for a number of other examples. A set $U \subset \mathbb{R}^n$ is open if and only if for every $\mathbf{x} \in U$ there exists $\epsilon > 0$ such that

$$I_\epsilon(\mathbf{x}) = \{\mathbf{y} \mid |y_i - x_i| < \epsilon, \text{ for all } i = 1, \ldots, n\} \subseteq U.$$

To show that vector addition ψ is continuous, it is necessary to show that $N = \psi^{-1}(I_\epsilon(\mathbf{x}))$ is an open subset of $\mathbb{R}^n \times \mathbb{R}^n$ for all $\mathbf{x} \in \mathbb{R}^n$, $\epsilon > 0$. If $\psi(\mathbf{u}, \mathbf{v}) = \mathbf{u} + \mathbf{v} = \mathbf{x}$ then for any $(\mathbf{u}', \mathbf{v}') \in I_{\epsilon/2}(\mathbf{u}) \times I_{\epsilon/2}(\mathbf{v})$, we have for all $i = 1, \ldots, n$

$$|(x_i - (u'_i + v'_i))| = |((u_i - u'_i) + (v_i - v'_i))|$$
$$\leq |((u_i - u'_i))| + |((v_i - v'_i))| \leq \frac{\epsilon}{2} + \frac{\epsilon}{2} = \epsilon.$$

Hence $I_{\epsilon/2}(\mathbf{u}) \times I_{\epsilon/2}(\mathbf{v}) \subset N$, and continuity of ψ is proved.

10.9 Topological vector spaces

For continuity of the scalar multiplication function τ, let $M = \tau^{-1}(I_\epsilon(\mathbf{x})) \subset \mathbb{R} \times \mathbb{R}^n$. If $\mathbf{x} = a\mathbf{u}$, let $\mathbf{v} \in I_\delta(\mathbf{u})$ and $b \in I_{\delta'}(a)$. Then, setting $A = \max_i |u_i|$, we have

$$|bv_i - au_i| = |bv_i - bu_i + bu_i - au_i|$$
$$\leq |b||v_i - u_i| + |b - a||u_i|$$
$$\leq (|a| + \delta')\delta + \delta' A$$
$$\leq \epsilon \quad \text{if } \delta' = \frac{\epsilon}{2A} \text{ and } \delta = \frac{\epsilon}{2|a|A + \epsilon}.$$

A similar proof may be used to show that the complex vector space $\mathbb{C}^n$ is a topological vector space.

Example 10.24 The vector space $\mathbb{R}^\infty$ consisting of all infinite sequences $\mathbf{x} = (x_1, x_2, \ldots)$ is an infinite dimensional vector space. We give it the product topology as described, whereby a set U is open if for every point $\mathbf{x} \in U$ there is a finite sequence of integers $\mathbf{i} = (i_1, i_2, \ldots, i_n)$ such that

$$I_{\mathbf{i},\epsilon}(\mathbf{x}) = \{\mathbf{y} \mid |y_{i_k} - x_{i_k}| < \epsilon \text{ for } k = 1, \ldots, n\} \subset U.$$

This neighbourhood of $\mathbf{x}$ is an infinite product of intervals of which all but a finite number consist of all of $\mathbb{R}$. To prove that ψ and τ are continuous functions, we again need only show that $\psi^{-1}(I_{\mathbf{i},\epsilon}(\mathbf{x}))$ and $\tau^{-1}(I_{\mathbf{i},\epsilon}(\mathbf{x}))$ are open sets. The argument follows along essentially identical lines to that in Example 10.23. To prove continuity of the scalar product τ, we set $A = \max_{i \in \mathbf{i}} |u_i|$, where $\mathbf{x} = a\mathbf{u}$ and continue as in the previous example.

Example 10.25 Let S be any set, and set $\mathcal{F}(S)$ to be the set of bounded real-valued functions on S. This is obviously a vector space, with vector addition defined by $(f + g)(x) = f(x) + g(x)$ and scalar multiplication by $(af)(x) = af(x)$. A metric can be defined on this space by setting $d(f, g)$ to be the least upper bound of $|f(x) - g(x)|$ on S. Conditions (Met1)–(Met4) are easy to verify. The vector space $\mathcal{F}(S)$ is a topological vector space with respect to the metric topology generated by this distance function. For example, let $f(x) = u(x) + v(x)$, then if $|u'(x) - u(x)| < \epsilon/2$ and $|v'(x) - v(x)| < \epsilon/2$ it follows at once that $|f(x) - (u'(x) + v'(x))| < \epsilon$. To prove continuity of scalar addition we again proceed as in Example 10.23. If $f(x) = au(x)$, let $|b - a| < \epsilon/2A$ where A is an upper bound of $u(x)$ in S and $|v(x) - u(x)| < \epsilon/(2|a|A + \epsilon)$ for all $x \in S$; then

$$|bv(x) - f(x)| < \epsilon \qquad \text{for all } x \in S.$$

Banach spaces

A **norm** on a vector space V is a map $\|\cdot\| : V \to \mathbb{R}$, associating a real number $\|v\|$ with every vector $v \in V$, such that

(Norm1) $\|v\| \geq 0$, and $\|v\| = 0$ if and only if $v = 0$.
(Norm2) $\|\lambda v\| = |\lambda| \, \|v\|$.
(Norm3) $\|u + v\| \leq \|u\| + \|v\|$.

In most cases the field of scalars is taken to be the complex numbers, $\mathbb{K} = \mathbb{C}$, although much of what we say also applies to real normed spaces. We have met this concept earlier, in the context of a complex inner product space (see Section 5.2).

A norm defines a distance function $d : V \times V \to \mathbb{R}$ by

$$d(u, v) = \|u - v\|.$$

The properties (Met1)–(Met3) are trivial to verify, while the triangle inequality

$$d(u, v) \leq d(u, w) + d(w, v) \qquad (10.8)$$

is an immediate consequence of (Norm3),

$$\|u - v\| = \|u - w + w - v\| \leq \|u - w\| + \|w - v\|.$$

We give V the standard metric topology generated by open balls $B_a(v) = \{u \mid d(u, v) < a\}$ as in Section 10.3. This makes it into a topological vector space. To show that the function $(u, v) \mapsto u + v$ is continuous with respect to this topology,

$$\|u' - u\| < \frac{\epsilon}{2} \text{ and } \|v' - v\| < \frac{\epsilon}{2} \implies \|u' + v' - (u + v)\| < \epsilon$$

on using the triangle inequality. The proof that $(\lambda, v) \mapsto \lambda v$ is continuous follows the lines of Example 10.23.

Exercise: Show that the 'norm' is a continuous function $\|\cdot\| : V \to \mathbb{R}$.

Example 10.26 The vector space $\mathcal{F}(S)$ of bounded real-valued functions on a set S defined in Example 10.25 has a norm

$$\|f\| = \sup_{x \in S} |f(x)|,$$

giving rise to the distance function $d(f, g)$ of Example 10.25. This is called the *supremum norm*.

Convergence of sequences is defined on V by

$$u_n \to u \text{ if } d(u_n, u) = \|u_n - u\| \to 0.$$

As in Section 10.3, every convergent sequence $u_i \to u$ is a Cauchy sequence

$$\|u_i - u_j\| \leq \|u - u_i\| + \|u - u_j\| \to 0 \text{ as } i, j \to \infty,$$

but the converse need not always hold. We say a normed vector space $(V, \|\cdot\|)$ is **complete**, or is a **Banach space**, if every Cauchy sequence converges,

$$\|u_i - u_j\| \to 0 \text{ as } i, j \to \infty \implies u_i \to u \text{ for some } u \in V.$$

Exercise: Give an example of a vector subspace of $\mathbb{C}^\infty$ that is an incomplete normed vector space.

Example 10.27 On the vector space $\mathbb{C}^n$ define the standard norm

$$\|\mathbf{x}\| = \sqrt{|x_1|^2 + |x_2|^2 + \cdots + |x_n|^2}.$$

10.9 Topological vector spaces

Conditions (Norm1) and (Norm2) are trivial, while (Norm3) follows from Theorem 5.6, since this norm is precisely that defined in Eq. (5.11) from the inner product $\langle x | y \rangle = \sum_{i=1}^{n} \overline{x_i} y_i$. If $\|\mathbf{x}_n - \mathbf{x}_m\| \to 0$ for a sequence of vectors $\mathbf{x}_n$, then each component is a Cauchy sequence $|x_{ni} - x_{mi}| \to 0$, and therefore has a limit $x_{ni} \to x_i$. It is straightforward to show that $\|\mathbf{x}_n - \mathbf{x}\| = 0$ where $\mathbf{x} = (x_1, x_2, \ldots, x_n)$. Hence this normed vector space is complete.

Example 10.28 Let $\mathcal{D}([-1, 1])$ be the vector space of bounded differentiable complex-valued functions on the closed interval $[-1, 1]$. As in Example 10.26 we adopt the supremum norm $\|f\| = \sup_{x \in [-1,1]} |f(x)|$. This normed vector space is not complete, for consider the sequence of functions

$$f_n(x) = |x|^{1+1/n} \quad (n = 1, 2, \ldots).$$

These functions are all differentiable on $[-1, 1]$ and have zero derivative at $x = 0$ from both the left and the right. Since they approach the bounded function $|x|$ as $n \to \infty$ this is necessarily a Cauchy sequence. However, the limit function $|x|$ is not differentiable at $x = 0$, and the norm is incomplete.

By a **linear functional** on a Banach space V we always mean a *continuous* linear map $\varphi : V \to \mathbb{C}$. The vector space of all linear functionals on V is called the **dual space** V' of V. If V is finite dimensional then V' and V^* coincide, since all linear functionals are continuous with respect to the norm

$$\|\mathbf{u}\| = \sqrt{|u_1|^2 + \cdots + |u_n|^2},$$

but for infinite dimensional spaces it is important to stipulate the continuity requirement. A linear map φ on a Banach space is said to be **bounded** if there exists $M > 0$ such that

$$|\varphi(x)| \le M \|x\| \quad \text{for all } x \in V.$$

The following theorem shows that the words 'bounded' and 'continuous' are interchangeable for linear functionals on a Banach space.

Theorem 10.23 *A linear functional $\varphi : V \to \mathbb{R}$ on a Banach space V is continuous if and only if it is bounded.*

Proof: If φ is bounded, let $M > 0$ be such that $|\varphi(x)| \le M\|x\|$ for all $x \in V$. Then for any pair of vectors $x, y \in V$

$$|\varphi(x - y)| \le M\|x - y\|,$$

and for any $\epsilon > 0$ we have

$$\|x - y\| < \frac{\epsilon}{M} \implies |\varphi(x) - \varphi(y)| = |\varphi(x - y)| \le \epsilon.$$

Hence φ is continuous.

Conversely, suppose φ is continuous. In particular, it is continuous at the origin $x = 0$ and there exists $\delta > 0$ such that

$$\|x\| < \delta \implies |\varphi(x)| < 1.$$

For any vector $y \in V$ we have

$$\left\| \frac{\delta y}{\|y\|} \right\| < \delta$$

whence

$$\left| \varphi\left(\frac{\delta y}{\|y\|} \right) \right| < 1.$$

Thus

$$|\varphi(y)| < \frac{\|y\|}{\delta},$$

showing that φ is bounded. ∎

Example 10.29 Let ℓ^1 be the vector space of all complex infinite sequences $\mathbf{x} = (x_1, x_2, \dots)$ that are absolutely convergent,

$$\|\mathbf{x}\| = \sum_{i=1}^{\infty} |x_i| < \infty.$$

If $c = (c_1, c_2, \dots)$ is a bounded infinite sequence of complex numbers, $|c_i| \leq C$ for all $i = 1, 2, \dots$, then

$$\varphi_c(\mathbf{x}) = c_1 x_1 + c_2 x_2 + \dots$$

is a continuous linear functional on ℓ^1. Linearity is obvious as long as the series converges, and convergence and boundedness are proved in one step,

$$\|\varphi_c(\mathbf{x})\| = \sum_{i=1}^{\infty} |c_i x_i| \leq |C| \sum_{i=1}^{\infty} |x_i| = |C| \, \|\mathbf{x}\| < \infty.$$

Hence φ_c is a continuous linear operator by Theorem 10.23.

Problems

Problem 10.23 Prove the properties (10.3)–(10.7).

Problem 10.24 Show that a linear map $T : V \to W$ between topological vector spaces is continuous everywhere on V if and only if it is continuous at the origin $0 \in V$.

Problem 10.25 Give an example of a linear map $T : V \to W$ between topological vector spaces V and W that is not continuous.

Problem 10.26 Complete the proof that a normed vector space is a topological vector space with respect to the metric topology induced by the norm.

Problem 10.27 Show that a real vector space V of dimension ≥ 1 is not a topological vector space with respect to either the discrete or indiscrete topology.

10.9 Topological vector spaces

Problem 10.28 Show that the following are all norms in the vector space $\mathbb{R}^2$:

$$\|\mathbf{u}\|_1 = \sqrt{(u_1)^2 + (u_2)^2},$$
$$\|\mathbf{u}\|_2 = \max\{|u_1|, |u_2|\},$$
$$\|\mathbf{u}\|_3 = |u_1| + |u_2|.$$

What are the shapes of the open balls $B_a(\mathbf{u})$? Show that the topologies generated by these norms are the same.

Problem 10.29 Show that if $x_n \to x$ in a normed vector space then

$$\frac{x_1 + x_2 + \cdots + x_n}{n} \to x.$$

Problem 10.30 Show that if x_n is a sequence in a normed vector space V such that every subsequence has a subsequence convergent to x, then $x_n \to x$.

Problem 10.31 Let V be a Banach space and W be a vector subspace of V. Define its *closure* $\overline{W}$ to be the union of W and all limits of Cauchy sequences of elements of W. Show that $\overline{W}$ is a closed vector subspace of V in the sense that the limit points of all Cauchy sequences in $\overline{W}$ lie in $\overline{W}$ (note that the Cauchy sequences may include the newly added limit points of W).

Problem 10.32 Show that every space $\mathcal{F}(S)$ is complete with respect to the supremum norm of Example 10.26. Hence show that the vector space ℓ_∞ of bounded infinite complex sequences is a Banach space with respect to the norm $\|\mathbf{x}\| = \sup(x_i)$.

Problem 10.33 Show that the set V' consisting of bounded linear functionals on a Banach space V is a normed vector space with respect to the norm

$$\|\varphi\| = \sup\{M \mid |\varphi(x)| \le M\|x\| \text{ for all } x \in V\}.$$

Show that this norm is complete on V'.

Problem 10.34 We say two norms $\|u\|_1$ and $\|u\|_2$ on a vector space V are *equivalent* if there exist constants A and B such that

$$\|u\|_1 \le A\|u\|_2 \quad \text{and} \quad \|u\|_2 \le B\|u\|_1$$

for all $u \in V$. If two norms are equivalent then show the following:

(a) If $u_n \to u$ with respect to one norm then this is also true for the other norm.
(b) Every linear functional that is continuous with respect to one norm is continuous with respect to the other norm.
(c) Let $V = C[0, 1]$ be the vector space of continuous complex functions on the interval $[0, 1]$. By considering the sequence of functions

$$f_n(x) = \frac{n}{\sqrt{\pi}} e^{-n^2 x^2}$$

show that the norms

$$\|f\|_1 = \sqrt{\int_0^1 |f|^2 \, dx} \quad \text{and} \quad \|f\|_2 = \max\{|f(x)| \mid 0 \le x \le 1\}$$

are not equivalent.

(d) Show that the linear functional defined by $F(f) = f(1)$ is continuous with respect to $\|\cdot\|_2$ but not with respect to $\|\cdot\|_1$.

References

[1] R. Geroch. *Mathematical Physics*. Chicago, The University of Chicago Press, 1985.
[2] J. Kelley. *General Topology*. New York, D. Van Nostrand Company, 1955.
[3] M. Nakahara. *Geometry, Topology and Physics*. Bristol, Adam Hilger, 1990.
[4] C. Nash and S. Sen. *Topology and Geometry for Physicists*. London, Academic Press, 1983.
[5] J. G. Hocking and G. S. Young. *Topology*. Reading, Mass., Addison-Wesley, 1961.
[6] D. W. Kahn. *Topology*. New York, Dover Publications, 1995.
[7] E. M. Patterson. *Topology*. Edinburgh, Oliver and Boyd, 1959.
[8] I. M. Singer and J. A. Thorpe. *Lecture Notes on Elementary Topology and Geometry*. Glenview, Ill., Scott Foresman, 1967.
[9] L. H. Loomis and S. Sternberg. *Advanced Calculus*. Reading, Mass., Addison-Wesley, 1968.
[10] T. Apostol. *Mathematical Analysis*. Reading, Mass., Addison-Wesley, 1957.
[11] L. Debnath and P. Mikusiński. *Introduction to Hilbert Spaces with Applications*. San Diego, Academic Press, 1990.
[12] N. B. Haaser and J. A. Sullivan. *Real Analysis*. New York, Van Nostrand Reinhold Company, 1971.

11 Measure theory and integration

Topology does not depend on the notion of 'size'. We do not need to know the length, area or volume of subsets of a given set to understand the topological structure. *Measure theory* is that area of mathematics concerned with the attribution of precisely these sorts of properties. The structure that tells us which subsets are *measurable* is called a *measure space*. It is somewhat analogous with a topological structure, telling us which sets are *open*, and indeed there is a certain amount of interaction between measure theory and topology. A measure space requires firstly an algebraic structure known as a σ-*algebra* imposed on the power set of the underlying space. A *measure* is a positive-valued real function on the σ-algebra that is *countably additive*, whereby the measure of a union of disjoint measurable sets is the sum of their measures. The measure of a set may well be zero or infinite. Full introductions to this subject are given in [1–5], while the flavour of the subject can be found in [6–8].

It is important that measure be not just finitely additive, else it is not far-reaching enough, yet to allow it to be additive on arbitrary unions of disjoint sets would lead to certain contradictions – either all sets would have to be assigned zero measure, or the measure of a set would not be well-defined. By general reckoning the broadest useful measure on the real line or its cartesian products is that due to Lebesgue (1875–1941), and Lebesgue's theory of *integration* based on this theory is in most ways the best definition of integration available.

Use will frequently be made in this chapter of the **extended real line** $\overline{\mathbb{R}}$ consisting of $\mathbb{R} \cup \{\infty\} \cup \{-\infty\}$, having rules of addition $a + \infty = \infty$, $a + (-\infty) = -\infty$ for all $a \in \mathbb{R}$, but no value is given to $\infty + (-\infty)$. The natural order on the real line is supplemented by the inequalities $-\infty < a < \infty$ for all real numbers a. Multiplication can also be extended in some cases, such as $a\infty = \infty$ if $a > 0$, but it is best to avoid the product 0∞ unless a clear convention can be adopted. The natural order topology on $\mathbb{R}$, generated by open intervals (a, b) is readily extended to $\overline{\mathbb{R}}$.

Exercise: Show $\overline{\mathbb{R}}$ is a compact topological space with respect to the order topology.

11.1 Measurable spaces and functions

Measurable spaces

Given a set X, a σ-**algebra** $\mathcal{M}$ on X consists of a collection of subsets, known as **measurable sets**, satisfying

Measure theory and integration

(Meas1) The empty set $\emptyset$ is a measurable set, $\emptyset \in \mathcal{M}$.
(Meas2) If A is measurable then so is its complement:
$$A \in \mathcal{M} \implies A^c = X - A \in \mathcal{M}.$$

(Meas3) $\mathcal{M}$ is closed under countable unions:
$$A_1, A_2, A_3, \ldots \in \mathcal{M} \implies \bigcup_i A_i \in \mathcal{M}.$$

The pair $(X, \mathcal{M})$ is known as a **measurable space**. Although there are similarities between these axioms and (Top1)–(Top3) for a topological space, (Meas2) is distinctly different in that the complement of an open set is a closed set and is rarely open. It follows from (Meas1) and (Meas2) that the whole space $X = \emptyset^c$ is measurable. The intersection of any pair of measurable sets is measurable, for
$$A \cap B = (A^c \cup B^c)^c.$$

Also, $\mathcal{M}$ is closed with respect to taking differences,
$$A - B = A \cap B^c = (A^c \cup B)^c.$$

Exercise: Show that any countable intersection of measurable sets is measurable.

Example 11.1 Given any set X, the collection $\mathcal{M} = \{\emptyset, X\}$ is obviously a σ-algebra. This is the smallest σ-algebra possible. By contrast, the largest σ-algebra is the set of all subsets 2^X. All interesting examples fall somewhere between these two extremes.

It is trivial to see that the intersection of any two σ-algebras $\mathcal{M} \cap \mathcal{M}'$ is another σ-algebra – check that properties (Meas1)–(Meas3) are satisfied by the sets common to the two σ-algebras. This statement extends to the intersection of an arbitrary family of σ-algebras, $\bigcap_{i \in I} \mathcal{M}_i$. Hence, given any collection of subsets $\mathcal{A} \subset 2^X$, there is a unique 'smallest' σ-algebra $\mathcal{S} \supseteq \mathcal{A}$. This is the intersection of all σ-algebras that contain $\mathcal{A}$. It is called the σ-**algebra generated by** $\mathcal{A}$. For a topological space X, the σ-algebra generated by the open sets are called **Borel sets** on X. They include all open and all closed sets and, in general, many more that are neither open nor closed.

Example 11.2 In the standard topology on the real line $\mathbb{R}$ every open set is a countable union of open intervals. Hence the Borel sets are generated by the set of all open intervals $\{(a, b) \mid a < b\}$. Infinite left-open intervals such as $(a, \infty) = (a, a+1) \cup (a, a+2) \cup (a, a+3) \cup \ldots$ are Borel sets by (Meas3), and similarly all intervals $(-\infty, a)$ are Borel. The complements of these sets are the infinite right or left-closed intervals $(-\infty, a]$ and $[a, \infty)$. Hence all closed intervals $[a, b] = (-\infty, b] \cap [a, \infty)$ are Borel sets.

Exercise: Prove that the σ-algebra of Borel sets on $\mathbb{R}$ is generated by (a) the infinite left-open intervals (a, ∞), (b) the closed intervals $[a, b]$.

Exercise: Prove that all singletons $\{a\}$ are Borel sets on $\mathbb{R}$.

11.1 Measurable spaces and functions

If $(X, \mathcal{M})$ and $(Y, \mathcal{N})$ are two measurable spaces then we define the **product measurable space** $(X \times Y, \mathcal{M} \otimes \mathcal{N})$, by setting the σ-algebra $\mathcal{M} \otimes \mathcal{N}$ on $X \times Y$ to be the σ-algebra generated by all sets of the form $A \times B$ where $A \in \mathcal{M}$ and $B \in \mathcal{N}$.

Measurable functions

Given two measurable spaces $(X, \mathcal{M})$ and $(Y, \mathcal{N})$, a map $f : X \to Y$ is said to be a **measurable function** if the inverse image of every measurable set is measurable:

$$A \in \mathcal{N} \implies f^{-1}(A) \in \mathcal{M}.$$

This definition mirrors that for a continuous function in topological spaces.

Theorem 11.1 *If X and Y are topological spaces and $\mathcal{M}$ and $\mathcal{N}$ are the σ-algebras of Borel sets, then every continuous function $f : X \to Y$ is Borel measurable.*

Proof: Let $\mathcal{O}_X$ and $\mathcal{O}_Y$ be the families of open sets in X and Y respectively. We adopt the notation $f^{-1}(\mathcal{A}) \equiv \{f^{-1}(A) \mid A \in \mathcal{A}\}$ for any family of sets $\mathcal{A} \subseteq 2^Y$. Since f is continuous, $f^{-1}(\mathcal{O}_Y) \subseteq \mathcal{O}_X$. The σ-algebras of Borel sets on the two spaces are $\mathcal{M} = \mathcal{S}(\mathcal{O}_X)$ and $\mathcal{N} = \mathcal{S}(\mathcal{O}_Y)$. To prove f is Borel measurable we must show that $f^{-1}(\mathcal{N}) \subseteq \mathcal{M}$. Let

$$\mathcal{N}' = \{B \subseteq Y \mid f^{-1}(B) \in \mathcal{M}\} \subset 2^Y.$$

This is a σ-algebra on Y, for $f^{-1}(\emptyset) = \emptyset$ and

$$f^{-1}(B^c) = (f^{-1}(B))^c, \qquad f^{-1}\left(\bigcup_i B_i\right) = \bigcup_i f^{-1}(B_i).$$

Hence $\mathcal{N}' \supseteq \mathcal{O}_Y$ for $f^{-1}(\mathcal{O}_Y) \subseteq \mathcal{O}_X \subset \mathcal{M}$. Since $\mathcal{N}$ is the σ-algebra generated by $\mathcal{O}_Y$ we must have that $\mathcal{N}' \supseteq \mathcal{N}$. Hence $f^{-1}(\mathcal{N}) \subseteq f^{-1}(\mathcal{N}') \subseteq \mathcal{M}$ as required. ∎

Exercise: If $f : X \to Y$ and $g : Y \to Z$ are measurable functions between measure spaces, show that the composition $g \circ f : X \to Z$ is a measurable function.

If $f : X \to \mathbb{R}$ is a measurable real-valued function on a measurable space $(X, \mathcal{M})$, where $\mathbb{R}$ is assumed given the Borel structure of Example 11.2, it follows that the set

$$\{x \mid f(x) > a\} = f^{-1}((a, \infty))$$

is measurable in X. Since the family of Borel sets $\mathcal{B}$ on the real line is generated by the intervals (a, ∞) (see the exercise following Example 11.2), this can actually be used as a criterion for measurability: $f : X \to \mathbb{R}$ is a measurable function iff for any $a \in \mathbb{R}$ the set $\{x \mid f(x) > a\}$ is measurable.

Exercise: Prove the sufficiency of this condition [refer to the proof of Theorem 11.1].

Example 11.3 If $(X, \mathcal{M})$ is a measurable space then the characteristic function $\chi_A : X \to \mathbb{R}$ of a set $A \subset X$ is measurable if and only if $A \in \mathcal{M}$, since for any $a \in \mathbb{R}$

$$\{x \mid \chi_A(x) > a\} = \begin{cases} X & \text{if } a < 0, \\ A & \text{if } 0 \leq a < 1, \\ \emptyset & \text{if } a \geq 1. \end{cases}$$

Exercise: Show that for any $a \in \mathbb{R}$, the set $\{x \in X \mid f(x) = a\}$ is a measurable set of X.

If $f : X \to \mathbb{R}$ is a measurable function then so is its modulus $|f|$, since the continuous function $x \mapsto |x|$ on $\mathbb{R}$ is necessarily Borel measurable, and $|f|$ is the composition function $|\cdot| \circ f$. Similarly the function f^a for $a > 0$ is measurable, and $1/f$ is measurable if $f(x) \neq 0$ for all $x \in X$. If $g : X \to \mathbb{R}$ is another measurable function then the function $f + g$ is measurable since it can be written as the composition $f = \rho \circ F$ where $F : X \to \mathbb{R}^2 = \mathbb{R} \times \mathbb{R}$ and $\rho : \mathbb{R}^2 \to \mathbb{R}$ are the maps

$$F : x \mapsto (f(x), g(x)) \quad \text{and} \quad \rho(a, b) = a + b.$$

The function F is measurable since the inverse image of any product of intervals I_1, I_2 is

$$F^{-1}(I_1 \times I_2) = f^{-1}(I_1) \cap g^{-1}(I_2),$$

which is a measurable set in X since f and g are assumed measurable functions, while the map ρ is evidently continuous on $\mathbb{R}^2$.

Exercise: Show that for measurable functions $f, g : X \to \mathbb{R}$, the function fg is measurable.

An important class of functions are the **simple functions**: measurable functions $h : X \to \overline{\mathbb{R}}$ that take on only a finite set of extended real values $a_1, a_2, \ldots, a_n$. Since $A_i = h^{-1}(\{a_i\})$ is a measurable subset of X for each a_i, we can write a simple function as a linear combination of measurable characteristic functions

$$h = a_1 \chi_{A_1} + a_2 \chi_{A_2} + \cdots + a_n \chi_{A_n} \quad \text{where all } a_i \neq 0.$$

Some authors use the term *step function* instead of simple function, but the common convention is to preserve this term for simple functions $h : \mathbb{R} \to \mathbb{R}$ in which each set A_i is a union of disjoint *intervals* (Fig. 11.1).

Figure 11.1 Simple (step) function

11.1 Measurable spaces and functions

Let f and g be any pair of measurable functions from X into the extended reals $\overline{R}$. The function $h = \sup(f, g)$ defined by

$$h(x) = \begin{cases} f(x) & \text{if } f(x) \geq g(x) \\ g(x) & \text{if } g(x) > f(x) \end{cases}$$

is measurable, since

$$\{x \mid h(x) > a\} = \{x \mid f(x) > a \text{ or } g(x) > a\} = \{x \mid f(x) > a\} \cup \{x \mid g(x) > a\}$$

is measurable. Similarly, $\inf(f, g)$ is a measurable function. In particular, if f is a measurable function, its positive and negative parts

$$f^+ = \sup(f, 0) \quad \text{and} \quad f^- = -\inf(f, 0)$$

are measurable functions.

Exercise: Show that $f = f^+ - f^-$ and $|f| = f^+ + f^-$.

A simple extension of the above argument shows that $\sup f_n$ is a measurable function for any countable set of measurable functions $f_1, f_2, f_3, \ldots$ with values in the extended real numbers $\overline{R}$. We define the lim sup as

$$\limsup f_n(x) = \inf_{n \geq 1} F_n(x) \quad \text{where} \quad F_n(x) = \sup_{k \geq n} f_k(x).$$

The lim sup always exists since the functions F_n are everywhere monotone decreasing,

$$F_1(x) \geq F_2(x) \geq F_3(x) \geq \ldots$$

and therefore have a limit if they are bounded below or approach $-\infty$ if unbounded below. Similarly we can define

$$\liminf f_n(x) = \sup_{n \geq 1} G_n(x) \quad \text{where} \quad G_n(x) = \inf_{k \geq n} f_k(x).$$

It follows that if f_n is a sequence of measurable functions then $\limsup f_n$ and $\liminf f_n$ are also measurable. By standard arguments in analysis $f_n(x)$ is a convergent sequence if and only if $\limsup f_n(x) = \liminf f_n(x) = \lim f_n(x)$. Hence the limit of any convergent sequence of measurable functions $f_n(x) \to f(x)$ is measurable. Note that the convergence need only be 'pointwise convergence', not uniform convergence as is required in many theorems in Riemann integration.

Theorem 11.2 *Any measurable function $f : X \to \mathbb{R}$ is the limit of a sequence of simple functions. The sequence can be chosen to be monotone increasing at all positive values of f, and monotone decreasing at negative values.*

Proof: Suppose f is positive and bounded above, $0 \leq f(x) \leq M$. For each integer $n = 1, 2, \ldots$ let h_n be the simple function

$$h_n = \sum_{k=0}^{n} \frac{kM}{2^n} \chi_{A_k}$$

where A_k is the measurable set

$$A_k = \left\{ x \mid \frac{kM}{2^n} \leq f(x) < \frac{k+1}{2^n} \right\}.$$

These simple functions are increasing, $0 \leq h_1(x) \leq h_2(x) \leq \ldots$ and $|f(x) - h_n(x)| < M/2^n$. Hence $h_n(x) \to f(x)$ for all $x \in X$ as $n \to \infty$.

If f is any positive function, possibly unbounded, the functions $g_n = \inf(h_n, n)$ are positive and bounded above. Hence for each n there exists a simple function h'_n such that $|g_n - h'_n| < 1/n$. The sequence of simple functions h'_n clearly converges everywhere to $f(x)$. To obtain a monotone increasing sequence of simple functions that converge to f, set

$$f_n = \sup(h'_1, h'_2, \ldots, h'_n).$$

If f is not positive, construct simple function sequences approaching the positive and negative parts and use $f = f^+ - f^-$. ∎

Problems

Problem 11.1 If $(X, \mathcal{M})$ and $(Y, \mathcal{N})$ are measurable spaces, show that the projection maps $\text{pr}_1 : X \times Y \to X$ and $\text{pr}_2 : X \times Y \to Y$ defined by $\text{pr}_1(x, y) = x$ and $\text{pr}_2(x, y) = y$ are measurable functions.

Problem 11.2 Find a step function $s(x)$ that approximates $f(x) = x^2$ uniformly to within $\varepsilon > 0$ on $[0, 1]$, in the sense that $|f(x) - s(x)| < \varepsilon$ everywhere in $[0, 1]$.

Problem 11.3 Let $f : X \to \mathbb{R}$ and $g : X \to \mathbb{R}$ be measurable functions and $E \subset X$ a measurable set. Show that

$$h(x) = \begin{cases} f(x) & \text{if } x \in E \\ g(x) & \text{if } x \notin E \end{cases}$$

is a measurable function on X.

Problem 11.4 If $f, g : \mathbb{R} \to \mathbb{R}$ are Borel measurable real functions show that $h(x, y) = f(x)g(y)$ is a measurable function $h : \mathbb{R}^2 \to \mathbb{R}$ with respect to the product measure on $\mathbb{R}^2$.

11.2 Measure spaces

Given a measurable space $(X, \mathcal{M})$, a **measure** μ on X is a function $\mu : \mathcal{M} \to \overline{\mathbb{R}}$ such that

(Meas4) $\mu(A) \geq 0$ for all measurable sets A, and $\mu(\emptyset) = 0$.
(Meas5) If $A_1, A_2, \ldots \in \mathcal{M}$ is any mutually disjoint sequence (finite or countably infinite) of measurable sets such that $A_i \cap A_j = \emptyset$ if $i \neq j$, then

$$\mu(A_1 \cup A_2 \cup A_3 \cup \ldots) = \sum_i \mu(A_i).$$

A function μ satisfying property (Meas5) is often referred to as being *countably additive*. If the series on the right-hand side is not convergent, it is given the value ∞. A **measure space** is a triple $(X, \mathcal{M}, \mu)$ consisting of a measurable space $(X, \mathcal{M})$ together with a measure μ.

11.2 Measure spaces

Exercise: Show that if $B \subset A$ are measurable sets, then $\mu B \leq \mu A$.

Example 11.4 An occasionally useful measure on a σ-algebra $\mathcal{M}$ defined on a set X is the *Dirac measure*. Let a be any fixed point of X, and set

$$\delta_a(A) = \begin{cases} 1 & \text{if } a \in A, \\ 0 & \text{if } a \notin A. \end{cases}$$

(Meas4) holds trivially for $\mu = \delta_a$, and (Meas5) follows from the obvious fact that the union of a disjoint family of sets $\{A_i\}$ can contain a if and only if $a \in A_j$ for precisely one member A_j. This measure has applications in distribution theory, Chapter 12.

Example 11.5 The branch of mathematics known as *probability theory* is best expressed in terms of measure theory. A **probability space** is a measure space $(\Omega, \mathcal{M}, P)$, where $P(\Omega) = 1$. Sets $A \in \Omega$ are known as **events** and Ω is sometimes referred to as the *universe*. This 'universe' is usually thought of as the set of all possible outcomes of a specific experiment. Note that events are not outcomes of the experiment, but *sets* of possible outcomes. The measure function P is known as the **probability measure** on Ω, and $P(A)$ is the **probability of the event** A. All events have probability in the range $0 \leq P(A) \leq 1$. The element $\emptyset$ has probability 0, $P(\emptyset) = 0$, and is called the **impossible event**. The entire space Ω has probability 1; it can be thought of as the **certainty event**.

The event $A \cup B$ is referred to as either A or B, while $A \cap B$ is A and B. Since P is additive on disjoint sets and $A \cup B = (A - (A \cap B)) \cup (B - (A \cap B)) \cup (A \cap B)$, the probabilities are related by

$$P(A \cup B) = P(A) + P(B) - P(A \cap B).$$

The two events are said to be **independent** if $P(A \cap B) = P(A)P(B)$. This is by no means always the case.

We think of the probability of event B after knowing that A has occurred as the **conditional probability** $P(B|A)$, defined by

$$P(B|A) = \frac{P(A \cap B)}{P(A)}.$$

Events A and B are independent if and only if $P(B|A) = P(B)$ – in other words, the probability of B in no way depends on the occurrence of A.

For a finite or countably infinite set of disjoint events $H_1, H_2, \ldots$ partitioning Ω (sometimes called a **hypothesis**), $\Omega = \bigcup_i H_i$, we have for any event B

$$B = \bigcup_{i=1}^{\infty}(H_i \cap B).$$

Since the sets in this countable union are mutually disjoint, the probability of B is

$$P(B) = \sum_{i=1}^{\infty} P(H_i \cap B) = \sum_{i=1}^{\infty} P(B|H_i)P(H_i).$$

Measure theory and integration

This leads to **Bayes' formula** for the conditional probability of the hypothesis H_i given the outcome event B,

$$P(H_i|B) = \frac{P(B \cap H_i)}{P(B)}$$
$$= \frac{P(H_i)P(B|H_i)}{P(B)}$$
$$= \frac{P(H_i)P(B|H_i)}{\sum_{k=1}^{\infty} P(B|H_k)P(H_k)}.$$

Theorem 11.3 Let $E_1, E_2, \ldots$ be a sequence of measurable sets, which is increasing in the sense that $E_n \subset E_{n+1}$ for all $n = 1, 2, \ldots$ Then

$$E = \bigcup_{n=1}^{\infty} E_n \in \mathcal{M} \quad \text{and} \quad \mu E = \lim_{n \to \infty} \mu E_n.$$

Proof: E is measurable by condition (Meas3). Set

$$F_1 = E_1, \quad F_2 = E_2 - E_1, \ldots, \quad F_n = E_n - E_{n-1}, \ldots.$$

The sets F_n are all measurable and disjoint, $F_n \cap F_m = \emptyset$ if $n \neq m$. Since $E_n = F_1 \cup F_2 \cup \cdots \cup F_n$ we have by (Meas5)

$$\lim_{n \to \infty} \mu E_n = \sum_{i=1}^{\infty} \mu(F_n) = \mu\left(\bigcup_{k=1}^{\infty} F_k\right) = \mu(E). \quad \blacksquare$$

Lebesgue measure

Every open set U on the real line is a countable union of disjoint open intervals. This follows from the fact that every rational point $r \in U$ lies in a maximal open interval $(a, b) \subseteq U$ where $a < r < b$. These intervals must be disjoint, else they would not be maximal, and there are countably many of them since the rational numbers are countably infinite. On the real line $\mathbb{R}$ set $\mu(I) = b - a$ for all open intervals $I = (a, b)$. This extends uniquely by countable additivity (Meas5) to all open sets of $\mathbb{R}$. By finite additivity μ must take the value $b - a$ on all intervals, open, closed or half-open. For example, for any $\epsilon > 0$

$$(a - \epsilon, b) = (a - \epsilon, a) \cup [a, b),$$

and since the right-hand side is the union of two disjoint Borel sets we have

$$b - a + \epsilon = a - a + \epsilon + \mu\big([a, b)\big).$$

Hence the measure of the left-closed interval $[a, b)$ is

$$\mu\big([a, b)\big) = b - a.$$

Using

$$[a, b) = \{a\} \cup (a, b)$$

11.2 Measure spaces

we see that every singleton has zero measure, $\mu(\{a\}) = 0$, and

$$(a, b] = (a, b) \cup \{b\}, [a, b] = (a, b) \cup \{a\} \cup \{b\} \implies \mu((a, b]) = \mu([a, b]) = b - a.$$

Exercise: Show that if $A \subset \mathbb{R}$ is a countable set, then $\mu(A) = 0$.

Exercise: Show that the set of finite unions of left-closed intervals $[a, b)$ is closed with respect to the operation of taking differences of sets.

For any set $A \subset \mathbb{R}$ we define its **outer measure**

$$\mu^*(A) = \inf\{\mu(U) \mid U \text{ is an open set in } \mathbb{R} \text{ and } A \subseteq U\}. \tag{11.1}$$

While outer measure can be defined for arbitrary sets A of real numbers it is not really a measure at all, for it does not satisfy the countably additive property (Meas5). The best we can do is a property known as *countable subadditivity*: if $A_1, A_2, \ldots \in 2^X$ is any mutually disjoint sequence of sets then

$$\mu^*\left(\bigcup_i A_i\right) \leq \sum_{i=1}^{\infty} \mu^*(A_i). \tag{11.2}$$

Proof: Let $\epsilon > 0$. For each $n = 1, 2, \ldots$ there exists an open set $U_n \supseteq A_n$ such that

$$\mu(U_n) \leq \mu^*(A_n) + \frac{\epsilon}{2^n}.$$

Since $U = U_1 \cup U_2 \cup \ldots$ covers $A = A_1 \cup A_2 \cup \ldots$,

$$\mu^*(A) \leq \mu(U) \leq \mu(U_1) + \mu(U_2) + \cdots \leq \sum_{i=1}^{\infty} \mu^*(A_n) + \epsilon.$$

As this is true for arbitrary $\epsilon > 0$, the inequality (11.2) follows immediately. ∎

Exercise: Show that outer measure satisfies (Meas4), $\mu^*(\emptyset) = 0$.

Exercise: For an open interval $I = (a, b)$ show that $\mu^*(I) = \mu(I) = b - a$.

Exercise: Show that

$$A \subseteq B \implies \mu^*(A) \leq \mu^*(B). \tag{11.3}$$

Following Carathéodory, a set E is said to be **Lebesgue measurable** if for any open interval $I = (a, b)$

$$\mu(I) = \mu^*(I \cap E) + \mu^*(I \cap E^c). \tag{11.4}$$

At first sight this may not seem a very intuitive notion.

What it is saying is that when we try to cover the mutually disjoint sets $I \cap E$ and $I - E = I \cap E^c$ with open intervals, the overlap of the two sets of intervals can be made 'arbitrarily small' (see Fig. 11.2). From now on we will often refer to Lebesgue measurable sets simply as *measurable*.

Measure theory and integration

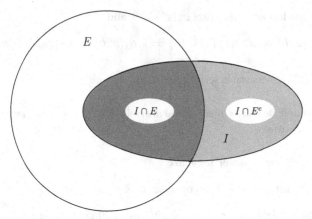

Figure 11.2 Lebesgue measurable set

Theorem 11.4 *If E is measurable then for any set A, measurable or not,*
$$\mu^*(A) = \mu^*(A \cap E) + \mu^*(A \cap E^c).$$

Proof: Given $\epsilon > 0$, let U be an open set such that $A \subseteq U$ and
$$\mu(U) < \mu^*(A) + \epsilon. \tag{11.5}$$

Since $A = (A \cap E) \cup (A \cap E^c)$ is a union of two disjoint sets, Eq. (11.2) gives
$$\mu^*(A) \leq \mu^*(A \cap E) + \mu^*(A \cap E^c).$$

Setting $U = \bigcup_n I_n$ where I_n are a finite or countable collection of disjoint open intervals, we have by (11.3)
$$\begin{aligned}
\mu^*(A) &\leq \mu^*(U \cap E) + \mu^*(U \cap E^c) \\
&\leq \sum_n \mu^*(I_n \cap E) + \mu^*(I_n \cap E^c) \\
&= \sum_n \mu^*(I_n) \quad \text{since } E \text{ is measurable.}
\end{aligned}$$

Using the inequality (11.5)
$$\mu^*(A) \leq \mu^*(U \cap E) + \mu^*(U \cap E^c) < \mu^*(A) + \epsilon$$
for arbitrary $\epsilon > 0$. This proves the desired result. ∎

Corollary 11.5 *If $E_1, E_2, \ldots, E_n$ is any family of disjoint measurable sets then*
$$\mu^*(E_1 \cup E_2 \cup \cdots \cup E_n) = \sum_{i=1}^n \mu^*(E_i).$$

Proof: If E and F are disjoint measurable sets, setting $A = E \cup F$ in (11.4) gives
$$\mu^*(E \cup F) = \mu^*\big((E \cup F) \cap E\big) + \mu^*\big((E \cup F) \cap E^c\big) = \mu^*(E) + \mu^*(F).$$

The result follows by induction on n. ∎

11.2 Measure spaces

Theorem 11.6 *The set of all Lebesgue measurable sets $\mathcal{L}$ is a σ-algebra, and μ^* is a measure on $\mathcal{L}$.*

Proof: The empty set is Lebesgue measurable, since for any open interval I,

$$\mu^*(I \cap \emptyset) + \mu^*(I \cap \mathbb{R}) = \mu^*(\emptyset) + \mu^*(I) = \mu^*(I) = \mu(I).$$

If E is a measurable set then so is E^c, on substituting in Eq. (11.4) and using $(E^c)^c = E$. Hence conditions (Meas1) and (Meas2) are satisfied. We prove (Meas3) in several stages.

Firstly, if E and F are measurable sets then $E \cup F$ is measurable. For, $I = \big(I \cap (E \cup F)\big) \cup \big(I \cap (E \cup F)^c\big)$ and using Eq. (11.2) gives

$$\mu^*(I) \leq \mu^*\big(I \cap (E \cup F)\big) + \mu^*\big(I \cap (E \cup F)^c\big). \tag{11.6}$$

The sets in the two arguments on the right-hand side can be decomposed as

$$I \cap (E \cup F) = \big((I \cap F) \cap E\big) \cup \big((I \cap F) \cap E^c\big) \cup \big((I \cap F^c) \cap E\big)$$

and

$$I \cap (E \cup F)^c = I \cap (E^c \cap F^c) = (I \cap F^c) \cap E^c.$$

Again using Eq. (11.2) gives

$$\mu^*\big(I \cap (E \cup F)\big) + \mu^*\big(I \cap (E \cup F)^c\big)$$
$$\leq \mu^*\big((I \cap F) \cap E\big) + \mu^*\big((I \cap F) \cap E^c\big) + \mu^*\big((I \cap F^c) \cap E\big) + \mu^*\big((I \cap F^c) \cap E^c\big)$$

On setting $A = I \cap F$ and $A = I \cap F^c$ respectively in Theorem 11.4 we have

$$\mu^*\big(I \cap (E \cup F)\big) + \mu^*\big(I \cap (E \cup F)^c\big) \leq \mu^*(I \cap F) + \mu^*(I \cap F^c) = \mu^*(I),$$

since F is measurable. Combining this with the inequality (11.6) we conclude that for all intervals I

$$\mu^*\big(I \cap (E \cup F)\big) + \mu^*\big(I \cap (E \cup F)^c\big) = \mu^*(I) = \mu(I),$$

which shows that $E \cup F$ is measurable. Incidentally it also follows that the intersection $E \cap F = (E^c \cup F^c)^c$ and the difference $F - E = F \cap E^c$ is measurable. Simple induction shows that any finite union of measurable sets $E_1 \cup E_2 \cup \cdots \cup E_n$ is measurable.

Let $E_1, E_2, \ldots$ be any sequence of disjoint measurable sets. Set

$$S_n = \bigcup_{i=1}^{n} E_i \quad \text{and} \quad S = \bigcup_{i=1}^{\infty} E_i.$$

By subadditivity (11.2),

$$\mu^*(S) \leq \sum_{i=1}^{\infty} \mu^*(E_i)$$

297

and since $S \supset S_n$ we have, using Corollary 11.5,

$$\mu^*(S) \geq \mu^*(S_n) = \sum_{i=1}^{n} \mu^*(E_i).$$

Since this holds for all integers n and the right-hand side is a monotone increasing series,

$$\mu^*(S) = \sum_{i=1}^{\infty} \mu^*(E_i). \tag{11.7}$$

If the series does not converge, the right-hand side is assigned the value ∞.

Since the E_i are disjoint sets, so are the sets $I \cap E_i$. Furthermore by Corollary 11.5

$$\sum_{i=1}^{n} \mu^*(I \cap E_i) = \mu^*\left(\bigcup_{i=1}^{n}(I \cap E_i)\right) = \mu^*\left(I \cap \bigcup_{i=1}^{n} E_i\right) \leq \mu^*(I).$$

Hence the series $\sum_{i=1}^{\infty} \mu^*(I \cap E_i)$ is convergent and for any $\epsilon > 0$ there exists an integer n such that

$$\sum_{i=n}^{\infty} \mu^*(I \cap E_i) < \epsilon.$$

Now since $S_n \subset S$,

$$I \cap S = I \cap (S_n \cup (S - S_n)) = (I \cap S_n) \cup (I \cap (S - S_n))$$

and by subadditivity (11.2),

$$\mu(I) = \mu^*\big((I \cap S) \cup (I \cap S^c)\big)$$
$$\leq \mu^*(I \cap S) + \mu^*(I \cap S^c)$$
$$\leq \mu^*\big((I \cap S_n)\big) + \mu^*\big(I \cap (S - S_n)\big) + \mu^*\big(I \cap (S_n)^c\big)$$
$$= \mu(I) + \mu^*\left(I \cap \bigcup_{i=n+1}^{\infty} E_i\right)$$
$$= \mu(I) + \sum_{i=n+1}^{\infty} \mu^*(I \cap E_i)$$
$$< \mu(I) + \epsilon.$$

Since $\epsilon > 0$ is arbitrary

$$\mu(I) = \mu^*(I \cap S) + \mu^*(I \cap S^c),$$

which proves that S is measurable.

If $E_1, E_2, \ldots$ are a sequence of measurable sets, not necessarily disjoint, then set

$$F_1 = E_1, \quad F_2 = E_2 - E_1, \ldots, \quad F_n = E_n - (E_1 \cup E_2 \cup \cdots \cup E_{n-1}), \ldots$$

These sets are all measurable and disjoint, and

$$\bigcup_{i=1}^{\infty} E_i = \bigcup_{i=1}^{\infty} F_i.$$

11.2 Measure spaces

The union of any countable collection $\{E_n\}$ of measurable sets is therefore measurable, proving that $\mathcal{L}$ is a σ-algebra. The outer measure μ^* is a measure on $\mathcal{L}$ since it satisfies $\mu^*(\emptyset) = 0$ and is countably additive by (11.7). ∎

Theorem 11.6 shows that $(\mathbb{R}, \mathcal{L}, \mu = \mu^*|_{\mathcal{L}})$ is a measure space. The notation μ for Lebesgue measure agrees with the earlier convention $\mu\big((a, b)\big) = b - a$ on open intervals. All open sets are measurable as they are disjoint unions of open intervals. Since the Borel sets form the σ-algebra generated by all open sets, they are included in $\mathcal{L}$. Hence every Borel set is Lebesgue measurable. It is not true however that every Lebesgue measurable set is a Borel set.

A property that holds everywhere except on a set of measure zero is said to hold **almost everywhere**, often abbreviated to 'a.e.'. For example, two functions f and g are said to be equal a.e. if the set of points where $f(x) \ne g(x)$ is a set of measure zero. It is sufficient for the set to have outer measure zero, $\mu^*(A) = 0$, in order for it to have measure zero (see Problem 11.8).

Lebesgue measure is defined on cartesian product spaces $\mathbb{R}^n$ in a similar manner. We give the construction for $n = 2$. We have already seen that the σ-algebra of measurable sets on the product space $\mathbb{R}^2 = \mathbb{R} \times \mathbb{R}$ is defined as that generated by products of measurable sets $E \times E'$ where E and E' are Lebesgue measurable on $\mathbb{R}$. The outer measure of any set $A \subset \mathbb{R}^2$ is defined as

$$\mu^*(A) = \sup \sum_i \mu(I_i)\mu(I_i')$$

where I_i and I_i' are any finite or countable family of open intervals of $\mathbb{R}$ such that the union $\bigcup_i I_i \times I_i'$ covers A. The outer measure of any product of open intervals is clearly the product of their measures, $\mu^*(I \times I') = \mu(I)\mu(I')$. We say a set $E \subset \mathbb{R}^2$ is **Lebesgue measurable** if, for any pair of open intervals I, I'

$$\mu(I)\mu(I') = \mu^*\big((I \times I') \cap E\big) + \mu^*\big((I \times I') \cap E^c\big).$$

As for the real line, outer measure μ^* is then a measure on $\mathbb{R}^2$. Lebesgue measure on higher dimensional products $\mathbb{R}^n$ is completely analogous. We sometimes denote this measure by μ^n.

Example 11.6 The Cantor set, Example 1.11, is a closed set since it is formed by taking the complement of a sequence of open intervals. It is therefore a Borel set and is Lebesgue measurable. The Cantor set is an uncountable set of measure 0 since the length remaining after the nth step in its construction is

$$1 - \tfrac{1}{3} - 2\big(\tfrac{1}{3}\big)^2 - 2^2\big(\tfrac{1}{3}\big)^3 - \cdots - 2^{n-1}\big(\tfrac{1}{3}\big)^n = \big(\tfrac{2}{3}\big)^n \to 0.$$

Its complement is an open subset of $[0, 1]$ with measure 1 – that is, having 'no gaps' between its component open intervals.

Example 11.7 Not every set is Lebegue measurable, but the sets that fail are non-constructive in character and invariably make use of the axiom of choice. A classic example is the following. For any pair of real numbers $x, y \in I = (0, 1)$ set $x Q y$ if and only if $x - y$

Measure theory and integration

is a rational number. This is an equivalence relation on I and it partitions this set into disjoint equivalence classes $Q_x = \{y \mid y - x = r \in \mathbb{Q}\}$ where $\mathbb{Q}$ is the set of rational numbers. Assuming the axiom of choice, there exists a set T consisting of exactly one representative from each equivalence class Q_x. Suppose it has Lebesgue measure $\mu(T)$. For each rational number $r \in (-1, 1)$ let $T_r = \{x + r \mid x \in T\}$. Every real number $y \in I$ belongs to some T_r since it differs by a rational number r from some member of T. Hence, since $|r| < 1$ for each such T_r, we must have

$$(-1, 2) \supset \bigcup_r T_r \supset (0, 1).$$

The sets T_r are mutually disjoint and all have measure equal to $\mu(T)$. If the rational numbers are displayed as a sequence $r_1, r_2, \ldots$ then

$$3 \geq \mu(T_1) + \mu(T_2) + \cdots = \sum_{i=1}^{\infty} \mu(T) \geq 1.$$

This yields a contradiction either for $\mu(T) = 0$ or $\mu(T) > 0$; in the first case the sum is 0, in the second it is ∞.

Problems

Problem 11.5 Show that every countable subset of $\mathbb{R}$ is measurable and has Lebesgue measure zero.

Problem 11.6 Show that the union of a sequence of sets of measure zero is a set of Lebesgue measure zero.

Problem 11.7 If $\mu^*(N) = 0$ show that for any set E, $\mu^*(E \cup N) = \mu^*(E - N) = \mu^*(E)$. Hence show that $E \cup N$ and $E - N$ are Lebesgue measurable if and only if E is measurable.

Problem 11.8 A measure is said to be **complete** if every subset of a set of measure zero is measurable. Show that if $A \subset \mathbb{R}$ is a set of outer measure zero, $\mu^*(A) = 0$, then A is Lebesgue measurable and has measure zero. Hence show that Lebesgue measure is complete.

Problem 11.9 Show that a subset E of $\mathbb{R}$ is measurable if for all $\epsilon > 0$ there exists an open set $U \supset E$ such that $\mu^*(U - E) < \epsilon$.

Problem 11.10 If E is bounded and there exists an interval $I \supset E$ such that

$$\mu^*(I) = \mu^*(I \cap E) + \mu^*(I - E)$$

then this holds for all intervals, possibly even those overlapping E.

Problem 11.11 The *inner measure* $\mu_*(E)$ of a set E is defined as the least upper bound of the measures of all measurable subsets of E. Show that $\mu_*(E) \leq \mu^*(E)$.
For any open set $U \supset E$, show that

$$\mu(U) = \mu_*(U \cap E) + \mu^*(U - E)$$

and that E is measurable with finite measure if and only if $\mu_*(E) = \mu^*(E) < \infty$.

11.3 Lebesgue integration

Let h be a simple function on a measure space $(X, \mathcal{M}, \mu)$,

$$h = \sum_{i=1}^{n} a_i \chi_{A_i} \quad (a_i \in \dot{\mathbb{R}})$$

where $A_i = h^{-1}(a_i)$ are measurable subsets of X. We define its **integral** to be

$$\int h \, d\mu = \sum_{i=1}^{n} a_i \mu(A_i).$$

This integral only gives a finite answer if the measure of all sets A_i is finite, and in some cases it may not have a sensible value at all. For example $h : \mathbb{R} \to \mathbb{R}$ defined by

$$h(x) = \begin{cases} 1 & \text{if } x > 0 \\ -1 & \text{if } x < 0 \end{cases}$$

has integral $\infty + (-\infty)$ which is not well-defined.

If $h : X \to \mathbb{R}$ is a simple function, then for any constant b we have

$$\int bf \, d\mu = b \int f \, d\mu. \tag{11.8}$$

If $g = \sum_{j=1}^{m} b_j \chi_{B_j}$ is another simple function then $f + g$ is a simple function

$$f + g = \sum_{i=1}^{n} \sum_{j=1}^{m} (a_i + b_j) \chi_{A_i \cap B_j},$$

and has integral

$$\int (f + g) \, d\mu = \int f \, d\mu + \int g \, d\mu. \tag{11.9}$$

It is best to omit any term from the double sum where $a_i + b_j = 0$, else we may face the awkward problem of assigning a value to the product 0∞.

Exercise: Prove (11.8).

For any pair of functions $f, g : X \to \mathbb{R}$ we write $f \leq g$ to mean $f(x) \leq g(x)$ for all $x \in X$. If h and h' are simple functions such that $h \leq h'$ then

$$\int h \, d\mu \leq \int h' \, d\mu.$$

This follows immediately from the fact that $h - h'$ is a simple function that is non-negative everywhere, and therefore has an integral ≥ 0.

Taking the measure on $\mathbb{R}$ to be Lebesgue measure, the **integral of a non-negative measurable function** $f : X \to \mathbb{R}$ is defined as

$$\int f \, d\mu = \sup \int h \, d\mu$$

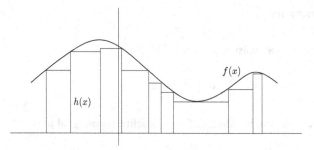

Figure 11.3 Integral of a non-negative measurable function

where the supremum is taken over all non-negative simple functions $h : X \to \mathbb{R}$ such that $h \leq f$ (see Fig. 11.3). If $E \subset X$ is a measurable set then $f\chi_E$ is a measurable function on X that vanishes outside E. We define the **integral of f over E** to be

$$\int_E f \, d\mu = \int f\chi_E \, d\mu.$$

Exercise: Show that for any pair of non-negative measurable functions f and g

$$f \geq g \implies \int f \, d\mu \geq \int g \, d\mu. \tag{11.10}$$

The following theorem is often known as the **monotone convergence theorem**:

Theorem 11.7 (Beppo Levi) *If f_n is an increasing sequence of non-negative measurable real-valued functions on X, $f_{n+1} \geq f_n$, such that $f_n(x) \to f(x)$ for all $x \in X$ then*

$$\lim_{n \to \infty} \int f_n \, d\mu = \int f \, d\mu.$$

Proof: From the comments before Theorem 11.2 we know that f is a measurable function, as it is the limit of a sequence of measurable functions. If f has a finite integral then, by definition, for any $\epsilon > 0$ there exists a simple function $h : X \to \mathbb{R}$ such that $0 \leq h \leq f$ and $\int f \, d\mu - \int h \, d\mu < \epsilon$. For any real number $0 < c < 1$ let

$$E_n = \{x \in X \mid f_n(x) \geq ch(x)\} = (f_n - ch)^{-1}([0, \infty)),$$

clearly a measurable set for each positive integer n. Furthermore, since f_n is an increasing sequence of functions we have

$$E_n \subset E_{n+1} \quad \text{and} \quad X = \bigcup_{n=1}^{\infty} E_n,$$

since every point $x \in X$ lies in some E_n for n big enough. Hence

$$\int f \, d\mu \geq \int f_n \, d\mu \geq c \int h\chi_{E_n} \, d\mu.$$

If $h = \sum_i a_i \chi_{A_i}$ then

$$\int h\chi_{E_n} \, d\mu = \int \sum_i a_i \chi_{A_i \cap E_n} \, d\mu = \sum_i a_i \mu(A_i \cap E_n).$$

11.3 Lebesgue integration

Hence, by Theorem 11.3,

$$\lim_{n \to \infty} \int h \chi_{E_n} \, d\mu = \sum_i a_i \mu(A_i \cap X) = \sum_i a_i \mu(A_i) = \int h \, d\mu,$$

so that

$$\int f \, d\mu \geq \lim_{n \to \infty} \int f_n \, d\mu \geq c \int h \, d\mu \geq c \int f \, d\mu - c\epsilon.$$

Since c can be chosen arbitrarily close to 1 and ϵ arbitrarily close to 0, we have

$$\int f \, d\mu \geq \lim_{n \to \infty} \int f_n \, d\mu \geq \int f \, d\mu,$$

which proves the required result. ∎

Exercise: How does this proof change if $\int f \, d\mu = \infty$?

Using the result from Theorem 11.2, that every positive measurable function is a limit of increasing simple functions, it follows from Theorem 11.7 that simple functions can be replaced by arbitrary measurable functions in Eqs. (11.8) and (11.9).

Theorem 11.8 *The integral of a non-negative measurable function $f \geq 0$ vanishes if and only if $f(x) = 0$ almost everywhere.*

Proof: If $f(x) = 0$ a.e., let $h = \sum_i a_i \chi_{A_i} \geq 0$ be a simple function such that $h \leq f$. Every set A_i must have measure zero, else $f(x) > 0$ on a set of positive measure. Hence $\int f \, d\mu = \sup \int h \, d\mu = 0$.

Conversely, suppose $\int f \, d\mu = 0$. Let $E_n = \{x \mid f(x) \geq 1/n\}$. These are an increasing sequence of measurable sets, $E_{n+1} \supset E_n$, and

$$f \geq \frac{1}{n} \chi_{A_n}.$$

Hence

$$\int \frac{1}{n} \chi_{A_n} \, d\mu = \frac{1}{n} \mu(A_n) \leq \int f \, d\mu = 0,$$

which is only possible if $\mu(A_n) = 0$. By Theorem 11.3 it follows that

$$\mu(\{x \mid f(x) > 0\}) = \mu\left(\bigcup_{n=1}^{\infty} A_n\right) = \lim_{n \to \infty} \mu(A_n) = 0.$$

Hence $f(x) = 0$ almost everywhere. ∎

Integration may be extended to real-valued functions that take on positive or negative values. We say a measurable function f is **integrable with respect to the measure** μ if both its positive and negative parts, f^+ and f^-, are integrable. The **integral of** f is then defined as

$$\int f \, d\mu = \int f^+ \, d\mu - \int f^- \, d\mu.$$

Measure theory and integration

If f is integrable then so is its modulus $|f| = f^+ + f^-$, and

$$\left|\int f\,d\mu\right| \le \left|\int f^+\,d\mu - \int f^-\,d\mu\right| \le \left|\int f^+\,d\mu + \int f^-\,d\mu\right| \le \int |f|\,d\mu. \quad (11.11)$$

Hence a measurable function f is integrable if and only if $|f|$ is integrable.

A function $f : \mathbb{R} \to \mathbb{R}$ is said to be **Lebesgue integrable** if it is measurable and integrable with respect to the Lebesgue measure on $\mathbb{R}$. As for Riemann integration it is common to use the notations

$$\int f(x)\,dx = \int_{-\infty}^{\infty} f(x)\,dx \equiv \int f\,d\mu,$$

and for integration over an interval $I = [a, b]$,

$$\int_a^b f(x)\,dx \equiv \int_I f\,d\mu.$$

Riemann integrable functions on an interval $[a, b]$ are Lebesgue integrable on that interval. A function f is Riemann integrable if for any $\epsilon > 0$ there exist step functions h_1 and h_2 – simple functions that are constant on intervals – such that $h_1 \le f \le h_2$ and

$$\int_a^b h_2(x)\,dx - \int_a^b h_1(x)\,dx < \epsilon.$$

By taking $H_n = \sup(h_{i1})$ for the sequence of functions (h_{i1}, h_{i2}) $(i = 1, \dots, n)$ defined by $\epsilon = 1, \frac{1}{2}, \dots, \frac{1}{n}$, it is straightforward to show that the H_n are simple functions the supremum of whose integrals is the Riemann integral of f. Hence f is Lebesgue integrable, and its Lebesgue integral is equal to its Riemann integral. The difference between the two concepts of integration is that for Lebesgue integration the simple functions used to approximate a function f need not be step functions, but can be constant on arbitrary measurable sets. For example, the function on $[0, 1]$ defined by

$$f(x) = \begin{cases} 1 & \text{if } x \text{ is irrational} \\ 0 & \text{if } x \text{ is rational} \end{cases}$$

is certainly Lebesgue integrable, and since $f = 1$ a.e. its Lebesgue integral is 1. It cannot, however, be approximated in the required way by step functions, and is not Riemann integrable.

Exercise: Prove the last statement.

Theorem 11.9 *If f and g are Lebesgue integrable real functions, then for any $a, b \in \mathbb{R}$ the function $af + bg$ is Lebesgue integrable and for any measurable set E*

$$\int_E (af + bg)\,d\mu = a \int_E f\,d\mu + b \int_E g\,d\mu.$$

The proof is straightforward and is left as an exercise (see problems at end of chapter).

11.3 Lebesgue integration

Lebesgue's dominated convergence theorem

One of the most important results of Lebesgue integration is that, under certain general circumstances, the limit of a sequence of integrable functions is integrable. First we need a lemma, relating to the concept of lim sup of a sequence of functions, defined in the paragraph prior to Theorem 11.2.

Lemma 11.10 (Fatou) *If (f_n) is any sequence of non-negative measurable functions defined on the measure space $(X, \mathcal{M}, \mu)$, then*

$$\int \liminf_{n \to \infty} f_n \, d\mu \leq \liminf_{n \to \infty} \int f_n \, d\mu.$$

Proof: The functions $G_n(x) = \inf_{k \geq n} f_k(x)$ form an increasing sequence of non-negative measurable functions such that $G_n \leq f_n$ for all n. Hence the lim inf of the sequence (f_n) is the limit of the sequence G_n,

$$\liminf_{n \to \infty} f_n = \sup_n G_n = \lim_{n \to \infty} G_n.$$

By the monotone convergence theorem 11.7,

$$\lim_{n \to \infty} \int G_n \, d\mu = \int \liminf_{n \to \infty} f_n \, d\mu$$

while the inequality $G_n \leq f_n$ implies that

$$\int G_n \, d\mu \leq \int f_n \, d\mu.$$

Hence

$$\int \liminf_{n \to \infty} f_n \, d\mu = \lim_{n \to \infty} \int G_n \, d\mu \leq \liminf_{n \to \infty} \int f_n \, d\mu. \quad \blacksquare$$

Theorem 11.11 (Lebesgue) *Let (f_n) be any sequence of real-valued measurable functions defined on the measure space $(X, \mathcal{M}, \mu)$ that converges almost everywhere to a function f. If there exists a positive integrable function $g : X \to \mathbb{R}$ such that $|f_n| < g$ for all n then*

$$\lim_{n \to \infty} \int f_n \, d\mu = \int f \, d\mu.$$

Proof: The function f is measurable since it is the limit a.e. of a sequence of measurable functions, and as $|f_n| < g$ all functions f_n and f are integrable with respect to the measure μ. Apply Fatou's lemma 11.10 to the sequence of positive measurable functions $g_n = 2g - |f_n - f| > 0$,

$$\int \liminf_{n \to \infty} (2g - |f_n - f|) \, d\mu \leq \liminf_{n \to \infty} \int (2g - |f_n - f|) \, d\mu.$$

Since $\int g \, d\mu < \infty$ and $\liminf |f_n - f| = \lim |f_n - f| = 0$, we have

$$0 \leq \liminf_{n \to \infty} \int -|f_n - f| \, d\mu = -\limsup_{n \to \infty} \int |f_n - f| \, d\mu.$$

Since $|f_n - f| > 0$ this is only possible if

$$\lim_{n \to \infty} \int |f_n - f| \, d\mu = 0.$$

Hence

$$\left| \int (f_n - f) \, d\mu \right| \leq \int |f_n - f| \, d\mu \to 0,$$

so that $\int f_n \, d\mu \to \int f \, d\mu$, as required. ∎

The convergence in this theorem is said to be **dominated convergence,** g being the dominating function. An attractive feature of Lebesgue integration is that an integral over an unbounded set is defined exactly as for a bounded set. The same is true of unbounded integrands. This contrasts sharply with Riemann integration where such integrals are not defined directly, but must be defined as 'improper integrals' that are limits of bounded functions over a succession of bounded intervals. The concept of an improper integral is not needed at all in Lebesgue theory. However, Lebesgue's dominated convergence theorem can be used to evaluate such integrals as limits of finite integrands over finite regions.

Example 11.8 The importance of a dominating function is shown by the following example. The sequence of functions $(\chi_{[n,n+1]})$ consists of a 'unit hump' drifting steadily to the right and clearly has the limit $f(x) = 0$ everywhere. However it has no dominating function and the integrals do not converge

$$\int \chi_{[n,n+1]} \, d\mu = \int_n^{n+1} 1 \, dx = 1 \nrightarrow \int 0 \, dx = 0.$$

We mention, without proof, the following theorem relating Lebesgue integration on higher dimensional Euclidean spaces to multiple integration.

Theorem 11.12 (Fubini) *If $f : \mathbb{R}^2 \to \mathbb{R}$ is a Lebesgue measurable function, then for each $x \in \mathbb{R}$ the function $f_x(y) = f(x, y)$ is measurable. Similarly for each $y \in \mathbb{R}$ the function $f'_y(x) = f(x, y)$ is measurable on $\mathbb{R}$. It is common to write*

$$\int f(x, y) \, dy \equiv \int f_x \, d\mu \quad \text{and} \quad \int f(x, y) \, dx \equiv \int f'_y \, d\mu.$$

Then

$$\int \int f(x, y) \, dx \, dy \equiv \int f \, d\mu^2 = \int \left(\int f(x, y) \, dy \right) dx$$
$$= \int \left(\int f(x, y) \, dx \right) dy.$$

The result generalizes to a product of an arbitrary pair of measure spaces. For a proof see, for example, [1, 2].

Problems

Problem 11.12 Show that if f and g are Lebesgue integrable on $E \subset \mathbb{R}$ and $f \geq g$ a.e., then
$$\int_E f \, d\mu \geq \int_E g \, d\mu.$$

Problem 11.13 Prove Theorem 11.9.

Problem 11.14 If f is a Lebesgue integrable function on $E \subset \mathbb{R}$ then show that the function ψ defined by
$$\psi(a) = \mu(\{x \in E \mid |f(x)| > a\}) = O(a^{-1}) \quad \text{as } a \to \infty.$$

References

[1] N. Boccara. *Functional Analysis*. San Diego, Academic Press, 1990.
[2] B. D. Craven. *Lebesgue Measure and Integral*. Marshfield, Pitman Publishing Company, 1982.
[3] L. Debnath and P. Mikusiński. *Introduction to Hilbert Spaces with Applications*. San Diego, Academic Press, 1990.
[4] N. B. Haaser and J. A. Sullivan. *Real Analysis*. New York, Van Nostrand Reinhold Company, 1971.
[5] P. R. Halmos. *Measure Theory*. New York, Van Nostrand Reinhold Company, 1950.
[6] R. Geroch. *Mathematical Physics*. Chicago, The University of Chicago Press, 1985.
[7] C. de Witt-Morette, Y. Choquet-Bruhat and M. Dillard-Bleick. *Analysis, Manifolds and Physics*. Amsterdam, North-Holland, 1977.
[8] F. Riesz and B. Sz.-Nagy. *Functional Analysis*. New York, F. Ungar Publishing Company, 1955.

12 Distributions

In physics and some areas of engineering it has become common to make use of certain 'functions' such as the *Dirac delta function* $\delta(x)$, having the property

$$\int_{-\infty}^{\infty} f(x)\delta(x)\,\mathrm{d}x = f(0)$$

for all continuous functions $f(x)$. If we set $f(x)$ to be a continuous function that is everywhere zero except on a small interval $(a-\epsilon, a+\epsilon)$ on which $f > 0$, it follows that $\delta(a) = 0$ for all $a \neq 0$. However, setting $f(x) = 1$ implies $\int \delta(x)\,\mathrm{d}x = 1$, so we must assign an infinite value to $\delta(0)$,

$$\delta(x) = \begin{cases} 0 & \text{if } x \neq 0, \\ \infty & \text{if } x = 0. \end{cases} \tag{12.1}$$

As it stands this really won't do, since the δ-function vanishes a.e. and should therefore be assigned Lebesgue integral zero. Our aim in this chapter is to give a rigorous definition of such 'generalized functions', which avoids these contradictions.

In an intuitive sense we might think of the Dirac delta function as being the 'limit' of a sequence of functions (see Fig. 12.1) such as

$$\varphi_n(x) = \begin{cases} 2n & \text{if } |x| \leq 1/n \\ 0 & \text{if } |x| > 1/n \end{cases}$$

or of Gaussian functions

$$\psi_n(x) = \frac{1}{n\sqrt{\pi}} e^{-x^2/n^2}.$$

Lebesgue's dominated convergence theorem does not apply to these sequences, yet the limit of the integrals is clearly 1, and for any continuous function $f(x)$ it is not difficult to show that

$$\lim_{n\to\infty} \int_{-\infty}^{\infty} f(x)\varphi_n(x)\,\mathrm{d}x = \lim_{n\to\infty} \int_{-\infty}^{\infty} f(x)\psi_n(x)\,\mathrm{d}x = f(0).$$

However, we will not attempt to define Dirac-like functions as limiting functions in some sense. Rather, following Laurent Schwartz, we define them as continuous linear functionals on a suitably defined space of regular test functions. This method is called the theory of *distributions* [1–6].

12.1 Test functions and distributions

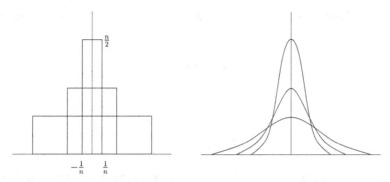

Figure 12.1 Dirac delta function as a limit of functions

12.1 Test functions and distributions

Spaces of test functions

The **support** of a function $f : \mathbb{R}^n \to \mathbb{R}$ is the closure of the region of $\mathbb{R}^n$ where $f(\mathbf{x}) \neq 0$. We will say a real-valued function f on $\mathbb{R}^n$ has **compact support** if the support is a closed bounded set; that is, there exists $R > 0$ such that $f(x_1, \ldots, x_n) = 0$ for $|\mathbf{x}| \geq R$. A function f is said to be C^m if all partial derivatives of order m,

$$D_{\underline{m}} f = \frac{\partial^m f}{\partial x_1^{m_1}, \ldots, \partial x_n^{m_n}},$$

exist and are continuous, where $\underline{m} = (m_1, \ldots, m_n)$ and $m = |\underline{m}| \equiv \sum_{i=1}^n m_i$. We adopt the convention that $D_{(0,0,\ldots,0)} f = f$. A function f is said to be C^∞, or **infinitely differentiable**, if it is C^m to all orders $m = 1, 2, \ldots$ We set $\mathcal{D}^m(\mathbb{R}^n)$ to be the vector space of all C^m functions on $\mathbb{R}$ with compact support, called the space of **test functions of order** m.

Exercise: Show that $\mathcal{D}^m(\mathbb{R}^n)$ is a real vector space.

The space of infinitely differentiable test functions, $\mathcal{D}^\infty(\mathbb{R}^n)$, is often denoted simply as $\mathcal{D}(\mathbb{R}^n)$ and is called the space of **test functions**,

$$\mathcal{D}(\mathbb{R}^n) = \bigcap_{n=1}^\infty \mathcal{D}^m(\mathbb{R}^n).$$

To satisfy ourselves that this space is not empty, consider the function $f : \mathbb{R} \to \mathbb{R}$ defined by

$$f(x) = \begin{cases} e^{-1/x} & \text{if } x > 0, \\ 0 & \text{if } x \leq 0. \end{cases}$$

This function is infinitely differentiable everywhere, including the point $x = 0$ where all derivatives vanish both from the left and the right. Hence the function $\varphi : \mathbb{R} \to \mathbb{R}$

$$\varphi(x) = f(-(x-a))f(x+a) = \begin{cases} \exp\left(\frac{2a}{x^2 - a^2}\right) & \text{if } |x| < a, \\ 0 & \text{if } |x| \geq a, \end{cases}$$

is everywhere differentiable and has compact support $[-a, a]$. There is a counterpart in $\mathbb{R}^n$,

$$\varphi(\mathbf{x}) = \begin{cases} \exp\left(\frac{2a}{|\mathbf{x}|^2 - a^2}\right) & \text{if } |\mathbf{x}| < a, \\ 0 & \text{if } |\mathbf{x}| \geq a, \end{cases}$$

where

$$|\mathbf{x}| = \sqrt{(x_1)^2 + (x_2)^2 + \cdots + (x_n)^2}.$$

A sequence of functions $\varphi_n \in \mathcal{D}(\mathbb{R}^n)$ is said to **converge to order** m to a function $\varphi \in \mathcal{D}(\mathbb{R}^n)$ if the functions φ_n and φ all have supports within a common bounded set and

$$D_{\underline{k}} \varphi_n(\mathbf{x}) \to D_{\underline{k}} \varphi(\mathbf{x})$$

uniformly for all $\mathbf{x}$, for all $\underline{k}$ of orders $k = 0, 1, \ldots, m$. If we have convergence to order m for all $m = 0, 1, 2, \ldots$ then we simply say φ_n **converges** to φ, written $\varphi_n \to \varphi$.

Example 12.1 Let $\varphi : \mathbb{R} \to \mathbb{R}$ be any differentiable function having compact support on K in $\mathbb{R}$. The sequence of functions

$$\varphi_n(x) = \frac{1}{n} \varphi(x) \sin nx$$

are all differentiable and have common compact support K. Since $|\varphi(x) \sin nx| < 1$ it is evident that these functions approach the zero function uniformly as $n \to \infty$, but their derivatives

$$\varphi'_n(x) = \frac{1}{n} \varphi'(x) \sin nx + \varphi(x) \cos nx \not\to 0.$$

This is an example of a sequence of functions that converge to order 0 to the zero function, but not to order 1.

To define this convergence topologically, we can proceed in the following manner. For every compact set $K \subset \mathbb{R}^n$ let $\mathcal{D}^m(K)$ be the space of C^m functions of compact support within K. This space is made into a topological space as in Example 10.25, by defining a norm

$$\|f\|_{K, m} = \sup_{\mathbf{x} \in K} \sum_{|\underline{k}| \leq m} |D_{\underline{k}} f(\mathbf{x})|.$$

On $\mathcal{D}^m(\mathbb{R}^n)$ we define a set U to be open if for every $f \in U$ there exists a compact set K and a real $a > 0$ such that $f \in K$ and

$$\{g \in K \mid \|g - f\|_{K, m} < a\} \subseteq U.$$

It then follows that a sequence $f_k \in \mathcal{D}^m(\mathbb{R}^n)$ converges to order m to a function $f \in \mathcal{D}^m(\mathbb{R}^n)$

12.1 Test functions and distributions

if and only if $f_n \to f$ with respect to this topology. A similar treatment gives a topology on $\mathcal{D}(\mathbb{R}^n)$ leading to convergence in all orders (see Problem 12.2).

Distributions

In this chapter, when we refer to 'continuity' of a functional S on a space such as $\mathcal{D}(\mathbb{R}^n)$, we will mean that whenever $f_n \to f$ in some specified sense on $\mathcal{D}(\mathbb{R}^n)$ we have $S(f_n) \to S(f)$. A **distribution of order** m on $\mathbb{R}^n$ is a linear functional T on $\mathcal{D}(\mathbb{R}^n)$,

$$T(a\varphi + b\psi) = aT(\varphi) + bT(\psi),$$

which is continuous to order m; that is, if $\varphi_k \to \varphi$ is any sequence of functions in $\mathcal{D}(\mathbb{R}^n)$ convergent to order m then $T(\varphi_k) \to T(\varphi)$. A linear functional T on $\mathcal{D}(\mathbb{R}^n)$ that is continuous with respect to sequences φ_i in $\mathcal{D}(\mathbb{R}^n)$ that are convergent to all orders will simply be referred to as a **distribution** on $\mathbb{R}^n$. In this sense of continuity, the space of distributions of order m on $\mathbb{R}^n$ is the dual space of $\mathcal{D}^m(\mathbb{R}^n)$ (see Section 10.9), and the space of distributions is the dual space of $\mathcal{D}(\mathbb{R}^n)$. Accordingly, these are denoted $\mathcal{D}'^m(\mathbb{R}^n)$ and $\mathcal{D}'(\mathbb{R}^n)$ respectively.

Note that a distribution T of order m is also a distribution of order m' for all $m' > m$. For, if φ_i is a convergent sequence of functions in $\mathcal{D}^{m'}(\mathbb{R}^n)$, then φ_i and all its derivatives up to order m' converge uniformly to a function $\varphi \in \mathcal{D}^{m'}(\mathbb{R}^n)$. In particular, it is also a sequence in $\mathcal{D}^m(\mathbb{R}^n)$ converging to order $m < m'$ to φ. Therefore a linear functional T of order m, having the property $T(\varphi_i) \to T(\varphi)$ for all convergent sequences φ_i in $\mathcal{D}^m(\mathbb{R}^n)$, automatically has this property for all convergent sequences in $\mathcal{D}^{m'}(\mathbb{R}^n)$. This is a curious feature, characteristic of dual spaces: given a function that is C^m we can only conclude that it is $C^{m'}$ for $m' \le m$, yet given a distribution of order m we are guaranteed that it is a distribution of order m' for all $m' \ge m$.

Regular distributions

A function $f : \mathbb{R}^n \to \mathbb{R}$ is said to be **locally integrable** if it is integrable on every compact subset $K \subset \mathbb{R}^n$. Set $T_f : \mathcal{D}(\mathbb{R}^n) \to \mathbb{R}$ to be the continuous linear functional defined by

$$T_f(\varphi) = \int_{\mathbb{R}^n} \varphi f \, \mathrm{d}\mu^n = \int \cdots \int_{\mathbb{R}^n} \varphi(\mathbf{x}) f(\mathbf{x}) \, \mathrm{d}x_1 \ldots \mathrm{d}x_n.$$

The integral always exists, since every test function φ vanishes outside some compact set. Linearity is straightforward by elementary properties of the integral operator,

$$T_f(a\varphi + b\psi) = aT_f(\varphi) + bT_f(\psi).$$

Continuity of T_f follows from the inequality (11.11) and Lebesgue's dominated convergence theorem 11.11,

$$|T_f(\varphi_i) - T_f(\varphi)| = \left| \int \cdots \int_{\mathbb{R}^n} f(\mathbf{x})(\varphi_i(\mathbf{x}) - \varphi(\mathbf{x})) \, d^n x \right|$$

$$\leq \int \cdots \int_{\mathbb{R}^n} |f(\mathbf{x})(\varphi_i(\mathbf{x}) - \varphi(\mathbf{x}))| \, d^n x$$

$$\leq \int \cdots \int_{\mathbb{R}^n} |f(\mathbf{x})||(\varphi_i(\mathbf{x}) - \varphi(\mathbf{x}))| \, d^n x$$

$$\to 0,$$

since the sequence of integrable functions $f\varphi_i$ is dominated by the integrable function $(\sup_i |\varphi_i|)|f|$. Hence T_f is a distribution and the function f is called its **density**. In fact T_f is a distribution of order 0, since only convergence to order 0 is needed in its definition.

Two locally integrable functions f and g that are equal almost everywhere give rise to the same distribution, $T_f = T_g$. Conversely, if $T_f(\varphi) = T_g(\varphi)$ for all test functions φ then the density functions f and g are equal a.e. An outline proof is as follows: let I^n be any product of closed intervals $I^n = I_1 \times I_2 \times \cdots \times I_n$, and choose a test function φ arbitrarily close to the unit step function χ_{I^n}. Then $\int_{I^n} (f - g) \, d\mu^n = 0$, which is impossible for all I^n if $f - g$ has non-vanishing positive part, $(f - g)^+ > 0$, on a set of positive measure. This argument may readily be refined to show that $f - g = 0$ a.e. Hence the density f is uniquely determined by T_f except on a set of measure zero. By identifying f with T_f, locally integrable functions can be thought of as distributions. Not all distributions, however, arise in this way; distributions having a density $T = T_f$ are sometimes referred to as **regular distributions**, while those not corresponding to any locally integrable function are called **singular**.

Example 12.2 Define the distribution δ_a on $\mathcal{D}(\mathbb{R})$ by

$$\delta_a(\varphi) = \varphi(a).$$

In particular, we write δ for δ_0, so that $\delta(\varphi) = \varphi(0)$. The map $\delta_a : \mathcal{D}(\mathbb{R}) \to \mathbb{R}$ is obviously linear, $\delta_a(b\varphi + c\psi) = b\varphi(a) + c\psi(a) = b\delta_a(\varphi) + c\delta_a(\psi)$, and is continuous since $\varphi_n \to \varphi \implies \varphi_n(a) \to \varphi(a)$. Hence δ_a is a distribution, but by the reasoning at the beginning of this chapter it cannot correspond to any locally integrable function. It is therefore a singular distribution. Nevertheless, physicists and engineers often maintain the density notation and write

$$\delta(\varphi) \equiv \int_{-\infty}^{\infty} \varphi(x)\delta(x) \, dx = \varphi(0). \qquad (12.2)$$

In writing such an equation, the distribution δ is imagined to have the form T_δ for a density function $\delta(x)$ concentrated at the point $x = 0$ and having an infinite value there as in Eq. (12.1), such that

$$\int_{-\infty}^{\infty} \delta(x) \, dx = 1.$$

12.1 Test functions and distributions

Using a similar convention, the distribution δ_a may be thought of as representing the density function $\delta_a(x)$ such that

$$\int_{-\infty}^{\infty} \varphi(x)\delta_a(x)\,dx = \varphi(a)$$

for all test functions φ. It is common to write $\delta_a(x) = \delta(x - a)$, for on performing the 'change of variable' $x = y + a$,

$$\int_{-\infty}^{\infty} \varphi(x)\delta(x-a)\,dx = \int_{-\infty}^{\infty} \varphi(y+a)\delta(y)\,dy = \varphi(a).$$

The n-dimensional delta function may be similarly defined by

$$\delta_{\mathbf{a}}^n(\varphi) = \varphi(\mathbf{a}) = \varphi(a_1, \dots, a_n)$$

and can be written

$$\delta_{\mathbf{a}}^n(\varphi) \equiv T_{\delta_{\mathbf{a}}^n}(\varphi) = \int \cdots \int_{\mathbb{R}^n} \varphi(\mathbf{x})\delta^n(\mathbf{x} - \mathbf{a})\,d^n x = \varphi(\mathbf{a})$$

where

$$\delta^n(\mathbf{x} - \mathbf{a}) = \delta(x_1 - a_1)\delta(x_2 - a_2)\dots\delta(x_n - a_n).$$

Although it is not in general possible to define the product of distributions, no problems arise in this instance because the delta functions on the right-hand side depend on separate and independent variables.

Problems

Problem 12.1 Construct a test function such that $\phi(x) = 1$ for $|x| \le 1$ and $\phi(x) = 0$ for $|x| \ge 2$.

Problem 12.2 For every compact set $K \subset \mathbb{R}^n$ let $\mathcal{D}(K)$ be the space of C^∞ functions of compact support within K. Show that if all integer vectors $\underline{k}$ are set out in a sequence where $N(\underline{k})$ denotes the position of $\underline{k}$ in the sequence, then

$$\|f\|_K = \sup_{x \in K} \sum_{|\underline{k}|} \frac{1}{2^{N(\underline{k})}} \frac{|D_{\underline{k}} f(\mathbf{x})|}{1 + |D_{\underline{k}} f(\mathbf{x})|}$$

is a norm on $\mathcal{D}(K)$. Let a set U be defined as open in $\mathcal{D}(\mathbb{R}^n)$ if it is a union of open balls $\{g \in K \mid \|g - f\|_K < a\}$. Show that this is a topology and sequence convergence with respect to this topology is identical with convergence of sequences of functions of compact support to all orders.

Problem 12.3 Which of the following is a distribution?

(a) $T(\phi) = \displaystyle\sum_{n=1}^{m} \lambda_n \phi^{(n)}(0)$ $(\lambda_n \in \mathbb{R})$.

(b) $T(\phi) = \displaystyle\sum_{n=1}^{m} \lambda_n \phi(x_n)$ $(\lambda_n, x_n \in \mathbb{R})$.

(c) $T(\phi) = (\phi(0))^2$.

(d) $T(\phi) = \sup \phi$.

(e) $T(\phi) = \int_{-\infty}^{\infty} |\phi(x)| \, dx$.

Problem 12.4 We say a sequence of distributions T_n converges to a distribution T, written $T_n \to T$, if $T_n(\phi) \to T(\phi)$ for all test functions $\phi \in \mathcal{D}$ (this is sometimes called *weak convergence*). If a sequence of continuous functions f_n converges uniformly to a function $f(x)$ on every compact subset of $\mathbb{R}$, show that the associated regular distributions $T_{f_n} \to T_f$.

In the distributional sense, show that we have the following convergences:

$$f_n(x) = \frac{n}{\pi(1+n^2x^2)} \to \delta(x),$$

$$g_n(x) = \frac{n}{\sqrt{\pi}} e^{-n^2 x^2} \to \delta(x).$$

12.2 Operations on distributions

If T and S are distributions of order m on $\mathbb{R}^n$, then clearly $T + S$ and aT are distributions of this order for all $a \in \mathbb{R}$. Thus $\mathcal{D}'^m(\mathbb{R}^n)$ is a vector space.

Exercise: Prove that $T + S$ is linear and continuous. Similarly for aT.

The product ST of two distributions is not a distribution. For example, if we were to define $(ST)(\varphi) = S(\varphi)T(\varphi)$, this is not linear in φ. However, if α is a C^m function on $\mathbb{R}^n$ and T is a distribution of order m then αT can be defined as a distribution of order m, by setting

$$(\alpha T)(\varphi) = T(\alpha \varphi),$$

since $\alpha \varphi \in \mathcal{D}^m(\mathbb{R}^n)$ for all $\varphi \in \mathcal{D}^m(\mathbb{R}^n)$. Note that α need not be a test function for this construction – it works even if the function α does not have compact support.

If T is a regular distribution on $\mathbb{R}^n$, $T = T_f$, then $\alpha T_f = T_{\alpha f}$. For

$$\alpha T_f(\varphi) = T_f(\alpha \varphi)$$
$$= \int \cdots \int_{\mathbb{R}^n} \varphi \alpha f \, d^n x$$
$$= T_{\alpha f}(\varphi).$$

The operation of multiplying the regular distribution T_f by α is equivalent to simply multiplying the corresponding density function f by α. In this case α need only be a locally integrable function.

Example 12.3 The distribution δ defined in Example 12.2 is a distribution of order zero, since it is well-defined on the space of continuous test functions, $\mathcal{D}^0(\mathbb{R})$. For any continuous function $\alpha(x)$ we have

$$\alpha \delta(\varphi) = \delta(\alpha \varphi) = \alpha(0)\varphi(0) = \alpha(0)\delta(\varphi).$$

12.2 Operations on distributions

Thus
$$\alpha\delta = \alpha(0)\delta. \tag{12.3}$$

In terms of the 'delta function' this identity is commonly written as
$$\alpha(x)\delta(x) = \alpha(0)\delta(x),$$

since
$$\int_{-\infty}^{\infty} \alpha(x)\delta(x)\varphi(x)\,dx = \alpha(0)\varphi(0) = \int_{-\infty}^{\infty} \alpha(0)\delta(x)\varphi(x)\,dx.$$

For the delta function at an arbitrary point a, these identities are replaced by
$$\alpha\delta_a = \alpha(a)\delta_a, \qquad \alpha(x)\delta(x-a) = \alpha(a)\delta(x-a). \tag{12.4}$$

Setting $\alpha(x) = x$ results in the useful identities
$$x\delta = 0, \qquad x\delta(x) = 0. \tag{12.5}$$

Exercise: Extend these identities to the n-dimensional delta function,
$$\alpha\delta_{\mathbf{a}} = \alpha(\mathbf{a})\delta_{\mathbf{a}}, \qquad \alpha(\mathbf{x})\delta^n(\mathbf{x}-\mathbf{a}) = \alpha(\mathbf{a})\delta(\mathbf{x}-\mathbf{a}).$$

Differentiation of distributions

Let T_f be a regular distribution where f is a differentiable function. Standard results in real analysis ensure that the derivative $f' = df/dx$ is a locally integrable function. Let φ be any test function from $\mathcal{D}^1(\mathbb{R})$. Using integration by parts

$$\begin{aligned}T_{f'}(\varphi) &= \int_{-\infty}^{\infty} \varphi(x)\frac{df}{dx}\,dx \\ &= \left[\varphi f\right]_{-\infty}^{\infty} - \int_{-\infty}^{\infty} \frac{d\varphi}{dx} f(x)\,dx \\ &= T_f(-\varphi'),\end{aligned}$$

since $\varphi(\pm\infty) = 0$. We can extend this identity to general distributions, by defining the **derivative of a distribution** T of order $m \geq 0$ on $\mathbb{R}$ to be the distribution T' of order $m+1$ given by
$$T'(\varphi) = T(-\varphi') = -T(\varphi'). \tag{12.6}$$

The derivative of a regular distribution then corresponds to taking the derivative of the density function. Note that the order of the distribution *increases* on differentiation, for $\varphi' \in \mathcal{D}^m(\mathbb{R})$ implies that $\varphi \in \mathcal{D}^{m+1}(\mathbb{R})$. In particular, if T is a distribution of order 0 then T' is a distribution of order 1.

To prove that T' is continuous (linearity is obvious), we use the fact that in the definition of convergence to order $m+1$ of a sequence of functions $\varphi_n \to \varphi$ it is required that all derivatives up to and including order $m+1$ converge uniformly on a compact subset K of

Distributions

$\mathbb{R}$. In particular, $\varphi'_n(x) \to \varphi'(x)$ for all $x \in K$, and

$$T'(\varphi_n) = T(-\varphi'_n) \to T(-\varphi') = T'(\varphi).$$

It follows that every distribution of any order is infinitely differentiable.

If T is a distribution of order greater or equal to 0 on $\mathbb{R}^n$, we may define its partial derivatives in a similar way,

$$\frac{\partial T}{\partial x_k}(\varphi) = -T\left(\frac{\partial \varphi}{\partial x_k}\right).$$

As for distributions on $\mathbb{R}$, any such distribution is infinitely differentiable. For higher derivatives it follows that

$$D_{\underline{m}}T(\varphi) = (-1)^m T(D_{\underline{m}}\varphi) \quad \text{where} \quad m = |\underline{m}| = \sum_i m_i.$$

Exercise: Show that

$$\frac{\partial^2 T}{\partial x_i \partial x_j} = \frac{\partial^2 T}{\partial x_j \partial x_j}.$$

Example 12.4 Set $\theta(x)$ to be the **Heaviside step function**

$$\theta(x) = \begin{cases} 1 & \text{if } x \geq 0, \\ 0 & \text{if } x < 0. \end{cases}$$

This is evidently a locally integrable function, and generates a regular distribution T_θ. For any test function $\varphi \in \mathcal{D}^1(\mathbb{R})$

$$\begin{aligned}
T_{\theta'}(\varphi) = T_\theta(-\varphi') &= -\int_{-\infty}^{\infty} \varphi'(x)\theta(x)\,dx \\
&= -\int_0^\infty \frac{d\varphi}{dx}\,dx \\
&= \varphi(0) \quad \text{since } \varphi(\infty) = 0 \\
&= \delta(\varphi).
\end{aligned}$$

Thus we have the distributional equation, valid only over $\mathcal{D}^1(\mathbb{R})$,

$$T'_\theta = \delta.$$

This is commonly written in terms of 'functions' as

$$\delta(x) = \theta'(x) = \frac{d\theta(x)}{dx}.$$

Intuitively, the step at $x = 0$ is 'infinitely steep'.

Example 12.5 The derivative of the delta distribution is defined as the distribution δ' of order 1, which may be applied to any test function $\varphi \in \mathcal{D}^1(\mathbb{R})$:

$$\delta'(\varphi) = \delta(-\varphi') = -\varphi'(0).$$

12.2 Operations on distributions

Expressed in terms of the delta function, this reads

$$\int_{-\infty}^{\infty} \delta'(x)\varphi(x)\,dx = -\varphi'(0),$$

for an arbitrary function differentiable on a neighbourhood of the origin $x = 0$. Continuing to higher derivatives, we have

$$\delta''(\varphi) = \varphi''(0)$$

or in Dirac's notation

$$\int_{-\infty}^{\infty} \delta''(x)\varphi(x)\,dx = \varphi''(0).$$

For the mth derivative,

$$\delta^{(m)}(\varphi) = (-1)^m \varphi^{(m)}(0), \qquad \int_{-\infty}^{\infty} \delta^{(m)}(x)\varphi(x)\,dx = (-1)^m \varphi^{(m)}(0).$$

For the product of a differentiable function α and a distribution T we obtain the usual Leibnitz rule,

$$(\alpha T)' = \alpha T' + \alpha' T,$$

for

$$\begin{aligned}(\alpha T)'(\varphi) &= \alpha T(-\varphi') \\ &= T(-\alpha\varphi') \\ &= T((-\alpha\varphi)' + \alpha'\varphi) \\ &= T'(\alpha\varphi) + \alpha' T(\varphi) \\ &= \alpha T'(\varphi) + \alpha' T(\varphi).\end{aligned}$$

Example 12.6 From Examples 12.3 and 12.5 we have that

$$(x\delta)' = 0' = 0$$

and

$$(x\delta)' = x\delta' + x'\delta = x\delta' + \delta.$$

Hence

$$x\delta' = -\delta.$$

We can also derive this equation by manipulating the delta function in natural ways,

$$x\delta'(x) = \big(x\delta(x)\big)' - x'\delta(x) = 0' - 1.\delta(x) = -\delta(x).$$

Exercise: Verify the identity $x\delta' = -\delta$ by applying both sides as distributions to an arbitrary test function $\phi(x)$.

Distributions

Change of variable in δ-functions

In applications of the mathematics of delta functions it is common to consider 'functions' such as $\delta(f(x))$. While this is not an operation that generalizes to all distributions, there is a sense in which we can define this concept for the delta distribution for many functions f. Firstly, if $f : \mathbb{R} \to \mathbb{R}$ is a continuous monotone increasing function such that $f(\pm\infty) = \pm\infty$ and we adopt Dirac's notation then, assuming integrals can be manipulated by the standard rules for change of variable,

$$\int_{-\infty}^{\infty} \varphi(x)\delta(f(x))\,dx = \int_{-\infty}^{\infty} \varphi(x)\delta(y)\frac{dy}{f'(x)} \quad \text{where } y = f(x)$$

$$= \int_{-\infty}^{\infty} \frac{\varphi(f^{-1}(y))}{f'(f^{-1}(y))}\delta(y)\,dy$$

$$= \frac{\varphi(a)}{f'(a)} \quad \text{where } f(a) = 0.$$

If $f(x)$ is monotone decreasing then the range of integration is inverted to $\int_{-\infty}^{\infty}$ resulting in a sign change. The general formula for a monotone function f of either direction, having a unique zero at $x = a$, is

$$\int_{-\infty}^{\infty} \varphi(x)\delta(f(x))\,dx = \frac{\varphi(a)}{|f'(a)|}. \tag{12.7}$$

Symbolically, we may write

$$\delta(f(x)) = \frac{1}{|f'(a)|}\delta(x-a),$$

or in terms of distributions,

$$\delta \circ f = \frac{1}{|f'(a)|}\delta_a. \tag{12.8}$$

Essentially this equation can be taken as the definition of the distribution $\delta \circ f$. Setting $f(x) = -x$, it follows that $\delta(x)$ is an even function, $\delta(-x) = \delta(x)$.

If two test functions φ and ψ agree on an arbitrary neighbourhood $[-\epsilon, \epsilon]$ of the origin $x = 0$ then

$$\delta(\varphi) = \delta(\psi) = \varphi(0) = \varphi(\psi).$$

Hence the distribution δ can be regarded as being a distribution on the space of functions $\mathcal{D}([-\epsilon, \epsilon])$, since essentially it only samples values of any test function φ in a neighbourhood of the origin. Thus it is completely consistent to write

$$\delta(\varphi) = \int_{-\epsilon}^{\epsilon} \varphi(x)\delta(x)\,dx.$$

This just reiterates the idea that $\delta(x) = 0$ for all $x \neq 0$.

If $f(x)$ has zeros at $x = a_1, a_2, \ldots$ and f is a monotone function in the neighbourhood of each a_i, then a change of variable to $y = f(x)$ gives, on restricting integration to a small

12.2 Operations on distributions

neighbourhood of each zero,

$$\int_{-\infty}^{\infty} \varphi(x)\delta(f(x))\,dx = \sum_i \frac{\varphi(a_i)}{|f'(a_i)|}.$$

Hence

$$\delta(f(x)) = \sum_i \frac{1}{|f'(a_i)|} \delta(x - a_i), \tag{12.9}$$

or equivalently

$$\delta \circ f = \sum_i \frac{1}{|f'(a_i)|} \delta_{a_i}.$$

Example 12.7 The function $f = x^2 - a^2 = (x-a)(x+a)$ is locally monotone at both its zeros $x = \pm a$, provided $a \neq 0$. In a small neighbourhood of $x = a$ the function f may be approximated by the monotone increasing function $2a(x-a)$, while in a neighbourhood of $x = -a$ it is monotone decreasing and approximated by $-2a(x+a)$. Thus

$$\delta(x^2 - a^2) = \delta(2a(x-a)) + \delta(-2a(x+a)) = \frac{1}{2a}(\delta(x-a) + \delta(x+a)),$$

in agreement with Eq. (12.9).

Problems

Problem 12.5 In the sense of convergence defined in Problem 12.4 show that if $T_n \to T$ then $T'_n \to T'$.

In the distributional sense, show that we have the following convergences:

$$f_n(x) = -\frac{2n^3 x}{\sqrt{\pi}} e^{-n^2 x^2} \to \delta'(x).$$

Problem 12.6 Evaluate

(a) $\displaystyle\int_{-\infty}^{\infty} e^{at} \sin bt\, \delta^{(n)}(t)\,dt$ for $n = 0, 1, 2$.

(b) $\displaystyle\int_{-\infty}^{\infty} (\cos t + \sin t)\delta^{(n)}(t^3 + t^2 + t)\,dt$ for $n = 0, 1$.

Problem 12.7 Show the following identities:

(a) $\delta((x-a)(x-b)) = \dfrac{1}{b-a}(\delta(x-a) + \delta(x-b)).$

(b) $\dfrac{d}{dx}\theta(x^2 - 1) = \delta(x-1) - \delta(x+1) = 2x\delta(x^2 - 1).$

(c) $\dfrac{d}{dx}\delta(x^2 - 1) = \tfrac{1}{2}(\delta'(x-1) + \delta'(x+1)).$

(d) $\delta'(x^2 - 1) = \tfrac{1}{4}(\delta'(x-1) - \delta'(x+1) + \delta(x-1) + \delta(x+1)).$

Problem 12.8 Show that for a monotone function $f(x)$ such that $f(\pm\infty) = \pm\infty$ with $f(a) = 0$

$$\int_{-\infty}^{\infty} \varphi(x)\delta'(f(x))\,dx = -\frac{1}{f'(x)}\frac{d}{dx}\left(\frac{\varphi(x)}{|f'(x)|}\right)\bigg|_{x=a}.$$

Distributions

For a general function $f(x)$ that is monotone on a neighbourhood of all its zeros, find a general formula for the distribution $\delta' \circ f$.

Problem 12.9 Show the identities

$$\frac{d}{dx}(\delta(f(x))) = f'(x)\delta'(f(x))$$

and

$$\delta(f(x)) + f(x)\delta'(f(x)) = 0.$$

Hence show that $\phi(x, y) = \delta(x^2 - y^2)$ is a solution of the partial differential equation

$$x\frac{\partial \phi}{\partial x} + y\frac{\partial \phi}{\partial y} + 2\phi(x, y) = 0.$$

12.3 Fourier transforms

For any function $\varphi(x)$ its **Fourier transform** is the function $\mathcal{F}\varphi$ defined by

$$\mathcal{F}\varphi(y) = \frac{1}{\sqrt{2\pi}} \int_{-\infty}^{\infty} e^{-ixy} \varphi(x)\, dx.$$

The **inverse Fourier transform** is defined by

$$\mathcal{F}^{-1}\varphi(y) = \frac{1}{\sqrt{2\pi}} \int_{-\infty}^{\infty} e^{ixy} \varphi(x)\, dx.$$

Fourier's integral theorem, applicable to all functions φ such that $|\varphi|$ is integrable over $[-\infty, \infty]$ and is of bounded variation, says that $\mathcal{F}^{-1}\mathcal{F}\varphi = \varphi$, expressed in integral form as

$$\varphi(a) = \frac{1}{2\pi} \int_{-\infty}^{\infty} dy\, e^{iay} \int_{-\infty}^{\infty} e^{-iyx} \varphi(x)\, dx$$

$$= \frac{1}{2\pi} \int_{-\infty}^{\infty} dx\, \varphi(x) \int_{-\infty}^{\infty} e^{iy(a-x)}\, dy.$$

The proof of this theorem can be found in many books on real analysis. The reader is referred to [6, chap. 7] or [2, p. 88].

Applying standard rules of integrals applied to delta functions, we expect

$$\delta_a(x) = \delta(x - a) = \frac{1}{2\pi} \int_{-\infty}^{\infty} e^{iy(a-x)}\, dy, \qquad (12.10)$$

or, on setting $a = 0$ and using $\delta(x) = \delta(-x)$,

$$\delta(x) = \frac{1}{2\pi} \int_{-\infty}^{\infty} e^{-iyx}\, dy = \frac{1}{2\pi} \int_{-\infty}^{\infty} e^{iyx}\, dy. \qquad (12.11)$$

Similarly, the Fourier transform of the delta function should be

$$\mathcal{F}\delta(y) = \frac{1}{\sqrt{2\pi}} \int_{-\infty}^{\infty} e^{-ixy} \delta(x)\, dx = \frac{1}{\sqrt{2\pi}} \qquad (12.12)$$

12.3 Fourier transforms

and Eq. (12.11) agrees with

$$\delta(x) = \mathcal{F}^{-1}\frac{1}{\sqrt{2\pi}} = \frac{1}{2\pi}\int_{-\infty}^{\infty} e^{ixy}\,dy. \tag{12.13}$$

Mathematical consistency can be achieved by defining the **Fourier transform of a distribution** T to be the distribution $\mathcal{F}T$ given by

$$\mathcal{F}T(\varphi) = T(\mathcal{F}\varphi), \tag{12.14}$$

for all test functions φ. For regular distributions we then have the desired result,

$$T_{\mathcal{F}f}(\varphi) = \mathcal{F}T_f(\varphi),$$

since

$$\mathcal{F}T_f(\varphi) = T_f(\mathcal{F}\varphi)$$
$$= \frac{1}{\sqrt{2\pi}}\int_{-\infty}^{\infty}\left(\int_{-\infty}^{\infty} e^{-iyx}\varphi(x)\,dx\right)f(y)\,dy$$
$$= \frac{1}{\sqrt{2\pi}}\int_{-\infty}^{\infty}\varphi(x)\left(\int_{-\infty}^{\infty} e^{-iyx}f(y)\,dy\right)dx$$
$$= \int_{-\infty}^{\infty}\varphi(x)\mathcal{F}f(x)\,dx$$
$$= T_{\mathcal{F}f}(\varphi).$$

If the inverse Fourier transform is defined on distributions by $\mathcal{F}^{-1}T(\varphi) = T(\mathcal{F}^{-1}\varphi)$, then

$$\mathcal{F}^{-1}\mathcal{F}T = T,$$

for

$$\mathcal{F}^{-1}\mathcal{F}T(\varphi) = \mathcal{F}T(\mathcal{F}^{-1}\varphi) = T(\mathcal{F}\mathcal{F}^{-1}\varphi) = T(\varphi).$$

There is, however, a serious problem with these definitions. If φ is a function of bounded support then $\mathcal{F}\varphi$ is generally an entire analytic function and cannot be of bounded support, since an entire function that vanishes on any open set must vanish everywhere. Hence the right-hand side of (12.14) is not in general well-defined. A way around this is to define a more general space of test functions $\mathcal{S}(\mathbb{R})$ called the space of **rapidly decreasing functions** – functions that approach 0 as $|x| \to \infty$ faster than any inverse power $|x|^{-n}$,

$$\mathcal{S}(\mathbb{R}) = \{\varphi \mid \sup_{x \in \mathbb{R}} |x^m \varphi^{(p)}(x)| < \infty \text{ for all integers } m, p > 0\}.$$

Convergence in $\mathcal{S}(\mathbb{R})$ is defined by $\varphi_n \to \varphi$ if and only if

$$\lim_{n\to\infty}\sup_{x\in\mathbb{R}}|x^m(\varphi^{(p)}(x) - \varphi^{(p)})| = 0 \text{ for all integers } m, p > 0.$$

The space of continuous linear functions on $\mathcal{S}(\mathbb{R})$ is denoted $\mathcal{S}'(\mathbb{R})$, and they are called **tempered distributions**. Since every test function is obviously a rapidly decreasing function, $\mathcal{D}(\mathbb{R}) \subset \mathcal{S}(\mathbb{R})$. If T is a tempered distribution in Eq. (12.14), the Fourier transform $\mathcal{F}T$ is well-defined, since the Fourier transform of any rapidly decreasing function may be shown to be a function of rapid decrease.

Distributions

Example 12.8 The Fourier transform of the delta distribution is defined by

$$\mathcal{F}\delta_a(\varphi) = \delta_a(\mathcal{F}\varphi)$$
$$= \mathcal{F}\varphi(a)$$
$$= \frac{1}{\sqrt{2\pi}} \int_{-\infty}^{\infty} e^{-iax} \varphi(x) \, dx$$
$$= T_{(2\pi)^{-1/2}e^{iax}}(\varphi).$$

Similarly

$$\mathcal{F}^{-1} T_{e^{-iax}} = \sqrt{2\pi}\, \delta_a.$$

The delta function versions of these distributional equations are

$$\mathcal{F}\delta_a(y) = \frac{1}{\sqrt{2\pi}} \int_{-\infty}^{\infty} e^{-iyx} \delta(x-a) \, dx = \frac{e^{-iay}}{\sqrt{2\pi}}$$

and

$$\mathcal{F}^{-1} e^{-iax} = \frac{1}{\sqrt{2\pi}} \int_{-\infty}^{\infty} e^{ixy} e^{-iax} \, dx = \sqrt{2\pi}\, \delta(x-a),$$

in agreement with Eqs. (12.10)–(12.13) above.

Problems

Problem 12.10 Find the Fourier transforms of the functions

$$f(x) = \begin{cases} 1 & \text{if } -a \leq x \leq a \\ 0 & \text{otherwise} \end{cases}$$

and

$$g(x) = \begin{cases} 1 - \frac{|x|}{2} & \text{if } -a \leq x \leq a \\ 0 & \text{otherwise.} \end{cases}$$

Problem 12.11 Show that

$$\mathcal{F}(e^{-a^2 x^2/2}) = \frac{1}{|a|} e^{-k^2/2a^2}.$$

Problem 12.12 Evaluate Fourier transforms of the following distributional functions:

(a) $\delta(x-a)$.
(b) $\delta'(x-a)$.
(c) $\delta^{(n)}(x-a)$.
(d) $\delta(x^2 - a^2)$.
(e) $\delta'(x^2 - a^2)$.

Problem 12.13 Prove that

$$x^m \delta^{(n)}(x) = (-1)^m \frac{n!}{(n-m)!} \delta^{(n-m)}(x) \qquad \text{for } n \geq m.$$

12.4 Green's functions

Hence show that the Fourier transform of the distribution
$$\sqrt{2\pi}\frac{k!}{(m+k)!}x^m\delta^{(m+k)}(-x)\quad (m,k\geq 0)$$
is $(-iy)^k$.

Problem 12.14 Show that the Fourier transform of the distribution
$$\delta_0+\delta_a+\delta_{2a}+\cdots+\delta_{(2n-1)a}$$
is a distribution with density
$$\frac{1}{\sqrt{2\pi}}\frac{\sin(nay)}{\sin(\tfrac{1}{2}ay)}e^{-(n-\tfrac{1}{2})iay}.$$

Show that
$$\mathcal{F}^{-1}(f(y)e^{iby})=(\mathcal{F}^{-1}f)(x+b).$$

Hence find the inverse Fourier transform of
$$g(y)=\frac{\sin nay}{\sin(\tfrac{1}{2}ay)}.$$

12.4 Green's functions

Distribution theory may often be used to find solutions of inhomogeneous linear partial differential equations by the technique of Green's functions. We give here two important standard examples.

Poisson's equation

To solve an inhomogeneous equation such as Poisson's equation
$$\nabla^2\phi=-4\pi\rho \tag{12.15}$$
we seek a solution to the distributional equation
$$\nabla^2 G(\mathbf{x}-\mathbf{x}')=\delta^3(\mathbf{x}-\mathbf{x}')=\delta(x-x')\delta(y-y')\delta(z-z'). \tag{12.16}$$
A solution of Poisson's equation (12.15) is then
$$\phi(\mathbf{x})=-\iiint 4\pi\rho(\mathbf{x}')G(\mathbf{x}-\mathbf{x}')\,d^3x',$$
for
$$\nabla^2\phi=-\iiint 4\pi\rho(\mathbf{x}')\nabla^2 G(\mathbf{x}-\mathbf{x}')\,d^3x'$$
$$=-\iiint 4\pi\rho(\mathbf{x}')\delta^3(\mathbf{x}-\mathbf{x}')\,d^3x'$$
$$=-4\pi\rho(\mathbf{x}).$$

323

Distributions

To solve, set

$$g(\mathbf{k}) = \mathcal{F}G = \frac{1}{(2\pi)^{3/2}} \iiint_{-\infty}^{\infty} e^{-i\mathbf{k}\cdot\mathbf{y}} G(\mathbf{y}) \, d^3y.$$

By Fourier's theorem

$$G(\mathbf{y}) = \frac{1}{(2\pi)^{3/2}} \iiint_{-\infty}^{\infty} e^{i\mathbf{k}\cdot\mathbf{y}} g(\mathbf{k}) \, d^3k,$$

which implies that

$$\nabla^2 G(\mathbf{x} - \mathbf{x}') = \frac{1}{(2\pi)^{3/2}} \iiint_{-\infty}^{\infty} -k^2 e^{i\mathbf{k}\cdot(\mathbf{x}-\mathbf{x}')} g(\mathbf{k}) \, d^3k.$$

But

$$\delta(\mathbf{y}) = \frac{1}{(2\pi)^3} \int_{-\infty}^{\infty} e^{ik_1 y_1} \, dk_1 \int_{-\infty}^{\infty} e^{ik_2 y_2} \, dk_2 \int_{-\infty}^{\infty} e^{ik_3 y_3} \, dk_3$$

$$= \frac{1}{(2\pi)^3} \iiint_{-\infty}^{\infty} e^{i\mathbf{k}\cdot\mathbf{y}} \, d^3k,$$

so

$$\delta^3(\mathbf{x} - \mathbf{x}') = \frac{1}{2\pi^3} \iiint_{-\infty}^{\infty} e^{i\mathbf{k}\cdot(\mathbf{x}-\mathbf{x}')} \, d^3k.$$

Substituting in Eq. (12.16) gives

$$g(\mathbf{k}) = -\frac{1}{(2\pi)^{3/2} k^2},$$

and

$$G(\mathbf{x} - \mathbf{x}') = -\frac{1}{(2\pi)^3} \iiint_{-\infty}^{\infty} \frac{e^{i\mathbf{k}\cdot\mathbf{y}}}{k^2} \, d^3k. \tag{12.17}$$

The integration in $\mathbf{k}$-space is best performed using polar coordinates (k, θ, ϕ) with the k_3-axis pointing along the direction $\mathbf{R} = \mathbf{x} - \mathbf{x}'$ (see Fig. 12.2). Then

$$\mathbf{k} \cdot (\mathbf{x} - \mathbf{x}') = kR\cos\theta \quad (k = \sqrt{\mathbf{k}\cdot\mathbf{k}})$$

and

$$d^3k = k^2 \sin\theta \, dk \, d\theta \, d\phi.$$

12.4 Green's functions

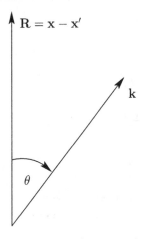

Figure 12.2 Change to polar coordinates in **k**-space.

This results in

$$
\begin{aligned}
G(\mathbf{R}) &= -\frac{1}{(2\pi)^3} \int_0^\infty dk \int_0^\pi d\theta \int_0^{2\pi} d\phi \frac{e^{ikR\cos\theta}}{k^2} k^2 \sin\theta \\
&= -\frac{2\pi}{(2\pi)^3} \int_0^\infty dk \int_0^\pi d\theta \frac{d}{d\theta}\left(\frac{-e^{ikR\cos\theta}}{ikR}\right) \\
&= -\frac{1}{(2\pi)^2 R} \int_0^\infty dk \frac{e^{ikR} - e^{-ikR}}{ik} \\
&= -\frac{1}{(2\pi)^2 R} \int_0^\infty dk\, 2\frac{\sin kR}{k} \\
&= -\frac{1}{4\pi R},
\end{aligned}
$$

on making use of the well-known definite integral

$$\int_0^\infty \frac{\sin x}{x}\, dx = \frac{\pi}{2}.$$

Hence

$$G(\mathbf{x} - \mathbf{x}') = -\frac{1}{4\pi |\mathbf{x} - \mathbf{x}'|}, \qquad (12.18)$$

and a solution of Poisson's equation (12.15) is

$$\phi(\mathbf{x}) = \iiint \frac{\rho(\mathbf{x}')}{|\mathbf{x} - \mathbf{x}'|}\, d^3 x',$$

where the integral is taken over all of the space, $-\infty < x', y', z' < \infty$. For a point charge, $\rho(\mathbf{x}) = q\delta^3(\mathbf{x} - \mathbf{a})$ the solution reduces to the standard coulomb solution

$$\phi(\mathbf{x}) = \frac{q}{|\mathbf{x} - \mathbf{a}|}.$$

Green's function for the wave equation

To solve the inhomogeneous wave equation

$$\Box \psi = -\frac{\partial^2}{c^2 \partial t^2}\psi + \nabla^2 \psi = f(\mathbf{x}, t) \qquad (12.19)$$

it is best to adopt a relativistic 4-vector notation, setting $x^4 = ct$. The wave equation can then be written as in Section 9.4,

$$\Box \psi = g^{\mu\nu}\frac{\partial}{\partial x^\mu}\frac{\partial}{\partial x^\nu}\psi = f(x)$$

where μ and ν range from 1 to 4, $g^{\mu\nu}$ is the diagonal metric tensor having diagonal components 1, 1, 1, -1 and the argument x in the last term is shorthand for $(\mathbf{x}, x^4)$.
Again we look for a solution of the equation

$$\Box G(x - x') = \delta^4(x - x') \equiv \delta(x^1 - x'^1)\delta(x^2 - x'^2)\delta(x^3 - x'^3)\delta(x^4 - x'^4). \qquad (12.20)$$

Every Green's function G generates a solution $\psi_G(x)$ of Eq. (12.19),

$$\psi_G(x) = \iiiint G(x - x') f(x') \, d^4 x'$$

for

$$\Box \psi_G = \iiiint \Box G(x - x') f(x') \, d^4 x' = \iiiint \delta^4(x - x') f(x') \, d^4 x' = f(x).$$

Exercise: Show that the general solution of the inhomogeneous wave equation (12.19) has the form $\psi_G(x) + \phi(x)$ where $\Box \phi = 0$.

Set

$$G(x - x') = \frac{1}{(2\pi)^2} \iiiint g(k) e^{ik.(x-x')} \, d^4 k$$

where $k = (k_1, k_2, k_3, k_4)$, and

$$k.(x - x') = k_\mu(x^\mu - x'^\mu) = k_4(x^4 - x'^4) + \mathbf{k} \cdot (\mathbf{x} - \mathbf{x}')$$

and $d^4 k = dk_1 \, dk_2 \, dk_3 \, dk_4$. Writing the four-dimensional δ function as a Fourier transform we have

$$\Box G(x - x') = \frac{1}{(2\pi)^2} \iiiint -k^2 g(k) e^{ik.(x-x')} \, d^4 k$$

$$= \delta^4(x - x') = \frac{1}{(2\pi)^4} \iiiint e^{ik.(x-x')} \, d^4 k,$$

whence

$$g(k) = -\frac{1}{4\pi^2 k^2}$$

12.4 Green's functions

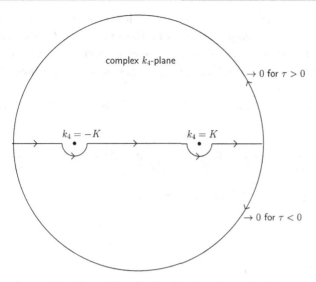

Figure 12.3 Green's function for the three-dimensional wave equation

where $k^2 \equiv k.k = k_\mu k^\mu$. The Fourier transform expression of the Green's function is thus

$$G(x - x') = -\frac{1}{(2\pi)^4} \iiiint \frac{e^{ik.(x-x')}}{k^2} d^4k. \qquad (12.21)$$

To evaluate this integral set

$$\tau = x^4 - x'^4, \qquad \mathbf{R} = \mathbf{x} - \mathbf{x}', \qquad K = |\mathbf{k}| = \sqrt{\mathbf{k} \cdot \mathbf{k}},$$

whence $k^2 = K^2 - k_4^2$ and

$$G(x - x') = \frac{1}{(2\pi)^4} \int_{-\infty}^{\infty} dk_4 \frac{e^{ik_4\tau}}{k_4^2 - K^2} \iiint d^3k\, e^{i\mathbf{k}\cdot\mathbf{R}}.$$

Deform the path in the complex k_4-plane to avoid the pole singularities at $k_4 = \pm K$ as shown in Fig. 12.3 – convince yourself, however, that this has no effect on G satisfying Eq. (12.20).

For $\tau > 0$ the contour is completed in a counterclockwise sense by the upper half semi-circle and

$$\int_{-\infty}^{\infty} \frac{e^{ik_4\tau}}{k_4^2 - K^2} dk_4 = 2\pi i \times \text{sum of residues}$$

$$= 2\pi i \left(\frac{e^{iK\tau}}{2K} - \frac{e^{-iK\tau}}{2K} \right).$$

For $\tau < 0$ we complete the contour with the lower semicircle in a clockwise direction; no poles are enclosed and the integral vanishes. Hence

$$\int_{-\infty}^{\infty} \frac{e^{ik_4\tau}}{k_4^2 - K^2} dk_4 = -\frac{2\pi}{K} \theta(\tau) \sin K\tau$$

where $\theta(\tau)$ is the Heaviside step function.

327

Distributions

This particular contour gives rise to a Green's function that vanishes for $\tau < 0$; that is, for $x^4 < x'^4$. It is therefore called the **outgoing wave condition** or **retarded Green's function**, for a source switched on at $(\mathbf{x}', x'^4)$ only affects field points at later times. If the contour had been chosen to lie above the poles, then the **ingoing wave condition** or **advanced Green's function** would have resulted.

To complete the calculation of G, use polar coodinates in k-space with the k_3-axis parallel to $\mathbf{R}$. This gives

$$G(x - x') = -\frac{1}{(2\pi)^3}\theta(\tau) \int_0^{2\pi} d\phi \int_0^\infty dK \int_0^\pi d\theta\, K^2 \sin\theta\, e^{iKR\cos\theta}\frac{\sin K\tau}{K}$$

$$= -\frac{\theta(\tau)}{2\pi^2 R}\int_0^\infty dK\, \sin K\tau\, \sin KR$$

$$= -\frac{\theta(\tau)}{2\pi^2 R}\int_0^\infty dK\, \frac{(e^{iK\tau} - e^{-iK\tau})}{2i}\frac{(e^{iKR} - e^{-iKR})}{2i}$$

$$= \frac{\theta(\tau)}{4\pi R}(\delta(\tau + R) - \delta(\tau - R))$$

$$= -\frac{\delta(\tau - R)}{4\pi R}.$$

The last step follows because the whole expression vanishes for $\tau < 0$ on account of the $\theta(\tau)$ factor, while for $\tau > 0$ we have $\delta(\tau + R) = 0$. Hence the Green's function may be written

$$G(x - x') = -\frac{1}{4\pi|\mathbf{x} - \mathbf{x}'|}\delta(x^4 - x'^4 - |\mathbf{x} - \mathbf{x}'|), \tag{12.22}$$

which is non-vanishing only on the future light cone of x'.

The solution of the inhomogeneous wave equation (12.19) generated by this Green's function is

$$\psi(\mathbf{x}, t) = \iiiint G(x - x')f(x')\, d^4x'$$

$$= -\frac{1}{4\pi}\iiint \frac{[f(\mathbf{x}', t')]_{\text{ret}}}{|\mathbf{x} - \mathbf{x}'|}\, d^3x' \tag{12.23}$$

where $[f(\mathbf{x}', t')]_{\text{ret}}$ means f evaluated at the retarded time

$$t' = t - \frac{|\mathbf{x} - \mathbf{x}'|}{c}.$$

Problems

Problem 12.15 Show that the Green's function for the time-independent Klein–Gordon equation

$$(\nabla^2 - m^2)\phi = \rho(\mathbf{r})$$

can be expressed as the Fourier integral

$$G(\mathbf{x} - \mathbf{x}') = -\frac{1}{(2\pi)^3}\iiint d^3k\, \frac{e^{i\mathbf{k}.(\mathbf{x}-\mathbf{x}')}}{k^2 + m^2}.$$

Evaluate this integral and show that it results in

$$G(\mathbf{R}) = -\frac{e^{-mR}}{4\pi R} \quad \text{where} \quad \mathbf{R} = \mathbf{x} - \mathbf{x}', \quad R = |\mathbf{R}|.$$

Find the solution ϕ corresponding to a point source

$$\rho(\mathbf{r}) = q\delta^3(\mathbf{r}).$$

Problem 12.16 Show that the Green's function for the one-dimensional diffusion equation,

$$\frac{\partial^2 G(x,t)}{\partial x^2} - \frac{1}{\kappa}\frac{\partial G(x,t)}{\partial t} = \delta(x-x')\delta(t-t')$$

is given by

$$G(x-x', t-t') = -\theta(t-t')\sqrt{\frac{\kappa}{4\pi(t-t')}}e^{-(x-x')^2/4\kappa(t-t')},$$

and write out the corresponding solution of the inhomogeneous equation

$$\frac{\partial^2 \psi(x,t)}{\partial x^2} - \frac{1}{\kappa}\frac{\partial \psi(x,t)}{\partial t} = F(x,t).$$

Do the same for the two- and three-dimensional diffusion equations

$$\nabla^2 G(x,t) - \frac{1}{\kappa}\frac{\partial G(x,t)}{\partial t} = \delta^n(\mathbf{x}-\mathbf{x}')\delta(t-t') \quad (n=2,3).$$

References

[1] J. Barros-Neto. *An Introduction to the Theory of Distributions*. New York, Marcel Dekker, 1973.
[2] N. Boccara. *Functional Analysis*. San Diego, Academic Press, 1990.
[3] R. Geroch. *Mathematical Physics*. Chicago, The University of Chicago Press, 1985.
[4] R. F. Hoskins. *Generalized Functions*. Chichester, Ellis Horwood, 1979.
[5] C. de Witt-Morette Y. Choquet-Bruhat and M. Dillard-Bleick. *Analysis, Manifolds and Physics*. Amsterdam, North-Holland, 1977.
[6] A. H. Zemanian. *Distribution Theory and Transform Analysis*. New York, Dover Publications, 1965.

13 Hilbert spaces

13.1 Definitions and examples

Let V be a complex vector space with an inner product $\langle \cdot | \cdot \rangle : V \times V \to \mathbb{C}$ satisfying (IP1)–(IP3) of Section 5.2. Such a space is sometimes called a **pre-Hilbert space**. As in Eq. (5.11) define a norm on an inner product space by

$$\|u\| = \sqrt{\langle u | u \rangle}. \tag{13.1}$$

The properties (Norm1)–(Norm3) of Section 10.9 hold for this choice of $\|\cdot\|$. Condition (Norm1) is equivalent to (IP3), and (Norm2) is an immediate consequence of (IP1) and (IP2), for

$$\|\lambda v\| = \sqrt{\langle \lambda v | \lambda v \rangle} = \sqrt{\lambda \bar{\lambda} \langle v | v \rangle} = |\lambda| \, \|v\|.$$

The triangle inequality (Norm3) is a consequence of Theorem 5.6. These properties hold equally in finite or infinite dimensional vector spaces. A **Hilbert space** $(\mathcal{H}, \langle \cdot | \cdot \rangle)$ is an inner product space that is complete in the induced norm; that is, $(\mathcal{H}, \|\cdot\|)$ is a Banach space. An introduction to Hilbert spaces at the level of this chapter may be found in [1–6], while more advanced topics are dealt with in [7–11].

The **parallelogram law**

$$\|x+y\|^2 + \|x-y\|^2 = 2\|x\|^2 + 2\|y\|^2 \tag{13.2}$$

holds for all pairs of vectors x, y in an inner product space $\mathcal{H}$. The proof is straightforward, by substituting $\|x+y\|^2 = \langle x+y | x+y \rangle = \|x\|^2 + \|y\|^2 + 2\text{Re}(\langle x | y \rangle)$, etc. It immediately gives rise to the inequality

$$\|x+y\|^2 \le 2\|x\|^2 + 2\|y\|^2. \tag{13.3}$$

For complex numbers (13.2) and (13.3) hold with norm replaced by modulus.

Example 13.1 The typical inner product defined on $\mathbb{C}^n$ in Example 5.4 by

$$\langle (u_1, \ldots, u_n) | (v_1, \ldots, v_n) \rangle = \sum_{i=1}^{n} \overline{u_i} v_i$$

makes it into a Hilbert space. The norm is

$$\|\mathbf{v}\| = \sqrt{|v_1|^2 + |v_2|^2 + \cdots + |v_n|^2},$$

13.1 Definitions and examples

which was shown to be complete in Example 10.27. In any finite dimensional inner product space the Schmidt orthonormalization procedure creates an orthonormal basis for which the inner product takes this form (see Section 5.2). Thus every finite dimensional Hilbert space is isomorphic to $\mathbb{C}^n$ with the above inner product. The only thing that distinguishes finite dimensional Hilbert spaces is their dimension.

Example 13.2 Let ℓ^2 be the set of all complex sequences $u = (u_1, u_2, \dots)$ where $u_i \in \mathbb{C}$ such that

$$\sum_{i=1}^{\infty} |u_i|^2 < \infty.$$

This space is a complex vector, for if u, v are any pair of sequences in ℓ^2, then $u + v \in \ell^2$. For, using the complex number version of the inequality (13.3),

$$\sum_{i=1}^{\infty} |u_i + v_i|^2 \le 2 \sum_{i=1}^{\infty} |u_i|^2 + 2 \sum_{i=1}^{\infty} |v_i|^2 < \infty.$$

It is trivial that $u \in \ell^2$ implies $\lambda u \in \ell^2$ for any complex number λ.

Let the inner product be defined by

$$\langle u \mid v \rangle = \sum_{i=1}^{\infty} \overline{u_i} v_i.$$

This is well-defined for any pair of sequences $u, v \in \ell^2$, for

$$\left| \sum_{i=1}^{\infty} \overline{u_i} v_i \right| \le \sum_{i=1}^{\infty} |\overline{u_i} v_i|$$
$$\le \frac{1}{2} \sum_{i=1}^{\infty} (|u_i|^2 + |v_i|^2) < \infty.$$

The last step follows from

$$2|\overline{a}b|^2 = 2|a|^2|b|^2 = |a|^2 + |b|^2 - (|a| - |b|)^2 \le |a|^2 + |b|^2.$$

The norm defined by this inner product is

$$\|u\| = \sqrt{\sum_{i=1}^{\infty} |u_i|^2} \le \infty.$$

For any integer M and $n, m > N$

$$\sum_{i=1}^{M} |u_i^{(m)} - u_i^{(n)}|^2 \le \sum_{i=1}^{\infty} |u_i^{(m)} - u_i^{(n)}|^2 = \|u^{(m)} - u^{(n)}\|^2 < \epsilon^2,$$

and taking the limit $n \to \infty$ we have

$$\sum_{i=1}^{M} |u_i^{(m)} - u_i|^2 \le \epsilon^2.$$

In the limit $M \to \infty$

$$\sum_{i=1}^{\infty} |u_i^{(m)} - u_i|^2 \leq \epsilon^2$$

so that $u^{(m)} - u \in \ell^2$. Hence $u = u^{(m)} - (u^{(m)} - u)$ belongs to ℓ^2 since it is the difference of two vectors from ℓ^2 and it is the limit of the sequence $u^{(m)}$ since $\|u^{(m)} - u\| < \epsilon$ for all $m > N$. It turns out, as we shall see, that ℓ^2 is isomorphic to most Hilbert spaces of interest – the so-called *separable Hilbert spaces*.

Example 13.3 On $C[0, 1]$, the continuous complex functions on $[0, 1]$, set

$$\langle f | g \rangle = \int_0^1 \overline{f} g \, dx.$$

This is a pre-Hilbert space, but fails to be a Hilbert space since a sequence of continuous functions may have a discontinuous limit.

Exercise: Find a sequence of functions in $C[0, 1]$ that have a discontinuous step function as their limit.

Example 13.4 Let $(X, \mathcal{M}, \mu)$ be a measure space, and $\mathcal{L}^2(X)$ be the set of all square integrable complex-valued functions $f : X \to \mathbb{C}$, such that

$$\int_X |f|^2 \, d\mu < \infty.$$

This space is a complex vector space, for if f and g are square integrable then

$$\int_X |f + \lambda g|^2 \, d\mu \leq 2 \int_X |f|^2 \, d\mu + 2|\lambda|^2 \int_X |g|^2 \, d\mu$$

by (13.3) applied to complex numbers.

Write $f \sim f'$ iff $f(x) = f'(x)$ almost everywhere on X; this is clearly an equivalence relation on X. We set $L^2(X)$ to be the factor space $\mathcal{L}^2(X)/\sim$. Its elements are equivalence classes $\tilde{f}$ of functions that differ at most on a set of measure zero. Define the inner product of two classes by

$$\langle \tilde{f} | \tilde{g} \rangle = \int_X \overline{f} g \, d\mu,$$

which is well-defined (see Example 5.6) and independent of the choice of representatives. For, if $f' \sim f$ and $g' \sim g$ then let A_f and A_g be the sets on which $f(x) \neq f'(x)$ and $g'(x) \neq g(x)$, respectively. These sets have measure zero, $\mu(A_f) = \mu(A_g) = 0$. The set on which $\overline{f'(x)} g'(x) \neq \overline{f(x)} g(x)$ is a subset of $A_f \cup A_g$ and therefore must also have measure zero, so that $\int_X \overline{f} g \, d\mu = \int_X \overline{f'} g' \, d\mu$.

The inner product axioms (IP1) and (IP2) are trivial, and (IP3) follows from

$$\|\tilde{f}\| = 0 \implies \int_X |f|^2 \, d\mu = 0 \implies f = 0 \text{ a.e.}$$

13.1 Definitions and examples

It is common to replace an equivalence class of functions $\tilde{f} \in \mathcal{L}^2(X)$ simply by a representative function f when there is no danger of confusion.

It turns out that the inner product space $\mathcal{L}^2(X)$ is in fact a Hilbert space. The following theorem is needed in order to show completeness.

Theorem 13.1 (Riesz–Fischer) *If $f_1, f_2, \ldots$ is a Cauchy sequence of functions in $\mathcal{L}^2(X)$, there exists a function $f \in \mathcal{L}^2(X)$ such that $\|f - f_n\| \to 0$ as $n \to \infty$.*

Proof: The Cauchy sequence condition $\|f_n - f_m\| \to 0$ implies that for any $\epsilon > 0$ there exists N such that

$$\int_X |f_n - f_m|^2 \, d\mu < \epsilon \qquad \text{for all } m, n > N.$$

We may, with some relabelling, pick a subsequence such that $f_0 = 0$ and

$$\|f_n - f_{n-1}\| < 2^{-n}.$$

Setting

$$h(x) = \sum_{n=1}^{\infty} |f_n(x) - f_{n-1}(x)|$$

we have from (Norm3),

$$\|h\| < \sum_{n=1}^{\infty} \|f_n - f_{n-1}\| < \sum_{n=1}^{\infty} 2^{-n} = 1.$$

The function $x \mapsto h^2(x)$ is thus a positive real integrable function on X, and the set of points where its defining sequence diverges, $E = \{x \mid h(x) = \infty\}$, is a set of measure zero, $\mu(E) = 0$. Let g_n be the sequence of functions

$$g_n(x) = \begin{cases} f_n - f_{n-1} & \text{if } x \notin E, \\ 0 & \text{if } x \in E. \end{cases}$$

Since $g_n = f_n - f_{n-1}$ a.e. these functions are measurable and $\|g_n\| = \|f_n - f_{n-1}\| < 2^{-n}$. The function

$$f(x) = \sum_{n=1}^{\infty} g_n(x)$$

is defined almost everywhere, since the series is absolutely convergent to $h(x)$ almost everywhere. Furthermore it belongs to $\mathcal{L}^2(X)$, for

$$|f(x)|^2 = \left|\sum g_n(x)\right|^2 \leq \left(\sum |g_n(x)|\right)^2 \leq (h(x))^2.$$

Since

$$f_n = \sum_{k=1}^{n} (f_k - f_{k-1}) = \sum_{k=1}^{n} g_k \quad \text{a.e.}$$

it follows that

$$\|f - f_n\| = \left\|f - \sum_{k=1}^{n} g_k\right\|$$

$$= \left\|\sum_{k=n+1}^{\infty} g_k\right\|$$

$$\leq \sum_{k=n+1}^{\infty} \|g_k\|$$

$$< \sum_{k=n+1}^{\infty} 2^{-k} = 2^{-n}.$$

Hence $\|f - f_n\| \to 0$ as $n \to \infty$ and the result is proved. ∎

Problems

Problem 13.1 Let E be a Banach space in which the norm satisfies the parallelogram law (13.2). Show that it is a Hilbert space with inner product given by

$$\langle x | y \rangle = \tfrac{1}{4}(\|x+y\|^2 - \|x-y\|^2 + i\|x-iy\|^2 - i\|x+iy\|^2).$$

Problem 13.2 On the vector space $\mathcal{F}^1[a, b]$ of complex continuous differentiable functions on the interval $[a, b]$, set

$$\langle f | g \rangle = \int_a^b \overline{f'(x)} g'(x)\, dx \quad \text{where} \quad f' = \frac{df}{dx}, \quad g' = \frac{dg}{dx}.$$

Show that this is not an inner product, but becomes one if restricted to the space of functions $f \in \mathcal{F}^1[a, b]$ having $f(c) = 0$ for some fixed $a \leq c \leq b$. Is it a Hilbert space?

Give a similar analysis for the case $a = -\infty$, $b = \infty$, and restricting functions to those of compact support.

Problem 13.3 In the space $L^2([0, 1])$ which of the following sequences of functions (i) is a Cauchy sequence, (ii) converges to 0, (iii) converges everywhere to 0, (iv) converges almost everywhere to 0, and (v) converges almost nowhere to 0?

(a) $f_n(x) = \sin^n(x)$, $n = 1, 2, \ldots$

(b) $f_n(x) = \begin{cases} 0 & \text{for } x < 1 - \frac{1}{n}, \\ nx + 1 - n & \text{for } 1 - \frac{1}{n} \leq x \leq 1. \end{cases}$

(c) $f_n(x) = \sin^n(nx)$.

(d) $f_n(x) = \chi_{U_n}(x)$, the characteristic function of the set

$$U_n = \left[\frac{k}{2^m}, \frac{k+1}{2^m}\right] \quad \text{where} \quad n = 2^m + k, \ m = 0, 1, \ldots \text{ and } k = 0, \ldots, 2^m - 1.$$

13.2 Expansion theorems

Subspaces

A **subspace** V of a Hilbert space $\mathcal{H}$ is a vector subspace that is **closed** with respect to the norm topology. For a vector subspace to be closed we require the limit of any sequence of vectors in V to belong to V,

$$u_1, u_2, \ldots \to u \text{ and all } u_n \in V \implies u \in V.$$

If V is any vector subspace of $\mathcal{H}$, its **closure** $\overline{V}$ is the smallest subspace containing V. It is the intersection of all subspaces containing V.

If K is any subset of $\mathcal{H}$ then, as in Chapter 3, the *vector subspace* generated by K is

$$L(K) = \left\{ \sum_{i=1}^{n} \alpha^i u_i \,|\, \alpha^i \in \mathbb{C}, u_i \in K \right\},$$

but the **subspace generated** by K will always refer to the *closed* subspace $\overline{L(K)}$ generated by K. A Hilbert space $\mathcal{H}$ is called **separable** if there is a countable set $K = \{u_1, u_2, \ldots\}$ such that $\mathcal{H}$ is generated by K,

$$\mathcal{H} = \overline{L(K)} = \overline{L(u_1, u_2, \ldots)}.$$

Orthonormal bases

If the Hilbert space $\mathcal{H}$ is separable and is generated by $\{u_1, u_2, \ldots\}$, we may use the **Schmidt orthonormalization** procedure (see Section 5.2) to produce an orthonormal set $\{e_1, e_2, \ldots, e_n\}$,

$$\langle e_i | e_j \rangle = \delta_{ij} = \begin{cases} 1 & \text{if } i = j, \\ 0 & \text{if } i \neq j. \end{cases}$$

The steps of the procedure are

$$f_1 = u_1 \qquad\qquad e_1 = f_1/\|f_1\|$$
$$f_2 = u_2 - \langle e_1 | u_2 \rangle e_1 \qquad\qquad e_2 = f_2/\|f_2\|$$
$$f_3 = u_3 - \langle e_1 | u_3 \rangle e_1 - \langle e_2 | u_3 \rangle e_2 \qquad e_3 = f_3/\|f_3\|, \text{ etc.}$$

from which it can be seen that each u_n is a linear combination of $\{e_1, e_2, \ldots, e_n\}$. Hence $\mathcal{H} = \overline{L(\{e_1, e_2, \ldots\})}$ and the set $\{e_n \,|\, n = 1, 2, \ldots\}$ is called a **complete orthonormal set** or **orthonormal basis** of $\mathcal{H}$.

Theorem 13.2 *If $\mathcal{H}$ is a separable Hilbert space and $\{e_1, e_2, \ldots\}$ is a complete orthonormal set, then any vector $u \in \mathcal{H}$ has a unique expansion*

$$u = \sum_{n=1}^{\infty} c_n e_n \quad \text{where} \quad c_n = \langle e_n | u \rangle. \tag{13.4}$$

Hilbert spaces

The meaning of the sum in this theorem is

$$\left\| u - \sum_{n=1}^{N} c_n e_n \right\| \to 0 \quad \text{as} \quad N \to \infty.$$

A critical part of the proof is **Bessel's inequality**:

$$\sum_{n=1}^{N} |\langle e_n | u \rangle|^2 \leq \|u\|^2. \tag{13.5}$$

Proof: For any $N > 1$

$$0 \leq \left\| u - \sum_{n=1}^{N} \langle e_n | u \rangle e_n \right\|^2$$

$$= \langle u - \sum_n \langle e_n | u \rangle e_n | u - \sum_m \langle e_m | u \rangle e_m \rangle$$

$$= \|u\|^2 - 2 \sum_{n=1}^{N} \overline{\langle e_n | u \rangle} \langle e_n | u \rangle + \sum_{n=1}^{N} \sum_{m=1}^{N} \overline{\langle e_n | u \rangle} \delta_{mn} \langle e_m | u \rangle$$

$$= \|u\|^2 - \sum_{n=1}^{N} |\langle e_n | u \rangle|^2,$$

which gives the desired inequality. ∎

Taking the limit $N \to \infty$ in Bessel's inequality (13.5) shows that the series

$$\sum_{n=1}^{\infty} |\langle e_n | u \rangle|^2$$

is bounded above and therefore convergent since it consists entirely of non-negative terms. To prove the expansion theorem 13.2, we first show two lemmas.

Lemma 13.3 *If $v_n \to v$ in a Hilbert space $\mathcal{H}$, then for all vectors $u \in \mathcal{H}$*

$$\langle u | v_n \rangle \to \langle u | v \rangle.$$

Proof: By the Cauchy–Schwarz inequality (5.13)

$$|\langle u | v_n \rangle - \langle u | v \rangle| = |\langle u | v_n - v \rangle|$$
$$\leq \|u\| \, \|v_n - v\|$$
$$\to 0$$
∎

Lemma 13.4 *If $\{e_1, e_2, \ldots\}$ is a complete orthonormal set and $\langle v | e_n \rangle = 0$ for $n = 1, 2, \ldots$ then $v = 0$.*

Proof: Since $\{e_n\}$ is a complete o.n. set, every vector $v \in \mathcal{H}$ is the limit of a sequence of vectors spanned by the vectors $\{e_1, e_2, \ldots\}$,

$$v = \lim_{n \to \infty} v_n \quad \text{where} \quad v_n = \sum_{i=1}^{N} v_{ni} e_i.$$

13.2 Expansion theorems

Setting $u = v$ in Lemma 13.3, we have

$$\|v\|^2 = \langle v | v \rangle = \lim_{n \to \infty} \langle v | v_n \rangle = 0.$$

Hence $v = 0$ by the condition (Norm1). ∎

We now return to the proof of the expansion theorem.

Proof of Theorem 13.2: Set

$$u_N = \sum_{n=1}^{N} \langle e_n | u \rangle e_n.$$

This is a Cauchy sequence,

$$\|u_N - u_M\|^2 = \sum_{n=M}^{N} |\langle e_n | u \rangle|^2 \to 0 \quad \text{as} \quad M, N \to \infty$$

since the series $\sum_n |\langle e_n | u \rangle|^2$ is absolutely convergent by Bessel's inequality (13.5). By completeness of the Hilbert space $\mathcal{H}$, $u_N \to u'$ for some vector $u' \in \mathcal{H}$. But

$$\langle e_k | u - u' \rangle = \lim_{N \to \infty} \langle e_k | u - u_N \rangle = \langle e_k | u \rangle - e_k u = 0$$

since $\langle e_k | u_N \rangle = \langle e_k | u \rangle$ for all $N \geq k$. Hence, by Lemma 13.4,

$$u = u' = \lim_{N \to \infty} u_N,$$

and Theorem 13.2 is proved. ∎

Exercise: Show that every separable Hilbert space is either a finite dimensional inner product space, or is isomorphic with ℓ^2.

Example 13.5 For any real numbers $a < b$ the Hilbert space $L^2([a, b])$ is separable. The following is an outline proof; details may be found in [1]. By Theorem 11.2 any positive measurable function $f \geq 0$ on $[a, b]$ may be approximated by an increasing sequence of positive simple functions $0 < s_n(x) \to f(x)$. If $f \in \mathcal{L}^2([a, b])$ then by the dominated convergence, Theorem 11.11, $\|f - s_n\| \to 0$. By a straightforward, but slightly technical, argument these simple functions may be approximated with continuous functions, and prove that for any $\epsilon > 0$ there exists a positive continuous function $h(x)$ such that $\|f - h\| < \epsilon$. Using a famous theorem of Weierstrass that any continuous function on a closed interval can be arbitrarily closely approximated by polynomials, it is possible to find a complex-valued polynomial $p(x)$ such that $\|f - p\| < \epsilon$. Since all polynomials are of the form $p(x) = c_0 + c_1 x + c_2 x^2 + \cdots + c_n x^n$ where $c \in \mathbb{C}$, the functions $1, x, x^2, \ldots$ form a countable sequence of functions on $[a, b]$ that generate $L^2([a, b])$. This proves separability of $L^2([a, b])$.

Separability of $L^2(\mathbb{R})$ is proved by showing the restricted polynomial functions $f_{n,N} = x^n \chi_{[-N, N]}$ are a countable set that generates $L^2(\mathbb{R})$.

Hilbert spaces

Example 13.6 On $L^2([-\pi, \pi])$ the functions

$$\phi_n(x) = \frac{e^{inx}}{\sqrt{2\pi}}$$

form an orthonormal basis,

$$\langle \phi_m | \phi_n \rangle = \frac{1}{2\pi} \int_{-\pi}^{\pi} e^{i(n-m)x} \, dx = \delta_{mn}$$

as is easily calculated for the two separate cases $n \neq m$ and $n = m$. These generate the Fourier series of an arbitrary square integrable function f on $[-\pi, \pi]$

$$f = \sum_{n=-\infty}^{\infty} c_n \phi_n \quad \text{a.e.}$$

where c_n are the Fourier coefficients

$$c_n = \langle \phi_n | f \rangle = \frac{1}{\sqrt{2\pi}} \int_{-\pi}^{\pi} e^{-inx} f(x) \, dx.$$

Example 13.7 The *hermite polynomials* $H_n(x)$ ($n = 0, 1, 2, \ldots$) are defined by

$$H_n(x) = (-1)^n e^{x^2} \frac{d^n e^{-x^2}}{dx^n}.$$

The first few are

$$H_0(x) = 1, \quad H_1(x) = 2x, \quad H_2(x)4x^2 - 2, \quad H_3(x) = 8x^3 - 12x, \ldots$$

The nth polynomial is clearly of degree n with leading term $(-2x)^n$. The functions $\psi_n(x) = e^{-(1/2)x^2} H_n(x)$ form an orthogonal system in $L^2(\mathbb{R})$:

$$\langle \psi_m | \psi_n \rangle = (-1)^{n+m} \int_{-\infty}^{\infty} e^{x^2} \frac{d^m e^{-x^2}}{dx^m} \frac{d^n e^{-x^2}}{dx^n} \, dx$$

$$= (-1)^{n+m} \left(\left[e^{x^2} \frac{d^m e^{-x^2}}{dx^m} \frac{d^{n-1} e^{-x^2}}{dx^{n-1}} \right]_{-\infty}^{\infty} \right.$$

$$\left. - \int_{-\infty}^{\infty} \frac{d}{dx}\left(e^{x^2} \frac{d^m e^{-x^2}}{dx^m} \right) \frac{d^{n-1} e^{-x^2}}{dx^{n-1}} \, dx \right)$$

on integration by parts. The first expression on the right-hand side of this equation vanishes since it involves terms of order $e^{-x^2} x^k$ that approach 0 as $x \to \pm\infty$. We may repeat the integration by parts on the remaining integral, until we arrive at

$$\langle \psi_m | \psi_n \rangle = (-1)^m \int_{-\infty}^{\infty} e^{-x^2} \frac{d^n}{dx^n} \left(e^{x^2} \frac{d^m e^{-x^2}}{dx^m} \right) dx,$$

which vanishes if $n > m$ since the expression in the brackets is a polynomial of degree m. A similar argument for $n < m$ yields

$$\langle \psi_m | \psi_n \rangle = 0 \quad \text{for } n \neq m.$$

13.2 Expansion theorems

For $n = m$ we have, from the leading term in the hermite polynomials,

$$\|\psi_n\|^2 = \langle \psi_n | \psi_n \rangle = (-1)^n \int_{-\infty}^{\infty} e^{-x^2} \frac{d^n}{dx^n} \left(e^{x^2} \frac{d^n e^{-x^2}}{dx^n} \right) dx$$

$$= (-1)^n \int_{-\infty}^{\infty} e^{-x^2} \frac{d^n}{dx^n} \left((-2x)^n \right) dx$$

$$= 2^n n! \int_{-\infty}^{\infty} e^{-x^2} dx$$

$$= 2^n n! \sqrt{\pi}.$$

Thus the functions

$$\phi_n(x) = \frac{e^{-(1/2)x^2}}{\sqrt{2^n n! \sqrt{\pi}}} H_n(x) \qquad (13.6)$$

form an orthonormal set. From Weierstrass's theorem they form a complete o.n. basis for $L^2(\mathbb{R})$.

The following generalization of Lemma 13.3 is sometimes useful.

Lemma 13.5 *If* $u_n \to u$ *and* $v_n \to v$ *then* $\langle u_n | v_n \rangle \to \langle u | v \rangle$.

Proof: Using the Cauchy–Schwarz inequality (5.13)

$$\left| \langle u_n | v_n \rangle - \langle u | v \rangle \right| = \left| \langle u_n | v_n \rangle - \langle u_n | v \rangle + \langle u_n | v \rangle - \langle u | v \rangle \right|$$
$$\leq \left| \langle u_n | v_n \rangle - \langle u_n | v \rangle \right| + \left| \langle u_n | v \rangle - \langle u | v \rangle \right|$$
$$\leq \|u_n\| \|v_n - v\| + \|u_n - u\| \|v\|$$
$$\to \|u\|.0 + 0.\|v\| \to 0.$$

■

Exercise: If $u_n \to u$ show that $\|u_n\| \to \|u\|$, used in the last step of the above proof.

The following identity has widespread application in quantum mechanics.

Theorem 13.6 (Parseval's identity)

$$\langle u | v \rangle = \sum_{i=1}^{\infty} \langle u | e_i \rangle \langle e_i | v \rangle. \qquad (13.7)$$

Proof: Set

$$u_n = \sum_{i=1}^{n} \langle e_i | u \rangle e_i \quad \text{and} \quad v_n = \sum_{i=1}^{n} \langle e_i | v \rangle e_i.$$

339

Hilbert spaces

By Theorem 13.2, $u_n \to u$ and $v_n \to v$ as $n \to \infty$. Now using Lemma 13.5,

$$\langle u \mid v \rangle = \lim_{n \to \infty} \langle u_n \mid v_n \rangle$$

$$= \lim_{n \to \infty} \sum_{i=1}^{n} \sum_{j=1}^{n} \overline{\langle e_i \mid u \rangle} \langle e_j \mid v \rangle \langle e_i \mid e_j \rangle$$

$$= \lim_{n \to \infty} \sum_{i=1}^{n} \sum_{j=1}^{n} \langle u \mid e_i \rangle \langle e_j \mid v \rangle \delta_{ij}$$

$$= \lim_{n \to \infty} \sum_{i=1}^{n} \langle u \mid e_i \rangle \langle e_i \mid v \rangle$$

$$= \sum_{i=1}^{\infty} \langle u \mid e_i \rangle \langle e_i \mid v \rangle. \quad \blacksquare$$

For a function $f(x) = \sum_{n=-\infty}^{\infty} c_n \phi_n$ on $[-\pi, \pi]$, where $\phi_n(x)$ are the standard Fourier functions given in Example 13.6, Parseval's identity becomes the well-known formula

$$\|f\|^2 = \int_{-\pi}^{\pi} |f(x)|^2 \, dx = \sum_{n=-\infty}^{\infty} |c_n|^2.$$

Problems

Problem 13.4 Show that a vector subspace is a closed subset of $\mathcal{H}$ with respect to the norm topology iff the limit of every sequence of vectors in V belongs to V.

Problem 13.5 Let ℓ_0 be the subset of ℓ^2 consisting of sequences with only finitely many terms different from zero. Show that ℓ_0 is a vector subspace of ℓ^2, but that it is not closed. What is its closure $\overline{\ell_0}$?

Problem 13.6 We say a sequence $\{x_n\}$ *converges weakly* to a point x in a Hilbert space $\mathcal{H}$, written $x_n \rightharpoonup x$ if $\langle x_n \mid y \rangle \to \langle x \mid y \rangle$ for all $y \in \mathcal{H}$. Show that every strongly convergent sequence, $\|x_n - x\| \to 0$ is weakly convergent to x. In finite dimensional Hilbert spaces show that every weakly convergent sequence is strongly convergent.

Give an example where $x_n \rightharpoonup x$ but $\|x_n\| \not\to \|x\|$. Is it true in general that the weak limit of a sequence is unique?

Show that if $x_n \rightharpoonup x$ and $\|x_n\| \not\to \|x\|$ then $x_n \not\to x$.

Problem 13.7 In the Hilbert space $L^2([-1, 1])$ let $\{f_n(x)\}$ be the sequence of functions $1, x, x^2, \ldots, f_n(x) = x^n, \ldots$

(a) Apply Schmidt orthonormalization to this sequence, writing down the first three polynomials so obtained.

(b) The nth Legendre polynomial $P_n(x)$ is defined as

$$P_n(x) = \frac{1}{2^n n!} \frac{d^n}{dx^n} (x^2 - 1)^n.$$

Prove that

$$\int_{-1}^{1} P_m(x) P_n(x) \, dx = \frac{2}{2n+1} \delta_{mn}.$$

(c) Show that the nth member of the o.n. sequence obtained in (a) is $\sqrt{n + \tfrac{1}{2}} P_n(x)$.

13.3 Linear functionals

Problem 13.8 Show that Schmidt orthonormalization in $L^2(\mathbb{R})$, applied to the sequence of functions

$$f_n(x) = x^n e^{-x^2/2},$$

leads to the normalized hermite functions (13.6) of Example 13.7.

Problem 13.9 Show that applying Schmidt orthonormalization in $L^2([0, \infty])$ to the sequence of functions

$$f_n(x) = x^n e^{-x/2}$$

leads to a normalized sequence of functions involving the *Laguerre polynomials*

$$L_n(x) = e^x \frac{d^n}{dx^n}\left(x^n e^{-x}\right).$$

13.3 Linear functionals

Orthogonal subspaces

Two vectors $u, v \in \mathcal{H}$ are said to be **orthogonal** if $\langle u|v \rangle = 0$, written $u \perp v$. If V is a subspace of $\mathcal{H}$ we denote its **orthogonal complement** by

$$V^\perp = \{u \,|\, u \perp v \text{ for all } v \in V\}.$$

Theorem 13.7 *If V is a subspace of $\mathcal{H}$ then $V^\perp$ is also a subspace.*

Proof: $V^\perp$ is clearly a vector subspace, for $v, v' \in V^\perp$ since

$$\langle \alpha v + \beta v' | u \rangle = \overline{\alpha} \langle v|u \rangle + \overline{\beta} \langle v'|u \rangle = 0$$

for all $u \in V$. The space $V^\perp$ is closed, for if $v_n \to v$ where $v_n \in V^\perp$, then

$$\langle v|u \rangle = \lim_{n \to \infty} \langle v_n | u \rangle = \lim_{n \to \infty} 0 = 0$$

for all $u \in V$. Hence $v \in V$. ∎

Theorem 13.8 *If V is a subspace of a Hilbert space $\mathcal{H}$ then every $u \in \mathcal{H}$ has a unique decomposition*

$$u = u' + u'' \quad \text{where} \quad u' \in V, \ u'' \in V^\perp.$$

Proof: The idea behind the proof of this theorem is to find the element of V that is 'nearest' to u. Just as in Euclidean space, this is the orthogonal projection of the vector u onto the subspace V. Let

$$d = \inf\{\|u - v\| \,|\, v \in V\}$$

and $v_n \in V$ a sequence of vectors such that $\|u - v_n\| \to d$. The sequence $\{v_n\}$ is Cauchy, for if we set $x = u - \frac{1}{2}(v_n + v_m)$ and $y = \frac{1}{2}(v_n - v_m)$ in the parallelogram law (13.2), then

$$\|v_n - v_m\|^2 = 2\|u - v_n\|^2 + 2\|u - v_m\|^2 - 4\|u - \tfrac{1}{2}(v_n + v_m)\|^2. \tag{13.8}$$

Hilbert spaces

For any $\epsilon > 0$ let $N > 0$ be such that for all $k > N$, $\|u - v_k\|^2 \leq d^2 + \frac{1}{4}\epsilon$. Setting n, m both $> N$ in Eq. (13.8) we find $\|v_n - v_m\|^2 \leq \epsilon$. Hence v_n is a Cauchy sequence.

Since $\mathcal{H}$ is complete and V is a closed subspace, there exists a vector $u' \in V$ such that $v_n \to u'$. Setting $u'' = u - u'$, it follows from the exercise after Lemma 13.5 that

$$\|u''\| = \lim_{n \to \infty} \|u - v_n\| = d.$$

For any $v \in V$ set $v_0 = v/\|v\|$, so that $\|v_0\| = 1$. Then

$$d^2 \leq \|u - (u' + \langle v_0 | u'' \rangle v_0)\|^2$$
$$= \|u'' - \langle v_0 | u'' \rangle v_0\|^2$$
$$= \langle u'' - \langle v_0 | u'' \rangle v_0 | u'' - \langle v_0 | u'' \rangle v_0 \rangle$$
$$= d^2 - |\langle v_0 | u'' \rangle|^2.$$

Hence $\langle v_0 | u'' \rangle = 0$, so that $\langle v | u'' \rangle = 0$. Since v is an arbitrary vector in V, we have $u'' \in V^\perp$.

A subspace and its orthogonal complement can only have the zero vector in common, $V \cap V^\perp = \{0\}$, for if $w \in V \cap V^\perp$ then $\langle w | w \rangle = 0$, which implies that $w = 0$. If $u = u' + u'' = v' + v''$, with $u', v' \in V$ and $u'', v'' \in V^\perp$, then the vector $u' - v' \in V$ is equal to $v'' - u'' \in V^\perp$. Hence $u' = v'$ and $u'' = v''$, the decomposition is unique. ∎

Corollary 13.9 *For any subspace V, $V^{\perp\perp} = V$.*

Proof: $V \subseteq V^{\perp\perp}$ for if $v \in V$ then $\langle v | u \rangle = 0$ for all $u \in V^\perp$. Conversely, let $v \in V^{\perp\perp}$. By Theorem 13.8 v has a unique decomposition $v = v' + v''$ where $v' \in V \subseteq V^{\perp\perp}$ and $v'' \in V^\perp$. Using Theorem 13.8 again but with V replaced by $V^\perp$, it follows that $v'' = 0$. Hence $v = v' \in V$. ∎

Riesz representation theorem

For every $v \in \mathcal{H}$ the map $\varphi_v : u \mapsto \langle v | u \rangle$ is a linear functional on $\mathcal{H}$. Linearity is obvious and continuity follows from Lemma 13.3. The following theorem shows that all (continuous) linear functionals on a Hilbert space are of this form, a result of considerable significance in quantum mechanics, as it motivates Dirac's *bra-ket notation*.

Theorem 13.10 (Riesz representation theorem) *If φ is a linear functional on a Hilbert space $\mathcal{H}$, then there is a unique vector $v \in \mathcal{H}$ such that*

$$\varphi(u) = \varphi_v(u) = \langle v | u \rangle \quad \text{for all } u \in \mathcal{H}.$$

Proof: Since a linear functional $\varphi : \mathcal{H} \to \mathbb{C}$ is required to be continuous, we always have

$$|\varphi(x_n) - \varphi(x)| \to 0 \quad \text{whenever} \quad \|x - x_n\| \to 0.$$

Let V be the null space of φ,

$$V = \{x \mid \varphi(x) = 0\}.$$

This is a closed subspace, for if $x_n \to x$ and $\varphi(x_n) = 0$ for all n, then $\varphi(x) = 0$ by continuity. If $V = \mathcal{H}$ then φ vanishes on $\mathcal{H}$ and one can set $v = 0$. Assume therefore that $V \neq \mathcal{H}$, and let

13.3 Linear functionals

w be a non-zero vector such that $w \notin V$. By Theorem 13.8, there is a unique decomposition

$$w = w' + w'' \quad \text{where} \quad w' \in V, \quad w'' \in V^\perp.$$

Then $\varphi(w'') = \varphi(w) - \varphi(w') = \varphi(w) \neq 0$ since $w \notin V$. For any $u \in \mathcal{H}$ we may write

$$u = \left(u - \frac{\varphi(u)}{\varphi(w'')} w''\right) + \frac{\varphi(u)}{\varphi(w'')} w'',$$

where the first term on the right-hand side belongs to V since the linear functional φ gives the value 0 when applied to it, while the second term belongs to $V^\perp$ as it is proportional to w''. For any $v \in V^\perp$ we have then

$$\langle v | u \rangle = \frac{\overline{\varphi(u)}}{\overline{\varphi(w'')}} \langle v | w'' \rangle.$$

In particular, setting

$$v = \frac{\overline{\varphi(w'')}}{\|w''\|^2} w'' \in V^\perp$$

gives

$$\langle v | u \rangle = \frac{\overline{\varphi(u)}}{\overline{\varphi(w'')}} \frac{\varphi(w'')}{\|w''\|^2} \langle w'' | w'' \rangle = \overline{\varphi(u)}.$$

Hence this v is the vector required for the theorem. It is the unique vector with this property, for if $\langle v - v' | u \rangle = 0$ for all $u \in \mathcal{H}$ then $v = v'$, on setting $u = v - v'$. ∎

Problems

Problem 13.10 If S is any subset of $\mathcal{H}$, and V the closed subspace generated by S, $V = \overline{L(S)}$, show that $S^\perp = \{u \in \mathcal{H} \mid \langle u | x \rangle = 0 \text{ for all } x \in S\} = V^\perp$.

Problem 13.11 Which of the following is a vector subspace of ℓ^2, and which are closed? In each case find the space of vectors orthogonal to the set.

(a) $V_N = \{(x_1, x_2, \ldots) \in \ell^2 \mid x_i = 0 \text{ for } i > N\}$.
(b) $V = \bigcup_{N=1}^{\infty} V_N = \{(x_1, x_2, \ldots) \in \ell^2 \mid x_i = 0 \text{ for } i > \text{some } N\}$.
(c) $U = \{(x_1, x_2, \ldots) \in \ell^2 \mid x_i = 0 \text{ for } i = 2n\}$.
(d) $W = \{(x_1, x_2, \ldots) \in \ell^2 \mid x_i = 0 \text{ for some } i\}$.

Problem 13.12 Show that the real Banach space $\mathbb{R}^2$ with the norm $\|(x, y)\| = \max\{|x|, |y|\}$ does not have the closest point property of Theorem 13.8. Namely for a given point $\mathbf{x}$ and one-dimensional subspace L, there does not in general exist a unique point in L that is closest to $\mathbf{x}$.

Problem 13.13 If $A : \mathcal{H} \to \mathcal{H}$ is an operator such that $Au \perp u$ for all $u \in \mathcal{H}$, show that $A = 0$.

343

13.4 Bounded linear operators

Let V be any normed vector space. A linear operator $A : V \to V$ is said to be **bounded** if

$$\|Au\| \le K\|u\|$$

for some constant $K \ge 0$ and all $u \in V$.

Theorem 13.11 *A linear operator on a normed vector space is bounded if and only if it is continuous with respect to the norm topology.*

Proof: If A is bounded then it is continuous, for if $\epsilon > 0$ then for any pair of vectors u, v such that $\|u - v\| < \epsilon/K$

$$\|Au - Av\| = \|A(u - v)\| \le K\|u - v\| < \epsilon.$$

Conversely, let A be a continuous operator on V. If A is *not* bounded, then for each $N > 0$ there exists u_N such that $\|Au_N\| \ge N\|u_N\|$. Set

$$w_N = \frac{u_N}{N\|u_N\|},$$

so that

$$\|w_N\| = \frac{1}{N} \to 0.$$

Hence $w_N \to 0$, but $\|Aw_N\| \ge 1$, so that Aw_n definitely does *not* $\to 0$, contradicting the assumption that A is continuous. ∎

The **norm** of a bounded operator A is defined as

$$\|A\| = \sup\{\|Au\| \mid \|u\| \le 1\}.$$

By Theorem 13.11, A is continuous at $x = 0$. Hence there exists $\epsilon > 0$ such that $\|Ax\| \le 1$ for all $\|x\| \le \epsilon$. For any u with $\|u\| \le 1$ let $v = \epsilon u$ so that $\|v\| \le \epsilon$ and

$$\|Au\| = \frac{1}{\epsilon}\|Av\| \le \frac{1}{\epsilon}.$$

This shows $\|A\|$ always exists for a bounded operator.

Example 13.8 On ℓ^2 define the two *shift operators* S and S' by

$$S\big((x_1, x_2, x_3, \dots)\big) = (0, x_1, x_2, \dots)$$

and

$$S'\big((x_1, x_2, x_3, \dots)\big) = (x_2, x_3, \dots).$$

These operators are clearly linear, and satisfy

$$\|Sx\| = \|x\| \quad \text{and} \quad \|S'x\| \le \|x\|.$$

Hence the norm of the operator S is 1, while $\|S'\|$ is also 1 since equality holds for $x_1 = 0$.

13.4 Bounded linear operators

Example 13.9 Let α be any bounded measurable function on the Hilbert space $L^2(X)$ of square integrable functions on a measure space X. The *multiplication operator* $A_\alpha : L^2(X) \to L^2(X)$ defined by $A_\alpha(f) = \alpha f$ is a bounded linear operator, for αf is measurable for every $f \in L^2(X)$, and it is square integrable since

$$|\alpha f|^2 \le M^2 |f|^2 \quad \text{where} \quad M = \sup_{x \in X} |\alpha(x)|.$$

The multiplication operator is well-defined on $L^2(X)$, for if f and f' are equal almost everywhere, $f \sim f'$, then $\alpha f \sim \alpha f'$; thus there is no ambiguity in writing $A_\alpha f$ for $A_\alpha[f]$. Linearity is trivial, while boundedness follows from

$$\|A_\alpha f\|^2 = \int_X |\alpha f|^2 \, d\mu \le M^2 \int_X |f|^2 \, d\mu = M^2 \|f\|^2.$$

Exercise: If A and B are bounded linear operators on a normed vector space, show that $A + \lambda B$ and AB are also bounded.

A bounded operator $A : V \to V$ is said to be **invertible** if there exists a bounded operator $A^{-1} : V \to V$ such that $AA^{-1} = A^{-1}A = I \equiv \mathrm{id}_V$. A^{-1} is called the **inverse** of A. It is clearly unique, for if $BA = CA$ then $B = BI = BAA^{-1} = CAA^{-1} = C$. It is important that we specify A^{-1} to be both a right and left inverse. For example, in ℓ^2, the shift operator S defined in Example 13.8 has left inverse S', since $S'S = I$, but it is not a right inverse for $SS'(x_1, x_2, \dots) = (0, x_2, x_3, \dots)$. Thus S is not an invertible operator, despite the fact that it is injective and an isometry, $\|Sx\| = \|x\|$. For a finite dimensional space these conditions would be enough to guarantee invertibility.

Theorem 13.12 *If A is a bounded operator on a Banach space V, with $\|A\| < 1$, then the operator $I - A$ is invertible and*

$$(I - A)^{-1} = \sum_{n=0}^{\infty} A^n.$$

Proof: Let x be any vector in V. Since $\|A^k x\| \le \|A\|(A^{k-1}x)$ it follows by simple induction that A^k is bounded and has norm $\|A^k\| \le (\|A\|)^k$. The vectors $u_n = (I + A + A^2 + \cdots + A^n)x$ form a Cauchy sequence, since

$$\begin{aligned}
\|u_n - u_m\| &= \|(A^{m+1} + \cdots + A^n)x\| \\
&\le \left(\|A\|^{m+1} + \cdots + \|A\|^n\right)\|x\| \\
&\le \frac{\|A\|^{m+1}}{1 - \|A\|} \|x\| \\
&\to 0 \quad \text{as } m \to \infty.
\end{aligned}$$

Since V is a Banach space, $u_m \to u$ for some $u \in V$, so there is a linear operator $T : V \to V$ such that $u = Tx$. Furthermore, since $T - (I + A + \cdots + A^n)$ is a bounded linear operator, it follows that T is bounded. Writing $T = \sum_{k=1}^{\infty} A^k$, in the sense that

$$\lim_{m \to \infty} \left(T - \sum_{k=1}^{m} A^k\right)x = 0,$$

Hilbert spaces

it is straightforward to verify that $(I - A)Tx = T(I - A)x = x$, which shows that $T = (I - A)^{-1}$. ∎

Adjoint operators

Let $A : \mathcal{H} \to \mathcal{H}$ be a bounded linear operator on a Hilbert space $\mathcal{H}$. We define its **adjoint** to be the operator $A^* : \mathcal{H} \to \mathcal{H}$ that has the property

$$\langle u \mid Av \rangle = \langle A^*u \mid v \rangle \quad \text{for all } u, v \in \mathcal{H}. \tag{13.9}$$

This operator is well-defined, linear and bounded.

Proof: For fixed u, the map $\varphi_u : v \mapsto \langle u \mid Av \rangle$ is clearly linear and continuous, on using Lemma 13.3. Hence φ_u is a linear functional, and by the Riesz representation theorem there exists a unique element $A * u \in \mathcal{H}$ such that

$$\langle A^*u \mid v \rangle = \varphi_u(v) = \langle u \mid Av \rangle.$$

The map $u \mapsto A * u$ is linear, since for an arbitrary vector v

$$\langle A^*(u + \lambda w) \mid v \rangle = \langle u + \lambda w \mid Av \rangle$$
$$= \langle u \mid Av \rangle + \overline{\lambda} \langle w \mid Av \rangle$$
$$= \langle A^*u + \lambda A^*w \mid v \rangle.$$

To show that the linear operator A^* is bounded, let u be any vector,

$$\|A^*u\|^2 = |\langle A^*u \mid A^*u \rangle|$$
$$= |\langle u \mid AA^*u \rangle|$$
$$\le \|u\| \, \|AA^*u\|$$
$$\le \|A\| \, \|u\| \, \|A^*u\|.$$

Hence, either $A^*u = 0$ or $\|A^*u\| \le \|A\| \, \|u\|$. In either case $\|A^*u\| \le \|A\| \, \|u\|$. ∎

Theorem 13.13 *The adjoint satisfies the following properties:*

(i) $(A + B)^* = A^* + B^*$,
(ii) $(\lambda A)^* = \overline{\lambda} A^*$,
(iii) $(AB)^* = B^* A^*$,
(iv) $A^{**} = A$,
(v) *if A is invertible then* $(A^{-1})^* = (A^*)^{-1}$.

Proof: We provide proofs of (i) and (ii), leaving the others as exercises.

(i) For arbitrary $u, v \in \mathcal{H}$

$$\langle (A + B)^*u \mid v \rangle = \langle u \mid (A + B)v \rangle = \langle u \mid Av + Bv \rangle$$
$$= \langle u \mid Av \rangle + \langle u \mid Bv \rangle = \langle A^*u \mid v \rangle + \langle B^*u \mid v \rangle = \langle A^*u + B^*u \mid v \rangle. \tag{13.10}$$

As $\langle w \mid v \rangle = \langle w' \mid v \rangle$ for all $v \in \mathcal{H}$ implies $w = w'$, we have

$$(A + B)^*u = A^*u + B^*u.$$

13.4 Bounded linear operators

(ii) For any pair of vectors $u, v \in \mathcal{H}$

$$\langle (\lambda A)^* u \,|\, v \rangle = \langle u \,|\, \lambda A v \rangle = \lambda \langle u \,|\, A v \rangle$$
$$= \lambda \langle A^* u \,|\, v \rangle = \langle \overline{\lambda} A^* u \,|\, v \rangle. \qquad (13.11)$$

The proofs of (iii)–(v) are on similar lines.

Example 13.10 The right shift operator S on ℓ^2 (see Example 13.8) induces the inner product

$$\langle x \,|\, Sy \rangle = \overline{x_1}.0 + \overline{x_2} y_1 + \overline{x_3} y_2 + \cdots = \langle S'x \,|\, y \rangle,$$

where S' is the left shift. Hence $S^* = S'$. Similarly $S'^* = S$, since

$$\langle x \,|\, S'y \rangle = \overline{x_1} y_2 + \overline{x_2} y_3 + \cdots = \langle S'x \,|\, y \rangle.$$

Example 13.11 Let α be a bounded measurable function on the Hilbert space $L^2(X)$ of square integrable functions on a measure space X, and A_α the multiplication operator defined in Example 13.9. For any pair of functions f, g square integrable on X, the equation $\langle A_\alpha^* f \,|\, g \rangle = \langle f \,|\, A_\alpha g \rangle$ reads

$$\int_X \overline{A_\alpha^* f} g \, d\mu = \int_X \overline{f} A_\alpha g \, d\mu = \int_X \overline{f} \alpha g \, d\mu.$$

Since g is an arbitrary function from $L^2(X)$, we have $\overline{A_\alpha^* f} = \alpha \overline{f}$ a.e., and in terms of the equivalence classes of functions in $L^2(X)$ the adjoint operator reads

$$A_\alpha^*[f] = [\overline{\alpha} f].$$

The adjoint operator of a multiplication operator is the multiplication operator by the complex conjugate function.

We define the **matrix element of the operator** A **between the vectors** u **and** v **in** $\mathcal{H}$ to be $\langle u \,|\, Av \rangle$. If the Hilbert space is separable and e_i is an o.n. basis then, by Theorem 13.2, we may write

$$A e_j = \sum_i a_{ij} e_i \quad \text{where} \quad a_{ij} = \langle e_i \,|\, A e_j \rangle.$$

Thus the matrix elements of the operator between the basis vectors are identical with the components of the matrix of the operator with respect to this basis, $\mathsf{A} = [a_{ij}]$. The adjoint operator has decomposition

$$A^* e_j = \sum_i a_{ij}^* e_i \quad \text{where} \quad a_{ij}^* = \langle e_i \,|\, A^* e_j \rangle.$$

The relation between the matrix elements $[a_{ij}^*]$ and $[a_{ij}]$ is determined by

$$a_{ij}^* = \langle e_i \,|\, A^* e_j \rangle = \langle A e_i \,|\, e_j \rangle = \overline{\langle e_j \,|\, A e_i \rangle} = \overline{a_{ji}},$$

or, in matrix notation,

$$\mathsf{A}^* \equiv [a_{ij}^*] = [\overline{a_{ji}}] = \overline{\mathsf{A}^T} = \mathsf{A}^\dagger.$$

Hilbert spaces

In quantum mechanics it is common to use the conjugate transpose notation $A^\dagger$ for the adjoint operator, but the equivalence with the complex adjoint matrix only holds for orthonormal bases.

Exercise: Show that in an o.n. basis $Au = \sum_i u'_i e_i$ where $u = \sum_i u_i e_i$ and $u'_i = \sum_j a_{ij} u_j$.

Hermitian operators

An operator A is called **hermitian** if $A = A^*$, so that

$$\langle u | Av \rangle = \langle A^* u | v \rangle = \overline{\langle v | A^* u \rangle} = \langle Au | v \rangle.$$

If $\mathcal{H}$ is separable and $e_1, e_2, \ldots$ a complete orthonormal set, then the matrix elements in this basis, $a_{ij} = \langle e_i | A e_j \rangle$, have the hermitian property

$$a_{ij} = \overline{a_{ji}}.$$

In other words, a bounded operator A is hermitian if and only if its matrix with respect to any o.n. basis is hermitian,

$$\mathsf{A} = [a_{ij}] = \overline{\mathsf{A}^T} = \mathsf{A}^\dagger.$$

These operators are sometimes referred to as *self-adjoint*, but in line with modern usage we will use this term for a more general concept defined in Section 13.6.

Let M be a closed subspace of $\mathcal{H}$ then, by Theorem 13.8, any $u \in \mathcal{H}$ has a unique decomposition

$$u = u' + u'' \quad \text{where} \quad u' \in M, \ u'' \in M^\perp.$$

We define the **projection operator** $P_M : \mathcal{H} \to \mathcal{H}$ by $P_M(u) = u'$, which maps every vector of $\mathcal{H}$ onto its orthogonal projection in the subspace M.

Theorem 13.14 *For every subspace M, the projection operator P_M is a bounded hermitian operator and satisfies $P_M^2 = P_M$ (called an **idempotent operator**). Conversely any idempotent hermitian operator P is a projection operator into some subspace.*

Proof: 1. P_M is hermitian. For any two vectors from $u, v \in \mathcal{H}$

$$\langle u | P_M v \rangle = \langle u | v' \rangle = \langle u' + u'' | v' \rangle = \langle u' | v' \rangle$$

since $\langle u'' | v' \rangle = 0$. Similarly,

$$\langle P_M u | v \rangle = \langle u' | v \rangle = \langle u' | v' + v'' \rangle = \langle u' | v' \rangle.$$

Thus $P_M = P_M^*$.

2. P_M is bounded, for $\| P_M u \|^2 \le \| u \|^2$ since

$$\| u \|^2 = \langle u | u \rangle = \langle u' + u'' | u' + u'' \rangle = \langle u' | u' \rangle + \langle u'' | u'' \rangle \ge \| u' \|^2.$$

3. P_M is idempotent, for $P_M^2 u = P_M u' = u'$ since $u' \in M$. Hence $P_M^2 = P_M$.

13.4 Bounded linear operators

4. Suppose P is hermitian and idempotent, $P^2 = P$. The operator P is bounded and therefore continuous, for by the Cauchy–Schwarz inequality (5.13),

$$\|Pu\|^2 = |\langle Pu | Pu \rangle| = |\langle u | P^2 u \rangle| = |\langle u | Pu \rangle| \le \|u\| \|Pu\|.$$

Hence either $\|Pu\| = 0$ or $\|Pu\| \le \|u\|$.

Let $M = \{u \mid u = Pu\}$. This is obviously a vector subspace of $\mathcal{H}$. It is closed by continuity of P, for if $u_n \to u$ and $Pu_n = u_n$, then $Pu_n \to Pu = \lim_{n \to \infty} u_n = u$. Thus M is a subspace of $\mathcal{H}$. For any vector $v \in \mathcal{H}$, set $v' = Pv$ and $v'' = (I - P)v = v - v'$. Then $v = v' + v''$ and $v' \in M$, $v'' \in M^\perp$, for

$$Pv' = P(Pv) = P^2 v = Pv = v',$$

and for all $w \in M$

$$\langle v'' | w \rangle = \langle (I - P)v | w \rangle = \langle v | w \rangle - \langle Pv | w \rangle = \langle v | w \rangle - \langle v | Pw \rangle = \langle v | w \rangle - \langle v | w \rangle = 0.$$ ∎

Unitary operators

An operator $U : \mathcal{H} \to \mathcal{H}$ is called **unitary** if

$$\langle Uu | Uv \rangle = \langle u | v \rangle \qquad \text{for all } u, v \in \mathcal{H}.$$

Since this implies $\langle U^* U u | v \rangle = \langle u | v \rangle$, an operator U is unitary if and only if $U^{-1} = U^*$. Every unitary operator is **isometric**, $\|Uu\| = \|u\|$ for all $u \in \mathcal{H}$ – it preserves the distance $d(u, v) = \|u - v\|$ between any two vectors. Conversely, every isometric operator is unitary, for if U is isometric then

$$\langle U(u+v) | U(u+v) \rangle - i \langle U(u+iv) | U(u+iv) \rangle = \langle u + v | u + v \rangle - i \langle u + iv | u + iv \rangle.$$

Expanding both sides and using $\langle Uu | Uu \rangle = \langle u | u \rangle$ and $\langle Uv | Uv \rangle = \langle v | v \rangle$, gives

$$2 \langle Uu | Uv \rangle = 2 \langle u | v \rangle.$$

If $\{e_1, e_2, \ldots\}$ is an orthonormal basis then so is

$$e'_1 = Ue_1, \ e'_2 = Ue_2, \ldots,$$

for

$$\langle e'_i | e'_j \rangle = \langle Ue_i | Ue_j \rangle = \langle U^* U e_i | e_j \rangle = \langle e_i | e_j \rangle = \delta_{ij}.$$

Conversely for any pair of complete orthonormal sets $\{e_1, e_2, \ldots\}$ and $\{e'_1, e'_2, \ldots\}$ the operator defined by $Ue_i = e'_i$ is unitary, for if u is any vector then, by Theorem 13.2,

$$u = \sum_i u_i e_i \quad \text{where} \quad u_i = \langle e_i | u \rangle.$$

Hence

$$Uu = \sum_i u_i Ue_i = \sum_i u_i e'_i,$$

Hilbert spaces

which gives
$$u_i = \langle e_i | u \rangle = \langle e'_i | Uu \rangle.$$

Parseval's identity (13.7) can be applied in the primed basis,
$$\begin{aligned}\langle Uu | Uv \rangle &= \sum_i \langle Uu | e'_i \rangle \langle e'_i | Uv \rangle \\ &= \sum_i \overline{u_i} v_i \\ &= \sum_i \langle u | e_i \rangle \langle e_i | v \rangle \\ &= \langle u | v \rangle,\end{aligned}$$

which shows that U is a unitary operator.

Exercise: Show that if U is a unitary operator then $\|U\| = 1$.

Exercise: Show that the multiplication operator A_α on $L^2(X)$ is unitary iff $|\alpha(x)| = 1$ for all $x \in X$.

Problems

Problem 13.14 The norm $\|\phi\|$ of a bounded linear operator $\phi : \mathcal{H} \to \mathbb{C}$ is defined as the greatest lower bound of all M such that $|\phi(u)| \le M\|u\|$ for all $u \in \mathcal{H}$. If $\phi(u) = \langle v | u \rangle$ show that $\|\phi\| = \|v\|$. Hence show that the bounded linear functional norm satisfies the parallelogram law
$$\|\phi + \psi\|^2 + \|\phi - \psi\|^2 = 2\|\phi\|^2 + 2\|\psi\|^2.$$

Problem 13.15 If $\{e_n\}$ is a complete o.n. set in a Hilbert space $\mathcal{H}$, and α_n a bounded sequence of scalars, show that there exists a unique bounded operator A such that $Ae_n = \alpha_n e_n$. Find the norm of A.

Problem 13.16 For bounded linear operators A, B on a normed vector space V show that
$$\|\lambda A\| = |\lambda| \, \|A\|, \qquad \|A + B\| \le \|A\| + \|B\|, \qquad \|AB\| \le \|A\| \, \|B\|.$$
Hence show that $\|A\|$ is a genuine norm on the set of bounded linear operators on V.

Problem 13.17 Prove properties (iii)–(v) of Theorem 13.13. Show that $\|A^*\| = \|A\|$.

Problem 13.18 Let A be a bounded operator on a Hilbert space $\mathcal{H}$ with a one-dimensional range.

(a) Show that there exist vectors u, v such that $Ax = \langle v | x \rangle u$ for all $x \in \mathcal{H}$.
(b) Show that $A^2 = \lambda A$ for some scalar λ, and that $\|A\| = \|u\| \|v\|$.
(c) Prove that A is hermitian, $A^* = A$, if and only if there exists a real number a such that $v = au$.

Problem 13.19 For every bounded operator A on a Hilbert space $\mathcal{H}$ show that the exponential operator
$$e^A = \sum_{n=0}^{\infty} \frac{A^n}{n!}$$

13.5 Spectral theory

is well-defined and bounded on $\mathcal{H}$. Show that

(a) $e^0 = I$.
(b) For all positive integers n, $(e^A)^n = e^{nA}$.
(c) e^A is invertible for all bounded operators A (even if A is not invertible) and $e^{-A} = (e^A)^{-1}$.
(d) If A and B are commuting operators then $e^{A+B} = e^A e^B$.
(e) If A is hermitian then e^{iA} is unitary.

Problem 13.20 Show that the sum of two projection operators $P_M + P_N$ is a projection operator iff $P_M P_N = 0$. Show that this condition is equivalent to $M \perp N$.

Problem 13.21 Verify that the operator on three-dimensional Hilbert space, having matrix representation in an o.n. basis

$$\begin{pmatrix} \frac{1}{2} & 0 & \frac{i}{2} \\ 0 & 1 & 0 \\ -\frac{i}{2} & 0 & \frac{1}{2} \end{pmatrix}$$

is a projection operator, and find a basis of the subspace it projects onto.

Problem 13.22 Let $\omega = e^{2\pi i/3}$. Show that $1 + \omega + \omega^2 = 0$.

(a) In Hilbert space of three dimensions let V be the subspace spanned by the vectors $(1, \omega, \omega^2)$ and $(1, \omega^2, \omega)$. Find the vector u_0 in this subspace that is closest to the vector $u = (1, -1, 1)$.
(b) Verify that $u - u_0$ is orthogonal to V.
(c) Find the matrix representing the projection operator P_V into the subspace V.

Problem 13.23 An operator A is called **normal** if it is bounded and commutes with its adjoint, $A^*A = AA^*$. Show that the operator

$$A\psi(x) = c\psi(x) + i \int_a^b K(x, y)\psi(y)\,dy$$

on $L^2([a, b])$, where c is a real number and $K(x, y) = \overline{K(y, x)}$, is normal.

(a) Show that an operator A is normal if and only if $\|Au\| = \|A^*u\|$ for all vectors $u \in \mathcal{H}$.

(b) Show that if A and B are commuting normal operators, AB and $A + \lambda B$ are normal for all $\lambda \in \mathbb{C}$.

13.5 Spectral theory

Eigenvectors

As in Chapter 4 a complex number α is an **eigenvalue** of a bounded linear operator $A : \mathcal{H} \to \mathcal{H}$ if there exists a non-zero vector $u \in \mathcal{H}$ such that

$$Au = \alpha u.$$

u is called the **eigenvector** of A corresponding to the eigenvalue α.

Theorem 13.15 *All eigenvalues of a hermitian operator A are real, and eigenvectors corresponding to different eigenvalues are orthogonal.*

Hilbert spaces

Proof: If $Au = \alpha u$ then
$$\langle u \mid Au \rangle = \langle u \mid \alpha u \rangle = \alpha \|u\|^2.$$

Since A is hermitian
$$\langle u \mid Au \rangle = \langle Au \mid u \rangle = \langle \alpha u \mid u \rangle = \bar{\alpha}\|u\|^2.$$

For a non-zero vector $\|u\| \neq 0$, we have $\alpha = \bar{\alpha}$; the eigenvalue α is real.

If $Av = \beta v$ then
$$\langle u \mid Av \rangle = \langle u \mid \beta v \rangle = \beta \langle u \mid v \rangle$$

and
$$\langle u \mid Av \rangle = \langle Au \mid v \rangle = \langle \alpha u \mid v \rangle = \bar{\alpha}\langle u \mid v \rangle = \alpha \langle u \mid v \rangle.$$

If $\beta \neq \alpha$ then $\langle u \mid v \rangle = 0$. ∎

A hermitian operator is said to be **complete** if its eigenvectors form a complete o.n. set.

Example 13.12 The eigenvalues of a projection operator P are always 0 or 1, for
$$Pu = \alpha u \implies P^2 u = P(\alpha u) = \alpha Pu = \alpha^2 u$$

and since P is idempotent,
$$P^2 u = Pu = \alpha u.$$

Hence $\alpha^2 = \alpha$, so that $\alpha = 0$ or 1. If $P = P_M$ then the eigenvectors corresponding to eigenvalue 1 are the vectors belonging to the subspace M, while those having eigenvalue 0 belong to its orthogonal complement $M^\perp$. Combining Theorems 13.8 and 13.2, we see that every projection operator is complete.

Theorem 13.16 *The eigenvalues of a unitary operator U are of the form $\alpha = e^{ia}$ where a is a real number, and eigenvectors corresponding to different eigenvalues are orthogonal.*

Proof: Since U is an isometry, if $Uu = \alpha u$ where $u \neq 0$, then
$$\|u\|^2 = \langle u \mid u \rangle = \langle Uu \mid Uu \rangle = \langle \alpha u \mid \alpha u \rangle = \bar{\alpha}\alpha \|u\|^2.$$

Hence $\bar{\alpha}\alpha = |\alpha|^2 = 1$, and there exists a real a such that $\alpha = e^{ia}$.

If $Uu = \alpha u$ and $Uv = \beta v$, then
$$\langle u \mid Uv \rangle = \beta uv.$$

But $U^*U = I$ implies $u = U^*Uu = \alpha U^* u$, so that
$$U^* u = \alpha^{-1} u = \bar{\alpha} u \quad \text{since} \quad |\alpha|^2 = 1.$$

Therefore
$$\langle u \mid Uv \rangle = \langle U^* u \mid v \rangle = \langle \bar{\alpha} u \mid v \rangle = \alpha \langle u \mid v \rangle.$$

Hence $(\alpha - \beta)\langle u \mid v \rangle = 0$. If $\alpha \neq \beta$ then u and v are orthogonal, $\langle u \mid v \rangle = 0$. ∎

13.5 Spectral theory

Spectrum of a bounded operator

In the case of a finite dimensional space, the set of eigenvalues of an operator is known as its *spectrum*. The spectrum is non-empty (see Chapter 4), and forms the diagonal elements in the Jordan canonical form. In infinite dimensional spaces, however, operators may have no eigenvalues at all.

Example 13.13 In ℓ^2, the right shift operator S has no eigenvalues, for suppose

$$S(x_1, x_2, \ldots) = (0, x_1, x_2, \ldots) = \lambda(x_1, x_2, \ldots).$$

If $\lambda \neq 0$ then $x_1 = 0, x_2 = 0, \ldots$, hence λ is not an eigenvalue. But $\lambda = 0$ also implies $x_1 = x_2 = \cdots = 0$, so this operator has no eigenvalues at all.

Exercise: Show that every λ such that $|\lambda| < 1$ is an eigenvalue of the left shift operator $S' = S^*$. Note that the spectrum of S and its adjoint S^* may be unrelated in the infinite dimensional case.

Example 13.14 Let $\alpha(x)$ be a bounded integrable function on a measure space X, and let $A_\alpha : g \mapsto \alpha g$ be the multiplication operator defined in Example 13.9. There is no normalizable function $g \in L^2(X)$ such that $\alpha(x)g(x) = \lambda g(x)$ unless $\alpha(x)$ has the constant value λ on an interval E of non-zero measure. For example, if $\alpha(x) = x$ on $X = [a, b]$, then $f(x)$ is an eigenvector of A_x iff there exists $\lambda \in \mathbb{C}$ such that

$$xf(x) = \lambda f(x),$$

which is only possible through $[a, b]$ if $f(x) = 0$. In quantum mechanics (see Chapter 14) this problem is sometimes overcome by treating the eigenvalue equation as a distributional equation. Then the Dirac delta function $\delta(x - x_0)$ acts as a distributional eigenfunction, with eigenvalue $a < \lambda = x_0 < b$,

$$x\delta(x - x_0) = x_0 \delta(x - x_0).$$

Examples such as 13.14 lead us to consider a new definition for the spectrum of an operator. Every operator A has a degeneracy at an eigenvalue λ, in that $A - \lambda I$ is not an invertible operator. For, if $(A - \lambda I)^{-1}$ exists then $Au \neq \lambda u$, for if $Au = \lambda u$ then

$$u = (A - \lambda I)^{-1}(A - \lambda I)u = (A - \lambda I)^{-1}0 = 0.$$

We say a complex number λ is a **regular value** of a bounded operator A on a Hilbert space $\mathcal{H}$ if $A - \lambda I$ is invertible – that is, $(A - \lambda I)^{-1}$ exists and is bounded. The **spectrum** $\Sigma(A)$ of A is defined to be the set of $\lambda \in \mathbb{C}$ that are not regular values of A. If λ is an eigenvalue of A then, as shown above, it is in the spectrum of A but the converse is not true. The eigenvalues are often called the **point spectrum**. The other points of the spectrum are called the **continuous spectrum**. At such points it is conceivable that the inverse exists but is not bounded. More commonly, the inverse only exists on a dense domain of $\mathcal{H}$ and is unbounded on that domain. We will leave discussion of this to Section 13.6.

Example 13.15 If $\alpha(x) = x$ then the multiplication operator A_α on $L^2([0, 1])$ has spectrum consisting of all real numbers λ such that $0 \leq \lambda \leq 1$. If $\lambda > 1$ or $\lambda < 0$ or has non-zero

Hilbert spaces

imaginary part then the function $\beta = x - \lambda$ is clearly invertible and bounded on the interval $[0, 1]$. Hence all these are regular values of the operator A_x. The real values $0 \le \lambda \le 1$ form the spectrum of A_x. From Example 13.14 none of these numbers are eigenvalues, but they do lie in the spectrum of A_x since the function β is not invertible. The operator A_β is then defined, but unbounded, on the dense set $[0, 1] - \{\lambda\}$.

Theorem 13.17 *Let A be a bounded operator on a Hilbert space $\mathcal{H}$.*

(i) *Every complex number $\lambda \in \Sigma(A)$ has magnitude $|\lambda| \le \|A\|$.*
(ii) *The set of regular values of A is an open subset of $\mathbb{C}$.*
(iii) *The spectrum of A is a compact subset of $\mathbb{C}$.*

Proof: (i) Let $|\lambda| > \|A\|$. The operator A/λ then has norm < 1 and by Theorem 13.12 the operator $I - A/\lambda$ is invertible and

$$(A - \lambda I)^{-1} = -\lambda^{-1}\left(I - \frac{A}{\lambda}\right)^{-1} = -\lambda^{-1}\sum_{n=0}^{\infty}\left(\frac{A}{\lambda}\right)^n.$$

Hence λ is a regular value. Spectral values must therefore have $|\lambda| \le \|A\|$.

(ii) If λ_0 is a regular value, then for any other complex number λ

$$I - (A - \lambda_0 I)^{-1}(A - \lambda I) = (A - \lambda_0 I)^{-1}\big((A - \lambda_0 I) - (A - \lambda I)\big)$$
$$= (A - \lambda_0 I)^{-1}(\lambda - \lambda_0).$$

Hence

$$\|I - (A - \lambda_0 I)^{-1}(A - \lambda I)\| = |\lambda - \lambda_0|\,\|(A - \lambda_0 I)^{-1}\| < 1$$

if

$$|\lambda - \lambda_0| < \frac{1}{\|(A - \lambda_0 I)^{-1}\|}.$$

By Theorem 13.12, for λ in a small enough neighbourhood of λ the operator $I - \big(I - (A - \lambda_0 I)^{-1}(A - \lambda)\big) = (A - \lambda_0 I)^{-1}(A - \lambda I)$ is invertible. If B is its inverse, then

$$B(A - \lambda_0 I)^{-1}(A - \lambda) = I$$

and $A - \lambda I$ is invertible with inverse $B(A - \lambda_0 I)^{-1}$. Hence the regular values form an open set.

(iii) The spectrum $\Sigma(A)$ is a closed set since it is the complement of an open set (the regular values). By part (i), it is a subset of a bounded set $|\lambda| \le \|A\|$, and is therefore a compact set. ∎

Spectral theory of hermitian operators

Of greatest interest is the spectral theory of hermitian operators. This theory can become quite difficult, and we will only sketch some of the proofs.

13.5 Spectral theory

Theorem 13.18 *The spectrum $\Sigma(A)$ of a hermitian operator A consists entirely of real numbers.*

Proof: Suppose $\lambda = a + ib$ is a complex number with $b \neq 0$. Then $\|(A - \lambda I)u\|^2 = \|(A - aI)u\|^2 + b^2\|u\|^2$, and

$$\|u\| \leq \frac{1}{|b|}\|(A - \lambda I)u\|. \tag{13.12}$$

The operator $A - \lambda I$ is therefore one-to-one, for if $(A - \lambda I)u = 0$ then $u = 0$.

The set $V = \{(A - \lambda I)u \mid u \in \mathcal{H}\}$ is a subspace of $\mathcal{H}$. To show closure (the vector subspace property is trivial), let $v_n = (A - \lambda I)u_n \to v$ be a convergent sequence of vectors in V. From the fact that it is a Cauchy sequence and the inequality (13.12), it follows that u_n is also a Cauchy sequence, having limit u. By continuity of the operator $A - \lambda I$, it follows that V is closed, for

$$(A - \lambda I)u = \lim_{n \to \infty}(A - \lambda I)u_n = \lim_{n \to \infty} v_n = v.$$

Finally, $V = \mathcal{H}$, for if $w \in V^\perp$, then $\langle (A - \lambda I)u \mid w \rangle = \langle u \mid (A - \overline{\lambda}I)w \rangle = 0$ for all $u \in \mathcal{H}$. Setting $u = (A - \overline{\lambda}I)w$ gives $(A - \overline{\lambda}I)w = 0$. Since $A - \overline{\lambda}I$ is one-to-one, $w = 0$. Hence $V^\perp = \{0\}$, the subspace $V = \mathcal{H}$ and every vector $u \in \mathcal{H}$ can be written in the form $u = (A - \lambda I)v$. Thus $A - \lambda I$ is invertible, and the inequality (13.12) can be used to show it is bounded. ∎

The full spectral theory of a hermitian operator involves reconstructing the operator from its spectrum. In the finite dimensional case, the spectrum consists entirely of eigenvalues, making up the point spectrum. From Theorem 13.15 the eigenvalues may be written as a non-empty ordered set of real numbers $\lambda_1 < \lambda_2 < \cdots < \lambda_k$. For each eigenvalue λ_i there corresponds an eigenspace M_i of eigenvectors, and different spaces are orthogonal to each other. A standard inductive argument can be used to show that every hermitian operator on a finite dimensional Hilbert space is complete, so the eigenspaces span the entire Hilbert space. In terms of projection operators into these eigenspaces $P_i = P_{M_i}$, these statements can be summarized as

$$A = \lambda_1 P_1 + \lambda_2 P_2 + \cdots + \lambda_k P_k$$

where

$$P_1 + P_2 + \cdots + P_k = I, \qquad P_i P_j = P_j P_i = \delta_{ij} P_i.$$

Essentially, this is the familiar statement that a hermitian matrix can be 'diagonalized' with its eigenvalues along the diagonal. If we write, for any two projection operators, $P_M \leq P_N$ iff $M \subseteq N$, we can replace the operators P_i with an increasing family of projection operators $E_i = P_1 + P_2 + \cdots + P_i$. These are projection operators since they are clearly hermitian and idempotent, $(E_i)^2 = E_i$, and project into an increasing family of subspaces, $V_i = L(M_1 \cup M_2 \cup \cdots \cup M_i)$, having the property $V_i \subset V_j$ if $i < j$. Since $P_i = E_i - E_{i-1}$,

Hilbert spaces

where $E_0 = 0$, we can write the spectral theorem in the form

$$A = \sum_{i=1}^{n} \lambda_i (E_i - E_{i-1}).$$

For infinite dimensional Hilbert spaces, the situation is considerably more complicated, but the projection operator language can again be used to effect. The full spectral theorem in arbitrary dimensions is as follows:

Theorem 13.19 *Let A be a hermitian operator on a Hilbert space $\mathcal{H}$, with spectrum $\Sigma(A)$. By Theorem 13.17 this is a closed bounded subset of $\mathbb{R}$. There exists an increasing family of projection operators E_λ ($\lambda \in \mathbb{R}$), with $E_\lambda \leq P_{\lambda'}$ for $\lambda \leq \lambda'$, such that*

$$E_\lambda = 0 \text{ for } \lambda < \inf(\Sigma(A)), \qquad E_\lambda = I \text{ for } \lambda > \sup(\Sigma(A))$$

and

$$A = \int_{-\infty}^{\infty} \lambda \, dE_\lambda.$$

The integral in this theorem is defined in the **Lebesgue–Stieltjes** sense. Essentially it means that if $f(x)$ is a measurable function, and $g(x)$ is of the form

$$g(x) = c + \int_0^x h(x) \, dx$$

for some complex constant c and integrable function $h(x)$, then

$$\int_a^b f(x) \, d(g(x)) = \int_a^b f(x) h(x) \, dx.$$

A function g of this form is said to be **absolutely continuous**; the function h is uniquely defined almost everywhere by g and we may write it as a kind of derivative of g, $h(x) = g'(x)$. For the finite dimensional case this theorem reduces to the statement above, on setting E_λ to have discrete jumps by P_i at each of the eigenvalues λ_i. The proof of this result is not easy. The interested reader is referred to [3, 6] for details.

Problems

Problem 13.24 Show that a non-zero vector u is an eigenvector of an operator A if and only if $|\langle u | Au \rangle| = \|Au\| \|u\|$.

Problem 13.25 For any projection operator P_M show that every value $\lambda \neq 0, 1$ is a regular value, by showing that $(P_M - \lambda I)$ has a bounded inverse.

Problem 13.26 Show that every complex number λ in the spectrum of a unitary operator has $|\lambda| = 1$.

Problem 13.27 Prove that every hermitian operator A on a finite dimensional Hilbert space can be written as

$$A = \sum_{i=1}^{k} \lambda_i P_i \quad \text{where} \quad \sum_{i=1}^{k} P_i = I, \quad P_i P_j = P_j P_i = \delta_{ij} P_i.$$

13.6 Unbounded operators

Problem 13.28 For any pair of hermitian operators A and B on a Hilbert space $\mathcal{H}$, define $A \leq B$ iff $\langle u | Au \rangle \leq \langle u | Bu \rangle$ for all $u \in \mathcal{H}$. Show that this is a partial order on the set of hermitian operators – pay particular attention to the symmetry property, $A \leq B$ and $B \leq A$ implies $A = B$.

(a) For multiplication operators on $L^2(X)$ show that $A_\alpha \leq A_\beta$ iff $\alpha(x) \leq \beta(x)$ a.e. on X.
(b) For projection operators show that the definition given here reduces to that given in the text, $P_M \leq P_N$ iff $M \subseteq N$.

13.6 Unbounded operators

A linear operator A on a Hilbert space $\mathcal{H}$ is **unbounded** if for any $M > 0$ there exists a vector u such that $\|Au\| \geq M\|u\|$. Very few interesting examples of unbounded operators are defined on all of $\mathcal{H}$ – for self-adjoint operators, there are none at all. It is therefore usual to consider an unbounded operator A as not being necessarily defined over all of $\mathcal{H}$ but only on some vector subspace $D_A \subseteq \mathcal{H}$ called the **domain** of A. Its **range** is defined as the set of vectors that are mapped onto, $R_A = A(D_A)$. In general we will refer to a pair (A, D_A), where D_A is a vector subspace of $\mathcal{H}$ and $A : D_A \to R_A \subseteq \mathcal{H}$ is a linear map, as being an **operator in** $\mathcal{H}$. Often we will simply refer to the operator A when the domain D_A is understood.

We say the domain D_A is a **dense** subspace of $\mathcal{H}$ if for every vector $u \in \mathcal{H}$ and any $\epsilon > 0$ there exists a vector $v \in D_A$ such that $\|u - v\| < \epsilon$. The operator A is then said to be **densely defined**.

We say A is an **extension of** B, written $B \subseteq A$, if $D_B \subseteq D_A$ and $A|_{D_B} = B$. Two operators (A, D_A) and (B, D_B) in $\mathcal{H}$ are called **equal** if and only if they are extensions of each other – their domains are equal, $D_A = D_B$ and $Au = Bu$ for all $u \in D_A$.

For any two operators in $\mathcal{H}$ we must be careful about simple operations such as addition $A + B$ and multiplication AB. The former only exists on the domain $D_{A+B} = D_A \cap D_B$, while the latter only exists on the set $B^{-1}(R_B \cap D_A)$. Thus operators in $\mathcal{H}$ do not form a vector space or algebra in any natural sense.

Example 13.16 In $\mathcal{H} = \ell^2$ let $A : \mathcal{H} \to \mathcal{H}$ be the operator defined by

$$(Ax)_n = \frac{1}{n} x_n.$$

This operator is bounded, hermitian and has domain $D_A = \mathcal{H}$ since

$$\sum_{n=1}^\infty |x_n|^2 < \infty \implies \sum_{n=1}^\infty \left|\frac{x_n}{n}\right|^2 < \infty.$$

The range of this operator is

$$R_A = \left\{ y \;\Big|\; \sum_{n=1}^\infty n^2 |y_n|^2 < \infty \right\},$$

which is dense in ℓ^2 – since every $x \in \ell^2$ can be approximated arbitrarily closely by, for example, a finite sum $\sum_{n=1}^N x_n e_n$ where e_n are the standard basis vectors having components

$(e_n)_m = \delta_{nm}$. The inverse operator A^{-1}, defined on the dense domain $D_{A^{-1}} = R_A$, is unbounded since

$$\|A^{-1}e_n\| = \|ne_n\| = n \to \infty.$$

Example 13.17 In the Hilbert space $L^2(\mathbb{R})$ of equivalence classes of square integrable functions (see Example 13.4), set D to be the vector subspace of elements $\widetilde{\varphi}$ having a representative φ from the C^∞ functions on $\mathbb{R}$ of compact support. This is essentially the space of test functions $\mathcal{D}^\infty(\mathbb{R})$ defined in Chapter 12. An argument similar to that outlined in Example 13.5 shows that D is a dense subspace of $L^2(\mathbb{R})$. We define the *position operator* $Q : D \to D \subset L^2(\mathbb{R})$ by $Q\widetilde{\varphi} = \widetilde{x\varphi}$. We may write this more informally as

$$(Q\varphi)(x) = x\varphi(x).$$

Similarly the *momentum operator* $P : D \to D$ is defined by

$$P\varphi(x) = -i\frac{d}{dx}\varphi(x).$$

Both these operators are evidently linear on their domains.

Exercise: Show that the position and momentum operators in $L^2(\mathbb{R})$ are unbounded.

If A is a bounded operator defined on a dense domain D_A, it has a unique extension to all of $\mathcal{H}$ (see Problem 13.30). We may always assume then that a bounded operator is defined on all of $\mathcal{H}$, and when we refer to a densely defined operator whose domain is a proper subspace of $\mathcal{H}$ we implicitly assume it to be an unbounded operator.

Self-adjoint and symmetric operators

Lemma 13.20 *If D_A is a dense domain and u a vector in $\mathcal{H}$ such that $\langle u | v \rangle = 0$ for all $v \in D_A$, then $u = 0$.*

Proof: Let w be any vector in $\mathcal{H}$ and $\epsilon > 0$. Since D_A is dense there exists a vector $v \in D_A$ such that $\|w - v\| < \epsilon$. By the Cauchy–Schwarz inequality

$$|\langle u | w \rangle| = |\langle u | w - v \rangle| \le \|u\| \|w - v\| < \epsilon \|u\|.$$

Since ϵ is an arbitrary positive number, $\langle u | w \rangle = 0$ for all $w \in \mathcal{H}$; hence $u = 0$. ∎

If (A, D_A) is an operator in $\mathcal{H}$ with dense domain D_A, then let D_{A^*} be defined by

$$u \in D_{A^*} \iff \exists u^* \text{ such that } \langle u | Av \rangle = \langle u^* | v \rangle, \quad \forall v \in D_A.$$

If $u \in D_{A^*}$ we set $A^*u = u^*$. This is uniquely defined, for if $\langle u_1^* - u_2^* | v \rangle = 0$ for all $v \in D_A$ then $u_1^* = u_2^*$ by Lemma 13.20. The operator (A^*, D_{A^*}) is called the **adjoint of** (A, D_A).

We say a densely defined operator (A, D_A) in $\mathcal{H}$ is **closed** if for every sequence $u_n \in D_A$ such that $u_n \to u$ and $Au_n \to v$ it follows that $u \in D_A$ and $Au = v$. Another way of expressing this is to say that an operator is closed if and only if its graph $G_A = \{(x, Ax) \,|\, x \in D_A\}$ is a closed subset of the product set $\mathcal{H} \times \mathcal{H}$. The notion of closedness is similar to

13.6 Unbounded operators

continuity, but differs in that we must *assert* the limit $Au_n \to v$, while for continuity it is deduced. Clearly every continuous operator is closed, but the converse does not hold in general.

Theorem 13.21 *If A is a densely defined operator then its adjoint A^* is closed.*

Proof: Let y_n be any sequence of vectors in D_{A^*} such that $y_n \to y$ and $A^*y_n \to z$. Then for all $x \in D_A$

$$\langle y | Ax \rangle = \lim_{n \to \infty} \langle y_n | Ax \rangle = \lim_{n \to \infty} \langle A^*y_n | x \rangle = \langle z | x \rangle.$$

Since D_A is a dense domain, it follows from Lemma 13.20 that $y \in D_{A^*}$ and $A^*y = z$. ∎

Example 13.18 Let $\mathcal{H}$ be a separable Hilbert space with complete orthonormal basis e_n ($n = 0, 1, 2, \dots$). Let the operators a and a^* be defined by

$$a\,e_n = \sqrt{n}\,e_{n-1}, \qquad a^*\,e_n = \sqrt{n+1}\,e_{n+1}.$$

The effect on a typical vector $x = \sum_{n=0}^{\infty} x_n e_n$, where $x_n = \langle x | e_n \rangle$, is

$$a\,x = \sum_{n=0}^{\infty} x_{n+1} \sqrt{n+1}\,e_n, \qquad a^*\,x = \sum_{n=1}^{\infty} x_{n-1} \sqrt{n}\,e_n.$$

The operator a^* is the adjoint of a since

$$\langle a^* y | x \rangle = \langle y | ax \rangle = \sum_{n=1}^{\infty} \overline{y_n} \sqrt{n+1}\,x_{n+1}$$

and both operators have domain of definition

$$D = D_a = D_{a^*} = \left\{ y \,\Big|\, \sum_{n=1}^{\infty} |y_n|^2 n < \infty \right\},$$

which is dense in $\mathcal{H}$ (see Example 13.16). In physics, $\mathcal{H}$ is the symmetric *Fock space*, in which e_n represents n identical (bosonic) particles in a given state, and a^* and a are interpreted as *creation* and *annihilation operators*, respectively.

Exercise: Show that $N = a^*a$ is the particle number operator, $Ne_n = ne_n$, and the commutator is $[a, a^*] = aa^* - a^*a = I$. What are the domains of validity of these equations?

Theorem 13.22 *If (A, D_A) and (B, D_B) are densely defined operators in $\mathcal{H}$ then $A \subseteq B \implies B^* \subseteq A^*$.*

Proof: If $A \subseteq B$ then for any vectors $u \in D_A$ and $v \in D_{B^*}$

$$\langle v | Au \rangle = \langle v | Bu \rangle = \langle B^*v | u \rangle.$$

Hence $v \in D_{A^*}$, so that $D_{B^*} \subseteq D_{A^*}$ and

$$\langle v | Au \rangle = \langle A^*v | u \rangle = \langle B^*v | u \rangle.$$

By Lemma 13.20 $A^*v = B^*v$, hence $B^* \subseteq A^*$. ∎

Hilbert spaces

An operator (A, D_A) on a dense domain is said to be **self-adjoint** if $A = A^*$. This means that not only is $Au = A^*u$ wherever both sides are defined, but also that the domains are equal, $D_A = D_{A^*}$. By Theorem 13.21 every self-adjoint operator is closed. This is not the only definition that generalizes the concept of a hermitian operator to unbounded operators. The following related definition is also useful. A densely defined operator (A, D_A) in $\mathcal{H}$ is called a **symmetric operator** if $\langle Au | v\rangle = \langle u | Av\rangle$ for all $u, v \in D_A$.

Theorem 13.23 *An operator (A, D_A) on a dense domain in $\mathcal{H}$ is symmetric if and only if A^* is an extension of A, $A \subseteq A^*$.*

Proof: If $A \subseteq A^*$ then for all $u, v \in D_A \subseteq D_{A^*}$

$$\langle u | Av\rangle = \langle A^*u | v\rangle.$$

Furthermore, since $A^*u = Au$ for all $u \in D_A$, we have the symmetry condition $\langle u | Av\rangle = \langle Au | v\rangle$.

Conversely, if A is symmetric then

$$\langle u | Av\rangle = \langle Au | v\rangle \quad \text{for all } u, v \in D_A.$$

On the other hand, the definition of adjoint gives

$$\langle u | Av\rangle = \langle A^*u | v\rangle \quad \text{for all } u \in D_{A^*}, v \in D_A.$$

Hence if $u \in D_A$ then $u \in D_{A^*}$ and $Au = A^*u$, which two conditions are equivalent to $A \subseteq A^*$. ∎

From this theorem it is immediate that every self-adjoint operator is symmetric, since $A = A^* \implies A \subseteq A^*$.

Exercise: Show that the operators A and A^{-1} of Example 13.16 are both self-adjoint.

Example 13.19 In Example 13.17 we defined the position operator (Q, D) having domain D, the space of C^∞ functions of compact support on $\mathbb{R}$. This operator is symmetric in $L^2(\mathbb{R})$, since

$$\langle \varphi | Q\psi\rangle = \int_{-\infty}^{\infty} \overline{\varphi(x)} x \psi(x) \, dx = \int_{-\infty}^{\infty} \overline{x\varphi(x)} \psi(x) \, dx = \langle Q\varphi | \psi\rangle$$

for all functions $\varphi, \psi \in D$. However it is not self-adjoint, since there are many functions $\varphi \notin D$ for which there exists a function φ^* such that $\langle \varphi | Q\psi\rangle = \langle \varphi^* | \psi\rangle$ for all $\psi \in D$. For example, the function

$$\varphi(x) = \begin{cases} 1 & \text{for } -1 \leq x \leq 1 \\ 0 & \text{for } |x| > 1 \end{cases}$$

is not in D since it is not C^∞, yet

$$\langle \varphi | Q\psi\rangle = \langle \varphi^* | \psi\rangle, \quad \forall \psi \in D \quad \text{where} \quad \varphi^*(x) = x\varphi(x).$$

Similarly, the function $\varphi = 1/(1+x^2)$ does not have compact support, yet satisfies the same equation. Thus the domain D_{Q^*} of the adjoint operator Q^* is larger than the domain D, and (Q, D) is not self-adjoint.

13.6 Unbounded operators

To rectify the situation, let D_Q be the subspace of $L^2(\mathbb{R})$ of functions φ such that $x\varphi \in L^2(\mathbb{R})$,

$$\int_{-\infty}^{\infty} |x\varphi(x)|^2 \, dx < \infty.$$

Functions φ and φ' are always to be identified, of course, if they are equal almost everywhere. The operator (Q, D_Q) is symmetric since

$$\langle \varphi | Q\psi \rangle = \int_{-\infty}^{\infty} \overline{\varphi(x)} x \psi(x) \, dx = \langle Q\varphi | \psi \rangle$$

for all $\varphi, \psi \in D_Q$. The domain D_Q is dense in $L^2(\mathbb{R})$, for if φ is any square integrable function then the sequence of functions

$$\varphi_n(x) = \begin{cases} \varphi(x) & \text{for } -n \leq x \leq n \\ 0 & \text{for } |x| > n \end{cases}$$

all belong to D_Q and $\varphi_n \to \varphi$ as $n \to \infty$ since

$$\|\varphi - \varphi_n\|^2 = \int_{-\infty}^{-n} |\varphi(x)|^2 \, dx + \int_n^{\infty} |\varphi(x)|^2 \, dx \to 0.$$

By Theorem 13.23, Q^* is an extension of Q since the operator (Q, D_Q) is symmetric. It only remains to show that $D_{Q^*} \subseteq D_Q$. The domain D_{Q^*} is the set of functions $\varphi \in L^2(\mathbb{R})$ such that there exists a function φ^* such that

$$\langle \varphi | Q\psi \rangle = \langle \varphi^* | \psi \rangle, \quad \forall \psi \in D_Q.$$

The function φ^* has the property

$$\int_{-\infty}^{\infty} \left(\overline{x\varphi(x)} - \overline{\varphi^*}\right) \psi(x) \, dx = 0, \quad \forall \psi \in D_Q.$$

Since D_Q is a dense domain this is only possible if $\varphi^*(x) = x\varphi(x)$ a.e. Since $\varphi^* \in L^2(\mathbb{R})$ it must be true that $x\varphi(x) \in L^2(\mathbb{R})$, whence $\varphi(x) \in D_Q$. This proves that $D_{Q^*} \subseteq D_Q$. Hence $D_{Q^*} = D_Q$, and since $\varphi^*(x) = x\varphi(x)$ a.e., we have $\varphi^* = Q\varphi$. The position operator is therefore self-adjoint, $Q = Q^*$.

Example 13.20 The momentum operator defined in Example 13.17 on the domain D of differentiable functions of compact support is symmetric, for

$$\langle \varphi | P\psi \rangle = \int_{-\infty}^{\infty} \overline{-i\varphi(x)} \frac{d\psi}{dx} \, dx$$

$$= \left[-i\overline{\varphi(x)}\psi(x)\right]_{-\infty}^{\infty} + \int_{-\infty}^{\infty} i\overline{\frac{d\varphi(x)}{dx}} \psi(x) \, dx$$

$$= \int_{-\infty}^{\infty} i\overline{\frac{d\varphi(x)}{dx}} \psi(x) \, dx$$

$$= \langle P\varphi | \psi \rangle$$

for all $\varphi, \psi \in D$. Again, it is not hard to find functions φ outside D that satisfy this relation for all ψ, so this operator is not self-adjoint. Extending the domain so that the momentum

Hilbert spaces

operator becomes self-adjoint is rather trickier than for the position operator. We only give the result; details may be found in [3, 7]. Recall from the discussion following Theorem 13.19 that a function $\varphi : \mathbb{R} \to \mathbb{C}$ is said to be *absolutely continuous* if there exists a measurable function ρ on $\mathbb{R}$ such that

$$\varphi(x) = c + \int_0^x \rho(x)\,dx.$$

We may then set $D\varphi = \varphi' = \rho$. When ρ is a continuous function, $\varphi(x)$ is differentiable and $D\varphi = d\varphi(x)/dx$. Let D_P consist of those absolutely continuous functions such that φ and $D\varphi$ are square integrable. It may be shown that D_P is a dense vector subspace of $L^2(\mathbb{R})$ and that the operator (P, D_P) where $P\varphi = -iD\varphi$ is a self-adjoint extension of the momentum operator P defined in Example 13.17.

Spectral theory of unbounded operators

As for hermitian operators, the eigenvalues of a self-adjoint operator (A, D_A) are real and eigenvectors corresponding to different eigenvalues are orthogonal. If $Au = \lambda u$, then λ is real since

$$\lambda = \frac{\langle u \,|\, Au \rangle}{\|u\|^2} = \frac{\langle Au \,|\, u \rangle}{\|u\|^2} = \overline{\frac{\langle u \,|\, Au \rangle}{\|u\|^2}} = \bar{\lambda}.$$

If $Au = \lambda u$ and $Av = \mu v$, then

$$0 = \langle Au \,|\, v \rangle - \langle u \,|\, Av \rangle = (\lambda - \mu)\langle u \,|\, v \rangle$$

whence $\langle u \,|\, v \rangle = 0$ whenever $\lambda \neq \mu$.

For each complex number define Δ_λ to be the domain of the *resolvent operator* $(A - \lambda I)^{-1}$,

$$\Delta_\lambda = D_{(A-\lambda I)^{-1}} = R_{A-\lambda I}.$$

The operator $(A - \lambda I)^{-1}$ is well-defined with domain Δ_λ provided λ is not an eigenvalue. For, if λ is not an eigenvalue then $\ker(A - \lambda I) = \{0\}$ and for every $y \in R_{A-\lambda I}$ there exists a unique $x \in D_A$ such that $y = (A - \lambda I)x$.

Exercise: Show that for all complex numbers λ, the operator $A - \lambda I$ is closed.

As for bounded operators a complex number λ is said to be a **regular value** for A if $\Delta_\lambda = \mathcal{H}$. The resolvent operator $(A - \lambda I)^{-1}$ can then be shown to be a bounded (continuous) operator. The set of complex numbers that are not regular are again known as the **spectrum** of A.

Theorem 13.24 λ *is an eigenvalue of a self-adjoint operator* (A, D_A) *if and only if the resolvent set* Δ_λ *is not dense in* $\mathcal{H}$.

Proof: If $Ax = \lambda x$ where $x \neq 0$, then

$$0 = \langle (A - \lambda I)x \,|\, u \rangle = \langle x \,|\, (A - \lambda I)u \rangle$$

13.6 Unbounded operators

for all $u \in D_A$. Hence $\langle x \,|\, v \rangle = 0$ for all $v \in \Delta_\lambda = R_{A-\lambda I}$. If Δ_λ is dense in $\mathcal{H}$ then, by Lemma 13.20, this can only be true for $x = 0$, contrary to assumption.

Conversely if Δ_λ is not dense then by Theorem 13.8 there exists a non-zero vector $x \in (\overline{\Delta_\lambda})^\perp$. This vector has the property

$$0 = \langle x \,|\, (A - \lambda I)u \rangle = \langle (A - \lambda I)x \,|\, u \rangle$$

for all $u \in D_A$. Since D_A is a dense domain, x must be an eigenvector, $Ax = \lambda x$. ∎

It is natural to classify the spectrum into two parts – the **point spectrum** consisting of eigenvalues, where the resolvent set Δ_λ is not dense in $\mathcal{H}$, and the **continuous spectrum** consisting of those values λ for which Δ_λ is not closed. Note that these are not mutually exclusive; it is possible to have eigenvalues λ for which the resolvent set is neither closed nor dense. The entire spectrum of a self-adjoint operator can, however, be shown to consist of real numbers. The spectral theorem 13.19 generalizes for self-adjoint operators as follows:

Theorem 13.25 *Let A be a self-adjoint operator on a Hilbert space $\mathcal{H}$. There exists an increasing family of projection operators E_λ ($\lambda \in \mathbb{R}$), with $E_\lambda \le P_{\lambda'}$ for $\lambda \le \lambda'$, such that*

$$E_{-\infty} = 0 \quad \text{and} \quad E_\infty = I$$

such that

$$A = \int_{-\infty}^{\infty} \lambda \, dE_\lambda,$$

where the integral is interpreted as the Lebesgue–Stieltjes integral

$$\langle u \,|\, Au \rangle = \int_{-\infty}^{\infty} \lambda \, d\langle u \,|\, E_\lambda u \rangle$$

valid for all $u \in D_A$

The proof is difficult and can be found in [7]. Its main use is that it permits us to define functions $f(A)$ of a self-adjoint operator A for a very wide class of functions. For example if $f : \mathbb{R} \to \mathbb{C}$ is a Lebesgue integrable function then we set

$$f(A) = \int_{-\infty}^{\infty} f(\lambda) \, dE_\lambda.$$

This is shorthand for

$$\langle u \,|\, f(A)v \rangle = \int_{-\infty}^{\infty} f(\lambda) \, d\langle u \,|\, E_\lambda v \rangle$$

for arbitrary vectors $u \in \mathcal{H}$, $v \in D_A$. One of the most useful of such functions is $f = e^{ix}$, giving rise to a unitary transformation

$$U = e^{iA} = \int_{-\infty}^{\infty} e^{i\lambda} \, dE_\lambda.$$

This relation between unitary and self-adjoint operators has its main expression in Stone's theorem, which generalizes the result for finite dimensional vector spaces, discussed in Example 6.12 and Problem 6.12.

Theorem 13.26 *Every one-parameter unitary group of transformations U_t on a Hilbert space, such that $U_t U_s = U_{t+s}$, can be expressed in the form*

$$U_t = e^{iAt} = \int_{-\infty}^{\infty} e^{i\lambda t} \, dE_\lambda.$$

Problems

Problem 13.29 For unbounded operators, show that

(a) $(AB)C = A(BC)$.
(b) $(A+B)C = AC + BC$.
(c) $AB + AC \subseteq A(B+C)$. Give an example where $A(B+C) \neq AB + AC$.

Problem 13.30 Show that a densely defined bounded operator A in $\mathcal{H}$ has a unique extension to an operator $\hat{A}$ defined on all of $\mathcal{H}$. Show that $\|\hat{A}\| = \|A\|$.

Problem 13.31 If A is self-adjoint and B a bounded operator, show that B^*AB is self-adjoint.

Problem 13.32 Show that if (A, D_A) and (B, D_B) are operators on dense domains in $\mathcal{H}$ then $B^*A^* \subseteq (AB)^*$.

Problem 13.33 For unbounded operators, show that $A^* + B^* \subseteq (A+B)^*$.

Problem 13.34 If (A, D_A) is a densely defined operator and D_{A^*} is dense in $\mathcal{H}$, show that $A \subseteq A^{**}$.

Problem 13.35 If A is a symmetric operator, show that A^* is symmetric if and only if it is self-adjoint, $A^* = A^{**}$.

Problem 13.36 If $A_1, A_2, \ldots, A_n$ are operators on a dense domain such that

$$\sum_{i=1}^{n} A_i^* A_i = 0,$$

show that $A_1 = A_2 = \cdots = A_n = 0$.

Problem 13.37 If A is a self-adjoint operator show that

$$\|(A+iI)u\|^2 = \|Au\|^2 + \|u\|^2$$

and that the operator $A + iI$ is invertible. Show that the operator $U = (A - iI)(A + iI)^{-1}$ is unitary (called the *Cayley transform* of A).

References

[1] N. Boccara. *Functional Analysis*. San Diego, Academic Press, 1990.
[2] L. Debnath and P. Mikusiński. *Introduction to Hilbert Spaces with Applications*. San Diego, Academic Press, 1990.
[3] R. Geroch. *Mathematical Physics*. Chicago, The University of Chicago Press, 1985.
[4] P. R. Halmos. *Introduction to Hilbert Space*. New York, Chelsea Publishing Company, 1951.

References

[5] J. M. Jauch. *Foundations of Quantum Mechanics*. Reading, Mass., Addison-Wesley, 1968.

[6] J. von Neumann. *Mathematical Foundations of Quantum Mechanics*. Princeton, N. J., Princeton University Press, 1955.

[7] N. I. Akhiezer and I. M. Glazman. *Theory of Linear Operators in Hilbert Space*. New York, F. Ungar Publishing Company, 1961.

[8] F. Riesz and B. Sz.-Nagy. *Functional Analysis*. New York, F. Ungar Publishing Company, 1955.

[9] E. Zeidler. *Applied Functional Analysis*. New York, Springer-Verlag, 1995.

[10] R. D. Richtmyer. *Principles of Advanced Mathematical Physics, Vol. 1*. New York, Springer-Verlag, 1978.

[11] M. Reed and B. Simon. *Methods of Modern Mathematical Physics, Vol. I: Functional Analysis*. New York, Academic Press, 1972.

14 Quantum mechanics

Our purpose in this chapter is to present the key concepts of quantum mechanics in the language of Hilbert spaces. The reader who has not previously met the physical ideas motivating quantum mechanics, and some of the more elementary applications of Schrödinger's equation, is encouraged to read any of a number of excellent texts on the subject such as [1–4]. Otherwise, the statements given here must to a large extent be taken on trust – not an altogether easy thing to do, since the basic assertions of quantum theory are frequently counterintuitive to anyone steeped in the classical view of physics. Quantum mechanics is frequently presented in the form of several postulates, as though it were an axiomatic system such as Euclidean geometry. As often presented, these postulates may not meet the standards of mathematical rigour required for a strictly logical set of axioms, so that little is gained by such an approach. We will do things a little more informally here. For those only interested in the mathematical aspects of quantum mechanics and the role of Hilbert space see [5–8].

Many of the standard applications, such as the hydrogen atom, will be omitted here as they can be found in all standard textbooks, and we leave aside the enormous topic of measurement theory and interpretations of quantum mechanics. This is not to say that we need be totally comfortable with quantum theory as it stands. Undoubtedly, there are some philosophically disquieting features in the theory, often expressed in the form of so-called paradoxes. However, to attempt an 'interpretation' of the theory in order to resolve these apparent paradoxes assumes that there are natural metaphysical concepts. Suitable introductions to this topic can be found in [3, chap. 11] or [4, chap. 5].

14.1 Basic concepts

Photon polarization experiments

To understand how quantum mechanics works, we look at the outcome of a number of *Gedanken experiments* involving polarized light beams. Typically, a monochromatic plane wave solution or Maxwell equations (see Problem 9.23) has electric field

$$\mathbf{E} \propto \mathrm{Re}\,(\alpha \mathbf{e}_x + \beta \mathbf{e}_y) e^{i(kz - \omega t)}$$

where $\mathbf{e}_x$ and $\mathbf{e}_y$ are the unit vectors in the x- and y-directions and α and β are complex numbers such that

$$|\alpha|^2 + |\beta|^2 = 1.$$

14.1 Basic concepts

When α/β is real we have a linearly polarized wave, as for example

$$\mathbf{E} \propto \frac{1}{\sqrt{2}}(\mathbf{e}_x + \mathbf{e}_y)e^{i(kz-\omega t)}.$$

If $b = \pm ia = \pm i/\sqrt{2}$ the wave is circularly polarized; the $+$ sign is said to be right circularly polarized, and the $-$ sign is left circularly polarized. In all other cases it is said to be elliptically polarized. If we pass a polarized beam through a polarizer with axis of polarization $\mathbf{e}_x$, then the beam is reduced in intensity by the factor $|a|^2$ and the emergent beam is $\mathbf{e}_x$-polarized. Thus, if the resultant beam is passed through another $\mathbf{e}_x$-polarizer it will be 100% transmitted, while if it is passed through an $\mathbf{e}_y$-polarizer it will be totally absorbed and nothing will come through. This is the classical situation.

As was discovered by Planck and Einstein at the turn of the twentieth century, light beams come in discrete packets called *photons*, having energy $E = h\nu = \hbar\omega$ where $h \approx 6.625 \times 10^{-27}$ g cm^2 s^{-1} is Planck's constant and $\hbar = h/2\pi$. What happens if we send the beams through the polarizers one photon at a time? Since the frequency of each photon is unchanged it emerges with the same energy, and since the intensity of the beam is related to the energy, it must mean that the number of photons is reduced. However, the most obvious conclusion that the beam consists of a mixture of photons consisting of a fraction $|\alpha|^2$ polarized in the $\mathbf{e}_x$-direction and $|\beta|^2$ in the $\mathbf{e}_y$-direction will not stand up to scrutiny. For, if a beam with $\alpha = \beta = 1/\sqrt{2}$ were passed through a polarizer designed to only transmit waves linearly polarized in the $(\mathbf{e}_x + \mathbf{e}_y)/\sqrt{2}$ direction, then it should be 100% transmitted. However, on the mixture hypothesis only half the $\mathbf{e}_x$-polarized photons should get through, and half the $\mathbf{e}_y$-polarized photons, leaving a total fraction $\frac{1}{2} \cdot \frac{1}{2} + \frac{1}{2} \cdot \frac{1}{2} = \frac{1}{2}$ being transmitted.

In quantum mechanics it is proposed that each photon is a 'complex superposition' $\alpha \mathbf{e}_x + \beta \mathbf{e}_y$ of the two polarization states $\mathbf{e}_x$ and $\mathbf{e}_y$. The probability of transmission by an $\mathbf{e}_x$-polarizer is given by $|\alpha|^2$, while the probability of transmission by an $\mathbf{e}_y$-polarizer is $|\beta|^2$. The effect of the $\mathbf{e}_x$-polarizer is essentially to 'collapse' the photon into an $\mathbf{e}_x$-polarized state. The polarizer can be regarded both as a measuring device or equally as a device for preparing photons in an $\mathbf{e}_x$-polarized state. If used as a measuring device it returns the value 1 if the photon is transmitted, or 0 if not – in either case the act of measurement has changed the state of the photon being measured.

An interesting arrangement to illustrate the second point of view is shown in Fig. 14.1. Consider a beam of photons incident on an $\mathbf{e}_x$-polarizer, followed by an $\mathbf{e}_y$-polarizer; the net result is that no photons come out of the second polarizer. Now introduce a polarizer for the direction $(1/\sqrt{2})(\mathbf{e}_x + \mathbf{e}_y)$ – in other words a device that should block some of the photons between the two initial polarizers. If the mixture theory were correct, it is inconceivable that this could increase transmission. Yet the reality is that half the photons emerge from this intermediary polarizer with polarization $(1/\sqrt{2})(\mathbf{e}_x + \mathbf{e}_y)$, and a further half of these, namely a quarter in all, are now transmitted by the $\mathbf{e}_y$-polarizer.

How can we find the transmission probability of a polarized state $\mathbf{A}$ with respect to an arbitrary polarization direction $\mathbf{B}$? The following argument is designed to be motivational rather than rigorous. Let $\mathbf{A} = \alpha \mathbf{e}_x + \beta \mathbf{e}_y$, where α and β are complex numbers subject to $|\alpha|^2 + |\beta|^2 = 1$. We write $\alpha = \langle \mathbf{e}_x | \mathbf{A} \rangle$, called the *amplitude* for $\mathbf{e}_x$-transmission of an

Quantum mechanics

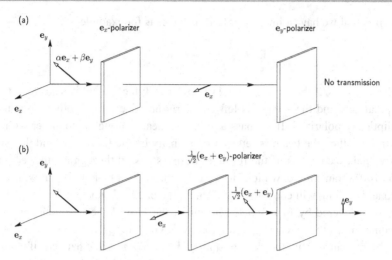

Figure 14.1 Photon polarization experiment

A-polarized photon. It is a complex number having no obvious physical interpretation of itself, but its magnitude square $|\alpha|^2$ is the probability of transmission by an $\mathbf{e}_x$-polarizer. Similarly $\beta = \langle \mathbf{e}_y | \mathbf{A} \rangle$ is the amplitude for $\mathbf{e}_y$-transmission, and $|\beta|^2$ the probability of transmission by an $\mathbf{e}_y$-polarizer. What is the polarization $A^\perp$ such that a polarizer of this type allows for no transmission, $\langle \mathbf{A}^\perp | \mathbf{A} \rangle = 0$? For linearly polarized waves with α and β both real we expect it to be geometrically orthogonal, $\mathbf{A}^\perp = \beta \mathbf{e}_x - \alpha \mathbf{e}_y$. For circularly polarized waves, the orthogonal 'direction' is the opposite circular sense. Hence

$$\frac{1}{\sqrt{2}}(\mathbf{e}_x \pm i\mathbf{e}_y)^\perp = \frac{1}{\sqrt{2}}(\mathbf{e}_x \mp i\mathbf{e}_y) \equiv \mp \frac{1}{\sqrt{2}}(i\mathbf{e}_x - \mathbf{e}_y)$$

since phase factors such as $\pm i$ are irrelevant. In the general elliptical case we might guess that $\mathbf{A}^\perp = \overline{\beta}\mathbf{e}_x - \overline{\alpha}\mathbf{e}_y$, since it reduces to the correct answer for linear and circular polarization. Solving for $\mathbf{e}_x$ and $\mathbf{e}_y$ we have

$$\mathbf{e}_x = \overline{\alpha}\mathbf{A} + \beta \mathbf{A}^\perp, \qquad \mathbf{e}_y = \overline{\beta}\mathbf{A} - \alpha \mathbf{A}^\perp.$$

Let $\mathbf{B} = \gamma \mathbf{e}_x + \delta \mathbf{e}_y$ be any other polarization, then substituting for $\mathbf{e}_x$ and $\mathbf{e}_y$ gives

$$\mathbf{B} = (\gamma \overline{\alpha} + \delta \overline{\beta})\mathbf{A} + (\gamma \beta - \alpha \delta)\mathbf{A}^\perp.$$

Setting $\mathbf{B} = \mathbf{A}$ gives the normalization condition $|\alpha|^2 + |\beta|^2 = 1$. Hence, since $\langle \mathbf{A} | \mathbf{A} \rangle = 1$ (transmission probability of 1),

$$\langle \mathbf{B} | \mathbf{A} \rangle = (\gamma \overline{\alpha} + \delta \overline{\beta}) = \overline{\langle \mathbf{A} | \mathbf{B} \rangle}.$$

Other systems such as the Stern–Gerlach experiment, in which an electron of magnetic moment μ is always deflected in a magnetic field $\mathbf{H}$ in just two directions, exhibit a completely analogous formalism. The conclusion is that the quantum mechanical states of a system form a complex vector space with inner product $\langle \phi | \psi \rangle$ satisfying the usual

14.1 Basic concepts

conditions

$$\langle\phi|\alpha\psi_1+\beta\psi_2\rangle = \alpha\langle\phi|\psi_1\rangle + \beta\langle\phi|\psi_2\rangle \quad \text{and} \quad \langle\phi|\psi\rangle = \overline{\langle\psi|\phi\rangle}.$$

The probability of obtaining a value corresponding to ϕ in a measurement is

$$P(\phi,\psi) = |\langle\phi|\psi\rangle|^2.$$

As will be seen, states are in fact normalized to $\langle\psi|\psi\rangle = 1$, so that only linear combinations $\alpha\psi_1 + \beta\psi_2$ with $|\alpha|^2 + |\beta|^2 = 1$ are permitted.

The Hilbert space of states

We will now assume that every physical system corresponds to a separable Hilbert space $\mathcal{H}$, representing all possible states of the system. The Hilbert space may be finite dimensional, as for example the states of polarization of a photon or electron, but often it is infinite dimensional. A **state** of the system is represented by a non-zero vector $\psi \in \mathcal{H}$, but this correspondence is not one-to-one, as any two vectors ψ and ψ' that are proportional through a non-zero complex factor, $\psi' = \lambda\psi$ where $\lambda \in \mathbb{C}$, will be assumed to represent identical states. In other words, a state is an equivalence class or *ray* of vectors $[\psi]$ all related by proportionality. A state may be represented by any vector from the class, and it is standard to select a representative having unit norm $\|\psi\| = 1$. Even this restriction does not uniquely define a vector to represent the state, as any other vector $\psi' = \lambda\psi$ with $|\lambda| = 1$ will also satisfy the unit norm condition. The angular freedom, $\lambda = e^{ic}$, is sometimes referred to as the *phase* of the state vector. Phase is only significant in a relative sense; for example, $\psi + e^{ic}\phi$ is in general a different state to $\psi + \phi$, but $e^{ic}(\psi + \phi)$ is not.

In this chapter we will adopt Dirac's bra-ket notation which, though slightly quirky, has largely become the convention of choice among physicists. Vectors $\psi \in \mathcal{H}$ are written as **kets** $|\psi\rangle$ and one makes the identification $|\lambda\psi\rangle = \lambda|\psi\rangle$. By the Riesz representation theorem 13.10, to each linear functional $f : \mathcal{H} \to \mathbb{C}$ there corresponds a unique vector $\phi \equiv |\phi\rangle \in \mathcal{H}$ such that $f(\psi) = \langle\phi|\psi\rangle$. In Dirac's terminology the linear functional is referred to as a **bra**, written $\langle\phi|$. The relation between bras and kets is antilinear,

$$\langle\lambda\psi + \phi| = \bar{\lambda}\langle\psi| + \langle\phi|.$$

In Dirac's notation it is common to think of a linear operator (A, D_A) as acting to the left on kets (vectors), while acting to the right on bras (linear functionals):

$$A|\psi\rangle \equiv |A\psi\rangle,$$

and if $\phi \in D_{A^*}$

$$\langle\phi|A \equiv \langle A^*\phi|.$$

The following notational usages for the matrix elements of an operator between two vectors are all equivalent:

$$\langle\phi|A|\psi\rangle \equiv \langle\phi|A\psi\rangle = \langle A^*\phi|\psi\rangle = \overline{\langle\psi|A^*\phi\rangle} = \overline{\langle\psi|A^*|\phi\rangle}.$$

If $|e_i\rangle$ is an o.n. basis of kets in a separable Hilbert space then we may write

$$A|e_j\rangle = \sum_i a_{ij}|e_i\rangle \quad \text{where} \quad a_{ij} = \langle e_i|A|e_j\rangle.$$

Observables

In classical mechanics, *physical observables* refer to quantities such as position, momentum, energy or angular momentum, which are real numbers or real multicomponented objects. In quantum mechanics **observables** are represented by self-adjoint operators on the Hilbert space of states. We first consider the case where A is a hermitian operator (bounded and continuous). Such an observable is said to be **complete** if the corresponding hermitian operator A is complete, so that there is an orthonormal basis made up of eigenvectors $|\psi_1\rangle, |\psi_2\rangle, \ldots$ such that

$$A|\psi_n\rangle = \alpha_n|\psi_n\rangle \quad \text{where} \quad \langle\psi_m|\psi_n\rangle = \delta_{mn}. \tag{14.1}$$

The result of measuring a complete observable is always one of the eigenvalues α_n, and the fact that these are real numbers provides a connection with classical physics. By Theorem 13.2 every state $|\psi\rangle$ can be written uniquely in the form

$$|\psi\rangle = \sum_{n=1}^{\infty} c_n|\psi_n\rangle \quad \text{where} \quad c_n = \langle\psi_n|\psi\rangle, \tag{14.2}$$

or, since the vector $|\psi\rangle$ is arbitrary, we can write

$$I \equiv \text{id}_{\mathcal{H}} = \sum_{n=1}^{\infty} |\psi_n\rangle\langle\psi_n|. \tag{14.3}$$

Exercise: Show that the operator A can be written in the form

$$A = \sum_{n=1}^{\infty} \alpha_n|\psi_n\rangle\langle\psi_n|.$$

The matrix element of the identity operator between two states $|\phi\rangle$ and $|\psi\rangle$ is

$$\langle\phi|\psi\rangle = \langle\phi|I|\psi\rangle = \sum_{n=1}^{\infty} \langle\phi|\psi_n\rangle\langle\psi_n|\psi\rangle.$$

Its physical interpretation is that $|\langle\phi|\psi\rangle|^2$ is the probability of realizing a state $|\psi\rangle$ when the system is in the state $|\phi\rangle$. Since both state vectors are unit vectors, the Cauchy–Schwarz inequality ensures that the probability is less than one,

$$|\langle\phi|\psi\rangle|^2 \le \|\phi\|^2 \|\psi\|^2 = 1.$$

If A is a complete hermitian operator with eigenstates $|\psi_n\rangle$ satisfying Eq. (14.1) then, according to this assumption, the probability that the eigenstate $|\psi_n\rangle$ is realized when the system is in the state $|\psi\rangle$ is given by $|c_n|^2 = |\langle\psi_n|\psi\rangle|^2$ where the c_n are the coefficients in the expansion (14.2). Thus $|c_n|^2$ is the probability that the value α_n be obtained on measuring

14.1 Basic concepts

the observable A when the system is in the state $|\psi\rangle$. By Parseval's identity (13.7) we have

$$\sum_{n=1}^{\infty} |c_n|^2 = \|\psi\|^2 = 1,$$

and the **expectation value** of the observable A in a given state $|\psi\rangle$ is given by

$$\langle A \rangle \equiv \langle A \rangle_\psi = \sum_{n=1}^{\infty} |c_n|^2 \alpha_n = \langle \psi | A | \psi \rangle. \tag{14.4}$$

The act of measuring the observable A 'collapses' the system into one of the eigenstates $|\psi_n\rangle$, with probability $|c_n|^2 = |\langle \psi_n | \psi \rangle|^2$. This feature of quantum mechanics, that the result of a measurement can only be known to within a probability, and that the system is no longer in the same state after a measurement as before, is one of the key differences between quantum and classical physics, where a measurement is always made delicately enough so as to minimally disturb the system. Quantum mechanics asserts that this is impossible, even in principle.

The **root mean square deviation** ΔA of an observable A in a state $|\psi\rangle$ is defined by

$$\Delta A = \sqrt{\langle (A - \langle A \rangle I)^2 \rangle}.$$

The quantity under the square root is positive, for

$$\langle (A - \langle A \rangle I)^2 \rangle = \langle \psi | (A - \langle A \rangle I)^2 \psi \rangle = \|(A - \langle A \rangle I)\psi\|^2 \geq 0$$

since A is hermitian and $\langle A \rangle$ is real. A useful formula for the RMS deviation is

$$\begin{aligned}(\Delta A)^2 &= \langle \psi | A^2 - 2A\langle A \rangle + \langle A \rangle^2 I | \psi \rangle \\ &= \langle A^2 \rangle - \langle A \rangle^2.\end{aligned} \tag{14.5}$$

Hence $|\psi\rangle$ is an eigenstate of A if and only if it is **dispersion-free**, $\Delta A = 0$. For, by (14.5), if $A|\psi\rangle = \alpha|\psi\rangle$ then $\langle A \rangle = \alpha$ and $\langle A^2 \rangle = \alpha^2$ immediately results in $\Delta A = 0$, and conversely if $\Delta A = 0$ then $\|(A - \langle A \rangle I)\psi\|^2$, which is only possible if $A|\psi\rangle = \langle A \rangle|\psi\rangle$. Dispersion-free states are sometimes referred to as **pure states** with respect to the observable A.

Theorem 14.1 (Heisenberg) *Let A and B be two hermitian operators, then for any state $|\psi\rangle$*

$$\Delta A \Delta B \geq \tfrac{1}{2} |\langle [A, B] \rangle| \tag{14.6}$$

*where $[A, B] = AB - BA$ is the **commutator** of the two operators.*

Proof: Let

$$|\psi_1\rangle = (A - \langle A \rangle I)|\psi\rangle, \qquad |\psi_2\rangle = (B - \langle B \rangle I)|\psi\rangle,$$

so that $\Delta A = \|\psi_1\|$ and $\Delta B = \|\psi_2\|$. Using the Cauchy–Schwarz inequality,

$$\begin{aligned}\Delta A \, \Delta B = \|\psi_1\| \, \|\psi_2\| &\geq |\langle \psi_1 | \psi_2 \rangle| \\ &\geq |\mathrm{Im}\, \langle \psi_1 | \psi_2 \rangle| \\ &= \left| \frac{1}{2i} (\langle \psi_1 | \psi_2 \rangle - \langle \psi_2 | \psi_1 \rangle) \right|.\end{aligned}$$

Now
$$\langle\psi_1|\psi_2\rangle = \langle(A - \langle A\rangle I)\psi \,|\, (B - \langle B\rangle I)\psi\rangle$$
$$= \langle\psi|AB|\psi\rangle - \langle A\rangle\langle B\rangle.$$

Hence
$$\Delta A \,\Delta B \geq \tfrac{1}{2}|\langle\psi|AB - BA|\psi\rangle| = \tfrac{1}{2}|\langle[A, B]\rangle|.$$
∎

Exercise: Show that for any two hermitian operators A and B, the operator $i[A, B]$ is hermitian.

Exercise: Show that $\langle[A, B]\rangle = 0$ for any state $|\psi\rangle$ that is an eigenvector of either A or B.

A particularly interesting case of Theorem 14.1 occurs when A and B satisfy the *canonical commutation relations*,

$$[A, B] = i\hbar I, \qquad (14.7)$$

where $\hbar = h/2\pi$ is Planck's constant divided by 2π. Such a pair of observables are said to be **complementary**. With some restriction on admissible domains they hold for the position operator $Q = A_x$ and the momentum operator $P = -i\hbar \mathrm{d}/\mathrm{d}x$ discussed in Examples 13.19 and 13.20. For, let f be a function in the intersection of their domains, then

$$[Q, P] = x\left(-i\hbar\frac{\mathrm{d}f}{\mathrm{d}x}\right) + i\hbar\frac{\mathrm{d}(xf)}{\mathrm{d}x} = i\hbar f,$$

whence

$$[Q, P] = i\hbar I. \qquad (14.8)$$

Theorem 14.1 results in the classic **Heisenberg uncertainty relation**

$$\Delta Q \,\Delta P \geq \frac{\hbar}{2}.$$

Sometimes it is claimed that this relation has no effect at a macroscopic level because Planck's constant h is so 'small' ($h \approx 6.625 \times 10^{-27} \mathrm{g\ cm^2\ s^{-1}}$). Little could be further from the truth. The fact that we are supported by a solid Earth, and not collapse in towards its centre, can be traced to this and similar relations.

Exercise: Show that Eq. (14.7) cannot possibly hold in a finite dimensional space. [*Hint*: Take the trace of both sides.]

Theorem 14.2 *A pair of complete hermitian observables A and B commute, $[A, B] = 0$, if and only if there exists a complete set of common eigenvectors. Such observables are said to be* **compatible**.

Proof: If there exists a basis of common eigenvectors $|\psi_1\rangle, |\psi_2\rangle, \ldots$ such that

$$A|\psi_n\rangle = \alpha_n|\psi_n\rangle, \quad B|\psi_n\rangle = \beta_n|\psi_n\rangle,$$

14.1 Basic concepts

then $AB|\psi_n\rangle = \alpha_n\beta_n|\psi_n\rangle = BA|\psi_n\rangle$ for each n. Hence for arbitrary vectors ψ we have from (14.2)

$$[A, B]|\psi\rangle = (AB - BA) \sum_n |\psi_n\rangle\langle\psi_n|\psi\rangle = 0.$$

Conversely, suppose that A and B commute. Let α be an eigenvalue of A with eigenspace $M_\alpha = \{|\psi\rangle \mid A|\psi\rangle = \alpha|\psi\rangle\}$, and set P_α to be the projection operator into this subspace. If $|\psi\rangle \in M_\alpha$ then $B|\psi\rangle \in M_\alpha$, since

$$AB|\psi\rangle = BA|\psi\rangle = B\alpha|\psi\rangle = \alpha B|\psi\rangle.$$

For any $|\phi\rangle \in \mathcal{H}$ we therefore have $BP_\alpha|\phi\rangle \in M_\alpha$. Hence

$$P_\alpha B P_\alpha |\phi\rangle = B P_\alpha |\phi\rangle$$

and since $|\phi\rangle$ is an arbitrary vector,

$$P_\alpha B P_\alpha = B P_\alpha.$$

Taking the adjoint of this equation, and using $B^* = B$, $P_\alpha = P_\alpha^*$, gives

$$P_\alpha B P_\alpha = P_\alpha^* B^* P_\alpha^* = P_\alpha^* B^* = P_\alpha B$$

and it follows that $P_\alpha B = B P_\alpha$, the operator B commutes with all projection operators P_α.

If β is any eigenvalue of B with projection map P_β, then since P_α is a hermitian operator that commutes with B the above argument shows that it commutes with P_β,

$$P_\alpha P_\beta = P_\beta P_\alpha.$$

Hence, the operator $P_{\alpha\beta} = P_\alpha P_\beta$ is hermitian and idempotent, and using Theorem 13.14 it is a projection operator. The space it projects into is $M_{\alpha\beta} = M_\alpha \cap M_\beta$. Two such spaces $M_{\alpha\beta}$ and $M_{\alpha'\beta'}$ are clearly orthogonal unless $\alpha = \alpha'$ and $\beta = \beta'$. Choose an orthonormal basis for each $M_{\alpha\beta}$. The collection of these vectors is a complete o.n. set consisting entirely of common eigenvectors to A and B. For, if $|\phi\rangle \neq 0$ is any non-zero vector orthogonal to all $M_{\alpha\beta}$, then $P_\alpha P_\beta |\phi\rangle \neq 0$ for all α, β. Since A is complete this implies $P_\beta|\phi\rangle = 0$ for all eigenvalues β of B, and since B is complete we must have $|\phi\rangle = 0$. ∎

Example 14.1 Consider spin $\frac{1}{2}$ electrons in a Stern–Gerlach device for measuring spin in the z-direction. Let σ_z be the operator for the observable 'spin in the z-direction'. It can only take on two values – up or down. This results in two eigenvalues ± 1, and the eigenvectors are written

$$\sigma_z|+z\rangle = |+z\rangle, \qquad \sigma_z|-z\rangle = -|-z\rangle.$$

Thus

$$\sigma_z = |+z\rangle\langle+z| - |-z\rangle\langle-z|, \qquad I = |+z\rangle\langle+z| + |-z\rangle\langle-z|,$$

and setting $|e_1\rangle = |+z\rangle$, $|e_2\rangle = |-z\rangle$ results in the matrix components

$$(\sigma_z)_{ij} = \langle e_i|\sigma_z|e_j\rangle = \begin{pmatrix} 1 & 0 \\ 0 & -1 \end{pmatrix}. \tag{14.9}$$

Quantum mechanics

Every state of the system can be written

$$|\psi\rangle = \psi_1|+z\rangle + \psi_2|-z\rangle \quad \text{where} \quad \psi_i = \langle e_i|\psi\rangle.$$

The operator $\sigma_\mathbf{n}$ representing spin in an arbitrary direction

$$\mathbf{n} = \cos\theta\, \mathbf{e}_z + \sin\theta\cos\phi\, \mathbf{e}_x + \sin\theta\sin\phi\, \mathbf{e}_y$$

has expectation values in different directions given by the classical values

$$\langle +z|\sigma_\mathbf{n}|+z\rangle = \cos\theta,$$
$$\langle +x|\sigma_\mathbf{n}|+x\rangle = \sin\theta\cos\phi,$$
$$\langle +y|\sigma_\mathbf{n}|+y\rangle = \sin\theta\sin\phi,$$

where $|\pm x\rangle$ refers to the pure states in the x direction, $\sigma_x|\pm x\rangle = \pm|\pm x\rangle$, etc.

Since $\sigma_\mathbf{n}$ is hermitian with eigenvalues $\lambda_i = \pm 1$ its matrix with respect to any o.n. basis has the form

$$(\sigma_\mathbf{n})_{ij} = \begin{pmatrix} \alpha & \beta \\ \bar{\beta} & \delta \end{pmatrix}$$

where

$$\alpha + \delta = \lambda_1 + \lambda_2 = 0, \qquad \alpha\delta - \beta\bar{\beta} = \lambda_1\lambda_2 = -1.$$

Hence $\delta = -\alpha$ and $\alpha^2 = 1 - |\beta|^2$. The expectation value of $\sigma_\mathbf{n}$ in the $|+z\rangle$ state is given by

$$\langle +z|\sigma_\mathbf{n}|+z\rangle = (\sigma_\mathbf{n})_{11} = \alpha = \cos\theta$$

so that $\beta = \sin\theta\, e^{-ic}$ where c is a real number. For $\mathbf{n} = \mathbf{e}_x$ and $\mathbf{n} = \mathbf{e}_y$ we have $\cos\theta = 0$,

$$\sigma_x = \begin{pmatrix} 0 & e^{-ia} \\ e^{ia} & 0 \end{pmatrix}, \qquad \sigma_y = \begin{pmatrix} 0 & e^{-ib} \\ e^{ib} & 0 \end{pmatrix}. \tag{14.10}$$

The states $|\pm x\rangle$ and $|\pm y\rangle$ are the eigenstates of σ_x and σ_y with normalized components

$$|\pm x\rangle = \frac{1}{\sqrt{2}} \begin{pmatrix} e^{-ia} \\ \pm 1 \end{pmatrix}, \qquad |\pm y\rangle = \frac{1}{\sqrt{2}} \begin{pmatrix} e^{-ib} \\ \pm 1 \end{pmatrix}$$

and as the expection values of σ_x in the orthogonal states $|\pm y\rangle$ vanish,

$$\langle +y|\sigma_x|+y\rangle = \tfrac{1}{2}\left(e^{i(b-a)} + e^{i(a-b)}\right) = \cos(b-a) = 0.$$

Hence $b = a + \pi/2$. Applying the unitary operator U

$$U = \begin{pmatrix} e^{ia} & 0 \\ 0 & 1 \end{pmatrix}$$

results in $a = 0$, and the spin operators are given by the *Pauli representation*

$$\sigma_x = \sigma_1 = \begin{pmatrix} 0 & 1 \\ 1 & 0 \end{pmatrix}, \quad \sigma_y = \sigma_2 = \begin{pmatrix} 0 & -i \\ i & 0 \end{pmatrix}, \quad \sigma_z = \sigma_3 = \begin{pmatrix} 1 & 0 \\ 0 & -1 \end{pmatrix}. \tag{14.11}$$

14.1 Basic concepts

For a spin operator in an arbitrary direction the expectation values are $\langle +x|\sigma_n|+x\rangle = \sin\theta\cos\phi$, etc., from which it is straightforward to verify that

$$\sigma_n = \begin{pmatrix} \cos\theta & \sin\theta e^{-i\phi} \\ \sin\theta e^{i\phi} & -\cos\theta \end{pmatrix} = \sin\theta\cos\phi\,\sigma_x + \sin\theta\sin\phi\,\sigma_y + \cos\theta\,\sigma_z.$$

Exercise: Find the eigenstates $|+n\rangle$ and $|-n\rangle$ of σ_n.

Unbounded operators in quantum mechanics

An important part of the framework of quantum mechanics is the *correspondence principle*, which asserts that to every classical dynamical variable there corresponds a quantum mechanical observable. This is at best a sort of guide – for example, as there is no natural way of defining general functions $f(Q, P)$ for a pair of non-commuting operators such as Q and P, it is not clear what operators correspond to generalized position and momentum in classical canonical coordinates. For rectangular cartesian coordinates x, y, z and momenta $p_x = m\dot{x}$, etc. experience has taught that the Hilbert space of states corresponding to a one-dimensional dynamical system is $\mathcal{H} = L^2(\mathbb{R})$, and the position and momentum operators are given by

$$Q\psi(x) = x\psi(x) \quad \text{and} \quad P\psi(x) = -i\hbar\frac{d\psi}{dx}.$$

These operators are unbounded operators and have been discussed in Examples 13.17, 13.19 and 13.20 of Chapter 13.

As these operators are not defined on all of $\mathcal{H}$ it is most common to take domains

$$D_Q = \left\{\psi(x) \in L^2(\mathbb{R}) \,\Big|\, \int_{-\infty}^{\infty} x^2|\psi(x)|^2\,dx < \infty\right\},$$

$$D_P = \left\{\psi(x) \in L^2(\mathbb{R}) \,\Big|\, \psi(x) \text{ is differentiable and } \int_{-\infty}^{\infty}\left\|\frac{d\psi(x)}{dx}\right\|^2 dx < \infty\right\}.$$

These domains are dense in $L^2(\mathbb{R})$ since the basis of functions $\phi_n(x)$ constructed from hermite polynomials in Example 13.7 (see Eq. (13.6)) belong to both of them. As shown in Example 13.19 the operator (Q, D_Q) is self-adjoint, but (P, D_P) is a symmetric operator that is not self-adjoint (see Example 13.20). To make it self-adjoint it must be extended to the domain of absolutely continuous functions.

Example 14.2 The position operator Q has no eigenvalues and eigenfunctions in $L^2(\mathbb{R})$ (see Example 13.14). For the momentum operator P the eigenvalue equation reads

$$\frac{d\psi}{dx} = i\lambda\psi(x) \implies \psi(x) = e^{i\lambda x},$$

and even when λ is a real number the function $\psi(x)$ does not belong to D_P,

$$\int_{-\infty}^{\infty}|e^{i\lambda x}|^2\,dx = \int_{-\infty}^{\infty} 1\,dx = \infty.$$

Quantum mechanics

For each real number k set $\varepsilon_k(x) = e^{ikx}$, and

$$\langle \varepsilon_k | \psi \rangle = \int_{-\infty}^{\infty} e^{-ikx} \psi(x) \, dx$$

is a linear functional on $L^2(\mathbb{R})$ – in fact, it is the Fourier transform of Section 12.3. This linear functional can be thought of as a tempered distribution $\langle \varepsilon_k |$ on the space of test functions of rapid decrease D_P. It is a bra that corresponds to no ket vector $|\varepsilon_k\rangle$ (this does not violate the Riesz representation theorem 13.10 since the domain D_P is not a closed subspace of $L^2(\mathbb{R})$). In quantum theory it is common to write equations that may be interpreted as

$$\langle \varepsilon_k | P = k \langle \varepsilon_k |,$$

which hold in the distributional sense,

$$\langle \varepsilon_k | P | \psi \rangle = k \langle \varepsilon_k | \psi \rangle \qquad \text{for all } \psi \in D_P.$$

Its integral version holds if we permit integration by parts, as for distributions,

$$\int_{-\infty}^{\infty} e^{-ikx} \left(-i \frac{d\psi(x)}{dx} \right) dx = \int_{-\infty}^{\infty} i \frac{d e^{-ikx}}{dx} \psi(x) \, dx = k \int_{-\infty}^{\infty} e^{-ikx} \psi(x) \, dx.$$

Similarly, for each real a define the linear functional $\langle \delta_a |$ by

$$\langle \delta_a | \psi \rangle = \psi(a)$$

for all kets $|\psi\rangle \equiv \psi(x) \in D_Q$. These too can be thought of as distributions on a set of test functions of rapid decrease. They behave as 'eigenbras' of the position operator Q,

$$\langle \delta_a | Q = a \langle \delta_a |$$

since

$$\langle \delta_a | Q | \psi \rangle = \langle \delta_a | x \psi(x) \rangle$$
$$= a \psi(a) = a \langle \delta_a | \psi \rangle$$

for all $|\psi\rangle \in D_Q$. While there is no function $\delta_a(x)$ in $L^2(\mathbb{R})$ having $\langle \delta_a |$, the Dirac delta function $\delta_a(x) = \delta(x-a)$ can be thought of as fulfilling this role in a distributional sense (see Chapter 12).

For a self-adjoint operator A we may apply Theorem 13.25. Let E_λ be the spectral family of increasing projection operators defined by A, such that

$$A = \int_{-\infty}^{\infty} \lambda \, dE_\lambda \quad \text{and} \quad I = \int_{-\infty}^{\infty} dE_\lambda.$$

The latter relation follows from

$$1 = \langle \psi | \psi \rangle = \int_{-\infty}^{\infty} d\langle \psi | E_\lambda \psi \rangle \qquad (14.12)$$

for all $\psi \in D_A$.

Exercise: Prove Eq. (14.12).

14.1 Basic concepts

If S is any measurable subset of $\mathbb{R}$ then the probability of the measured value of A lying in S, when the system is in a state $|\psi\rangle$, is given by

$$P_S(A) = \int_S d\langle\psi | E_\lambda \psi\rangle.$$

The expectation value and RMS deviation are given by

$$\langle A\rangle = \int_{-\infty}^{\infty} \lambda\, d\langle\psi | E_\lambda \psi\rangle$$

and

$$(\Delta A)^2 = \int_{-\infty}^{\infty} (\lambda - \langle A\rangle)^2 d\langle\psi | E_\lambda \psi\rangle.$$

Example 14.3 The spectral family for the position operator is defined as the 'cut-off' operators

$$(E_\lambda \psi)(x) = \begin{cases} \psi(x) & \text{if } x \leq \lambda, \\ 0 & \text{if } x > \lambda. \end{cases}$$

Firstly, these operators are projection operators since they are idempotent ($E_\lambda^2 = E_\lambda$) and hermitian:

$$\langle\phi | E_\lambda \psi\rangle = \int_{-\infty}^{\lambda} \overline{\phi(x)}\psi(x)\, dx = \langle E_\lambda \phi | \psi\rangle$$

for all $\phi, \psi \in L^2(\mathbb{R})$. They are an increasing family since the image spaces are clearly increasing, and $\mathbb{E}_{-\infty} = O$, $E_\infty = I$. The function $\lambda \mapsto \langle\phi | E_\lambda \psi\rangle$ is absolutely continuous, since

$$\langle\phi | E_\lambda \psi\rangle = \int_{-\infty}^{\lambda} \overline{\phi(x)}\psi(x)\, dx$$

and has generalized derivative $'$ with respect to λ given by

$$\langle\phi | E_\lambda \psi\rangle' = \overline{\phi(\lambda)}\psi(\lambda).$$

Hence

$$\langle\phi | Q\psi\rangle = \int_{-\infty}^{\infty} \overline{\phi(\lambda)}\lambda\psi(\lambda)\, d\lambda = \int_{-\infty}^{\infty} \lambda\, d\langle\phi | E_\lambda \psi\rangle,$$

which is equivalent to the required spectral decomposition

$$Q = \int_{-\infty}^{\infty} \lambda\, dE_\lambda.$$

Exercise: Show that for any $-\infty \leq a < b \leq \infty$ for the spectral family of the previous example

$$\int_a^b d\langle\psi | E_\lambda \psi\rangle = \int_a^b |\psi(x)|^2 dx.$$

Quantum mechanics

Problems

Problem 14.1 Verify for each direction

$$\mathbf{n} = \sin\theta\cos\phi\,\mathbf{e}_x + \sin\theta\sin\phi\,\mathbf{e}_y + \cos\theta\,\mathbf{e}_z$$

the spin operator

$$\sigma_\mathbf{n} = \begin{pmatrix} \cos\theta & \sin\theta\,e^{-i\phi} \\ \sin\theta\,e^{i\phi} & -\cos\theta \end{pmatrix}$$

has eigenvalues ± 1. Show that up to phase, the eigenvectors can be expressed as

$$|+\mathbf{n}\rangle = \begin{pmatrix} \cos\tfrac{1}{2}\theta\, e^{-i\phi} \\ \sin\tfrac{1}{2}\theta \end{pmatrix}, \quad |-\mathbf{n}\rangle = \begin{pmatrix} -\sin\tfrac{1}{2}\theta\, e^{-i\phi} \\ \cos\tfrac{1}{2}\theta \end{pmatrix}$$

and compute the expectation values for spin in the direction of the various axes

$$\langle\sigma_i\rangle_{\pm\mathbf{n}} = \langle\pm\mathbf{n}|\sigma_i|\pm\mathbf{n}\rangle.$$

For a beam of particles in a pure state $|+\mathbf{n}\rangle$ show that after a measurement of spin in the $+x$ direction the probability that the spin is in this direction is $\tfrac{1}{2}(1+\sin\theta\cos\phi)$.

Problem 14.2 If **A** and **B** are vector observables that commute with the Pauli spin matrices, $[\sigma_i, A_j] = [\sigma_i, B_j] = 0$ (but $[A_i, B_j] \neq 0$ in general) show that

$$(\sigma\cdot\mathbf{A})(\sigma\cdot\mathbf{B}) = \mathbf{A}\cdot\mathbf{B} + i(\mathbf{A}\times\mathbf{B})\cdot\sigma$$

where $\sigma = (\sigma_1, \sigma_2, \sigma_3)$.

Problem 14.3 Prove the following commutator identities:

$$[A, [B, C]] + [B, [C, A]] + [C, [A, B]] = 0 \quad \text{(Jacobi identity)}$$
$$[AB, C] = A[B, C] + [A, C]B$$
$$[A, BC] = [A, B]C + B[A, C]$$

Problem 14.4 Using the identities of Problem 14.3 show the following identities:

$$[Q^n, P] = ni\hbar Q^{n-1},$$
$$[Q, P^m] = mi\hbar P^{m-1},$$
$$[Q^n, P^2] = 2ni\hbar Q^{n-1}P + n(n-1)\hbar^2 Q^{n-2},$$
$$[L_m, Q_k] = i\hbar\epsilon_{mkj} Q_j, \quad [L_m, P_k] = i\hbar\epsilon_{mkj} P_j,$$

where $L_m = \epsilon_{mij} Q_i P_j$ are the **angular momentum operators**.

Problem 14.5 Consider a one-dimensional wave packet

$$\psi(x,t) = \frac{1}{\sqrt{2\pi\hbar}} \int_{-\infty}^{\infty} e^{i(xp - p^2 t/2m)/\hbar} \Psi(p)\, dp$$

where

$$\Psi(p) \propto e^{-(p-p_0)^2/2(\Delta p)^2}.$$

Show that $|\psi(x,t)|^2$ is a Gaussian normal distribution whose peak moves with velocity p/m and whose spread Δx increases with time, always satisfying $\Delta x\, \Delta p \geq \hbar/\sqrt{2}$.

If an electron ($m = 9 \times 10^{-28}$ g) is initially within an atomic radius $\Delta x_0 = 10^{-8}$ cm, after how long will Δx be (a) 2×10^{-8} cm, (b) the size of the solar system (about 10^{14} cm)?

14.2 Quantum dynamics

The discussion of Section 14.1 refers only to *quantum statics* – the essential framework in which quantum descriptions are to be set. The dynamical evolution of quantum systems is determined by a hermitian operator H, possibly but not usually a function of time $H = H(t)$, such that the time development of any state $|\psi(t)\rangle$ of the system is given by **Schrödinger's equation**

$$i\hbar \frac{d}{dt}|\psi\rangle = H|\psi\rangle. \tag{14.13}$$

The operator H is known as the **Hamiltonian** or **energy operator**. Equation (14.13) guarantees that all inner products are preserved for, taking the adjoint gives

$$-i\hbar \frac{d}{dt}\langle\psi| = \langle\psi|H^* = \langle\psi|H$$

and for any pair of states $|\psi\rangle$ and $|\phi\rangle$,

$$i\hbar \frac{d}{dt}\langle\psi|\phi\rangle = i\hbar\left(\left(\frac{d}{dt}\langle\psi|\right)|\phi\rangle + \langle\psi|\left(\frac{d}{dt}|\phi\rangle\right)\right)$$
$$= -\langle\psi|H|\phi\rangle + \langle\psi|H|\phi\rangle = 0.$$

In particular the normalization $\|\psi(t)\| = \|\phi(t)\| = 1$ is preserved by Schrödinger's equation. Since

$$\langle\psi(t)|\phi(t)\rangle = \langle\psi(0)|\phi(0)\rangle$$

for all pairs of states, there exists a unitary operator $U(t)$ such that

$$|\psi(t)\rangle = U(t)|\psi(0)\rangle. \tag{14.14}$$

If H is independent of t then

$$U(t) = e^{(-i/\hbar)Ht} \tag{14.15}$$

where the exponential function can be defined as in the comments prior to Theorem 13.26 at the end of Chapter 13. If H is a complete hermitian operator $H = \sum_n \lambda_n |\psi_n\rangle\langle\psi_n|$ then

$$e^{(-i/\hbar)Ht} = \sum_n e^{(-i/\hbar)\lambda t}|\psi_n\rangle\langle\psi_n|$$

and for a self-adjoint operator

$$H = \int_{-\infty}^{\infty} \lambda \, dE_\lambda \implies e^{(-i/\hbar)Ht} = \int_{-\infty}^{\infty} e^{(-i/\hbar)\lambda t} dE_\lambda.$$

To prove Eq. (14.15) substitute (14.14) in Schrödinger's equation

$$i\hbar \frac{dU}{dt}|\psi(0)\rangle = HU|\psi(0)\rangle,$$

Quantum mechanics

and since $|\psi(0)\rangle$ is an arbitrary initial state vector,

$$i\hbar \frac{dU}{dt} = HU.$$

Setting $U(t) = e^{(-i/\hbar)Ht} V(t)$ (always possible since the operator $e^{(-i/\hbar)Ht}$ is invertible with inverse $e^{(i/\hbar)Ht}$) we obtain

$$e^{(-i/\hbar)Ht} HV(t) + i\hbar e^{(-i/\hbar)Ht} \frac{dV(t)}{dt} = H e^{(-i/\hbar)Ht} V(t).$$

As H and $e^{(-i/\hbar)Ht}$ commute it follows that $V(t) = \text{const.} = V(0) = I$ since $U(0) = I$ on setting $t = 0$ in Eq. (14.14).

The Heisenberg picture

The above description of the evolution of a quantum mechanical system is called the **Schrödinger picture**. There is an equivalent version called the **Heisenberg picture** in which the states are treated as constant, but observables undergo a dynamic evolution. The idea is to perform a unitary transformation on $\mathcal{H}$, simultaneously on states and operators:

$$|\psi\rangle \mapsto |\psi'\rangle = U^*|\psi\rangle \qquad (|\psi\rangle = U|\psi'\rangle)$$
$$A \mapsto A' = U^* A U,$$

where U is given by (14.15). This transformation has the effect of bringing every solution of Schrödinger's equation to rest, for if $|\psi(t)\rangle$ is a solution of Eq. (14.13) then

$$|\psi'\rangle = U^*|\psi(t)\rangle = U^*U|\psi(0)\rangle = |\psi(0)\rangle \implies \frac{d}{dt}|\psi'(t)\rangle = 0.$$

It preserves all matrix elements, and in particular all expectation values:

$$\langle A'\rangle_{\psi'} = \langle \psi'|A'|\psi'\rangle = \langle \psi|UU^*AUU^*|\psi\rangle = \langle \psi|A|\psi\rangle = \langle A\rangle_\psi.$$

Thus the states and observables are physically equivalent in the two pictures.

We derive a dynamical equation for the Heisenberg operator A':

$$\frac{d}{dt}A' = \frac{d}{dt}(U^* A U)$$
$$= \frac{dU^*}{dt} A U + U^* \frac{dA}{dt} U + U^* A \frac{dU}{dt}$$
$$= \frac{1}{-i\hbar} U^* H A U + U^* \frac{dA}{dt} U + \frac{1}{i\hbar} U^* A H U$$

since, by Eq. (14.15),

$$\frac{dU}{dt} = \frac{1}{i\hbar} HU, \qquad \frac{dU^*}{dt} = \frac{1}{-i\hbar} U^* H^* = \frac{1}{-i\hbar} U^* H.$$

Hence

$$\frac{d}{dt} A' = \frac{1}{i\hbar}[A', H'] + \frac{\partial A'}{\partial t} \qquad (14.16)$$

14.2 Quantum dynamics

where

$$\frac{\partial A'}{\partial t} = U^* \frac{\mathrm{d}A}{\mathrm{d}t} U = \left(\frac{\mathrm{d}A}{\mathrm{d}t}\right)'.$$

The motivation for this identification is the following: if $|\psi_i\rangle$ is a rest basis of $\mathcal{H}$ in the Schrödinger picture, so that $\mathrm{d}|\psi_i\rangle/\mathrm{d}t = 0$, and if $|\psi_i'\rangle = U^*|\psi_i\rangle$ is the 'moving basis' obtained from it, then

$$\langle \psi_i' | A' | \psi_j' \rangle = \langle \psi_i | A | \psi_j \rangle$$

and

$$\langle \psi_i' | \frac{\partial A'}{\partial t} | \psi_j' \rangle = \langle \psi_i | U \frac{\partial A'}{\partial t} U^* | \psi_j \rangle = \langle \psi_i | \frac{\mathrm{d}A}{\mathrm{d}t} | \psi_j \rangle = \frac{\mathrm{d}}{\mathrm{d}t} \langle \psi_i | A | \psi_j \rangle.$$

Thus the matrix elements of $\partial A'/\partial t$ measure the *explicit* time rate of change of the matrix elements of the operator A in the Schrödinger representation.

If A is an operator having no explicit time dependence, so that $\partial A'/\partial t = 0$, then A' is a constant of the motion if and only if it commutes with the Hamiltonian, $[A', H'] = 0$,

$$\frac{\mathrm{d}A'}{\mathrm{d}t} = 0 \iff [A' H'] = [A, H] = 0.$$

In particular, since every operator commutes with itself, the Hamiltonian H' is a constant of the motion if and only if it is time independent, $\partial H'/\partial t = 0$.

Example 14.4 For an electron of charge e, mass m and spin $\frac{1}{2}$, notation as in Example 14.1, the Hamiltonian in a magnetic field **B** is given by

$$H = \frac{-e\hbar}{2mc} \sigma \cdot \mathbf{B}.$$

If **B** is parallel to the z-axis then $H = -(e\hbar/2mc)\sigma_z B$ and setting $|\psi\rangle = \psi_1(t)|+z\rangle + \psi_2(t)|-z\rangle$, Schrödinger's equation (14.13) can be written as the two differential equations

$$i\hbar \dot{\psi}_1 = -\frac{e\hbar}{2mc} B \psi_1, \qquad i\hbar \dot{\psi}_2 = \frac{e\hbar}{2mc} B \psi_2$$

with solutions

$$\psi_1(t) = \psi_{10} e^{i(\omega/2)t}, \qquad \psi_2(t) = \psi_{20} e^{-i(\omega/2)t},$$

where $\omega = eB/mc$. Substituting in the expectation values

$$\langle \psi(t) | \sigma | \psi(t) \rangle = \begin{pmatrix} \sin\theta \cos\phi(t) \\ \sin\theta \sin\phi(t) \\ \cos\theta(t) \end{pmatrix}$$

results in $\theta(t) = \theta_0 = \mathrm{const.}$ and

$$\cos\phi(t) = \frac{e^{i(\phi_0 - \omega t)} + e^{-i(\phi_0 - \omega t)}}{2} = \cos(\phi_0 - \omega t).$$

Hence $\phi(t) = \phi_0 - \omega t$, and the motion is a precession with angular velocity ω about the direction of the magnetic field.

Quantum mechanics

In the Heisenberg picture, set $\sigma_x = \sigma_x(t)$, etc., where $\sigma_x(0) = \sigma_1, \sigma_y(0) = \sigma_2, \sigma_z(0) = \sigma_3$ are the Pauli values, (14.11). From the commutation relations

$$[\sigma_1, \sigma_2] = 2i\sigma_3, \quad [\sigma_2, \sigma_3] = 2i\sigma_1, \quad [\sigma_3, \sigma_1] = 2i\sigma_2 \qquad (14.17)$$

and $\sigma_x = U^*\sigma_1 U$, etc., it follows that

$$[\sigma_x, \sigma_y] = 2i\sigma_z, \text{ etc.}$$

Heisenberg equations of motion are

$$\dot{\sigma}_x = \frac{1}{i\hbar}[\sigma_x, H] = \omega\sigma_y,$$

$$\dot{\sigma}_y = -\omega\sigma_x,$$

$$\dot{\sigma}_z = 0.$$

Hence $\ddot{\sigma}_x = \omega\dot{\sigma}_y = -\omega^2\sigma_x$ and the solution of Heisenberg's equation is

$$\sigma_x = Ae^{i\omega t} + Be^{-i\omega t}, \qquad \sigma_y = \frac{1}{\omega}\dot{\sigma}_x.$$

The 2×2 matrices **A**, **B** are evaluated by initial values at $t = 0$, resulting in

$$\sigma_x(t) = \cos\omega t\, \sigma_1 + \sin\omega t\, \sigma_2,$$

$$\sigma_y(t) = -\sin\omega t\, \sigma_1 + \cos\omega t\, \sigma_2$$

$$\sigma_z = \sigma_3 = \text{const.}$$

Correspondence with classical mechanics and wave mechanics

For readers familiar with Hamiltonian mechanics (see Section 16.5), the following correspondence can be set up between classical and quantum mechanics:

	Quantum mechanics	Classical mechanics
State space	Hilbert space $\mathcal{H}$	Phase space Γ
States	Normalized kets $\|\psi\rangle \in \mathcal{H}$	Points $(q_i, p_j) \in \Gamma$
Observables	Self-adjoint operators in $\mathcal{H}$; multiple values $\lambda_b i$ in each state $\|\psi\rangle$ with probability $P = \langle\psi_i\|\psi\rangle^2$	Real functions $f(q_i, p_j)$ on phase space; one value for each state
Commutators	Bracket commutators $[A, B]$	Poisson brackets (f, g)
Dynamics	1. Schrödinger picture $i\hbar\frac{d}{dt}\|\psi\rangle = H\|\psi\rangle$ 2. Heisenberg picture $\dot{A} = \frac{1}{i\hbar}[A, H] + \frac{\partial A}{\partial t}$	1. Hamilton's equations $\dot{q}_i = \frac{\partial H}{\partial p_i}, \dot{p}_i = -\frac{\partial H}{\partial q_i}$ 2. Poisson bracket form $\dot{f} = (f, H) + \frac{\partial f}{\partial t}$

If f and g are classical observables with quantum mechanical equivalents F and G then, from Heisenberg's equation of motion, the proposal is that the commutator $[F, G]$

14.2 Quantum dynamics

corresponds to the $i\hbar$ times the Poisson bracket,

$$[F, G] \longleftrightarrow i\hbar(f, g) = i\hbar\left(\frac{\partial f}{\partial q_i}\frac{\partial g}{\partial p_i} - \frac{\partial f}{\partial p_i}\frac{\partial g}{\partial q_i}\right).$$

For example if Q_i are position operators representing classical variable q_i and $P_i = -i\hbar\partial/\partial q_i$ the momentum operators, then the classical canonical commutation relations imply

$$(q_i, q_j) = 0 \implies [Q_i, Q_j] = 0,$$
$$(p_i, p_j) = 0 \implies [P_i, P_j] = 0,$$
$$(q_i, p_j) = \delta_{ij} \implies [Q_i, P_j] = i\hbar\delta_{ij}I.$$

Generalizing from the one-dimensional case, we assume $\mathcal{H}$ is the set of differentiable functions in $L^2(\mathbb{R}^n)$ such that $x_i\psi(x_1, \ldots, x_n)$ belongs to $L^2(\mathbb{R}^n)$ for each x_i. The above commutation relations are satisfied by the standard operators:

$$Q_i\psi(x_1, \ldots, x_n) = x_i\psi(x_1, \ldots, x_n), \qquad P_i\psi(x_1, \ldots, x_n) = -i\hbar\frac{\partial \psi}{\partial x_i}.$$

For a particle in a potential $V(x, y, z)$ the Hamiltonian is $H = \mathbf{p}^2/2m + V(x, y, z)$, which corresponds to the quantum mechanical Schrödinger equation

$$i\hbar\frac{\partial \psi}{\partial t} = -\frac{\hbar^2}{2m}\nabla^2\psi + V(\mathbf{r})\psi. \tag{14.18}$$

Exercise: Show that the probability density $P(\mathbf{r}, t) = \psi\overline{\psi}$ satisfies the conservation equation

$$\frac{\partial P}{\partial t} = -\nabla\mathbf{J} \quad \text{where} \quad \mathbf{J} = \frac{i\hbar}{2m}(\psi\nabla\overline{\psi} - \overline{\psi}\nabla\psi).$$

A trial solution of Eq. (14.18) by separation of variables, $\psi = T(t)\phi(\mathbf{x})$, results in

$$\psi = e^{-i\omega t}\phi(\mathbf{x})$$

where $\phi(\mathbf{x})$ satisfies the **time-independent Schrödinger equation**

$$H\phi(\mathbf{x}) = -\frac{\hbar^2}{2m}\nabla^2\phi(\mathbf{x}) + V(\mathbf{r})\phi(\mathbf{x}) = E\phi(\mathbf{x})$$

where E is given by Planck's relation, $E = \hbar\omega = h\nu$. From its classical analogue, the eigenvalue E of the Hamiltonian is interpreted as the energy of the system, and if the Hamiltonian is a complete operator with discrete spectrum E_n then the general solution of the Schrödinger equation is given by

$$\psi(\mathbf{x}, t) = \sum_n c_n\phi_n(\mathbf{x})e^{-iE_nt/\hbar}$$

where

$$H\phi_n(\mathbf{x}) = E_n\phi_n(\mathbf{x}).$$

Quantum mechanics

Harmonic oscillator

The classical one-dimensional harmonic oscillator has Hamiltonian

$$H_{cl} = \frac{1}{2m}p^2 + \frac{k}{2}q^2.$$

Its quantum mechanical equivalent should have energy operator

$$H = \frac{1}{2m}P^2 + \frac{k}{2}Q^2 \tag{14.19}$$

where

$$P = -i\hbar\frac{d}{dx}, \quad [Q, P] = i\hbar I.$$

Set

$$A = \frac{1}{\sqrt{\omega\hbar}}\left(\frac{1}{\sqrt{2m}}P + i\sqrt{\frac{k}{2}}Q\right)$$

where $\omega = \sqrt{k/m}$ and we find

$$H = \omega\hbar(N + \tfrac{1}{2}I) \tag{14.20}$$

where N is the self-adjoint operator $N = AA^* = N^*$.

It is not hard to show that

$$[A, A^*] = AA^* - A^*A = -I, \tag{14.21}$$

and from the identities in Problem 14.3 it follows that

$$[N, A] = [AA^*, A] = A[A^*, A] + [A, A]A^* = A, \tag{14.22}$$
$$[N, A^*] = [A, N^*]^* = -[N, A]^* = -A^*. \tag{14.23}$$

All eigenvalues of N are non-negative, $n \geq 0$, for if $N|\psi_n\rangle = n|\psi_n\rangle$ then

$$0 \leq \|A^*\psi_n\|^2 = \langle\psi_n|AA^*|\psi_n\rangle = \langle\psi_n|N|\psi_n\rangle = n\langle\psi_n|\psi_n\rangle. \tag{14.24}$$

Let $n_0 \geq 0$ be the lowest eigenvalue. Using (14.23), the state $A^*|\psi_n\rangle$ is an eigenstate of N with eigenvalue $(n-1)$

$$NA^*|\psi_n\rangle = (A^*N - A^*)|\psi_n\rangle = (n-1)A^*|\psi_n\rangle.$$

Hence $A^*|\psi_{n_0}\rangle = 0$, else $n_0 - 1$ would be an eigenvalue, contradicting n_0 being lowest, and setting $n = n_0$ in Eq. (14.24) gives $n_0 = 0$. Furthermore, if n is an eigenvalue then $A|\psi_n\rangle \neq 0$ is an eigenstate with eigenvalue $(n+1)$ for, by Eqs. (14.22) and (14.21)

$$NA|\psi_n\rangle = (AN + A)|\psi_n\rangle = (n+1)|\psi_n\rangle,$$
$$\|A|\psi_n\rangle\|^2 = \langle\psi_n|A^*A|\psi_n\rangle = \langle\psi_n|AA^* + I|\psi_n\rangle = (n+1)\langle\psi_n|\psi_n\rangle > 0.$$

The eigenvalues of N are therefore $n = 0, 1, 2, 3, \ldots$ and the eigenvalues of H are $\tfrac{1}{2}\hbar\omega$, $\tfrac{3}{2}\hbar\omega, \ldots, (n+\tfrac{1}{2})\hbar\omega, \ldots$

14.2 Quantum dynamics

Angular momentum

A similar analysis can be used to find the eigenvalues of the angular momentum operators $L_i = \epsilon_{ijk} Q_j P_k$. Using the identities in Problems 14.3 and 14.4 it is straightforward to derive the commutation relations of angular momentum

$$[L_i, L_j] = i\hbar \epsilon_{ijk} L_k \tag{14.25}$$

and

$$[L^2, L_i] = 0 \tag{14.26}$$

where $L^2 = (L_1)^2 + (L_2)^2 + (L_3)^2$ is the **total angular momentum**.

Exercise: Prove the identities (14.25) and (14.26).

Any set of three operators L_1, L_2, L_3 satisfying the commutation relations (14.25) are said to be **angular momentum operators**. If they are of the form $L_i = \epsilon_{ijk} Q_j P_k$ then we term them **orbital angular momentum**, else they are called **spin angular momentum**, or a combination thereof. If we set $J_i = L_i/\hbar$ and $J^2 = (J_1)^2 + (J_2)^2 + (J_3)^2$ then

$$[J_i, J_j] = i\epsilon_{ijk} J_k, \qquad [J^2, J_i] = 0$$

and since J_3 and J^2 commute there exist, by Theorem 14.2, a common set of eigenvectors $|j^2 m\rangle$ such that

$$J^2 |j^2 m\rangle = j^2 |j^2 m\rangle, \qquad J_3 |j^2 m\rangle = m |j^2 m\rangle.$$

Thus

$$\langle j^2 m | J^2 | j^2 m \rangle = \sum_{k=1}^{3} \langle j^2 m | (J_k)^2 | j^2 m \rangle = \sum_{k=1}^{3} \| J_k | j^2 m \rangle \|^2 \geq m^2 \| |j^2 m\rangle \|^2.$$

Since the left-hand side is equal to $j^2 \| |j^2 m\rangle \|^2$ we have $j^2 \geq m^2 \geq 0$, and there is an upper and lower bound to the eigenvalue m for any fixed j^2. Let this upper bound be l.

If we set $J_\pm = J_1 \pm iJ_2$, then it is simple to show the identities

$$[J_3, J_\pm] = \pm J_\pm, \qquad J_\pm J_\mp = J^2 - (J_3)^2 \pm J_3. \tag{14.27}$$

Exercise: Prove the identities (14.27).

Hence $J_\pm$ are raising and lowering operators for the eigenvalues of J_3,

$$J_3 J_\pm |j^2 m\rangle = (J_\pm J_3 \pm J_\pm) |j^2 m\rangle = (m \pm 1) J_\pm |j^2 m\rangle$$

while they leave the eigenvalue of J^2 alone,

$$J^2 (J_\pm |j^2 m\rangle) = J_\pm J^2 |j^2 m\rangle = j^2 J_\pm J^2 |j^2 m\rangle.$$

Since l is the maximum possible value of m, we must have $J_\pm |j^2 l\rangle = 0$ and using the second identity of (14.27) we have

$$J_- J_+ |j^2 l\rangle = (J^2 - (J_3)^2 - J_3) |j^2 l\rangle = (j^2 - l^2 - l) |j^2 l\rangle = 0,$$

whence $j^2 = l(l+1)$. Since for each integer n, $(J_-)^n |j^2 l\rangle$ is an eigenket of J_3 with eigenvalue $(l-n)$ and the eigenvalues of J_3 are bounded below, there exists an integer n such that $(J_-)^n |j^2 l\rangle \neq 0$ and $(J_-)^{n+1} |j^2 l\rangle = 0$. Using (14.27) we deduce

$$0 = J_+ J_- (J_-)^n |j^2 l\rangle = \big(J^2 - (J_3)^2 \pm J_3\big)(J_-)^n |j^2 l\rangle$$
$$= \big(j^2 - (l-n)^2 + (l-n)\big)(J_-)^n |j^2 l\rangle$$

so that

$$j^2 = (l-n)(l-n-1) = l^2 + l$$

from which it follows that $l = \frac{1}{2}n$. Thus the eigenvalues of total angular momentum L^2 are of the form $l(l+1)\hbar^2$ where l has integral or half integral values. The eigenspaces are $(2l+1)$-degenerate, and the simultaneous eigenstates of L_3 have eigenvalues $m\hbar$ where $-l \le m \le m$. For orbital angular momentum it turns out that the value of l is always integral, but spin eigenstates may have all possible eigenvalues, $l = 0, \frac{1}{2}, 1, \frac{3}{2}, \ldots$ depending on the particle in question.

Problems

Problem 14.6 In the Heisenberg picture show that the time evolution of the expection value of an operator A is given by

$$\frac{d}{dt}\langle A' \rangle_{\psi'} = \frac{1}{i\hbar}\langle [A', H'] \rangle_{\psi'} + \langle \frac{\partial A'}{\partial t} \rangle_{\psi'}.$$

Convert this to an equation in the Schrödinger picture for the time evolution of $\langle A \rangle_\psi$.

Problem 14.7 For a particle of spin half in a magnetic field with Hamiltonian given in Example 14.4, show that in the Heisenberg picture

$$\langle \sigma_x(t) \rangle_\mathbf{n} = \sin\theta \cos(\phi - \omega t),$$
$$\langle \sigma_y(t) \rangle_\mathbf{n} = \sin\theta \sin(\phi - \omega t),$$
$$\langle \sigma_z(t) \rangle_\mathbf{n} = \cos\theta.$$

Problem 14.8 A particle of mass m is confined by an infinite potential barrier to remain within a box $0 \le x, y, z \le a$, so that the wave function vanishes on the boundary of the box. Show that the energy levels are

$$E = \frac{1}{2m}\frac{\pi^2 \hbar^2}{a^2}(n_1^2 + n_2^2 + n_3^2),$$

where n_1, n_2, n_3 are positive integers, and calculate the stationary wave functions $\psi_E(\mathbf{x})$. Verify that the lowest energy state is non-degenerate, but the next highest is triply degenerate.

Problem 14.9 For a particle with Hamiltonian

$$H = \frac{\mathbf{p}^2}{2m} + V(\mathbf{x})$$

show from the equation of motion in the Heisenberg picture that

$$\frac{d}{dt}\langle \mathbf{r} \cdot \mathbf{p} \rangle = \langle \frac{\mathbf{p}^2}{m} \rangle - \langle \mathbf{r} \cdot \nabla V \rangle.$$

14.3 Symmetry transformations

This is called the *Virial theorem*. For stationary states, show that

$$2\langle T \rangle = \langle \mathbf{r} \cdot \nabla V \rangle$$

where T is the kinetic energy. If $V \propto r^n$ this reduces to the classical result $\langle 2T + nV \rangle = 0$.

Problem 14.10 Show that the nth normalized eigenstate of the harmonic oscillator is given by

$$|\psi_n\rangle = \frac{1}{(n!)^{1/2}} A^n |\psi_0\rangle.$$

Show from $A^* \psi_0 = 0$ that

$$\psi_0 = c e^{-\sqrt{km} x^2 / 2\hbar} \quad \text{where} \quad c = \left(\frac{m\omega}{\pi\hbar}\right)^{1/4}$$

and the nth eigenfunction is

$$\psi_n(x) = \frac{i^n}{(2^n n!)^{1/2}} \left(\frac{m\omega}{\pi\hbar}\right)^{1/4} e^{-m\omega x^2 / 2\hbar} H_n\left(\sqrt{\frac{m\omega}{\hbar}} x\right),$$

where $H_n(y)$ is the nth hermite polynomial (see Example 13.7).

Problem 14.11 For the two-dimensional harmonic oscillator define operators A_1, A_2 such that

$$[A_i, A_j] = [A_i^*, A_j^*] = 0, \quad [A_i, A_j^*] = -\delta_{ij}, \quad H = \hbar\omega(2N + I)$$

where $i, j = 1, 2$ and N is the number operator

$$N = \tfrac{1}{2}(A_1 A_1^* + A_2 A_2^*).$$

Let J_1, J_2 and J_3 be the operators

$$J_1 = \tfrac{1}{2}(A_2 A_1^* + A_1 A_2^*), \quad J_2 = \tfrac{1}{2}i(A_2 A_1^* - A_1 A_2^*), \quad J_3 = \tfrac{1}{2}(A_1 A_1^* - A_2 A_2^*),$$

and show that:

(a) The J_i satisfy the angular momentum commutation relations $[J_1, J_2] = i J_3$, etc.
(b) $J^2 = J_1^2 + J_2^2 + J_3^2 = N(N + 1)$.
(c) $[J^2, N] = 0$, $[J_3, N] = 0$.

From the properties of angular momentum deduce the energy levels and their degeneracies for the two-dimensional harmonic oscillator.

Problem 14.12 Show that the eigenvalues of the three-dimensional harmonic oscillator have the form $(n + \tfrac{3}{2})\hbar\omega$ where n is a non-negative integer. Show that the degeneracy of the nth eigenvalue is $\tfrac{1}{2}(n^2 + 3n + 2)$. Find the corresponding eigenfunctions.

14.3 Symmetry transformations

Consider two observers O and O' related by a symmetry transformation such as a translation $\mathbf{x}' = \mathbf{x} - \mathbf{a}$, or rotation $\mathbf{x}' = A\mathbf{x}$ where $AA^T = I$, etc. For any state corresponding to a ray $[|\psi\rangle]$ according to O let O' assign the ray $[|\psi'\rangle]$, and for any observable assigned the self-adjoint operator A by O let O' assign the operator A'. Since the physical elements are determined by the modulus squared of the matrix elements between states (being the probability of transition between states) and the expectation values of observables, this correspondence is said to be a **symmetry transformation** if

$$|\langle \phi' | \psi' \rangle|^2 = |\langle \phi | \psi \rangle|^2 \tag{14.28}$$

$$\langle A' \rangle_{\psi'} = \langle \psi' | A' | \psi' \rangle = \langle A \rangle_\psi = \langle \psi | A | \psi \rangle \tag{14.29}$$

for all states $|\psi\rangle$ and observables A.

Theorem 14.3 (Wigner) *A ray correspondence that satisfies (14.28) for all rays is generated up to a phase by a transformation $U : \mathcal{H} \to \mathcal{H}$ that is either unitary or anti-unitary.*

A unitary transformation was defined in Chapter 13 as a linear transformation, $U(|\psi\rangle + \alpha|\phi\rangle) = U|\psi\rangle + \alpha U|\phi\rangle$, which preserves inner products

$$\langle U\psi | U\phi \rangle = \langle \psi | \phi \rangle \iff UU^* = U^*U = I.$$

A transformation V is said to be **antilinear** if

$$V(|\psi\rangle + \alpha|\phi\rangle) = V|\psi\rangle + \bar{\alpha} V|\phi\rangle.$$

The adjoint V^* is defined by

$$\langle V^*\psi | \phi \rangle = \overline{\langle \psi | V\phi \rangle} = \langle V\phi | \psi \rangle,$$

in order that it too will be antilinear (we pay no attention to domains here). An operator V is called **anti-unitary** if it is antilinear and $V^*V = VV^* = I$. In this case

$$\langle V\psi | V\phi \rangle = \overline{\langle \psi | \phi \rangle} = \langle \phi | \psi \rangle$$

for all vectors $|\psi\rangle$ and $|\phi\rangle$.

Outline proof of Theorem 14.3: We prove Wigner's theorem in the case of a two-dimensional Hilbert space. The full proof is along similar lines. If $|e_1\rangle$, $|e_2\rangle$ is an orthonormal basis then, up to phases, so is $|e'_1\rangle$, $|e'_2\rangle$,

$$\delta_{ij} = |\langle e_i | e_j \rangle|^2 = |\langle e'_i | e'_j \rangle|^2.$$

Let $|\psi\rangle = a_1|e_1\rangle + a_2|e_2\rangle$ be any unit vector, $\langle \psi | \psi \rangle = |a_1|^2 + |a_2|^2 = 1$. Set

$$|\psi'\rangle = a'_1|e'_1\rangle + a'_2|e'_2\rangle$$

and we have, from Eq. (14.28),

$$|a'_1|^2 = |\langle e'_1 | \psi' \rangle|^2 = |\langle e_1 | \psi \rangle|^2 = |a_1|^2,$$

and similarly $|a'_2|^2 = |a_2|^2$. Hence we can set real angles α, θ, φ, etc. such that

$$a_1 = \cos\alpha e^{i\theta}, \quad a'_1 = \cos\alpha e^{i\theta'},$$
$$a_2 = \sin\alpha e^{i\varphi}, \quad a'_2 = \sin\alpha e^{i\varphi'}.$$

Let $|\psi_1\rangle$ and $|\psi_2\rangle$ be an arbitrary pair of unit vectors,

$$|\psi_i\rangle = \cos\alpha_i e^{i\theta_i} |e_1\rangle + \sin\alpha_i e^{i\varphi_i} |e_2\rangle,$$

14.3 Symmetry transformations

then $|\langle\psi_1'|\psi_2'\rangle|^2 = |\langle\psi_1|\psi_2\rangle|^2$ implies

$$\cos(\theta_2' - \varphi_2' - \theta_1' + \varphi_1') = \cos(\theta_2 - \varphi_2 - \theta_1 + \varphi_1).$$

Hence

$$\theta_2' - \varphi_2' - (\theta_1' - \varphi_1') = \pm(\theta_2 - \varphi_2 - (\theta_1 - \varphi_1)). \tag{14.30}$$

Define an angle δ by

$$\theta_1' - \varphi_1' = \delta \pm (\theta_1 - \varphi_1),$$

and it follows from (14.30) that

$$\theta_2' - \varphi_2' = \delta \pm (\theta_2 - \varphi_2).$$

Hence for an arbitrary vector $|\psi\rangle$,

$$\theta' - \varphi' = \delta \pm (\theta - \varphi).$$

For the + sign this results in the transformation

$$|\psi\rangle \mapsto |\psi'\rangle = e^{i(\varphi' - \varphi)}\left(a_1 e^{i\delta}|e_1'\rangle + a_2|e_2'\rangle\right)$$

while for the − sign it is

$$|\psi\rangle \mapsto |\psi'\rangle = e^{i(\varphi' + \varphi)}\left(\overline{a_1} e^{i\delta}|e_1'\rangle + \overline{a_2}|e_2'\rangle\right).$$

These transformations are, up to a phase e^{if}, unitary and anti-unitary respectively. This is Wigner's theorem. ∎

Exercise: Show that the phase $f = \varphi' \pm \varphi$ is independent of the state.

If U is a unitary transformation and A is an observable, then

$$\langle A'\rangle_{\psi'} = \langle\psi'|A'|\psi'\rangle = \langle\psi|U^* e^{-if} A' e^{if} U|\psi\rangle = \langle U^* A' U\rangle_\psi$$

and the requirement (14.29) implies that this holds for arbitrary vectors $|\psi\rangle$ if and only if

$$A' = UAU^*.$$

Performing two symmetries g and h in succession results in a symmetry transformation of $\mathcal{H}$ satisfying

$$U(g)U(h) = e^{i\varphi} U(gh)$$

where the phase φ may depend on g and h. This is called a **projective** or **ray representation** of the group G on the Hilbert space $\mathcal{H}$. It is not in general possible to choose the phases such that all $e^{i\varphi} = 1$, giving a genuine representation. For a continuous group (see Section 10.8), elements in the component of the identity G_0 must be unitary since they are connected continuously with the identity element, which is definitely unitary. Anti-unitary transformations can only correspond to group elements in components that are not continuously connected with the identity.

389

Quantum mechanics

Infinitesimal generators

If G is a Lie group, the elements of which are unitary transformations characterized as in Section 6.5 by a set of real parameters $U = U(a_1, \ldots, a_n)$ such that $U(0, \ldots, 0) = I$, we define the **infinitesimal generators** by

$$S_j = -i \frac{\partial U}{\partial a_j}\bigg|_{a=0}. \tag{14.31}$$

These are self-adjoint operators since $UU^* = I$ implies that

$$0 = \frac{\partial U}{\partial a_j}\bigg|_{a=0} + \frac{\partial U^*}{\partial a_j}\bigg|_{a=0} = i(S_j - S_j^*).$$

Note that self-adjoint operators do not form a Lie algebra, since their commutator is not in general self-adjoint. However, as seen in Section 6.5, Problem 6.12, the operators iS_j do form a Lie algebra,

$$[iS_i, iS_j] = \sum_{k=1}^n C_{ij}^k iS_k.$$

Exercise: Show that an operator S satisfies $S^* = -S$ iff it is of the form $S = iA$ where A is self-adjoint. Show that the commutator product preserves this property.

Example 14.5 If S is a hermitian operator the set of unitary transformations $U(a) = e^{iaS}$ where $-\infty < a < \infty$ is a one-parameter group of unitary transformations,

$$U(a)U^*(a) = e^{iaS}e^{-iaS} = I, \qquad U(a)U(b) = e^{iaS}e^{ibS} = e^{i(a+b)S} = U(a+b).$$

Its infinitesimal generator is

$$-i\frac{\partial}{\partial a}e^{iaS}\bigg|_{a=0} = S.$$

Example 14.6 Let O and O' be two observers related by a displacement of the origin through the vector $\mathbf{a}$. Let the state vectors be related by

$$|\psi'\rangle = T(\mathbf{a})|\psi\rangle.$$

If $\mathbf{Q}$ is the position operator then

$$\mathbf{q}' = \langle \mathbf{Q}\rangle_{\psi'} = \langle \psi'|\mathbf{Q}|\psi'\rangle = \mathbf{q} - \mathbf{a} = \langle\psi|\mathbf{Q}|\psi\rangle - \mathbf{a}\langle\psi|I|\psi\rangle,$$

whence

$$T^*(\mathbf{a})\mathbf{Q}T(\mathbf{a}) = \mathbf{Q} - \mathbf{a}I. \tag{14.32}$$

Taking the partial derivative with respect to a_i at $\mathbf{a} = 0$ we find

$$-iS_i Q_j + iQ_j S_i = -\delta_{ij} \quad \text{where} \quad S_j = -i\frac{\partial T}{\partial a_j}\bigg|_{a=0}.$$

Hence $[S_i, Q_j] = -i\delta_{ij}$, and we may expect

$$S_i = \frac{1}{\hbar}P_i$$

14.3 Symmetry transformations

where P_i are the momentum operators

$$P_i = -i\hbar \frac{\partial}{\partial q_i}, \qquad [Q_i, P_j] = i\hbar \delta_{ij}.$$

This is consistent with $\mathbf{P}' = \mathbf{P}$, since

$$T^*(\mathbf{a})P_i T(\mathbf{a}) = P_i \implies [S_j, P_i] \propto [P_j, P_i] = 0.$$

To find the translation operators $T(\mathbf{a})$, we use the group property $T(\mathbf{a})T(\mathbf{b}) = T(\mathbf{a} + \mathbf{b})$, and take the derivative with respect to b_i at $\mathbf{b} = 0$:

$$iT(\mathbf{a})S_i = T(\mathbf{a})\frac{\partial T}{\partial b_i}\bigg|_{\mathbf{b}=0} = \frac{\partial T}{\partial a_i}(\mathbf{a}).$$

The solution of this operator equation may be written

$$T(\mathbf{a}) = e^{i \sum_i a_i S_i} = e^{i \mathbf{a} \cdot \mathbf{S}}$$
$$= e^{i \mathbf{a} \cdot \mathbf{P}/\hbar} = e^{i a_1 P_1/\hbar} e^{i a_2 P_2/\hbar} e^{i a_3 P_3/\hbar}$$

since the P_i commute with each other. It is left as an exercise (Problem 14.14) to verify Eq. (14.32).

Example 14.7 Two observers related by a rotation through an angle θ about the z-axis

$$q_1' = q_1 \cos\theta + q_2 \sin\theta, \qquad q_2' = -q_1 \sin\theta + q_2 \cos\theta, \qquad q_3' = q_3$$

are related by a unitary operator $R(\theta)$ such that

$$|\psi'\rangle = R(\theta)|\psi\rangle, \qquad R^*(\theta)R(\theta) = I.$$

In order to arrive at the correct transformation of expectation values we require that

$$R^*(\theta)Q_1 R(\theta) = Q_1 \cos\theta + Q_2 \sin\theta, \qquad (14.33)$$
$$R^*(\theta)Q_2 R(\theta) = -Q_1 \sin\theta + Q_2 \cos\theta, \qquad (14.34)$$
$$R^*(\theta)Q_3 R(\theta) = Q_3. \qquad (14.35)$$

Setting

$$J = -i\frac{dR}{d\theta}\bigg|_{\theta=0}, \qquad J^* = J$$

we find on taking derivatives at $\theta = 0$ of Eqs. (14.33)–(14.35)

$$[J, Q_1] = iQ_2, \qquad [J, Q_2] = -iQ_1, \qquad [J, Q_3] = 0.$$

A solution is the z-component of angular momentum,

$$J = \frac{1}{\hbar}L_3 = \frac{1}{\hbar}(Q_1 P_2 - Q_2 P_1),$$

since $[J, Q_1] = iQ_2$, etc. (see Problem 14.4).

It is again easy to verify the group property $R(\theta_1)R(\theta_2) = R(\theta_1 + \theta_2)$ and as in the translational example above,

$$iR(\theta)J = \frac{dR}{d\theta} \implies R\theta = e^{i\theta J} = e^{i\theta L_3/\hbar}.$$

It is again left as an exercise to show that this operator satisfies Eqs. (14.33)–(14.35). For a rotation of magnitude θ about an axis $\mathbf{n}$ the rotation operator is

$$R(\mathbf{n}, \theta) = e^{i\theta \mathbf{n} \cdot \mathbf{L}/\hbar}$$

where $\mathbf{L}$ is the angular momentum operator having components $L_i = \epsilon_{ijk} Q_j P_k$. Since these operators do not commute, satisfying the commutation relations (14.25), we have in general

$$R(\mathbf{n}, \theta) \neq e^{i\theta_1 L_1/\hbar} e^{i\theta_2 L_2/\hbar} e^{i\theta_3 L_3/\hbar}.$$

Exercise: Show that the transformation of momentum components under a rotation with infinitesimal generator J is

$$[J, P_1] = iP_2, \qquad [J, P_2] = -iP_1, \qquad [J, P_3] = 0.$$

Example 14.8 Under a time translation $t' = t - \tau$, we have $|\psi'(t')\rangle = |\psi(t)\rangle$, so that

$$|\psi'(t)\rangle = |\psi(t + \tau)\rangle = T(\tau)|\psi(t)\rangle.$$

Hence, by Schrödinger's equation

$$iS|\psi(t)\rangle = \left.\frac{\partial T}{\partial \tau}\right|_{\tau=0} |\psi(t)\rangle = \frac{d}{dt}|\psi(t)\rangle = \frac{1}{i\hbar} H|\psi(t)\rangle.$$

The infinitesimal generator of the time translation is essentially the Hamiltonian, $S = -H/\hbar$. If the Hamiltonian H is time-independent,

$$T(\tau) = e^{-iH\tau/\hbar}. \tag{14.36}$$

Conserved quantities

Under a time-dependent unitary transformation

$$|\psi'\rangle = U(t)|\psi\rangle.$$

Schrödinger's equation (14.13) results in

$$i\hbar \frac{d}{dt}|\psi'\rangle = i\hbar \frac{\partial U}{\partial t}|\psi\rangle + i\hbar U(t)\frac{d}{dt}|\psi\rangle$$
$$= H'|\psi'\rangle$$

where

$$H' = UHU^* + i\hbar \frac{\partial U}{\partial t} U^*. \tag{14.37}$$

Exercise: Show that under an anti-unitary transformation

$$H' = -UHU^* + i\hbar \frac{\partial U}{\partial t} U^*.$$

14.3 Symmetry transformations

U is called a **Hamiltonian symmetry** if $H' = H$. Then, multiplying Eq. (14.37) by U on the right gives

$$[U, H] + i\hbar \frac{\partial U}{\partial t} = 0. \tag{14.38}$$

If U is independent of time then U commutes with the Hamiltonian, $[U, H] = 0$.

If G is an n-parameter Lie group of unitary Hamiltonian symmetries $U(t, a_1, a_2, \ldots, a_n)$, having infinitesimal generators

$$S_i(t) = -i \frac{\partial U}{\partial a_i}\bigg|_{a=0},$$

then differentiating Eq. (14.38) with respect to a_i gives

$$[S_i, H] + i\hbar \frac{\partial S_i}{\partial t} = 0. \tag{14.39}$$

Any hermitian operator $S(t)$ satisfying this equation is said to be a **constant of the motion** or **conserved quantity**, for Schrödinger's equation implies that its expection values are constant:

$$\frac{d}{dt}\langle S \rangle_\psi = \frac{d}{dt}\langle \psi | S | \psi \rangle = \langle \frac{d\psi}{dt} | S | \psi \rangle + \langle \psi | S \frac{d}{dt} | \psi \rangle + \langle \psi | \frac{\partial S}{\partial t} | \psi \rangle$$

$$= \frac{-1}{i\hbar} \langle \psi | HS | \psi \rangle + \frac{1}{i\hbar} \langle \psi | SH | \psi \rangle + \langle \psi | \frac{\partial S}{\partial t} | \psi \rangle$$

$$= \frac{1}{i\hbar} \langle \psi | [S, H] + i\hbar \frac{\partial S}{\partial t} | \psi \rangle = 0.$$

Exercise: Show that in the Heisenberg picture, this is equivalent to

$$\frac{dS_H}{dt} = 0 \quad \text{where} \quad S_H = e^{-iHt/\hbar} S e^{iHt/\hbar}.$$

From Examples 14.6 and 14.7 it follows that invariance of the Hamiltonian under translations and rotations is equivalent to conservation of momentum and angular momentum respectively. In both cases the infinitesimal generators are time-independent. If the Hamiltonian is invariant under time translations, having generator $S = -H/\hbar$ (see Example 14.8), then Eq. (14.39) reduces to

$$-\frac{1}{\hbar}[H, H] - i\frac{\partial H}{\partial t} = 0,$$

which is true if and only if H has no explicit time dependence, $\partial H/\partial t = 0$.

Discrete symmetries

There are a number of important symmetries of a more discrete nature, illustrated in the following examples.

Example 14.9 Consider a spatial inversion $\mathbf{r} \mapsto \mathbf{r}' = -\mathbf{r}$, which can be thought of as a rotation by $180°$ about the z-axis followed by a reflection $x' = x$, $y' = y$, $z' = -z$. Let Π

Quantum mechanics

be the operator on $\mathcal{H}$ induced by such an inversion, satisfying

$$\Pi^* Q_i \Pi = -Q_i, \qquad \Pi^* P_i \Pi = -P_i.$$

By Wigner's theorem,

$$\Pi^*[Q_i, P_j]\Pi = \Pi^* i\hbar \delta_{ij} \Pi = \begin{cases} i\hbar \delta_{ij} & \text{if } \Pi \text{ is unitary,} \\ -i\hbar \delta_{ij} & \text{if } \Pi \text{ is anti-unitary.} \end{cases}$$

It turns out that Π must be a unitary operator, for

$$\Pi^*[Q_i, P_j]\Pi = [\Pi^* Q_i \Pi, \Pi^* P_j \Pi] = [-Q_i, -P_j] = [Q_i, P_j] = i\hbar \delta_{ij}.$$

Note also that angular momentum operators are invariant under spatial reflections,

$$\Pi^* L_i \Pi = L_i \quad \text{where} \quad \mathbf{L} = \mathbf{Q} \times \mathbf{P}.$$

Since successive reflections result in the identity $\Pi^2 = I$, we have $\Pi^* = \Pi$. Hence Π is a hermitian operator, corresponding to an observable called **parity**, having eigenvalues ± 1. States of eigenvalue 1, $\Pi|\psi\rangle = |\psi\rangle$, are said to be of **even parity**, while those of eigenvalue -1 are called **odd parity**, $\Pi|\psi\rangle = -|\psi\rangle$. Every state can be decomposed as a sum of an even and an odd parity state,

$$|\psi\rangle = \tfrac{1}{2}(I + \Pi)|\psi\rangle + \tfrac{1}{2}(I - \Pi)|\psi\rangle = |\psi_+\rangle + |\psi_-\rangle.$$

Exercise: Show that if $[\Pi, H] = 0$, the parity of any state is preserved throughout its motion, and eigenstates of H with non-degenerate eigenvalue have definite parity.

Example 14.10 In classical physics, if $\mathbf{q}(t)$ is a solution of Newton's equations then so is the reverse motion $\mathbf{q}_{\text{rev}}(t) = \mathbf{q}(t)$ having opposite momentum $\mathbf{p}_{\text{rev}}(t) = -\mathbf{p}(-t)$. If O' is an observer having time in the reversed direction $t' = -t$ to that of an observer O, let the time-reversed states be

$$|\psi'\rangle = \Theta|\psi\rangle,$$

where Θ is the time-reversal operator. Since we require

$$\Theta^* Q_i \Theta = Q_i, \qquad \Theta^* P_i \Theta = -P_i$$

a similar discussion to that in Example 14.9 gives

$$\Theta^*[Q, P]\Theta = \Theta^* i\hbar I \Theta = \pm i\hbar I$$
$$= [Q, -P] = -i\hbar I.$$

Hence time-reversal Θ is an anti-unitary operator.

If the Hamiltonian H is invariant under time reversal, $[H, \Theta] = 0$, then applying Θ to Schrödinger's equation (14.13) gives

$$-i\hbar \frac{d}{dt} \Theta|\psi(t)\rangle = \Theta i\hbar \frac{d}{dt}|\psi\rangle = \Theta H|\psi\rangle = H\Theta|\psi(t)\rangle.$$

14.3 Symmetry transformations

Changing the time variable t to $-t$,

$$i\hbar \frac{d}{dt} \Theta |\psi(-t)\rangle = H\Theta |\psi(-t)\rangle.$$

It follows that $|\psi_{\text{rev}}(t)\rangle = \Theta |\psi(-t)\rangle$ is a solution of Schrödinger's equation, which may be thought of as the time-reversed solution. In this sense, the dynamics of quantum mechanics is time-reversable, but note that because of the anitilinear nature of the operator Θ, a complex conjugation is required in addition to time inversion. For example in the position representation, if $\psi(\mathbf{r}, t)$ is a solution of Schrödinger's wave equation (14.18), then $\psi(\mathbf{r}, -t)$ is not in general a solution. However, taking the complex conjugate shows that $\psi_{\text{rev}}(t) = \overline{\psi(\mathbf{r}, -t)}$ is a solution of (14.18),

$$i\hbar \frac{\partial}{\partial t} \overline{\psi(\mathbf{r}, -t)} = -\left(\frac{\hbar^2}{2m}\nabla^2 + V(\mathbf{r})\right) \overline{\psi(\mathbf{r}, -t)}.$$

Identical particles

Consider a system consisting of N indistinguishable particles. If the Hilbert space of each individual particle is $\mathcal{H}$ we take the Hilbert space of the entire system to be the tensor product

$$\mathcal{H}^N = \mathcal{H} \otimes \mathcal{H} \otimes \cdots \otimes \mathcal{H}.$$

As in Chapter 7 this may be regarded as the tensor space spanned by free formal products

$$|\psi_1 \psi_2 \ldots \psi_N\rangle \equiv |\psi_1\rangle |\psi_2\rangle \ldots |\psi_N\rangle$$

where each $|\psi_i\rangle \in \mathcal{H}$, subject to identifications

$$(\alpha|\psi_1\rangle + \beta|\phi\rangle)|\psi_2\rangle \cdots = \alpha|\psi_1\rangle |\psi_2\rangle \cdots + \beta|\phi\rangle|\psi_2\rangle \ldots, \text{ etc.}$$

The inner product on $\mathcal{H}^N$ is defined by

$$\langle \psi_1 \psi_2 \ldots \psi_N | \phi_1 \phi_2 \ldots \phi_N \rangle = \langle \psi_1 | \phi_1 \rangle \langle \psi_2 | \phi_2 \rangle \ldots \langle \psi_N | \phi_N \rangle.$$

For each pair $1 \le i < j \le N$ define the permutation operator $P_{ij} : \mathcal{H}^N \to \mathcal{H}^N$ by

$$\langle \phi_1 \ldots \phi_i \ldots \phi_j \ldots \phi_N | P_{ij} \psi \rangle = \langle \phi_1 \ldots \phi_j \ldots \phi_i \ldots \phi_N | \psi \rangle$$

for all $|\phi_i\rangle \in \mathcal{H}$. This is a linear operator that 'interchanges particles' i and j,

$$P_{ij}|\psi_1 \ldots \psi_i \ldots \psi_j \ldots \psi_N\rangle = |\psi_1 \ldots \psi_j \ldots \psi_i \ldots \psi_N\rangle.$$

If $|\psi_a\rangle$ ($a = 1, 2, \ldots$) is an o.n. basis of the Hilbert space $\mathcal{H}$ then the set of all vectors

$$|\psi_{a_1} \psi_{a_2} \ldots \psi_{a_N}\rangle \equiv |\psi_{a_1}\rangle |\psi_{a_2}\rangle \ldots |\psi_{a_N}\rangle$$

forms an o.n. basis of $\mathcal{H}^N$. Since P_{ij} transforms any such o.n. basis to an o.n. basis it must be a unitary operator. These statements extend to a general permutation P, since it can be written as a product of interchanges

$$P = P_{ij} P_{kl} \ldots$$

Quantum mechanics

As there is no dynamical way of detecting an interchange of identical particles, the expection values of the Hamiltonian must be invariant under permutations, $\langle H \rangle_{P\psi} = \langle H \rangle_\psi$, so that

$$\langle \psi | P^* H P | \psi \rangle = \langle \psi | H | \psi \rangle$$

for all $|\psi\rangle \in \mathcal{H}$. Hence $P^* H P = H$ and as P is unitary, $PP^* = I$,

$$[H, P] = 0.$$

This is yet another example of a discrete symmetry. In classical mechanics it is taken as given that all particles have an individuality and are in principle distinguishable. It is basic to the philosophy of quantum mechanics, however, that since no physical procedure exists for 'marking' identical particles such as a pair of electrons in order to keep track of them, there can be no method even in principle of distinguishing between them.

All interchanges have the property $(P_{ij})^2 = I$, from which they are necessarily hermitian, $P_{ij} = P_{ij}^*$. Every interchange therefore corresponds to an observable. It has eigenvalues $\epsilon \pm 1$ and, since P_{ij} is a constant of the motion, any eigenstate

$$P_{ij}|\psi\rangle = \epsilon|\psi\rangle$$

remains an eigenstate corresponding to the same eigenvalue, for

$$\epsilon = \langle \psi | P_{ij} | \psi \rangle = \langle P_{ij} \rangle_\psi = \text{const.}$$

Since no physical observable can distinguish between states related by a permutation, a similar argument to that used for the Hamiltonian shows that every observable A commutes with all permutation operators,

$$[A, P] = 0.$$

Hence, if $|\psi\rangle$ is a non-degenerate eigenstate of A, it is an eigenstate of every permutation operator P,

$$A|\psi\rangle = a|\psi\rangle \implies AP|\psi\rangle = PA|\psi\rangle = aP|\psi\rangle$$
$$\implies P|\psi\rangle = p|\psi\rangle$$

for some factor $p = \pm 1$. If, as is commonly assumed, every state is representable as a sum of non-degenerate common eigenvectors of a commuting set of complete observables $A, B, C, \ldots$ we must then assume that every physical state $|\psi\rangle$ of the system is a common eigenvector of all permutation operators. In particular, for all interchanges P_{ij}

$$P_{ij}|\psi\rangle = p_{ij}|\psi\rangle \quad \text{where} \quad p_{ij} = \pm 1.$$

All p_{ij} are equal for the state $|\psi\rangle$, since for any pair k, l

$$P_{ij}|\psi\rangle = P_{ki} P_{lj} P_{kl} P_{lj} P_{ki} |\psi\rangle,$$

from which it follows that $p_{ij} = p_{kl}$ since

$$p_{ij}|\psi\rangle = (p_{ki})^2 (p_{lj})^2 p_{kl}|\psi\rangle = p_{kl}|\psi\rangle.$$

14.4 Quantum statistical mechanics

Thus for all permutations either $P|\psi\rangle = |\psi\rangle$ or $P|\psi\rangle = (-1)^P|\psi\rangle$. In the first case, $p_{ij} = 1$, the state is said to be **symmetrical** and the particles are called **bosons** or to obey **Bose–Einstein statistics**. If $p_{ij} = -1$ the state is **antisymmetrical**, the particles are said to be **fermions** and obey **Fermi–Dirac statistics**. It turns out that bosons are always particles of integral spin such as photons or mesons, while fermions such as electrons or protons always have half-integral spin. This is known as the **spin-statistics theorem**, but lies beyond the scope of this book (see, for example, [9]).

The celebrated **Pauli exclusion principle** asserts that two identical fermions cannot occupy the same state, for if

$$|\psi\rangle = |\psi_1\rangle \ldots |\psi_i\rangle \ldots |\psi_j\rangle \ldots |\psi_N\rangle$$

and $|\psi_i\rangle = |\psi_j\rangle$ then

$$P_{ij}|\psi\rangle = |\psi\rangle = -|\psi\rangle$$

since every state has eigenvalue -1. Hence $|\psi\rangle = 0$.

Problems

Problem 14.13 If the operator K is complex conjugation with respect to a complete o.n. set,

$$K\left(\sum_i \alpha|e_i\rangle\right) = \sum_i \overline{\alpha_i}|e_i\rangle,$$

show that every anti-unitary operator V can be written in the form $V = UK$, where U is a unitary operator.

Problem 14.14 For any pair of operators A and B show by induction on the coefficients that

$$e^{aB} A e^{-aB} = A + a[B, A] + \frac{a^2}{2!}[B, [B, A]] + \frac{a^3}{3!}[B, [B, [B, A]]] + \ldots$$

Hence show the relation (14.32) holds for $T(\mathbf{a}) = e^{i\mathbf{a} \cdot \mathbf{P}/\hbar}$.

Problem 14.15 Using the expansion in Problem 14.14 show that $R(\theta) = e^{i\theta L_3/\hbar}$ satisfies Eqs. (14.33)–(14.35).

Problem 14.16 Show that the time reversal of angular momentum $\mathbf{L} = \mathbf{Q} \times \mathbf{P}$ is $\Theta^* L_i \Theta = -L_i$, and that the commutation relations $[L_i, L_j] = i\hbar \epsilon_{ijk} L_k$ are only preserved if Θ is anti-unitary.

14.4 Quantum statistical mechanics

Statistical mechanics is the physics of large systems of particles, which are usually identical. The systems are generally so large that only averages of physical quantities can be accurately dealt with. This section will give only the briefest introduction to this enormous and far-ranging subject.

Quantum mechanics

Density operator

Let a quantum system have a complete o.n. basis $|\psi_i\rangle$. If we imagine the rest of the universe (taken in a somewhat restricted sense) to be spanned by an o.n. set $|\theta_a\rangle$, then the general state of the combined system can be written

$$|\Psi\rangle = \sum_i \sum_a c_{ia} |\psi_i\rangle |\theta_a\rangle.$$

An operator A acting on the system only acts on the vectors $|\psi_i\rangle$, hence

$$\begin{aligned}\langle A\rangle_\Psi &= \langle \Psi | A | \Psi \rangle \\ &= \sum_i \sum_a \sum_j \sum_b \overline{c_{ia}} \langle \theta_a | \langle \psi_i | A | \psi_j \rangle | \theta_b \rangle c_{jb} \\ &= \sum_i \sum_a \sum_j \sum_b A_{ij} \overline{c_{ia}} c_{jb} \delta_{ab} \\ &= \sum_i \sum_j A_{ij} \rho_{ji}\end{aligned}$$

where

$$A_{ij} = \langle \psi_i | A | \psi_j \rangle, \qquad \rho_{ji} = \sum_a c_{ja} \overline{c_{ia}}.$$

The operator A can be written

$$A = \sum_i \sum_j A_{ij} |\psi_i\rangle \langle \psi_j|.$$

Exercise: Verify that for any $|\phi\rangle \in \mathcal{H}$,

$$A|\phi\rangle = \sum_i \sum_j A_{ij} |\psi_i\rangle \langle \psi_j | \phi \rangle.$$

Define the **density operator** ρ as that having components ρ_{ij},

$$\rho = \sum_i \sum_j \rho_{ij} |\psi_i\rangle \langle \psi_j|,$$

which is hermitian since

$$\overline{\rho_{ji}} = \sum_a \overline{c_{ia}} c_{ja} = \rho_{ij}.$$

A useful expression for the expectation value of A is

$$\langle A \rangle = \operatorname{tr}(A\rho) = \operatorname{tr}(\rho A) \tag{14.40}$$

where the trace of an operator is given by

$$\operatorname{tr} B = \sum_i \langle \psi_i | B | \psi_i \rangle = \sum_i B_{ii}.$$

Exercise: Show that the trace of an operator is independent of the o.n. basis $|\psi_i\rangle$.

14.4 Quantum statistical mechanics

Setting $A = I$ we have

$$\langle I \rangle = \langle \Psi | \Psi \rangle = \|\Psi\|^2 = 1 \implies \text{tr}(\rho) = 1,$$

and setting $A = |\psi_k\rangle\langle\psi_k|$ gives

$$\langle A \rangle = \langle \Psi | \psi_k \rangle \langle \psi_k | \Psi \rangle = \|\langle \Psi | \psi_k \rangle\|^2 \geq 0.$$

On the other hand

$$\text{tr}(A\rho) = \sum_i \langle \psi_i | A\rho | \psi_i \rangle = \sum_i \langle \psi_i | \psi_k \rangle \langle \psi_k | \rho | \psi_i \rangle$$
$$= \sum_i \delta_{ik} \langle \psi_k | \rho | \psi_i \rangle = \langle \psi_k | \rho | \psi_k \rangle = \rho_{kk}.$$

Hence all diagonal elements of the density matrix are positive, $\rho_{kk} = \langle \psi_k | \rho | \psi_k \rangle \geq 0$. Assuming ρ is a complete operator, select $|\psi_i\rangle$ to be eigenvectors, $\rho|\psi_i\rangle = w_i|\psi_i\rangle$, so that ρ is diagonalized

$$\rho = \sum_i w_i |\psi_i\rangle\langle\psi_i|. \tag{14.41}$$

We then have

$$\sum_i w_i = 1, \quad w_i \geq 0.$$

The interpretation of the density operator ρ, or its related state $|\Psi\rangle$, is as a **mixed state** of the system, with the ith eigenstate $|\psi_i\rangle$ having probability w_i. A **pure state** occurs when there exists k such that $w_k = 1$ and $w_i = 0$ for all $i \neq k$. In this case $\rho^2 = \rho$ and the density operator is idempotent – it acts as a projection operator into the one-dimensional subspace spanned by the associated eigenstate ψ_k.

Exercise: Show the converse: if $\rho^2 = \rho$ then all $w_i = 1$ or 0, and there exists k such that $\rho = |\psi_k\rangle\langle\psi_k|$.

Exercise: Show that the probability of finding the system in a state $|\chi\rangle$ is $\text{tr}\,\rho|\chi\rangle\langle\chi|$.

Example 14.11 Consider a beam of photons in the z-direction. Let $|\psi_1\rangle$ be the state of a photon polarized in the x-direction, and $|\psi_2\rangle$ be the state of a photon polarized in the y-direction. The general state is a linear sum of these two,

$$|\psi\rangle = a|\psi_1\rangle + b|\psi_2\rangle \quad \text{where} \quad |a|^2 + |b|^2 = 1.$$

The pure state represented by this vector has density operator $\rho = |\psi\rangle\langle\psi|$, having components

$$\rho_{ij} = \langle \psi_i | \rho | \psi_j \rangle = \langle \psi_i | \psi \rangle \overline{\langle \psi_j | \psi \rangle} = \begin{pmatrix} a\bar{a} & a\bar{b} \\ b\bar{a} & b\bar{b} \end{pmatrix}.$$

For example, the pure states corresponding to $45°$-polarization $(a = b = \frac{1}{\sqrt{2}})$ and

135°-polarization ($a = -b = -\frac{1}{\sqrt{2}}$) have respective density operators

$$\rho = \begin{pmatrix} \frac{1}{2} & \frac{1}{2} \\ \frac{1}{2} & \frac{1}{2} \end{pmatrix} \quad \text{and} \quad \rho = \begin{pmatrix} \frac{1}{2} & -\frac{1}{2} \\ -\frac{1}{2} & \frac{1}{2} \end{pmatrix}.$$

A half–half mixture of 45°- and 135°-polarized photons is indistinguishable from an equal mixture of x-polarized photons and y-polarized photons, since

$$\left(\frac{1}{\sqrt{2}}|\psi_1\rangle + \frac{1}{\sqrt{2}}|\psi_2\rangle\right)\left(\frac{1}{\sqrt{2}}\langle\psi_1| + \frac{1}{\sqrt{2}}\langle\psi_2|\right) + \left(\frac{1}{\sqrt{2}}|\psi_1\rangle - \frac{1}{\sqrt{2}}|\psi_2\rangle\right)\left(\frac{1}{\sqrt{2}}\langle\psi_1| - \frac{1}{\sqrt{2}}\langle\psi_2|\right)$$
$$= \tfrac{1}{2}|\psi_1\rangle\langle\psi_1| + \tfrac{1}{2}|\psi_2\rangle\langle\psi_2|.$$

From Schrödinger's equation (14.13),

$$i\hbar \frac{d}{dt}|\psi_i\rangle = H|\psi_i\rangle, \qquad -i\hbar \frac{d}{dt}\langle\psi_i| = \langle\psi_i|H.$$

It follows that the density operator satisfies the evolution equation

$$\frac{d\rho}{dt} = \frac{-i}{\hbar}[H, \rho]. \tag{14.42}$$

From the solution (14.14), (14.15) of the Schrödinger equation, the solution of (14.42) is

$$\rho(t) = e^{(-i/\hbar)Ht} \rho(0) e^{(i/\hbar)Ht}. \tag{14.43}$$

Hence for any function $f(\rho) = \sum_i f(w_i)|\psi_i\rangle\langle\psi_i|$ the trace is constant,

$$\text{tr}(f(\rho)) = \text{tr}\!\left(e^{(-i/\hbar)Ht} f(\rho(0)) e^{(i/\hbar)Ht}\right) = \text{tr}(f(\rho(0))) = \text{const}.$$

A mixed state is said to be **stationary** if $d\rho/dt = 0$. From Eq. (14.42) this implies $[H, \rho] = 0$, and for any pair of energy eigenvectors

$$H|E_j\rangle = E_j|E_j\rangle, \qquad H|E_k\rangle = E_k|E_k\rangle$$

we have

$$0 = \langle E_j|\rho H - H\rho|E_k\rangle = (E_k - E_j)\langle E_j|\rho|E_k\rangle.$$

Hence, if $E_j \neq E_k$ then $\langle E_j|\rho|E_k\rangle = \rho_{jk} = 0$, and if H has no degenerate energy levels then

$$\rho = \sum_i w_i |E_i\rangle\langle E_i|,$$

which is equivalent to the assertion that the density operator is a function of the Hamiltonian, $\rho = \rho(H)$. If H has degenerate energy levels then ρ and H can be simultaneously diagonalized, and it is possible to treat this as a limiting case of non-degenerate levels. It is reasonable therefore to assume that $\rho = \rho(H)$ in all cases.

14.4 Quantum statistical mechanics

Ensembles

An **ensemble** of physical systems is another way of talking about the density operator. Essentially, we consider a large number of copies of the same system, within certain constraints, to represent a statistical system of particles. Each member of the ensemble is a possible state of the system; it is an eigenstate of the Hamiltonian and the density operator tells us its probability within the ensemble.

One of the simplest examples is the **microcanonical ensemble**, where ρ is constant for energy values in a narrow range, $E < E_k < E + \Delta E$, and $w_j = 0$ for all energy values E_j outside this range. For those energy values within the allowed range, we set $w_k = w = 1/s$ where s is the number of energy values in the range $(E, E + \Delta E)$. Let $j(E)$ be the number of states with energy $E_k < E$, then

$$s = j(E + \Delta E) - j(E) = \Sigma(E)\Delta E,$$

where

$$\Sigma(E) = \frac{dj(E)}{dE} = \text{the density of states}.$$

For the microcanonical ensemble all $w_k = 0$ for $E_k \leq E$ or $E_k \geq E + \Delta E$, while

$$w_k = \frac{1}{\Sigma(E)\Delta E} \quad \text{if} \quad E < E_k < E + \Delta E.$$

The **canonical ensemble** can be thought of as a system embedded in a heat reservoir consisting of the external world. Let H be the Hamiltonian of the system and H_R that of the reservoir. The total Hamiltonian of the universe is $H_U = H_R + H$. Suppose the system is in the eigenstate $|\psi_m\rangle$ of energy E_m

$$H|\psi_m\rangle = E_m|\psi_m\rangle$$

and let $|\Psi\rangle = |\theta\rangle|\psi_m\rangle$ be the total state of the universe. If we assume the universe to be in a microcanonical ensemble, then

$$H_U|\Psi\rangle = E_U|\Psi\rangle \quad \text{where} \quad E < E_U < E + \Delta E.$$

Using the decomposition $H_U = H_R + H$ we have

$$H_R|\theta\rangle|\psi_m\rangle + |\theta\rangle E_m|\psi_m\rangle = E_U|\theta\rangle|\psi_m\rangle,$$

whence

$$H_R|\theta\rangle = (E_U - E_m)|\theta\rangle.$$

Thus $|\theta\rangle$ is an eigenstate of H_R with energy $E_U - E_m$. If $\Sigma_R(E_R)$ is the density of states in the reservoir, then

$$w_m \Sigma_U(E_U)\Delta E = \Sigma_R(E_U - E_m)\Delta E,$$

Quantum mechanics

whence

$$w_m = \frac{\Sigma_R(E_U - E_m)}{\Sigma_U(E_U)}.$$

For $E_m \ll E_U$, as expected of a system in a much larger reservoir,

$$\ln w_m = \text{const.} - \beta E_m,$$

most commonly written in the form

$$w_m = \frac{1}{Z} e^{-\beta E_m} \quad \text{where} \quad Z = \sum_{m=0}^{\infty} e^{-\beta E_m}, \qquad (14.44)$$

where the last identity follows from $\sum_m w_m = 1$. The density operator for the canonical ensemble is thus

$$\rho = \frac{1}{Z} e^{-\beta H} \qquad (14.45)$$

where, by the identity $\text{tr}\,\rho = 1$,

$$Z = \text{tr}\, e^{-\beta H} = \sum_{m=0}^{\infty} e^{-\beta E_m}, \qquad (14.46)$$

known as the **canonical partition function**. The average energy is

$$U = \langle E \rangle = \text{tr}(\rho H) = \frac{1}{Z} \sum_k E_k e^{-\beta E_k} = -\frac{\partial \ln Z}{\partial \beta}. \qquad (14.47)$$

Example 14.12 Consider a linear harmonic oscillator having Hamiltonian given by Eq. (14.19). The energy eigenvalues are

$$E_m = \tfrac{1}{2}\hbar\omega + m\hbar\omega$$

and the partition function is

$$Z = \sum_{m=0}^{\infty} e^{-\beta E_m}$$

$$= e^{-\frac{1}{2}\beta\hbar\omega} \sum_{m=0}^{\infty} (e^{-\beta\hbar\omega})^m.$$

$$= \frac{e^{\beta\hbar\omega/2}}{e^{\beta\hbar\omega} - 1}.$$

From Eq. (14.47) the average energy is

$$U = -\frac{\partial \ln Z}{\partial \beta}$$

$$= \frac{1}{2}\hbar\omega + \frac{\hbar\omega}{e^{\beta\hbar\omega} - 1}.$$

As $\beta \to 0$ we have $U \approx \beta^{-1}$. This is the classical limit $U = kT$, where T is the temperature,

14.4 Quantum statistical mechanics

and is an indication of the identity $\beta = 1/kT$. As $\beta \to \infty$ we arrive at the low temperature limit,

$$U \to \tfrac{1}{2}\hbar\omega + \hbar\omega e^{-\beta\hbar\omega} \approx \tfrac{1}{2}\hbar\omega.$$

The **entropy** is defined as

$$S = -k\,\mathrm{tr}(\rho \ln \rho) = -k \sum_i w_i \ln w_i. \tag{14.48}$$

For a pure state, $w_i = 1$ or 0, we have $S = 0$. This is interpreted as a state of maximum order. For a completely random state, $w_i = \mathrm{const.} = 1/N$ where N is the total number of states in the ensemble (assumed finite here), the entropy is

$$S = k \ln N.$$

This state of maximal disorder corresponds to a maximum value of S, as may be seen by using the method of Lagrange multipliers: the maximum of S occurs where $dS = 0$ subject to the constraint $\sum_i w_i = 1$,

$$dS = 0 \implies d\sum_i w_i \ln w_i - \lambda \sum_i dw_i = 0$$
$$\implies \sum_i (1 + \ln w_i - \lambda)\,dw_i = 0.$$

Since dw_i is arbitrary the Lagrange multiplier is $\lambda = 1 + \ln w_i$, so that

$$w_1 = w_2 = \cdots = e^{\lambda - 1}.$$

Exercise: Two systems may be said to be independent if their combined density operator is $\rho = \rho_1 \rho_2 = \rho_2 \rho_1$. Show that the entropy has the additive property for independent systems, $S = S_1 + S_2$.

If the Hamiltonian depends on a parameter a, we define 'generalized force' A conjugate to a by

$$A = \left\langle -\frac{\partial H}{\partial a} \right\rangle = -\mathrm{tr}\left(\rho \frac{\partial H}{\partial a}\right). \tag{14.49}$$

For example, for a gas in a volume V, the **pressure** p is defined as

$$p = \left\langle -\frac{\partial H}{\partial V} \right\rangle = -\mathrm{tr}\left(\rho \frac{\partial H}{\partial V}\right).$$

If $H|\psi_k\rangle = E_k|\psi_k\rangle$ where $E_k = E_k(a)$ and $|\psi_k\rangle = |\psi_k(a)\rangle$, then

$$\frac{\partial H}{\partial a}|\psi_k\rangle + H\frac{\partial}{\partial a}|\psi_k\rangle = \frac{\partial E_k}{\partial a}|\psi_k\rangle + E_k \frac{\partial}{\partial a}|\psi_k\rangle$$

Quantum mechanics

so that

$$A = -\left\langle \frac{\partial H}{\partial a} \right\rangle = -\mathrm{tr}\left(\rho \frac{\partial H}{\partial a}\right)$$

$$= -\sum_k w_k \langle \psi_k | \frac{\partial H}{\partial a} | \psi_k \rangle$$

$$= -\sum_k w_k \left(\langle \psi_k | \frac{\partial E_k}{\partial a} | \psi_k \rangle + \langle \psi_k | E_k - H | \frac{\partial}{\partial a} \psi_k \rangle \right)$$

$$= -\sum_k w_k \left(\frac{\partial E_k}{\partial a} \| \psi_k \|^2 + \langle \psi_k | E_k - E_k | \frac{\partial}{\partial a} \psi_k \rangle \right)$$

$$= -\sum_k w_k \frac{\partial E_k}{\partial a}.$$

The total work done under a change of parameter da is defined to be

$$dW = -dU = -\sum_k w_k \frac{\partial E_k}{\partial a} = A\, da.$$

For a change in volume this gives the classical formula $dW = p\,dV$.

For the canonical ensemble we have, by Eq. (14.44),

$$A = -\sum_k w_k \frac{\partial E_k}{\partial a} = \frac{1}{\beta} \frac{\partial \ln Z}{\partial a}, \tag{14.50}$$

and as the entropy is given by

$$S = -k \sum_k w_k \ln w_k = k \sum_k w_k (\ln Z + \beta E_k) = k(\ln Z + \beta U)$$

we have

$$dS = k(d \ln Z + \beta\, dU) = \frac{1}{T}(dU + A\, da) \tag{14.51}$$

where

$$\beta = \frac{1}{kT}.$$

This relation forms the basic connection between statistical mechanics and thermodynamics (see Section 16.4); the quantity T is known as the **temperature** of the system.

Systems of identical particles

For a system of N identical particles, bosons or fermions, let h be the Hamiltonian of each individual particle, having eigenstates

$$h|\varphi_a\rangle = \varepsilon_a |\varphi_a\rangle \quad (a = 0, 1, \ldots).$$

14.4 Quantum statistical mechanics

The Hamiltonian of the entire system is $H : \mathcal{H}^N \to \mathcal{H}^N$ given by

$$H = h_1 + h_2 + \cdots + h_N$$

where

$$h_i |\psi_1\rangle \ldots |\psi_i\rangle \ldots |\psi_N\rangle = |\psi_1\rangle \ldots h|\psi_i\rangle \ldots |\psi_N\rangle.$$

The eigenstates of the total Hamiltonian,

$$H|\Phi_k\rangle = E_k|\Phi_k\rangle,$$

are linear combinations of state vectors

$$|\psi_1\rangle |\psi_2\rangle \ldots |\psi_N\rangle$$

such that

$$E_k = \varepsilon_0 + \varepsilon_1 + \cdots + \varepsilon_N.$$

If n_0 particles are in state $|\varphi_0\rangle$, n_1 particles in state $|\varphi_1\rangle$, etc. then the energy eigenstates are determined by the set of occupation numbers $(n_0, n_1, n_2, \ldots)$ such that

$$E_k = \sum_{a=0}^{\infty} n_a \varepsilon_a.$$

If we are looking for eigenstates that are simultaneously eigenstates of the permutation operators P, then they must be symmetric states $P|\Psi_k\rangle = |\Psi_k\rangle$ for bosons, and antisymmetric states $P|\Psi_k\rangle = (-1)^P |\Psi_k\rangle$ in the case of fermions. Let S be the symmetrization operator and A the antisymmetrization operator

$$S = \frac{1}{N!} \sum_P P, \qquad A = \frac{1}{N!} \sum_P (-1)^P P.$$

Both are hermitian and idempotent

$$S^* = \frac{1}{N!} \sum_P P^* = \frac{1}{N!} \sum_P P^{-1} = S, \qquad A^* = A \text{ since } (-1)^{P^*} = (-1)^P$$

$$S^2 = S, \qquad A^2 = A, \qquad AS = SA = 0.$$

Thus S and A are orthogonal projection operators, and for any state $|\Psi\rangle$

$$PS|\Psi\rangle = S|\Psi\rangle, \qquad PA|\Psi\rangle = (-1)^P A|\Psi\rangle$$

for all permutations P. For bosons the eigenstates are of the form $S|\varphi_{a_1}\rangle|\varphi_{a_2}\rangle \ldots |\varphi_{a_N}\rangle$, while for fermions they are $A|\varphi_{a_1}\rangle|\varphi_{a_2}\rangle \ldots |\varphi_{a_N}\rangle$. In either case the state of the system is completely determined by the occupation numbers $n_0, n_1, \ldots$ For bosons the occupation numbers run from 0 to N, while the Pauli exclusion principle implies that fermionic occupation numbers only take on values 0 or 1. Thus, for the canonical distribution

$$w(n_0, n_1, \ldots) = \frac{1}{Z} e^{-\beta \sum_a n_a \varepsilon_a}, \qquad \sum_a n_a = N,$$

where

$$Z_{Bose} = \sum_{n_0=0}^{N}\sum_{n_1=0}^{N}\ldots e^{-\beta \sum_a n_a \varepsilon_a},$$

$$Z_{Fermi} = \sum_{n_0=0}^{1}\sum_{n_1=0}^{1}\ldots e^{-\beta \sum_a n_a \varepsilon_a}.$$

The constraint $\sum_a n_a = N$ makes these sums quite difficult to calculate directly.

In the classical version where particles are distinguishable, all ways of realizing a configuration are counted separately,

$$\begin{aligned}Z_{Boltzmann} &= \sum_{n_0=0}^{N}\sum_{n_1=0}^{N}\ldots e^{-\beta \sum_a n_a \varepsilon_a}\frac{N!}{n_0! n_1! \ldots}\\ &= \left(e^{-\beta \varepsilon_0} + e^{-\beta \varepsilon_1} + \ldots\right)^N \\ &= (Z_1)^N,\end{aligned}$$

where Z_1 is the one-particle partition function

$$Z_1 = \sum_a e^{-\beta \varepsilon_a}.$$

The average energy is

$$U = \langle E \rangle = -\frac{\partial \ln Z}{\partial \beta} = \frac{N \sum_a \varepsilon_a e^{-\beta \varepsilon_a}}{\sum_a e^{-\beta \varepsilon_a}} = N\langle \varepsilon \rangle.$$

It is generally accepted that $Z_{Boltzmann}$ should be divided by $N!$, discounting all possible permutations of particles, in order to avoid the Gibbs paradox.

In the quantum case, it is easier to consider an even larger distribution, wherein the number of particles is no longer fixed. Assuming an *open system*, allowing exchange of particles between the system and reservoir, an argument similar to that used to arrive at the canonical ensemble gives

$$w_{mN} = \frac{1}{Z_g}e^{\alpha N - \beta E_m},$$

where

$$Z_g = \sum_{N=0}^{\infty}\sum_m e^{\alpha N - \beta E_m}.$$

This is known as the partition function for the **grand canonical ensemble**. In terms of the density operator,

$$\rho = \frac{1}{Z_g}e^{-\beta H + \alpha N}, \quad Z_g = \text{tr}\, e^{-\beta H + \alpha N}.$$

For a system of identical particles,

$$w(n_0, n_1, \ldots) = \frac{1}{Z}e^{\alpha \sum_a n_a - \beta \sum_a n_a \varepsilon_a},$$

14.4 Quantum statistical mechanics

where $\sum_a n_a = N$ is no longer fixed. We can therefore write

$$Z_{\text{Bose}} = \sum_{n_0=0}^{\infty}\sum_{n_1=0}^{\infty}\ldots e^{\alpha \sum_a n_a - \beta \sum_a n_a \varepsilon_a}$$

$$= \sum_{n_0=0}^{\infty} e^{(\alpha n_0 - \beta \varepsilon_0)n_0} \sum_{n_1=0}^{\infty} e^{(\alpha n_1 - \beta \varepsilon_1)n_1}\ldots$$

$$= \frac{1}{1-e^{\alpha-\beta\varepsilon_0}}\frac{1}{1-e^{\alpha-\beta\varepsilon_1}}\ldots$$

and

$$\ln Z_{\text{Bose}} = -\sum_{a=0}^{\infty}\ln\left(1-\lambda e^{-\beta\varepsilon_a}\right), \quad \lambda = e^{\alpha}.$$

Similarly

$$Z_{\text{Fermi}} = \sum_{n_0=0}^{1}\sum_{n_1=0}^{1}\ldots e^{\alpha\sum_a n_a - \beta\sum_a n_a \varepsilon_a}\ldots$$

results in

$$\ln Z_{\text{Fermi}} = \sum_{a=0}^{\infty}\ln\left(1+\lambda e^{-\beta\varepsilon_a}\right), \quad \lambda = e^{\alpha}.$$

Summarizing, we have

$$\ln Z = \pm\sum_{a=0}^{\infty}\ln\left(1\pm\lambda e^{-\beta\varepsilon_a}\right) \tag{14.52}$$

where the + sign occurs for fermions, the − sign for bosons.

The average occupation numbers are

$$\langle n_a\rangle = \sum_{n_0}\sum_{n_1}\ldots n_a w(n_0, n_1, \ldots)$$

$$= \frac{1}{Z}\sum_{n_0}\sum_{n_1}\ldots n_a e^{\sum_a(\alpha-\beta\varepsilon_a)n_a}$$

$$= -\frac{1}{\beta}\frac{\partial}{\partial\varepsilon_a}\ln Z.$$

Using Eq. (14.52),

$$\langle n_a\rangle = -\frac{1}{\beta}\frac{\partial}{\partial\varepsilon_a}\left(\pm\ln\left(1\pm\lambda e^{-\beta\varepsilon_a}\right)\right) = \frac{1}{\lambda^{-1}e^{\beta\varepsilon_a}\pm 1}, \tag{14.53}$$

where the + sign applies to Fermi particles, and the − to Bose. The parameter λ is often written $\lambda = e^{\beta\mu}$, where μ is known as the **chemical potential**. It can be shown, using the method of steepest descent (see [10]) that the formulae (14.53) are also valid for the canonical ensemble. In this case the total particle number is fixed so that the chemical potential may be found from

$$\sum_a \langle n_a\rangle = \frac{1}{Z}\sum_{n_0}\sum_{n_1}\ldots\left(\sum_a n_a\right)e^{\sum_b(\alpha-\beta\varepsilon_b)n_b} = N\sum_{n_0}\sum_{n_1}\ldots w(n_0, n_1, \ldots) = N.$$

Quantum mechanics

That is,

$$N = \sum_a \frac{1}{e^{\beta(\varepsilon_a - \mu)} \pm 1}$$

and

$$U = \langle E \rangle = -\frac{\partial Z}{\partial \beta} = \sum_a \frac{\varepsilon_a}{e^{\beta(\varepsilon_a - \mu)} \pm 1} = \sum_a \varepsilon_a \langle n_a \rangle.$$

Application to perfect gases, black body radiation and other systems may be found in any standard book on statistical mechanics [1, 10–12].

Problems

Problem 14.17 Show that the correctly normalized fermion states are

$$\frac{1}{\sqrt{N!}} \sum_P (-1)^P P |\varphi_{a_1}\rangle |\varphi_{a_2}\rangle \ldots |\varphi_{a_N}\rangle$$

and normalized boson states are

$$\frac{1}{\sqrt{N!}\sqrt{n_0!}\sqrt{n_1!}\ldots} \sum_P P |\varphi_{a_1}\rangle |\varphi_{a_2}\rangle \ldots |\varphi_{a_N}\rangle.$$

Problem 14.18 Calculate the canonical partition function, mean energy U and entropy S, for a system having just two energy levels 0 and E. If $E = E(a)$ for a parameter a, calculate the force A and verify the thermodynamic relation $dS = \frac{1}{T}(dU + A\, da)$.

Problem 14.19 let $\rho = e^{-\beta H}$ be the unnormalized canonical distribution. For a free particle of mass m in one dimension show that its position representation form $\rho(x, x'; \beta) = \langle x|\rho|x'\rangle$ satisfies the diffusion equation

$$\frac{\partial \rho(x, x'; \beta)}{\partial \beta} = \frac{\hbar^2}{2m} \frac{\partial^2}{\partial x^2} \rho(x, x'; \beta)$$

with 'initial' condition $\rho(x, x'; 0) = \delta(x - x')$. Verify that the solution is

$$\rho(x, x'; \beta) = \left(\frac{m}{2\pi \hbar^2 \beta}\right)^{1/2} e^{-m(x-x')^2/2\hbar^2 \beta}.$$

Problem 14.20 A solid can be regarded as being made up of $3N$ independent quantum oscillators of angular frequency ω. Show that the canonical partition function is given by

$$Z = \left(\frac{e^{-\beta \hbar \omega / 2}}{1 - e^{-\beta \hbar \omega}}\right)^{3N},$$

and the specific heat is given by

$$C_V = \frac{dU}{dT} = 3Nk \left(\frac{T_0}{T}\right)^2 \frac{e^{T_0/T}}{(e^{T_0/T} - 1)^2} \quad \text{where} \quad kT_0 = \hbar \omega.$$

Show that the high temperature limit $T \gg T_0$ is the classical value $C_V = 3Nk$.

Problem 14.21 Show that the average occupation numbers for the classical distribution, $Z_{\text{Boltzmann}}$ are given by

$$\langle n_a \rangle = -\frac{1}{\beta} \frac{\partial}{\partial \varepsilon_a} \ln Z_{\text{Boltzmann}} = \lambda e^{-\beta \varepsilon_a}.$$

Hence show that

$$\langle n_a \rangle_{\text{Fermi}} < \langle n_a \rangle_{\text{Boltzmann}} < \langle n_a \rangle_{\text{Bose}}$$

and that all three types agree approximately for low occupation numbers $\langle n_a \rangle \ll 1$.

Problem 14.22 A *spin system* consists of N particles of magnetic moment μ in a magnetic field B. When n particles have spin up, $N - n$ spin down, the energy is $E_n = n\mu B - (N - n)\mu B = (2n - N)\mu B$. Show that the canonical partition function is

$$Z = \frac{\sinh\big((N+1)\beta\mu B\big)}{\sinh \beta\mu B}.$$

Evaluate the mean energy U and entropy S, sketching their dependence on the variable $x = \beta\mu B$.

References

[1] R. H. Dicke and J. P. Wittke. *Introduction to Quantum Mechanics*. Reading, Mass., Addison-Wesley, 1960.

[2] P. Dirac. *The Principles of Quantum Mechanics*. Oxford, Oxford University Press, 1958.

[3] J. M. Jauch. *Foundations of Quantum Mechanics*. Reading, Mass., Addison-Wesley, 1968.

[4] A. Sudbery. *Quantum Mechanics and the Particles of Nature*. Cambridge, Cambridge University Press, 1986.

[5] R. D. Richtmyer. *Principles of Advanced Mathematical Physics, Vol. 1*. New York, Springer-Verlag, 1978.

[6] M. Schlechter. *Operator Methods in Quantum Mechanics*. New York, Elsevier-North Holland, 1981.

[7] J. von Neumann. *Mathematical Foundations of Quantum Mechanics*. Princeton, N.J., Princeton University Press, 1955.

[8] E. Zeidler. *Applied Functional Analysis*. New York, Springer-Verlag, 1995.

[9] R. F. Streater and A. S. Wightman. *PCT, Spin and Statistics, and All That*. New York, W. A. Benjamin, 1964.

[10] E. Schrödinger. *Statistical Thermodynamics*. Cambridge, Cambridge University Press, 1962.

[11] L. D. Landau and E. M. Lifshitz. *Statistical Physics*. Oxford, Pergamon, 1980.

[12] F. Reif. *Statistical and Thermal Physics*. New York, McGraw-Hill, 1965.

15 Differential geometry

For much of physics and mathematics the concept of a continuous map, provided by topology, is not sufficient. What is often required is a notion of *differentiable* or *smooth* maps between spaces. For this, our spaces will need a structure something like that of a surface in Euclidean space $\mathbb{R}^n$. The key ingredient is the concept of a *differentiable manifold*, which can be thought of as topological space that is 'locally Euclidean' at every point. *Differential geometry* is the area of mathematics dealing with these structures. Of the many excellent books on the subject, the reader is referred in particular to [1–14].

Think of the surface of the Earth. Since it is a sphere, it is neither metrically nor topologically identical with the Euclidean plane $\mathbb{R}^2$. A typical atlas of the world consists of separate pages called *charts*, each representing different regions of the Earth. This representation is not metrically correct since the curved surface of the Earth must be flattened out to conform with a sheet of paper, but it is at least smoothly continuous. Each chart has regions where it connects with other charts – a part of France may find itself on a map of Germany, for example – and the correspondence between the charts in the overlapping regions should be continuous and smooth. Some charts may even find themselves entirely inside others; for example, a map of Italy will reappear on a separate page devoted entirely to Europe. Ideally, the entire surface of the Earth should be covered by the different charts of the atlas, although this may not strictly be the case for a real atlas, since the north and south poles are not always properly covered by some chart. We have here the archetype of a differentiable manifold.

Points of $\mathbb{R}^n$ will usually be denoted from now on by superscripted coordinates, $\mathbf{x} = (x^1, x^2, \ldots, x^n)$. In Chapter 12 we defined a function $f : \mathbb{R}^n \to \mathbb{R}$ to be C^r if all its partial derivatives

$$\frac{\partial^s f(x^1, x^2, \ldots, x^n)}{\partial x^{i_1} \partial x^{i_2} \ldots \partial x^{i_s}}$$

exist and are continuous for $s = 1, 2, \ldots, r$. A C^0 function is simply a continuous function, while a C^∞ function is one that is C^r for all values of $r = 0, 1, 2, \ldots$; such a function will generally be referred to simply as a **differentiable function**. A differentiable function need not be *analytic* (expandable as a power series in a neighbourhood of any point), as illustrated by the function

$$f(x) = \begin{cases} 0 & \text{if } x \leq 0, \\ e^{-1/x^2} & \text{if } x > 0, \end{cases}$$

410

15.1 Differentiable manifolds

which is differentiable but not analytic at $x = 0$ since its power series would have all coefficients zero at $x = 0$.

A map $\phi : \mathbb{R}^n \to \mathbb{R}^m$ is said to be C^r if, when expressed in coordinates

$$y^i = \phi_i(x^1, x^2, \ldots, x^n) \quad \text{where} \quad \phi_i = \mathrm{pr}_i \circ \phi \quad (i = 1, 2, \ldots, m),$$

each of the real-valued functions $\phi_i : \mathbb{R}^n \to \mathbb{R}$ is C^r. Similarly, the notion of differentiable and analytic functions can be extended to maps between Euclidean spaces of arbitrary dimensions.

15.1 Differentiable manifolds

A **locally Euclidean space** or **topological manifold** M of **dimension** n is a Hausdorff topological space M in which every point x has a neighbourhood homeomorphic to an open subset of $\mathbb{R}^n$. If p is any point of M then a **(coordinate) chart** at p is a pair (U, ϕ) where U is an open subset of M, called the **domain** of the chart and $\phi : U \to \phi(U) \subset \mathbb{R}^n$ is a homeomorphism between U and its image $\phi(U)$. The image $\phi(U)$ is an open subset of $\mathbb{R}^n$, given the relative topology in $\mathbb{R}^n$. It is also common to call U a **coordinate neighbourhood** of p and ϕ a **coordinate map**. The functions $x^i = \mathrm{pr}_i \circ \phi : U \to \mathbb{R}$ $(i = 1, \ldots, n)$, where $\mathrm{pr}_i : \mathbb{R}^n \to \mathbb{R}$ are the standard projection maps, are known as the **coordinate functions** determined by this chart, and the real numbers $x^i(p)$ are called the **coordinates** of p in this chart (see Fig. 15.1). Sometimes, when we wish to emphasize the symbols to be used for the coordinate functions, we denote the chart by $(U, \phi; x^i)$, or simply $(U; x^i)$. Occasionally the term **coordinate system at** p is used for a chart whose domain U covers p. The use of superscripts rather than subscripts for coordinate functions is not universal, but its advantages will become apparent as the tensor formalism on manifolds is developed.

For any pair of coordinate charts $(U, \phi; x^i)$ and $(U', \phi'; x'^j)$ such that $U \cap U' \ne \emptyset$, define the **transition functions**

$$\phi' \circ \phi^{-1} : \phi(U \cap U') \to \phi'(U \cap U'),$$
$$\phi \circ \phi'^{-1} : \phi'(U \cap U') \to \phi(U \cap U'),$$

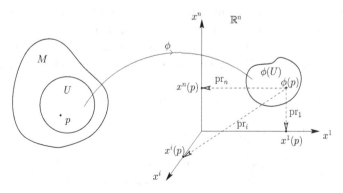

Figure 15.1 Chart at a point p

Differential geometry

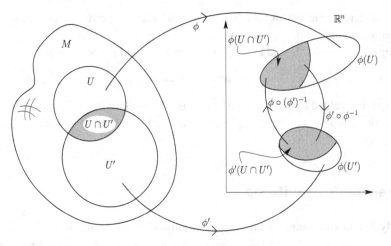

Figure 15.2 Transition functions on compatible charts

which are depicted in Fig. 15.2. The transition functions are often written

$$x'^j = x'^j(x^1, x^2, \ldots, x^n) \quad \text{and} \quad x^i = x^i(x'^1, x'^2, \ldots, x'^n) \quad (i, j = 1, \ldots, n) \quad (15.1)$$

which is an abbreviated form of the awkward, but technically correct,

$$x'^j(p) = \mathrm{pr}_i \circ \phi' \circ \phi^{-1}(x^1(p), x^2(p), \ldots, x^n(p)),$$
$$x^i(p) = \mathrm{pr}_i \circ \phi \circ \phi'^{-1}(x'^1(p), x'^2(p), \ldots, x'^n(p)).$$

The two charts are said to be C^r-**compatible** where r is a non-negative integer or ∞, if all the functions in (15.1) are C^r. For convenience we will generally assume that the charts are C^∞-compatible.

An **atlas** on M is a family of charts $\mathcal{A} = \{(U_\alpha, \phi_\alpha) \mid \alpha \in A\}$ such that the coordinate neighbourhoods U_α cover M, and any pair of charts from the family are C^∞-compatible. If $\mathcal{A}$ and $\mathcal{A}'$ are two atlases on M then so is their union $\mathcal{A} \cup \mathcal{A}'$.

Exercise: Prove this statement. [*Hint*: A differentiable function of a differentiable function is always differentiable.]

Any atlas $\mathcal{A}$ may thus be extended to a *maximal atlas* by adding to it all charts that are C^∞-compatible with the charts of $\mathcal{A}$. This maximal atlas is called a **differentiable structure** on M. A pair $(M, \mathcal{A})$, where M, is an n-dimensional topological manifold and $\mathcal{A}$ is a differentiable structure on M, is called a **differentiable manifold**; it is usually just denoted M.

The Jacobian matrix $\mathsf{J} = [\partial x'^k/\partial x^j]$ is non-singular since its inverse is $\mathsf{J}^{-1} = [\partial x^i/\partial x'^k]$,

$$\mathsf{J}^{-1}\mathsf{J} = \left[\frac{\partial x^i}{\partial x'^k}\frac{\partial x'^k}{\partial x^j}\right] = \left[\frac{\partial x^i}{\partial x^j}\right] = [\delta^i_j] = \mathsf{I}.$$

Similarly $\mathsf{J}\mathsf{J}^{-1} = \mathsf{I}$. Hence the Jacobian determinant is non-vanishing, $\det[\partial x'^j/\partial x^i] \neq 0$. We are making a return here and in the rest of this book to the summation convention of earlier chapters.

15.1 Differentiable manifolds

Example 15.1 Euclidean space $\mathbb{R}^n$ is trivially a manifold, since the single chart $(U = \mathbb{R}^n, \phi = \mathrm{id})$ covers it and generates a unique atlas consisting of all charts that are compatible with it. For example, in $\mathbb{R}^2$ it is permissible to use polar coordinates (r, θ) defined by

$$x = r\cos\theta, \qquad y = r\sin\theta,$$

which are compatible with (x, y) on the open set $U = \mathbb{R}^2 - \{(x, y) \mid x \geq 0, \ y = 0\}$. The inverse transformation is

$$r = \sqrt{x^2 + y^2}, \qquad \theta = \begin{cases} \arctan y/x & \text{if } y > 0, \\ \pi & \text{if } y = 0, \ x < 0, \\ \pi + \arctan y/x & \text{if } y < 0. \end{cases}$$

The image set $\phi(U)$ in the (r, θ)-plane is a semi-infinite open strip $r > 0$, $0 < \theta < 2\pi$.

Example 15.2 Any open region U of $\mathbb{R}^n$ is a differentiable manifold formed by giving it the relative topology and the differentiable structure generated by the single chart $(U, \mathrm{id}_U : U \to \mathbb{R}^n)$. Every chart on U is the restriction of a coordinate neighbourhood and coordinate map on $\mathbb{R}^n$ to the open region U and can be written $(U \cap V, \psi|_{U \cap V})$ where (V, ψ) is a chart on $\mathbb{R}^n$. Such a manifold is called an **open submanifold** of $\mathbb{R}^n$.

Exercise: Describe the open region of $\mathbb{R}^3$ and the image set in the (r, θ, ϕ) on which spherical polar coordinates are defined,

$$x = r\sin\theta\cos\phi, \qquad y = r\sin\theta\sin\phi, \qquad z = r\cos\theta. \tag{15.2}$$

Example 15.3 The unit circle $S^1 \subset \mathbb{R}^2$, defined by the equation $x^2 + y^2 = 1$, is a one-dimensional manifold. The coordinate x can be used on either the upper semicircle $y > 0$ or the lower semicircle $y < 0$, but not on all of S^1. Alternatively, setting $r = 1$ in polar coordinates as defined in Example 15.1, a possible chart is $(U, \phi; \theta)$ where $U = S^1 - \{(1, 0)\}$ and $\phi : U \to \mathbb{R}$ is defined by $\phi((x, y)) = \theta$. The image set $\phi(U)$ is the open interval $(0, 2\pi) \subset \mathbb{R}$. These charts are clearly compatible with each other. S^1 is the only one-dimensional manifold that is not homeomorphic to the real line $\mathbb{R}$.

Example 15.4 The 2-sphere S^2 defined as the subset of points (x, y, z) of $\mathbb{R}^3$ satisfying

$$x^2 + y^2 + z^2 = 1$$

is a two-dimensional differentiable manifold. Some possible charts on S^2 are:

(i) Rectangular coordinates (x, y), defined on the upper and lower hemisphere, $z > 0$ and $z < 0$, separately. These two charts are non-intersecting and do not cover the sphere since points on the central plane $z = 0$ are omitted.

(ii) Stereographic projection from the north pole, Eqs. (10.1) and (10.2), defines a chart $(S^2 - \{(0, 0, 1)\}, \mathrm{St}_N)$ where $\mathrm{St}_N : (x, y, z) \mapsto (X, Y)$ is given by

$$X = \frac{x}{1-z}, \qquad Y = \frac{y}{1-z}.$$

These coordinates are not defined on the sphere's north pole $N = (0, 0, 1)$, but a similar projection St_S from the south pole $S = (0, 0, -1)$ will cover N,

$$X' = \frac{x}{1+z}, \qquad Y' = \frac{y}{1+z}.$$

Both of these charts are evidently compatible with the rectangular coordinate charts (i) and therefore with each other in their region of overlap.

(iii) Spherical polar coordinates (θ, ϕ) defined by seting $r = 1$ in Eq. (15.2). Simple algebra shows that these are related to the stereographic coordinates (ii) by

$$X = \cot \tfrac{1}{2}\theta \, \cos\phi, \qquad Y = \cot \tfrac{1}{2}\theta \, \sin\phi,$$

and therefore form a compatible chart on their region of definition.

In a similar way the n-sphere S^n,

$$S^n = \{\mathbf{x} \in \mathbb{R}^{n+1} \mid (x^1)^2 + (x^2)^2 + \cdots + (x^{n+1})^2 = 1\}$$

is a differentiable manifold of dimension n. A set of charts providing an atlas is the set of rectangular coordinates on all hemispheres, $(U_i^+, \phi_i^\pm)$ and $(U_i^-, \phi_i^\pm)$, where

$$U_i^+ = \{\mathbf{x} \in S^n \mid x^i > 0\}, \qquad U_i^- = \{\mathbf{x} \in S^n \mid x^i < 0\},$$

and $\phi_i^+ : U_i^+ \to \mathbb{R}^n$ and $\phi_i^- : U_i^- \to \mathbb{R}^n$ are both defined by

$$\phi_i^\pm(\mathbf{x}) = (x^1, x^2, \ldots, x^{i-1}, x^{i+1}, \ldots, x^{n+1}).$$

Exercise: Prove that St_N and St_S are compatible, by showing they are related by

$$X' = \frac{X}{X^2 + Y^2}, \qquad Y' = \frac{Y}{X^2 + Y^2}.$$

Example 15.5 The set of $n \times n$ real matrices $M(n, \mathbb{R})$ can be put in one-to-one correspondence with points of $\mathbb{R}^{n^2}$, through the map $\phi : M(n, \mathbb{R}) \to \mathbb{R}^{n^2}$ defined by

$$\phi(A = [a_{ij}]) = (a_{11}, a_{12}, \ldots, a_{1n}, a_{21}, a_{22}, \ldots, a_{nn}).$$

This provides $M(n, \mathbb{R})$ with a Hausdorff topology inherited from R^{n^2} in the obvious way. The differentiable structure generated by the chart $(M(n, \mathbb{R}), \phi)$ converts $M(n, \mathbb{R})$ into a differentiable manifold of dimension n^2.

The group of $n \times n$ real non-singular matrices $GL(n, \mathbb{R})$ consists of $n \times n$ real matrices having non-zero determinant. The determinant map $\det : GL(n, \mathbb{R}) \to \mathbb{R}$ is continuous since it is made up purely of polynomial operations, so that $\phi(GL(n, \mathbb{R})) = \det^{-1}(\mathbb{R} - \{0\})$ is an open subset of $\mathbb{R}^{n^2}$. Thus $GL(n, \mathbb{R})$ is a differentiable manifold of dimension n^2, as it is in one-to-one correspondence with an open submanifold of $\mathbb{R}^{n^2}$.

From any two differentiable manifolds M and N of dimensions m and n respectively, it is possible to form their **product** $M \times N$, which is the topological space defined in Section 10.4. Let (U_α, ϕ_α) and (V_β, ψ_β) be any families of mutually compatible charts on M and

15.2 Differentiable maps and curves

N respectively, which generate the differentiable structures on these manifolds. The charts $(U_\alpha \times V_\beta, \phi_\alpha \times \psi_\beta)$, where $\phi_\alpha \times \psi_\beta : U_\alpha \times V_\beta \to \mathbb{R}^m \times \mathbb{R}^n = \mathbb{R}^{m+n}$ defined by

$$\phi_\alpha \times \psi_\beta((p, q)) = (\phi_\alpha(p), \psi_\beta(q)) = (x^1(p), \ldots, x^m(p), y^1(q), \ldots, y^n(q)),$$

manifestly cover $M \times N$, and are clearly compatible in their overlaps. The maximal atlas generated by these charts is a differentiable structure on $M \times N$ making it into a differentiable manifold of dimension $m + n$.

Example 15.6 The topological 2-torus $T^2 = S^1 \times S^1$ (see Example 10.13) can be given a differentiable structure as a product manifold in the obvious way from the manifold structure on S^1. Similarly, one can define the n-**torus** to be the product of n circles, $T^n = S^1 \times S^1 \times \cdots \times S^1 = (S^1)^n$.

Problems

Problem 15.1 Show that the group of unimodular matrices $SL(n, \mathbb{R}) = \{A \in GL(n, \mathbb{R}) \mid \det A = 1\}$ is a differentiable manifold.

Problem 15.2 On the n-sphere S^n find coordinates corresponding to (i) stereographic projection, (ii) spherical polars.

Problem 15.3 Show that the real projective n-space P^n defined in Example 10.15 as the set of straight lines through the origin in $\mathbb{R}^{n+1}$ is a differentiable manifold of dimension n, by finding an atlas of compatible charts that cover it.

Problem 15.4 Define the complex projective n-space CP^n in a similar way to Example 10.15 as lines in $\mathbb{C}^{n+1}$ of the form $\lambda(z^0, z^1, \ldots, z^n)$ where $\lambda, z^0, \ldots, z^n \in \mathbb{C}$. Show that CP^n is a differentiable (real) manifold of dimension $2n$.

15.2 Differentiable maps and curves

Let M be a differentiable manifold of dimension n. A map $f : M \to \mathbb{R}$ is said to be **differentiable** at a point $p \in M$ if for some coordinate chart $(U, \phi; x^i)$ at p the function $\hat{f} = f \circ \phi^{-1} : \phi(U) \to \mathbb{R}$ is differentiable at $\phi(p) = \mathbf{x}(p) = (x^1(p), x^2(p), \ldots, x^n(p))$. This definition is independent of the choice of chart at p, for if (U', ϕ') is a second chart at p that is compatible with (U, ϕ), then

$$\hat{f}' = f \circ \phi'^{-1} = \hat{f} \circ \phi \circ \phi'^{-1}$$

is C^∞ since it is a differentiable function of a differentiable function. We denote by $\mathcal{F}_p(M)$ the set of all real-valued functions on M that are differentiable at $p \in M$.

Given an open set $V \subseteq M$, a real-valued function $f : M \to \mathbb{R}$ is said to be **differentiable** or **smooth** between manifolds on V if it is differentiable at every point $p \in V$. Clearly, the function need only be defined on the open subset V for this definition. We will denote the set of all real-valued functions on M that are differentiable on an open subset V by the symbol $\mathcal{F}(V)$. Since the sum $f + g$ and product fg of any pair of differentiable functions f and

g are differentiable functions, $\mathcal{F}(V)$ is a ring. Furthermore, $\mathcal{F}(V)$ is closed with respect to taking linear combinations $f + ag$ where $a \in \mathbb{R}$ and is therefore also a real vector space that at the same time is a commutative algebra with respect to multiplication of functions fg. All functions in $\mathcal{F}_p(M)$ are differentiable on some open neighbourhood V of the point $p \in M$.

Exercise: Show that $\mathcal{F}_p(M)$ is a real commutative algebra with respect to multiplication of functions.

If M and N are differentiable manifolds, dimensions m and n respectively, then a map $\alpha : M \to N$ is **differentiable at** $p \in M$ if for any pair of coordinate charts $(U, \phi; x^i)$ and $(V, \psi; y^a)$ covering p and $\alpha(p)$ respectively, its coordinate representation

$$\hat{\alpha} = \psi \circ \alpha \circ \phi^{-1} : \phi(U) \to \psi(V)$$

is differentiable at $\phi(p)$. As for differentiable real-valued functions this definition is independent of the choice of charts. The map $\hat{\alpha}$ is represented by n differentiable real-valued functions

$$y^a = \alpha^a(x^1, x^2, \ldots, x^m) \quad (a = 1, 2, \ldots, n),$$

where $\alpha^a = \text{pr}_a \circ \hat{\alpha}$.

A **diffeomorphism** is a map $\alpha : M \to N$ that is one-to-one and both α and $\alpha^{-1} : N \to M$ are differentiable. Two manifolds M and N are said to be **diffeomorphic**, written $M \cong N$, if there exists a diffeomorphism $\alpha : M \to N$; the dimensions of the two manifolds must of course be equal, $m = n$. It is a curious and difficult fact that there exist topological manifolds with more than one inequivalent differentiable structure.

A **smooth parametrized curve** on an n-dimensional manifold M is a differentiable map $\gamma : (a, b) \to M$ from an open interval $(a, b) \subseteq \mathbb{R}$ of the real line into M. The curve is said to **pass through** p at $t = t_0$ if $\gamma(t_0) = p$, where $a < t_0 < b$. Note that a parametrized curve consists of a map, not the image points $\gamma(t) \in M$. Changing the parameter from t to $t' = f(t)$, where $f : \mathbb{R} \to \mathbb{R}$ is a monotone differentiable function and $a' = f(a) < t' < b' = f(b)$, changes the parametrized curve to $\gamma' = \gamma \circ f$, but has no effect on the image points in M. Given a chart $(U, \phi; x^i)$ at p, the inverse image of the open set U is an open subset $\gamma^{-1}(U) \subseteq \mathbb{R}$. Let (a_1, b_1) be the connected component of this set that contains the real number t_0 such that $p = \gamma(t_0)$. The 'coordinate representation' of the parametrized curve γ induced by this chart is the smooth curve $\hat{\gamma} = \phi \circ \gamma : (a_1, b_1) \to \mathbb{R}^n$, described by n real-valued functions $x^i = \gamma^i(t)$ where $\gamma^i = \text{pr}_i \circ \phi \circ \gamma$. We often write this simply as $x^i = x^i(t)$ when there is no danger of any misunderstanding (see Fig. 15.3). In another chart $(U'; x'^j)$ the n functions representing the curve change to $x'^j = \gamma'^j(t) = x'^j(\gamma(t))$. Assuming compatible charts, these new functions representing the curve are again smooth, although it is possible that the parameter range (a', b') is altered.

Problems

Problem 15.5 Let $\mathbb{R}'$ be the manifold consisting of $\mathbb{R}$ with differentiable structure generated by the chart $(\mathbb{R}; y = x^3)$. Show that the identity map $\text{id}_\mathbb{R} : \mathbb{R}' \to \mathbb{R}$ is a differentiable homeomorphism, which is not a diffeomorphism.

15.3 Tangent, cotangent and tensor spaces

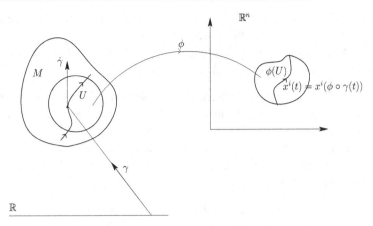

Figure 15.3 Parametrized curve on a differentiable manifold

Problem 15.6 Show that the set of real $m \times m$ matrices $M(m, n; \mathbb{R})$ is a manifold of dimension mn. Show that the matrix multiplication map $M(m, k; \mathbb{R}) \times M(k, n; \mathbb{R}) \to M(m, n; \mathbb{R})$ is differentiable.

15.3 Tangent, cotangent and tensor spaces

Tangent vectors

Let $x^i = x^i(t)$ be a curve in $\mathbb{R}^n$ passing through the point $\mathbf{x}_0 = \mathbf{x}(t_0)$. In elementary mathematics it is common to define the 'tangent' to the curve, or 'velocity', at $\mathbf{x}_0$ as the n-vector $\mathbf{v} = (\dot{\mathbf{x}}) = (\dot{x}^1, \dot{x}^2, \ldots, \dot{x}^n)$ where $\dot{x}^i = (dx^i/dt)_{t=t_0}$. In an n-dimensional manifold it is not satisfactory to define the tangent by its components, since general coordinate transformations are permitted. For example, by a rotation of axes in $\mathbb{R}^n$ it is possible to achieve that the tangent vector has components $\mathbf{v} = (v, 0, 0, \ldots, 0)$. A coordinate-independent, or *invariant*, approach revolves around the concept of the **directional derivative** of a differentiable function $f : \mathbb{R}^n \to \mathbb{R}$ along the curve at $\mathbf{x}_0$,

$$Xf = \frac{df(\mathbf{x}(t))}{dt}\bigg|_{t=t_0} = \frac{dx^i(t)}{dt}\bigg|_{t=t_0} \frac{\partial f(\mathbf{x})}{\partial x^i}\bigg|_{\mathbf{x}=\mathbf{x}_0},$$

where X is the linear differential operator

$$X = \frac{dx^i(t)}{dt}\bigg|_{t=t_0} \frac{\partial}{\partial x^i}\bigg|_{\mathbf{x}=\mathbf{x}_0}.$$

The value of the operator X when applied to a function f only depends on the values taken by the function in a neighbourhood of $\mathbf{x}_0$ along the curve in question, and is independent of coordinates chosen for the space $\mathbb{R}^n$. The above expansion demonstrates, however, that the components of the tangent vector in any coordinates on $\mathbb{R}^n$ can be extracted from the directional derivative operator from its coefficients of expansion in terms of coordinate partial derivatives.

The directional derivative operator X is a real-valued map on the algebra of differentiable functions at $\mathbf{x}_0$. Two important properties hold for the map $X : \mathcal{F}_{\mathbf{x}_0}(\mathbb{R}^n) \to \mathbb{R}$:

Differential geometry

(i) It is linear on the vector space $\mathcal{F}_{\mathbf{x}_0}(\mathbb{R}^n)$; that is, for any pair of functions f, g and real numbers a, b we have $X(af + bg) = aXf + bXg$.

(ii) The application of X on any product of functions fg in the algebra $\mathcal{F}_{\mathbf{x}_0}(\mathbb{R}^n)$ is determined by the **Leibnitz rule**, $X(fg) = f(\mathbf{x}_0)Xg + g(\mathbf{x}_0)Xf$.

These two properties completely characterize the class of directional derivative operators (see Theorem 15.1), and will be used to motivate the definition of a tangent vector at a point of a general manifold.

A **tangent vector** X_p at any point p of a differentiable manifold M is a linear map from the algebra of differentiable functions at p to the real numbers, $X_p : \mathcal{F}_p(M) \to \mathbb{R}$, which satisfies the Leibnitz rule for products:

$$X_p(af + bg) = aX_pf + bX_pg \quad \text{(linearity)}, \tag{15.3}$$

$$X_p(fg) = f(p)X_pg + g(p)X_pf \quad \text{(Leibnitz rule)}. \tag{15.4}$$

The set of tangent vectors at p form a vector space $T_p(M)$, since any linear combination $aX_p + bY_p$ of tangent vectors at p, defined by

$$(aX_p + bY_p)f = aX_pf + bY_pf,$$

is a tangent vector at p since it satisfies (15.3) and (15.4). It is called the **tangent space at** p. If (U, ϕ) is any chart at p with coordinate functions x^i, define the operators

$$\left(\partial_{x^i}\right)_p \equiv \left.\frac{\partial}{\partial x^i}\right|_p : \mathcal{F}_p(M) \to \mathbb{R}$$

by

$$\left(\partial_{x^i}\right)_p f \equiv \left.\frac{\partial}{\partial x^i}\right|_p f = \left.\frac{\partial \hat{f}(x^1,\ldots,x^n)}{\partial x^i}\right|_{\mathbf{x}=\phi(p)}, \tag{15.5}$$

where $\hat{f} = f \circ \phi^{-1} : \mathbb{R}^n \to \mathbb{R}$. These operators are clearly tangent vectors since they satisfy (15.3) and (15.4). Thus any linear combination

$$X_p = X^i \left.\frac{\partial}{\partial x^i}\right|_p \equiv \sum_{i=1}^n X^i \left.\frac{\partial}{\partial x^i}\right|_p \quad \text{where} \quad X^i \in \mathbb{R}$$

is a tangent vector. The coefficients X^j can be computed from the action of X on the coordinate functions x^j themselves:

$$X_p x^j = X^i \left.\frac{\partial x^j}{\partial x^i}\right|_{\mathbf{x}=\phi(p)} = X^i \delta_i^j = X^j.$$

Theorem 15.1 *If $(U, \phi; x^i)$ is a chart at $p \in M$, then the operators $\left(\partial_{x^i}\right)_p$ defined by (15.5) form a basis of the tangent space $T_p(M)$, and its dimension is $n = \dim M$.*

Proof: Let X_p be a tangent vector at the given fixed point p. Firstly, it follows by the Leibnitz rule (15.4) that X_p applied to a unit constant function $f = 1$ always results in zero, $X_p 1 = 0$, for

$$X_p 1 = X_p(1.1) = 1.X_p 1 + 1.X_p 1 = 2X_p 1.$$

15.3 Tangent, cotangent and tensor spaces

By linearity, X_p applied to any constant function $f = c$ results in zero, $Xc = X(c.1) = cX1 = 0$.

Set the coordinates of p to be $\phi(p) = \mathbf{a} = (a^1, a^2, \ldots, a^n)$, and let $\mathbf{y} = \phi(q)$ be any point in a neighbourhood ball $B_r(\mathbf{a}) \subseteq \phi(U)$. The function $\hat{f} = f \circ \phi^{-1}$ can be written as

$$\hat{f}(y^1, y^2, \ldots, y^n) = \hat{f}(y^1, y^2, \ldots, y^n) - \hat{f}(y^1, \ldots, y^{n-1}, a^n)$$
$$+ \hat{f}(y^1, \ldots, y^{n-1}, a^n) - \hat{f}(y^1, \ldots, y^{n-2}, a^{n-1}, a^n) + \ldots$$
$$+ \hat{f}(y^1, a^2, \ldots, a^n) - \hat{f}(a^1, a^2, \ldots, a^n) + \hat{f}(a^1, a^2, \ldots, a^n)$$
$$= \hat{f}(a^1, a^2, \ldots, a^n)$$
$$+ \sum_{i=1}^{n} \int_0^1 \frac{\partial \hat{f}(y^1, \ldots, y^{i-1}, a^i + t(y^i - a^i), a^{i+1}, \ldots, a^n)}{\partial t} dt$$
$$= \hat{f}(\mathbf{a}) + \sum_{i=1}^{n} \int_0^1 \frac{\partial \hat{f}}{\partial x^i}(y^1, \ldots, y^{i-1}, a^i + t(y^i - a^i), a^{i+1}, \ldots, a^n)$$
$$\times dt (y^i - a^i).$$

Hence, in a neighbourhood of $\mathbf{a} = \phi(p)$, any function $\hat{f}$ can be written in the form

$$\hat{f}(\mathbf{y}) = \hat{f}(\mathbf{a}) + \hat{f}_i(\mathbf{y})(y^i - a^i) \tag{15.6}$$

where the functions $\hat{f}_i(y^1, y^2, \ldots, y^n)$ are differentiable at $\mathbf{a}$. Thus, in a neighbourhood of p,

$$f(q) = \hat{f} \circ \phi = f(p) + f_i(q)(x^i(q) - a^i)$$

where $f_i = \hat{f}_i \circ \phi \in \mathcal{F}_p(M)$. Using the linear and Leibnitz properties of X_p,

$$X_p f = X_p f(p) + X_p f_i (x^i(p) - a^i) + f_i(p)(X_p x^i - X_p a^i) = f_i(p) X_p x^i$$

since $X_p c = 0$ for any constant c, and $x^i(p) = a^i$. Furthermore,

$$f_i(p) = \frac{\partial \hat{f}}{\partial x^i}(a^1, \ldots, a^n) \int_0^1 dt = \frac{\partial}{\partial x^i}\Big|_p f,$$

and the tangent vectors $(\partial_{x^i})_p$ span the tangent space $T_p(M)$,

$$X_p = X^i \frac{\partial}{\partial x^i}\Big|_p = X^i (\partial_{x^i})_p \quad \text{where} \quad X^i = X_p x^i. \tag{15.7}$$

To show that they form a basis, we need linear independence. Suppose

$$A^i \frac{\partial}{\partial x^i}\Big|_p = 0,$$

then the action on the coordinate functions $f = x^j$ gives

$$0 = A^i \frac{\partial}{\partial x^i}\Big|_p x^j = A^i \frac{\partial x^j}{\partial x^i}\Big|_\mathbf{a} = A^i \delta_i^j = A^j$$

as required. ∎

Differential geometry

This proof shows that, for every tangent vector X_p, the decomposition given by Eq. (15.7) is unique. The coefficients $X^i = X_p x^i$ are said to be the **components** of the tangent vector X_p in the chart $(U; x^i)$.

How does this definition of tangent vector relate to that given earlier for a curve in $\mathbb{R}^n$? Let $\gamma : (a, b) \to M$ be a smooth parametrized curve passing through the point $p \in M$ at $t = t_0$. Define the **tangent vector to the curve** at p to be the operator $\dot{\gamma}_p$ defined by the action on an arbitrary differentiable function f at p,

$$\dot{\gamma}_p f = \left.\frac{\mathrm{d} f \circ \gamma(t)}{\mathrm{d} t}\right|_{t=t_0}.$$

It is straightforward to verify that the $\dot{\gamma}_p$ is a tangent vector at p, as it satisfies Eqs. (15.3) and (15.4). In a chart with coordinate functions x^i at p, let the coordinate representation of the curve be $\hat{\gamma} = \phi \circ \gamma = (\gamma^1(t), \ldots, \gamma^n(t))$. Then

$$\dot{\gamma}_p f = \left.\frac{\mathrm{d}\hat{f} \circ \hat{\gamma}(t)}{\mathrm{d} t}\right|_{t=t_0} = \left.\frac{\partial \hat{f}}{\partial x^i}\right|_{\phi(p)} \left.\frac{\mathrm{d}\gamma^i(t)}{\mathrm{d} t}\right|_{t=t_0}$$

and

$$\dot{\gamma}_p = \dot{\gamma}^i(t_0) \left.\frac{\partial}{\partial x^i}\right|_p \quad \text{where} \quad \dot{\gamma}^i(t) = \frac{\mathrm{d}\gamma^i(t)}{\mathrm{d} t}.$$

In the case $M = \mathbb{R}^n$ the operator $\dot{\gamma}_p$ is precisely the directional derivative of the curve.

It is also true that every tangent vector is tangent to some curve. For example, the basis vectors $(\partial_{x^i})_p$ are tangent to the 'coordinate lines' at $p = \phi^{-1}(\mathbf{a})$,

$$\gamma^i : t \mapsto \phi^{-1}((a^1, a^2, \ldots, x^i = a^i + t - t_0, \ldots, a^n)).$$

An arbitrary tangent vector $X_p = X^i (\partial_{x^i})_p$ at p is tangent to the curve

$$\gamma : t \mapsto \phi^{-1}((a^1 + X^1(t - t_0), \ldots, x^i = a^i + X^i(t - t_0), \ldots, a^n + X^n(t - t_0))).$$

Example 15.7 The curves α, β and γ on $\mathbb{R}^2$ given respectively by

$$\alpha^1(t) = 1 + \sin t \cos t \qquad \alpha^2(t) = 1 + 3t \cos 2t,$$
$$\beta^1(t) = 1 + t \qquad \beta^2(t) = 1 + 3te^{3t},$$
$$\gamma^1(t) = e^t \qquad \gamma^2(t) = e^{3t},$$

all pass through the point $p = (1, 1)$ at $t = 0$ and are tangent to each other there,

$$\dot{\alpha}_p = \dot{\beta}_p = \dot{\gamma}_p = (\partial_{x^1})_p + 3(\partial_{x^2})_p.$$

Cotangent and tensor spaces

The dual space $T_p^*(M)$ associated with the tangent space at $p \in M$ is called the **cotangent space** at p. It consists of all linear functionals on $T_p(M)$, also called **covectors** or **1-forms** at p. The action of a covector ω_p at p on a tangent vector X_p will be denoted by $\omega_p(X_p)$, $\langle \omega_p, X_p \rangle$ or $\langle X_p, \omega_p \rangle$. From Section 3.7 we have that $\dim T_p^*(M) = n = \dim T_p(M) = \dim M$.

15.3 Tangent, cotangent and tensor spaces

If f is any function that is differentiable at p, we define its **differential** at p to be the covector $(df)_p$ whose action on any tangent vector X_p at p is given by

$$\langle (df)_p, X_p \rangle = X_p f. \tag{15.8}$$

This is a linear functional since, for any tangent vectors X_p, Y_p and scalars $a, b \in \mathbb{R}$,

$$\langle (df)_p, aX_p + bY_p \rangle = (aX_p + bY_p)f = aX_p f + bY_p f = a\langle (df)_p, X_p \rangle + b\langle (df)_p, Y_p \rangle.$$

Given a chart $(U, \phi; x^i)$ at p, the differentials of the coordinate functions have the property

$$\langle (dx^i)_p, X_p \rangle = X_p x^i = X^i$$

where X^i are the components of the tangent vector, $X_p = X^i (\partial_{x^i})_p$. Applying $(dx^i)_p$ to the basis tangent vectors, we have

$$\langle (dx^i)_p, (\partial_{x^j})_p \rangle = \frac{\partial}{\partial x^j}\bigg|_p x^i = \frac{\partial x^i}{\partial x^j}\bigg|_{\phi(p)} = \delta^i_j.$$

Hence the linear functionals $(dx^1)_p, (dx^2)_p, \ldots, (dx^n)_p$ are the dual basis, spanning the cotangent space, and every covector at p has a unique expansion

$$\omega_p = w_i (dx^i)_p \quad \text{where} \quad w_i = \langle \omega_p, (\partial_{x^i})_p \rangle.$$

The w_i are called the **components** of the linear functional ω_p in the chart $(U; x^i)$.

The differential of any function at p has a coordinate expansion

$$(df)_p = f_i (dx^i)_p$$

where

$$f_i = \langle (df)_p, (\partial_{x^i})_p \rangle = \frac{\partial}{\partial x^i}\bigg|_p f = \frac{\partial \hat{f}}{\partial x^i}\bigg|_{\phi(p)}.$$

A common way of writing this is the 'chain rule'

$$(df)_p = f_{,i}(p)(dx^i)_p \tag{15.9}$$

where

$$f_{,i} = \frac{\partial \hat{f}}{\partial x^i} \circ \phi.$$

These components are often referred to as the *gradient* of the function at p. Differentials have never found a comfortable place in calculus as non-vanishing quantities that are 'arbitrarily small'. The concept of differentials as linear functionals avoids these problems, yet has all the desired properties such as the chain rule of multivariable calculus.

As in Chapter 7, a **tensor of type** (r, s) at p is a multilinear functional

$$A_p : \underbrace{T_p^*(M) \times T_p^*(M) \times \cdots \times T_p^*(M)}_{r} \times \underbrace{T_p(M) \times \cdots \times T_p(M)}_{s} \to \mathbb{R},$$

We denote the tensor space of type (r, s) at p by $T_p^{(r,s)}(M)$. It is a vector space of dimension n^{r+s}.

Differential geometry

Vector and tensor fields

A **vector field** X is an assignment of a tangent vector X_p at each point $p \in M$. In other words, X is a map from M to the set $\bigcup_{p \in M} T_p(M)$ with the property that the image of every point, $X(p)$, belongs to the tangent space $T_p(M)$ at p. We may thus write X_p in place of $X(p)$. The vector field is said to be **differentiable** or **smooth** if for every differentiable function $f \in \mathcal{F}(M)$ the function Xf defined by

$$(Xf)(p) = X_p f$$

is differentiable, $Xf \in \mathcal{F}(M)$. The set of all differentiable vector fields on M is denoted $\mathcal{T}(M)$.

Exercise: Show that $\mathcal{T}(M)$ forms a module over the ring of functions $\mathcal{F}(M)$: if X and Y are vector fields, and $f \in \mathcal{F}(M)$ then $X + fY$ is a vector field.

Every smooth vector field defines a map $X : \mathcal{F}(M) \to \mathcal{F}(M)$, which is linear

$$X(af + bg) = aXf + bXg \qquad \text{for all } f, g \in \mathcal{F}(M) \text{ and all } a, b \in \mathbb{R},$$

and satisfies the Leibnitz rule for products

$$F(fg) = fXg + gXf.$$

Conversely, any map X with these properties defines a smooth vector field, since for each point p the map $X_p : \mathcal{F}_p(M) \to \mathcal{F}_p(M)$ defined by $X_p f = (Xf)(p)$ satisfies Eqs. (15.3) and (15.4) and is therefore a tangent vector at p.

We may also define vector fields on any open set U in a similar way as an assignment of a tangent vector at every point of U such that $Xf \in \mathcal{F}(U)$ for all $f \in \mathcal{F}(U)$. By the term **local basis of vector fields** at p we will mean an open neighbourhood U of p and a set of vector fields $\{e_1, e_2, \ldots, e_n\}$ on U such that the tangent vectors $(e_i)_q$ span the tangent space $T_q(M)$ at each point $q \in U$. For any chart $(U, \phi; x^i)$, define the vector fields on the domain U

$$\partial_{x^i} \equiv \frac{\partial}{\partial x^i} : \mathcal{F}(U) \to \mathcal{F}(U)$$

by

$$\partial_{x^i} f = \frac{\partial}{\partial x^i} f = \frac{\partial f \circ \phi^{-1}}{\partial x^i}.$$

These vector fields assign the basis tangent vectors $(\partial_{x^i})_p$ at each point $p \in U$, and form a local basis of vector fields at any point of U. When it is restricted to the coordinate domain U, every differentiable vector field X on M has a unique expansion in terms of these vector fields

$$X\big|_U = X^i \frac{\partial}{\partial x^i} = X^i \partial_{x^i}$$

where the components $X^i : U \in \mathbb{R}$ are differentiable functions on U. The local vector fields ∂_{x^i} form a module basis on U, but they are not a vector space basis since as a vector space $\mathcal{T}(U)$ is the direct sum of tangent spaces at all points $p \in U$, and is infinite dimensional.

15.3 Tangent, cotangent and tensor spaces

In a similar way we define a **covector field** or **differentiable 1-form** ω as an assignment of a covector ω_p at each point $p \in M$, such that the function $\langle \omega, X \rangle$ defined by $\langle \omega, X \rangle(p) = \langle \omega_p, X_p \rangle$ is differentiable for every smooth vector field X. The space of differentiable 1-forms will be denoted $T^*(M)$. Given any smooth function f, let df be the differentiable 1-form defined by assigning the differential df_p at each point p, so that

$$\langle df, X \rangle = Xf \quad \text{for all } X \in T(M).$$

We refer to this covector field simply as the **differential** of f. A local module basis on any chart $(U, \phi; x^i)$ consists of the 1-forms dx^i, which have the property

$$\langle dx^i, \partial_{x^j} \rangle = \frac{\partial x^i}{\partial x^j} = \delta^i{}_j.$$

Every differential can be expanded locally by the chain rule,

$$df = f_{,i} \, dx^i \quad \text{where} \quad f_{,i} = \frac{\partial}{\partial x^i} f. \tag{15.10}$$

Tensor fields are defined in a similar way, where the differentiable tensor field A of type (r, s) has a local expansion in any coordinate chart

$$A = A^{i_1 i_2 \ldots i_r}{}_{j_1 \ldots j_s} \frac{\partial}{\partial x^{i_1}} \otimes \frac{\partial}{\partial x^{i_2}} \otimes \cdots \otimes \frac{\partial}{\partial x^{i_r}} \otimes dx^{j_1} \otimes \cdots \otimes dx^{j_s}. \tag{15.11}$$

The components are differentiable functions over the coordinate domain U given by

$$A^{i_1 i_2 \ldots i_r}{}_{j_1 \ldots j_s} = A\left(dx^{i_1}, dx^{i_2}, \ldots, dx^{i_r}, \frac{\partial}{\partial x^{j_1}}, \ldots, \frac{\partial}{\partial x^{j_s}}\right).$$

Coordinate transformations

Let $(U, \phi; x^i)$ and $(U', \phi'; x'^j)$ be any two coordinate charts. From the chain rule of partial differentiation

$$\frac{\partial}{\partial x'^j} = \frac{\partial x^i}{\partial x'^j} \frac{\partial}{\partial x^i}, \quad \frac{\partial}{\partial x^i} = \frac{\partial x'^j}{\partial x^i} \frac{\partial}{\partial x'^j}. \tag{15.12}$$

Exercise: Show these equations by applying both sides to an arbitrary differentiable function f on M.

Substituting the transformations (15.12) into the expression of a tangent vector with respect to either of these bases

$$X = X^i \frac{\partial}{\partial x^i} = X'^j \frac{\partial}{\partial x'^j}$$

gives the *contravariant law of transformation* of components

$$X'^j = X^i \frac{\partial x'^j}{\partial x^i}. \tag{15.13}$$

The chain rule (15.10), written in coordinates x'^j and setting $f = x^i$, gives

$$dx^i = \frac{\partial x^i}{\partial x'^j} dx'^j.$$

Differential geometry

Expressing a differentiable 1-form ω in both coordinate bases,

$$\omega = w_i \, dx^i = w'_j \, dx'^j,$$

we obtain the *covariant transformation law* of components

$$w'_j = \frac{\partial x^i}{\partial x'^j} w_i. \tag{15.14}$$

The component transformation laws (15.13) and (15.14) can be identified with similar formulae in Chapter 3 on setting

$$A^j_i = \frac{\partial x'^j}{\partial x^i}, \qquad A'^i_k = \frac{\partial x^i}{\partial x'^k}.$$

The transformation law of a general tensor of type (r, s) follows from Eq. (7.30):

$$T'^{i'_1\ldots i'_r}{}_{j'_1\ldots j'_s} = T^{i_1\ldots i_r}{}_{j_1\ldots j_s} \frac{\partial x'^{i'_1}}{\partial x^{i_1}} \cdots \frac{\partial x'^{i'_r}}{\partial x^{i_r}} \frac{\partial x^{j_1}}{\partial x'^{j'_1}} \cdots \frac{\partial x^{j_s}}{\partial x'^{j'_s}}. \tag{15.15}$$

Tensor bundles

The **tangent bundle** TM on a manifold M consists of the set-theoretical union of all tangent spaces at all points

$$TM = \bigcup_{p \in M} T_p(M).$$

There is a natural **projection map** $\pi : TM \to M$ defined by $\pi(u) = p$ if $u \in T_p(M)$, and for each chart $(U, \phi; x^i)$ on M we can define a chart $(\pi^{-1}(U), \tilde{\phi})$ on TM where the coordinate map $\tilde{\phi} : \pi^{-1}(U) \to \mathbb{R}^{2n}$ is defined by

$$\tilde{\phi}(v) = \left(x^1(p), \ldots, x^n(p), v^1, \ldots, v^n\right)$$

$$\text{where} \quad p = \pi(v) \quad \text{and} \quad v = \sum_{i=1}^n v^i \frac{\partial}{\partial x^i}\bigg|_p.$$

The topology on TM is taken to be the coarsest topology such that all sets $\tilde{\phi}^{-1}(A)$ are open whenever A is an open subset of $\mathbb{R}^{2n}$. With this topology these charts generate a maximal atlas on the tangent bundle TM, making it into a differentiable manifold of dimension $2n$.

Given an open subset $U \subseteq M$, a smooth map $X : U \to TM$ is said to be a **smooth vector field on** U if $\pi \circ X = \mathrm{id}\big|_U$. This agrees with our earlier notion, since it assigns exactly one tangent vector from the tangent space $T_p(M)$ to the point $p \in U$. A similar idea may be used for a **smooth vector field along a parametrized curve** $\gamma : (a, b) \to M$, defined to be a smooth curve $X : (a, b) \to TM$ that *lifts* γ to the tangent bundle in the sense that $\pi \circ X = \gamma$. Essentially, this defines a tangent vector at each point of the curve, not necessarily tangent *to* the curve, in a differentiable manner.

The **cotangent bundle** T^*M is defined in an analogous way, as the union of all cotangent spaces $T^*_p(M)$ at all points $p \in M$. The generating charts have the form $(\pi^{-1}(U), \tilde{\phi})$ on

15.3 Tangent, cotangent and tensor spaces

T^*M where the coordinate map $\tilde{\phi} : \pi^{-1}(U) \to \mathbb{R}^{2n}$ is defined by

$$\tilde{\phi}(\omega_p) = (x^1(p), \ldots, x^n(p), w_1, \ldots, w_n)$$

$$\text{where } p = \pi(\omega) \text{ and } \omega = \sum_{i=1}^{n} w_i (\mathrm{d}x^i)_p,$$

making T^*M into a differentiable manifold of dimension $2n$. This process may be extended to produce the tensor bundle of type $T^{(r,s)}M$, a differentiable manifold of dimension $n + n^{r+s}$.

Problems

Problem 15.7 Let $\gamma : \mathbb{R} \to \mathbb{R}^2$ be the curve $x = 2t + 1$, $y = t^2 - 3t$. Show that at an arbitrary parameter value t the tangent vector to the curve is $X_{\gamma(t)} = \dot\gamma = 2\partial_x + (2t - 3)\partial_y = 2\partial_x + (x - 4)\partial_y$. If $f : \mathbb{R}^2 \to \mathbb{R}$ is the function $f = x^2 - y^2$, write f as a function of t along the curve and verify the identities

$$X_{\gamma(t)} f = \frac{\mathrm{d}f(t)}{\mathrm{d}t} = \langle (\mathrm{d}f)_{\gamma(t)}, X_{\gamma(t)} \rangle = C_1^1 (\mathrm{d}f)_{\gamma(t)} \otimes X_{\gamma(t)}.$$

Problem 15.8 Let $x^1 = x$, $x^2 = y$, $x^3 = z$ be ordinary rectangular cartesian coordinates in $\mathbb{R}^3$, and let $x'^1 = r$, $x'^2 = \theta$, $x'^3 = \phi$ be the usual transformation to polar coordinates.

(a) Calculate the Jacobian matrices $[\partial x^i / \partial x'^j]$ and $[\partial x'^i / \partial x^j]$.

(b) In polar coordinates, work out the components of the covariant vector fields having components in rectangular coordinates (i) $(0, 0, 1)$, (ii) $(1, 0, 0)$, (iii) (x, y, z).

(c) In polar coordinates, what are the components of the contravariant vector fields whose components in rectangular coordinates are (i) (x, y, z), (ii) $(0, 0, 1)$, (iii) $(-y, x, 0)$.

(d) If g_{ij} is the covariant tensor field whose components in rectangular coordinates are δ_{ij}, what are its components g'_{ij} in polar coordinates?

Problem 15.9 Show that the curve

$$2x^2 + 2y^2 + 2xy = 1$$

can be converted by a rotation of axes to the standard form for an ellipse

$$x'^2 + 3y'^2 = 1.$$

If $x' = \cos \psi$, $y' = \frac{1}{\sqrt{3}} \sin \psi$ is used as a parametrization of this curve, show that

$$x = \frac{1}{\sqrt{2}} \left(\cos \psi + \frac{1}{\sqrt{3}} \sin \psi \right), \qquad y = \frac{1}{\sqrt{2}} \left(-\cos \psi + \frac{1}{\sqrt{3}} \sin \psi \right).$$

Compute the components of the tangent vector

$$X = \frac{\mathrm{d}x}{\mathrm{d}\psi} \partial_x + \frac{\mathrm{d}y}{\mathrm{d}\psi} \partial_y.$$

Show that $X(f) = (2/\sqrt{3})(x^2 - y^2)$.

Problem 15.10 Show that the tangent space $T_{(p,q)}(M \times N)$ at any point (p, q) of a product manifold $M \times N$ is naturally isomorphic to the direct sum of tangent spaces $T_p(M) \oplus T_q(N)$.

Differential geometry

Problem 15.11 On the unit 2-sphere express the vector fields ∂_x and ∂_y in terms of the polar coordinate basis ∂_θ and ∂_ϕ. Again in polar coordinates, what are the dual forms to these vector fields?

Problem 15.12 Express the vector field ∂_ϕ in polar coordinates (θ, ϕ) on the unit 2-sphere in terms of stereographic coordinates X and Y.

15.4 Tangent map and submanifolds

The tangent map and pullback of a map

Let $\alpha : M \to N$ be a differentiable map between manifolds M and N, where $\dim M = m$, $\dim N = n$. This induces a map $\alpha_* : T_p(M) \to T_{\alpha(p)}(N)$, called the **tangent map** of α, whereby the tangent vector $Y_{\alpha(p)} = \alpha_* X_p$ is defined by

$$Y_{\alpha(p)} f = (\alpha_* X_p) f = X_p (f \circ \alpha)$$

for any function $f \in \mathcal{F}_{\alpha(p)}(N)$. This map is often called the *differential* of the map α, but this may cause confusion with our earlier use of this term.

Let $(U, \phi; x^i)$ and $(V, \psi; y^a)$ be charts at p and $\alpha(p)$, respectively. The map α has coordinate representation $\hat{\alpha} = \psi \circ \alpha \circ \phi^{-1} : \phi(U) \to \psi(V)$, written

$$y^a = \alpha^a(x^1, x^2, \ldots, x^m) \quad (a = 1, \ldots, n).$$

To compute the components Y^a of $Y_{\alpha(p)} = \alpha_* X_p = Y^a (\partial_{y^a})_{\alpha(p)}$, we perform the following steps:

$$Y_{\alpha(p)} f = Y^a \left.\frac{\partial f \circ \psi^{-1}}{\partial y^a}\right|_{\alpha(p)} = X^i \left.\frac{\partial}{\partial x^i}\right|_p f \circ \alpha$$

$$= X^i \left.\frac{\partial f \circ \alpha \circ \phi^{-1}}{\partial x^i}\right|_{\phi(p)}$$

$$= X^i \left.\frac{\partial f \circ \psi^{-1} \circ \hat{\alpha}}{\partial x^i}\right|_{\phi(p)}$$

$$= X^i \left.\frac{\partial f \circ \psi^{-1}(\alpha^1(\mathbf{x}), \alpha^2(\mathbf{x}), \ldots, \alpha^n(\mathbf{x}))}{\partial x^i}\right|_{\mathbf{x}=\phi(p)}$$

$$= X^i \left.\frac{\partial y^a}{\partial x^i}\right|_{\phi(p)} \left.\frac{\partial f \circ \psi^{-1}}{\partial y^a}\right|_{\hat{\alpha}(\phi(p))}$$

$$= X^i \left.\frac{\partial y^a}{\partial x^i}\right|_{\phi(p)} \left.\frac{\partial}{\partial y^a}\right|_{\alpha(p)} f.$$

Hence

$$Y^a = X^i \left.\frac{\partial y^a}{\partial x^i}\right|_{\phi(p)}. \tag{15.16}$$

Exercise: If $\alpha : M \to N$ and $\beta : K \to M$ are differentiable maps between manifolds, show that

$$(\alpha \circ \beta)_* = \alpha_* \circ \beta_*. \tag{15.17}$$

15.4 Tangent map and submanifolds

The map $\alpha : M \to N$ also induces a map α^* between cotangent spaces, but in this case it acts in the reverse direction, called the **pullback** induced by α,

$$\alpha^* : T^*_{\alpha(p)}(N) \to T^*_p(M).$$

The pullback of a 1-form $\omega_{\alpha(p)}$ is defined by requiring

$$\langle \alpha^* \omega_{\alpha(p)}, X_p \rangle = \langle \omega_{\alpha(p)}, \alpha_* X_p \rangle \qquad (15.18)$$

for arbitrary tangent vectors X_p.

Exercise: Show that this definition uniquely defines the pullback $\alpha^* \omega_{\alpha(p)}$.

Exercise: Show that the pullback of a functional composition of maps is given by

$$(\alpha \circ \beta)^* = \beta^* \circ \alpha^*. \qquad (15.19)$$

The notion of tangent map or pullback of a map can be extended to totally contravariant or totally covariant tensors, such as r-vectors or r-forms, but is only available for mixed tensors if the map is a diffeomorphism (see Problem 15.13). The tangent map does not in general apply to vector fields, for if it is not one-to-one the tangent vector may not be uniquely defined at the image point $\alpha(p)$ (see Example 15.8 below). However, no such ambiguity in the value of the pullback $\alpha^* \omega$ can ever arise at the inverse image point p, even if ω is a covector field, since its action on every tangent vector X_p is well-defined by Eq. (15.18). The pullback can therefore be applied to arbitrary differentiable 1-forms; the map α need not be either injective or surjective. This is one of the features that makes covector fields more attractive geometrical objects to deal with than vector fields. The following example should make this clear.

Example 15.8 Let $\alpha : \dot{\mathbb{R}}^3 = \mathbb{R}^3 - \{(0, 0, 0)\} \to \mathbb{R}^2$ be the differentiable map

$$\alpha : (x, y, z) \mapsto (u, v) \quad \text{where} \quad u = x + y + z, \quad v = \sqrt{x^2 + z^2}.$$

This map is neither surjective, since the whole lower half plane $v < 0$ is not mapped onto, nor injective, since, for example, the points $\mathbf{p} = (1, y, 0)$ and $\mathbf{q} = (0, y, 1)$ are both mapped to the point $(y + 1, 1)$. Consider a vector field

$$X = X^i \frac{\partial}{\partial x^i} = X^1 \frac{\partial}{\partial x} + X^2 \frac{\partial}{\partial y} + X^3 \frac{\partial}{\partial z}$$

and the action of the tangent map α_* at any point (x, y, z) is

$$\alpha_* X = \left(X^1 + X^2 + X^3 \right) \frac{\partial}{\partial u} + \left(X^1 \frac{x}{\sqrt{x^2 + z^2}} + X^3 \frac{z}{\sqrt{x^2 + z^2}} \right) \frac{\partial}{\partial v}.$$

While this map is well-defined on the tangent space at any point $\mathbf{x} = (x, y, z)$, it does not in general map the vector field X to a vector field on $\mathbb{R}^2$. For example, no tangent vector can be assigned at $\mathbf{u} = \alpha(\mathbf{p}) = \alpha(\mathbf{q})$ as we would need

$$(X^1 + X^2 + X^3)(\mathbf{p}) \frac{\partial}{\partial u}\bigg|_\mathbf{u} + X^1(\mathbf{p}) \frac{\partial}{\partial v}\bigg|_\mathbf{u} = (X^1 + X^2 + X^3)(\mathbf{q}) \frac{\partial}{\partial u}\bigg|_\mathbf{u} + X^3(\mathbf{q}) \frac{\partial}{\partial v}\bigg|_\mathbf{u}.$$

There is no reason to expect these two tangent vectors at $\mathbf{u}$ to be identical.

427

Differential geometry

However, if $\omega = w_1 \, du + w_2 \, dv$ is a differentiable 1-form on $\mathbb{R}^2$ it induces a differentiable 1-form on $\mathring{\mathbb{R}}^3$, on substituting $du = (\partial u/\partial x) \, dx + (\partial u/\partial y) \, dy + (\partial u/\partial z) \, dz$, etc.

$$\alpha^*\omega = \left(w_1 + w_2 \frac{x}{\sqrt{x^2 + z^2}}\right) dx + w_1 \, dy + \left(w_1 + w_2 \frac{z}{\sqrt{x^2 + z^2}}\right) dz,$$

which is uniquely determined at any point $(x, y, z) \neq (0, 0, 0)$ by the components $w_1(u, v)$ and $w_2(u, v)$ of the differentiable 1-form at $(u, v) = \alpha(x, y, z)$.

Example 15.9 If $\gamma : \mathbb{R} \to M$ is a curve on M and $p = \gamma(t_0)$, the tangent vector to the curve at p is the image under the tangent map induced by γ of the ordinary derivative on the real line,

$$\dot{\gamma}_p = \gamma_* \frac{d}{dt}\bigg|_{t_0},$$

for if $f : M \to \mathbb{R}$ is any function differentiable at p then

$$\gamma_* \frac{d}{dt}\bigg|_{t_0}(f) = \frac{df \circ \gamma}{dt}\bigg|_{t_0} = \dot{\gamma}_p(f).$$

By a *curve with endpoints* we shall mean the restriction of a parametrized curve $\gamma : (a, b) \to M$ to a closed subinterval of $\gamma : [t_1, t_2] \to M$ where $a < t_1 < t_2 < b$. The **integral** of a 1-form α on the curve with end points γ is defined as

$$\int_\gamma \alpha = \int_{t_1}^{t_2} \alpha(\dot{\gamma}) \, dt.$$

In a coordinate representation $x^i = \gamma^i(t)$ and $\alpha = \alpha_i \, dx^i$,

$$\int_\gamma \alpha = \int_{t_1}^{t_2} \alpha_i(x(t)) \frac{dx^i}{dt} dt.$$

Let $\gamma' = \gamma \circ f$ be the curve related to γ by a change of parametrization $t' = f(t)$ where $f : \mathbb{R} \to \mathbb{R}$ is a monotone function on the real line. Then

$$\int_\gamma \alpha = \int_{\gamma'} \alpha$$

for, by the standard change of variable formula for a definite integral,

$$\int_\gamma \alpha = \int_{t_1}^{t_2} \alpha_i(x(t)) \frac{dx^i(t(t'))}{dt'} \frac{dt'}{dt} dt$$

$$= \int_{t'_1}^{t'_2} \alpha_i(x(t(t'))) \frac{dx'^i(t')}{dt'} dt'$$

$$= \int_{\gamma'} \alpha.$$

Hence the integral of a 1-form is independent of the parametrization on the curve γ.

The integral of α along γ is zero if its pullback to the real line vanishes, $\gamma^*(\alpha) = 0$, for

$$\int_\gamma \alpha = \int_{t_1}^{t_2} \langle \alpha, \gamma_* \frac{d}{dt} \rangle dt = \int_{t_1}^{t_2} \langle \gamma^*(\alpha), \frac{d}{dt} \rangle dt = 0.$$

15.4 Tangent map and submanifolds

If α is the differential of a scalar field $\alpha = df$ it is called an **exact** 1-form. The integral of an exact 1-form is independent of the curve connecting two points $p_1 = \gamma(t_1)$ and $p_2 = \gamma(t_2)$, for

$$\int_\gamma df = \int_{t_1}^{t_2} \langle df, \dot{\gamma} \rangle \, dt = \int_{t_1}^{t_2} \frac{df(\gamma(t))}{dt} dt = f(\gamma(t_2)) - f(\gamma(t_1)),$$

which only depends on the value of f at the end points. In particular, the integral of an exact 1-form vanishes on any closed circuit, since $\gamma(t_1) = \gamma(t_2)$.

For general 1-forms, the integral is usually curve-dependent. For example, let $\alpha = x\, dy$ on the manifold $\mathbb{R}^2$ with coordinates (x, y). Consider the following two curves connecting $p_1 = (-1, 0)$ to $p_2 = (1, 0)$:

$$\begin{array}{llll} \gamma_1: x = t, & y = 0 & t_1 = -1, & t_2 = 1, \\ \gamma_2: x = \cos t, & y = \sin t & t_1 = -\pi, & t_2 = 0. \end{array}$$

The pullback of α to the first curve vanishes, $\gamma_1^* x\, dy = t\, d0 = 0$, while the pullback to γ_2 is given by

$$\gamma_2^* x\, dy = \cos t\, d(y \circ \gamma_2(t)) = \cos t\, d(\sin t) = \cos^2 t\, dt.$$

Hence

$$\int_{\gamma_2} x\, dy = \int_{-\pi}^{0} \cos^2 t\, dt = \frac{\pi}{2} \neq \int_{\gamma_1} x\, dy = 0.$$

Submanifolds

Let $\alpha: M \to N$ be a differentiable mapping where $m = \dim M \leq n \dim N$. The map is said to be an **immersion** if the tangent map $\alpha_*: T_p(M) \to T_{\alpha(p)}(N)$ is injective at every point $p \in M$; i.e., α_* is everywhere a non-degenerate linear map. From the inverse function theorem, it is straightforward to show that there exist charts at any point p and its image $\alpha(p)$ such that the map $\hat{\alpha}$ is represented as

$$\begin{array}{ll} y^i = \alpha^i(x^1, x^2, \ldots, x^m) = x^i & \text{for } i = 1, \ldots, m, \\ y^a = \alpha^a(x^1, x^2, \ldots, x^m) = 0 & \text{for } a > m. \end{array}$$

A detailed proof may be found in [11].

Example 15.10 In general the image $\alpha(M) \subset N$ of an immersion is not a genuine 'submanifold', since there is nothing to prevent self-intersections. For example the mapping $\alpha: \mathbb{R} \to \mathbb{R}^2$ defined by

$$x = \alpha^1(t) = t(t^2 - 1), \qquad y = \alpha^2(t) = t^2 - 1$$

is an immersion since its Jacobian matrix is everywhere non-degenerate,

$$\left(\frac{\partial x}{\partial t} \quad \frac{\partial y}{\partial t} \right) = (2t^2 - 1 \quad 2t) \neq (0 \quad 0) \qquad \text{for all } t \in \mathbb{R}.$$

The subset $\alpha(\mathbb{R}) \subset \mathbb{R}^2$ does not, however, inherit the manifold structure of $\mathbb{R}$ since there is a self-intersection at $t = \pm 1$, as shown in Fig. 15.4.

Differential geometry

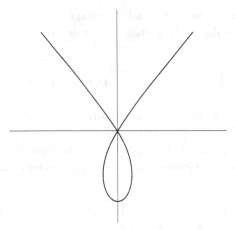

Figure 15.4 Immersion that is not a submanifold

In order to have a natural manifold structure on the subset $\alpha(M)$ we require that the map α is itself injective as well as its tangent map α_*. The map is then called an **embedding**, and the pair (M, α) an **embedded submanifold** of N.

Example 15.11 Let A be any open subset of a manifold M. As in Example 15.2, it inherits a manifold structure from M, whereby a chart is said to be admissible if it has the form $(U \cap A, \phi|_{U \cap A})$ for some chart (U, ϕ) on M. With this differentiable structure, A is said to be an **open submanifold** of M. It evidently has the same dimension as M. The pair $(A, \mathrm{id}|_A)$ is an embedded submanifold of M.

Example 15.12 Let $T^2 = S^1 \times S^1$ be the 2-torus (see Example 15.6). The space T^2 can also be viewed as the factor space $R^2/\mathrm{mod}\ 1$, where $(x, y) = (x', y')$ mod 1 if there exist integers k and l such that $x - x' = k$ and $y - y' = l$. Denote equivalence classes mod 1 by the symbol $[(x, y)]$. Consider the curve $\alpha : \mathbb{R} \to T^2$ defined by $\alpha(t) = [(at, bt)]$. This map is an immersion unless $a = b = 0$. If a/b is a rational number it is not an embedding since the curve eventually passes through $(1, 1) = (0, 0)$ for some t and α is not injective. For a/b irrational the curve never passes through any point twice and is therefore an embedding. Figure 15.5 illustrates these properties. When a/b is rational the image $C = \alpha(\mathbb{R})$ has the relative topology in T^2 of a circle. Hence there is an embedding $\beta : S^1 \to C$, making (S^1, β) an embedded submanifold of T^2. It is left to the reader to explicitly construct the map β. In this case the subset $C = \beta(S^1) \subset T^2$ is closed.

The set $\alpha(\mathbb{R})$ is dense in T^2 when a/b is irrational, since the curve eventually passes arbitrarily close to any point of T^2, and cannot be a closed subset. Hence the relative topology on $\alpha(\mathbb{R})$ induced on it as a subset of T^2 is much coarser than the topology it would obtain from $\mathbb{R}$ through the bijective map α. The embedding α is therefore not a homeomorphism from $\mathbb{R}$ to $\alpha(\mathbb{R})$ when the latter is given the relative topology.

In general, an embedding $\alpha : M \to N$ that is also a homeomorphism from M to $\alpha(M)$ when the latter is given the relative topology in N is called a **regular embedding**. A necessary and sufficient condition for this to hold is that there be a coordinate chart $(U, \phi; x^i)$

15.4 Tangent map and submanifolds

$$x = at, \quad y = bt$$

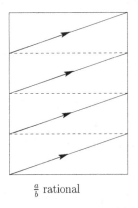
$\frac{a}{b}$ rational

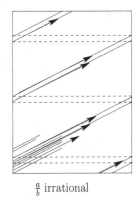
$\frac{a}{b}$ irrational

Figure 15.5 Submanifolds of the torus

at every point $p \in \alpha(M)$ such that $\alpha(M) \cap U$ is defined by the 'coordinate slice'

$$x^{m+1} = x^{m+2} = \cdots = x^n = 0.$$

It also follows that the set $\alpha(M)$ must be a closed subset of N for this to occur. The proofs of these statements can be found in [11]. The above embedded submanifold (S^1, β) is a regular embedding when a/b is rational.

Problems

Problem 15.13 Show that if $\rho_p = r_i(dx^i)_p = \alpha^* \omega_{\alpha(p)}$ then the components are given by

$$r_i = \left.\frac{\partial y^a}{\partial x^i}\right|_{\phi(p)} w_a$$

where $\omega_{\alpha(p)} = w_a(dy^a)_{\alpha(p)}$.

If α is a diffeomorphism, define a map $\alpha_* : T_p^{(1,1)}(M) \to T_{\alpha(p)}^{(1,1)}(N)$ by setting

$$\alpha_* T\left(\omega_{\alpha(p)}, X_{\alpha(p)}\right) = T\left(\alpha^* \omega_{\alpha(p)}, \alpha_*^{-1} X_{\alpha(p)}\right)$$

and show that the components transform as

$$(\alpha_* T)^a{}_b = T^i{}_j \frac{\partial y^a}{\partial x^i} \frac{\partial x^j}{\partial y^b}.$$

Problem 15.14 If $\gamma : \mathbb{R} \to M$ is a curve on M and $p = \gamma(t_0)$ and $\alpha : M \to N$ is a differentiable map show that

$$\alpha_* \dot\gamma_p = \dot\sigma_{\alpha(p)} \quad \text{where} \quad \sigma = \alpha \circ \gamma : \mathbb{R} \to N.$$

Problem 15.15 Is the map $\alpha : \mathbb{R} \to \mathbb{R}^2$ given by $x = \sin t, y = \sin 2t$ (i) an immersion, (ii) an embedded submanifold?

Problem 15.16 Show that the map $\alpha : \dot{\mathbb{R}}^2 \to \mathbb{R}^3$ defined by

$$u = x^2 + y^2, \quad v = 2xy, \quad w = x^2 - y^2$$

is an immersion. Is it an embedded submanifold?

Evaluate $\alpha^*(u\,du + v\,dv + w\,dw)$ and $\alpha_*(\partial_x)_{(a,b)}$. Find a vector field X on $\mathring{\mathbb{R}}^2$ for which $\alpha_* X$ is not a well-defined vector field.

15.5 Commutators, flows and Lie derivatives

Commutators

Let X and Y be smooth vector fields on an open subset U of a differentiable manifold M. We define their **commutator** or **Lie bracket** $[X, Y]$ as the vector field on U defined by

$$[X, Y]f = X(Yf) - Y(Xf) \tag{15.20}$$

for all differentiable functions f on U. This is a vector field since (i) it is linear

$$[X, Y](af + bg) = a[X, Y]f + b[x, y]g$$

for all $f, g \in \mathcal{F}(U)$ and $a, b \in \mathbb{R}$, and (ii) it satisfies the Leibnitz rule

$$[X, Y](fg) = f[X, Y]g + g[X, Y]f.$$

Linearity is trivial, while the Leibnitz rule follows from

$$\begin{aligned}[][X, Y](fg) &= X(fYg + gYf) - Y(fXg + gXf) \\ &= Xf\,Yg + fX(Yg) + Xg\,Yf + gX(Yf) \\ &\quad - Yf\,Xg - fY(Xg) - Yg\,Xf - gY(Xf) \\ &= f[X, Y]g + g[X, Y]f.\end{aligned}$$

A number of identities are easily verified for the Lie bracket:

$$[X, Y] = -[Y, X], \tag{15.21}$$
$$[X, aY + bZ] = a[X, Y] + b[X, Z], \tag{15.22}$$
$$[X, fY] = f[X, Y] + Xf\,Y, \tag{15.23}$$
$$[[X, Y], Z] + [[Y, Z], X] + [[Z, X], Y] = 0. \tag{15.24}$$

Equations (15.21) and (15.22) are trivial, and (15.23) follows from

$$\begin{aligned}[][X, fY]g &= X(fYg) - fY(Xg) = fX(Yg) + X(f)Yg - fY(Xg) \\ &= f[X, Y]g + Xf\,Yg.\end{aligned}$$

The **Jacobi identity** (15.24) is proved much as for commutators in matrix theory, Example 6.7.

Exercise: Show that for any functions f, g and vector fields X, Y

$$[fX, gY] = fg[X, Y] + fXg\,Y - gYf\,X.$$

To find a coordinate formula for the Lie product, let $X = X^i(x^1, \ldots, x^n)\partial_{x^i}$, $Y = Y^i(x^1, \ldots, x^n)\partial_{x^i}$. Then $[X, Y] = [X, Y]^k(x^1, \ldots, x^n)\partial_{x^k}$, where

$$[X, Y]^k = [X, Y](x^k) = X(Yx^k) - Y(Xx^k) = X^i \frac{\partial Y^k}{\partial x^i} - Y^i \frac{\partial X^k}{\partial x^i},$$

15.5 Commutators, flows and Lie derivatives

or in the comma derivative notation

$$[X, Y]^k = X^i Y^k{}_{,i} - Y^i X^k{}_{,i}. \tag{15.25}$$

If we regard the vector field X as acting on the vector field Y by the Lie bracket to produce a new vector field, $X : Y \mapsto \mathcal{L}_X Y = [X, Y]$, this action is remarkably 'derivative-like' in that it is both linear

$$\mathcal{L}_X(aY + bZ) = a\mathcal{L}_X Y + b\mathcal{L}_X Z$$

and has the property

$$\mathcal{L}_X(fY) = Xf\, Y + f\mathcal{L}_X Y. \tag{15.26}$$

These properties follow immediately from (15.22) and (15.23). A geometrical interpretation of this derivative will appear in terms of the concept of a *flow* induced by a vector field.

Integral curves and flows

Let X be a smooth vector field on a manifold M. An **integral curve** of X is a parametrized curve $\sigma : (a, b) \to M$ whose tangent vector $\dot{\sigma}(t)$ at each point $p = \sigma(t)$ on the curve is equal to the tangent vector X_p assigned to p,

$$\dot{\sigma}(t) = X_{\sigma(t)}.$$

In a local coordinate chart $(U; x^i)$ at p where the curve can be written as n real functions $x^i(t) = x^i(\sigma(t))$ and the vector field has the form $X = X^i(x^1, \ldots, x^n)\partial_{x^i}$, this requirement appears as n ordinary differential equations,

$$\frac{\mathrm{d}x^i}{\mathrm{d}t} = X^i\bigl(x^1(t), \ldots, x^n(t)\bigr). \tag{15.27}$$

The existence and uniqueness theorem of ordinary differential equations asserts that through each point $p \in M$ there exists a unique maximal integral curve $\gamma_p : (a, b) \to M$ such that $a = a(p) < 0 < b = b(p)$ and $p = \gamma_p(0)$ [15, 16]. Uniqueness means that if $\sigma : (c, d) \to M$ is any other integral curve passing through p at $t = 0$ then $a \le c < 0 < d \le b$ and $\sigma = \gamma_p|_{(c,d)}$.

By a **transformation** of the manifold M is meant a diffeomorphism $\varphi : M \to M$. A **one-parameter group of transformations** of M, or on M, is a map $\sigma : \mathbb{R} \times M \to M$ such that:

(i) for each $t \in \mathbb{R}$ the map $\sigma_t : M \to M$ defined by $\sigma_t(p) = \sigma(t, p)$ is a transformation of M;
(ii) for all $t, s \in \mathbb{R}$ we have the abelian group property, $\sigma_{t+s} = \sigma_t \circ \sigma_s$.

Since the maps σ_t are one-to-one and onto, every point $p \in M$ is the image of a unique point $q \in M$; that is, we can write $p = \sigma_t(q)$ where $q = \sigma_t^{-1}(p)$. Hence σ_0 is the identity transformation, $\sigma_0 = \mathrm{id}_M$ since $\sigma_0(p) = \sigma_0 \circ \sigma_t(q) = \sigma_t(q) = p$ for all $p \in M$. Furthermore, the inverse of each map σ_t^{-1} is σ_{-t} since $\sigma_t \circ \sigma_{-t} = \sigma_0 = \mathrm{id}_M$.

Differential geometry

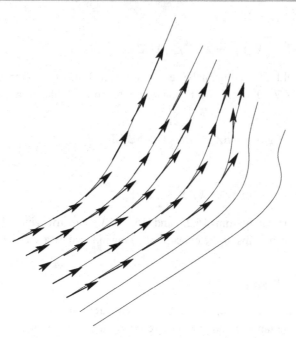

Figure 15.6 Streamlines representing the flow generated by a vector field

The curve $\gamma_p : \mathbb{R} \to M$ defined by $\gamma_p(t) = \sigma_t(p)$ clearly passes through p at $t = 0$. It is called the **orbit** of p under the flow σ and defines a tangent vector X_p at p by

$$X_p f = \left.\frac{\mathrm{d}f(\gamma_p(t))}{\mathrm{d}t}\right|_{t=0} = \left.\frac{\mathrm{d}f(\sigma_t(p))}{\mathrm{d}t}\right|_{t=0}.$$

Since p is an arbitrary point of M we have a vector field X on M, said to be the **vector field induced** by the flow σ. Any vector field X induced by a one-parameter group of transformations of M is said to be **complete**. The one-parameter group σ_t can be thought of as 'filling in' the vector field X with a set of curves, which play the role of streamlines for a fluid whose velocity is everywhere given by X (see Fig. 15.6).

Not every vector field is complete, but there is a local concept that is always applicable. A **local one-parameter group of transformations**, or **local flow**, consists of an open subset $U \subseteq M$ and a real interval $I_\epsilon = (-\epsilon, \epsilon)$, together with a map $\sigma : I_\epsilon \times U \to M$ such that:

(i') for each $t \in I_\epsilon$ the map $\sigma_t : U \to M$ defined by $\sigma_t(p) = \sigma(t, p)$ is a diffeomorphism of U onto $\sigma_t(U)$;
(ii') if t, s and $t + s \in I_\epsilon$ and $p, \sigma_s(p) \in U$ then $\sigma_{t+s}(p) = \sigma_t(\sigma_s(p))$.

A local flow induces a vector field X on U in a similar way to that described above for a flow:

$$X_p f = \left.\frac{\mathrm{d}f(\sigma_t(p))}{\mathrm{d}t}\right|_{t=0} \quad \text{for all } p \in U. \tag{15.28}$$

It now turns out that every vector field X corresponds to a local one-parameter group of transformations, which it may be said to **generate**.

15.5 Commutators, flows and Lie derivatives

Theorem 15.2 *If X is a vector field on M, and $p \in M$ then there exists an interval $I = (-\epsilon, \epsilon)$, a neighbourhood U of p, and a local flow $\sigma : I \times U \to M$ that induces the vector field $X|_U$ restricted to U.*

Proof: If $(U, \phi; x^i)$ is a coordinate chart at p we may set

$$X|_U = X^i \frac{\partial}{\partial x^i} \quad \text{where} \quad X^i : U \to \mathbb{R}.$$

The existence and uniqueness theorem of ordinary differential equations implies that for any $\mathbf{x} \in \phi(U)$ there exists a unique curve $\mathbf{y} = \mathbf{y}(t; x^1, \ldots, x^n)$ on some interval $I = (-\epsilon, \epsilon)$ such that

$$\frac{dy^i(t; x^1, \ldots, x^n)}{dt} = X^i \circ \phi^{-1}(y^1(t; \mathbf{x}), y^2(t; \mathbf{x}), \ldots, y^n(t; \mathbf{x}))$$

and

$$y^i(0; x^1, \ldots, x^n) = x^i.$$

As the solutions of a family of differential equations depend smoothly on the initial coordinates [15, 16], the functions $y(t; x^1, \ldots, x^n)$ are differentiable with respect to t and x^i.

For fixed s and fixed $\mathbf{x} \in \phi(U)$ the curves $t \to z^i(t, s; \mathbf{x}) = y^i(t; \mathbf{y}(s; \mathbf{x}))$ and $t \to z(t, s, \mathbf{x}) = y^i(t + s; \mathbf{x})$ satisfy the same differential equation

$$\frac{dz^i(t, s; x^1, \ldots, x^n)}{dt} = X^i(z^1(t, s; \mathbf{x}), \ldots, z^n(t, s; \mathbf{x}))$$

and have the same initial conditions at $t = 0$,

$$y^i(0; \mathbf{y}(s; \mathbf{x})) = y^i(s; \mathbf{x}) = y^i(0 + s; \mathbf{x}).$$

These solutions are therefore identical and the map $\sigma : I \times U \to U$ defined by $\sigma(t, p) = \phi^{-1}\mathbf{y}(t; \phi(p))$ satisfies the local one-parameter group condition

$$\sigma(t, \sigma(s, p)) = \sigma(t + s, p).$$

∎

A useful consequence of this theorem is the local existence of a coordinate system that 'straightens out' any given vector field X so that its components point along the 1-axis, $X^i = (1, 0, \ldots, 0)$. The local flow σ_t generated by X is then simply a translation in the 1-direction, $\sigma_t(x^1, x^2, \ldots, x^n) = (x^1 + t, x^2, \ldots, x^n)$.

Theorem 15.3 *If X is a vector field on a manifold M such that $X_p \neq 0$, then there exists a coordinate chart $(U, \phi; x^i)$ at p such that*

$$X = \frac{\partial}{\partial x^1}. \tag{15.29}$$

Outline proof: The idea behind the proof is not difficult. Pick any coordinate system $(U, \psi; y^i)$ at p such that $y^i(p) = 0$, and $X_p = (\partial_{y^1})_p$. Let $\sigma : I_\epsilon \times A \to M$ be a local flow that induces X on the open set A. In a neighbourhood of p consider a small $(n-1)$-dimensional 'open ball' of points through p that cuts across the flow, whose typical point q has coordinates $(0, y^2, \ldots, y^n)$, and assign coordinates $(x^1 = t, x^2 = y^2, \ldots, x^n = y^n)$

Differential geometry

to points on the streamline $\sigma_t(q)$ through q. The coordinates $x^2, \ldots, x^n$ are then constant along the curves $t \to \sigma_t(q)$, and the vector field X, being tangent to the streamlines, has coordinates $(1, 0, \ldots, 0)$ throughout a neighbourhood of p. A detailed proof may be found in [11, theorem 4.3] or [4, p. 124]. ∎

Example 15.13 Let X be the differentiable vector field $X = x^2 \partial_x$ on the real line manifold $\mathbb{R}$. To find a coordinate $y = y(x)$ such that $X = \partial_y$, we need to solve the differential equation

$$x^2 \frac{\partial y}{\partial x} = 1.$$

The solution is $y = C - 1/x$.

The local one-parameter group generated by X is found by solving the ordinary differential equation,

$$\frac{dx}{dt} = x^2.$$

The solution is

$$\sigma_t(x) = \frac{1}{x^{-1} - t} = \frac{x}{1 - tx}.$$

It is straightforward to verify the group property

$$\sigma_t(\sigma_s(x)) = \frac{x}{1 - (t+s)x} = \sigma_{t+s}(x).$$

Example 15.14 If X and Y are vector fields on M generating flows ϕ_t and ψ_t respectively, let σ be the curve through $p \in M$ defined by

$$\sigma(t) = \psi_{-t} \circ \phi_{-t} \circ \psi_t \circ \phi_t p.$$

Then $\sigma(\sqrt{t})$ is a curve whose tangent vector is the commutator $[X, Y]$ at p. The proof is to let f be any differentiable function at p and show that

$$[X, Y]_p f = \lim_{t \to 0} \frac{f[\sigma(\sqrt{t})] - f[\sigma(0)]}{t}.$$

Details may be found in [3, p. 130]. Some interesting geometrophysical applications of this result are discussed in [17].

Lie derivative

Let X be a smooth vector field on a manifold M, which generates a local one-parameter group of transformations σ_t on M. If Y is any differentiable vector field on M, we define its **Lie derivative** along X to be

$$\mathcal{L}_X Y = \lim_{t \to 0} \frac{Y - (\sigma_t)_* Y}{t}. \tag{15.30}$$

Figure 15.7 illustrates the situation. Essentially, the tangent map of the diffeomorphism σ_t is used to 'drag' the vector field Y forward along the integral curves from a point $\sigma_{-t}(p)$ to p and the result is compared with original value Y_p of the vector field. Equation (15.7)

15.5 Commutators, flows and Lie derivatives

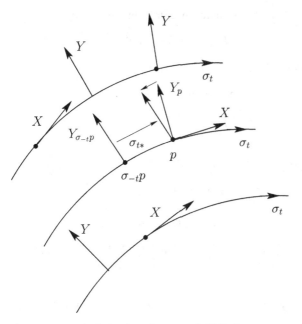

Figure 15.7 Lie derivative of a vector field Y along a vector field X

performs this operation for neighbouring points and takes the limit on dividing by t. We now show that this derivative is identical with the 'derivative-like' operation of taking the commutator of two vector fields.

Let $f : M \to \mathbb{R}$ be a differentiable function, and p any point of M. From (15.30) we have at p,

$$
\begin{aligned}
(\mathcal{L}_X Y)_p f &= \lim_{t \to 0} \frac{1}{t} \left(Y_p f - ((\sigma_t)_* Y)_p f \right) \\
&= \lim_{t \to 0} \frac{1}{t} \left(Y_p f - Y_{\sigma_{-t}(p)} (f \circ \sigma_t) \right) \\
&= \lim_{t \to 0} \frac{1}{t} \left(Y_p f - Y_{\sigma_{-t}(p)} f - Y_{\sigma_{-t}(p)} (f \circ \sigma_t - f) \right) \\
&= \lim_{t \to 0} \frac{1}{t} \left((Yf - (Yf) \circ \sigma_{-t})(p) - Y_{\sigma_{-t}(p)} (f \circ \sigma_t - f) \right).
\end{aligned}
$$

On setting $s = -t$ in the first term and using Eq. (15.28), the right-hand side reduces to $X(Yf)(p) - Y(Xf)(p) = [X, Y]_p f$, and we have the desired relation

$$\mathcal{L}_X Y = [X, Y]. \tag{15.31}$$

The concept of Lie derivative can be extended to all tensor fields. First, for any diffeomorphism $\varphi : M \to M$, we define the induced map $\tilde{\varphi} : T^{(r,s)}(M) \to T^{(r,s)}(M)$ in the following way:

(i) for vector fields set $\tilde{\varphi} = \varphi_*$;
(ii) for scalar fields $f : M \to \mathbb{R}$ set $\tilde{\varphi} f = f \circ \varphi^{-1}$;
(iii) for covector fields set $\tilde{\varphi} = (\varphi^{-1})^*$;

(iv) the map $\tilde{\varphi}$ is extended to all tensor fields by demanding linearity and

$$\tilde{\varphi}(T \otimes S) = \tilde{\varphi}T \otimes \tilde{\varphi}S$$

for arbitrary tensor fields T and S.

If ω and X are arbitrary covector and vector fields, then

$$\langle \tilde{\varphi}\omega, \tilde{\varphi}X \rangle = \tilde{\varphi}\langle \omega, X \rangle, \qquad (15.32)$$

since

$$\langle \tilde{\varphi}\omega, \tilde{\varphi}X \rangle(p) = \langle ((\varphi^{-1})^* \omega_{\varphi^{-1}(p)}, \varphi_* X_{\varphi^{-1}(p)} \rangle$$
$$= \langle \omega_{\varphi^{-1}(p)}, X_{\varphi^{-1}(p)} \rangle = \langle \omega, X \rangle (\varphi^{-1}(p)).$$

Exercise: For arbitrary vector fields X show from (ii) that $\tilde{\varphi}X(\tilde{\varphi}f) = \tilde{\varphi}(X(f))$.

Using Eq. (15.11), property (iv) provides a unique definition for the application of the map $\tilde{\varphi}$ to all higher order tensors. Alternatively, as for covector fields, the following is a characterization of the map $\tilde{\varphi}$:

$$(\tilde{\varphi}T)(\tilde{\varphi}\omega^1, \ldots, \tilde{\varphi}\omega^r, \tilde{\varphi}X_1, \ldots, \tilde{\varphi}X_s) = \tilde{\varphi}(T(\omega^1, \ldots, \omega^r, X_1, \ldots, X_s))$$

for all vector fields $X_1, \ldots X_s$ and covector fields $\omega^1, \ldots, \omega^r$.

The **Lie derivative** $\mathcal{L}_X T$ of a smooth tensor field T with respect to the vector field X is defined as

$$\mathcal{L}_X T = \lim_{t \to 0} \frac{1}{t}(T - \tilde{\sigma}_t T). \qquad (15.33)$$

Exercise: Show that for any tensor field T

$$\mathcal{L}_X T = -\frac{d\tilde{\sigma}_t T}{dt}\bigg|_{t=0} \qquad (15.34)$$

and prove the Leibnitz rule

$$\mathcal{L}_X(T \otimes S) = T \otimes (\mathcal{L}_X S) + (\mathcal{L}_X T) \otimes S. \qquad (15.35)$$

When T is a scalar field f, we find, on changing the limit variable to $s = -t$,

$$(\mathcal{L}_X f)_p = \frac{df \circ \sigma_s(p)}{ds}\bigg|_{s=0} = Xf(p),$$

and in a local coordinate chart $(U; x^i)$

$$\mathcal{L}_X f = Xf = f_{,i}X^i. \qquad (15.36)$$

Since for any pair i, j

$$\mathcal{L}_{\partial_{x^i}} \frac{\partial}{\partial x^j} = \left[\frac{\partial}{\partial x^i}, \frac{\partial}{\partial x^j}\right] = \frac{\partial^2}{\partial x^i \partial x^j} - \frac{\partial^2}{\partial x^j \partial x^i} = 0,$$

and $\mathcal{L}_X Y = [X, Y] = -[Y, X] = -\mathcal{L}_Y X$ for any pair of vector fields X, Y, we find

$$\mathcal{L}_X \frac{\partial}{\partial x^j} = -\mathcal{L}_{\partial_{x^j}}\left(X^i \frac{\partial}{\partial x^i}\right) = -X^i{}_{,j}\frac{\partial}{\partial x^i}.$$

15.5 Commutators, flows and Lie derivatives

Applying the Leibnitz rule (15.35) results in

$$\mathcal{L}_X Y = \mathcal{L}_X\left(Y^i \frac{\partial}{\partial x^i}\right) = Y^i{}_{,j} X^j \frac{\partial}{\partial x^i} - Y^j X^i{}_{,j} \frac{\partial}{\partial x^i},$$

in agreement with the component formula for the Lie bracket in Eq. (15.25),

$$(\mathcal{L}_X Y)^i = Y^i{}_{,j} X^j - Y^j X^i{}_{,j}. \tag{15.37}$$

To find the component formula for the Lie derivative of a 1-form $\omega = w_i dx^i$, we note that for any pair of vector fields X, Y

$$\mathcal{L}_X \langle \omega, Y \rangle = X \langle \omega, Y \rangle = \langle \mathcal{L}_X \omega, Y \rangle + \langle \omega, \mathcal{L}_X Y \rangle, \tag{15.38}$$

which follows from Eqs. (15.32) and (15.34),

$$\mathcal{L}_X \langle \omega, Y \rangle = X \langle \omega, Y \rangle = -\frac{d}{dt} \tilde{\sigma}_t \langle \omega, Y \rangle \Big|_{t=0}$$
$$= -\frac{d}{dt} \langle \tilde{\sigma}_t \omega, \tilde{\sigma}_t Y \rangle \Big|_{t=0}$$
$$= -\langle \frac{d}{dt} \tilde{\sigma}_t \omega, Y \rangle \Big|_{t=0} - \langle \omega, \frac{d}{dt} \tilde{\sigma}_t Y \rangle \Big|_{t=0}$$
$$= \langle \mathcal{L}_X \omega, Y \rangle + \langle \omega, \mathcal{L}_X Y \rangle.$$

If $\omega = w_i \, dx^i$ is a 1-form, then its Lie derivative $\mathcal{L}_X \omega$ with respect to the vector field X has components in a coordinate chart $(U; x^i)$ given by

$$(\mathcal{L}_X \omega)_j = \langle \mathcal{L}_X \omega, \frac{\partial}{\partial x^j} \rangle$$
$$= \mathcal{L}_X \langle \omega, \frac{\partial}{\partial x^j} \rangle - \langle \omega, \mathcal{L}_X \frac{\partial}{\partial x^j} \rangle$$
$$= \mathcal{L}_X w_j + \langle \omega, X^i{}_{,j} \frac{\partial}{\partial x^i} \rangle$$
$$= w_{j,i} X^i + w_i X^i{}_{,j}.$$

Extending this argument to a general tensor of type (r, s), we find

$$(\mathcal{L}_X T)^{ij\cdots}{}_{kl\cdots} = T^{ij\cdots}{}_{kl\cdots,m} X^m - T^{mj\cdots}{}_{kl\cdots} X^i{}_{,m} - T^{im\cdots}{}_{kl\cdots} X^j{}_{,m} - \cdots$$
$$+ T^{ij\cdots}{}_{ml\cdots} X^m{}_{,k} + T^{ij\cdots}{}_{km\cdots} X^m{}_{,l} + \cdots \tag{15.39}$$

Example 15.15 In local coordinates such that $X = \partial_{x^1}$ (see Theorem 15.3), all $X^i{}_{,j} = 0$ since the components X^i = consts. and the components of the Lie derivative are simply the derivatives in the 1-direction,

$$(\mathcal{L}_X T)^{ij\cdots}{}_{kl\cdots} = T^{ij\cdots}{}_{kl\cdots,1}.$$

Problems

Problem 15.17 Show that the components of the Lie product $[X, Y]^k$ given by Eq. (15.25) transform as a contravariant vector field under a coordinate transformation $x'^j(x^i)$.

Differential geometry

Problem 15.18 Show that the Jacobi identity can be written

$$\mathcal{L}_{[X,Y]}Z = \mathcal{L}_X\mathcal{L}_Y Z - \mathcal{L}_Y\mathcal{L}_X Z,$$

and this property extends to all tensors T:

$$\mathcal{L}_{[X,Y]}T = \mathcal{L}_X\mathcal{L}_Y T - \mathcal{L}_Y\mathcal{L}_X T.$$

Problem 15.19 Let $\alpha : M \to N$ be a diffeomorphism between manifolds M and N and X a vector field on M that generates a local one-parameter group of transformations σ_t on M. Show that the vector field $X' = \alpha_* X$ on N generates the local flow $\sigma'_t = \alpha \circ \sigma_t \circ \alpha^{-1}$.

Problem 15.20 For any real positive number n show that the vector field $X = x^n \partial_x$ is differentiable on the manifold $\mathbb{R}^+$ consisting of the positive real line $\{x \in \mathbb{R} \mid x > 0\}$. Why is this not true in general on the entire real line $\mathbb{R}$? As done for the case $n = 2$ in Example 15.13, find the maximal one-parameter subgroup σ_t generated by this vector field at any point $x > 0$.

Problem 15.21 On the manifold $\mathbb{R}^2$ with coordinates (x, y), let X be the vector field $X = -y\partial_x + x\partial_y$. Determine the integral curve through any point (x, y), and the one-parameter group generated by X. Find coordinates (x', y') such that $X = \partial_{x'}$.

Problem 15.22 Repeat the previous problem for the vector fields, $X = y\partial_x + x\partial_y$ and $X = x\partial_x + y\partial_y$.

Problem 15.23 On a compact manifold show that every vector field X is complete. [*Hint*: Let σ_t be a local flow generating X, and let ϵ be the least bound required on a finite open covering. Set $\sigma_t = (\sigma_{t/N})^N$ for N large enough that $|t| < \epsilon N$.]

Problem 15.24 Show that the Lie derivative $\mathcal{L}_X$ commutes with all operations of contraction C^i_j on a tensor field T,

$$\mathcal{L}_X C^i_j T = C^i_j \mathcal{L}_X T.$$

Problem 15.25 Prove the formula (15.39) for the Lie derivative of a general tensor.

15.6 Distributions and Frobenius theorem

A **k-dimensional distribution** D^k on a manifold M is an assignment of a k-dimensional subspace $D^k(p)$ of the tangent space $T_p(M)$ at every point $p \in M$. The distribution is said to be C^∞ or **smooth** if for all $p \in M$ there is an open neighbourhood U and k smooth vector fields $X_1, \ldots, X_k$ on U that span $D^k(q)$ at each point $q \in U$. A vector field X on an open domain A is said to **lie in** or **belong to** the distribution D^k if $X_p \in D^k(p)$ at each point $p \in A$. A one-dimensional distribution is equivalent to a vector field up to an arbitrary scalar factor at every point, and is sometimes called a *direction field*.

An **integral manifold** of a distribution D^k is a k-dimensional submanifold (K, ψ) of M such that all vector fields tangent to the submanifold belong to D^k,

$$\psi_*(T_p(K)) = D^k(\psi(p)).$$

15.6 Distributions and Frobenius theorem

Every one-dimensional distribution has integral manifolds, for if X is any vector field that spans a distribution D^1 then any family of integral curves of X act as integral manifolds of the distribution D^1. We will see, however, that not every distribution of higher dimension has integral manifolds.

A distribution D^k is said to be **involutive** if for any pair of vector fields X, Y lying in D^k, their Lie bracket $[X, Y]$ also belongs to D^k. If $\{e_1, \ldots, e_k\}$ is any local basis of vector fields spanning an involutive D^k on an open neighbourhood U, then

$$[e_\alpha, e_\beta] = \sum_{\gamma=1}^{k} C_{\alpha\beta}^\gamma e_\gamma \quad (\alpha, \beta = 1, \ldots, k) \tag{15.40}$$

where $C_{\alpha\beta}^\gamma = -C_{\beta\alpha}^\gamma$ are C^∞ functions on U. Conversely, if there exists a local basis $\{e_\alpha\}$ satisfying (15.40) for some scalar structure fields $C_{\alpha\beta}^\gamma$, the distribution is involutive, for if $X = X^\alpha e_\alpha$ and $Y = Y^\beta e_\beta$ then

$$[X, Y] = [X^\alpha e_\alpha, Y^\beta e_\beta] = \big(X(Y^\gamma) - Y(X^\gamma) + X^\alpha Y^\beta C_{\alpha\beta}^\gamma\big)e_\gamma,$$

which belongs to D^k as required. For example, if there exists a coordinate chart $(U; x^i)$ such that the distribution D^k is spanned by the first k coordinate basis vector fields

$$e_1 = \partial_{x^1}, \ e_2 = \partial_{x^2}, \ldots, \ e_k = \partial_{x^k}$$

then D^k is involutive on U since all $[e_\alpha, e_\beta] = 0$, a trivial instance of the relation (15.40). In this case we can restrict the chart to a cubical neighbourhood $U' = \{p \mid -a < x^i(p) < a\}$, and the 'slices' $x^a = $ const. $(a = k+1, \ldots, n)$ are local integral manifolds of the distribution D^k. The key result is the **Frobenius theorem**:

Theorem 15.4 *A smooth k-dimensional distribution D^k on a manifold M is involutive if and only if every point $p \in M$ lies in a coordinate chart $(U; x^i)$ such that the coordinate vector fields $\partial/\partial x^\alpha$ for $\alpha = 1, \ldots, k$ span D^k at each point of U.*

Proof: The *if* part follows from the above remarks. The converse will be shown by induction on the dimension k. The case $k = 1$ follows immediately from Theorem 15.3. Suppose now that the statement is true for all $(k-1)$-dimensional distributions, and let D^k be a k-dimensional involutive distribution spanned at all points of an open set A by vector fields $\{X_1, \ldots, X_k\}$. At any point $p \in A$ there exist coordinates $(V; y^i)$ such that $X_k = \partial_{y^k}$. Set

$$Y_\alpha = X_\alpha - (X_\alpha y^k) X_k, \quad Y_k = X_k$$

where Greek indices $\alpha, \beta, \ldots$ range from 1 to $k-1$. The vector fields $Y_1, Y_2, \ldots, Y_k$ clearly span D^k on V, and

$$Y_\alpha y^k = 0, \quad Y_k y^k = 1. \tag{15.41}$$

Since D^k is an involutive we can write

$$[Y_\alpha, Y_\beta] = C_{\alpha\beta}^\gamma Y_\gamma + a_{\alpha\beta} Y_k,$$
$$[Y_\alpha, Y_k] = C_\alpha^\gamma Y_\gamma + a_\alpha Y_k.$$

Differential geometry

Applying both sides of these equations to the coordinate function y^k and using (15.41), we find $a_{\alpha\beta} = a_\alpha = 0$, whence

$$[Y_\alpha, Y_\beta] = C^\gamma_{\alpha\beta} Y_\gamma, \tag{15.42}$$

$$[Y_\alpha, Y_k] = C^\gamma_\alpha Y_\gamma. \tag{15.43}$$

The distribution D^{k-1} spanned by $Y_1, Y_2, \ldots, Y_{k-1}$ is therefore involutive on V, and by the induction hypothesis there exists a coordinate chart $(W; z^i)$ such that D^{k-1} is spanned by $\{\partial_{z^1}, \ldots, \partial_{z^{k-1}}\}$. Set

$$\frac{\partial}{\partial z^\alpha} = A^\beta_\alpha Y_\beta$$

where $[A^\beta_\alpha]$ is a non-singular matrix of functions on W. The original distribution D^k is spanned on W by the set of vector fields

$$\{\partial_{z^1}, \partial_{z^2}, \ldots, \partial_{z^{k-1}}, Y_k\}.$$

It follows then from (15.43) that

$$[\partial_{z^\alpha}, Y_k] = K^\beta_\alpha \partial_{z^\beta} \tag{15.44}$$

for some functions K^β_α. If we write

$$Y_k = \sum_{\alpha=1}^{k-1} \xi^\alpha \partial_{z^\alpha} + \sum_{a=k}^{n} \xi^a \partial_{z^a}$$

and apply Eq. (15.44) to the coordinate functions z^a ($a = k, \ldots, n$), we find

$$\frac{\partial \xi^a}{\partial z^\alpha} = 0.$$

Hence $\xi^a = \xi^a(z^k, \ldots, z^n)$ for all $a \geq k$. Since Y_k is linearly independent of the vectors ∂_{z^α}, the distribution D^k is spanned by the set of vectors $\{\partial_{z^1}, \partial_{z^2}, \ldots, \partial_{z^{k-1}}, Z\}$, where

$$Z = Y_k - \xi^\alpha \partial_{z^\alpha} = \xi^a(z^k, \ldots, z^n) \partial_{z^a}.$$

By Theorem 15.3 there exists a coordinate transformation not involving the first $(k-1)$ coordinates,

$$x^k = x^k(z^k, \ldots, z^n), \quad x^{k+1} = x^{k+1}(z^k, \ldots, z^n), \ldots, \quad x^n = x^n(z^k, \ldots, z^n)$$

such that $Z = \partial_{x^k}$. Setting $x^1 = z^1, \ldots, x^{k-1} = z^{k-1}$, we have coordinates $(U; x^i)$ in which D^k is spanned by $\{\partial_{x^1}, \ldots, \partial_{x^{k-1}}, \partial_{x^k}\}$. ∎

Theorem 15.5 *A set of vector fields $\{X_1, X_2, \ldots, X_k\}$ is equal to the first k basis fields of a local coordinate system, $X_1 = \partial_{x^1}, \ldots, X_k = \partial_{x^k}$ if and only if they commute with each other, $[X_\alpha, X_\beta] = 0$.*

Proof: The vanishing of all commutators is clearly a necessary condition for the vector fields to be local basis fields of a coordinate system, for if $X_\alpha = \partial_{x^\alpha}$ ($\alpha = 1, \ldots, r$) then $[X_\alpha, X_\beta] = [\partial_{x^\alpha}, \partial_{x^\beta}] = 0$.

15.6 Distributions and Frobenius theorem

To prove sufficiency, we again use induction on k. The case $k = 1$ is essentially Theorem 15.3. By the induction hypothesis, there exists local coordinates $(U; x^i)$ such that $X_\alpha = \partial_{x^\alpha}$ for $\alpha = 1, \ldots, k-1$. Set $Y = X_k = Y^i(x^1, \ldots, x^n)\partial_{x^i}$, and by Example 15.15 $Y^i{}_{,\alpha} = 0$, so that we may write

$$Y = \sum_{\alpha=1}^{k-1} Y^\alpha(x^k, \ldots, x^n)\partial_{x^\alpha} + \sum_{a=k}^{n} Y^a(x^k, \ldots, x^n)\partial_{x^a}.$$

Using Theorem 15.3 we may perform a coordinate transformation on the last $n - k + 1$ coordinates such that

$$Y = \sum_{\alpha=1}^{k-1} Y^\alpha(x^k, \ldots, x^n)\partial_{x^\alpha} + \partial_{x^k}.$$

A coordinate transformation

$$x'^\alpha = x^\alpha + f^\alpha(x^k, \ldots, x^n) \quad (\alpha = 1, \ldots, k-1)$$
$$x'^a = x^a \quad (a = k, \ldots, n)$$

has the effect

$$Y = \sum_{\beta=1}^{k-1}\left(Y^\beta + \frac{\partial f^\beta}{\partial x^k}\right)\frac{\partial}{\partial x'^\beta} + \frac{\partial}{\partial x'^k}.$$

Solving the differential equations

$$\frac{\partial f^\beta}{\partial x^k} = -Y^\beta(x^k, \ldots, x^n)$$

by a straightforward integration leads to $Y = \partial_{x'^k}$ as required. ∎

Example 15.16 On $\mathring{\mathbb{R}}^3 = \mathbb{R}^3 - \{(0,0,0)\}$ let X_1, X_2, X_3 be the three vector fields

$$X_1 = y\partial_z - z\partial_y, \quad X_2 = z\partial_x - x\partial_z, \quad X_3 = x\partial_y - y\partial_x.$$

These three vector fields generate a two-dimensional distribution D^2, as they are not linearly independent

$$xX_1 + yX_2 + zX_3 = 0.$$

The Lie bracket of any pair of these vector fields is easily calculated,

$$[X_1, X_2]f = [y\partial_z - z\partial_y, z\partial_x - x\partial_z]f$$
$$= yz[\partial_z, \partial_x]f + yf_{,x} - yx[\partial_z, \partial_z]f - z^2[\partial_y, \partial_x]f + zx[\partial_y, \partial_z]f - xf_{,y}$$
$$= (-x\partial_y + y\partial_x)f = -X_3 f.$$

There are similar identities for the other commutators,

$$[X_1, X_2] = -X_3, \quad [X_2, X_3] = -X_1, \quad [X_3, X_1] = -X_2. \tag{15.45}$$

Differential geometry

Hence the distribution D^2 is involutive and by the Frobenius theorem it is possible to find a local transformation to coordinates y^1, y^2, y^3 such that ∂_{y^1} and ∂_{y^2} span all three vector fields X_1, X_2 and X_3.

The vector field $X = x\partial_x + y\partial_y + z\partial_z$ commutes with all X_i: for example,

$$[X_3, X]f = [x\partial_y - y\partial_x, xf_{,x} + yf_{,y} + zf_{,z}]f$$
$$= x^2[\partial_y, \partial_x]f - x\partial_y f + x\partial_y f + xy[\partial_y, \partial_y]f - y^2[\partial_x, \partial_y]f$$
$$+ xz[\partial_y, \partial_z]f - yz[\partial_x, \partial_z]f$$
$$= 0.$$

Hence the distribution E^2 generated by the pair of vector fields $\{X_3, X\}$ is also involutive. Let us consider spherical polar coordinates, Eq. (15.2), having inverse transformations

$$r = \sqrt{x^2 + y^2 + z^2}, \qquad \theta = \cos^{-1}\left(\frac{z}{r}\right), \qquad \phi = \tan^{-1}\left(\frac{y}{x}\right).$$

Express the basis vector fields in terms of these coordinates

$$\partial_x = \frac{\partial r}{\partial x}\partial_r + \frac{\partial \theta}{\partial x}\partial_\theta + \frac{\partial \phi}{\partial x}\partial_\phi = \sin\theta\cos\phi\,\partial_r + \frac{\cos\theta\cos\phi}{r}\partial_\theta - \frac{\sin\phi}{r\sin\theta}\partial_\phi,$$

$$\partial_y = \frac{\partial r}{\partial y}\partial_r + \frac{\partial \theta}{\partial y}\partial_\theta + \frac{\partial \phi}{\partial y}\partial_\phi = \sin\theta\sin\phi\,\partial_r + \frac{\cos\theta\sin\phi}{r}\partial_\theta + \frac{\cos\phi}{r\sin\theta}\partial_\phi,$$

$$\partial_z = \frac{\partial r}{\partial z}\partial_r + \frac{\partial \theta}{\partial z}\partial_\theta + \frac{\partial \phi}{\partial z}\partial_\phi = \cos\theta\,\partial_r - \frac{\sin\theta}{r}\partial_\theta,$$

and a simple calculation gives

$$X_1 = y\partial_z - z\partial_y = -\sin\phi\,\partial_\theta - \cot\theta\cos\phi\,\partial_\phi,$$
$$X_2 = z\partial_x - x\partial_z = -\cos\phi\,\partial_\theta - \cot\theta\sin\phi\,\partial_\phi,$$
$$X_3 = x\partial_y - y\partial_x = \partial_\phi,$$
$$X = x\partial_x + y\partial_y + z\partial_z = r\partial_r = \partial_{r'} \quad \text{where } r' = \ln r.$$

The distribution D^2 is spanned by the basis vector fields ∂_θ and ∂_ϕ, while the distribution E^2 is spanned by the vector fields ∂_r and ∂_ϕ in spherical polars.

Exercise: Find a chart, two of whose basis vector fields span the distribution generated by X_1 and X. Do the same for the distribution generated by X_2 and X.

Problems

Problem 15.26 Let D_k be an involutive distribution spanned locally by coordinate vector fields $e_\alpha = \partial/\partial x^\alpha$, where Greek indices α, β, etc. all range from 1 to k. If $X_\alpha = A^\beta{}_\alpha e_\beta$ is any local basis spanning a distribution D^k, show that the matrix of functions $[A^\beta{}_\alpha]$ is non-singular everywhere on its region of definition, and that $[X_\alpha, X_\beta] = C^\gamma_{\alpha\beta} X_\gamma$ where

$$C^\gamma_{\alpha\beta} = \left(A^\delta{}_\alpha A^\eta{}_{\beta,\delta} - A^\delta{}_\beta A^\eta{}_{\alpha,\delta}\right)\left(A^{-1}\right)^\gamma{}_\eta.$$

Problem 15.27 There is a classical version of the Frobenius theorem stating that a system of partial differential equations of the form

$$\frac{\partial f^\beta}{\partial x^j} = A^\beta_j(x^1, \ldots, x^k, f^1(x), \ldots, f^r(x))$$

where $i, j = 1, \ldots, k$ and $\alpha, \beta = 1, \ldots, r$ has a unique local solution through any point $(a^1, \ldots, a^k, b^1, \ldots, b^r)$ if and only if

$$\frac{\partial A^\beta_j}{\partial x^i} - \frac{\partial A^\beta_i}{\partial x^j} + A^\alpha_i \frac{\partial A^\beta_j}{\partial y^\alpha} - A^\alpha_j \frac{\partial A^\beta_i}{\partial y^\alpha} = 0$$

where $A^\beta_j = A^\beta_j(x^1, \ldots, x^k, y^1, \ldots, y^r)$. Show that this statement is equivalent to the version given in Theorem 15.4. [*Hint*: On $\mathbb{R}^n$ where $n = r + k$ consider the distribution spanned by vectors

$$Y_i = \frac{\partial}{\partial x^i} + A^\beta_i \frac{\partial}{\partial y^\beta} \quad (i = 1, \ldots, k)$$

and show that the integrability condition is precisely the involutive condition $[Y_i, Y_j] = 0$, while the condition for an integral submanifold of the form $y^\beta = f^\beta(x^1, \ldots, x^k)$ is $A^\beta_j = f^\beta_{,j}$.]

References

[1] L. Auslander and R. E. MacKenzie. *Introduction to Differentiable Manifolds*. New York, McGraw-Hill, 1963.

[2] R. W. R. Darling. *Differential Forms and Connections*. New York, Cambridge University Press, 1994.

[3] T. Frankel. *The Geometry of Physics*. New York, Cambridge University Press, 1997.

[4] N. J. Hicks. *Notes on Differential Geometry*. New York, D. Van Nostrand Company, 1965.

[5] S. Kobayashi and K. Nomizu. *Foundations of Differential Geometry*. New York, Interscience Publishers, 1963.

[6] L. H. Loomis and S. Sternberg. *Advanced Calculus*. Reading, Mass., Addison-Wesley, 1968.

[7] M. Nakahara. *Geometry, Topology and Physics*. Bristol, Adam Hilger, 1990.

[8] C. Nash and S. Sen. *Topology and Geometry for Physicists*. London, Academic Press, 1983.

[9] I. M. Singer and J. A. Thorpe. *Lecture Notes on Elementary Topology and Geometry*. Glenview, Ill., Scott Foresman, 1967.

[10] M. Spivak. *Differential Geometry, Vols. 1–5*. Boston, Publish or Perish Inc., 1979.

[11] W. H. Chen, S. S. Chern, and K. S. Lam. *Lectures on Differential Geometry*. Singapore, World Scientific, 1999.

[12] S. Sternberg. *Lectures on Differential Geometry*. Englewood Cliffs, N.J., Prentice-Hall, 1964.

[13] F. W. Warner. *Foundations of Differential Manifolds and Lie Groups*. New York, Springer-Verlag, 1983.

[14] C. de Witt-Morette, Y. Choquet-Bruhat and M. Dillard-Bleick. *Analysis, Manifolds and Physics*. Amsterdam, North-Holland, 1977.
[15] E. Coddington and N. Levinson. *Theory of Ordinary Differential Equations*. New York, McGraw-Hill, 1955.
[16] W. Hurewicz. *Lectures on Ordinary Differential Equations*. New York, John Wiley & Sons, 1958.
[17] E. Nelson. *Tensor Analysis*. Princeton, N.J., Princeton University Press, 1967.

16 Differentiable forms

16.1 Differential forms and exterior derivative

Let M be a differentiable manifold of dimension n. At any point $p \in M$ let $(\Lambda_r)_p(M) \equiv \Lambda^{*r}(T_p(M))$ be the space of totally antisymmetric tensors, or r-forms, generated by the tangent space $T_p(M)$ (see Chapter 8). Denote the associated exterior algebra

$$\Lambda_p(M) \equiv \Lambda^*(T_p(M)) = (\Lambda_0)_p(M) \oplus (\Lambda_1)_p(M) \oplus \cdots \oplus (\Lambda_n)_p(M)$$

with graded exterior product $\wedge : (\Lambda_r)_p(M) \times (\Lambda_s)_p(M) \to (\Lambda_{r+s})_p(M)$.

A **differential r-form** α on an open subset $U \subseteq M$ is an r-form field, or assignment of an r-form α_p at every point $p \in U$, such that the function $\alpha(X_1, X_2, \ldots, X_r)(p) \equiv \alpha_p((X_1)_p, (X_2)_p, \ldots, (X_r)_p)$ is differentiable for all smooth vector fields $X_1, X_2, \ldots, X_r$ on U. The set of all differential r-forms on U is denoted $\Lambda_r(U)$ and the **differential exterior algebra** on U is the direct sum

$$\Lambda(U) = \Lambda_0(U) \oplus \Lambda_1(U) \oplus \cdots \oplus \Lambda_n(U)$$

with exterior product defined by $(\alpha \wedge \beta)_p = \alpha_p \wedge \beta_p$. This product is linear, associative and obeys the usual anticommutative rule:

$$\alpha \wedge (\beta + \gamma) = \alpha \wedge \beta + \alpha \wedge \gamma$$
$$\alpha \wedge (\beta \wedge \gamma) = (\alpha \wedge \beta) \wedge \gamma$$
$$\alpha \wedge \beta = (-1)^{rs} \beta \wedge \alpha \quad \text{if } \alpha \in \Lambda_r(U), \ \beta \in \Lambda_s(U). \tag{16.1}$$

Differential 0-forms are simply scalar fields, smooth real-valued functions f on U, $\Lambda_0(U) = \mathcal{F}(U)$.

In a coordinate chart $(U, \phi; x^i)$ a basis of $\Lambda_r(U)$ is

$$dx^{i_1} \wedge dx^{i_2} \wedge \cdots \wedge dx^{i_r} = \mathcal{A}(dx^{i_1} \otimes dx^{i_2} \otimes \cdots \otimes dx^{i_r}),$$

and every differential r-form on U has a unique expansion

$$\alpha = \alpha_{i_1 i_2 \ldots i_r} \, dx^{i_1} \wedge dx^{i_2} \wedge \cdots \wedge dx^{i_r} \tag{16.2}$$

where the components $\alpha_{i_1 i_2 \ldots i_r}$ are smooth functions on U and are antisymmetric in all indices,

$$\alpha_{i_1 i_2 \ldots i_r} = \alpha_{[i_1 i_2 \ldots i_r]}.$$

Differentiable forms

If f is a scalar field then its gradient

$$f_{,i} = \frac{\partial f}{\partial x^i}$$

forms the components of a covariant vector field known as its differential df (see Chapter 15). This concept may be extended to a map on all differential forms, $d : \Lambda(M) \to \Lambda(M)$, called the **exterior derivative**, such that $d\Lambda_r(M) \subseteq \Lambda_{(r+1)}(M)$ and satisfying the following conditions:

(ED1) If f is a differential 0-form, then df is its differential, defined by $\langle df, X \rangle = Xf$ for any smooth vector field X.
(ED2) For any pair of differential forms, $\alpha, \beta \in \Lambda(M)$, $d(\alpha + \beta) = d\alpha + d\beta$.
(ED3) If f is a differential 0-form then $d^2 f \equiv d(df) = 0$.
(ED4) For any r-form α, and any $\beta \in \Lambda(M)$

$$d(\alpha \wedge \beta) = (d\alpha) \wedge \beta + (-1)^r \alpha \wedge d\beta. \tag{16.3}$$

Condition (ED4) says that it is an *anti-derivation* (see Section 8.4, Eq. (8.18)), and (ED3) will be shown to hold for differential forms of all orders. The general theory of differential forms and exterior derivative may be found in [1–9]. Our aim in the following discussion is to show that the operator d exists and is uniquely defined.

Lemma 16.1 *Let U be an open subset of a differentiable manifold M. For any point $p \in U$ there exist open sets W and W' where W has compact closure with $p \in \overline{W} \subseteq W' \subseteq U$, and a smooth function $h \geq 0$ such that $h = 1$ on W and $h = 0$ on $M - W'$.*

Proof: Let $f : \mathbb{R} \to \mathbb{R}$ be the smooth non-negative function, defined by

$$f(t) = \begin{cases} e^{-1/t} & \text{if } t > 0, \\ 0 & \text{if } t \leq 0. \end{cases}$$

For every $a > 0$ let $g_a : \mathbb{R} \to \mathbb{R}$ be the non-negative smooth function

$$g_a(t) = \frac{f(t)}{f(t) + f(a-t)} = \begin{cases} 0 & \text{if } t \leq 0, \\ > 0 & \text{if } 0 < t < a, \\ 1 & \text{if } t \geq a. \end{cases}$$

If $b > a$ the smooth function $h_{a,b} : \mathbb{R} \to \mathbb{R}$ defined by

$$h_{a,b}(t) = 1 - g_{b-a}(t-a)$$

has the value 1 for $t \leq a$ and is 0 for $t \geq b$. On the open interval (a, b) it is positive with values between 0 and 1. Let $(V, \phi; x^i)$ be a coordinate chart at p such that $x^i(p) = 0$ and $V \subseteq U$. Let $b > 0$ be any real number such that the open ball $B_b(0) \subset \phi(V)$, and let $0 < a < b$. Set $W' = \phi^{-1}(B_b(0))$ and $W = \phi^{-1}(B_a(0))$. The closure of W, being the homeomorphic image of a compact set, is compact and $\overline{W} \subset W'$. Let $\tilde{h} : \mathbb{R}^n \to \mathbb{R}$ be the smooth map

$$\tilde{h}(x^1, x^2, \ldots, x^n) = h_{a,b}(r) \quad \text{where} \quad r = \sqrt{(x^1)^2 + (x^2)^2 + \cdots + (x^n)^2},$$

16.1 Differential forms and exterior derivative

and the positive function $h : M \to \mathbb{R}$ defined by

$$h(p) = \begin{cases} \tilde{h} \circ \phi(p) & \text{for } p \in W \\ 0 & \text{for } p \in M - W' \end{cases}$$

has all the desired properties. ∎

If α is an r-form whose restriction to U vanishes, $\alpha|_U = 0$, then $h\alpha = 0$ on all of M where h is the function defined in Lemma 16.1, and by property (ED4),

$$d(h\alpha) = dh \wedge \alpha + h \, d\alpha = 0.$$

Restricting this equation to W we have $d\alpha|_W = 0$, and in particular $(d\alpha)|_p = 0$. Since p is an arbitrary point of U it follows that $d\alpha|_U = 0$. Hence, if α and β are any pair of r-forms such that $\alpha|_U = \beta|_U$, then $(d\alpha)|_U = (d\beta)|_U$. Thus if d exists, satisfying (ED1)–(ED4), then it has a local character and is uniquely defined everywhere.

To show the existence of the operator d, let $(U; x^i)$ be a coordinate chart at any point p. Expanding α according to Eq. (16.2) we have, using (ED1)–(ED4),

$$\begin{aligned} d\alpha &= d\alpha_{i_1 \ldots i_r} \wedge dx^{i_1} \wedge \cdots \wedge dx^{i_r} + \alpha_{i_1 \ldots i_r} d^2 x^{i_1} \wedge dx^{i_2} \wedge \cdots \wedge dx^{i_r} \\ &\quad - \alpha_{i_1 \ldots i_r} dx^{i_1} \wedge d^2 x^{i_2} \wedge \cdots \wedge dx^{i_r} + \cdots \\ &= d\alpha_{i_1 \ldots i_r} dx^{i_1} \wedge \cdots \wedge dx^{i_r} \\ &= \alpha_{i_1 \ldots i_r, j} \, dx^j \wedge dx^{i_1} \wedge \cdots \wedge dx^{i_r}. \end{aligned}$$

Performing a cyclic permutation of indices, and using the total antisymmetry of the wedge product,

$$d\alpha = (-1)^r \alpha_{[i_1 \ldots i_r, i_{r+1}]} \, dx^{i_1} \wedge \cdots \wedge dx^{i_r} \wedge dx^{i_{r+1}}. \tag{16.4}$$

It still remains to verify that conditions (ED1)–(ED4) hold for (16.4). Firstly, this formula reduces to $df = f_{,i} \, dx^i$ in the case of a 0-form, consistent with (ED1). Condition (ED2) follows trivially. To verify (ED3),

$$d^2 f = d(f_{,i} \, dx^i) = f_{,ij} \, dx^j \wedge dx^i = f_{,[ij]} \, dx^j \wedge dx^i = 0$$

since

$$f_{,[ij]} = \frac{1}{2}\left(\frac{\partial^2 f}{\partial x^i \partial x^j} - \frac{\partial^2 f}{\partial x^j \partial x^i}\right) = 0.$$

Finally Eq. (16.4) implies (ED4):

$$\begin{aligned} d(\alpha \wedge \beta) &= d\big(\alpha_{i_1 i_2 \ldots i_r} \beta_{j_1 j_2 \ldots j_s} dx^{i_1} \wedge \cdots \wedge dx^{i_r} dx^{j_1} \wedge \cdots \wedge dx^{j_s}\big) \\ &= d\big(\alpha_{i_1 i_2 \ldots i_r} \beta_{j_1 j_2 \ldots j_s}\big) \wedge dx^{i_1} \wedge \cdots \wedge dx^{i_r} dx^{j_1} \wedge \cdots \wedge dx^{j_s} \\ &= \big(d\alpha_{i_1 i_2 \ldots i_r} \, \beta_{j_1 j_2 \ldots j_s} + \alpha_{i_1 i_2 \ldots i_r} d\beta_{j_1 j_2 \ldots j_s}\big) \wedge dx^{i_1} \wedge \cdots \wedge dx^{i_r} dx^{j_1} \wedge \cdots \wedge dx^{j_s} \\ &= (d\alpha) \wedge \beta + (-1)^r \alpha \wedge d\beta. \end{aligned}$$

The last step follows on performing the r interchanges needed to bring the $d\beta_{j_1 j_2 \ldots j_s}$ term between dx^{i_r} and dx^{j_1}. This shows the existence and uniqueness of the operator d on every coordinate neighbourhood on M.

Differentiable forms

Exercise: For all differential forms α and β, and any pair of real numbers a and b, show that

$$d(a\alpha + b\beta) = a d(\alpha) + b d(\beta).$$

The property (ED3) extends to arbitrary differential forms

$$d^2\alpha = d(d\alpha) = 0, \tag{16.5}$$

for, applying the operator d to (16.2) and using (ED3) gives

$$d(d\alpha) = \left(d^2\alpha_{i_1\ldots i_r}\right) \wedge dx^{i_1} \wedge \cdots \wedge dx^{i_r} = 0.$$

Example 16.1 Let $x = x^1, y = x^2, z = x^3$ be coordinates on the three-dimensional manifold $M = \mathbb{R}^3$. The exterior derivative of any 0-form $\alpha = f$ is

$$df = f_{,i} dx^i = \frac{\partial f}{\partial x} dx + \frac{\partial f}{\partial y} dy + \frac{\partial f}{\partial z} dz.$$

The three components are commonly known as the *gradient* of the scalar field f.
If $\omega = w_i dx^i = A\, dx + B\, dy + C\, dz$ is a differential 1-form then

$$d\omega = \left(\frac{\partial C}{\partial y} - \frac{\partial B}{\partial z}\right) dy \wedge dz + \left(\frac{\partial A}{\partial z} - \frac{\partial C}{\partial x}\right) dz \wedge dx + \left(\frac{\partial B}{\partial x} - \frac{\partial A}{\partial y}\right) dx \wedge dy.$$

The components of the exterior derivative are traditionally written as components of a vector field, known as the *curl* of the three-component vector field (A, B, C). Notice, however, that the *tensor* components of $d\omega = -w_{[i,j]} dx^i \wedge dx^j$ are half the curl components,

$$(d\omega)_{ij} = -w_{[i,j]} = \tfrac{1}{2}\left(w_{j,i} - w_{i,j}\right).$$

If $\alpha = \alpha_{ij} dx^i \wedge dx^j = P\, dy \wedge dz + Q\, dz \wedge dx + R\, dx \wedge dy$ is a 2-form then

$$d\alpha = \left(\frac{\partial P}{\partial x} + \frac{\partial Q}{\partial y} + \frac{\partial R}{\partial z}\right) dx \wedge dy \wedge dz.$$

The single component of this 3-form is known as the *divergence* of the three-component vector field (P, Q, R). Equation (16.5) applied to the 0-form f and 1-form ω gives the following classical results:

$$d^2 f = 0 \implies \text{curl grad} = 0,$$
$$d^2 \omega = 0 \implies \text{div curl} = 0.$$

Exercise: If $\alpha = \alpha_{ij} dx^i \wedge dx^j$ is a 2-form on a manifold M, show that

$$(d\alpha)_{ijk} = \tfrac{1}{3}\left(\alpha_{ij,k} + \alpha_{jk,i} + \alpha_{ki,j}\right). \tag{16.6}$$

More generally, lumping together the permutations of the first r indices in Eq. (16.4) we obtain the following formula for the tensor components of the exterior derivative of an r-form α:

$$(d\alpha)_{i_1\ldots i_{r+1}} = \frac{(-1)^r}{r+1} \sum_{\text{cyclic } \pi} (-1)^\pi \alpha_{i_{\pi(1)}\ldots i_{\pi(r)}, i_{\pi(r+1)}}. \tag{16.7}$$

16.2 Properties of exterior derivative

Problems

Problem 16.1 Let $x^1 = x$, $x^2 = y$, $x^3 = z$ be coordinates on the manifold $\mathbb{R}^3$. Write out the components α_{ij} and $(d\alpha)_{ijk}$, etc. for each of the following 2-forms:

$$\alpha = dy \wedge dz + dx \wedge dy,$$
$$\beta = x\, dz \wedge dy + y\, dx \wedge dz + z\, dy \wedge dx,$$
$$\gamma = d(r^2(x\, dx + y\, dy + z\, dz)) \quad \text{where} \quad r^2 = x^2 + y^2 + z^2.$$

Problem 16.2 On the manifold $\mathbb{R}^n$ compute the exterior derivative d of the differential form

$$\alpha = \sum_{i=1}^{n}(-1)^{i-1} x^i dx^1 \wedge \cdots \wedge dx^{i-1} \wedge dx^{i+1} \wedge \cdots \wedge dx^n.$$

Do the same for $\beta = r^{-n}\alpha$ where $r^2 = (x^1)^2 + \cdots + (x^n)^2$.

Problem 16.3 Show that the right-hand side of Eq. (16.6) transforms as a tensor field of type (0, 3). Generalize this result to the right-hand side of Eq. (16.7), to show that this equation could be used as a local definition of exterior derivative independent of the choice of coordinate system.

16.2 Properties of exterior derivative

If $\varphi : M \to N$ is a smooth map between two differentiable manifolds M and N, we define the induced map $\varphi^* : \Lambda_r(N) \to \Lambda_r(M)$ in a similar way to the pullback map, Eq. (15.18):

$$(\varphi^*\alpha)_p\big((X_1)_p, (X_2)_p, \ldots, (X_r)_p\big) = \alpha_{\varphi(p)}\big(\varphi_*(X_1)_p, \varphi_*(X_2)_p, \ldots, \varphi_*(X_r)_p\big).$$

As for covector fields, this map is well-defined on all differential r-forms, $\varphi^*\alpha$. The pullback of a 0-form $f \in \mathcal{F}(N) = \Lambda_0(N)$ is defined by $\varphi^* f = f \circ \varphi$, and it preserves wedge products

$$\varphi^*(\alpha \wedge \beta) = \varphi^*\alpha \wedge \varphi^*\beta,$$

which follows immediately from the definition $\alpha \wedge \beta = \mathcal{A}(\alpha \otimes \beta)$.

Exercise: Show that the composition of two maps φ and ψ results in a reverse composition of pullbacks, as in Eq. (15.19), $(\varphi \circ \psi)^* = \psi^* \circ \varphi^*$.

Theorem 16.2 *For any differential form $\alpha \in \Lambda(N)$, the induced map φ^* commutes with the exterior derivative,*

$$d\varphi^*\alpha = \varphi^* d\alpha.$$

Proof: For a 0-form, $\alpha = f : N \to \mathbb{R}$, at any point $p \in M$ and any tangent vector X_p

$$\langle(\varphi^* df)_p, X_p\rangle = \langle(df)_{\varphi(p)}, \varphi_* X_p\rangle$$
$$= (\varphi_* X_p) f$$
$$= X_p(f \circ \varphi)$$
$$= \langle(d(f \circ \varphi))_p, X_p\rangle.$$

As this equation holds for all tangent vectors X_p, we have

$$\varphi^* df = d(f \circ \varphi) = d(\varphi^* f).$$

Differentiable forms

For a general r-form, it is only necessary to prove the result in any local coordinate chart $(U; x^i)$. If $\alpha = \alpha_{i_1 \ldots i_r} \, dx^{i_1} \wedge \cdots \wedge dx^{i_r}$, then

$$\varphi^* d\alpha = \varphi^*(d\alpha_{i_1 \ldots i_r} \wedge dx^{i_1} \wedge \ldots dx^{i_r})$$
$$= d(\varphi^* \alpha_{i_1 \ldots i_r}) \wedge d(\varphi^* x^{i_1}) \wedge \cdots \wedge d(\varphi^* x^{i_r})$$
$$= d(\varphi^* \alpha).$$

∎

Applying the definition (15.33) of Lie derivative to the tensor field α and using $\widetilde{\sigma}_t = (\sigma_{-t})^*$, where σ_t is a local one-parameter group generating a vector field X, it follows from Theorem 16.2 that the exterior derivative and Lie derivative commute,

$$\mathcal{L}_X \, d\alpha = d\mathcal{L}_X \alpha. \tag{16.8}$$

For any vector field X define the **interior product** $i_X : \Lambda_r(M) \to \Lambda_{(r-1)}(M)$ as in Section 8.4,

$$i_X \alpha = r C_1^1(X \otimes \alpha), \tag{16.9}$$

or equivalently, for arbitrary vector fields $X_1, X_2, \ldots, X_r$

$$(i_{X_1} \alpha)(X_2, \ldots, X_r) = r\alpha(X_1, X_2, \ldots, X_r). \tag{16.10}$$

By Eq. (8.20) i_X is an antiderivation – for any differential r-form α and arbitrary differential form β

$$i_X(\alpha \wedge \beta) = (i_X \alpha) \wedge \beta + (-1)^r \alpha \wedge (i_X \beta). \tag{16.11}$$

Exercise: Show that for any pair of vector fields X and Y, $i_X \circ i_Y = -i_Y \circ i_X$.

Theorem 16.3 (Cartan) *If X and Y are smooth vector fields on a differentiable manifold M and ω is a differential 1-form then*

$$i_{[X,Y]} = \mathcal{L}_X \circ i_Y - i_Y \circ \mathcal{L}_X, \tag{16.12}$$
$$\mathcal{L}_X = i_X \circ d + d \circ i_X, \tag{16.13}$$
$$d\omega(X, Y) = \tfrac{1}{2}\bigl(X(\langle Y, \omega \rangle) - Y(\langle X, \omega \rangle) - \langle [X, Y], \omega \rangle\bigr). \tag{16.14}$$

Proof: The first identity follows essentially from the fact that the Lie derivative $\mathcal{L}_X$ commutes with contraction operators, $\mathcal{L}_X C_j^i = C_j^i \mathcal{L}_X$ (see Problem 15.24). Thus for an arbitrary r-form α, using the Leibnitz rule (15.35) gives

$$\mathcal{L}_X(i_Y \alpha) = r C_1^1 \mathcal{L}_X(Y \otimes \alpha)$$
$$= r C_1^1 \bigl[(\mathcal{L}_X Y) \otimes \alpha + Y \otimes \mathcal{L}_X \alpha \bigr]$$
$$= i_{[X,Y]} \alpha + i_Y(\mathcal{L}_X \alpha)$$

as required.

To show (16.13) set K_X to be the operator $K_X = i_X \circ d + d \circ i_X : \Lambda_r(M) \to \Lambda_r(M)$. Using the fact that both i_X and d are antiderivations, Eqs. (16.11) and (16.3), it is straightforward to show that K_X is a derivation,

$$K_X(\alpha \wedge \beta) = K_X \alpha \wedge \beta + \alpha \wedge K_X \beta$$

16.2 Properties of exterior derivative

for all differential forms α and β. From $d^2 = 0$ the operator K_X commutes with d,

$$K_X \circ d = i_X \circ d^2 + d \circ i_X \circ d = d \circ i_X \circ d = d \circ K_X.$$

If α is a 0-form $\alpha = f$ then $i_X f = 0$ by definition, and

$$K_X f = i_X(df) + d i_X f = \langle df, X \rangle = Xf = \mathcal{L}_X f.$$

Hence, since K_X commutes both with d and $\mathcal{L}_X$,

$$K_X df = dK_X f = d(\mathcal{L}_X f) = \mathcal{L}_X df.$$

On applying the derivation property we obtain $K_X(g\,df) = \mathcal{L}_X(g\,df)$ and the required identity holds for any 1-form ω, as it can be expressed locally in a coordinate chart at any point as $\omega = w_i\,dx^i$. The argument may be generalized to higher order r-forms to show that the operators $\mathcal{L}_X$ and K_X are identical on all of $\Lambda_r(M)$.

The final identity (16.14) is proved on applying (16.13) to a 1-form ω,

$$\langle Y, \mathcal{L}_X \omega \rangle = \langle Y, i_X(d\omega) + d(i_X \omega) \rangle$$

and using the Leibnitz rule for the Lie derivative,

$$\mathcal{L}_X(\langle Y, \omega \rangle) - \langle \mathcal{L}_X Y, \omega \rangle = i_X\,d\omega(Y) + Y(i_X \omega).$$

Setting $r = 1$ and $\alpha = \omega$ in Eq. (16.10),

$$X(\langle Y, \omega \rangle) - \langle \mathcal{L}_X Y, \omega \rangle = 2\,d\omega(X, Y) + Y(\langle X, \omega \rangle)$$

from which (16.14) is immediate. ∎

If α is an r-form on M, a formula for $\alpha(X_1, X_2, \ldots, X_{r+1})$ that generalizes Eq. (16.14) is left to the reader (see Problem 16.5).

Problems

Problem 16.4 Let $\varphi : \mathbb{R}^2 \to \mathbb{R}^3$ be the map

$$(x, y) \to (u, v, w) \quad \text{where} \quad u = \sin(xy),\ v = x + y,\ w = 2.$$

For the 1-form $\omega = w_1 du + w_2 dv + w_3 dw$ on $\mathbb{R}^3$ evaluate $\varphi^*\omega$. For any function $f : \mathbb{R}^3 \to \mathbb{R}$ verify Theorem 16.2, that $d(\varphi^* f) = \varphi^* df$.

Problem 16.5 If α is an r-form on a differentiable manifold M, show that for any vector fields $X_1, X_2, \ldots X_{r+1}$

$$d\alpha(X_1, X_2, \ldots, X_{r+1}) = \frac{1}{r+1} \Big[\sum_{i=1}^{r+1}(-1)^{i+1} X_i \alpha(X_1, X_2, \ldots, \hat{X}_i, \ldots, X_{r+1})$$

$$+ \sum_{i=1}^{r}\sum_{j=i+1}^{r+1}(-1)^{i+j}\alpha([X_i, X_j], \ldots, \hat{X}_i, \ldots, \hat{X}_j, \ldots, X_{r+1})\Big]$$

where $\hat{X}_i$ signifies that the argument X_i is to be omitted. The case $r = 0$ simply asserts that $df(X) = Xf$, while Eq. (16.14) is the case $r = 1$. Proceed by induction, assuming the identity is true for all $(r-1)$-forms, and use the fact that any r-form can be written locally as a sum of tensors of the type $\omega \wedge \beta$ where ω is a 1-form and β an r-form.

Differentiable forms

Problem 16.6 Show that the Laplacian operator on $\mathbb{R}^3$ may be defined by

$$d * d\phi = \nabla^2 \phi \, dx \wedge dy \wedge dz = \left(\frac{\partial^2 \phi}{\partial x^2} + \frac{\partial^2 \phi}{\partial y^2} + \frac{\partial^2 \phi}{\partial z^2}\right) dx \wedge dy \wedge dz$$

where $*$ is the Hodge star operator of Section 8.6.

Use this to express the Laplacian operator in spherical polar coordinates (r, θ, ϕ).

16.3 Frobenius theorem: dual form

Let D^k be a k-dimensional distribution on a manifold M, assigning a k-dimensional subspace $D^k(p)$ of the tangent space at each point $p \in M$. Its **annihilator subspace** $(D^k)^\perp(p)$ (see Problem 3.16) consists of the set of covectors at p that vanish on $D^k(p)$,

$$(D^k)^\perp(p) = \{\omega_p \mid \langle \omega_p, X_p \rangle = 0 \text{ for all } X_p \in D^k(p)\}.$$

Since the distribution D^k is required to be C^∞, it follows from Theorem 3.7 that every point p has a neighbourhood U and a basis e_i of smooth vector fields on U, such that $e_{r+1}, \ldots, e_n$ span $D^k(q)$ at every point $q \in U$, where $r = n - k$. The dual basis of 1-forms ω^i defined by $\langle \omega^i, e_j \rangle = \delta^i_j$ has the property that the first r 1-forms $\omega^1, \omega^2, \ldots, \omega^r$ are linearly independent and span the annihilator subspace $(D^k)^\perp(q)$ at each $q \in U$.

The annihilator property is reciprocal: given r linearly independent 1-forms ω^a ($a = 1, \ldots, r$) on an open subset U of M, they span the annihilator subspace $(D^k)^\perp$ of the $k = (n-r)$-dimensional distribution

$$D^k = \{X \in \mathcal{T}(U) \mid \langle \omega^a, X \rangle = 0\}.$$

As shown at the end of Section 8.3, the simple differential k-form

$$\Omega = \omega^1 \wedge \omega^2 \wedge \cdots \wedge \omega^k$$

is uniquely defined up to a scalar field factor by the subspace $(D^k)^\perp$, and has the property that a 1-form ω belongs to $(D^k)^\perp$ if and only if $\omega \wedge \Omega = 0$.

Suppose the distribution D^k is involutive, so that $X, Y \in D^k \Rightarrow [X, Y] \in D^k$. From Eq. (16.14)

$$d\omega^a(X, Y) = \tfrac{1}{2}\bigl(X(\langle Y, \omega^a\rangle) - Y(\langle X, \omega^a\rangle) - \langle[X, Y], \omega^a\rangle\bigr) = 0$$

for any pair of vectors $X, Y \in D^k$. Conversely, if all ω^a and $d\omega^a$ vanish when restricted to the distribution D^k, then $\langle [X, Y], \omega^a \rangle = 0$ for all $X, Y \in D^k$. Thus, a necessary and sufficient condition for a distribution D^k to be involutive is that for all $\omega^a \in (D^k)^\perp$ the exterior derivative $d\omega^a$ vanishes on D^k.

Let $A^a_{ij} = -A^a_{ji}$ be scalar fields such that $d\omega^a = A^a_{ij} \omega^i \wedge \omega^j$. If $d\omega^a(e_\alpha, e_\beta) = 0$ for all $\alpha, \beta = r+1, \ldots, n$, then $A^a_{\alpha\beta} = 0$ and

$$d\omega^a = A^a_{bc} \omega^b \wedge \omega^c + A^a_{b\beta} \omega^b \wedge \omega^\beta + A^a_{\alpha c} \omega^\alpha \wedge \omega^c.$$

Thus, D^k is involutive if and only if for the 1-forms $d\omega^a$ there exist 1-forms θ^a_b such that

$$d\omega^a = \theta^a_b \wedge \omega^b.$$

16.3 Frobenius theorem: dual form

On the other hand, the Frobenius theorem 15.4 asserts that D^k is involutive if and only if there exist local coordinates $(U; x^i)$ at any point p such that $e_\alpha = B_\alpha^\beta \partial_{x^\beta}$ for an invertible matrix of scalar fields $[B_\alpha^\beta]$ on U. In these coordinates, set $\omega^a = A_b^a \, dx^b + W_\alpha^a \, dx^\alpha$, and using $\langle \omega^a, e_\alpha \rangle = 0$, we have $W_\alpha^a = 0$. Hence an alternative necessary and sufficient condition for D^k to be involutive is the existence of coordinates $(U; x^i)$ such that

$$\omega^a = A_b^a \, dx^b.$$

Theorem 16.4 *Let ω^a $(a = 1, \ldots, r)$ be a set of 1-forms on an open set U, linearly independent at every point $p \in U$. The following statements are all equivalent:*

(i) *There exist local coordinates $(U; x^i)$ at every point $p \in U$ such that $\omega^a = A_b^a \, dx^b$.*
(ii) *There exist 1-forms θ_b^a such that $d\omega^a = \theta_b^a \wedge \omega^b$.*
(iii) $d\omega^a \wedge \Omega = 0$ *where* $\Omega = \omega^1 \wedge \omega^2 \wedge \cdots \wedge \omega^r$.
(iv) $d\Omega \wedge \omega^a = 0$.
(v) *There exists a 1-form θ such that $d\Omega = \theta \wedge \Omega$.*

Proof: We have seen by the above remarks that (i) $\Leftrightarrow$ (ii) as both statements are equivalent to the statement that the distribution D^k that annihilates all ω^a is involutive. Condition (ii) $\Leftrightarrow$ (iii) since $\omega^a \wedge \Omega = 0$, while the converse follows on setting $d\omega^a = \theta_b^a \wedge \omega^b + A_{\alpha\beta}^a \omega^\alpha \wedge \omega^\beta$, where ω^i $(i = 1, \ldots, n)$ is any local basis of 1-forms completing the ω^a. The implication (iii) $\Leftrightarrow$ (iv) follows at once from Eq. (16.3), and (v) $\Rightarrow$ (iv) since

$$d\Omega = \theta \wedge \Omega \implies d\Omega \wedge \omega^a = \theta \wedge \Omega \wedge \omega^a = 0.$$

Finally, (ii) $\rightarrow$ (v), for if $d\omega^a = \theta_b^a \wedge \omega^b$ then

$$d\Omega = d\omega^1 \wedge \omega^2 \wedge \cdots \wedge \omega^r - \omega^1 \wedge d\omega^2 \wedge \cdots \wedge \omega^r + \ldots$$
$$= \theta_1^1 \wedge \omega^1 \wedge \omega^2 \wedge \cdots \wedge \omega^r - \omega^1 \wedge \theta_2^2 \wedge \omega^2 \wedge \cdots \wedge \omega^r + \ldots$$
$$= (\theta_1^1 + \theta_2^2 + \cdots + \theta_r^r) \wedge \omega^1 \wedge \omega^2 \wedge \cdots \wedge \omega^r$$
$$= \theta \wedge \Omega$$

where $\theta = \theta_a^a$. Hence (iv) $\Rightarrow$ (iii) $\Rightarrow$ (ii) $\Rightarrow$ (v) and the proof is completed. ∎

A system of linearly independent 1-forms $\omega^1, \ldots, \omega^r$ on an open set U, satisfying any of the conditions (i)–(v) of this theorem is said to be **completely integrable**. The equations defining the distribution D^k $(k = n - r)$ that annihilates these ω^a is given by the equations $\langle \omega^a, X \rangle = 0$, often written as a **Pfaffian system of equations**

$$\omega^a = 0 \quad (a = 1, \ldots, r).$$

Condition (i) says that locally there exist r functions $g^a(x^1, \ldots, x^n)$ on U such that

$$\omega^a = f_b^a \, dg^b$$

where the functions f_b^a form a non-singular $r \times r$ matrix at every point of U. The functions g^a are known as a **first integral of the system**. The r-dimensional submanifolds (N_c, ψ_c) defined by $g^a(x^1, \ldots, x^n) = c^a \equiv \text{const.}$ have the property

$$\psi_c^* \omega^a = f_b^a \circ \psi_c \, dc^b = 0,$$

and are known as **integral submanifolds** of the system.

Differentiable forms

Example 16.2 Consider a single Pfaffian equation in three dimensions,

$$\omega = P(x, y, z)\,dx + Q(x, y, z)\,dy + R(x, y, z)\,dz = 0.$$

If $\omega = f\,dg$ where $f(0, 0, 0) \neq 0$, the function $f(x, y, z)$ is said to be an *integrating factor*. It is immediate then that

$$d\omega = df \wedge dg = df \wedge \frac{1}{f}\omega = \theta \wedge \omega$$

where $\theta = d(\ln f)$. This is equivalent to conditions (ii) and (v) of Theorem 16.4. Conditions (iii) and (iv) are identical since $\Omega = \omega$, and follow at once from

$$d\omega \wedge \omega = \theta \wedge \omega \wedge \omega = 0,$$

which reduces to Euler's famous integrability condition for the existence of an integrating factor,

$$P\left(\frac{\partial R}{\partial y} - \frac{\partial Q}{\partial z}\right) + Q\left(\frac{\partial P}{\partial z} - \frac{\partial R}{\partial x}\right) + R\left(\frac{\partial Q}{\partial x} - \frac{\partial P}{\partial y}\right) = 0.$$

For example, if $\omega = dx + z\,dy + dz$ there is no integrating factor, $\omega = f\,dg$, since

$$d\omega \wedge \omega = dz \wedge dy \wedge dx = -dx \wedge dy \wedge dz \neq 0.$$

On the other hand, if $\omega = 2xz\,dx + 2yz\,dy + dz$, then

$$d\omega \wedge \omega = (2x\,dz \wedge dx + 2y\,dz \wedge dy) \wedge \omega = 4xyz(dz \wedge dx \wedge dy + dz \wedge dy \wedge dx) = 0.$$

It should therefore be possible locally to express ω in the form $f\,dg$. The functions f and g are not unique, for if $G(g)$ is an arbitrary function then $\omega = F\,dG$ where $F = f/(dG/dg)$. To find an integrating factor f we solve a system of three differential equations

$$f\frac{\partial g}{\partial x} = 2xz, \tag{a}$$

$$f\frac{\partial g}{\partial y} = 2yz, \tag{b}$$

$$f\frac{\partial g}{\partial z} = 1. \tag{c}$$

Eliminating f from (a) and (b) we have

$$\frac{1}{x}\frac{\partial g}{\partial x} = \frac{1}{y}\frac{\partial g}{\partial y},$$

which can be expressed as

$$\frac{\partial g}{\partial x^2} = \frac{\partial g}{\partial y^2}.$$

This equation has a general solution $g(z, u)$ where $u = x^2 + y^2$, and eliminating f from (b) and (c) results in

$$\frac{\partial g}{\partial z} = 2yz\frac{\partial g}{\partial y} \Longrightarrow \frac{\partial g}{\partial \ln z} = \frac{\partial g}{\partial y^2} = \frac{\partial g}{\partial u}.$$

Hence $g = G(\ln z + u)$, and since it is possible to pick an arbitrary function G we can set $g = ze^{x^2+y^2}$. From (c) it follows that $f = e^{-x^2-y^2}$, and it is easy to check that

$$\omega = e^{-x^2-y^2} d(ze^{x^2+y^2}) = 2xz\,dx + 2yz\,dy + dz.$$

Problems

Problem 16.7   Let $\omega = yz\,dx + xz + z^2\,dz$. Show that the Pfaffian system $\omega = 0$ has integral surfaces $g = z^3 e^{xy} = \text{const.}$, and express ω in the form $f\,dg$.

Problem 16.8   Given an $r \times r$ matrix of 1-forms Ω, show that the equation

$$dA = \Omega A - A\Omega$$

is soluble for an $r \times r$ matrix of functions A only if

$$\Theta A = A\Theta$$

where $\Theta = d\Omega - \Omega \wedge \Omega$.

If the equation has a solution for arbitrary initial values $A = A_0$ at any point $p \in M$, show that there exists a 2-form α such that $\Theta = \alpha I$ and $d\alpha = 0$.

16.4 Thermodynamics

Thermodynamics deals with the overall properties of systems such as a vessel of gas or mixture of gases, a block of ice, a magnetized iron bar, etc. While such systems may be impossibly complex at the microscopic level, their thermodynamic behaviour is governed by a very few number of variables. For example, the state of a simple gas is determined by two variables, its volume V and pressure p, while a mixture of gases also requires specification of the molar concentrations $n_1, n_2, \ldots$ representing the relative number of particles of each species of gas. An iron bar may need information from among variables such as its length ℓ, cross-section A, tensile strength f and Young's modulus Y, the magnetic field $\mathbf{H}$, magnetization μ, electric field $\mathbf{E}$ and conductivity σ. In any case, the number of variables needed for a thermodynamic description of the system is tiny compared to the 10^{24} or so variables required for a complete description of the microscopic state of the system (see Section 14.4).

The following treatment is similar to that given in [10]. Every thermodynamic system will be assumed to have a special class of states known as **equilibrium states**, forming an n-dimensional manifold K, and given locally by a set of **thermodynamic variables** $\mathbf{x} = (x^1, x^2, \ldots, x^n)$. The dimension n is called the number of **degrees of freedom** of the thermodynamic system. Physically, we think of an equilibrium state as one in which the system remains when all external forces are removed. For a perfect gas there are two degrees of freedom, usually set to be $x^1 = p$ and $x^2 = V$. The variable p is called an **internal** or **thermal** variable, characterized physically by the fact that no work is done on or by the system if we change p alone, leaving V unaltered. Variables such as volume V, a change in which results in work being done on the system, are called **external** or **deformation** variables.

Differentiable forms

A **quasi-static** or **reversible** process, resulting in a transition from one equilibrium state x_1 to another x_2, is a parametrized curve $\gamma : [t_1, t_2] \to K$ such that $\gamma(t_1) = x_1$ and $\gamma(t_2) = x_2$. Since the curve passes through a continuous succession of equilibrium states, it should be thought of as occurring *infinitely slowly*, and its parameter t is not to be identified with real time. For example, a gas in a cylinder with a piston attached will undergo a quasi-static transition if the piston is withdrawn so slowly that the effect on the gas is reversible. If the piston is withdrawn rapidly the action is irreversible, as non-equilibrium intermediate states arise in which the gas swirls and eddies, creating regions of non-uniform pressure and density throughout the container. The same can be said of the action of a 'stirrer' on a gas or liquid in an adiabatic container – you can never 'unstir' the milk or sugar added to a cup of tea. Irreversible transitions from one state of the system cannot be represented by parametrized curves in the manifold of equilibrium states K. Whether the transition be reversible or irreversible, we assume that there is always associated with it a well-defined quantity ΔW, known as the **work done by the system**. The work done **on the system** is defined to be the negative of this quantity, $-\Delta W$.

We will also think of thermodynamic systems as being confined to certain 'enclosures', to be thought of as closed regions of three-dimensional space. Most importantly, a system K is said to be in an **adiabatic enclosure** if equilibrium states can only be disturbed by doing work on the system through mechanical means (reversible or irreversible), such as the movement of a piston or the rotation of a stirrer. In all cases, transitions between states of a system in an adiabatic enclosure are called **adiabatic processes**.

The boundary of an adiabatic enclosure can be considered as being an *insulating wall* through which no 'heat transfer' is allowed; a precise meaning to the concept of *heat* will be given directly. A **diathermic wall** within an adiabatic enclosure is one that permits heat to be transferred across it without any work being done. Two systems K_A and K_B are said to be in **thermal contact** if both are enclosed in a common adiabatic enclosure, but are separated by a diathermic wall. The states x_A and x_B of the two systems are then said to be in **thermal equilibrium** with each other.

Zeroth law of thermodynamics: temperature *For every thermodynamic system K there exists a function $\tau : K \to \mathbb{R}$ called **empirical temperature** such that two systems K_A and K_B are in equilibrium with each other if and only if $\tau_A(x_A) = \tau_B(x_B)$.*

This law serves as little more than a definition of empirical temperature, but the fact that a single function of state achieves the definition of equilibrium is significant. Any set of states $\{x \mid \tau(x) = \text{const.}\}$ is called an **isotherm** of a system K.

Example 16.3 For an **ideal gas** we find $\tau = pV$ is an empirical temperature, and the isotherms are curves $pV = \text{const}$. Any monotone function $\tau' = \varphi \circ \tau$ will also do as empirical temperature. A system of ideal gases in equilibrium with each other, $(p_1, V_1), (p_2, V_2), \ldots, (p_n, V_n)$ have common empirical temperature $\tau = T_g$, called the **absolute gas temperature**, given by

$$T_g = \frac{p_1 V_1}{n_1 R} = \frac{p_2 V_2}{n_2 R} = \cdots = \frac{p_n V_n}{n_n R}$$

16.4 Thermodynamics

where n_i are the relative molar quantities of the gases involved and R is the universal gas constant. While an arbitrary function φ may still be applied to the absolute gas temperature, the same function must be applied equally to all component gases. In this example it is possible to eliminate all pressures except one, and the total system can be described by a single thermal variable, p_1 say, and n external deformation variables $V_1, V_2, \ldots, V_n$.

This example illustrates a common assumption made about thermodynamic systems of n degrees of freedom, that it is possible to pick coordinates $(x^1, x^2, \ldots, x^n)$ in a local neighbourhood of any point in K such that the first $n-1$ coordinates are external variables and x^n is an internal variable. We call this the **thermal variable assumption**.

First law of thermodynamics: energy. *For every thermodynamic system K there is a function $U : K \to \mathbb{R}$ known as **internal energy** and a 1-form $\omega \in \Lambda_1(K)$ known as the **work form** such that the work done by the system in any reversible process $\gamma : [t_1, t_2] \to K$ is given by*

$$\Delta W = \int_\gamma \omega$$

(see Example 15.9 for the definition of integral of ω along the curve γ). In every reversible adiabatic process

$$\gamma^*(\omega + dU) = 0.$$

From Example 15.9 the integral of $\omega + dU$ along the curve γ vanishes, since

$$\int_\gamma \omega + dU = \int_{t_1}^{t_2} \langle \gamma^*(\omega + dU), \frac{d}{dt} \rangle dt = 0.$$

Furthermore, since

$$\int_\gamma dU = \int_{t_1}^{t_2} \frac{dU}{dt} dt = U(t_2) - U(t_1) = \Delta U$$

the conservation law of energy holds for any reversible adiabatic process,

$$\Delta W + \Delta U = 0.$$

Thus the change of internal energy is equal to the work done on the system, $-\Delta W$. The work done in any reversible adiabatic transition from one equilibrium state of a system K to another is independent of the path. In particular, no work is done by the system in any cyclic adiabatic process, returning a system to its original state – commonly known as the *impossibility of a perpetual motion machine of the first kind*.

The **heat 1-form** is defined to be $\theta = \omega + dU$, and we refer to $\Delta Q = \int_\gamma \theta$ as the **heat added** to the system in any reversible process γ. The conservation of energy in the form

$$\Delta Q = \Delta W + \Delta U$$

is often referred to in the literature as the first law of thermodynamics. Adiabatic transitions are those with $\Delta Q = 0$.

Differentiable forms

If the thermal variable assumption holds, then it is generally assumed that the work form is a linear expansion of the external variables alone,

$$\omega = \sum_{k=1}^{n-1} P_k(x^1, \ldots, x^n) \, dx^k,$$

where the component function P_k is known as the k**th generalized force**. Since U is a thermal variable, it is always possible to choose the nth coordinate as $x^n = U$, in which case

$$\theta = \sum_{k=1}^{n-1} P_k(x^1, \ldots, x^n) \, dx^k + dU = \sum_{k=1}^{n-1} P_k(x^1, \ldots, x^n) \, dx^k + dx^n.$$

Second law of thermodynamics

Not every transition between equilibrium states is possible, even if conservation of energy holds. The second law of thermodynamics limits the possible transitions consistent with energy conservation, and has a number of equivalent formulations. For example, the version due to Clausis asserts that no machine can perform work, or mechanical energy, while at the same time having no other effect than to lower the temperature of a thermodynamic system. Such a machine is sometimes referred to as a *perpetual motion machine of the second kind* – if it were possible one could draw on the essentially infinite heat reservoir of the oceans to perform an unlimited amount of mechanical work.

An equivalent version is Kelvin's principle: no cyclic quasi-static thermodynamic process permits the conversion of heat *entirely* into mechanical energy. By this is meant that no quasi-static thermodynamic cycle γ exists, the first half of which consists of a quasi-static process γ_1 purely of heat transfer in which no work is done, $\gamma_1^* \omega = 0$, while the second half γ_2 is adiabatic and consists purely of mechanical work, $\gamma_2^* \theta = 0$. Since U is a function of state it follows, on separating the cycle into its two parts, that

$$0 = \Delta U = \oint_\gamma dU = \Delta_1 U + \Delta_2 U = \Delta_1 Q - \Delta_2 W.$$

Thus in any such cycle an amount of heat would be converted entirely into its mechanical equivalent of work.

Consider a quasi-static process γ_1 taking an equilibrium state $\mathbf{x}$ to another state $\mathbf{x}'$ along a curve of constant volume, $x^k = $ const. for $k = 1, \ldots, n-1$. Such a curve can be thought of as 'cooling at constant volume' and is achieved purely by heat transfer; no mechanical work is done,

$$\gamma_1^* \omega = \gamma_1^* \sum_{k=1}^{n-1} P_k \, dx^k = \sum_{k=1}^{n-1} P_k \, d(x^k \circ \gamma_1) = 0.$$

It then follows that no reversible adiabatic transition γ_2 such that $\gamma_2^* \theta = 0$ exists between these two states. Since processes such as γ_1 may always be assumed to be locally possible, it follows that every state $\mathbf{x}$ has equilibrium states in its neighbourhood that cannot be reached by quasi-static adiabatic paths. This leads to Carathéodory's more general version of the second law.

16.4 Thermodynamics

Second law of thermodynamics: entropy. *In a thermodynamic system K, every neighbourhood U of an arbitrary equilibrium state $\mathbf{x}$ contains a state $\mathbf{x}'$ that is inaccessible by a quasi-static adiabatic path from $\mathbf{x}$.*

Theorem 16.5 (Carathéodory) *The heat 1-form θ is integrable, $\theta \wedge d\theta = 0$, if and only if every neighbourhood of any state $\mathbf{x} \in K$ contains a state $\mathbf{x}'$ adiabatically inaccessible from $\mathbf{x}$.*

Outline proof: If θ is integrable, then by Theorem 16.4 it is possible to find local coordinates $(U; y^i)$ of any state $\mathbf{x}$ such that $\theta|_U = Q_n \, dy^n$. Adiabatics satisfy $\gamma^*\theta = 0$, or $y^n = $ const. Hence, if U' is an open neighbourhood of $\mathbf{x}$ such that $U' \subseteq U$, any state $\mathbf{x}' \in U'$ such that $y'^n = y^n(\mathbf{x}') \ne y^n(\mathbf{x})$ is adiabatically inaccessible from $\mathbf{x}$.

Conversely, if $\theta \wedge d\theta \ne 0$, then the 1-form θ is not integrable on an open subset U of every state $\mathbf{x} \in M$. Hence the distribution D^{n-1} such that $\theta \in (D^{n-1})^\perp$ is not involutive on an open neighbourhood U' of $\mathbf{x}$, so that $[D^{n-1}, D^{n-1}] = T(U')$. Let X and Y be vector fields in D^{n-1} such that $[X, Y]$ is not in the distribution. It may then be shown that every state $\mathbf{x}'$ is accessible by a curve of the form

$$t \mapsto \psi_{-\sqrt{t}} \circ \phi_{-\sqrt{t}} \circ \psi_{\sqrt{t}} \circ \phi_{\sqrt{t}} \mathbf{x}$$

where ψ_t and ϕ_t are local flows generated by the vector field X and Y (see Example 15.14). ∎

For a reversible adiabatic process γ at constant volume we have $\gamma^*\theta = 0$ and

$$\gamma^*\omega = \sum_{k=1}^{n-1} P_k \frac{dx^k(\gamma(t))}{dt} dt = 0.$$

Hence there is no change in internal energy for such processes, $\Delta U = 0$. On the other hand, for an irreversible adiabatic process at constant volume, such as stirring a gas in an adiabatic enclosure, there is always an increase in internal energy, $U' > U$. Hence all states with $U' < U$ are adiabatically inaccessible by adiabatic processes at constant volume, be they reversible or not. As remarked above, it is impossible to 'unstir' a gas. In general, for any two states $\mathbf{x}$ and $\mathbf{x}'$ either (i) $\mathbf{x}$ is adiabatically inaccessible to $\mathbf{x}'$, (ii) $\mathbf{x}'$ is adiabatically inaccessible to $\mathbf{x}$, or (iii) there exists a reversible quasi-static process from $\mathbf{x}$ and $\mathbf{x}'$.

From Theorem 16.5 and Carathéodory's statement of the second law, the heat form θ can be expressed as

$$\theta = f \, ds$$

where f and s are real-valued functions on K. Any function $s(x^1, \ldots, x^{n-1}, U)$ for which this holds is known as an **empirical entropy**. A reversible adiabatic process $\gamma : [a, b] \to K$ is clearly **isentropic**, $s = $ const., since $\gamma^*\theta = 0$ along the process, and the hypersurface $s = $ const. through any state $\mathbf{x}$ represents the local boundary between adiabatically accessible and inaccessible states from $\mathbf{x}$.

For most thermodynamic systems the function s is globally defined by the identity $\theta = f \, ds$. Since a path in K connecting adiabatically accessible states has $dU/dt \ge 0$, we can assume that s is a monotone increasing function of U for fixed volume coordinates

Differentiable forms

$x^1, \ldots, x^{n-1}$. For any path γ with $x^k =$ const. for $k = 1, \ldots, n-1$, such that $\gamma^*\omega = 0$, it follows that

$$\frac{dU}{dt} = \langle \dot{\gamma}, \theta \rangle = f\frac{ds}{dt}$$

and the function f must be everywhere positive.

Absolute entropy and temperature

Consider two systems A and B in an adiabatic enclosure and in equilibrium through mutual contact with a diathermic wall. In place of variables $x^1, \ldots, x^{n-1}$, U_A for states of system A let us use variables $x^1, \ldots, x^{n-2}$, s_A, τ_A where τ_A is the empirical temperature, and similarly use variables $y^1, \ldots, y^{m-2}$, s_B, $\tau_B = \tau_A$ for states of system B. The combined system then has coordinates $x^1, \ldots, x^{n-2}$, $y^1, \ldots, y^{m-2}$, $s_A, s_B, \tau = \tau_A = \tau_B$. Since work done in any reversible process is an additive quantity, $\Delta W = \Delta W_A + \Delta W_B$, we may assume from the first law of thermodynamics that U is an additive function, $U = U_A + U_B$. Hence the work 1-form may be assumed to be additive, $\omega = \omega_A + \omega_B$, and so is the heat 1-form

$$\theta = \omega + dU = \omega_A + \omega_B + dU_A + dU_B = \theta_A + \theta_B, \tag{16.15}$$

which can be written

$$f\,ds = f_A\,ds_A + f_B\,ds_B, \tag{16.16}$$

where $f_A = f_A(x^1, \ldots, x^{n-2}, s_A, \tau)$ and $f_B = f_B(y^1, \ldots, y^{m-2}, s_B, \tau)$. Since s is a function of all variables $s = s(x^1, \ldots, y^{m-2}, s_A, s_B, \tau)$, it follows that $s = s(s_A, s_B)$ and

$$\frac{f_A}{f} = \frac{\partial s}{\partial s_A}, \quad \frac{f_B}{f} = \frac{\partial s}{\partial s_B}. \tag{16.17}$$

Hence $f = f(s_A, s_B, \tau)$, $f_A = f_A(s_A, \tau)$, $f_B = f_B(s_B, \tau)$ and

$$\frac{\partial \ln f_A}{\partial \tau} = \frac{\partial \ln f_B}{\partial \tau} = \frac{\partial \ln f}{\partial \tau} = g(\tau)$$

for some function g. Setting $T(\tau) = \exp\left(\int g(\tau)\,d\tau\right)$,

$$f_A = T(\tau)F_A(s_A), \quad f_B = T(\tau)F_B(s_B), \quad f = T(\tau)F(s_A, s_B),$$

and Eq. (16.16) results in

$$F\,ds = F_A\,ds_A + F_B\,ds_B. \tag{16.18}$$

By setting $S_A = \int F_A(s_A)\,ds_A$ and $S_B = \int F_B(s_B)\,ds_B$, we have

$$F\,ds = dS_A + dS_B = dS$$

where $S = S_A + S_B$. Hence

$$\theta_A = f_A\,ds_A = TF_A\,ds_A = T\,dS_A, \qquad \theta_B = T\,dS_B$$

and

$$\theta = TF\,ds = T\,dS, \tag{16.19}$$

16.4 Thermodynamics

which is consistent with the earlier requirement of additivity of heat forms, Eq. (16.15). The particular choice of empirical temperature T and entropy S such that (16.19) holds, and which has the additivity property $S = S_A + S_B$, is called **absolute temperature** and **absolute entropy**. In the literature one often finds the formula $dQ = T dS$ in place of (16.19) but this notation is not good, for the right-hand side is not an exact differential as $d\theta \neq 0$ in general.

When $d\theta \neq 0$ the original variables τ and s are independent and only simple scaling freedoms are available for absolute temperature and entropy. For example, if

$$\theta = T \, dS = T' \, dS'$$

then

$$\frac{T'(\tau)}{T(\tau)} = \frac{dS'(s)}{dS(s)} = a = \text{const.},$$

where $a > 0$ if the rule $\Delta S > 0$ for adiabatically accessible states is to be preserved. Hence

$$T' = aT, \qquad S' = \frac{1}{a} S + b.$$

Only a positive scaling may be applied to absolute temperature and there is an **absolute zero** of temperature; absolute entropy permits an affine transformation, consisting of both a rescaling and change of origin.

Example 16.4 An ideal or perfect gas is determined by two variables, volume V and absolute temperature T. The heat 1-form is given by

$$\theta = dU + p \, dV = T \, dS.$$

Using

$$d\left(\frac{\theta}{T}\right) = d^2 S = 0$$

we have

$$-\frac{1}{T^2} dT \wedge dU + d\left(\frac{p}{T}\right) \wedge dV = 0$$

and setting $U = U(V, T)$, $p = p(V, T)$ results in

$$\left[-\frac{1}{T^2}\left(\frac{\partial U}{\partial V}\right)_T - \frac{p}{T^2} + \frac{1}{T}\left(\frac{\partial p}{\partial T}\right)_V \right] dT \wedge dV = 0.$$

Hence

$$T\left(\frac{\partial p}{\partial T}\right)_V = \left(\frac{\partial U}{\partial V}\right)_T + p. \tag{16.20}$$

For a gas in an adiabatic enclosure, classic experiments of Gay-Lussac and Joule have led to the conclusion that $U = U(T)$. Substituting into (16.20) results in

$$\frac{\partial \ln p}{\partial T} = \frac{1}{T},$$

Differentiable forms

which integrates to give a function $f(V)$ such that

$$f(V)p = T.$$

Comparing with the discussion in Example 16.3, we have for a single mole of gas

$$Vp = RT_g,$$

and since $T = T(T_g)$ it follows that after a suitable scaling of temperature we may set $f(V) = V$ and $T = T_g$. Thus for an ideal gas the absolute temperature is identical with absolute gas temperature.

From $\theta = T\,dS = dU + p\,dV$ we have

$$dS = \frac{1}{T}dU + \frac{R}{V}dV$$

and the formula for absolute entropy of an ideal gas is

$$S = \int \frac{1}{T}\frac{dU}{dT}dT + R\ln V.$$

Problem

Problem 16.9 For a reversible process $\sigma : T \to K$, using absolute temperature T as the parameter, set

$$\sigma^*\theta = c\,dT$$

where c is known as the **specific heat** for the process. For a perfect gas show that for a process at constant volume, $V = \text{const.}$, the specific heat is given by

$$c_V = \left(\frac{\partial U}{\partial T}\right)_V.$$

For a process at constant pressure show that

$$c_p = c_V + R,$$

while for an adiabatic process, $\sigma^*\theta = 0$,

$$pV^\gamma = \text{const.} \quad \text{where} \quad \gamma = \frac{c_p}{c_V}.$$

16.5 Classical mechanics

Classical analytic mechanics comes in two basic forms, Lagrangian or Hamiltonian. Both have natural formulations in the language of differential geometry, which we will outline in this section. More details may be found in [2, 10–14] and [4, chap. 13].

Calculus of variations

The reader should have at least a rudimentary acquaintance with the calculus of variations as found in standard texts on applied mathematics such as [15]. The following is a brief introduction to the subject, as it applies to parametrized curves on manifolds.

16.5 Classical mechanics

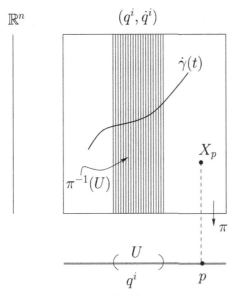

Figure 16.1 Tangent bundle

Let M be any differential manifold and TM its tangent bundle (refer to Section 15.3; see Fig. 16.1). If $\gamma : \mathbb{R} \to M$ is a smooth curve on M, define its **lift** to TM to be the curve $\dot\gamma : \mathbb{R} \to TM$ traced out by the tangent vector to the curve, so that $\dot\gamma(t)$ is the tangent vector to the curve at $\gamma(t)$ and $\pi(\dot\gamma(t)) = \gamma(t)$.

Exercise: Show that if $X(t)$ is the tangent to the curve $\dot\gamma(t)$ then $\pi_*X(t) = \dot\gamma(t) \in T_{\gamma(t)}(M)$.

A function $L : TM \to \mathbb{R}$ is called a **Lagrangian function**, and for any parametrized curve $\gamma : [t_0, t_1] \to M$ we define the corresponding **action** to be

$$S[\gamma] = \int_{t_0}^{t_1} L(\dot\gamma(t))\,dt. \tag{16.21}$$

If $(q^1, \ldots, q^n)$ are local coordinates on M let the induced local coordinates on the tangent bundle TM be written $(q^1, \ldots, q^n, \dot q^1, \ldots, \dot q^n)$. This notation may cause a little concern to the reader, but it is much loved by physicists – the quantities $\dot q^i$ are independent quantities, not to be thought of as 'derivatives' of q^i unless a specific curve γ having coordinate representation $q^i = q^i(t)$ is given. In that case, and only then, we find $\dot q^i(t) = dq^i(t)/dt$ along the lift $\dot\gamma(t)$ of the curve. Otherwise, the $\dot q^i$ refer to all possible components of tangent vectors at that point of M having coordinates q^j. A Lagrangian can be written as a function of $2n$ variables, $L(q^1, \ldots, q^n, \dot q^1, \ldots, \dot q^n)$.

By a **variation** of a given curve $\gamma : [t_0, t_1] \to M$ (see Fig. 16.2) is meant a one-parameter family of curves $\gamma : [t_0, t_1] \times [-a, a] \to M$ such that for all $\lambda \in [-a, a]$

$$\gamma(t_0, \lambda) = \gamma(t_0) \quad \text{and} \quad \gamma(t_1, \lambda) = \gamma(t_1)$$

465

Differentiable forms

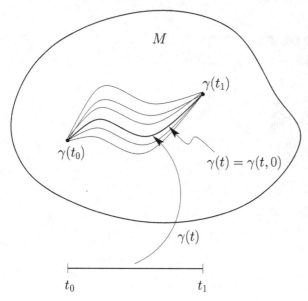

Figure 16.2   Variation of a curve

and the member of the family defined by $\lambda = 0$ is the given curve γ,

$$\gamma(t, 0) = \gamma(t) \qquad \text{for all } t_0 \le t \le t_1.$$

For each t in the range $[t_0, t_1]$ we define the *connection curve* $\gamma_t : [-a, a] \to M$ by $\gamma_t(\lambda) = \gamma(t, \lambda)$. Its tangent vector along the curve $\lambda = 0$ is written $\delta\gamma$, whose value at $t \in [t_0, t_1]$ is determined by the action on an arbitrary function $f : M \to \mathbb{R}$,

$$\delta\gamma_t(f) = \left.\frac{\partial f(\gamma_t(\lambda))}{\partial \lambda}\right|_{\lambda=0}. \tag{16.22}$$

This is referred to as the **variation field** along the curve. In traditional literature it is simply referred to as the 'variation of the curve'. Since all curves of the family meet at the end points $t = t_0, t_1$, the quantity on the right-hand side of Eq. (16.22) vanishes,

$$\delta\gamma_{t_0} = \delta\gamma_{t_1} = 0. \tag{16.23}$$

The lift of the variation field to the tangent bundle is a curve $\delta\dot\gamma : [t_0, t_1] \to TM$, which starts at the zero vector in the fibre above $\gamma(t_0)$ and ends at the zero vector in the fibre above $\gamma(t_1)$. In coordinates,

$$\delta\dot\gamma(t) = \big(\delta q^1(t), \ldots, \delta q^n(t), \delta\dot q^1(t), \ldots, \delta\dot q^n(t)\big)$$

where

$$\delta q^i(t) = \left.\frac{\partial q^i(t, \lambda)}{\partial \lambda}\right|_{\lambda=0}, \quad \delta\dot q^i(t) = \left.\frac{\partial \dot q^i(t, \lambda)}{\partial \lambda}\right|_{\lambda=0} = \frac{\partial \delta q^i(t)}{\partial t}.$$

Exercise: Justify the final identity in this equation.

16.5 Classical mechanics

The action $S[\gamma]$ becomes a function of λ if γ is replaced by its variation γ_λ. We say that a curve $\gamma : [t_0, t_1] \to M$ is an **extremal** if for every variation of the curve

$$\delta S \equiv \left.\frac{dS}{d\lambda}\right|_{\lambda=0} = \int_{t_0}^{t_1} \delta L \, dt = 0 \tag{16.24}$$

where

$$\delta L \equiv \left.\frac{\partial L(\dot{\gamma}_\lambda)}{\partial \lambda}\right|_{\lambda=0}$$
$$= \langle dL \mid \delta\dot{\gamma}\rangle$$
$$= \frac{\partial L}{\partial q^i}\delta q^i + \frac{\partial L}{\partial \dot{q}^i}\delta \dot{q}^i.$$

Substituting in (16.24) and performing an integration by parts results in

$$0 = \delta S = \int_{t_0}^{t_1}\left[\frac{\partial L}{\partial q^i} - \frac{d}{dt}\left(\frac{\partial L}{\partial \dot{q}^i}\right)\right]dt + \left.\frac{\partial L}{\partial \dot{q}^i}\delta q^i\right|_{t_0}^{t_1}.$$

The final term vanishes on account of $\delta q^i = 0$ at $t = t_0, t_1$, and since the $\delta q^i(t)$ are essentially arbitrary functions on the interval $[t_0, t_1]$ subject to the end-point constraints it may be shown that the term in the integrand must vanish,

$$\frac{\partial L}{\partial q^i} - \frac{d}{dt}\left(\frac{\partial L}{\partial \dot{q}^i}\right) = 0. \tag{16.25}$$

These are known as the **Euler–Lagrange equations**.

Example 16.5 In the plane, the shortest curve between two fixed points is a straight line. To prove this, use the length as action

$$S = \int_{t_0}^{t_1} \sqrt{\dot{x}^2 + \dot{y}^2}\, dt.$$

Setting $t = x$ and replacing $\dot{}$ with $'$ there is a single variable $q^1 = y$ and the Lagrangian is $L = \sqrt{1 + (y')^2}$. The Euler–Lagrange equation reads

$$\frac{\partial L}{\partial x} - \frac{d}{dx}\left(\frac{\partial L}{\partial y'}\right) = \frac{d}{dx}\left(\frac{y'}{\sqrt{1+(y')^2}}\right) = 0$$

with solution

$$\frac{y'}{\sqrt{1+(y')^2}} = \text{const.}$$

Hence $y' = a$ for some constant a, and the extremal curve is a straight line $y = ax + b$.

Lagrangian mechanics

In Newtonian mechanics a dynamical system of N particles is defined by positive real scalars $m_1, m_2, \ldots, m_N$ called the **masses** of the particles, and parametrized curves $t \mapsto \mathbf{r}_a = \mathbf{r}_a(t)$ $(a = 1, 2, \ldots, N)$ where each $\mathbf{r}_a \in \mathbb{R}^3$. The parameter t is interpreted as *time*.

Differentiable forms

The **kinetic energy** of the system is defined as

$$T = \frac{1}{2}\sum_{a=1}^{N} m_a \dot{\mathbf{r}}_a^2 \quad \text{where} \quad \dot{\mathbf{r}}_a = \frac{d\mathbf{r}_a}{dt} \quad \text{and} \quad \dot{\mathbf{r}}_a^2 = \dot{x}_a^2 + \dot{y}_a^2 + \dot{z}_a^2.$$

We will also assume **conservative systems** in which Newton's second law reads

$$m_a \ddot{\mathbf{r}}_a = -\nabla_a U \equiv -\frac{\partial U(\mathbf{r}_1, \mathbf{r}_2, \ldots, \mathbf{r}_N)}{\partial \mathbf{r}_a}, \tag{16.26}$$

where the given function $U : \mathbb{R}^{3N} \to r$ is known as the **potential energy** of the system.

A **constrained system** consists of a Newtonian dynamical system together with a manifold M of dimension $n \leq 3N$, and a map $C : M \to \mathbb{R}^{3N}$, called the **constraint**. In a local system of coordinates $(V; (q^1, q^2, \ldots, q^n))$ on M, the constraint can be written as a set of functions

$$\mathbf{r}_a = \mathbf{r}_a(q^1, q^2, \ldots, q^n)$$

and n is called the **number of degrees of freedom** of the constrained system. The coordinates q^i ($i = 1, \ldots, n$) are commonly called **generalized coordinates** for the constrained system. They may be used even for an unconstrained system, in which $n = 3N$ and V is an open submanifold of $\mathbb{R}^{3N}$; in this case we are essentially expressing the original Newtonian dynamical system in terms of general coordinates. It will always be assumed that (M, C) is an embedded submanifold of $\mathbb{R}^{3N}$, so that the tangent map C_* is injective everywhere. This implies that the matrix $[\partial \mathbf{r}_a / \partial q^i]$ has rank n everywhere (no critical points).

Using the chain rule

$$\dot{\mathbf{r}}_a = \frac{\partial \mathbf{r}_a}{\partial q^i}\dot{q}^i \quad \text{where} \quad \dot{q}^i = \frac{dq^i}{dt}$$

the kinetic energy for a constrained system may be written

$$T = \tfrac{1}{2} g_{ij} \dot{q}^i \dot{q}^j \tag{16.27}$$

where

$$g_{ij} = \sum_{a=1}^{N} m_a \frac{\partial \mathbf{r}_a}{\partial q^i} \cdot \frac{\partial \mathbf{r}_a}{\partial q^j}. \tag{16.28}$$

This is a tensor field of type (0, 2) over the coordinate neighbourhood V, since

$$g'_{i'j'} = \sum_{a=1}^{N} m_a \frac{\partial \mathbf{r}_a}{\partial q'^{i'}} \cdot \frac{\partial \mathbf{r}_a}{\partial q'^{j'}} = g_{ij} \frac{\partial q^i}{\partial q'^{i'}} \frac{\partial q^j}{\partial q'^{j'}}.$$

At each point of $q \in M$ we can define an inner product on the tangent space T_q,

$$g(u, v) \equiv u \cdot v = g_{ij} u^i v^j \quad \text{where} \quad u = u^i \frac{\partial}{\partial q^i}, \quad v = v^j \frac{\partial}{\partial q^j},$$

which is positive definite since

$$g(u, u) = \sum_{a=1}^{N} m_a \left(\frac{\partial \mathbf{r}_a}{\partial q^i} u^i\right)^2 \geq 0$$

16.5 Classical mechanics

and the value 0 is only possible if $u^i = 0$ since the constraint map is an embedding and has no critical points. A manifold M with a positive definite inner product defined everywhere is called a **Riemannian manifold**; further discussion of such manifolds will be found in Chapter 18. The associated symmetric tensor field $g = g_{ij}\, dq^i \otimes dq^j$ is called the **metric tensor**. These remarks serve as motivation for the following definition.

A **Lagrangian mechanical system** consists of an n-dimensional Riemannian manifold (M, g) called **configuration space**, together with a function $L : TM \to \mathbb{R}$ called the **Lagrangian** of the system. The Lagrangian will be assumed to have the form $L = T - U$ where, for any $u = (q^i, \dot{q}^j) \in TM$

$$T(u) = \tfrac{1}{2}g(u, u) = \tfrac{1}{2}g_{ij}(q^1, \ldots, q^n)\dot{q}^i \dot{q}^j$$

and

$$U(u) = U(\pi(u)) = U(q^1, \ldots, q^n).$$

As for the calculus of variations it will be common to write $L(q^1, \ldots, q^n, \dot{q}^1, \ldots, \dot{q}^n)$.

The previous discussion shows that every constrained system can be considered as a Lagrangian mechanical system with $U(q^1, \ldots, q^n) = U(\mathbf{r}_1(q^i), \ldots, \mathbf{r}_N(q^i))$. In place of Newton's law (16.26) we postulate **Hamilton's principle**, that every motion $t \mapsto \gamma(t) \equiv q^i(t)$ of the system is an extremal of the action determined by the Lagrangian L

$$\delta S = \int_{t_0}^{t_1} \delta L \, dt = 0.$$

The equations of motion are then the second-order differential equations (16.25),

$$\frac{d}{dt}\left(\frac{\partial L}{\partial \dot{q}^i}\right) - \frac{\partial L}{\partial q^i} = 0, \qquad L = T - U \tag{16.29}$$

known as **Lagrange's equations**.

Example 16.6 A Newtonian system of N unconstrained particles, $\mathbf{r}_a = (x_a, y_a, z_a)$ can be considered also as a Lagrangian system with $3N$ degrees of freedom if we set

$$q^1 = x_1,\ q^2 = y_1,\ q^3 = z_1,\ q^4 = x_2,\ \ldots,\ q^{3N} = z_N.$$

The metric tensor is diagonal with $g_{11} = m_1$, $g_{22} = m_1$, $\ldots$, $g_{44} = m_2$, $\ldots$, etc. Lagrange's equations (16.29) read, for $i = 3a - 2\ (a = 1, 2, \ldots, N)$

$$\frac{d}{dt}\left(\frac{\partial L}{\partial \dot{x}_a}\right) - \frac{\partial L}{\partial x_a} = \frac{d}{dt}(m_a \dot{x}_a) + \frac{\partial U}{\partial x_a} = 0,$$

that is,

$$m_a \ddot{x}_a = -\frac{\partial U}{\partial x_a}$$

and similarly

$$m_a \ddot{y}_a = -\frac{\partial U}{\partial y_a}, \qquad m_a \ddot{z}_a = -\frac{\partial U}{\partial z_a}$$

in agreement with Eq. (16.26).

Differentiable forms

For a single particle, in spherical polar coordinates, $q^1 = r > 0$, $0 < q^2 = \theta < \pi$, $0 < q^3 = \phi < 2\pi$,

$$x = r\sin\theta\cos\phi, \qquad y = r\sin\theta\sin\phi, \qquad z = r\cos\theta$$

the kinetic energy is

$$T = \frac{m}{2}(\dot{x}^2 + \dot{y}^2 + \dot{z}^2) = \frac{m}{2}(\dot{r}^2 + r^2\dot{\theta}^2 + r^2\sin^2\theta\,\dot{\phi}^2).$$

Hence the metric tensor g has components

$$[g_{ij}] = \begin{pmatrix} m\dot{r}^2 & 0 & 0 \\ 0 & mr^2\dot{\theta}^2 & 0 \\ 0 & 0 & mr^2\sin^2\theta\,\dot{\phi}^2 \end{pmatrix}$$

and Lagrange's equations for a *central potential* $U = U(r)$ read

$$m\ddot{r} - mr\dot{\theta}^2 - mr\sin^2\theta\,\dot{\phi}^2 + \frac{dU}{dr} = 0,$$

$$m\frac{d}{dt}(r^2\dot{\theta}) - mr^2\sin\theta\cos\theta\,\dot{\phi}^2 = 0,$$

$$m\frac{d}{dt}(r^2\sin^2\theta\,\dot{\phi}) = 0.$$

Exercise: Write out the equations of motion for a particle constrained to the plane $z = 0$ in polar coordinates, $x = r\cos\theta$, $y = r\sin\theta$.

Example 16.7 The plane pendulum has configuration space $M = S^1$, the one-dimensional circle, which can be covered with two charts, $0 < \phi_1 < 2\pi$ and $-\pi < \phi_2 < \pi$, such that on the overlaps they are related by

$$\theta_2 = \theta_1 \qquad \text{for } 0 < \theta_1 \le \pi$$
$$\theta_2 = \theta_1 - 2\pi \qquad \text{for } \pi \le \theta_1 < 2\pi$$

and constraint functions embedding this manifold in $\mathbb{R}^3$ are

$$x = 0, \qquad y = -a\sin\theta_1, \qquad z = -a\cos\theta_1;$$
$$x = 0, \qquad y = -a\sin\theta_2, \qquad z = -a\cos\theta_2.$$

For $\theta = \theta_1$ or $\theta = \theta_2$ we have

$$T = \frac{m}{2}a^2\dot{\theta}^2, \qquad U = mgz = -mga\cos\theta, \qquad L(\theta, \dot{\theta}) = T - U$$

and subsituting in Lagrange's equations (16.29) with $q^1 = \theta$ gives

$$\frac{d}{dt}\left(\frac{\partial L}{\partial \dot\theta}\right) - \frac{\partial L}{\partial \theta} = ma^2\ddot{\theta} + mga\sin\theta = 0.$$

For small values of θ, the pendulum hanging near vertical, the equation approximates the simple harmonic oscillator equation

$$\ddot{\theta} + \frac{g}{a}\theta = 0$$

with period $\tau = 2\pi\sqrt{g/a}$.

16.5 Classical mechanics

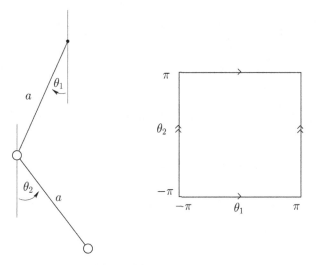

Figure 16.3 Double pendulum

Example 16.8 The spherical pendulum is similar to the plane pendulum, but the configuration manifold is the 2-sphere S^2. In spherical polars the constraint is

$$x = a \sin\theta \cos\phi, \qquad y = a \sin\theta \sin\phi, \qquad z = -a \cos\theta$$

and as for Example 16.6 we find

$$T = \frac{m}{2}a^2(\dot\theta^2 + \sin^2\theta \dot\phi^2), \qquad U = mgz = -mga \cos\theta.$$

Exercise: Write out Lagrange's equations for the spherical pendulum.

Example 16.9 The double pendulum consists of two plane pendula, of lengths a, b and equal mass m, one suspended from the end of the other (see Fig. 16.3). The configuration manifold is the 2-torus $M = S^1 \times S^1 = T^2$, and constraint functions are

$$x_1 = x_2 = 0, \qquad y_1 = -a \sin\theta_1, \qquad z_1 = -a \cos\theta_1,$$
$$y_2 = -a \sin\theta_1 - b \sin\theta_2, \qquad z_2 = -a \cos\theta_1 - b \cos\theta_2.$$

The kinetic energy is

$$T = \frac{m}{2}(\dot y_1^2 + \dot z_1^2 + \dot y_2^2 + \dot z_2^2)$$
$$= \frac{m}{2}(2a^2\dot\theta_1^2 + b^2\dot\theta_2^2 + 2ab \cos(\theta_1 - \theta - 2)\dot\theta_1\dot\theta_2)$$

and the potential energy is $U = -2mga \cos\theta_1 - mgb \cos\theta_2$.

Exercise: Write out Lagrange's equations for the double pendulum of this example.

Exercise: Write out the Lagrangian for a double pendulum with unequal masses, m_1 and m_2.

471

Differentiable forms

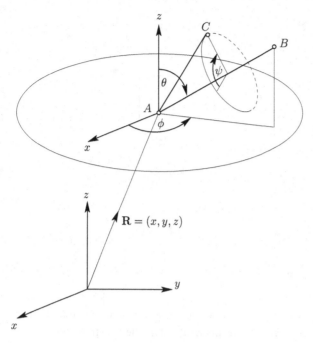

Figure 16.4 Degrees of freedom of a rigid body

Example 16.10 A rigid body is a system of particles subject to the constraint that all distances between particles are constant, $|\mathbf{r}_a - \mathbf{r}_b| = c_{ab} = $ const. These equations are not independent since their number is considerably greater in general than the number of components in the $\mathbf{r}_a$. The number of degrees of freedom is in general six, as can be seen from the following argument. Fix a point in the object A, such as its centre of mass, and assign to it three rectangular coordinates $\mathbf{R} = (X, Y, Z)$. Any other point B of the body is at a fixed distance from A and therefore is constrained to move on a sphere about A. It can be assigned two spherical angles θ, ϕ as for the spherical pendulum. The only remaining freedom is a rotation by an angle ψ, say, about the axis AB. Every point of the rigid body is now determined once these three angles are specified (see Fig. 16.4). Thus the configuration manifold of the rigid body is the six-dimensional manifold $\mathbb{R}^3 \times S^2 \times S^1$. Alternatively the freedom of the body about the point A may be determined by a member of the rotation group $SO(3)$, which can be specified by three Euler angles. These are the most commonly used generalized coordinates for a rigid body. Details may be found in [12, chap. 6].

Given a tangent vector $u = \dot{\gamma} = \dot{q}^i \partial_{q^i}$, the **momentum 1-form conjugate to** u is defined by

$$\langle \omega_u, v \rangle = g(u, v) = g_{ij} \dot{q}^i v^j.$$

Setting $\omega_u = p_i \, dq^i$ we see that

$$p_i = g_{ij} \dot{q}^j = \frac{\partial L}{\partial \dot{q}^i}. \qquad (16.30)$$

16.5 Classical mechanics

The last step follows either by direct differentiation of $L = \frac{1}{2}g_{ij}\dot{q}^i\dot{q}^j - U(q)$ or by applying Euler's theorem on homogeneous functions to $T(q^j, \lambda\dot{q}^i) = \lambda^2 T(q^j, \dot{q}^i)$. The components p_i of the momentum 1-form, given by Eq. (16.30), are called the **generalized momenta** conjugate to the generalized coordinates q^i.

Exercise: For a general Lagrangian L, not necessarily of the form $T - U$, show that $\omega = (\partial L/\partial \dot{q}^i)\, dq^i$ is a well-defined 1-form on M.

Example 16.11 The generalized momentum for an unconstrained particle $L = \frac{1}{2}m\dot{\mathbf{r}}^2 - U(\mathbf{r})$ given by

$$p_x = \frac{\partial L}{\partial \dot{x}} = m\dot{x}, \quad p_y = \frac{\partial L}{\partial \dot{y}} = m\dot{y}, \quad p_z = \frac{\partial L}{\partial \dot{z}} = m\dot{z},$$

which are the components of standard momentum $\mathbf{p} = (p_x, p_y, p_z) = m\dot{\mathbf{r}}$.

In spherical polar coordinates

$$L = \frac{m}{2}\left(\dot{r}^2 + r^2\dot{\theta}^2 + r^2\sin^2\theta\,\dot{\phi}^2\right) - U(r, \theta, \phi),$$

whence

$$p_\phi = \frac{\partial L}{\partial \dot{\phi}} = mr^2\dot{\phi}\sin^2\theta.$$

This can be identified with the z-component of angular momentum,

$$L_z = \mathbf{r} \times \mathbf{p} \cdot \hat{\mathbf{z}} = m(x\dot{y} - y\dot{x}) = m\dot{\phi}r^2\sin^2\theta.$$

It is a general result that the momentum conjugate to an angular coordinate about a fixed axis is the angular momentum about that axis.

Exercise: The angle θ in the previous example does not have a fixed axis of definition unless $\phi = \text{const}$. In this case show that $p_\theta = \mathbf{L} \cdot (-\cos\phi, \sin\phi, 0)$ and interpret geometrically.

Example 16.12 If the Lagrangian has no explicit dependence on a particular generalized coordinate q^k, so that $\partial L/\partial q^k = 0$, it is called an **ignorable** or **cyclic** coordinate, The corresponding generalized momentum p_k is then a constant of the motion, for the kth Lagrange's equation reads

$$0 = \frac{d}{dt}\frac{\partial L}{\partial \dot{q}^k} - \frac{\partial L}{\partial q_k} = \frac{dp_k}{dt}.$$

This is a particular instance of a more general statement, known as **Noether's theorem**.

Let $\varphi_s : M \to M$ be a local one-parameter group of motions on M, generating the vector field X by

$$X_q f = \left.\frac{\partial f(\varphi_s(q))}{\partial s}\right|_{s=0}.$$

The tangent map φ_{*s} induces a local flow on the tangent bundle, since $\varphi_{*s} \circ \varphi_{*t} = (\varphi_s \circ \varphi_t)_* = \varphi_{*(s+t)}$, and the Lagrangian is said to be **invariant** under this local one-parameter group if $L(\varphi_{*s}u) = L(u)$ for all $u \in TM$. Noether's theorem asserts that the quantity $\langle \omega_u, X \rangle$

Differentiable forms

is then a constant of the motion. The result is most easily proved in natural coordinates on TM.

Let $q^i(t)$ be any solution of Lagrange's equations, and set $\mathbf{q}(s,t) = \varphi_s \mathbf{q}(t)$. On differentiation with respect to t we have

$$\dot{\mathbf{q}}(s,t) = \varphi_s \dot{\mathbf{q}}(t) = \frac{\partial \mathbf{q}(s,t)}{\partial t}$$

and invariance of the Lagrangian implies, using Lagrange's equations at $s = 0$,

$$0 = \frac{\partial L}{\partial s}\Big|_{s=0} = \frac{\partial}{\partial s}\Big(L(\mathbf{q}(s,t), \dot{\mathbf{q}}(s,t))\Big)\Big|_{s=0}$$

$$= \Big(\frac{\partial L}{\partial q^i}\frac{\partial q^i}{\partial s} + \frac{\partial L}{\partial \dot{q}^i}\frac{\partial \dot{q}^i}{\partial s}\Big)\Big|_{s=0}$$

$$= \Big(\frac{\partial}{\partial t}\Big(\frac{\partial L}{\partial \dot{q}^i}\Big)\frac{\partial q^i}{\partial s} + \frac{\partial L}{\partial \dot{q}^i}\frac{\partial^2 q^i}{\partial s \partial t}\Big)\Big|_{s=0}$$

$$= \frac{\partial}{\partial t}\Big(\frac{\partial L}{\partial \dot{q}^i}\frac{\partial q^i}{\partial s}\Big)\Big|_{s=0}$$

$$= \frac{\partial}{\partial t}(p_i X^i).$$

Hence, along any solution of Lagrange's equations, we have an integral of the motion

$$\langle \omega_u, X \rangle = g_{ij}\dot{q}^i X^j = p_i X^j = \text{const.}$$

The one-parameter group is often called a **symmetry group** of the system, and Noether's theorem exhibits the relation between symmetries and conservation laws.

If q^k is an ignorable coordinate then the one-parameter group of motions

$$\varphi_s(\mathbf{q}) = (q^1, \ldots, q^{k-1}, q^k + s, q^{k+1}, \ldots, q^n)$$

is an invariance group of the Lagrangian. It generates the vector field $X^i = \partial(\varphi_s(\mathbf{q}))^i/\partial s\big|_{s=0} = \delta^i_k$, and the associated constant of the motion is the generalized momentum $p_i X^i = p_i \delta^i_k = p_k$ conjugate to the ignorable coordinate.

Hamiltonian mechanics

A 2-form Ω is said to be **non-degenerate** at $p \in M$ if

$$\Omega_p(X_p, Y_p) = 0 \quad \text{for all } Y_p \in T_p(M) \implies X_p = 0.$$

As for the concept of non-singularity for inner products (Chapter 5), this is true if and only if

$$\Omega_p = A_{ij}(\mathrm{d}x^i)_p \wedge (\mathrm{d}x^j)_p \quad \text{where} \quad A_{ij} = -A_{ji}, \ \det[A_{ij}] \neq 0.$$

Exercise: Prove this statement.

16.5 Classical mechanics

The manifold M must necessarily be of even dimension $m = 2n$ if there exists a non-degenerate 2-form, since $\det A = \det A^T = \det(-A) = (-1)^m \det A$. A **symplectic structure** on a $2n$-dimensional manifold M is a closed differentiable 2-form Ω that is everywhere non-degenerate. Recall that *closed* means that $d\Omega = 0$ everywhere. An even-dimensional manifold M with a symplectic structure is called a **symplectic manifold**.

As in Examples 7.6 and 7.7, a symplectic form Ω induces an isomorphic map $\bar{\Omega}$: $T_p(M) \to T_p^*(M)$ where the covector $\overline{X_p} = \bar{\Omega} X_p$ is defined by

$$\langle \overline{X_p}, Y_p \rangle \equiv \langle \bar{\Omega} X_p, Y_p \rangle = \Omega(X_p, Y_p). \tag{16.31}$$

We may naturally extend this correspondence to one between vector fields and differential 1-forms $X \leftrightarrow \overline{X}$ such that for any vector field Y

$$\langle \overline{X}, Y \rangle = \Omega(X, Y).$$

By Eq. (16.10), we find for any vector field

$$\overline{X} = \tfrac{1}{2} i_X \Omega. \tag{16.32}$$

In components $\overline{X}_i = A_{ji} X^j$.

We will write the vector field corresponding to a 1-form by the same notation $\overline{\omega} = \bar{\Omega}^{-1} \omega$, such that

$$\langle \omega, Y \rangle = \Omega(\overline{\omega}, Y) \tag{16.33}$$

for all vector fields Y. A vector field X is said to be a **Hamiltonian vector field** if there exists a function H on M such that $X = \overline{dH}$, or equivalently $\overline{X} = dH$. The function H is called the **Hamiltonian** generating this vector field. A function f is said to be a **first integral of the phase flow** generated by the Hamiltonian vector field $X = \overline{dH}$ if $Xf = 0$. The Hamiltonian H is a first integral of the phase flow, for

$$X(H) = \langle dH, X \rangle = \langle dH, \overline{dH} \rangle = \Omega(\overline{dH}, \overline{dH}) = 0$$

on setting $\omega = dH$ and $Y = \overline{dH}$ in Eq. (16.33) and using the antisymmetry of Ω.

Any function $f : M \to \mathbb{R}$ is known as a **dynamical variable**. For any dynamical variable f we set $X_f = \overline{df}$ to be the Hamiltonian vector field generated by f. Then for any vector field Y,

$$\Omega(X_f, Y) = \langle \overline{X_f}, Y \rangle = \langle df, Y \rangle = Y(f),$$

and we have the identity

$$i_{X_f} \Omega = 2\, df.$$

Define the **Poisson bracket** of two dynamical variables f and g to be

$$(f, g) = \Omega(X_f, X_g), \tag{16.34}$$

from which

$$(f, g) = \langle df, X_g \rangle = X_g f = -X_f g = -(f, g).$$

475

Differentiable forms

In these and other conventions, different authors adopt almost random sign conventions – so beware of any discrepencies between formulae given here and those in other books!
From Eq. (16.13) we have that $d\Omega = 0$ implies

$$\mathcal{L}_X\Omega = i_X\, d\Omega + d\circ i_X\Omega = d(i_X\Omega),$$

whence the Lie derivative of the symplectic form in any Hamiltonian direction vanishes,

$$\mathcal{L}_{X_f}\Omega = 2\, d(df) = 2\, d^2 f = 0.$$

Using Eq. (16.12) with $X = X_f$ and $Y = Y_g$ we obtain

$$i_{[X_f, X_g]}\Omega = \mathcal{L}_{X_f}(i_{X_g}\Omega).$$

By (16.8),

$$i_{[X_f, X_g]}\Omega = 2\mathcal{L}_{X_f} dg = 2\, d\mathcal{L}_{X_f} g = 2\, d(X_f g) = i_{X_{(g,f)}}\Omega,$$

whence

$$[X_f, X_g] = X_{(g,f)} = -X_{(f,g)}. \tag{16.35}$$

From the Jacobi identity (15.24) it then follows that

$$((f, g), h) + ((g, h), f) + ((h, f), g) = 0. \tag{16.36}$$

Exercise: Prove Eq. (16.36).

Exercise: Show that $(f, g) + (f, h) = (f, g + h)$ and $(f, gh) = g(f, h) + h(f, g)$.

The rate of change of a dynamical variable f along a Hamiltonian flow is given by

$$\dot f = \frac{df}{dt} = X_H f = (f, H). \tag{16.37}$$

Thus f is a first integral of the phase flow generated by the Hamiltonian vector field X_H if and only if it 'commutes' with the Hamiltonian, in the sense that its Poisson bracket with H vanishes, $(f, H) = 0$. The analogies with quantum mechanics (Chapter 14) are manifest.

Exercise: Show that if f and g are first integrals then so is (f, g).

Example 16.13 Let $M = \mathbb{R}^{2n}$ with coordinates labelled $(q^1, \dots, q^n, p_1, \dots, p_n)$. The 2-form $\Omega = 2dq^i \wedge dp_i = dq^i \otimes dp_i - dp_i \otimes dq^i$, having constant components

$$A = \begin{pmatrix} 0 & 1 \\ -1 & 0 \end{pmatrix}$$

is a symplectic structure, since $\det A = 1$ (a simple exercise!) and it is closed,

$$d\Omega = 2\, d^2 q^i \wedge dp_i - dq^i \wedge d^2 p_i = 0.$$

If X and Y are vector fields having components

$$X = \xi^i \frac{\partial}{\partial q^i} + \xi_j \frac{\partial}{\partial p_j}, \qquad Y = \eta^i \frac{\partial}{\partial q^i} + \eta_j \frac{\partial}{\partial p_j},$$

16.5 Classical mechanics

then
$$\langle \overline{X}, Y \rangle = \Omega(X, Y) = (dq^i \otimes dp_i - dp_i \otimes dq^i)(X, Y) = \xi^i \eta_i - \xi_i \eta^i$$
so that the 1-form $\overline{X}$ has components
$$\overline{X} = -\xi_j \, dq^j + \xi^i \, dp_i.$$
A Hamiltonian vector field X has $\xi_j = -\partial H/\partial q^j$ and $\xi^i = \partial H/\partial p_i$, so that
$$X = \overline{dH} = X_H = \frac{\partial H}{\partial p_i}\frac{\partial}{\partial q^i} - \frac{\partial H}{\partial q^j}\frac{\partial}{\partial p_j}.$$
A curve $\gamma : \mathbb{R} \to M$ is an integral curve of this vector field if the functions $q^i = q^i(t)$, $p_j = p_j(t)$ satisfy the differential equations known as **Hamilton's equations**:
$$\frac{dq^i}{dt} = \frac{\partial H}{\partial p_i}, \qquad \frac{dp_j}{dt} = -\frac{\partial H}{\partial q^j}. \tag{16.38}$$
The Poisson bracket is given by
$$(f, g) = X_g f = \left(\frac{\partial g}{\partial p_i}\frac{\partial}{\partial q^i} - \frac{\partial g}{\partial q^j}\frac{\partial}{\partial p_j}\right) f = \frac{\partial f}{\partial q^i}\frac{\partial g}{\partial p_i} - \frac{\partial f}{\partial p_j}\frac{\partial g}{\partial q^j}.$$
For any dynamical variable f it is straightforward to verify the Poisson bracket relations
$$(q^i, f) = \frac{\partial f}{\partial p_i}, \qquad (p_i, f) = -\frac{\partial f}{\partial q^i}, \tag{16.39}$$
from which the *canonical relations* are immediate
$$(q^i, q^j) = 0, \qquad (p_i, p_j) = 0, \qquad (q^i, p_j) = \delta^i{}_j.$$

Connection between Lagrangian and Hamiltonian mechanics

If M is a manifold of any dimension n its cotangent bundle T^*M, consisting of all covectors at all points, is a $2n$-dimensional manifold. If $(U; q^i)$ is any coordinate chart on M, a chart is generated on T^*M by assigning coordinates $(q^1, \ldots, q^n, p_1, \ldots, p_n)$ to any covector $\omega_q = p_i (dq^i)_q$ at $q \in M$. The natural projection map $\pi : T^*M \to M$ has the effect of sending any covector to its base point, $\pi(\omega_q) = q$. The tangent map corresponding to this projection map, $\pi_* : T_{\omega_q}(T^*M) \to T_q(M)$, maps every tangent vector $X_{\omega_q} \in T_{\omega_q}(T^*M)$ to a tangent vector $\pi_* X_{\omega_q} \in T_q(M)$. In canonical coordinates, set
$$X_{\omega_q} = \xi^i \frac{\partial}{\partial q^i} + \xi_j \frac{\partial}{\partial p_j}$$
and for any function $f : M \to \mathbb{R}$, written in coordinates as $f(q^1, \ldots, q^n)$, we have
$$(\pi_* X_{\omega_q}) f(\mathbf{q}) = X_{\omega_q}(f \circ \pi)(\mathbf{q}, \mathbf{p}) = \xi^i \frac{\partial f(\mathbf{q})}{\partial q^i} + \xi_i \frac{\partial f(\mathbf{q})}{\partial p_i}$$
so that
$$\pi_* X_{\omega_q} = \xi^i \frac{\partial}{\partial q^i}.$$

Differentiable forms

This defines a **canonical 1-form** θ on T^*M by setting

$$\theta_{\omega_q}(X_{\omega_q}) \equiv \langle \theta_{\omega_q}, X_{\omega_q} \rangle = \langle \omega_q, \pi_* X_{\omega_q} \rangle.$$

Alternatively, we can think of θ as the pullback $\theta_{\omega_q} = \pi^* \omega_q \in T^*_{\omega_q}(T^*M)$, for

$$\langle \pi^* \omega_q, X_{\omega_q} \rangle = \langle \omega_q, \pi_* X_{\omega_q} \rangle = \theta_{\omega_q}(X_{\omega_q})$$

for arbitrary $X_{\omega_q} \in T_{\omega_q}(T^*M)$. Writing $\omega_q = p_i\, dq^i$, we thus have $\langle \theta_{\omega_q}, X_{\omega_q} \rangle = p_i \xi^i$, so that in any canonical chart $(U \times \mathbb{R}^n; q^1, \ldots, q^n, p_1, \ldots, p_n)$

$$\theta = p_i\, dq^i. \tag{16.40}$$

The 2-form

$$\Omega = -2\, d\theta = 2\, dq^i \wedge dp_i \tag{16.41}$$

is of the same form as that in Example 16.13, and provides a natural symplectic structure on the cotangent bundle of any manifold M.

Given a Lagrangian system having configuration space (M, g) and Lagrangian function $L = T - U : TM \to \mathbb{R}$ where

$$T(\mathbf{q}, \dot{\mathbf{q}}) = \tfrac{1}{2} g_{ij}(\mathbf{q}) \dot{q}^i \dot{q}^j, \qquad U = U(\mathbf{q})$$

the cotangent bundle T^*M, consisting of momentum 1-forms on M, is known as the **phase space** of the system. The coordinates p_i and $\dot{q}^j$ are related by Eq. (16.30), so that velocity components can be expressed in terms of generalized momenta, $\dot{q}^j = g^{jk} p_k$ where $g^{jk} g_{ki} = \delta^j_i$, and Lagrange's equations (16.29) can be written

$$\dot{p}_i = \frac{\partial L}{\partial q^i}.$$

Our first task is to find a Hamiltonian function $H : T^*M \to \mathbb{R}$, written $H(q^1, \ldots, q^n, p_1, \ldots, p_n)$, such that the equations of motion of the system in phase space have the form of Hamiltonian equations (16.38) in Example 16.13. The Hamiltonian function H must then have exterior derivative

$$\begin{aligned}
dH &= \frac{\partial H}{\partial q^i} dq^i + H_{p_i}\, dp_i \\
&= -\dot{p}_i\, dq^i + \dot{q}^i\, dp_i \\
&= -\frac{\partial L}{\partial q^i} dq^i + d(\dot{q}^i p_i) - p_i\, d\dot{q}^i \\
&= d(\dot{q}^i p_i) - \left(\frac{\partial L}{\partial q^i} dq^i + \frac{\partial L}{\partial \dot{q}^j} d\dot{q}^j \right) \\
&= d(\dot{q}^i p_i - L)
\end{aligned}$$

whence, within an arbitrary constant

$$H = \dot{q}^i p_i - L = g_{ij} \dot{q}^i \dot{q}^j - L = 2T - (T - U) = T + U = E.$$

The Hamiltonian is thus the energy of the system expressed in terms of canonical coordinates on T^*M.

16.5 Classical mechanics

Apart from expressing the equations of mechanics as a first-order system of equations, one of the advantages of the Hamiltonian view is that coordinates in which the symplectic form takes the form given in Example 16.13 need not be restricted to the canonical coordinates generated by the tangent bundle construction. For example, let $(q^i, p_j) \to (\bar{q}^i, \bar{p}_j)$ be any coordinate transformation such that the canonical 1-forms $\theta = p_i \, dq^i$ and $\bar{\theta} = \bar{p}_i \, d\bar{q}^i$ generate the same symplectic form,

$$\bar{\Omega} = -2 \, d\bar{\theta} = \Omega = -2 \, d\theta$$

so that $\bar{\theta} = \theta - dF$ for some function F on T^*M.

Exercise: Show that $\bar{p}_i \dfrac{\partial \bar{q}^i}{\partial p_j} = -\dfrac{\partial F}{\partial p_j}$, $\bar{p}_i \dfrac{\partial \bar{q}^i}{\partial q^j} = p_j - \dfrac{\partial F}{\partial q^j}$.

Since $\bar{\Omega} = \Omega$ the Hamiltonian vector fields generated by any dynamical variable f, are identical for the two forms, $X_f = \bar{X}_f$, since for any vector field Y on T^*M,

$$\bar{\Omega}(\bar{X}_f, Y) = Yf = \Omega(X_f, Y).$$

Hence, Poisson brackets are invariant with respect to this change of coordinates, for

$$(f, g)_{\bar{q}, \bar{p}} = \bar{X}_g f = X_g f = (f, g)_{g, p}.$$

This result is easy to prove directly by change of variables, as is done in some standard books on analytic mechanics. Using Eqs. (16.37) and (16.39) we have then

$$\frac{d\bar{q}^i}{dt} = (\bar{q}^i, H)_{q,p} = (\bar{q}^i, H)_{\bar{q},\bar{p}} = \frac{\partial H}{\partial \bar{p}_i},$$

$$\frac{d\bar{p}_i}{dt} = (\bar{p}_i, H)_{q,p} = (\bar{p}_i, H)_{\bar{q},\bar{p}} = -\frac{\partial H}{\partial \bar{q}^i},$$

and Hamilton's equations are preserved under such transformations. These are called **homogeneous contact transformations**.

More generally, let $H(q^1, \ldots, q^n, p_1, \ldots, p_n, t)$ be a time-dependent Hamiltonian, defined on **extended phase space** $T^*M \times \mathbb{R}$, where $\mathbb{R}$ represents the time variable t, and let λ be the **contact 1-form**,

$$\lambda = p_i \, dq^i - H \, dt.$$

If $T^*\bar{M} \to \mathbb{R}$ is another extended phase of the same dimension with canonical coordinates $\bar{q}^i, \bar{p}_i$ and Hamiltonian $\bar{H}(\bar{q}, \bar{p}, t)$, then a diffeomorphism $\phi : T^*M \times \mathbb{R} \to T^*\bar{M} \to \mathbb{R}$ is called a **contact transformation** if $\phi^* \, d\bar{\lambda} = d\lambda$. Since $\phi^* \circ d = d \circ \phi$ there exists a function F on $T^*M \times \mathbb{R}$ in the neighbourhood of any point such that $\phi^*\bar{\lambda} = \lambda - dF$. If we write the function F as depending on the variables q^i and $\bar{q}^i$, which is generally possible locally,

$$\bar{p}_i \, d\bar{q}^i - \bar{H} \, dt = p_i \, dq^i - H \, dt - \frac{\partial F}{\partial q^i} dq^i - \frac{\partial F}{\partial \bar{q}^i} d\bar{q}^i - \frac{\partial F}{\partial t} dt$$

and we arrive at the classical canonical transformation equations

$$\bar{p}_i = -\frac{\partial F}{\partial \bar{q}^i}, \quad p_i = \frac{\partial F}{\partial q^i}, \quad \bar{H} = H + \frac{\partial F}{\partial t}.$$

If $\bar{H} = 0$ then the solution of the Hamilton equations trivially of the form $\bar{q}^i = $ const., $\bar{p}_i = $ const. To find the function F for a transformation to this system we seek the general solution of the first-order partial differential equation known as the **Hamilton–Jacobi equation**,

$$\frac{\partial S(q^1,\ldots,q^n,c^1,\ldots,c^n,t)}{\partial t} + H\left(q^1,\ldots,q^n,\frac{\partial S}{\partial q^1},\ldots,\frac{\partial S}{\partial q^n}\right) = 0, \qquad (16.42)$$

and set $F(q^1,\ldots,q^n,\bar{q}^1,\ldots,\bar{q}^n) = S(q^1,\ldots,q^n,\bar{q}^1,\ldots,\bar{q}^n)$.

References

[1] R. W. R. Darling. *Differential Forms and Connections*. New York, Cambridge University Press, 1994.
[2] H. Flanders. *Differential Forms*. New York, Dover Publications, 1989.
[3] S. I. Goldberg. *Curvature and Homology*. New York, Academic Press, 1962.
[4] L. H. Loomis and S. Sternberg. *Advanced Calculus*. Reading, Mass., Addison-Wesley, 1968.
[5] M. Spivak. *Calculus on Manifolds*. New York, W. A. Benjamin, 1965.
[6] W. H. Chen, S. S. Chern and K. S. Lam. *Lectures on Differential Geometry*. Singapore, World Scientific, 1999.
[7] S. Sternberg. *Lectures on Differential Geometry*. Englewood Cliffs, N.J., Prentice-Hall, 1964.
[8] F. W. Warner. *Foundations of Differential Manifolds and Lie Groups*. New York, Springer-Verlag, 1983.
[9] C. de Witt-Morette, Y. Choquet-Bruhat and M. Dillard-Bleick. *Analysis, Manifolds and Physics*. Amsterdam, North-Holland, 1977.
[10] T. Frankel. *The Geometry of Physics*. New York, Cambridge University Press, 1997.
[11] R. Abraham. *Foundations of Mechanics*. New York, W. A. Benjamin, 1967.
[12] V. I. Arnold. *Mathematical Methods of Classical Mechanics*. New York, Springer-Verlag, 1978.
[13] G. W. Mackey. *Mathematical Foundations of Quantum Mechanics*. New York, W. J. Benjamin, 1963.
[14] W. Thirring. *A Course in Mathematical Physics, Vol. 1: Classical Dynamical Systems*. New York, Springer-Verlag, 1978.
[15] F. P. Hildebrand *Methods of Applied Mathematics*. Englewood Cliffs, N.J., Prentice-Hall, 1965.

17 Integration on manifolds

The theory of integration over manifolds is only available for a restricted class known as *oriented manifolds*. The general theory can be found in [1–11]. An n-dimensional differentiable manifold M is called **orientable** if there exists a differential n-form ω that vanishes at no point $p \in M$. The n-form ω is called a **volume element** for M, and the pair (M, ω) is an **oriented manifold**. Since the space $(\Lambda_n)_p(M) \equiv (\Lambda^{*n})_p(M)$ is one-dimensional at each $p \in M$, any two volume elements are proportional to each other, $\omega' = f\omega$, where $f : M \to \mathbb{R}$ is a non-vanishing smooth function on M. If the manifold is a connected topological space it has the same sign everywhere; if $f(p) > 0$ for all $p \in M$, the two n-forms ω and ω' are said to assign the same **orientation** to M, otherwise they are **oppositely oriented**. Referring to Example 8.4, a manifold is orientable if each cotangent space $T_p^*(M)$ ($p \in M$) is oriented by assigning a non-zero n-form $\Omega = \omega_p$ at p and the orientations are assigned in a smooth and continuous way over the manifold.

With respect to a coordinate chart $(U, \phi; x^i)$ the volume element ω can be written

$$\omega = g(x^1, \ldots, x^n)\, dx^1 \wedge dx^2 \wedge \cdots \wedge dx^n = \frac{g}{n!} \epsilon_{i_1 i_2 \ldots i_n}\, dx^{i_1} \otimes dx^{i_2} \otimes \cdots \otimes dx^{i_n}.$$

If $(U', \phi'; x^{'i'})$ is a second coordinate chart then, in the overlap region $U \cap U'$,

$$\omega = g'(x^{'i_1}, \ldots, x^{'i_n})\, dx^{'1} \wedge dx^{'2} \wedge \cdots \wedge dx^{'n}$$

where

$$g'(\mathbf{x}') = g(\mathbf{x}) \det\left[\frac{\partial x^i}{\partial x^{'j'}}\right].$$

The sign of the component function g thus remains unchanged if and only if the Jacobian determinant of the coordinate transformation is positive throughout $U \cap V$, in which case the charts are said to have the **same orientation**. A differentiable manifold is in fact orientable if and only if there exists an atlas of charts (U_i, ϕ_i) covering M, such that any two charts (U_i, ϕ_i) and (U_j, ϕ_j) have the same orientation on their overlap $U_i \cap U_j$, but the proof requires the concept of a *partition of unity*.

Problems

Problem 17.1 Show that in spherical polar coordinates

$$dx \wedge dy \wedge dz = r^2 \sin\theta\, dr \wedge d\theta \wedge d\phi,$$

and that $a^2 \sin\theta\, d\theta \wedge d\phi$ is a volume element on the 2-sphere $x^2 + y^2 + z^2 = a^2$.

Integration on manifolds

Problem 17.2 Show that the 2-sphere $x^2 + y^2 + z^2 = 1$ is an orientable manifold.

17.1 Partitions of unity

Given an open covering $\{U_i \mid i \in I\}$ of a topological space S, an open covering $\{V_a\}$ of S is called a **refinement** of $\{U_i\}$ if each $V_a \subseteq U_i$ for some $i \in I$. The refinement is said to be **locally finite** if every point p belongs to at most a finite number of V_a's. The topological space S is said to be **paracompact** if for every open covering U_i ($i \in I$) there exists a locally finite refinement. As it may be shown that every locally compact Hausdorff second-countable space is paracompact [12], we will from now on restrict attention to manifolds that are paracompact topological spaces.

Given a locally finite open covering $\{V_a\}$ of a manifold M, a **partition of unity subordinate to the covering** $\{V_a\}$ consists of a family of differentiable functions $g_a : M \to \mathbb{R}$ such that

(1) $0 \le g_a \le 1$ on M for all α,
(2) $g_a(p) = 0$ for all $p \notin V_a$,
(3) $\sum_a g_a(p) = 1$ for all $p \in M$.

It is important that the covering be locally finite, so that the sum in (3) reduces to a finite sum.

Theorem 17.1 *For every locally finite covering $\{V_a\}$ of a paracompact manifold M there exists a partition of unity $\{g_a\}$ subordinate to this refinement.*

Proof: For each $p \in M$ let B_p be an open neighbourhood of p such that its closure $\overline{B_p}$ is compact and contained in some V_a – for example, take B_p to be the inverse image of a small coordinate ball in $\mathbb{R}^n$. As the sets $\{B_p\}$ form an open covering of M, they have a locally finite refinement $\{B'_\alpha\}$. For each a let V'_a be the union of all B'_α whose closure $\overline{B'_\alpha} \subset V_a$. Since every $\overline{B'_\alpha} \subset \overline{B_p} \subset V_a$ for some p and a, it follows that the sets V'_a are an open covering of M. For each a the closure of V'_a is compact and, by the local finiteness of the covering $\{B'_\alpha\}$,

$$\overline{V'_a} = \bigcup \overline{B'_\alpha} \subset V_a.$$

As seen in Lemma 16.1 for any point $p \in \overline{V'_a}$ it is possible to find a differentiable function $h_p : M \to \mathbb{R}$ such that $h_p(p) = 1$ and $h_p = 0$ on $M - V_a$. For each point $p \in \overline{V'_a}$ let U_p be the open neighbourhood $\{q \in V_a \mid f_p(q) > \frac{1}{2}\}$. Since $\overline{V'_a}$ is compact, there exists a finite subcover $\{U_{p_1}, \ldots, U_{p_k}\}$. The function $h_a = h_{p_1} + \cdots + h_{p_k}$ has the following three properties: (i) $h_a \ge 0$ on M, (ii) $h > 0$ on $\overline{V'_a}$, and (iii) $h_a = 0$ outside V_a. As $\{V_a\}$ is a locally finite covering of M, the function $h = \sum_a h_a$ is well-defined, and positive everywhere on M. The functions $g_a = h_a/h$ satisfy all requirements for a partition of unity subordinate to V_a. ∎

17.1 Partitions of unity

We can now construct a non-vanishing n-form on a manifold M from any atlas of charts $(U_\alpha, \phi_\alpha; x^i_\alpha)$ having positive Jacobian determinants on all overlaps. Let $\{V_a\}$ be a locally finite refinement of $\{U_\alpha\}$ and g_a a partition of unity subordinate to $\{V_a\}$. The charts $(V_a, \phi_a; x^i_a)$ where $\phi_a = \phi_\alpha|_{V_a}$ and $x^i_a = x^i_\alpha|_{V_a}$ form an atlas on M, and

$$\omega = \sum_a g_a \, dx_a^1 \wedge dx_a^2 \wedge \cdots \wedge dx_a^n$$

is a differential n-form on M that nowhere vanishes.

Example 17.1 The *Möbius band* can be thought of as a strip of paper with the two ends joined together after giving the strip a twist, as shown in Fig. 17.1. For example, let $M = \{(x, y) \mid -2 \le x \le 2, -1 < y < 1\}$ where the end edges are identified in opposing directions, $(2, y) \equiv (-2, -y)$. This manifold can be covered by two charts

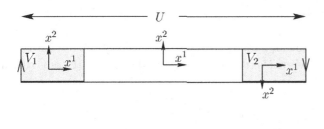

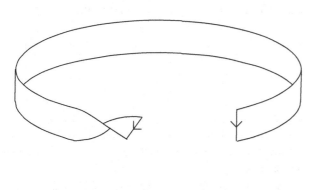

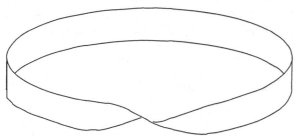

Figure 17.1 Möbius band

$(U = \{(x, y) \mid -2 < x < 2, -1 < y < 1\}, \phi = \mathrm{id}_U)$ and $(V = V_1 \cup V_2, \psi)$ where

$$V_1 = \{(x, y) \mid -2 \le x < -1, -1 < y < 1\},$$
$$V_2 = \{(x, y) \mid 1 \le x < 2, -1 < y < 1\},$$

$$(x', y') = \psi(x, y) = \begin{cases} (x + 2, y) & \text{if } (x, y) \in V_1, \\ (x - 2, -y) & \text{if } (x, y) \in V_2. \end{cases}$$

The Jacobian is $+1$ on $U \cap V_1$ and -1 on $U \cap V_2$, so these two charts do not have the same orientation everywhere. The Möbius band is non-orientable, for if there existed a non-vanishing 2-form ω, we would have $\omega = f \, dx \wedge dy$ with $f(x, y) > 0$ or $f(x, y) < 0$ everywhere on U. Setting $\omega = f' \, dx' \wedge dy'$ we have $f' = f$ on V_1 and $f' = -f$ on V_2. Hence $f'(x', y')$ must vanish on the line $x = \pm 2$, which contradicts ω being non-vanishing everywhere.

17.2 Integration of n-forms

There is no natural way to define the integral of a scalar function $f : M \to \mathbb{R}$ over a compact region D. For, if $D \subset U$ where U is the domain of a coordinate chart $(U, \phi; x^i)$ the multiple integral

$$\int_D f = \int_{\phi(D)} f \circ \phi^{-1}(x^1, \ldots, x^n) \, dx^1 \ldots dx^n$$

will have a different expression in a second chart $(V, \psi; y^i)$ such that $D \subset U \cap V$

$$\int_{\psi(D)} f \circ \psi^{-1}(y) \, dy^1 \ldots dy^n = \int_{\phi(D)} f \circ \phi^{-1}(x) \left| \det\left[\frac{\partial y^i}{\partial x^j}\right] \right| dx^1 \ldots dx^n$$

$$\ne \int_{\phi(D)} f \circ \phi^{-1}(x) \, dx^1 \ldots dx^n.$$

As seen above, n-forms absorb a Jacobian determinant in their coordinate transformation, and it turns out that these are the ideal objects for integration. However, it is necessary that the manifold be orientable so that the *absolute value* of the Jacobian occurring in the integral transformation law can be omitted.

Let (M, ω) be an n-dimensional oriented differentiable manifold, and $(U, \phi; x^i)$ a positively oriented chart. On U we can write $\omega = g(\mathbf{x}) \, dx^1 \wedge \cdots \wedge dx^n$ where $g > 0$. The **support** of an n-form α is defined as the closure of the set on which $\alpha \ne 0$,

$$\operatorname{supp} \alpha = \overline{\{p \in M \mid \alpha_p \ne 0\}}.$$

If α has compact support contained in U, and $\alpha = f \, dx^1 \wedge dx^2 \wedge \cdots \wedge dx^n$ on U we define its **integral** over M to be

$$\int_M \alpha = \int_{\phi(U)} f(x^1, \ldots, x^n) \, dx^1 \ldots dx^n = \int_{\phi(\operatorname{supp} \alpha)} f \, dx^1 \ldots dx^n$$

17.2 Integration of n-forms

where $f(x^1, \ldots, x^n)$ is commonly written in place of $\hat{f} = f \circ \phi^{-1}$. If $(V, \psi; x'^i)$ is a second positively oriented chart also containing the support of α and $f'(x'^1, \ldots, x'^n) \equiv f \circ \psi^{-1}$, we have by the change of variable formula in multiple integration

$$\int_{\psi(V)} f'(x'^1, \ldots, x'^n) \, dx'^1 \ldots dx'^n = \int_{\psi(\text{supp } \alpha)} f'(\mathbf{x}') \, dx'^1 \ldots dx'^n$$

$$= \int_{\phi(\text{supp } \alpha)} f'(\mathbf{x}') \left| \det\left[\frac{\partial x'^i}{\partial x^j} \right] \right| dx^1 \ldots dx^n$$

$$= \int_{\phi(U)} f(\mathbf{x}) \, dx^1 \ldots dx^n$$

since

$$\alpha = f \, dx^1 \wedge \cdots \wedge dx^n = f' \, dx'^1 \wedge \cdots \wedge'^n \quad \text{where} \quad f = f' \det\left[\frac{\partial x'^i}{\partial x^j} \right]$$

and the Jacobian determinant is everywhere positive. The definition of the integral is therefore independent of the coordinate chart, provided the support lies within the domain of the chart.

For an arbitrary n-form α with compact support and atlas (U_a, ϕ_a), assumed to be locally finite, let g_a be a partition of unity subordinate to the open covering $\{U_a\}$. Evidently

$$\alpha = \sum_a g_a \alpha$$

and each of the summands $g_a \alpha$ has compact support contained in U_a. We define the integral of α over M to be

$$\int_M \alpha = \sum_a \int_M g_a \alpha. \tag{17.1}$$

Exercise: Prove that $\int_M$ is a linear operator, $\int_M \alpha + c\beta = \int_M \alpha + c \int_M \beta$.

If α is a differential k-form with compact support on M and $\varphi : N \to M$ is a regular embedding of a k-dimensional manifold N in M (see Section 15.4), define the integral of α on $\varphi(N)$ to be

$$\int_{\varphi(N)} \alpha = \int_N \varphi^* \alpha.$$

The right-hand side is well-defined since $\varphi^* \alpha$ is a differential k-form on N with compact support, since $\varphi : M \to N$ is a homeomorphism from N to $\varphi(N)$ in the relative topology with respect to M.

Problems

Problem 17.3 Show that the definition of the integral of an n-form over a manifold M given in Eq. (17.1) is independent of the choice of partition of unity subordinate to $\{U_a\}$.

17.3 Stokes' theorem

Stokes' theorem requires the concept of a submanifold with boundary. This is not an easy notion in general, but for most practical purposes it is sufficient to restrict ourselves to regions made up of 'coordinate cubical regions'. Let Γ_k be the standard unit k-cube,

$$\Gamma_k = \{\mathbf{x} \in \mathbb{R}^k \mid 0 \le x^i \le 1 \, (i = 1, \ldots, k)\} \subset \mathbb{R}^k.$$

The unit 0-cube is taken to be the singleton $\Gamma_0 = \{0\} \subset \mathbb{R}$. A **$k$-cell** in a manifold M is a smooth map $\sigma : U \to M$ where U is an open neighbourhood U of Γ_k in $\mathbb{R}^k$ (see Fig. 17.2), and its **support** is defined as the image of the standard k-cube, $\sigma(\Gamma_k)$. A **cubical k-chain** in M consists of a formal sum

$$C = c^1 \sigma_1 + c^2 \sigma_2 + \cdots + c^r \sigma_r \quad \text{where} \quad c^i \in \mathbb{R}, \; r \text{ a positive integer.}$$

The set of all cubical k-chains is denoted $\mathcal{C}_k$; it forms an abelian group under addition of k-chains defined in the obvious way. It is also a vector space if we define scalar multiplication as $aC = \sum_i ac^i \sigma_i$ where $a \in \mathbb{R}$.

For each $i = 1, \ldots, k$ and $\epsilon = 0, 1$ define the maps $\varphi_i^\epsilon : \Gamma_{k-1} \to \Gamma_k$ by

$$\varphi_i^\epsilon(y^1, \ldots, y^{k-1}) = (y^1, \ldots, y^{i-1}, \epsilon, y^i, \ldots, y^{k-1}).$$

These maps can be thought of as the $(i, 0)$-face and $(i, 1)$-face respectively of Γ_k. If the interior of the standard k-cube Γ_k is oriented in the natural way, by assigning the k-form $dx^1 \wedge \cdots \wedge dx^k$ to be positively oriented over Γ_k, then for each face map $x^i = 0$ or 1 the orientation on Γ_{k-1} is assigned according to the following rule: set the k-form $dx^i \wedge dx^1 \wedge \cdots \wedge dx^{i-1} \wedge dx^{i+1} \wedge \cdots \wedge dx^k$ to be positively oriented if x^i is increasing outwards at the face, else it is negatively oriented. According to this rule the $(k-1)$-form $dy^1 \wedge dy^2 \wedge \cdots \wedge dy^{k-1}$ has orientation $(-1)^i$ on the $(i, 0)$-face, while on the $(i, 1)$-face it has orientation $(-1)^{i+1}$. This is sometimes called the *outward normal rule* – the orientation on the boundary surface must be chosen such that at every point there exist local positively oriented coordinates such that the first coordinate x^1 points outwards from the surface.

Exercise: On the two-dimensional square, verify that the outward normal rule implies that the direction of increasing x or y coordinate on each side proceeds in an anticlockwise fashion around the square.

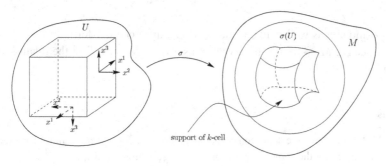

Figure 17.2 k-Cell on a manifold M

17.3 Stokes' theorem

This gives the rationale for the **boundary map** $\partial : C_k \to (C)_{k-1}$, defined by:
(i) for a k-cell σ $(k > 0)$, set

$$\partial \sigma = \sum_{i=1}^{k}(-1)^i (\sigma \circ \varphi_i^0 - \sigma \circ \varphi_i^1),$$

(ii) for a cubical k-chain $C = \sum c^i \sigma_i$ set

$$\partial C = \sum_i c^i \partial \sigma_i.$$

An important identity is $\partial^2 = 0$. For example if σ is a 2-cell, its boundary is given by

$$\partial \sigma = -\varphi_1^0 + \varphi_1^1 + \varphi_2^0 - \varphi_2^1.$$

The boundary of the face $\varphi_1^0(z) = (0, z)$ is $\partial \varphi_1^0 = -\rho^{01} + \rho^{00}$, where $\rho^{ab} : \{0\} \to \mathbb{R}$ is the map $\rho^{ab}(0) = (a, b)$. Hence

$$\partial \circ \partial \sigma = -\rho^{01} + \rho^{00} + \rho^{11} - \rho^{10} + \rho^{10} - \rho^{00} - \rho^{11} + \rho^{01} = 0.$$

For a k-cell,

$$\partial^2 \sigma = \sum_{\epsilon=0}^{1}\sum_{\epsilon'=0}^{1}\left[\sum_i \sum_{j<i}(-1)^{i+\epsilon+j+\epsilon'} \sigma \circ \varphi_{ij}^{\epsilon\epsilon'} + \sum_i \sum_{j\geq i}(-1)^{i+\epsilon+j+\epsilon'+1} \sigma \circ \varphi_{ij}^{\epsilon\epsilon'}\right]$$

where

$$\varphi_{ij}^{\epsilon\epsilon'}(z^1, \ldots, z^{k-2}) = \begin{cases} (z^1, \ldots, z^{j-1}, \epsilon', z^j, \ldots, z^{i-1}, \epsilon, z^i, \ldots, z^{k-2}) & \text{if } j < i, \\ (z^1, \ldots, z^{i-1}, \epsilon, z^i, \ldots, z^{j-1}, \epsilon', z^j, \ldots, z^{k-2}) & \text{if } j \geq i. \end{cases}$$

It follows from this equation that all terms cancel in pairs, so that $\partial^2 \sigma = 0$. The identity extends by linearity to all k-chains.

Exercise: Write out $\partial^2 \sigma$ for a 3-cube, and verify the cancellation property.

For a k-form α on M and a k-chain $C = \sum c^i \sigma_i$ we define the integral

$$\int_C \alpha = \sum c^i \int_{\sigma_i} \alpha = \sum c^i \int_{\Gamma_k} \sigma_i^* \alpha.$$

Theorem 17.2 (Stokes' theorem) *For any $(k+1)$-chain C, and differential k-form on M,*

$$\int_C d\alpha = \int_{\partial C} \alpha. \tag{17.2}$$

Proof: By linearity, it is only necessary to prove the theorem for a $(k+1)$-cell σ. The left-hand side of (17.2) can be written, using Theorem 16.2,

$$\int_\sigma d\alpha = \int_{\Gamma_{k+1}} \sigma^* d\alpha = \int_{\Gamma_{k+1}} d(\sigma^* \alpha)$$

Integration on manifolds

while the right-hand side is

$$\int_{\partial\sigma}\alpha = \sum_{\epsilon=0}^{1}\sum_{i=1}^{k+1}(-1)^{i+\epsilon}\int_{\sigma\circ\varphi_i^\epsilon}\alpha$$

$$= \sum_{\epsilon=0}^{1}\sum_{i=1}^{k+1}(-1)^{i+\epsilon}\int_{\Gamma_k}(\varphi_i^\epsilon)^*\circ\sigma^*\alpha.$$

Since $\sigma^*\alpha$ is a differential k-form on $\mathbb{R}^{k+1}$ it can be written as

$$\sigma^*\alpha = \sum_{i=1}^{k+1} A_i\,dx^1\wedge\cdots\wedge dx^{i-1}\wedge dx^{i+1}\wedge\cdots\wedge dx^{k+1}$$

where the A_i are differentiable functions $A_i : V \to \mathbb{R}$ on an open neighbourhood V of Γ_{k+1}. Hence

$$d(\sigma^*\alpha) = \sum_{i=1}^{k+1}\sum_{j=1}^{k+1}\frac{\partial A_i}{\partial x^j}dx^j\wedge dx^1\wedge\cdots\wedge dx^{i-1}\wedge dx^{i+1}\wedge\cdots\wedge dx^{k+1}$$

$$= \sum_{i=1}^{k+1}(-1)^{i+1}\frac{\partial A_i}{\partial x^i}dx^1\wedge\cdots\wedge dx^j\wedge\cdots\wedge dx^{k+1}.$$

Substituting in the left-hand integral of Eq. (17.2) we have

$$\int_{\Gamma_{k+1}} d(\sigma^*\alpha) = \sum_{i=1}^{k+1}(-1)^{i+1}\int_0^1\int_0^1\cdots\int_0^1\frac{\partial A_i}{\partial x^i}dx^1\,dx^2\ldots dx^{k+1}$$

$$= \sum_{i=1}^{k+1}(-1)^{i+1}\int_0^1\int_0^1\cdots\int_0^1 dx^1\ldots dx^{i-1}dx^{i+1}\ldots dx^{k+1}$$

$$\times\left[A_i(x^1,\ldots,x^{i-1},1,x^{i+1},\ldots,x^{k+1}) - A_i(x^1,\ldots,x^{i-1},0,x^{i+1},\ldots,x^{k+1})\right]$$

$$= \sum_{\epsilon=0}^{1}\sum_{i=1}^{k+1}(-1)^{i+\epsilon}\int_0^1\cdots\int_0^1 dx^1\ldots dx^{i-1}\,dx^{i+1}\ldots$$
$$dx^{k+1} A_i(x^1,\ldots,x^{i-1},\epsilon,x^{i+1},\ldots,x^{k+1})$$

$$= \sum_{\epsilon=0}^{1}\sum_{i=1}^{k+1}(-1)^{i+\epsilon}\int_{\Gamma_k}(\varphi_i^\epsilon)^*\circ\sigma^*\alpha$$

$$= \int_{\partial\sigma}\alpha,$$

as required. ∎

Example 17.2 In Example 15.9 we defined the integral of a differential 1-form ω over a curve with end points $\gamma : [t_1, t_2] \to M$ to be

$$\int_\gamma \omega = \int_{t_1}^{t_2}\langle\omega,\dot\gamma\rangle dt = \int_{t_1}^{t_2}\langle\gamma^*\omega,\frac{d}{dt}\rangle dt.$$

A 1-cell $\sigma : U \to M$ is a curve with parameter range $U = (-a, 1+a) \supset \Gamma_1 = [0, 1]$, and can be made to cover an arbitrary range $[t_1, t_2]$ by a change of parameter $t \to t' =$

17.3 Stokes' theorem

$t_1 + (t_2 - t_1)t$. The integral of $\omega = w_i(x^j)\,dx^i$ over the support of the 1-cell is

$$\int_\sigma \omega = \int_{\Gamma_1} \sigma * \omega = \int_0^1 w_i(\mathbf{x}(t))\frac{dx^i}{dt}\,dt,$$

which agrees with the definition of the integral given in Example 15.9.

The boundary of the 1-cell σ is

$$\partial\sigma = \sigma \circ \varphi_1^1 - \sigma \circ \varphi_1^0$$

where the two terms on the right-hand side are the 0-cells $\Gamma_0 = \{0\} \to \sigma(1)$ and $\Gamma_0 = \{0\} \to \sigma(0)$, respectively. Setting $\omega = df$ where f is a differentiable function on M, Stokes' theorem gives

$$\int_\sigma df = \int_{\partial\sigma} f = f(\sigma(1)) - f(\sigma(0)).$$

If $M = \mathbb{R}$ and the 1-cell σ is defined by $x = \sigma(t) = a + t(b-a)$, Stokes' theorem reduces to the fundamental theorem of calculus,

$$\int_a^b \frac{df}{dx}\,dx = \int_0^1 \frac{df}{dt}\,dt = \int_\sigma df = f(b) - f(a).$$

Regular domains

In the above discussion, the only requirement made concerning the cell maps σ was that they be differentiable on a neighbourhood of the unit k-cube Γ_k. For example, they could be completely degenerate and map the entire set Γ_k into a single point of M. For this reason, the chains are sometimes called **singular**. A **fundamental n-chain** on an n-dimensional manifold M has the form

$$C = \sigma_1 + \sigma_2 + \cdots + \sigma_N$$

where each $\sigma_i : U \to \sigma_i(U)$ is a diffeomorphism and the interiors of the supports $\sigma_i(\Gamma_k)$ of different cells are non-intersecting:

$$\sigma_i\big((\Gamma_n)^o\big) \cap \sigma_j\big((\Gamma_n)^o\big) = \emptyset \quad \text{for } i \neq j.$$

A **regular domain** $D \subseteq M$ is a closed set of the form

$$D = \bigcup_{i=1}^n \sigma_i(\Gamma_n)$$

where σ_i are the n-cells of a fundamental chain C (see Fig. 17.3). We may think of a regular domain as subdivided into cubical cells or a *region with boundary* – the 'boundary' consisting of boundary points of the chain that are not on the common faces of any pair of cells.

Theorem 17.3 *If D is a regular domain 'cubulated' in two different ways by fundamental chains $C = \sigma_i + \cdots + \sigma_N$ and $C' = \tau_1 + \cdots + \tau_M$ then $\int_C \omega = \int_{C'} \omega$ for every differential n-form ω.*

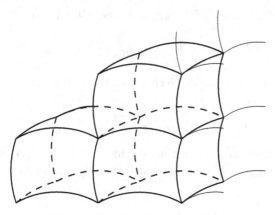

Figure 17.3 A regular domain on a manifold

Proof: Let $A_{ij} = \sigma_i(\Gamma_n) \cap \tau_j(\Gamma_n)$ and
$$B_{ij} = (\sigma_i)^{-1}(A_{ij}), \qquad C_{ij} = (\tau_j)^{-1}(A_{ij}).$$
The maps $\tau_j^{-1} \circ \sigma_i : B_{ij} \to C_{ij}$ are all diffeomorphisms and
$$\int_{B_{ij}} \sigma_i^* \omega = \int_{B_{ij}} \sigma_i^* \circ (\tau_j^{-1})^* \circ \tau_j^* \omega$$
$$= \int_{B_{ij}} \left((\tau_j^{-1}) \circ \sigma_i\right)^* \circ \tau_j^* \omega$$
$$= \int_{C_{ij}} \tau_j^* \omega.$$

Hence
$$\int_C \omega = \sum_i \int_{\sigma_i} \omega = \sum_{i,j} \int_{B_{ij}} \sigma_i^* \omega = \sum_{i,j} \int_{C_{ij}} \tau_j^* \omega = \int_{C'} \omega. \qquad\blacksquare$$

If α is a differential $(n-1)$-form,
$$\int_{\partial D} \alpha = \int_{\partial C} \alpha = \sum_{i=1}^N \sum_{j=1}^n \sum_{\epsilon=0}^1 \int_{\Gamma_{n-1}} (\varphi_j^\epsilon)^* \circ \sigma_i^* \alpha.$$

Since the outward normals on common faces of adjoining cells are oppositely directed, the faces will be oppositely oriented and the integrals will cancel, leaving only an integral on the 'free' parts of the boundary of D. This results in Stokes' theorem for a regular domain
$$\int_D d\alpha = \int_{\partial D} \alpha.$$

A **regular k-domain** D_k is defined as the image of a regular domain $D \subset K$ of a k-dimensional manifold K under a regular embedding $\varphi : K \to M$, and for any k-form

17.3 Stokes' theorem

β and $(k-1)$-form α we set

$$\int_{D_k} \beta = \int_{\varphi(D)} \beta = \int_D \varphi * \beta$$

$$\int_{\partial D_k} \alpha = \int_{\partial \varphi(D)} \alpha = \int_{\partial D} \varphi * \alpha.$$

The general form of Stokes' theorem asserts that for any $(k-1)$-form α and regular k-domain D_k,

$$\int_{D_k} d\alpha = \int_{\partial D_k} \alpha. \tag{17.3}$$

Example 17.3 In low dimensions, Stokes' theorem reduces to a variety of familiar forms. For example, let D be a regular 2-domain in $\mathbb{R}^2$ bounded by a circuit $C = \partial D$ having induced orientation according to the 'right hand rule'. By this we mean that if the outward normal is taken locally in the direction of the first coordinate x^1, and the tangent to C in the direction x^2, then $dx^1 \wedge dx^2$ is positively oriented. If $\alpha = P\,dx + Q\,dy$, then

$$d\alpha = \frac{\partial P}{\partial y} dy \wedge dx + \frac{\partial Q}{\partial x} dx \wedge dy = \left(\frac{\partial Q}{\partial x} - \frac{\partial P}{\partial y}\right) dx \wedge dy,$$

and Stokes' theorem is equivalent to **Green's theorem**

$$\iint_D \left(\frac{\partial Q}{\partial x} - \frac{\partial P}{\partial y}\right) dx\,dy = \oint_C P\,dx + Q\,dy.$$

If D is a bounded region in $\mathbb{R}^3$ whose boundary is a surface $S = \partial D$, the induced orientation is such that if $(\mathbf{e}, \mathbf{f})$ are a correctly ordered pair of tangent vectors to S and $\mathbf{n}$ the outward normal to S, then $(\mathbf{n}, \mathbf{e}, \mathbf{f})$ is a positively oriented basis of vectors in $\mathbb{R}^3$. Let $\alpha = A_1\,dy \wedge dz + A_2\,dz \wedge dx + A_3\,dx \wedge dz$ be a 2-form, then

$$d\alpha = (A_{1,1} + A_{2,2} + A_{3,3})dx \wedge dy \wedge dz = \nabla \cdot \mathbf{A}\,dx \wedge dy \wedge dz$$

where $\mathbf{A} = (A_1, A_2, A_3)$. Stokes' theorem reads

$$\int_D d\alpha = \iiint_D \nabla \cdot \mathbf{A}\,dx\,dy\,dz = \int_{\partial D} \alpha = \iint_S A_1\,dy\,dz + A_2\,dz\,dx + A_3\,dx\,dz.$$

If the bounding surface is locally parametrized by two parameters, $x = x(\lambda_1, \lambda_2)$, $y = y(\lambda_1, \lambda_2)$, $z = z(\lambda_1, \lambda_2)$, then we can write

$$\iint_S \alpha = \iint_S \epsilon_{ijk} A_i \frac{\partial x^j}{\partial \lambda_1} \frac{\partial x^k}{\partial \lambda_2} d\lambda_1 \wedge d\lambda_2$$

and it is common to write Stokes' theorem in the standard **Gauss theorem** form

$$\iiint_D \nabla \cdot \mathbf{A}\,dx\,dy\,dz = \iint_S \mathbf{A} \cdot d\mathbf{S},$$

where **S** is the vector area normal to S, having components

$$dS_i = \epsilon_{ijk} \frac{\partial x^j}{\partial \lambda_1} \frac{\partial x^k}{\partial \lambda_2} d\lambda_1 \, d\lambda_2.$$

Let Σ be an oriented 2-surface in $\mathbb{R}^3$, with boundary an appropriately oriented circuit $C = \partial \Sigma$, and α a differential 1-form $\alpha = A_1 \, dx + A_2 \, dy + A_3 \, dz = A_i \, dx^i$. Then

$$\int_\Sigma d\alpha = \int_\Sigma A_{i,j} \, dx^j \wedge dx^i$$

$$= \iint_\Sigma A_{k,j} \left(\frac{\partial x^j}{\partial \lambda_1} \frac{\partial x^k}{\partial \lambda_2} - \frac{\partial x^k}{\partial \lambda_1} \frac{\partial x^j}{\partial \lambda_2} \right) d\lambda_1 \wedge d\lambda_2$$

$$= \iint_\Sigma \epsilon_{ijk} A_{k,j} \, dS_i$$

and

$$\int_{\partial \Sigma} \alpha = \oint_C A_i \, dx^i = \oint_C A_1 \, dx + A_2 \, dy + A_3 \, dz.$$

This can be expressed in the familiar form of Stokes' theorem

$$\iint_\Sigma (\nabla \times \mathbf{A}) \cdot d\mathbf{S} = \oint_C \mathbf{A} \cdot d\mathbf{r}.$$

Exercise: Show that dS_i is 'normal' to the surface S in the sense that $dS_i \, \partial x^i / \partial \lambda_a = 0$ for $a = 1, 2$.

Problems

Problem 17.4 Let $\alpha = y^2 \, dx + x^2 \, dy$. If γ_1 is the stretch of y-axis from $(x = 0, y = -1)$ to $(x = 0, y = 1)$, and γ_2 the unit right semicircle connecting these points, evaluate

$$\int_{\gamma_1} \alpha, \quad \int_{\gamma_2} \alpha \quad \text{and} \quad \int_{S^1} \alpha.$$

Verify Stokes' theorem for the unit circle and the unit right semicircular region encompassed by γ_1 and γ_2.

Problem 17.5 If $\alpha = x \, dy \wedge dz + y \, dz \wedge dx + z \, dx \wedge dy$ compute $\int_{\partial \Omega} \alpha$ where Ω is (i) the unit cube, (ii) the unit ball in $\mathbb{R}^3$. In each case verify Stokes' theorem,

$$\int_{\partial \Omega} \alpha = \int_\Omega d\alpha.$$

Problem 17.6 Let S be the surface of a cylinder of elliptical cross-section and height $2h$ given by

$$x = a \cos \theta, \quad y = b \sin \theta \quad (0 \le \theta < 2\pi), \quad -h \le z \le h.$$

(a) Compute $\int_S \alpha$ where $\alpha = x \, dy \wedge dz + y \, dz \wedge dx - 2z \, dx \wedge dy$.
(b) Show $d\alpha = 0$, and find a 1-form ω such that $\alpha = d\omega$.
(c) Verify Stokes' theorem $\int_S \alpha = \int_{\partial S} \omega$.

17.4 Homology and cohomology

Problem 17.7 A torus in $\mathbb{R}^3$ may be represented parametrically by

$$x = \cos\phi(a + b\cos\psi), \qquad y = \sin\phi(a + b\cos\psi), \qquad z = b\sin\psi$$

where $0 \le \phi < 2\pi, 0 \le \psi < 2\pi$. If b is replaced by a variable ρ that ranges from 0 to b, show that

$$dx \wedge dy \wedge dz = \rho(a + \rho\cos\psi)\, d\phi \wedge d\psi \wedge d\rho.$$

By integrating this 3-form over the region enclosed by the torus, show that the volume of the solid torus is $2\pi^2 ab^2$. Can you see this by a simple geometrical argument?

Evaluate the volume by performing the integral of the 2-form $\alpha = x\, dy \wedge dz$ over the surface of the torus and using Stokes' theorem.

Problem 17.8 Show that in n dimensions, if V is a regular n-domain with boundary $S = \partial V$, and we set α to be an $(n-1)$-form with components

$$\alpha = \sum_{i=1}^{n} (-1)^{i+1} A^i\, dx^1 \wedge \cdots \wedge dx^{i-1} \wedge dx^{i+1} \wedge \cdots \wedge dx^n,$$

Stokes' theorem can be reduced to the n-dimensional Gauss theorem

$$\int \cdots \int_V A^i{}_{,i}\, dx^1 \ldots dx^n = \int \cdots \int_S A^i\, dS_i$$

where $dS_i = dx^1 \ldots dx^{i-1}\, dx^{i+1} \ldots dx^n$ is a 'vector volume element' normal to S.

17.4 Homology and cohomology

In the previous section we considered regions that could be subdivided into 'cubical' parts. While this has practical advantages when it comes to integration, and makes the proof of Stokes' theorem relatively straightforward, the subject of *homology* is more standardly based on triangular cells. There is no essential difference in this change since any k-cube is readily triangulated, as well as the converse. For example, a triangle in two dimensions is easily divided into squares (see Fig. 17.4). Dividing a tetrahedron into four cubical regions is harder to visualize, and is left as an excercise for the reader.

Ordered simplices and chains in Euclidean space

A set of points $\{x_0, x_1, \ldots, x_p\}$ in Euclidean space $\mathbb{R}^n$ is said to be **independent** if the p vectors $x_i - x_0$ ($i = 1, \ldots, p$) are linearly independent. The **ordered** p-**simplex** with these points as **vertices** consists of their convex hull,

$$\langle x_0, x_1, \ldots, x_p\rangle = \left\{ y = \sum_{i=0}^{p} t^i x_i \,\middle|\, \text{all } t^j \ge 0 \text{ and } \sum_{i=0}^{p} t^i = 1 \right\},$$

together with a specific ordering $(x_0, x_1, \ldots, x_p)$ of the vertex points. We will often denote an ordered r-simplex by a symbol such as Δ or Δ_p. Two ordered p-simplices with the

Integration on manifolds

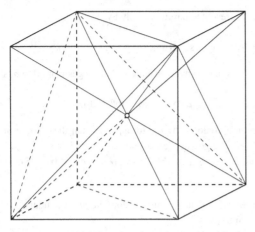

Figure 17.4 Dividing the triangular into 'cubical' cells

same set of vertices will be taken to be identical if the orderings are related by an even permutation, else they are given the opposite sign. For example,

$$\langle \mathbf{x}, \mathbf{y}, \mathbf{z}\rangle = -\langle \mathbf{x}, \mathbf{z}, \mathbf{y}\rangle = \langle \mathbf{y}, \mathbf{z}, \mathbf{x}\rangle.$$

The **standard n-simplex** on $\mathbb{R}^n$ is $\tilde{\Delta}_n = \langle \mathbf{0}, \mathbf{e}_1, \ldots, \mathbf{e}_n\rangle$ where $\mathbf{e}_i$ is the ith basis vector $\mathbf{e}_i = (0, 0, \ldots, 1, \ldots, 0)$, i.e.

$$\tilde{\Delta}_n = \{(t_1, t_2, \ldots, t_n) \mid \sum_{j=1}^n t_j = 1 \text{ and } 0 \le t_i \le 1 \ (i = 1, \ldots, n)\} \subset \mathbb{R}^n.$$

A 0-simplex $\langle \mathbf{x}_0\rangle$ is a single point $\mathbf{x}_0$ together with a plus or minus sign.

A 1-simplex $\langle \mathbf{x}_0, \mathbf{x}_1\rangle$ is a closed directed line from $\mathbf{x}_0$ to $\mathbf{x}_1$.

A 2-simplex $\langle \mathbf{x}_0, \mathbf{x}_1, \mathbf{x}_2\rangle$ is an oriented triangle, where the vertices are taken in a definite order.

A 3-simplex is a tetrahedron in which the vertices are again given a specific order up to even permutations. These examples are depicted in Fig. 17.5.

A *p*-**chain** is a formal sum $C = \sum_{\mu=1}^M c^\mu \Delta_\mu$ where a_μ are real numbers and Δ_μ are p-simplices in $\mathbb{R}^n$. The set of all p-chains on $\mathbb{R}^n$ is obviously a vector space, denoted $\mathcal{C}_p(\mathbb{R}^n)$.

The ith face of a p-simplex $\Delta = \langle \mathbf{x}_0, \mathbf{x}_1, \ldots, \mathbf{x}_p\rangle$ is defined as the ordered $(p-1)$-simplex

$$\Delta_i = (-1)^i \langle \mathbf{x}_0, \ldots, \widehat{\mathbf{x}_i}, \ldots, \mathbf{x}_p\rangle \equiv (-1)^i \langle \mathbf{x}_0, \mathbf{x}_1, \ldots, \mathbf{x}_{i-1}, \mathbf{x}_{i+1}, \ldots, \mathbf{x}_p\rangle,$$

and the **boundary** of a p-simplex Δ is defined as the $(p-1)$-chain

$$\partial \Delta = \sum_{i=0}^p \Delta_i = \sum_{i=0}^p (-1)^i \langle \mathbf{x}_0, \ldots, \widehat{\mathbf{x}_i}, \ldots, \mathbf{x}_p\rangle,$$

17.4 Homology and cohomology

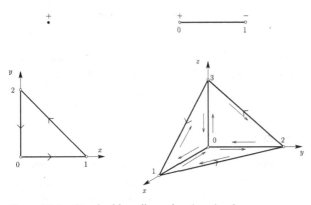

Figure 17.5 Standard low dimensional p-simplexes

and extends to all p-chains by linearity. For example,

$$\partial \langle x_0, x_1 \rangle = \langle x_1 \rangle - \langle x_0 \rangle,$$
$$\partial \langle x_0, x_1, x_2 \rangle = \langle x_1, x_2 \rangle - \langle x_0, x_2 \rangle + \langle x_0, x_1 \rangle,$$
$$\partial \langle x_0, x_1, x_2, x_3 \rangle = \langle x_1, x_2, x_3 \rangle - \langle x_0, x_2, x_3 \rangle + \langle x_0, x_1, x_3 \rangle - \langle x_0, x_1, x_2 \rangle.$$

For each $p = 0, \ldots, n$ the boundary operator generates a linear map $\partial : C_p(\mathbb{R}^n) \to C_{p-1}(\mathbb{R}^n)$ by setting

$$\partial \left(\sum_\mu c^\mu \Delta_\mu \right) = \sum_\mu c^\mu \partial \Delta_\mu.$$

If we set the boundary of any 0-form to be the zero chain, $\partial \langle x \rangle = 0$ it is trivial to see that two successive applications of the boundary operator on any 1-simplex vanishes,

$$\partial^2 \langle x_0, x_1 \rangle = \partial(\langle x_1 \rangle - \langle x_0 \rangle) = 0 - 0 = 0.$$

This identity generalizes to arbitrary p-simplices, for

$$\partial \partial \langle x_0, x_1, \ldots, x_p \rangle = \sum_{i=0}^{p} \partial (-1)^i \langle x_0, \ldots, \widehat{x}_i, \ldots, x_p \rangle$$
$$= \sum_{i=0}^{p} (-1)^i \left[\sum_{j<i} (-1)^j \langle x_0, \ldots, \widehat{x}_j, \ldots, \widehat{x}_i, \ldots, x_p \rangle \right.$$
$$\left. + \sum_{j>i} (-1)^{j+1} \langle x_0, \ldots, \widehat{x}_i, \ldots, \widehat{x}_j, \ldots, x_p \rangle \right]$$
$$= 0$$

since all terms cancel in pairs. The identity $\partial^2 = 0$ follows from the linearity of the boundary operator ∂ on $C_p(\mathbb{R}^n)$.

Integration on manifolds

Exercise: Write out the cancellation of terms in this argument explicitly for a 2-simplex $\langle x_0, x_1, x_2 \rangle$ and 3-simplex $\langle x_0, x_1, x_2, x_3 \rangle$.

A p-chain C is said to be a **cycle** if it has no boundary, $\partial C = 0$. It is said to be a **boundary** if there exists a $(p+1)$-chain C' such that $C = \partial C'$. Clearly every boundary is a p-chain since $\partial C = \partial^2 C' = 0$, but the converse need not be true.

Simplicial homology on manifolds

Let M be an n-dimensional differentiable manifold. A **(singular)** p-**simplex** σ_p on M is a smooth map $\phi : U \to M$ where U is an open subset of $\mathbb{R}^p$ containing the standard p-simplex $\bar{\Delta}_p$. A p-**chain on** M is a formal linear combination of p-simplices on M,

$$C = \sum_{\mu=1}^{M} c^{\mu} \sigma_{p\mu} \quad (c^{\mu} \in \mathbb{R}),$$

and let $\mathcal{C}_p(M)$ be the real vector space generated by all p-simplices on M.

For each $i = 1, 2, \ldots, p-1$ denote by $\varphi_i : \mathbb{R}^{p-1} \to \mathbb{R}^p$ the map that embeds $\mathbb{R}^{p-1}$ into the plane $x^i = 0$ of $\mathbb{R}^p$,

$$\varphi_i(x^1, \ldots, x^{p-1}) = (x^1, \ldots, x^i, 0, x^{i+1}, \ldots, x^{p-1}),$$

and for $i = 0$ set

$$\varphi_0(x^1, \ldots, x^{p-1}) = \left(1 - \sum_{i=1}^{p-1} x^i, x^1, \ldots, x^{p-1}\right).$$

The maps $\varphi_0, \varphi_1, \ldots, \varphi_{p-1}$ are $(p-1)$-simplices in $\mathbb{R}^p$, whose supports are the various faces of the standard p-simplex, $\bar{\Delta}_p$,

$$\varphi_i(\bar{\Delta}_{p-1}) = \bar{\Delta}_{pi} \quad (i = 0, 1, \ldots, p-1).$$

If σ is a p-simplex in M, define its ith face to be the $(p-1)$-simplex

$$\sigma_i = \sigma \circ \varphi_i : \bar{\Delta}_{p-1} \to M,$$

and its **boundary** to be the $(p-1)$-chain

$$\partial_p \sigma = \sum_{i=0}^{p-1} (-1)^i \sigma_i.$$

Extend by linearity to all chains $C \in \mathcal{C}_p(M)$,

$$\partial_p C = \partial \sum_{\mu=1}^{M} c^{\mu} \sigma_{p\mu} = \sum_{\mu=1}^{M} c^{\mu} \partial \sigma_{p\mu}.$$

A p-**boundary** B is a singular p-cycle on M that is the boundary of a $(p+1)$-chain, $B = \partial_{p+1} C$. A p-**cycle** C is a singular p-chain on M whose boundary vanishes, $\partial_p C = 0$. Since $\partial^2 = 0$ it is clear that every p-boundary is a p-cycle.

17.4 Homology and cohomology

If we let $B_p(M)$ be the set of all p-boundaries on M, and $Z_p(M)$ all p-cycles, these are both vector subspaces of $C_p(M)$:

$$B_p(M) = \operatorname{im} \partial_{p+1},$$
$$Z_p(M) = \ker \partial_p \supseteq B_p(M).$$

We define the pth **homology space** to be the factor space

$$H_p(M) = Z_p(M)/B_p(M).$$

Commonly this is called the pth *homology group*, only the abelian group property being relevant. Two cycles C_1 and C_2 are said to be **homologous** if they belong to the same homology class – that is, if there exists a chain C such that $C_1 - C_2 = \partial C$. The dimension of the pth homology space is known as the pth **Betti number**,

$$b_p = \dim H_p(M),$$

and the quantity

$$\chi(M) = \sum_{p=0}^{n} (-1)^p b_p$$

is known as the **Euler characteristic** of the manifold M. A non-trivial result that we shall not attempt to prove is that the Betti numbers are topological invariants – two manifolds that are topologically homeomorphic have the same Betti numbers and Euler characteristic [13].

Example 17.4 Since $\partial \langle 0 \rangle = 0$ every 0-simplex in a manifold M has boundary 0. Hence every 0-chain in M is a 0-cycle, and $Z_0(M) = C_0(M)$. The zeroth homology space $H_0(M) = Z_0(M)/B_0(M)$ counts the number of 0-chains that are not boundaries of 1-chains. Since a 1-simplex is essentially a smooth curve $\sigma : [0, 1] \to M$ it has boundary $\sigma(1) - \sigma(0)$, where we represent the 0-simplex map $0 \to p \in M$ simply by its image point p. Two 0-simplices p and q are homologous if $p - q$ is a boundary; that is, if they are the end points of a smooth curve connecting them. This is true if and only if they belong to the same connected component of M. Thus $H_0(M)$ is spanned by a set of simplices $\{p_0, p_1, \ldots\}$, one from each connected component of M, and the zeroth Betti number β_0 is the number of connected components of the topological space M.

De Rham cohomology groups and duality

Let $C^r(M) = \Lambda_r(M)$ be the real vector space consisting of all differential r-forms on M. Its elements are also known as r-**cochains** on M. The exterior derivative d is a linear operator $d : C^r(M) \to C^{r+1}(M)$ for each $r = 0, 1, \ldots, n$, with the property $d^2 = 0$. We write its restriction to $C^r(M)$ as d_r.

A differential r-form α is said to be **closed** if $d\alpha = 0$, and it is said to be **exact** if there exists an $(r-1)$-form β such that $\alpha = d\beta$. Clearly every exact r-form is closed since $d^2\beta = 0$. In the language of cochains these definitions can be expressed as follows: an r-cochain α is called an r-**cocycle** if it is a closed differential form, while it is an r-**coboundary** if it is exact. We denoted the vector subspace of r-cocycles by $Z^r(M)$, and the subspace of r-coboundaries by $B^r(M)$.

$$Z^r(M) = \{\alpha \in C^r(M) \mid d\alpha = 0\} = \ker d_r \subset C^r(M)$$
$$B^r(M) = \{\alpha \in C^r(M) \mid \alpha = d\beta, \ \beta \in C^{r-1}(M)\} = \operatorname{im} d_{r-1} \subset Z^r(M).$$

The rth de Rham cohomology space (group) is defined as the factor space $H^r(M) = Z^r(M)/B^r(M)$, and any two r-cocycles α and β are said to be **cohomologous** if they belong to the same coset, $\alpha - \beta = d\gamma$ for some $(r-1)$-cochain γ. The dimensions of the vector spaces $H^r(M)$ are denoted b^r.

Example 17.5 Since there are no differential forms of degree -1 we always set $B^0(M) = 0$. Hence $H^0(M) = Z^0(M)$. A 0-form f is closed and belongs to $Z^0(M)$ if and only if $df = 0$. Hence $f = \text{const.}$ on each connected component of M, and $H^0(M) = \mathbb{R} + \mathbb{R} + \cdots + \mathbb{R}$, one contribution from each such component. Hence $b^0 = b_0$ is the number of connected components of M (see Example 17.4).

Example 17.6 If $M = \mathbb{R}$, then from the previous example $H^0(\mathbb{R}) = \mathbb{R}$. A 1-form $\omega \in C^1(M)$ is closed if $d\omega = 0$. Setting $\omega = f(x)\,dx$ we can clearly always write

$$\omega = df = \frac{dF}{x} \quad \text{where} \quad F(x) = \int_0^x f(y)\,dy.$$

Hence every closed 1-form is exact and $H^1(\mathbb{R}) = 0$, $b^1 = 0$. It is not difficult to verify that this is also the value of the Betti numbers, $b_1 = 0$.

Define a bracket $\langle\,,\,\rangle : C_r(M) \times C^r(M) \to \mathbb{R}$ by setting

$$\langle C, \alpha \rangle = \int_C \alpha.$$

for every r-chain $C \in C_r(M)$ and r-cochain $\alpha \in C^r(M) = \Lambda_r(M)$. For every α the map $C \mapsto \langle C, \alpha \rangle$ is evidently linear on $C_r(M)$ and for every $C \in C_r(M)$ the map $\alpha \mapsto \langle C, \alpha \rangle$ is linear on $C^r(M)$. By Stokes' theorem the exterior derivative d is the adjoint of the boundary operator ∂ with respect to this bracket, in the sense that

$$\langle C, d\alpha \rangle = \langle \partial C, \alpha \rangle.$$

The bracket induces a bracket $\langle\,,\,\rangle$ on $H_r(M) \times H^r(M)$ by setting

$$\langle [C], [\alpha] \rangle = \langle C, \alpha \rangle = \int_C \alpha$$

for any pair $[C] \in H_r(M)$ and $[\alpha] \in H^r(M)$. It is independent of the choice of representative from the homology and cohomology classes, for if $C' = C + \partial C_1$ and $\alpha' = \alpha + d\alpha_1$, where

17.4 Homology and cohomology

$\partial C = 0$ and $d\alpha = 0$, then

$$\begin{aligned}
\langle C', \alpha' \rangle &= \int_{C+\partial C_1} \alpha + d\alpha_1 \\
&= \int_C \alpha + \int_{\partial C_1} \alpha + \int_C d\alpha_1 + \int_{\partial C_1} d\alpha_1 \\
&= \int_C \alpha + \int_{C_1} d\alpha + \int_{\partial C} \alpha_1 + \int_{\partial^2 C_1} \alpha_1 \quad \text{by Stokes' theorem} \\
&= \int_C \alpha = \langle C, \alpha \rangle.
\end{aligned}$$

For any fixed r-cohomology class $[\alpha]$, the map $f_{[\alpha]} : H_p(M) \to \mathbb{R}$ given by

$$f_{[\alpha]}([C]) = \langle [C], [\alpha] \rangle$$

is a well-defined linear functional on $H_r(M)$. De Rham's theorem asserts that this correspondence between linear functionals on $H_r(M)$ and cohomology classes $[\alpha] \in H^r(M)$ is bijective.

Theorem 17.4 (de Rham) *The bilinear map on $H_r(M) \times H^r(M) \to \mathbb{R}$ defined by $([C], [\alpha]) \mapsto \langle [C], [\alpha] \rangle$ is non-degenerate in both arguments. That is, every linear functional on $H_r(M)$ has the form $f_{[\alpha]}$ for a uniquely defined r-cohomology class $[\alpha]$.*

The proof lies beyond the scope of this book, and may be found in [6, 10]. There are a variety of ways of expressing de Rham's theorem. Essentially it says that the rth cohomology group is isomorphic with the dual space of the rth homology group,

$$H^r(M) \cong \left(H_r(M)\right)^*.$$

If the Betti numbers are finite then $b^r = b_r$.

The integral of a closed r-form α over an r-cycle C

$$\langle C, \alpha \rangle = \int_C \alpha$$

is sometimes called a **period** of α. By Stokes' theorem all periods of α vanish if α is an exact form, and the period of any closed r-form vanishes over a boundary r-cycle $C = \partial C'$. Let $C_1, \ldots, C_k$ be $k = b_r$ linearly independent cycles in $Z_r(M)$, such that $[C_i] \ne [C_j]$ for $i \ne j$. De Rham's theorem implies that an r-form α is exact if and only if all the periods $\int_{C_i} \alpha = 0$. If α is exact then we have already remarked that all its periods vanish. The converse follows from the fact that $\langle [C], [\alpha] \rangle = 0$ for every $[C] \in H_r(M)$, since $[C]$ can be expanded to $[C] = \sum_{i=1}^k [C_i]$. By non-degeneracy of the product $\langle \,,\, \rangle$ we must have $[\alpha] = 0$, so that $\alpha = d\beta$ for some $(r-1)$-form β.

Problems

Problem 17.9 Show that any tetrahedron may be divided into 'cubical' regions. Describe a procedure for achieving the same result for a general k-simplex.

Integration on manifolds

Problem 17.10 For any pair of subspaces H and K of the exterior algebra $\Lambda^*(M)$, set $H \wedge K$ to be the vector subspace spanned by all $\alpha \wedge \beta$ where $\alpha \in H$, $\beta \in K$. Show that

(a) $Z^p(M) \wedge Z^q(M) \subseteq Z^{p+q}(M)$,

(b) $Z^p(M) \wedge B^q(M) \subseteq B^{p+q}(M)$,

(c) $B^p(M) \wedge B^q(M) \subseteq B^{p+q}(M)$.

Problem 17.11 Show that for any set of real numbers $a_1, \ldots, a_k$ there exists a closed r-form α whose periods $\int_{C_i} \alpha = a_i$.

Problem 17.12 If S^1 is the unit circle, show that $b^0 = b^1 = 1$.

17.5 The Poincaré lemma

The fact that every exact differential form is closed has a kind of local converse.

Theorem 17.5 (Poincaré lemma) *On any open set $U \subseteq M$ homeomorphic to $\mathbb{R}^n$, every closed differential form of degree $k \geq 1$ is exact: if $d\alpha = 0$ on U where $\alpha \in \Lambda_k(U)$, then there exists a $(k-1)$-form β on U such that $\alpha = d\beta$.*

Proof: We prove the theorem on $\mathbb{R}^n$ itself, with coordinates $x^1, \ldots, x^n$, and set $\alpha = \alpha_{i_1 i_2 \ldots i_k}(\mathbf{x}) \, dx^{i_1} \wedge dx^{i_2} \wedge \cdots \wedge dx^{i_k}$. Let α^t $(0 \leq t \leq 1)$ be the one-parameter family of k-forms

$$\alpha^t = \alpha_{i_1 \ldots i_k}(t\mathbf{x}) \, dx^{i_1} \wedge \cdots \wedge dx^{i_k}.$$

The map $h_k : \Lambda_k(\mathbb{R}^n) \to \Lambda_{k-1}(\mathbb{R}^n)$ defined by

$$h_k \alpha = \int_0^1 t^{k-1} i_X \alpha^t \, dt \quad \text{where} \quad X = x^i \frac{\partial}{\partial x^i}$$

satisfies the key identity

$$(d \circ h_k + h_{k+1} \circ d)\alpha = \alpha \tag{17.4}$$

for any k-form α on $\mathbb{R}^n$. To prove (17.4) write out the left-hand side,

$$(d \circ h_k + h_{k+1} \circ d)\alpha = \int_0^1 t^{k-1} d i_X \alpha^t + t^k i_X (d\alpha)^t \, dt,$$

and

$$d\alpha^t = \frac{\partial \alpha_{i_1 \ldots i_k}(t\mathbf{x})}{\partial x^j} dx^j \wedge dx^{i_1} \wedge \cdots \wedge dx^{i_k}$$

$$= t(d\alpha)^t.$$

Using the Cartan identity, Eq. (16.13),

$$(d \circ h_k + h_{k+1} \circ d)\alpha = \int_0^1 t^{k-1} (d \circ i_X + i_X \circ d) \alpha^t \, dt$$

$$= \int_0^1 t^{k-1} \mathcal{L}_X \alpha^t \, dt$$

17.5 The Poincaré lemma

and from the component formula for the Lie derivative (15.39),

$$\begin{aligned}(\mathcal{L}_X \alpha^t)_{i_1\ldots i_k} &= \frac{\partial \alpha_{i_1\ldots i_k}(t\mathbf{x})}{\partial x^j} x^j + \frac{\partial x^j}{\partial x^{i_1}} \alpha_{j i_2\ldots i_k}(t\mathbf{x}) + \cdots + \frac{\partial x^j}{\partial x^{i_k}} \alpha_{i_1\ldots j}(t\mathbf{x}) \\ &= t \frac{\partial \alpha_{i_1\ldots i_k}(t\mathbf{x})}{\partial t x^j} \frac{dt x^j}{dt} + \delta^j_{i_1} \alpha_{j\ldots i_k}(t\mathbf{x}) + \cdots + \delta^j_{x^{i_k}} \alpha_{i_1\ldots j}(t\mathbf{x}) \\ &= t \frac{d\alpha_{i_1\ldots i_k}(t\mathbf{x})}{dt} + k\alpha_{i_1\ldots i_k}(t\mathbf{x}).\end{aligned}$$

Hence

$$\mathcal{L}_X \alpha^t = t \frac{d\alpha^t}{dt} + k\alpha^t,$$

and Eq. (17.4) follows from

$$(\mathrm{d} \circ h_k + h_{k+1} \circ \mathrm{d})\alpha = \int_0^1 t^k \frac{d\alpha^t}{dt} + kt^{k-1}\alpha^t \, dt = \int_0^1 \frac{dt^k \alpha^t}{dt} dt = \alpha.$$

If $d\alpha = 0$ we have $\alpha = d\beta$ where $\beta = h_k \alpha$, and the theorem is proved. ∎

An immediate corollary of this theorem and de Rham's theorem is that all homology groups $H_k(\mathbb{R}^n)$ are trivial for $k \geq 1$; that is, all Betti numbers for $k \geq 1$ vanish in Euclidean space, $b_k = b^k = 0$. Of course $b_0 = b^0 = 1$ since there is a single connected component.

Example 17.7 In $\mathbb{R}^3$ let α be the 1-form $\alpha = A_1 \, dx^1 + A_2 \, dx^2 + A_3 \, dx^3$. Its exterior derivative is

$$d\alpha = (A_{2,1} - A_{1,2}) \, dx^1 \wedge dx^2 + (A_{1,3} - A_{3,1}) \, dx^3 \wedge dx^1 + (A_{3,2} - A_{2,3}) \, dx^2 \wedge dx^3$$

and Poincaré's lemma asserts that $d\alpha = 0$ if and only if there exists a function f on $\mathbb{R}^3$ such that $\alpha = df$. In components,

$$A_{2,1} - A_{1,2} = A_{1,3} - A_{3,1} = A_{3,2} - A_{2,3} = 0 \iff A_1 = f_{,1}, \quad A_2 = f_{,2}, \quad A_3 = f_{,3},$$

or in standard 3-vector language, with $\mathbf{A} = (A_1, A_2, A_3)$,

$$\nabla \times \mathbf{A} = 0 \iff \mathbf{A} = \nabla f.$$

If α is the differential 2-form $\alpha = A_3 \, dx^1 \wedge dx^2 + A_2 \, dx^3 \wedge dx^1 + A_1 \, dx^2 \wedge dx^3$, then

$$d\alpha = (A_{1,1} + A_{2,2} + A_{3,3}) \, dx^1 \wedge dx^2 \wedge dx^3.$$

The Poincaré lemma says

$$d\alpha = 0 \iff \alpha = d\beta \quad \text{where} \quad \beta = B_1 \, dx^1 + B_2 \, dx^2 + B_3 \, dx^3,$$

or in components

$$A_{1,1} + A_{2,2} + A_{3,3} = 0 \iff A_1 = B_{3,2} - B_{2,3}, \quad A_2 = B_{1,3} - B_{3,1}, \quad A_3 = B_{2,1} - B_{1,2},$$

which reduces to the familiar 3-vector statement

$$\nabla \cdot \mathbf{A} = 0 \iff \text{there exists } \mathbf{B} \text{ such that } \mathbf{A} = \nabla \times \mathbf{B}.$$

Integration on manifolds

Example 17.8 On the manifold $M = \mathbb{R}^2 - \{0\}$ with coordinates $x^1 = x$, $x^2 = y$, let ω be the differential 1-form

$$\omega = \frac{-y\,dx + x\,dy}{x^2 + y^2},$$

which cannot be extended smoothly to a 1-form on all of $\mathbb{R}^2$ because of the singular behaviour at the origin. On M, however, it is closed since

$$d\omega = \frac{-dy \wedge dx}{x^2 + y^2} + \frac{2y^2\,dy \wedge dx}{(x^2 + y^2)^2} + \frac{dx \wedge dy}{x^2 + y^2} - \frac{2x^2\,dx \wedge dy}{(x^2 + y^2)^2}$$

$$= \frac{2\,dx \wedge dy}{x^2 + y^2} - \frac{2(x^2 + y^2)}{(x^2 + y^2)^2}dx \wedge dy = 0.$$

Locally it is possible everywhere to find a function f such that $\omega = df$. For example, it is straightforward to verify that the pair of differential equations

$$\frac{\partial f}{\partial x} = \frac{-y}{x^2 + y^2}, \qquad \frac{\partial f}{\partial y} = \frac{x}{x^2 + y^2}$$

has a solution $f = \arctan(y/x)$. However f is not globally defined on M, since it is essentially the polar angle given by $x = r \cos f$, $y = r \sin f$ and increases by 2π on any circuit of the origin beginning at the positive branch of the y-axis. This demonstrates that Poincaré's lemma does not in general hold on manifolds not homeomorphic with $\mathbb{R}^n$.

Electrodynamics

An electromagnetic field is represented by an antisymmetric 4-tensor field F in Minkowski space, having components $F_{\mu\nu}(x^\alpha)$ ($\mu, \nu = 1, \ldots, 4$) (see Chapter 9). Define the **Maxwell 2-form** φ as having components $F_{\mu\nu}$,

$$\varphi = F_{\mu\nu}\,dx^\mu\,dx^\nu$$
$$= 2\big(B_3\,dx^1 \wedge dx^2 + B_2\,dx^3 \wedge dx^1 + B_1\,dx^2 \wedge dx^3$$
$$+ E_1\,dx^1 \wedge dx^4 + E_2\,dx^2 \wedge dx^4 + E_3\,dx^3 \wedge dx^4\big)$$

where $\mathbf{E} = (E_1, E_2, E_3)$ is the electric field, $\mathbf{B} = (B_1, B_2, B_3)$ the magnetic field and $x^1 = x$, $x^2 = y$, $x^3 = z$, $x^4 = ct$ are inertial coordinates. The source-free Maxwell equations (9.37) can be written

$$d\varphi = 0 \iff F_{\mu\nu,\rho} + F_{\nu\rho,\mu} + F_{\rho\mu,\nu} = 0$$
$$\iff \nabla \cdot \mathbf{B} = 0, \quad \nabla \times \mathbf{E} + \frac{1}{c}\frac{\partial \mathbf{B}}{\partial t} = 0.$$

By the Poincaré lemma, there exists a 1-form α, known as the 4-vector potential, such that $\varphi = d\alpha$. Writing the components of α as $(A_1, A_2, A_3, -\phi)$ this equation reads

$$\mathbf{B} = \nabla \times \mathbf{A}, \qquad \mathbf{E} = -\frac{1}{c}\frac{\partial \mathbf{A}}{\partial t} - \nabla \phi.$$

To express the equations relating the electromagnetic field to its sources in terms of differential forms we must define the **dual Maxwell 2-form** $*\varphi = *F_{\mu\nu}\,dx^\mu \wedge dx^\nu$ where

17.5 The Poincaré lemma

$*F_{\mu\nu}$ is defined as in Example 8.8,

$$*\varphi = 2\big(-E_3\,dx^1 \wedge dx^2 - E_2\,dx^3 \wedge dx^1 - E_1\,dx^2 \wedge dx^3 \\ + B_1\,dx^1 \wedge dx^4 + B_2\,dx^2 \wedge dx^4 + B_3\,dx^3 \wedge dx^4\big).$$

The distribution of electric charge present is represented by a 4-current vector field $J = J^\mu e_\mu$ having components $J^\mu = (\mathbf{j}, \rho c)$ where $\rho(\mathbf{r}, t)$ is the charge density and $\mathbf{j}$ the charge flux density (see Section 9.4).

$$\vartheta = *J_{\mu\nu\rho}\,dx^\mu \wedge dx^\nu \wedge dx^\rho$$
$$= -\frac{1}{3!}\epsilon_{\mu\nu\rho\sigma}J^\sigma\,dx^\mu \wedge dx^\nu \wedge dx^\rho$$
$$= -c\rho\,dx^1 \wedge dx^2 \wedge dx^3 + J^1 dx^2 \wedge dx^3 \wedge dx^4$$
$$+ J^2\,dx^3 \wedge dx^1 \wedge dx^4 + J^3 dx^1 \wedge dx^2 \wedge dx^4.$$

Equations (9.38) may then be written as

$$d*\varphi = -\vartheta \iff \nabla \cdot \mathbf{E} = 4\pi\rho, \qquad -\frac{1}{c}\frac{\partial \mathbf{E}}{\partial t} + \nabla \times \mathbf{B} = \frac{4\pi}{c}\mathbf{j}.$$

Charge conservation follows from

$$d\vartheta = -d^2 *\varphi = 0 \iff \nabla \cdot \mathbf{j} + \frac{1}{c}\frac{\partial \rho}{\partial t} = 0.$$

Example 17.9 Although Maxwell's vacuum equations take on the deceptively symmetrical form

$$d\varphi = 0, \qquad d*\varphi = 0$$

we cannot assume that $*\varphi = d\beta$ for a globally defined 1-form β. For example, the coulomb field

$$\mathbf{B} = \mathbf{0}, \qquad \mathbf{E} = \left(\frac{qx}{r^3}, \frac{qy}{r^3}, \frac{qz}{r^3}\right)$$

corresponds to the Maxwell 2-form

$$\varphi = \frac{2q}{r^3}\big(x\,dx \wedge dx^4 + y\,dy \wedge dx^4 + z\,dz \wedge dx^4\big) = \frac{2q}{r^2}dr \wedge dx^4$$

where $r^2 = x^2 + y^2 + z^2$, with dual 2-form

$$*\varphi = -\frac{2q}{r^3}\big(z\,dx \wedge dy + y\,dz \wedge dx + x\,dy \wedge dz\big).$$

This 2-form is, however, only defined on the subspace $M = \mathbb{R}^3 - \{0\}$. A short calculation in spherical polar coordinates results in

$$*\varphi = -2q\sin\theta\,d\theta \wedge d\phi = d(2q\cos\theta\,d\phi) = d(2q(\cos\theta - 1)\,d\phi).$$

Either of the choices $\beta = 2q\cos\theta\,d\phi$ or $\beta' = 2q(\cos\theta - 1)d\phi$ will act as a potential 1-form for $*\varphi$, but neither is defined on all of M since the angular coordinate ϕ is not well-defined on the z-axis where $\theta = 0$ or π. The 1-form β is not well-defined on the entire z-axis, but the potential 1-form β' vanishes on the positive z-axis and has a singularity along the negative

503

Integration on manifolds

z-axis. It is sometimes called a **Dirac string** – a term commonly reserved for solutions representing magnetic monopoles.

The impossibility of a global potential β can be seen by integrating $*\varphi$ over the unit 2-sphere

$$\int_{S^2} *\varphi = \int_0^\pi \int_0^{2\pi} -2q \sin\theta \, d\theta \, d\phi = -8\pi q,$$

and using Stokes' theorem (note that S^2 has no boundary)

$$\int_{S^2} *\varphi = \int_{S^2} d\beta = \int_{\partial S^2} \beta = 0.$$

Problems

Problem 17.13 Let

$$\alpha = \frac{x \, dy - y \, dx}{x^2 + y^2}.$$

Show that α is a closed 1-form on $\mathbb{R}^2 - \{0\}$. Compute its integral over the unit circle S^1 and show that it is not exact. What does this tell us of the de Rham cohomology of $\mathbb{R}^2 - \{0\}$ and S^1?

Problem 17.14 Prove that every closed 1-form on S^2 is exact. Show that this statement does not extend to 2-forms by showing that the 2-form

$$\alpha = r^{-3/2}(x \, dy \wedge dz + y \, dz \wedge dx + z \, dx \wedge dy)$$

is closed, but has non-vanishing integral on S^2.

Problem 17.15 Show that the Maxwell 2-form satisfies the identities

$$\varphi \wedge *\varphi = *\varphi \wedge \varphi = 4(\mathbf{B}^2 - \mathbf{E}^2)\Omega$$

$$\varphi \wedge \varphi = -*\varphi \wedge *\varphi = 8\mathbf{B} \cdot \mathbf{E} \Omega$$

where $\Omega = dx^1 \wedge dx^2 \wedge dx^3 \wedge dx^4$.

References

[1] R. W. R. Darling. *Differential Forms and Connections*. New York, Cambridge University Press, 1994.

[2] S. I. Goldberg. *Curvature and Homology*. New York, Academic Press, 1962.

[3] L. H. Loomis and S. Sternberg. *Advanced Calculus*. Reading, Mass., Addison-Wesley, 1968.

[4] M. Nakahara. *Geometry, Topology and Physics*. Bristol, Adam Hilger, 1990.

[5] C. Nash and S. Sen. *Topology and Geometry for Physicists*. London, Academic Press, 1983.

[6] I. M. Singer and J. A. Thorpe. *Lecture Notes on Elementary Topology and Geometry*. Glenview, Ill., Scott Foresman, 1967.

[7] M. Spivak. *Calculus on Manifolds*. New York, W. A. Benjamin, 1965.

References

[8] W. H. Chen, S. S. Chern and K. S. Lam. *Lectures on Differential Geometry*. Singapore, World Scientific, 1999.

[9] S. Sternberg. *Lectures on Differential Geometry*. Englewood Cliffs, N.J., Prentice-Hall, 1964.

[10] F. W. Warner. *Foundations of Differential Manifolds and Lie Groups*. New York, Springer-Verlag, 1983.

[11] C. de Witt-Morette, Y. Choquet-Bruhat and M. Dillard-Bleick. *Analysis, Manifolds and Physics*. Amsterdam, North-Holland, 1977.

[12] J. Kelley. *General Topology*. New York, D. Van Nostrand Company, 1955.

[13] J. G. Hocking and G. S. Young. *Topology*. Reading, Mass., Addison-Wesley, 1961.

18 Connections and curvature

18.1 Linear connections and geodesics

There is no natural way of comparing tangent vectors Y_p and Y_q at p and q, for if they had identical components in one coordinate system this will not generally be true in a different coordinate chart covering the two points. In a slightly different light, consider the partial derivatives $Y^i{}_{,j} = \partial Y^i / \partial x^j$ of a vector field Y in a coordinate chart $(U; x^i)$. On performing a transformation to coordinates $(U'; x'^{i'})$, we have from Eq. (15.13)

$$Y'^{i'}{}_{,j'} = \frac{\partial Y'^{i'}}{\partial x'^{j'}} = \frac{\partial x^j}{\partial x'^{j'}} \frac{\partial}{\partial x^j}\left(Y^i \frac{\partial x'^{i'}}{\partial x^i}\right),$$

whence

$$Y'^{i'}{}_{,j'} = \frac{\partial x'^{i'}}{\partial x^i} \frac{\partial x^j}{\partial x'^{j'}} Y^i{}_{,j} + Y^i \frac{\partial x^j}{\partial x'^{j'}} \frac{\partial^2 x'^{i'}}{\partial x^i \partial x^j}. \tag{18.1}$$

The first term on the right-hand side has the form of a tensor transformation term, as in Eq. (15.15), but the second term is definitely not tensorial in character. Thus, if Y^i has constant components in the chart $(U; x^i)$ (so that $Y^i{}_{,j} = 0$) this will not be true in the chart $(U'; x'^{i'})$ unless the coordinate transformation functions are linear,

$$x'^{i'} = A^{i'}{}_j x^j \iff \frac{\partial^2 x'^{i'}}{\partial x^i \partial x^j} = 0. \tag{18.2}$$

Suppose we had a well-defined notion of 'directional derivative' of a vector field Y with respect to a tangent vector X_p at p, rather like the concept of directional derivative of a function f with respect to X_p. It would then be possible to define a 'constant' vector field $Y(t)$ along a parametrized curve γ connecting p and q by requiring that the directional derivative with respect to the tangent vector $\dot\gamma$ to the curve be zero for all t in the parameter range. The resulting tangent vector Y_q at q may, however, be dependent on the choice of connecting curve γ. If we write the action of a vector field X on a scalar field f as $D_X f \equiv Xf$, then for any real-valued function g

$$D_{gX} f = gXf = g D_X f. \tag{18.3}$$

This property is essential for D_X to be a *local action*, as the action of D_{gX} at p only depends on the value $g(p)$, not on the behaviour of the function g in an entire neighbourhood of p,

$$(D_{gX} f)(p) = g(p) D_X f(p).$$

18.1 Linear connections and geodesics

We may therefore write $D_{X_p} f = (D_X f)(p)$ without ambiguity, for if $X_p = Y_p$ then $(D_X f)(p) = (D_Y f)(p)$.

Exercise: Show that the last assertion follows from Eq. (18.3).

Extending this idea to vector fields Y, we seek a derivative $D_X Y$ having the property

$$D_{gX} Y = g D_X Y$$

for any function $g : M \to \mathbb{R}$. We will show that the derivative of Y relative to a tangent vector X_p at p can then be defined by setting $D_{X_p} Y = (D_X Y)_p$ – the result will be independent of the choice of vector field X reducing to X_p at p.

The Lie derivative $\mathcal{L}_X Y = [X, Y]$ (see Section 15.5) is not a derivative in the sense required here since, for a general function $g \in \mathcal{F}(M)$,

$$\mathcal{L}_{gX} Y = [gX, Y] = g[X, Y] - Y(g)X \neq g \mathcal{L}_X Y.$$

To calculate the Lie derivative $[X, Y]_p$ at a point p it is not sufficient to know the tangent vector X_p at p – we must know the behaviour of the vector field X in an entire neighbourhood of a point p.

A **connection**, also called a **linear** or **affine connection**, on a differentiable manifold M is a map $D : \mathcal{T}(M) \times \mathcal{T}(M) \to \mathcal{T}(M)$, where $\mathcal{T}(M)$ is the module of differentiable vector fields on M, such that the map $D_X : \mathcal{T}(M) \to \mathcal{T}(M)$ defined by $D_X Y = D(X, Y)$ satisfies the following conditions for arbitrary vector fields X, Y, Z and scalar fields f, g:

(Con1) $D_{X+Y} Z = D_X Z + D_Y Z$,
(Con2) $D_{gX} Y = g D_X Y$,
(Con3) $D_X (Y + Z) = D_X Y + D_X Z$,
(Con4) $D_X (fY) = (Xf)Y + f D_X Y = (D_X f)Y + f D_X Y$.

A linear connection is not inherent in the original manifold structure – it must be imposed as an extra structure on the manifold. Given a linear connection D on a manifold M, for every vector field Y there exists a tensor field DY of type $(1, 1)$ defined by

$$DY(\omega, X) = \langle \omega, D_X Y \rangle \tag{18.4}$$

for every 1-form ω and vector field X. The tensor nature of DY follows from linearity in both arguments. Linearity in ω is trivial, while linearity in X follows immediately from (Con1) and (Con2). The tensor field DY is called the **covariant derivative** of the vector field Y. The theory of connections as described here is called a *Koszul connection* [1–6], while the 'old-fashioned' coordinate version that will be deduced below appears in texts such as [7, 8].

A connection can be restricted to any open submanifold $U \subset M$ in a natural way. For example, if $(U; x^i)$ is a coordinate chart on M and ∂_{x^i} the associated local basis of vector fields, we may set $D_k = D_{\partial_{x^k}}$. Expanding the vector fields $D_k \partial_{x^j}$ in terms of the local basis,

$$D_k \partial_{x^j} = \Gamma^i_{jk} \partial_{x^i} \tag{18.5}$$

where Γ^i_{jk} are real-valued functions on U, known as the **components of the connection** D with respect to the coordinates $\{x^i\}$. Using (Con3) and (Con4) we can compute the covariant derivative of any vector field $Y = Y^i \partial_{x^i}$ on U:

$$D_k Y = D_k(Y^j \partial_{x^j}) = (\partial_{x^k} Y^j) \partial_{x^j} + Y^j \Gamma^i_{jk} \partial_{x^i}$$
$$= Y^i_{;k} \partial_{x^i}$$

where

$$Y^i_{;k} = Y^i_{,k} + \Gamma^i_{jk} Y^j. \qquad (18.6)$$

The coefficients $Y^i_{;k}$ are the components of the covariant derivative with respect to these coordinates since, by Eq. (18.4),

$$(DY)^i_{\ k} = DY(dx^i, \partial_{x^k}) = \langle dx^i, D_k Y \rangle$$
$$= \langle dx^i, Y^j_{;k} \rangle \partial_{x^j}$$
$$= Y^j_{;k} \delta^i_j = Y^i_{;k}.$$

Thus

$$DY = Y^i_{;k} \partial_{x^i} \otimes dx^k$$

and the components of $D_X Y$ with respect to the coordinates x^i are

$$(D_X Y)^i = DY(dx^i, X) = Y^i_{;k} X^k. \qquad (18.7)$$

As anticipated above, it is possible to define the covariant derivative of a vector field Y with respect to a tangent vector X_p at a point p as $D_{X_p} Y = (D_X Y)_p \in T_p(M)$, where X is any vector field that 'reduces' to X_p at p. For this definition to make sense, we must show that it is independent of the choice of vector field X. Suppose X' is a second vector field such that $X'_p = X_p$. The vector field $Z = X - X'$ vanishes at p, and we have

$$(D_X Y)_p - (D_{X'} Y)_p = (D_{X-X'} Y)_p$$
$$= (D_Z Y)_p$$
$$= Y^i_{;k}(p)(Z_p)^k \partial_{x^i} = 0 \quad \text{by Eq. (18.7)}.$$

The **covariant derivative of a vector field** Y **along a curve** $\gamma : \mathbb{R} \to M$ is defined to be

$$\frac{DY}{dt} = D_{\dot{\gamma}} Y$$

where $X = \dot{\gamma}$ is the tangent vector to the curve. By (18.7), the components are

$$\frac{DY^i}{dt} \equiv \left(\frac{DY}{dt}\right)^i = \frac{dx^k}{dt} \left(\frac{\partial Y^i}{\partial x^k} + \Gamma^i_{jk} Y^j\right)$$
$$= \frac{dY^i(t)}{dt} + \Gamma^i_{jk} Y^j \frac{dx^k}{dt}. \qquad (18.8)$$

We will say the vector field Y is **parallel along the curve** γ if $DY(t)/dt = 0$ for all t in the curve's parameter range. A curve will be called a **geodesic** if its tangent vector is everywhere parallel along the curve, $D_{\dot{\gamma}} \dot{\gamma} = 0$ – note that the expression on the right-hand side of

18.1 Linear connections and geodesics

Eq. (18.8) depends only on the values of the components $Y^i(t) \equiv Y^i(\gamma(t))$ along the curve. By (18.8) a geodesic can be written locally as a set of differential equations

$$\frac{D}{dt}\frac{dx^i}{dt} = \frac{d^2x^i}{dt^2} + \Gamma^i_{jk}\frac{dx^j}{dt}\frac{dx^k}{dt} = 0. \tag{18.9}$$

The above discussion can also be reversed. Let p be any point of M and $\gamma : [a, b] \to M$ a curve such that $p = \gamma(a)$. In local coordinates $(U; x^i)$ the equations for a vector field to be parallel along the curve are a linear set of differential equations

$$\frac{dY^i(t)}{dt} + \Gamma^i_{jk}Y^j(t)\frac{dx^k}{dt} = 0. \tag{18.10}$$

By the existence and uniqueness theorem of differential equations, for any tangent vector Y_p at p there exists a unique vector field $Y(t)$ parallel along $\gamma \cap U$ such that $Y(a) = Y_p$. The curve segment is a compact set and can be covered by a finite family of charts, so that existence and uniqueness extends over the entire curve $a \le t \le b$. Furthermore, as the differential equations are linear the map $P_t : T_p(M) \to T_{\gamma(t)}(M)$ such that $P_t(Y_p) = Y(t)$ is a linear map, called **parallel transport** along γ from $p = \gamma(a)$ to $\gamma(t)$. Since the parallel transport map can be reversed by changing the parameter to $t' = -t$, the map P_t is one-to-one and must be a linear isomorphism.

The uniqueness of a maximal solution to a set of differential equations also shows that if p is any point of M there exists a unique maximal geodesic $\sigma : [0, a)$ where $a \le \infty$ starting with any specified tangent vector $\dot\sigma(0) = X_p$ at $p = \sigma(0)$. The parameter t such that a geodesic satisfies Eq. (18.9) is called an **affine parameter**. Under a parameter transformation $t' = f(t)$ the tangent vector becomes

$$\dot\sigma' = \frac{dx^i}{dt'}\partial_{x^i} = \frac{1}{f'(t)}\frac{dx^i}{dt}\partial_{x^i}$$

where $f'(t) = df/dt$ and, using Eq. (18.9), we have

$$\frac{d^2x^i}{dt'^2} + \Gamma^i_{jk}\frac{dx^j}{dt'}\frac{dx^k}{dt'} = \frac{1}{f'(t)}\frac{d}{dt}\left(\frac{1}{f'(t)}\right)\frac{dx^i}{dt}$$

$$= -\frac{f''(t)}{(f'(t))^2}\frac{dx^i}{dt'}.$$

The new parameter t' is an affine parameter if and only if $f''(t) = 0$ – that is, an affine transformation, $t' = at + b$. Herein lies the reason behind the term *affine parameter*.

Coordinate transformations

Consider a coordinate transformation from a chart $(U; x^i)$ to a chart $(U'; x'^{i'})$. In the overlap $U \cap U'$ we have, using the transformations between coordinate bases given by

Connections and curvature

Eq. (15.12),

$$D_{\partial_{x'^{k'}}} \partial_{x'^{j'}} = \frac{\partial x^k}{\partial x'^{k'}} D_k \left(\frac{\partial x^j}{\partial x'^{j'}} \partial_{x^j} \right)$$

$$= \frac{\partial x^k}{\partial x'^{k'}} \frac{\partial x^j}{\partial x'^{j'}} \Gamma^i_{jk} \frac{\partial x'^{i'}}{\partial x^i} \partial_{x'^{i'}} + \frac{\partial^2 x^j}{\partial x'^{k'} \partial x'^{j'}} \partial_{x^j}$$

$$= \Gamma'^{i'}_{j'k'} \partial_{x'^{i'}}$$

where

$$\Gamma'^{i'}_{j'k'} = \frac{\partial x^k}{\partial x'^{k'}} \frac{\partial x^j}{\partial x'^{j'}} \frac{\partial x'^{i'}}{\partial x^i} \Gamma^i_{jk} + \frac{\partial^2 x^i}{\partial x'^{k'} \partial x'^{j'}} \frac{\partial x'^{i'}}{\partial x^i}. \qquad (18.11)$$

This is the law of transformation of components of a connection.

The first term on the right-hand side of (18.11) is tensorial in nature, but the second term adds a complication that only vanishes for linear transformations. It is precisely the expression needed to counteract the non-tensorial part of the transformation of the derivative of a vector field given in Eq. (18.1) – see Problem 18.1.

Problems

Problem 18.1 Show directly from the transformation laws (18.1) and (18.11) that the components of the covariant derivative (18.6) of a vector field transform as a tensor of type (1, 1).

Problem 18.2 Show that the transformation law (18.11) can be written in the form

$$\Gamma'^{i'}_{j'k'} = \frac{\partial x^k}{\partial x'^{k'}} \frac{\partial x^j}{\partial x'^{j'}} \frac{\partial x'^{i'}}{\partial x^i} \Gamma^i_{jk} - \frac{\partial^2 x'^{k'}}{\partial x^i \partial x^j} \frac{\partial x^i}{\partial x'^{i'}} \frac{\partial x^j}{\partial x'^{j'}}.$$

18.2 Covariant derivative of tensor fields

For every vector field X we define a map $D_X : T^{(r,s)}(M) \to T^{(r,s)}(M)$ that extends the covariant derivative to general tensor fields by requiring:

(Cov1) For scalar fields, $f \in \mathcal{F}(M) = T^{(0,0)}(M)$, we set $D_X f = X f$.
(Cov2) For 1-forms $\omega \in T^{(0,1)}(M)$ assume a Leibnitz rule for $\langle \, , \, \rangle$,

$$D_X \langle \omega, Y \rangle = \langle D_X \omega, Y \rangle + \langle \omega, D_X Y \rangle.$$

(Cov3) $D_X(T + S) = D_X T + D_X S$ for any pair of tensor fields $T, S \in T^{(r,s)}(M)$.
(Cov4) The Leibnitz rule holds with respect to tensor products

$$D_X(T \otimes S) = (D_X T) \otimes S + T \otimes D_X S.$$

These requirements define a unique tensor field $D_X T$ for any smooth tensor field T. Firstly, let $\omega = w_i dx^i$ be any 1-form defined on a coordinate chart $(U; x^i)$ covering p. Setting $Y = \partial_{x^i}$, condition (Cov1) gives $D_X \langle \omega, \partial_{x^i} \rangle = D_X(w_i) = X^k w_{i,k}$, while (Cov2) implies

$$D_X \langle \omega, \partial_{x^i} \rangle = \langle D_X \omega, \partial_{x^i} \rangle + \langle \omega, X^k D_k \partial_{x^i} \rangle.$$

18.2 Covariant derivative of tensor fields

Hence, using (18.5) and $(D_X\omega)_i = \langle D_X\omega, \partial_{x^i}\rangle$, we find

$$(D_X\omega)_i = w_{i;k}X^k \tag{18.12}$$

where

$$w_{i;k} = w_{i,k} - \Gamma^j_{ik}w_j. \tag{18.13}$$

Exercise: Verify that Eq. (18.13) implies the coordinate expression for condition (Cov2),

$$(w_i Y^i)_{,k} X^k = w_{i;k} X^k Y^i + w_i Y^i_{;k} X^k.$$

For a general tensor T, expand in terms of the basis consisting of tensor products of the ∂_{x^i} and dx^j and use (Cov3) and (Cov4). For example, if

$$T = T^{ij\cdots}{}_{kl\cdots}\,\partial_{x^i} \otimes \partial_{x^j} \otimes \cdots \otimes dx^k \otimes dx^l \otimes \cdots$$

a straightforward calculation results in

$$D_X T = T^{ij\cdots}{}_{kl\cdots;p}\, X^p\, \partial_{x^i} \otimes \partial_{x^j} \otimes \cdots \otimes dx^k \otimes dx^l \otimes \cdots$$

where

$$T^{ij\cdots}{}_{kl\cdots;p} = T^{ij\cdots}{}_{kl\cdots,p} + \Gamma^i_{ap} T^{aj\cdots}{}_{kl\cdots} + \Gamma^j_{ap} T^{ia\cdots}{}_{kl\cdots} + \cdots$$
$$- \Gamma^a_{kp} T^{ij\cdots}{}_{al\cdots} - \Gamma^a_{lp} T^{ij\cdots}{}_{ka\cdots} - \cdots \tag{18.14}$$

This demonstrates that (Cov1)–(Cov4) can be used to compute the components of $D_X T$ at any point $p \in M$ with respect to a coordinate chart $(U; x^i)$ covering p, and thus uniquely define the tensor field DXT throughout M.

For every tensor field T of type (r, s) its **covariant derivative** DT is the tensor field of type $(r, s + 1)$ defined by

$$DT(\omega^1, \omega^2, \ldots, \omega^r, Y_1, Y_2, \ldots, Y_s, X) = D_X T(\omega^1, \omega^2, \ldots, \omega^r, Y_1, Y_2, \ldots, Y_s).$$

Exercise: Show that, with respect to any local coordinates, the tensor field DT has components $T^{ij\cdots}{}_{kl\cdots;p}$ defined by Eq. (18.14).

Exercise: Show that

$$\delta^i{}_{j;k} = 0. \tag{18.15}$$

The covariant derivative commutes with all contractions on a tensor field

$$D(C^k_l T) = C^k_l DT. \tag{18.16}$$

This relation is most easily shown in local coordinates, where it reads

$$\left(T^{i_1\ldots a\ldots i_r}{}_{j_1\ldots a\ldots j_s}\right)_{;k} = T^{i_1\ldots a\ldots i_r}{}_{j_1\ldots a\ldots j_s;k}, \tag{18.17}$$

the upper index a being in the kth position, the lower in the lth. If we expand the right-hand side according to Eq. (18.14) the terms corresponding to these indices are

$$\Gamma^a_{bk} T^{i_1\ldots b\ldots i_r}{}_{j_1\ldots a\ldots j_s} - \Gamma^b_{ak} T^{i_1\ldots a\ldots i_r}{}_{j_1\ldots b\ldots j_s} = 0$$

Connections and curvature

and what remains reduces to the expression formed by expanding the left-hand side of (18.17).

A useful corollary of this property is the following relation:

$$X(T(\omega^1,\ldots,\omega^r,Y_1,\ldots,Y_s)) = D_X T(\omega^1,\ldots,\omega^r,Y_1,\ldots,Y_s)$$
$$+ T(D_X\omega^1,\ldots,\omega^r,Y_1,\ldots,Y_s) + \cdots + T(\omega^1,\ldots,D_X\omega^r,Y_1,\ldots,Y_s)$$
$$+ T(\omega^1,\ldots,\omega^r,D_X Y_1,\ldots,Y_s) + \cdots + T(\omega^1,\ldots,\omega^r,Y_1,\ldots,D_X Y_s). \quad (18.18)$$

Problems

Problem 18.3 Show directly from (Cov1)–(Cov4) that $D_{fX}T = f D_X T$ for all vector fields X, tensor fields T and scalar functions $f : M \to \mathbb{R}$.

Problem 18.4 Verify from the coordinate transformation rule (18.11) for Γ^i_{jk} that the components of the covariant derivative of an arbitrary tensor field, defined in Eq. (18.14), transform as components of a tensor field.

Problem 18.5 Show that the identity (18.18) follows from Eq. (18.16).

18.3 Curvature and torsion

Torsion tensor

In the transformation law of the components Γ^i_{jk}, Eq. (18.11), the term involving second derivatives of the transformation functions is symmetric in the indices jk. It follows that the antisymmetrized quantity $T^i_{jk} = \Gamma^i_{kj} - \Gamma^i_{jk}$ does transform as a tensor, since the non-tensorial parts of the transformation law (18.11) cancel out.

To express this idea in an invariant non-coordinate way, observe that for any vector field Y the tensor field DY defined in Eq. (18.4) satisfies the identities

$$DY(f\omega, X) = f DY(\omega, X), \qquad DY(\omega, fX) = f DY(\omega, X)$$

for all functions $f \in \mathcal{F}(M)$. These are called $\mathcal{F}$-**linearity** in the respective arguments. On the other hand, the map $D' : T^*(M) \times T(M) \times T(M) \to \mathcal{F}(M)$ defined by

$$D'(\omega, X, Y) = \langle \omega, D_X Y \rangle$$

is not a tensor field of type (1, 2), since $\mathcal{F}$-linearity fails for the third argument by (Con4),

$$D'(\omega, X, fY) = \langle \omega, D_X(fY) \rangle = \langle \omega, (Xf)Y + f D_X Y \rangle$$
$$= (Xf)\langle \omega, Y \rangle + f D'(\omega, X, Y) \neq f D'(\omega, X, Y).$$

Exercise: Show that the 'components' of D' are $D'(dx^i, \partial_{x^j}, \partial_{x^k}) = \Gamma^i_{jk}$.

Now let the *torsion map* $\tau : T(M) \times T(M) \to T(M)$ be defined by

$$\tau(X, Y) = D_X Y - D_Y X - [X, Y] = -\tau(Y, X). \quad (18.19)$$

18.3 Curvature and torsion

This map is $\mathcal{F}$-linear in the first argument,

$$\begin{aligned}\tau(fX, Y) &= D_{fX}Y - D_Y(fX) - [fX, Y]\\ &= fD_XY - (Yf)X - fD_YX - f[X, Y] + (Yf)X\\ &= f\tau(X, Y),\end{aligned}$$

and by antisymmetry it is also $\mathcal{F}$-linear in the second argument Y. Hence τ gives rise to a tensor field T of type $(1, 2)$ by

$$T(\omega, X, Y) = \langle \omega, \tau(X, Y)\rangle, \tag{18.20}$$

known as the **torsion tensor** of the connection D. In a local coordinate chart $(U; x^i)$ its components are precisely the antisymmetrized connection components:

$$T^i_{jk} = \Gamma^i_{kj} - \Gamma^i_{jk}. \tag{18.21}$$

The proof follows from setting $T^i_{jk} = \langle dx^i, \tau(\partial_{x^j}, \partial_{x^k})\rangle$ and sustituting Eq. (18.19). We call a connection **torsion-free** or **symmetric** if its torsion tensor vanishes, $T = 0$; equivalently, its components are symmetric with respect to all coordinates, $\Gamma^i_{jk} = \Gamma^i_{kj}$.

Exercise: Prove Eq. (18.21).

Curvature tensor

A similar problem occurs when commuting repeated covariant derivatives on a vector or tensor field. If X, Y and Z are any vector fields on M then $D_X D_Y Z - D_Y D_X Z$ is obviously a vector field, but the map $P : T^*(M) \times T(M) \times T(M) \times T(M) \to \mathcal{F}(M)$ defined by $P(\omega, X, Y, Z) = \langle \omega, D_X D_Y Z - D_Y D_X Z\rangle$ fails to be a tensor field of type $(1, 3)$ as it is not $\mathcal{F}$-linear in the three vector field arguments. The remedy is similar to that for creating the torsion tensor.

For any pair of vector fields X and Y define the operator $\rho_{X,Y} : T(M) \to T(M)$ by

$$\rho_{X,Y}Z = D_X D_Y Z - D_Y D_X Z - D_{[X,Y]}Z = -\rho_{Y,X}Z. \tag{18.22}$$

This operator is $\mathcal{F}$-linear with respect to X, and therefore Y, since

$$\begin{aligned}\rho_{fX,Y}Z &= D_{fX}D_Y Z - D_Y D_{fX} Z - D_{[fX,Y]}Z\\ &= fD_X D_Y Z - (Yf)D_X Z - fD_Y D_X Z - fD_{[X,Y]}Z + (Yf)D_X Z\\ &= f\rho_{X,Y}Z.\end{aligned}$$

$\mathcal{F}$-linearity with respect to Z follows from

$$\begin{aligned}\rho_{X,Y}fZ &= D_X D_Y(fZ) - D_Y D_X(fZ) - D_{[X,Y]}(fZ)\\ &= D_X\big((Yf)Z + fD_Y Z\big) - D_Y\big((Xf)Z + fD_X Z\big) - ([X, Y]f)Z - fD_{[X,Y]}Z\\ &= (Yf)D_X Z + X(Yf)Z + (Xf)D_Y Z + fD_X D_Y Z - Y(Xf)Z - (Xf)D_Y Z\\ &\quad - (Yf)D_X Z - fD_Y D_X Z - ([X, Y]f)Z - fD_{[X,Y]}Z\\ &= f\rho_{X,Y}Z.\end{aligned}$$

We can therefore define a tensor field R of type $(1, 3)$, by setting

$$R(\omega, Z, X, Y) = \langle \omega, \rho_{X,Y} Z \rangle = -R(\omega, Z, Y, X), \qquad (18.23)$$

called the **curvature tensor** of the connection D.

For a torsion-free connection, $D_X Y - D_Y X = [X, Y]$, there is a cyclic identity:

$$\begin{aligned}
\rho_{X,Y} Z + \rho_{Y,Z} X + \rho_{Z,X} Y &= D_X D_Y Z - D_Y D_X Z - D_{[X,Y]} Z \\
&\quad + D_Y D_Z X - D_Z D_Y X - D_{[Y,Z]} X + D_Z D_X Y \\
&\quad - D_X D_Z Y - D_{[Z,X]} Y \\
&= D_X [Y, Z] + D_Y [Z, X] + D_Z [X, Y] - D_{[Y,Z]} X \\
&\quad - D_{[Z,X]} Y - D_{[X,Y]} Z \\
&= [X, [Y, Z]] + [Y, [Z, X]] + [Z, [X, Y]] = 0
\end{aligned}$$

using $\tau(X, [Y, Z]) = 0$, etc. and the Jacobi identity (15.24). For $T = 0$ we thus have the so-called **first Bianchi identity**,

$$R(\omega, Z, X, Y) + R(\omega, X, Y, Z) + R(\omega, Y, Z, X) = 0. \qquad (18.24)$$

In a coordinate system $(U; x^i)$, using Eq. (18.5) and $[\partial_{x^k}, \partial_{x^l}] = 0$ for all k, l, the components of the curvature tensor are

$$\begin{aligned}
R^i{}_{jkl} &= R(dx^i, \partial_{x^j}, \partial_{x^k}, \partial_{x^l}) \\
&= \langle dx^i, D_k D_l \partial_{x^j} - D_l D_k \partial_{x^j} - D_{[\partial_{x^k}, \partial_{x^l}]} \partial_{x^j} \rangle \\
&= \langle dx^i, D_k(\Gamma^m_{jl} \partial_{x^m}) - D_l(\Gamma^m_{jk} \partial_{x^m}) \rangle \\
&= \langle dx^i, \Gamma^m_{jl,k} \partial_{x^m} + \Gamma^m_{jl} \Gamma^p_{mk} \partial_{x^p} - \Gamma^m_{jk,l} \partial_{x^m} - \Gamma^m_{jk} \Gamma^p_{ml} \partial_{x^p} \rangle
\end{aligned}$$

where $\Gamma^m_{jk,l} = \partial \Gamma^k_{jk}/\partial x^l$. Hence

$$R^i{}_{jkl} = \Gamma^i_{jl,k} - \Gamma^i_{jk,l} + \Gamma^m_{jl} \Gamma^i_{mk} - \Gamma^m_{jk} \Gamma^i_{ml} = -R^i{}_{jlk}. \qquad (18.25)$$

Setting $\omega = dx^i$, $X = \partial_{x^k}$, etc. in the first Bianchi identity (18.24) gives

$$R^i{}_{jkl} + R^i{}_{klj} + R^i{}_{ljk} = 0. \qquad (18.26)$$

Another class of identities, known as **Ricci identities**, are sometimes used to define the torsion and curvature tensors in a coordinate region $(U; x^i)$. For any smooth function f on U, set $f_{;ij} \equiv (f_{;i})_{;j} = (f_{,i})_{;j}$. Then, by Eq. (18.13),

$$f_{;ij} - f_{;ji} = f_{,ij} - \Gamma^k_{ij} f_{,k} - f_{,ji} + \Gamma^k_{ji} f_{,k} = T^k_{ij} f_{,k}. \qquad (18.27)$$

Similarly, for a smooth vector field $X = X^k \partial_{x^k}$, Eq. (18.25) gives rise to

$$X^k{}_{;ij} - X^k{}_{;ji} = X^a R^k{}_{aji} + T^a_{ij} X^k{}_{;a}, \qquad (18.28)$$

and for a 1-form $\omega = w_i \, dx^i$,

$$w_{k;ij} - w_{k;ji} = w_a R^a{}_{kij} + T^a_{ij} w_{k;a}. \qquad (18.29)$$

18.3 Curvature and torsion

Problems

Problem 18.6 Let f be a smooth function, $X = X^i \partial_{x^i}$ a smooth vector field and $\omega = w_i\, dx^i$ a differential 1-form. Show that

$$(D_j D_i - D_i D_j)f = 0,$$
$$(D_j D_i - D_i D_j)X = X^a R^k{}_{aji} \partial_{x^k},$$
$$(D_j D_i - D_i D_j)\omega = w_a R^a{}_{kji}\, dx^k.$$

Why does the torsion tensor not appear in these formulae, in contrast with the Ricci identities (18.27)–(18.29)?

Problem 18.7 Show that the coordinate expression for the Lie derivative of a vector field may be written

$$(\mathcal{L}_X Y)^i = [X, Y]^i = Y^i{}_{;j} X^j - X^i{}_{;j} Y^j + T^i_{jk} X^k Y^j. \qquad (18.30)$$

For a torsion-free connection show that the Lie derivative (15.39) of a general tensor field S of type (r, s) may be expressed by

$$(\mathcal{L}_X S)^{ij\ldots}{}_{kl\ldots} = S^{ij\ldots}{}_{kl\ldots;m} X^m - S^{mj\ldots}{}_{kl\ldots} X^i{}_{;m} - S^{im\ldots}{}_{kl\ldots} X^j{}_{;m} - \ldots$$
$$+ S^{ij\ldots}{}_{ml\ldots} X^m{}_{;k} + S^{ij\ldots}{}_{km\ldots} X^m{}_{;l} + \ldots \qquad (18.31)$$

Write down the full version of this equation for a general connection with torsion.

Problem 18.8 Prove the Ricci identities (18.28) and (18.29).

Problem 18.9 For a torsion-free connection prove the generalized Ricci identities

$$S^{kl\ldots}{}_{mn\ldots;ij} - S^{kl\ldots}{}_{mn\ldots;ji} = S^{al\ldots}{}_{mn\ldots} R^k{}_{aji} + S^{ka\ldots}{}_{mn\ldots} R^l{}_{aji} + \ldots$$
$$+ S^{kl\ldots}{}_{an\ldots} R^a{}_{mij} + S^{kl\ldots}{}_{ma\ldots} R^a{}_{nij} + \ldots$$

How is this equation modified in the case of torsion?

Problem 18.10 For arbitrary vector fields Y, Z and W show that the operator $\Sigma_{Y,Z,W} : T(M) \to T(M)$ defined by

$$\Sigma_{Y,Z,W} X = D_W(\rho_{Y,Z} X) - \rho_{Z,[Y,W]} X - \rho_{Y,Z}(D_W X)$$

has the cyclic symmetry

$$\Sigma_{Y,Z,W} X + \Sigma_{Z,W,Y} X + \Sigma_{W,Y,Z} X = 0.$$

Express this equation in components with respect to a local coordinate chart and show that it is equivalent to the **(second) Bianchi identity**

$$R^i{}_{jkl;m} + R^i{}_{jlm;k} + R^i{}_{jmk;l} = R^i{}_{jpk} T^p_{ml} + R^i{}_{jpl} T^p_{km} + R^i{}_{jpm} T^p_{lk}. \qquad (18.32)$$

Problem 18.11 Let $Y^i(t)$ be a vector that is parallel propagated along a curve having coordinate representation $x^j = \overset{0}{x}{}^j + A^j t$. Show that for $t \ll 1$

$$Y^i(t) = \overset{0}{Y}{}^i - \overset{0}{\Gamma}{}^i_{ja} \overset{0}{Y}{}^j A^a t + \frac{t^2}{2}(\overset{0}{\Gamma}{}^i_{ka}\overset{0}{\Gamma}{}^k_{ja} - \overset{0}{\Gamma}{}^i_{ja,b}) A^a A^b \overset{0}{Y}{}^j + O(t^3)$$

where $\overset{0}{\Gamma}{}^i_{jk} = \Gamma^i_{jk}(\overset{0}{x}{}^a)$ and $\overset{0}{Y}{}^i = Y^i(0)$. From the point P, having coordinates $\overset{0}{x}{}^i$, parallel transport the tangent vector $\overset{0}{Y}{}^i$ around a coordinate rectangle $PQRSP$ whose sides are each of parameter

Connections and curvature

length t and are along the a- and b-axes successively through these points. For example, the a-axis through P is the curve $x^j = \overset{0}{x}{}^j + \delta^j_a t$. Show that to order t^2, the final vector at P has components

$$Y^i = \overset{0}{Y}{}^i + t^2 \overset{0}{R}{}^i{}_{jba} \overset{0}{Y}{}^j$$

where $\overset{0}{R}{}^i{}_{jba}$ are the curvature tensor components at P.

18.4 Pseudo-Riemannian manifolds

A tensor field g of type $(0, 2)$ on a manifold M is said to be **non-singular** if $g_p \in T^{(0,2)}$ is a non-singular tensor at every point $p \in M$. A **pseudo-Riemannian manifold** (M, g) consists of a differentiable manifold M together with a symmetric non-singular tensor field g of type $(0, 2)$, called a **metric tensor**. This is equivalent to defining an inner product $X_p \cdot Y_p = g_p(X_p, Y_p)$ on the tangent space $T_p(M)$ at every point $p \in M$ (see Chapters 5 and 7). We will assume g is a differentiable tensor field, so that for every pair of smooth vector fields $X, Y \in \mathcal{T}(M)$ the inner product is a differentiable function,

$$g(X, Y) = g(Y, X) \in \mathcal{F}(M).$$

In any coordinate chart $(U; x^i)$ we can write

$$g = g_{ij}\, dx^i \otimes dx^j \quad \text{where} \quad g_{ij} = g_{ji} = g(\partial_{x^i}, \partial_{x^j}),$$

and $\mathbf{G} = [g_{ij}]$ is a non-singular matrix at each point $p \in M$. As in Example 7.7 there exists a smooth **inverse metric tensor** g^{-1} on M, a symmetric tensor field of type $(2, 0)$, such that in any coordinate chart $(U; x^i)$

$$g^{-1} = g^{ij} \frac{\partial}{\partial x^i} \otimes \frac{\partial}{\partial x^j} \quad \text{where} \quad g^{ik} g_{kj} = \delta^i{}_j.$$

It is always possible to find a set of orthonormal vector fields $e_1, \ldots, e_n$ on a neighbourhood of any given point p, spanning the tangent space at each point of the neighbourhood, such that

$$g(e_i, e_j) = \eta_{ij} = \begin{cases} \eta_i & \text{if } i = j \\ 0 & \text{if } i \neq j \end{cases}$$

where $\eta_i = \pm 1$. At p one can set up coordinates such that $e_i(p) = (\partial_{x^i})_p$, so that

$$g_p = \eta_{ij}(dx^i)_p \otimes (dx^j)_p,$$

but in general it is not possible to achieve that $e_i = \partial_{x^i}$ over an entire coordinate chart unless all Lie brackets of the orthonormal fields vanish, $[e_i, e_j] = 0$. We say (M, g) is a **Riemannian manifold** if the metric tensor is everywhere positive definite,

$$g_p(X_p, X_p) > 0 \quad \text{for all } X_p \neq 0 \in T_p(M),$$

or equivalently, $g(X, X) \geq 0$ for all vector fields $X \in \mathcal{T}(M)$. In this case all $\eta_i = 1$ in the above expansion. The word *Riemannian* is also applied to the negative definite case,

18.4 Pseudo-Riemannian manifolds

all $\eta_i = -1$. If the inner product defined by g_p on every tangent space is Minkowskian, as defined in Section 5.1, we say (M, g) is a **Minkowskian** or **hyperbolic** manifold. In this case there exists a local orthonormal set of vector fields e_i such that the associated coefficients are $\eta_1 = \epsilon, \eta_2 = \cdots = \eta_n = -\epsilon$ where $\epsilon = \pm 1$.

If $\gamma : [a, b] \to M$ is a parametrized curve on a Riemannian manifold, its **length** between t_0 and t is defined to be

$$s = \int_{t_0}^{t} \sqrt{g(\dot{\gamma}(u), \dot{\gamma}(u))}\, du. \tag{18.33}$$

If the curve is contained in a coordinate chart $(U; x^i)$ and is written $x^i = x^i(t)$ we have

$$s = \int_{t_0}^{t} \sqrt{g_{ij} \frac{dx^i}{du} \frac{dx^j}{u}}\, du.$$

Exercise: Verify that the length of the curve is independent of parametrization; i.e., s is unaltered under a change of parameter $u' = f(u)$ in the integral on the right-hand side of (18.33).

Let the value of t_0 in (18.33) be fixed. Then $ds/dt = \sqrt{g(\dot{\gamma}(t), \dot{\gamma}(t))}$, and

$$\left(\frac{ds}{dt}\right)^2 = g_{ij} \frac{dx^i}{dt} \frac{dx^j}{dt}. \tag{18.34}$$

If the parameter along the curve is set to be the distance parameter s, the tangent vector is a unit vector along the curve,

$$g(\dot{\gamma}(s), \dot{\gamma}(s)) = g_{ij} \frac{dx^i}{ds} \frac{dx^j}{ds} = 1. \tag{18.35}$$

Sometimes Eq. (18.34) is written symbolically in the form

$$ds^2 = g_{ij}\, dx^i\, dx^j, \tag{18.36}$$

commonly called the *metric* of the space. It is to be thought of as a symbolic expression for displaying the components of the metric tensor and replaces the more correct $g = g_{ij}\, dx^i \otimes dx^j$. This may be done even in the case of an indefinite metric where, strictly speaking, we can have $ds^2 < 0$.

The Riemannian space R^n with metric

$$ds^2 = (dx^1)^2 + (dx^2)^2 + \cdots + (dx^n)^2$$

is called **Euclidean space** and is denoted by the symbol $\mathbb{E}^n$. Of course other coordinates such as polar coordinates may be used, but when we use the symbol $\mathbb{E}^n$ we shall usually assume that the rectilinear system is being adopted unless otherwise specified.

Example 18.1 If $(M, \varphi : M \to \mathbb{E}^n)$ is any submanifold of $\mathbb{E}^n$, it has a naturally induced metric tensor

$$g = \varphi^*(dx^1 \otimes dx^1 + dx^2 \otimes dx^2 + \cdots + dx^n \otimes dx^n).$$

Let M be the 2-sphere of radius a, $x^2 + y^2 + z^2 = a^2$, and adopt polar coordinates

$$x = a \sin\theta \cos\phi, \qquad y = a \sin\theta \sin\phi, \qquad z = a \cos\theta.$$

Connections and curvature

It is straightforward to evaluate the induced metric tensor on S^2,

$$g = \varphi^*(dx \otimes dx + dy \otimes dy + dz \otimes dz)$$
$$= (\cos\theta \cos\phi \, d\theta - \sin\theta \sin\phi \, d\phi) \otimes (\cos\theta \cos\phi \, d\theta - \sin\theta \sin\phi \, d\phi) + \cdots$$
$$= a^2(d\theta \otimes d\theta + \sin^2\phi \, d\phi \otimes d\phi).$$

Alternatively, let $\theta = \theta(t)$, $\phi = \phi(t)$ be any curve lying in M. The components of its tangent vector in $\mathbb{E}^3$ are

$$\frac{dx}{dt} = \cos\theta(t)\cos\phi(t)\frac{d\theta}{dt} - \sin\theta(t)\sin\phi(t)\frac{d\phi}{dt}, \text{ etc.}$$

and the length of the curve in $\mathbb{E}^3$ is

$$\left(\frac{ds}{dt}\right)^2 = \left(\frac{dx}{dt}\right)^2 + \left(\frac{dy}{dt}\right)^2 + \left(\frac{dz}{dt}\right)^2 = a^2\left(\left(\frac{d\theta}{dt}\right)^2 + \sin^2\theta\left(\frac{d\phi}{dt}\right)^2\right).$$

The metric induced on M from $\mathbb{E}^3$ may thus be written

$$ds^2 = a^2 \, d\theta^2 + a^2 \cos^2\theta \, d\phi^2.$$

Riemannian connection

A pseudo-Riemannian manifold (M, g) has a natural connection D defined on it that is subject to the following two requirements:

(i) D is torsion-free.
(ii) The covariant derivative of the metric tensor field vanishes, $Dg = 0$.

This connection is called the **Riemannian connection** defined by the metric tensor g. An interesting example of a physical theory that does not impose condition (i) is the *Einstein–Cartan theory* where torsion represents spin [9]. Condition (ii) has the following consequence. Let γ be a curve with tangent vector $X(t) = \dot\gamma(t)$, and let Y and Z be vector fields parallel transported along γ, so that $D_X Y = D_X Z = 0$. By Eq. (18.18) it follows that their inner product $g(Y, Z)$ is constant along the curve:

$$\frac{d}{dt}g(Y, Z) = \frac{D}{dt}g(Y, Z) = D_X(g(Y, Z))$$
$$= (D_X g)(Y, Z) + g(D_X Y, Z) + g(Y, D_X Z)$$
$$= Dg(Y, Z, X) = 0.$$

In particular every vector field Y parallel transported along γ has constant magnitude $g(Y, Y)$ along the curve, a condition that is in fact necessary and sufficient for condition (ii) to hold.

Exercise: Prove the last statement.

Conditions (i) and (ii) define a unique connection, for let $(U; x^i)$ be any local coordinate chart, and $\Gamma^i_{jk} = \Gamma^i_{kj}$ the components of the connection with respect to this chart. Condition (ii) can be written using Eq. (18.14)

$$g_{ij;k} = g_{ij,k} - \Gamma^m_{ik}g_{mj} - \Gamma^m_{jk}g_{im} = 0. \tag{18.37}$$

18.4 Pseudo-Riemannian manifolds

Interchanging pairs of indices i, k and j, k results in

$$g_{kj;i} = g_{kj,i} - \Gamma^m_{ki} g_{mj} - \Gamma^m_{ji} g_{km} = 0, \tag{18.38}$$

$$g_{ik;j} = g_{ik,j} - \Gamma^m_{ij} g_{mk} - \Gamma^m_{kj} g_{im} = 0. \tag{18.39}$$

The combination $(18.37) + (18.38) - (18.39)$ gives, on using the symmetry of g_{ij} and Γ^m_{ij},

$$g_{ij,k} + g_{kj,i} - g_{ik,j} = 2 g_{mj} \Gamma^m_{ik}.$$

Multiply through by g^{jl} and, after a change of indices, we have

$$\Gamma^i_{jk} = \tfrac{1}{2} g^{im} \left(g_{mj,k} + g_{mk,j} - g_{jk,m} \right). \tag{18.40}$$

These expressions are called **Christoffel symbols**; they are the explicit expression for the components of the Riemannian connection in any coordinate system.

Exercise: Show that

$$g^{ij}{}_{;k} = 0. \tag{18.41}$$

Let $\gamma : \mathbb{R} \to M$ be a geodesic with affine parameter t. As the tangent vector $X(t) = \dot{\gamma}$ is parallel propagated along the curve,

$$\frac{DX}{dt} = D_X X = 0,$$

it has constant magnitude,

$$g(X, X) = g_{ij} \frac{dx^i}{dt} \frac{dx^j}{dt} = \text{const.}$$

A scaling transformation can be applied to the affine parameter such that

$$g(X, X) = g_{ij} \frac{dx^i}{dt} \frac{dx^j}{dt} = \pm 1 \text{ or } 0.$$

In Minkowskian manifolds, the latter case is called a **null geodesic**. If (M, g) is a Riemannian space and $p = \gamma(0)$ then $g(X, X) = 1$ and the affine parameter t is identical with the distance parameter along the geodesic.

Exercise: Show directly from the geodesic equation (18.9) and the Christoffel symbols (18.40) that

$$\frac{d}{dt} \left(g_{ij} \frac{dx^i}{dt} \frac{dx^j}{dt} \right) = 0.$$

Example 18.2 In a pseudo-Riemannian manifold the geodesic equations may be derived from a variation principle (see Section 16.5). Geodesics can be thought of as curves of *stationary length*,

$$\delta s = \delta \int_{t_1}^{t_2} \sqrt{|g(\dot{\gamma}(t), \dot{\gamma}(t))|} \, dt = 0.$$

Let $\gamma : [t_1, t_2] \times [-a, a] \to M$ be a variation of the given curve $\gamma : [t_1, t_2] \to M$, such that $\gamma(t, 0) = \gamma(t)$ and the end points of all members of the variation are fixed, $\gamma(t_1, \lambda) =$

Connections and curvature

$\gamma(t_1), \gamma(t_1, \lambda) = \gamma(t_1)$ for all $\lambda \in [-a, a]$. Set the Lagrangian $L : TM \to \mathbb{R}$ to be $L(X_p) = \sqrt{|g(X_p, X_p)|}$, and we follow the argument leading to the Euler–Lagrange equations (16.25):

$$0 = \delta s = \int_{t_1}^{t_2} \delta L(\dot{\gamma}(t))\, dt$$

$$= \int_{t_1}^{t_2} \frac{d}{d\lambda} \sqrt{|g(\dot{\gamma}(t, \lambda), \dot{\gamma}(t, \lambda))|}\bigg|_{\lambda=0} dt$$

$$= \pm \int_{t_1}^{t_2} \frac{1}{2L} \left(\delta g_{ij} \dot{x}^i \dot{x}^j + 2 g_{ij} \dot{x}^i \delta \dot{x}^j\right) dt$$

$$= \pm \int_{t_1}^{t_2} \left\{ \frac{1}{2L} \delta g_{ij,k} \dot{x}^i \dot{x}^j \delta x^k - \frac{d}{dt}\left(\frac{g_{ik}\dot{x}^i}{L}\right) \delta x^k \right\} dt + \left[\frac{g_{ij}\dot{x}^i}{L} \delta \dot{x}^j\right]_{t_1}^{t_2}$$

$$= \pm \int_{t_1}^{t_2} \left\{ \frac{1}{2L} \delta g_{ij,k} \dot{x}^i \dot{x}^j - \frac{d}{dt}\left(\frac{g_{ik}\dot{x}^i}{L}\right) \right\} \delta x^k\, dt$$

since $\delta x^k = 0$ at the end points $t = t_1$ and $t = t_2$. Since δx^k is arbitrary,

$$\frac{1}{2L} \delta g_{ij,k} \dot{x}^i \dot{x}^j - \frac{d}{dt}\left(\frac{g_{ik}\dot{x}^i}{L}\right) = 0$$

and expanding the second term on the left and multiplying the resulting equation by Lg^{km}, we find

$$\frac{d^2 x^m}{dt^2} + \Gamma^m_{ij} \frac{dx^i}{dt} \frac{dx^j}{dt} = \frac{1}{L} \frac{dL}{dt} \frac{dx^m}{dt}, \quad (18.42)$$

where Γ^m_{ij} are the Christoffel symbols given by Eq. (18.40). If we set t to be the distance parameter $t = s$, then $L = 1$ so that $dL/ds = 0$ and Eq. (18.42) reduces to the standard geodesic equation with affine parameter (18.9).

While we might think of this as telling us that geodesics are curves of 'shortest distance' connecting any pair of points, this is by no means true in general. More usually there is a critical point along any geodesic emanating from a given point, past which the geodesic is 'point of inflection' with respect to distance along neighbouring curves. In pseudo-Riemannian manifolds some geodesics may even be curves of 'longest length'. For timelike geodesics in Minkowski space this is essentially the time dilatation effect – a clock carried on an arbitrary path between two events will indicate less elapsed time than an inertial clock between the two events.

Geodesic coordinates

In cartesian coordinates for Euclidean space we have $g_{ij} = \delta_{ij}$ and by Eq. (18.40) all components of the Riemannian connection vanish, $\Gamma^i_{jk} = 0$. It therefore follows from Eq. (18.25) that all components of the curvature tensor R vanish. Conversely, if all components of the connection vanish in a coordinate chart $(U; x^i)$, we have $g_{ij,k} = 0$ by Eq. (18.37) and the metric tensor components g_{ij} are constant through the coordinate region U.

In Section 18.7 we will show that a necessary and sufficent condition for $\Gamma^i_{jk} = 0$ in a coordinate chart $(V; y^i)$ is that the curvature tensor vanish throughout an open region of

18.4 Pseudo-Riemannian manifolds

the manifold. However, as long as the torsion tensor vanishes, it is *always* possible to find coordinates such that $\Gamma^i_{jk}(p) = 0$ at any given point $p \in M$. For simplicity assume that p has coordinates $x^i(p) = 0$. We attempt a local coordinate transformation of the form

$$x^i = B^i_{i'} y^{i'} + A^i_{j'k'} y^{j'} y^{k'}$$

where $B = [B^i_{i'}]$ and $A^i_{j'k'} = A^i_{k'j'}$ are constant coefficients. Since

$$\left.\frac{\partial x^i}{\partial y^{i'}}\right|_p = B^i_{i'}$$

the transformation is invertible in a neighbourhood of p only if $B = [B^i_{i'}]$ is a non-singular matrix. The new coordinates of p are again zero, $y^{i'}(p) = 0$, and using the transformation formula (18.11), we have

$$\Gamma'^{i'}_{j'k'}(p) = B^j_{j'} B^k_{k'} (B^{-1})^{i'}_i \Gamma^i_{jk}(p) + 2 A^i_{j'k'} (B^{-1})^{i'}_i = 0$$

if we set

$$A^i_{j'k'} = -\tfrac{1}{2} B^j_{j'} B^k_{k'} \Gamma^i_{jk}(p).$$

Any such coordinates $(V; y^{j'})$ are called **geodesic coordinates**, or **normal coordinates**, at p. Their effect is to make geodesics appear locally 'straight' in a vanishingly small neighbourhood of p.

Exercise: Why does this procedure fail if the connection is not torsion free?

In the case of a pseudo-Riemannian manifold all derivatives of the metric tensor vanish in geodesic coordinates at p, $g_{ij,k}(p) = 0$. The constant coefficients $B^j_{j'}$ in the above may be chosen to send the metric tensor into standard diagonal form at p, such that $g'_{i'j'}(p) = g_{ij}(p) B^i_{i'} B^j_{j'} = \eta_{i'j'}$ has values ± 1 along the diagonal. Higher than first derivatives of g_{ij} will not in general vanish at p. For example, in normal coordinates at p the components of the curvature tensor can be expressed, using (18.40) and (18.25), in terms of the second derivatives $g_{ij,kl}(p)$:

$$\begin{aligned} R^i{}_{jkl}(p) &= \Gamma^i_{jl,k}(p) - \Gamma^i_{jk,l}(p) \\ &= \tfrac{1}{2} \eta^{im} \left(g_{ml,jk} + g_{jk,ml} - g_{mk,jl} - g_{jl,mk} \right)\big|_p . \end{aligned} \quad (18.43)$$

Problems

Problem 18.12 (a) Show that in a pseudo-Riemannian space the action principle

$$\delta \int_{t_1}^{t_2} L \, dt = 0$$

where $L = g_{\mu\nu} \dot{x}^\mu \dot{x}^\nu$ gives rise to geodesic equations with affine parameter t.
(b) For the sphere of radius a in polar coordinates,

$$ds^2 = a^2 (d\theta^2 + \sin^2 \theta \, d\phi^2),$$

use this variation principle to write out the equations of geodesics, and read off from them the Christoffel symbols $\Gamma^\mu_{\nu\rho}$.

(c) Verify by direct substitution in the geodesic equations that $L = \dot{\theta}^2 + \sin^2\theta\, \dot{\phi}^2$ is a constant along the geodesics and use this to show that the general solution of the geodesic equations is given by

$$b \cot\theta = -\cos(\phi - \phi_0) \quad \text{where} \quad b, \phi_0 = \text{const.}$$

(d) Show that these curves are great circles on the sphere.

Problem 18.13 Show directly from the tensor transformation laws of g_{ij} and g^{ij} that the Christoffel symbols

$$\Gamma^i_{jk} = \tfrac{1}{2} g^{ia}(g_{aj,k} + g_{ak,j} - g_{jk,a})$$

transform as components of an affine connection.

18.5 Equation of geodesic deviation

We now give a geometrical interpretation of the curvature tensor, which will subsequently be used in the measurement of the gravitational field (see Section 18.8). Let $\gamma : I \times J \to M$, where $I = [t_1, t_2]$ and $J = [\lambda_1, \lambda_2]$ are closed intervals of the real line, be a one-parameter family of curves on M. We will assume that the restriction of the map γ to $I' \times J'$, where I' and J' are the open intervals (t_1, t_2) and (λ_1, λ_2) respectively, is an embedded two-dimensional submanifold of M. We think of each map $\gamma_\lambda : I \to M$ defined by $\gamma_\lambda(t) = \gamma(t, \lambda)$ as being the curve represented by $\lambda = $ const. and t as the parameter along the curve. The one-parameter family of curves γ will be said to be **from** p **to** q if

$$p = \gamma(t_1, \lambda), \quad q = \gamma(t_2, \lambda)$$

for all $\lambda_1 \le \lambda \le \lambda_2$.

The tangent vectors to the curves of a one-parameter family constitute a vector field X on the two-dimensional submanifold $\gamma(I' \times J')$. If the curves are all covered by a single coordinate chart $(U; x^i)$, then

$$X = \frac{\partial \gamma^i(t, \lambda)}{\partial t} \frac{\partial}{\partial x^i} \quad \text{where} \quad \gamma^i(t, \lambda) = x^i\bigl(\gamma(t, \lambda)\bigr).$$

The **connection vector field** Y defined by

$$Y = \frac{\partial \gamma^i(t, \lambda)}{\partial \lambda} \frac{\partial}{\partial x^i}$$

is the tangent vector field to the curves connecting points having the same parameter value, $t = $ const. (see Fig. 18.1). The covariant derivative of the vector field Y along the curves γ_λ is given by

$$\frac{DY^i}{\partial t} \equiv \frac{\partial Y^i}{\partial t} + \Gamma^i_{jk} Y^j X^k$$

$$= \frac{\partial^2 \gamma^i}{\partial t \partial \lambda} + \Gamma^i_{jk} \frac{\partial \gamma^j}{\partial \lambda} \frac{\partial \gamma^k}{\partial t}$$

$$= \frac{\partial^2 \gamma^i}{\partial \lambda \partial t} + \Gamma^i_{jk} \frac{\partial \gamma^k}{\partial t} \frac{\partial \gamma^j}{\partial \lambda}.$$

18.5 Equation of geodesic deviation

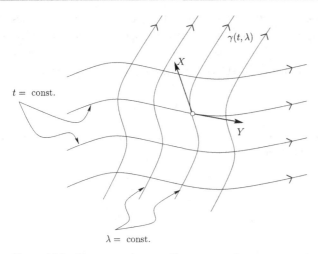

Figure 18.1 Tangent and connection vectors of a one-parameter family of geodesics

Hence
$$\frac{DY^i}{\partial t} = \frac{DX^i}{\partial \lambda}. \tag{18.44}$$

Alternatively, we can write
$$D_X Y = D_Y X. \tag{18.45}$$

If $A = A^i \partial_{x^i}$ is any vector field on U then
$$\left(\frac{D}{\partial t}\frac{D}{\partial \lambda} - \frac{D}{\partial \lambda}\frac{D}{\partial t}\right) A^i = \frac{D}{\partial t}\left(A^i_{;j} Y^j\right) - \frac{D}{\partial \lambda}\left(A^i_{;j} X^j\right)$$
$$= A^i_{;jk} Y^j X^k + A^i_{;j}\frac{DY^j}{\partial t} - A^i_{;jk} X^j Y^k - A^i_{;j}\frac{DX^j}{\partial \lambda}.$$

From Eq. (18.44) and the Ricci identity (18.28)
$$\left(\frac{D}{\partial t}\frac{D}{\partial \lambda} - \frac{D}{\partial \lambda}\frac{D}{\partial t}\right) A^i = R^i{}_{ajk} A^a X^j Y^k. \tag{18.46}$$

Let M be a pseudo-Riemannian manifold and $\gamma(\lambda, t)$ a one-parameter family of geodesics, such that the geodesics $\lambda = $ const. all have t as an affine parameter,
$$D_X X = \frac{DX^i}{\partial t} = 0,$$

the parametrization chosen to have the same normalization on all geodesics, $g(X, X) = \pm 1$ or 0. It then follows that $g(X, Y)$ is constant along each geodesic, since
$$\frac{\partial}{\partial t}\big(g(X, Y)\big) = D_X\big(g(X, Y)\big) = (D_X g)(X, Y) + g(D_X X, Y) + g(X, D_X Y)$$
$$= g(X, D_Y X) \quad \text{by (18.45)}$$
$$= \frac{1}{2} D_Y\big(g(X, X)\big) = \frac{1}{2}\frac{\partial}{\partial \lambda} g(X, X) = 0.$$

Connections and curvature

Thus, if the tangent and connection vector are initially orthogonal on a geodesic of the one-parameter family, $g(X_p, Y_p) = 0$, then they are orthogonal all along the geodesic.

In Eq. (18.46) set $A^i = X^i$ – this is possible since it is only necessary to have A^i defined in terms of t and λ (see Problem 18.14). With the help of Eq. (18.44) we have

$$\frac{D}{\partial t}\frac{DY^i}{\partial t} = R^i{}_{ajk} X^a X^j Y^k, \qquad (18.47)$$

known as the **equation of geodesic deviation**. For two geodesics, labelled by constants λ and $\lambda + \delta\lambda$, let δx^i be the tangent vector

$$\delta x^i \frac{\partial \gamma^i}{\partial \lambda}\delta\lambda = Y^i \delta\lambda.$$

For vanishingly small $\Delta\lambda$ it is usual to think of δx^i as an 'infinitesimal separation vector'. Since $\delta\lambda$ is constant along the geodesic we have

$$\delta\ddot{x}^i = R^i{}_{ajk} X^a X^j \delta x^k \qquad (18.48)$$

where $\cdot \equiv D/\partial t$. Thus $R^i{}_{jkl}$ measures the relative 'acceleration' between geodesics.

Problem

Problem 18.14 Equation (18.46) has strictly only been proved for a vector field A. Show that it holds equally for a vector field whose components $A^i(t, \lambda)$ are only defined on the one-parameter family of curves γ.

18.6 The Riemann tensor and its symmetries

In a pseudo-Riemannian manifold (M, g) it is possible to lower the contravariant index of the curvature tensor to form a tensor $\overline{R}$ of type $(0, 4)$,

$$\overline{R}(W, Z, X, Y) = R(\omega, Z, X, Y) \quad \text{where} \quad g(W, A) = \langle \omega, A \rangle.$$

This tensor will be referred to as the **Riemann curvature tensor** or simply the **Riemann tensor**. Setting $W = \partial_{x^i}$, $Z = \partial_{x^j}$, etc. then $\omega = g_{ia}\, dx^a$, whence

$$\overline{R}_{ijkl} = g_{ia} R^a{}_{jkl}.$$

In line with the standard index lowering convention, we denote the components of $\overline{R}$ by R_{ijkl}.

The following symmetries apply to the Riemann tensor:

$$R_{ijkl} = -R_{ijlk}, \qquad (18.49)$$

$$R_{ijkl} = -R_{jikl}, \qquad (18.50)$$

$$R_{ijkl} + R_{iklj} + R_{iljk} = 0, \qquad (18.51)$$

$$R_{ijkl} = R_{klij}. \qquad (18.52)$$

18.6 The Riemann tensor and its symmetries

Proof: Antisymmetry in the second pair of indices, Eq. (18.49), follows immediately from the definition of the curvature tensor (18.23) – it is not changed by the act of lowering the first index. Similarly, (18.51) follows immediately from Eq. (18.26). The remaining symmetries (18.50) and (18.52) may be proved by adopting geodesic coordinates at any given point p and using the expression (18.43) for the components of $R^i{}_{jkl}$:

$$R_{ijkl}(p) = \tfrac{1}{2}\left(g_{ml,jk} + g_{jk,ml} - g_{mk,jl} - g_{jl,mk}\right)\big|_p. \tag{18.53}$$

A more 'invariant' proof of (18.50) is to apply the generalized Ricci identities, Problem 18.9, to the metric tensor g,

$$0 = g_{ij;kl} - g_{ij;lk} = g_{aj} R^a{}_{ikl} + g_{ia} R^a{}_{jkl} = R_{jikl} + R_{ijkl}.$$

The symmetry (18.52) is actually a consequence of the first three symmetries, as may be shown by performing cyclic permutations on all four indices of Eq. (18.51):

$$R_{jkli} + R_{jlik} + R_{jikl} = 0, \tag{18.51a}$$
$$R_{klij} + R_{kijl} + R_{kjli} = 0, \tag{18.51b}$$
$$R_{lijk} + R_{ljki} + R_{lkij} = 0. \tag{18.51c}$$

The combination (18.51) – (18.51a) – (18.51b) + (18.51c) gives, after several cancellations using the symmetries (18.49) and (18.50),

$$2R_{ijkl} - 2R_{klij} = 0.$$

This is obviously equivalent to Eq. (18.52). ∎

Exercise: Prove from these symmetries that the cyclic symmetry also holds for any three indices; for example

$$R_{ijkl} + R_{jkil} + R_{kijl} = 0.$$

These symmetries permit us to count the number of independent components of the Riemann tensor. Since a skew symmetric tensor of type $(0, 2)$ on an n-dimensional vector space has $\tfrac{1}{2}n(n-1)$ independent components, a tensor of type $(0, 4)$ subject to symmetries (18.49) and (18.50) will have $\tfrac{1}{4}n^2(n-1)^2$ independent components. For fixed $i = 1, 2, \ldots, n$ only unequal triples $j \neq k \neq l$ need be considered in the symmetry (18.52), for if some pair are equal nothing new is added, by the cyclic identity: for example, if $k = l = 1$ then $R_{ij11} + R_{i11j} + R_{i1j1} = 0$ merely reiterates the skew symmetry on the second pair of indices, $R_{i11j} = -R_{i1j1}$. For every triple $j \neq k \neq l$ the total number of relations generated by (18.52) that are independent of (18.49) and (18.50) is therefore the number of such triples of numbers $j \neq k \neq l$ in the range $1, \ldots, n$, namely $n\binom{n}{3} = n^2(n-1)(n-2)/6$. By the above proof we need not consider the symmetry (18.52), and the total number of independent components of the Riemann tensor is

$$N = \frac{n^2(n-1)^2}{4} - \frac{n^2(n-1)(n-2)}{6}$$
$$= \frac{n^2(n^2-1)}{12}. \tag{18.54}$$

Connections and curvature

For low dimensions the number of independent components of the Riemann tensor is

$$n = 1: \quad N = 0,$$
$$n = 2: \quad N = 1,$$
$$n = 3: \quad N = 6,$$
$$n = 4: \quad N = 20.$$

A tensor of great interest in general relativity is the **Ricci tensor**, defined by

$$\text{Ric} = C_2^1 R.$$

It is common to write the components of $\text{Ric}(\partial_{x^i}, \partial_{x^j})$ in any chart $(U; x^i)$ as R_{ij}:

$$R_{ij} = R^a{}_{iaj} = g^{ab} R_{aibj}. \tag{18.55}$$

This tensor is symmetric since, by symmetry (18.52),

$$R_{ij} = g^{ab} R_{aibj} = g^{ab} R_{bjai} = R^a{}_{jai} = R_{ji}.$$

Contracting again gives the quantity known as the **Ricci scalar**,

$$R = R^i{}_i = g^{ij} R_{ij}. \tag{18.56}$$

Bianchi identities

For a torsion-free connection we have, on setting $T^p_{km} = 0$ in Eq. (18.32) of Problem 18.10, the **second Bianchi identity**

$$R^i{}_{jkl;m} + R^i{}_{jlm;k} + R^i{}_{jmk;l} = 0. \tag{18.57}$$

These are often referred to simply as the **Bianchi identities**. An alternative demonstration is to use normal coordinates at any point $p \in M$, such that $\Gamma^i_{jk}(p) = 0$. Making use of Eqs. (18.14) and (18.25) we have

$$R^i{}_{jkl;m}(p) = R^i{}_{jkl,m}(p) + \left(\Gamma^i_{am} R^a{}_{jkl} - \Gamma^a_{jm} R^i{}_{akl} - \Gamma^a_{km} R^i{}_{jal} - \Gamma^a_{lm} R^i{}_{jka} \right)(p)$$
$$= R^i{}_{jkl,m}(p)$$
$$= \Gamma^i_{jl,km}(p) - \Gamma^i_{jk,lm}(p) + \Gamma^a_{jl,m} \Gamma^i_{ak}(p) + \Gamma^a_{jl}(p)\Gamma^i_{ak,m} - \Gamma^a_{jk,m}\Gamma^i_{al}(p)$$
$$\quad - \Gamma^a_{jk}(p)\Gamma^i_{al,m}$$
$$= \Gamma^i_{jl,km}(p) - \Gamma^i_{jk,lm}(p).$$

If we substitute this expression in the left-hand side of (18.57) and use $\Gamma^i_{jk} = \Gamma^i_{kj}$, all terms cancel out.

Contracting Eq. (18.57) over i and m gives

$$R^i{}_{jkl;i} - R_{jl;k} + R_{jk;l} = 0. \tag{18.58}$$

Contracting Eq. (18.58) again by multiplying through by g^{jl} and using Eq. (18.41) we find

$$R^i{}_{k;i} - R_{;k} + R^j{}_{k;j} = 0,$$

or equivalently, the **contracted Bianchi identities**

$$R^j_{k;j} - \tfrac{1}{2} R_{,k} = 0. \tag{18.59}$$

A useful way of writing (18.59) is

$$G^j{}_{k;j} = 0, \tag{18.60}$$

where $G^i{}_j$ is the **Einstein tensor**,

$$G^i{}_j = R^i{}_j - \tfrac{1}{2} R \delta^i_j. \tag{18.61}$$

This tensor is symmetric when its indices are lowered,

$$G_{ij} = g_{ia} G^a{}_j = R_{ij} - \tfrac{1}{2} R g_{ij} = G_{ji}. \tag{18.62}$$

18.7 Cartan formalism

Cartan's approach to curvature is expressed entirely in terms of differential forms. Let e_i ($i = 1, \ldots, n$) be a local basis of vector fields, spanning $T(U)$ over an open set $U \subseteq M$ and $\{\varepsilon^i\}$ the dual basis of $T^*(U)$, such that $\langle e_i, \varepsilon^j \rangle = \delta^j_i$. For example, in a coordinate chart $(U; x^i)$ we may set $e_i = \partial_{x^i}$ and $\varepsilon^j = dx^j$, but such a coordinate system will exist for an arbitrary basis $\{e_i\}$ if and only if $[e_i, e_j] = 0$ for all $i, j = 1, \ldots, n$. We define the **connection 1-forms** $\omega^i{}_j : T(U) \to \mathcal{F}(U)$ by

$$D_X e_j = \omega^i{}_j(X) e_i. \tag{18.63}$$

These maps are differential 1-forms on U since they are clearly $\mathcal{F}$-linear,

$$D_{X+fY} e_j = D_X e_j + f D_Y e_j \implies \omega^i{}_j(X + fY) = \omega^i{}_j(X) + f \omega^i{}_j(Y).$$

If $\tau : T(U) \times T(U) \to T(U)$ is the torsion operator defined in Eq. (18.19), set

$$\tau(X, Y) = \tau^i(X, Y) e_i.$$

The maps $\tau^i : T(U) \times T(U) \to \mathcal{F}(U)$ are $\mathcal{F}$-linear in both arguments by the $\mathcal{F}$-linearity of τ, and are antisymmetric $\tau^i(X, Y) = -\tau^i(Y, X)$. They are therefore differential 2-forms on U, known as the **torsion 2-forms**.

Exercise: Show that $\tau^i = T^i_{jk} \varepsilon^j \wedge \varepsilon^k$ where $T^i_{jk} = \langle \tau(e_j, e_k), \varepsilon^i \rangle$.

From the identity $Z = \langle \varepsilon^i, Z \rangle e_i$ for any vector field Z on U, we have

$$\begin{aligned}\tau(X, Y) &= D_X(\langle \varepsilon^i, Y \rangle e_i) - D_Y(\langle \varepsilon^i, X \rangle e_i) - \langle \varepsilon^i, [X, Y] \rangle e_i \\ &= X(\langle \varepsilon^i, Y \rangle) e_i + \langle \varepsilon^i, Y \rangle \omega^k_i(X) e_k - Y(\langle \varepsilon^i, X \rangle) e_i - \langle \varepsilon^i, X \rangle \omega^k_i(X) e_k \\ &\quad - \langle \varepsilon^i, [X, Y] \rangle e_i \\ &= 2 d\varepsilon^i(X, Y) e_i + 2 \omega^k_i \wedge \varepsilon^i(X, Y) e_k,\end{aligned}$$

using the Cartan identity (16.14). Thus

$$\tau^i(X, Y) e_i = 2 \big(d\varepsilon^i(X, Y) + 2 \omega^i{}_k \wedge \varepsilon^k(X, Y)\big) e_i,$$

Connections and curvature

and by $\mathcal{F}$-linear independence of the vector fields e_i we have **Cartan's first structural equation**

$$d\varepsilon^i = -\omega^i{}_k \wedge \varepsilon^k + \tfrac{1}{2}\tau^i. \tag{18.64}$$

Define the **curvature 2-forms** $\rho^i{}_j$ by

$$\rho_{X,Y}e_j = \rho^i{}_j(X,Y)e_i, \tag{18.65}$$

where $\rho_{X,Y} : T(M) \to T(M)$ is the curvature operator in Eq. (18.22).

Exercise: Show that the $\rho^i{}_j$ are differential 2-forms on U; namely, they are $\mathcal{F}$-linear with respect to X and Y and $\rho^i{}_j(X,Y) = -\rho^i{}_j(Y,X)$.

Changing the dummy suffix on the right-hand side of Eq. (18.65) from i to k and applying $\langle \varepsilon^i, . \rangle$ to both sides of the equation we have, with the help of Eq. (18.23),

$$\begin{aligned}
\rho^i{}_j(X,Y) &= \langle \varepsilon^i, \rho_{X,Y}e_j \rangle \\
&= R(\varepsilon^i, e_j, X, Y) \\
&= R^i{}_{jkl}X^k Y^l \quad \text{where } R^i{}_{jkl} = R(\varepsilon^i, e_j, e_k, e_l) \\
&= \tfrac{1}{2}R^i{}_{jkl}(\varepsilon^k \otimes \varepsilon^l - \varepsilon^l \otimes \varepsilon^k)(X,Y).
\end{aligned}$$

Hence

$$\rho^i{}_j = R^i{}_{jkl}\varepsilon^k \wedge \varepsilon^l. \tag{18.66}$$

A similar analysis to that for the torsion operator results in

$$\begin{aligned}
\rho_{X,Y}e_j &= D_X D_Y e_j - D_Y D_X e_j - D_{[X,Y]}e_j \\
&= D_X(\langle \omega^i{}_j, Y \rangle e_i) - D_Y(\langle \omega^i{}_j, X \rangle e_i) - \langle \omega^i{}_j, [X,Y] \rangle e_i \\
&= 2[d\omega^i{}_j(X,Y) + \omega^i{}_k \wedge \omega^k{}_j(X,Y)]e_i,
\end{aligned}$$

and **Cartan's second structural equation**

$$d\omega^i{}_j = -\omega^i{}_k \wedge \omega^k{}_j + \tfrac{1}{2}\rho^i{}_j. \tag{18.67}$$

Example 18.3 With respect to a coordinate basis $e_i = \partial_{x^i}$, the Cartan structural equations reduce to formulae found earlier in this chapter. For any vector field $X = X^k \partial_{x^k}$,

$$D_X \partial_{x^j} = X^k D_k \partial_{x^j} = X^k \Gamma^i{}_{jk} \partial_{x^i}.$$

Hence, by Eq. (18.63), we have $\omega^i{}_j(X) = X^k \Gamma^i{}_{jk}$, so that

$$\omega^i{}_j = \Gamma^i{}_{jk}\, dx^k. \tag{18.68}$$

Thus the components of the connection 1-forms with respect to a coordinate basis are precisely the components of the connection.

Setting $\varepsilon^i = dx^i$ in Cartan's first structural equation (18.64), we have

$$d\varepsilon^i = d^2 x^i = 0 = -\omega^i{}_j \wedge dx^j + \tfrac{1}{2}\tau^i.$$

18.7 Cartan formalism

Hence $\tau^i = 2\Gamma^i_{jk}\, dx^k \wedge dx^j$, and it follows that the components of the torsion 2-forms are identical with those of the torsion tensor T in Eq. (18.21),

$$\tau^i = T^i_{jk}\, dx^j \wedge dx^k \quad \text{where} \quad T^i_{jk} = \Gamma^i_{kj} - \Gamma^i_{jk}. \tag{18.69}$$

Finally, Cartan's second structural equation (18.67) reduces in a coordinate basis to

$$d\Gamma^i_{jk} \wedge dx^k = -\Gamma^i_{kl}\, dx^l \wedge \Gamma^k_{jm}\, dx^m + \tfrac{1}{2}\rho^i_j,$$

whence, on using the decomposition (18.66),

$$R^i{}_{jkl}\, dx^k \wedge dx^l = 2\Gamma^i_{jk,l}\, dx^l \wedge dx^k + 2\Gamma^i_{ml}\Gamma^m_{jk}\, dx^l \wedge dx^k.$$

We thus find, in agreement with Eq. (18.25),

$$R^i{}_{jkl} = \Gamma^i_{jl,k} - \Gamma^i_{jk,l} + \Gamma^m_{jl}\Gamma^i_{mk} - \Gamma^m_{jk}\Gamma^i_{ml} = -R^i{}_{jlk}.$$

The big advantage of Cartan's structural equations over these various coordinate expressions is that they give expressions for torsion and curvature for arbitrary vector field bases.

Bianchi identities

Taking the exterior derivative of (18.64) gives, with the help of (18.67),

$$d^2\varepsilon^i = 0 = -d\omega^i_k \wedge \varepsilon^k + \omega^i_k \wedge d\varepsilon^k + \tfrac{1}{2}d\tau^i$$
$$= \omega^i_j \wedge \omega^j_k \wedge \varepsilon^k - \tfrac{1}{2}\rho^i_k \wedge \varepsilon^k - \omega^i_k \wedge \omega^k_j \wedge \varepsilon^j + \tfrac{1}{2}\omega^i_k \wedge \tau^k + \tfrac{1}{2}d\tau^i$$
$$= \tfrac{1}{2}(-\rho^i_k \wedge \varepsilon^k + \omega^i_k \wedge \tau^k + d\tau^i).$$

Hence, we obtain the *first Bianchi identity*

$$d\tau^i = \rho^i_k \wedge \varepsilon^k - \omega^i_k \wedge \tau^k. \tag{18.70}$$

Its relation to the earlier identity (18.24) is left as an exercise (see Problem 18.15).
Similarly

$$d^2\omega^i_j = 0 = -d\omega^i_k \wedge \omega^k_j + \omega^i_k \wedge d\omega^k_j + \tfrac{1}{2}d\rho^i_j$$
$$= \tfrac{1}{2}(-\rho^i_k \wedge \omega^k_j + \omega^i_k \wedge \rho^k_j + d\rho^i_j),$$

resulting in the *second Bianchi identity*

$$d\rho^i_j = \rho^i_k \wedge \omega^k_j - \omega^i_k \wedge \rho^k_j. \tag{18.71}$$

Pseudo-Riemannian spaces in Cartan formalism

In a pseudo-Riemannian manifold with metric tensor g, set $g_{ij} = g(e_i, e_j)$. Since g_{ij} is a scalar field for each $i, j = 1, \ldots, n$ and $D_X g = 0$, we have

$$\langle X, dg_{ij}\rangle = X(g_{ij}) = D_X\big(g(e_i, e_j)\big)$$
$$= g(D_X e_i, e_j) + g(e_i, D_X e_j)$$
$$= g\big(\omega^k_i(X)e_k, e_j\big) + g\big(e_i, \omega^k_j(X)e_k\big)$$
$$= g_{kj}\omega^k_i(X) + g_{ik}\omega^k_j(X)$$
$$= \langle X, \omega_{ji}\rangle + \langle X, \omega_{ij}\rangle$$

where
$$\omega_{ij} = g_{ki}\omega^k{}_j.$$

As X is an arbitrary vector field,
$$dg_{ij} = \omega_{ij} + \omega_{ji}. \tag{18.72}$$

For an orthonormal basis e_i, such that $g_{ij} = \eta_{ij}$, we have $dg_{ij} = 0$ and
$$\omega_{ij} = -\omega_{ji}. \tag{18.73}$$

In particular, all diagonals vanish, $g_{ii} = 0$ for $i = 1, \ldots, n$.

Lowering the first index on the curvature 2-forms $\rho_{ij} = g_{ik}\rho^k{}_j$, we have from the second Cartan structural equation (18.67),

$$\begin{aligned}\rho_{ij} &= 2\big(d\omega_{ij} + \omega_{ik} \wedge \omega^k{}_j\big) \\ &= 2\big(-d\omega_{ji} + \omega^k{}_j \wedge \omega_{ki}\big) \\ &= 2\big(-d\omega_{ji} - \omega_{jk} \wedge \omega^k{}_i\big),\end{aligned}$$

whence
$$\rho_{ij} = -\rho_{ji}. \tag{18.74}$$

Exercise: Show that (18.74) is equivalent to the symmetry $R_{ijkl} = -R_{jikl}$.

Example 18.4 The 3-sphere of radius a is the submanifold of $\mathbb{R}^4$,
$$S^3(a) = \{(x, y, z, w) \,|\, x^2 + y^2 + z^2 + w^2 = a^2\} \subset \mathbb{R}^4.$$

Spherical polar coordinates χ, θ, ϕ are defined by
$$\begin{aligned}x &= a \sin\chi \sin\theta \cos\phi \\ y &= a \sin\chi \sin\theta \sin\phi \\ z &= a \sin\chi \cos\theta \\ w &= a \cos\chi\end{aligned}$$

where $0 < \chi, \theta < \pi$ and $0 < \phi < 2\pi$. These coordinates cover all of $S^3(a)$ apart from the points $y = 0$, $x \geq 0$. The Euclidean metric on $\mathbb{R}^4$
$$ds^2 = dx^2 + dy^2 + dz^2 + dw^2$$

induces a metric on $S^3(a)$ as for the 2-sphere in Example 18.1:
$$ds^2 = a^2\big[d\chi^2 + \sin^2\chi(d\theta^2 + \sin^2\theta\, d\phi^2)\big].$$

An orthonormal frame is
$$e_1 = \frac{1}{a}\frac{\partial}{\partial\chi}, \qquad e_2 = \frac{1}{a\sin\chi}\frac{\partial}{\partial\theta}, \qquad e_3 = \frac{1}{a\sin\chi\sin\theta}\frac{\partial}{\partial\phi}$$
$$\varepsilon^1 = a\, d\chi, \qquad \varepsilon^2 = a\sin\chi\, d\theta, \qquad \varepsilon^3 = a\sin\chi\sin\theta\, d\phi$$

18.7 Cartan formalism

where

$$g_{ij} = g(e_i, e_j) = \delta_{ij}, \qquad g = \varepsilon^1 \otimes \varepsilon^1 + \varepsilon^2 \otimes \varepsilon^2 + \varepsilon^3 \otimes \varepsilon^3.$$

Since the metric connection is torsion-free, $\tau^i = 0$, the first structural equation reads

$$d\varepsilon^i = -\omega^i{}_k \wedge \varepsilon^k = -\omega_{ik} \wedge \varepsilon^k$$

setting

$$\omega_{ij} = A_{ijk}\varepsilon^k$$

where $A_{ijk} = -A_{jik}$. By interchanging dummy suffixes j and k we may also write

$$d\varepsilon^i = -A_{ikj}\varepsilon^j \wedge \varepsilon^k = A_{ikj}\varepsilon^k \wedge \varepsilon^j = A_{ijk}\varepsilon^j \wedge \varepsilon^k.$$

For $i = 1$, using $A_{11k} = 0$,

$$d\varepsilon^1 = ad^2\chi = 0 = A_{121}\varepsilon^2 \wedge \varepsilon^1 + A_{131}\varepsilon^3 \wedge \varepsilon^1 + (A_{123} - A_{132})\varepsilon^2 \wedge \varepsilon^3,$$

whence

$$A_{121} = A_{131} = 0, \qquad A_{123} = A_{132}.$$

For $i = 2$,

$$d\varepsilon^2 = a\cos\chi\, d\chi \wedge d\theta = a^{-1}\cot\chi \varepsilon^1 \wedge \varepsilon^2$$
$$= A_{212}\varepsilon^1 \wedge \varepsilon^2 + A_{232}\varepsilon^3 \wedge \varepsilon^2 + (A_{213} - A_{231})\varepsilon^1 \wedge \varepsilon^3,$$

which implies

$$A_{212} = a^{-1}\cot\chi, \qquad A_{213} = A_{231}, \qquad A_{232} = 0.$$

Similarly the $i = 3$ equation gives

$$A_{312} = A_{321}, \qquad A_{313} = a^{-1}\cot\chi, \qquad A_{323} = a^{-1}\frac{\cot\theta}{\sin\chi}.$$

All coefficients having all three indices different, such as A_{123}, must vanish since

$$A_{123} = A_{132} = -A_{312} = -A_{321} = A_{231} = A_{213} = -A_{123} \implies A_{123} = 0.$$

There is enough information now to write out the connection 1-forms:

$$\omega_{12} = -\omega_{21} = -a^{-1}\cot\chi \varepsilon^2,$$
$$\omega_{13} = -\omega_{31} = -a^{-1}\cot\chi \varepsilon^3,$$
$$\omega_{23} = -\omega_{32} = -a^{-1}\frac{\cot\theta}{\sin\chi}\varepsilon^3.$$

The second structural relations (18.67) can now be used to calculate the curvature 2-forms:

$$\rho_{12} = 2(d\omega_{12} + \omega_{1k} \wedge \omega^k{}_2)$$
$$= 2(a^{-1}\operatorname{cosec}^2\chi\, d\chi \wedge \varepsilon^2 - a^{-1}\cot\chi\, d\varepsilon^2 + \omega_{13} \wedge \omega_{32})$$
$$= 2a^{-2}(\operatorname{cosec}^2\chi \varepsilon^1 \wedge \varepsilon^2 - \cot^2\chi \varepsilon^1 \wedge \varepsilon^2)$$
$$= 2a^{-2}\varepsilon^1 \wedge \varepsilon^2,$$

Connections and curvature

and similarly

$$\rho_{13} = 2(d\omega_{13} + \omega_{12} \wedge \omega_{23}) = 2a^{-2}\varepsilon^1 \wedge \varepsilon^3,$$
$$\rho_{23} = 2(d\omega_{23} + \omega_{21} \wedge \omega_{13}) = 2a^{-2}\varepsilon^2 \wedge \varepsilon^3.$$

The components of the Riemann curvature tensor can be read off using Eq. (18.66):

$$R_{1212} = R_{1313} = R_{2323} = a^{-2},$$

and all other components of the Riemann tensor are simply related to these components by symmetries; for example, $R_{2121} = -R_{1221} = a^{-2}$, $R_{1223} = 0$, etc. It is straightforward to verify the relation

$$R_{ijkl} = \frac{1}{a^2}(g_{ik}g_{jl} - g_{il}g_{jk}).$$

For any Riemannian space (M, g) the **sectional curvature** of the vector 2-space spanned by a pair of tangent vectors X and Y at any point p is defined to be

$$K(X, Y) = \frac{R(X, Y, X, Y)}{A(X, Y)}$$

where $A(X, Y)$ is the 'area' of the parallelogram spanned by X and Y,

$$A(X, Y) = g(X, Y)g(X, Y) - g(X, X)g(Y, Y).$$

For the 3-sphere

$$K(X, Y) = \frac{1}{a^2} \frac{X_i Y^i X_k Y^k - X_i X^i Y_k Y^k}{X_i Y^i X_k Y^k - X_i X^i Y_k Y^k} = \frac{1}{a^2}$$

independent of the point $p \in S^3(a)$ and the choice of tangent vectors X, Y. For this reason the 3-sphere is said to be a **space of constant curvature**.

Locally flat spaces

A manifold M with affine connection is said to be **locally flat** if for every point $p \in M$ there is a chart $(U; x^i)$ such that all components of the connection vanish throughout U. This implies of course that both torsion tensor and curvature tensor vanish throughout U, but more interesting is that these conditions are both necessary and sufficient. This result is most easily proved in Cartan's formalism, and requires the transformation of the connection 1-forms under a change of basis

$$\varepsilon'^i = A^i{}_j \varepsilon^j \quad \text{where} \quad A^i{}_j = A^i{}_j(x^1, \ldots, x^n).$$

Evaluating $d\varepsilon'^i$, using Eq. (18.64), gives

$$d\varepsilon'^i = dA^i{}_j \wedge \varepsilon^j + A^i{}_j d\varepsilon^j = -\omega'^i{}_k \wedge \varepsilon'^k + \tfrac{1}{2}\tau'^i$$

where $\tau'^i = A^i{}_j \tau^j$ and

$$A^k{}_j \omega'^i{}_k = dA^i{}_j - A^i{}_k \omega^k{}_j. \tag{18.75}$$

18.7 Cartan formalism

Exercise: Show from this equation and Eq. (18.67) that if a transformation exists such that $\omega'^i_k = 0$ then the curvature 2-forms vanish, $\rho^k_j = 0$.

Theorem 18.1 *A manifold with symmetric connection is locally flat everywhere if and only if the curvature 2-forms vanish, $\rho^k_j = 0$.*

Proof: The *only if* part follows from the above comments. For the converse, we suppose $\tau^i = \rho^i_j = 0$ everywhere. If $(U; x^i)$ is any chart on M, let $N = U \times \mathbb{R}^{n^2}$ and denote coordinates on $\mathbb{R}^{n^2}$ by z^j_k. Using the second structural formula (18.67) with $\rho^i_j = 0$, the 1-forms $\alpha^i_j = \mathrm{d}z^i_j - z^i_k \omega^k_j$ on N satisfy

$$\begin{aligned}
\mathrm{d}\alpha^i_j &= -\mathrm{d}z^i_k \wedge \omega^k_j - z^i_k \, \mathrm{d}\omega^k_j \\
&= -(\alpha^i_k + z^i_m \, \mathrm{d}\omega^m_k) \wedge \omega^k_j + z^i_k \omega^k_m \wedge \omega^m_j \\
&= \omega^k_j \wedge \alpha^i_k.
\end{aligned} \quad (18.76)$$

By the Frobenius theorem 16.4, $\mathrm{d}\alpha^i_j = 0$ is an integrable system on N, and has a local integral submanifold through any point x^i_0, A^j_{0k} where $\det[A^j_{0k}] \neq 0$, that may be assumed to be of the form

$$z^j_i = A^j_i(x^1, \ldots, x^n) \quad \text{where} \quad A^j_i(x^1_0, \ldots, x^n_0) = A^j_{0i}.$$

We may assume that $\det[A^j_k]$ is non-singular in a neighbourhood of $\mathbf{x}_0$. Hence

$$\alpha^i_j = 0 \implies \mathrm{d}A^i_j - A^i_k \omega^k_j = 0$$

and substituting in Eq. (18.75) results in

$$\omega'^i_k = 0 \quad \text{if} \quad \varepsilon'^i = A^i_j \varepsilon^j.$$

Finally, the structural equation (18.64) gives $\mathrm{d}\varepsilon'^i = 0$, and the Poincaré lemma 17.5 implies that there exist local coordinates y^i such that $\varepsilon'^i = \mathrm{d}y^i$. ∎

In the case of a pseudo-Riemannian space, locally flat coordinates such that $\Gamma^i_{jk} = 0$ imply that $g_{ij,k} = 0$ by Eq. (18.37). Hence $g_{ij} = $ const. throughout the coordinate region, and a linear transformation can be used to diagonalize the metric into standard diagonal form $g_{ij} = \eta_{ij}$ with ± 1 along the diagonal.

Problems

Problem 18.15 Let $e_i = \partial_{x^i}$ be a coordinate basis.
(a) Show that the first Bianchi identity reads

$$R^i_{[jkl]} = T^i_{[jk;l]} - T^a_{[jk} T^i_{l]a},$$

and reduces to the cyclic identity (18.26) in the case of a torsion-free connection.
(b) Show that the second Bianchi identity becomes

$$R^i_{j[kl;m]} = R^i_{ja[k} T^a_{ml]},$$

which is identical with Eq. (18.32) of Problem 18.10.

Connections and curvature

Problem 18.16 In a Riemannian manifold (M, g) show that the sectional curvature $K(X, Y)$ at a point p, defined in Example 18.4, is independent of the choice of basis of the 2-space; i.e., $K(X', Y') = K(X, Y)$ if $X' = aX + bY$, $Y' = cX + dY$ where $ad - bc \neq 0$.

The space is said to be *isotropic* at $p \in M$ if $K(X, Y)$ is independent of the choice of tangent vectors X and Y at p. If the space is isotropic at each point p show that

$$R_{ijkl} = f(g_{ik}g_{jl} - g_{il}g_{jk})$$

where f is a scalar field on M. If the dimension of the manifold is greater than 2, show *Schur's theorem*: a Riemannian manifold that is everywhere isotropic is a space of constant curvature, $f = \text{const.}$ [*Hint*: Use the contracted Bianchi identity (18.59).]

Problem 18.17 Show that a space is locally flat if and only if there exists a local basis of vector fields $\{e_i\}$ that are *absolutely parallel*, $De_i = 0$.

Problem 18.18 Let (M, φ) be a *surface of revolution* defined as a submanifold of $\mathbb{E}^3$ of the form

$$x = g(u)\cos\theta, \qquad y = g(u)\sin\theta, \qquad z = h(u).$$

Show that the induced metric (see Example 18.1) is

$$ds^2 = (g'(u)^2 + h'(u)^2)\,du^2 + g^2(u)\,d\theta^2.$$

Picking the parameter u such that $g'(u)^2 + h'(u)^2 = 1$ (interpret this choice!), and setting the basis 1-forms to be $\varepsilon^1 = du$, $\varepsilon^2 = g\,d\theta$, calculate the connection 1-forms $\omega^i{}_j$, the curvature 1-forms $\rho^i{}_j$, and the curvature tensor component R_{1212}.

Problem 18.19 For the ellipsoid

$$\frac{x^2}{a^2} + \frac{y^2}{b^2} + \frac{z^2}{c^2} = 1$$

show that the sectional curvature is given by

$$K = \left(\frac{x^2 bc}{a^3} + \frac{y^2 ac}{b^3} + \frac{z^2 ab}{c^3}\right)^{-2}.$$

18.8 General relativity

The principle of equivalence

The Newtonian gravitational force on a particle of mass m is $\mathbf{F} = -m\nabla\phi$ where the scalar potential ϕ satisfies *Poisson's equation*

$$\nabla^2\phi = 4\pi G\rho. \tag{18.77}$$

Here $G = 6.672 \times 10^{-8}$ g^{-1} cm^3 s^{-2} is Newton's gravitational constant and ρ is the density of matter present. While it is in principle possible to generalize this theory by postulating a relativistically invariant equation such as

$$\Box\phi = \nabla^2\phi - \frac{\partial^2}{\partial t^2}\phi = 4\pi G\rho,$$

18.8 General relativity

there are a number of problems with this theory, not least of which is that it does not accord with observations.

Key to the formulation of a correct theory is the **principle of equivalence** that, in its simplest version, states that all particles fall with equal acceleration in a gravitational field, a fact first attributed to Galileo in an improbable tale concerning the leaning tower of Pisa. While the derivation is by simple cancellation of m from both sides of the Newtonian equation of motion

$$m\ddot{x} = mg,$$

the masses appearing on the two sides of the equation could conceivably be different. On the left we really have *inertial mass*, m_i, which measures the particle's resistance to *any* force, while on the right the mass should be identified as *gravitational mass*, measuring the particle's response to gravitational fields – its 'gravitational charge' so to speak. The principle of equivalence can be expressed as saying that the ratio of inertial to gravitational mass will be the same for all bodies irrespective of the material from which they are composed. This was tested to one part in 10^8 for a vast variety of materials in 1890 by Eötvös using a torsion balance, and repeated in the 1960s to one part in 10^{11} by Dicke using a solar balancing mechanism (see C. M. Will's article in [10]).

The principle of equivalence essentially says that it is impossible to distinguish inertial forces such as centrifugal or Coriolis forces from gravitational ones. A nice example of the equivalence of such forces is the *Einstein elevator*. An observer in an elevator at rest sees objects fall to the ground with acceleration g. However if the elevator is set in free fall, objects around the observer will no longer appear to be subject to forces, much as if he were in an inertial frame in outer space. It has in effect been possible to *transform away* the gravitational field by going to a freely falling laboratory. Conversely as all bodies 'fall' to the floor in an accelerated rocket with the same acceleration, an observer will experience an 'apparent' gravitational field.

The effect of a non-inertial frame in special relativity should be essentially indistinguishable from the effects of gravity. The metric interval of Minkowski space (see Chapter 9) in a general coordinate system $x'^{\mu'} = x'^{\mu'}(x^1, x^2, x^3, x^4)$, where x^ν are inertial coordinates, becomes

$$ds^2 = g_{\mu\nu} \, dx^\mu \, dx^\nu = g'_{\mu'\nu'} \, dx'^{\mu'} \, dx'^{\nu'},$$

where

$$g'_{\mu'\nu'} = g_{\mu\nu} \frac{\partial x^\mu}{\partial x'^{\mu'}} \frac{\partial x^\nu}{\partial x'^{\nu'}}.$$

Expressed in general coordinates, the (geodesic) equations of motion of an inertial particle, $d^2 x^\mu/ds^2 = 0$, are

$$\frac{d^2 x'^{\mu'}}{ds^2} + \Gamma'^{\mu'}_{\alpha'\beta'} \frac{dx'^{\alpha'}}{ds} \frac{dx'^{\beta'}}{ds} = 0,$$

where

$$\Gamma'^{\mu'}_{\alpha'\beta'} = -\frac{\partial^2 x'^{\mu'}}{\partial x^\mu \partial x^\nu} \frac{\partial x^\mu}{\partial x'^{\alpha'}} \frac{\partial x^\nu}{\partial x'^{\beta'}}.$$

Connections and curvature

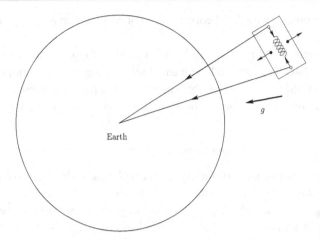

Figure 18.2 Tidal effects in a freely falling laboratory

The principle of equivalence is a purely local idea, and only applies to vanishingly small laboratories. A *real* gravitational field such as that due to the Earth cannot be totally transformed away in general. For example, if the freely falling Einstein elevator has significant size compared to the scale on which there is variation in the Earth's gravitational field, then particles at different positions in the lift will undergo different accelerations. Particles near the floor of the elevator will have larger accelerations than particles released from the ceiling, while particles released from the sides of the elevator will have a small horizontal acceleration relative to the central observer because the direction to the centre of the Earth is not everywhere parallel. These mutual accelerations or *tidal* forces can be measured in principle by connecting pairs of freely falling particles with springs (see Fig. 18.2).

Postulates of general relativity

The basic proposition of general relativity (Einstein, 1916) is the following: the world is a four-dimensional Minkowskian manifold (pseudo-Riemannian of index +2) called **space-time**. Its points are called **events**. The **world-line**, or space-time history of a material particle, is a parametrized curve $\gamma : \mathbb{R} \to M$ whose tangent $\dot{\gamma}$ is everywhere timelike. The **proper time**, or time as measured by a clock carried by the particle between parameter values $\lambda = \lambda_1$ and λ_2, is given by

$$\tau = \frac{1}{c} \int_{\lambda_1}^{\lambda_2} \sqrt{-g(\dot{\gamma}, \dot{\gamma})}\, d\lambda = \frac{1}{c} \int_{\lambda_1}^{\lambda_2} \sqrt{-g_{\mu\nu} \frac{dx^\mu}{d\lambda} \frac{dx^\nu}{d\lambda}}\, d\lambda = \frac{1}{c} \int_{s_1}^{s_2} ds. \tag{18.78}$$

A **test particle** is a particle of very small mass compared to the major masses in its neighbourhood and 'freely falling' in the sense that it is subject to no external forces. The world-line of a test particle is assumed to be a timelike geodesic. The world-line of a photon of small energy is a null geodesic. Both satisfy equations

$$\frac{d^2 x^\mu}{ds^2} + \Gamma^\mu_{\nu\rho} \frac{dx^\nu}{ds} \frac{dx^\rho}{ds} = 0,$$

18.8 General relativity

where the Christoffel symbols $\Gamma^\mu_{\nu\rho}$ are given by Eq. (18.40) with Greek indices subsituted. The affine parameter s is determined by

$$g_{\mu\nu}\frac{dx^\mu}{ds}\frac{dx^\nu}{ds} = \begin{cases} -1 & \text{for test particles,} \\ 0 & \text{for photons.} \end{cases}$$

An introduction to the theory of general relativity, together with many of its developments, can be found in [9, 11–16].

The principle of equivalence has a natural place in this postulate, since all particles have the same geodesic motion independent of their mass. Also, at each event p, geodesic normal coordinates may be found such that components of the metric tensor $g = g_{\mu\nu}\,dx^\mu \otimes dx^\nu$ have the Minkowski values $g_{\mu\nu}(p) = \eta_{\mu\nu}$ and the Christoffel symbols vanish, $\Gamma^\mu_{\nu\rho}(p) = 0$. In such coordinates the space-time appears locally to be Minkowski space, and gravitational forces have been locally transformed away since any geodesic at p reduces to a 'local rectilinear motion' $d^2x^\mu/ds^2 = 0$. When it is possible to find coordinates that transform the metric to constant values on an entire chart, the metric is locally flat and all gravitational fields are 'fictitious' since they arise entirely from non-inertial effects. By Theorem 18.1 such coordinate transformations are possible if and only if the curvature tensor vanishes. Hence it is natural to identify the 'real' gravitational field with the curvature tensor $R^\mu_{\nu\rho\sigma}$.

The equations determining the gravitational field are **Einstein's field equations**

$$G_{\mu\nu} = R_{\mu\nu} - \tfrac{1}{2}R g_{\mu\nu} = \kappa T_{\mu\nu}, \tag{18.79}$$

where $G_{\mu\nu}$ is the Einstein tensor defined above in Eqs. (18.61) and (18.62),

$$R_{\mu\nu} = R^\alpha_{\mu\alpha\nu} \quad \text{and} \quad R = R^\alpha_\alpha = g^{\mu\nu}R_{\mu\nu},$$

and $T_{\mu\nu}$ is the energy–stress tensor of all the matter fields present. The constant κ is known as **Einstein's gravitational constant**; we shall relate it to Newton's gravitational constant directly. These equations have the property that for weak fields, $g_{\mu\nu} \approx \eta_{\mu\nu}$, they reduce to the Poisson equation when appropriate identifications are made (see the discussion of the weak field approximation below), and by the contracted Bianchi identity (18.60) they guarantee a 'covariant' version of the conservation identities

$$T^{\mu\nu}{}_{;\nu} \equiv T^{\mu\nu}{}_{,\nu} + \Gamma^\mu_{\alpha\nu}T^{\alpha\nu} + \Gamma^\nu_{\alpha\nu}T^{\mu\alpha} = 0.$$

Measurement of the curvature tensor

The equation of geodesic deviation (18.47) can be used to give a physical interpretation of the curvature tensor. Consider a one-parameter family of timelike geodesics $\gamma(s, \lambda)$ with tangent vectors $U = U^\mu \partial_{x^\mu}$ when expressed in a coordinate chart $(A; x^\mu)$, where

$$U^\mu(s) = \left.\frac{\partial x^\mu}{\partial s}\right|_{\lambda=0}, \quad U^\mu U_\mu = -1, \quad \frac{DU^\mu}{\partial s} = 0.$$

Suppose the connection vector $Y = Y^\mu \partial_{x^\mu}$, where $Y^\mu = \partial x^\mu/\partial \lambda$, is initially orthogonal to U at $s = s_0$,

$$Y^\mu U_\mu\big|_{s=s_0} = g(U, Y)(s_0) = 0.$$

537

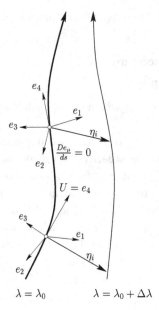

Figure 18.3 Physical measurement of curvature tensor

We then have $g(U, Y) = 0$ for all s, since $g(U, Y)$ is constant along each geodesic (see Section 18.5). Thus if e_1, e_2, e_3 are three mutually orthogonal spacelike vectors at $s = s_0$ on the central geodesic $\lambda = 0$ that are orthogonal to U, and they are parallel propagated along this geodesic, $De_i/\partial s = 0$, then they remain orthogonal to each other and U along this geodesic,

$$g(e_i, e_j) = \delta_{ij}, \qquad g(e_i, u) = 0.$$

In summary, if we set $e_4 = U$ then the four vectors $e_1, \ldots, e_4$ are an orthonormal tetrad of vectors along $\lambda = \lambda_0$,

$$g(e_\mu, e_\nu) = \eta_{\mu\nu}.$$

The situation is depicted in Fig. 18.3. Let $\lambda = \lambda_0 + \delta\lambda$ be any neighbouring geodesic from the family, then since we are assuming δx^μ is orthogonal to U^μ, the equation of geodesic deviation in the form (18.48) can be written

$$\frac{D^2 \delta x^\mu}{ds^2} = \delta\lambda \frac{D^2 Y^\mu}{ds^2} = R^\mu{}_{\alpha\rho\nu} U^\alpha U^\rho \delta x^\nu. \tag{18.80}$$

Expanding δx^μ in terms of the basis e_μ we have, adopting a cartesian tensor summation convention,

$$\delta x^\mu = \eta_j e_j^\mu \equiv \sum_{j=1}^{3} \eta_j e_j^\mu,$$

18.8 General relativity

where $\eta_j = \eta_j(s)$, so that

$$\frac{D\delta x^\mu}{ds} = \frac{d\eta_j}{ds}e_j^\mu + \eta_j\frac{e_i^\mu}{ds} = \frac{d\eta_j}{ds}e_j^\mu,$$

$$\frac{D^2\delta x^\mu}{ds^2} = \frac{d^2\eta_j}{ds^2}e_j^\mu + \frac{d\eta_j}{ds}\frac{e_i^\mu}{ds} = \frac{d^2\eta_j}{ds^2}e_j^\mu.$$

Substituting (18.80) results in

$$\frac{d^2\eta_i}{ds^2} = e_{i\mu}\frac{D^2\delta x^\mu}{ds^2} = R^\mu{}_{\alpha\rho\nu}e_{i\mu}e_4^\alpha e_j^\rho e_4^\nu \eta_j,$$

which reads in any local coordinates at any point on $\lambda = \lambda_0$ such that $e_\alpha^\mu = \delta_\alpha^\mu$,

$$\frac{d^2\eta_i}{ds^2} = -R_{i4j4}\eta_j. \tag{18.81}$$

Thus $R_{i4j4} \equiv R_{\mu\alpha\rho\nu}e_i^\mu e_4^\alpha e_j^\rho e_4^\nu$ measures the *relative accelerations* between neighbouring freely falling particles in the gravitational field. Essentially these are what are termed **tidal forces** in Newtonian physics, and could be measured by the strain on a spring connecting the two particles (see Fig. 18.2 and [17]).

The linearized approximation

Consider a one-parameter family of Minkowskian metrics having components $g_{\mu\nu} = g_{\mu\nu}(x^\alpha, \epsilon)$ such that $\epsilon = 0$ reduces to flat Minkowski space, $g_{\mu\nu}(0) = \eta_{\mu\nu}$. Such a family is known as a **linearized approximation** of general relativity. If we set

$$h_{\mu\nu} = \left.\frac{\partial g_{\mu\nu}}{\partial \epsilon}\right|_{\epsilon=0} \tag{18.82}$$

then for $|\epsilon| \ll 1$ we have 'weak gravitational fields' in the sense that the metric is only slightly different from Minkowski space,

$$g_{\mu\nu} \approx \eta_{\mu\nu} + \epsilon h_{\mu\nu}.$$

From $g^{\mu\rho}g_{\rho\nu} = \delta^\mu_\nu$ it follows by differentiating with respect to ϵ at $\epsilon = 0$ that

$$\left.\frac{\partial g^{\mu\rho}}{\partial \epsilon}\right|_{\epsilon=0}\eta_{\rho\nu} + \eta^{\mu\rho}h_{\rho\nu} = 0,$$

whence

$$\left.\frac{\partial g^{\mu\nu}}{\partial \epsilon}\right|_{\epsilon=0} = -h^{\mu\nu} \equiv -\eta^{\mu\alpha}\eta^{\nu\beta}. \tag{18.83}$$

In this equation and throughout the present discussion indices are raised and lowered with respect to the Minkowski metric, $\eta_{\mu\nu}, \eta^{\mu\nu}$. For $\epsilon \ll 1$ we evidently have $g^{\mu\nu} \approx \eta^{\mu\nu} - \epsilon h^{\mu\nu}$.

Assuming that partial derivatives with respect to x^μ and ϵ commute, it is straightforward to compute the linearization of the Christoffel symbols,

$$\bar\Gamma^\mu_{\nu\rho} \equiv \left.\frac{\partial}{\partial\epsilon}\Gamma^\mu_{\nu\rho}\right|_{\epsilon=0} = -\tfrac{1}{2}h^{\mu\alpha}(\eta_{\alpha\nu,\rho}+\eta_{\alpha\rho,\nu}-\eta_{\nu\rho,\alpha}) + \tfrac{1}{2}\eta^{\mu\alpha}(h_{\alpha\nu,\rho}+h_{\alpha\rho,\nu}-h_{\nu\rho,\alpha})$$
$$= \tfrac{1}{2}\eta^{\mu\alpha}(h_{\alpha\nu,\rho}+h_{\alpha\rho,\nu}-h_{\nu\rho,\alpha})$$

Connections and curvature

and
$$\Gamma^{\mu}_{\nu\rho,\sigma} = \tfrac{1}{2}\eta^{\mu\alpha}(h_{\alpha\nu,\rho\sigma} + h_{\alpha\rho,\nu\sigma} - h_{\nu\rho,\alpha\sigma}).$$

Thus, from the component expansion of the curvature tensor (18.25), we have
$$r^{\mu}{}_{\nu\rho\sigma} \equiv \frac{\partial}{\partial\epsilon}R^{\mu}{}_{\nu\rho\sigma}\bigg|_{\epsilon=0} = \Gamma^{\mu}_{\nu\sigma,\rho} - \gamma^{\mu}_{\nu\rho,\sigma}$$

since
$$\frac{\partial}{\partial\epsilon}\Gamma^{\alpha}_{\nu\sigma}\Gamma^{\mu}_{\alpha\rho}\bigg|_{\epsilon=0} = \Gamma^{\alpha}_{\nu\sigma}\Gamma^{\mu}_{\alpha\rho}\bigg|_{\epsilon=0} + \Gamma^{\alpha}_{\nu\sigma}\bigg|_{\epsilon=0}\Gamma^{\mu}_{\alpha\rho} = 0, \text{ etc.}$$

since $\Gamma^{\alpha}_{\nu\sigma}|_{\epsilon=0} = 0$. Thus
$$r^{\mu}{}_{\nu\rho\sigma} = \tfrac{1}{2}\eta^{\mu\alpha}\left(h_{\alpha\nu,\sigma\rho} + h_{\alpha\sigma,\nu\rho} - h_{\nu\sigma,\alpha\rho} - h_{\alpha\nu,\rho\sigma} - h_{\alpha\rho,\nu\sigma} + h_{\nu\rho,\alpha\sigma}\right)$$

and for small values of the parameter ϵ the Riemann curvature tensor is
$$R_{\mu\nu\rho\sigma} \approx \epsilon r_{\mu\nu\rho\sigma} = \frac{\epsilon}{2}\left(h_{\mu\sigma,\nu\rho} + h_{\nu\rho,\mu\sigma} - h_{\nu\sigma,\mu\rho} - h_{\mu\rho,\nu\sigma}\right). \tag{18.84}$$

It is interesting to compare this equation with the expression in geodesic normal coordinates, Eq. (18.43).

The Newtonian tidal equation is derived by considering the motion of two neighbouring particles
$$\ddot{x}_i = -\phi_{,i} \quad \text{and} \quad \ddot{x}_i + \ddot{\eta}_i = -\phi_{,i}(\mathbf{x}+\boldsymbol{\eta}).$$

Since $\phi_{,i}(\mathbf{x}+\boldsymbol{\eta}) = \phi_{,i}(\mathbf{x}) + \phi_{,ij}(\mathbf{x})\eta_j$ we have
$$\ddot{\eta}_i = -\phi_{,ij}(\mathbf{x})\eta_j.$$

Compare with the equation of geodesic deviation (18.81) with s replaced by ct, which is approximately correct for velocities $|\dot{\mathbf{x}}| \ll c$,
$$\ddot{\eta}_i = -c^2 R_{i4j4}\eta_j,$$

and we should have, by Eq. (18.84),
$$R_{i4j4} = \frac{\epsilon}{2}(h_{i4,4j} + h_{4j,i4} - h_{ij,44} - h_{44,ij}) = \frac{\phi_{,ij}}{c^2}.$$

This equation can only hold in a general way if
$$\epsilon h_{44} \approx -\frac{2\phi}{c^2} \quad \text{and} \quad h_{i4,4j}, h_{ij,44} \ll h_{44,ij},$$

and the Newtonian approximation implies that
$$g_{44} \approx -1 + \epsilon h_{44} \approx -1 - \frac{2\phi}{c^2} \quad (\phi \ll c^2). \tag{18.85}$$

Note that the Newtonian potential ϕ has the dimensions of a velocity square – the weak field slow motion approximation of general relativity arises when this velocity is small compared to the velocity of light c.

18.8 General relativity

Multiplying Eq. (18.79) through by $g^{\mu\nu}$ we find

$$R - 2R = -R = \kappa T \quad \text{where} \quad T = T^\mu_\mu = g^{\mu\nu} T_{\mu\nu},$$

and Einstein's field equations can be written in the 'Ricci tensor form'

$$R_{\mu\nu} = \kappa \left(T_{\mu\nu} - \tfrac{1}{2} T g_{\mu\nu} \right). \tag{18.86}$$

Hence

$$R_{44} = R^i{}_{4i4} \approx \frac{1}{c^2} \nabla^2 \phi = \kappa \left(T_{44} + \tfrac{1}{2} T \right).$$

If we assume a perfect fluid, Example 9.4, for low velocities compared to c we have

$$T_{\mu\nu} = \left(\rho + \frac{P}{c^2} \right) V_\mu V_\nu + P g_{\mu\nu} \quad \text{where} \quad V_\mu \approx (v_1, v_2, v_3, -c),$$

so that

$$T = -c^2 \left(\rho + \frac{P}{c^2} \right) + 4P = -\rho c^2 + 3P \approx -\rho c^2,$$

and

$$T_{44} \approx \left(\rho + \frac{P}{c^2} \right) c^2 + P \left(-1 - \frac{\phi}{c^2} P \right) \approx \rho c^2.$$

Substituting in the Ricci form of Einstein's equations we find

$$\frac{1}{c^2} \nabla^2 \phi \approx \frac{1}{2} \kappa \rho c^2,$$

which is in agreement with the Newtonian equation (18.77) provided Einstein's gravitational constant has the form

$$\kappa = \frac{8\pi G}{c^4}. \tag{18.87}$$

Exercise: Show that the contracted Bianchi identity (18.60) implies that in geodesic coordinates at any point representing a local freely falling frame, the conservation identities (9.56) hold, $T^\mu{}_{\nu,\mu} = 0$.

Exercise: Show that if we had assumed field equations of the form $R_{\mu\nu} = \lambda T_{\mu\nu}$, there would have resulted the physically unsavoury result $T = \text{const}$.

Consider now the effect of a one-parameter family of coordinate transformations $x^\mu = x^\mu(y^\alpha, \epsilon)$ on a linearized approximation $g_{\mu\nu} = g_{\mu\nu}(\epsilon)$ and set

$$\xi^\mu = \frac{\partial x^\mu}{\partial \epsilon}\bigg|_{\epsilon=0}.$$

The transformation of components of the metric tensor results in

$$g_{\mu\nu}(x, \epsilon) \to g'_{\mu\nu}(y, \epsilon) = g_{\alpha\beta}(x(y, \epsilon), \epsilon) \frac{\partial x^\alpha}{\partial y^\mu} \frac{\partial x^\beta}{\partial y^\nu}$$

and taking $\partial/\partial\epsilon$ at $\epsilon = 0$ gives

$$h'_{\mu\nu}(y) = h_{\mu\nu}(y) + \xi_{\mu,\nu} + \xi_{\nu,\mu}. \tag{18.88}$$

Connections and curvature

These may be thought of as 'gauge transformations' for the weak fields $h_{\mu\nu}$, comparable with the gauge transformations (9.49), $A'_\mu = A_\mu + \psi_{,\mu}$, which leave the electromagnetic field $F_{\mu\nu}$ unchanged. In the present case, it is straightforward to verify that the transformations (18.88) leave the linearized Riemann tensor (18.84), or *real gravitational field*, invariant.

We define the quantities $\varphi_{\mu\nu}$ by

$$\varphi_{\mu\nu} = h_{\mu\nu} - h\eta_{\mu\nu} \quad \text{where} \quad h = h^\alpha_\alpha = \eta^{\alpha\beta} h_{\alpha\beta}.$$

The transformation of $\varphi^\nu_{\mu,\nu}$ under a gauge transformation is then

$$\varphi'^\nu_{\mu,\nu} = \varphi^\nu_{\mu,\nu} + \xi_{\mu,\nu}{}^\nu = \varphi^\nu_{\mu,\nu} + \Box\xi_\mu,$$

where indices are raised and lowered with the Minkowski metric, $\eta^{\mu\nu}$, $\eta_{\mu\nu}$. Just as done for the Lorentz gauge (9.51), it is possible (after dropping primes) to find ξ such that

$$\varphi^\nu_{\mu,\nu} = 0. \tag{18.89}$$

Such a gauge is commonly known as a **harmonic gauge**. There are still available gauge freedoms ξ_μ subject to solutions of the wave equation $\Box\xi_\mu = 0$.

A computation of the linearized Ricci tensor $r_{\mu\nu} = \eta^{\rho\sigma} r_{\rho\mu\sigma\nu}$ using Eq. (18.84) gives

$$r_{\mu\nu} = \tfrac{1}{2}\left(-\Box h_{\mu\nu} + \varphi^\rho_{\nu,\rho\mu} + \varphi^\rho_{\mu,\rho\nu}\right) = -\tfrac{1}{2}\Box h_{\mu\nu}$$

in a harmonic gauge. The Einstein tensor is thus $G_{\mu\nu} \approx -(\epsilon/2)\Box\varphi_{\mu\nu}$, and the linearized Einstein equation is

$$\epsilon\Box\varphi_{\mu\nu} = -\kappa T_{\mu\nu} = -\frac{16\pi G}{c^4} T_{\mu\nu},$$

having solution in terms of retarded Green's functions (12.23)

$$\epsilon\varphi_{\mu\nu}(\mathbf{x}, t) = -\frac{4G}{c^4} \iiint \frac{[T_{\mu\nu}(\mathbf{x}', t')]_{\text{ret}}}{|\mathbf{x}-\mathbf{x}'|} d^3 x'.$$

In vacuo, $T_{\mu\nu}$, Einstein's field equations can be written $R_{\mu\nu} = 0$, so that in the linearized approximation we have $\Box h_{\mu\nu} = 0$; these solutions are known as *gravitational waves* (see Problem 18.20 for further details).

The Schwarzschild solution

The vacuum Einstein field equations, $R_{\mu\nu} = 0$ are a non-linear set of 10 second-order equations for 10 unknowns $g_{\mu\nu}$ that can only be solved in a handful of special cases. The most important is that of *spherical symmetry* which, as we shall see in the next chapter, implies that the metric has the form in a set of coordinates $x^1 = r$, $x^2 = \theta$, $x^3 = \phi$, $x^4 = ct$,

$$ds^2 = e^\lambda dr^2 + r^2(d\theta^2 + \sin^2\theta \, d\phi^2) - e^\nu c^2 \, dt^2 \tag{18.90}$$

where θ and ϕ take the normal ranges of polar coordinates (r does not necessarily range from 0 to ∞), and λ and ν are functions of r and t. We will assume for simplicity that the solutions are *static* so that they are functions of the radial coordinate r alone, $\lambda = \lambda(r)$, $\nu = \nu(r)$. A remarkable theorem of Birkhoff assures us that all spherically symmetric vacuum solutions

18.8 General relativity

are in fact static for an appropriate choice of the coordinate t; a proof may be found in Synge [15].

We will perform calculations using Cartan's formalism. Many books prefer to calculate Christoffel symbols and do all computations in the coordinate system of Eq. (18.90). Let $e_1, \ldots, e_4$ be the orthonormal basis

$$e_1 = e^{-\frac{1}{2}\lambda}\partial_r, \quad e_2 = \frac{1}{r}\partial_\theta, \quad e_3 = \frac{1}{r\sin\theta}\partial_\phi, \quad e_4 = e^{-\frac{1}{2}\nu}c^{-1}\partial_t$$

such that

$$g_{\mu\nu} = g(e_\mu, e_\nu) = \eta = \mathrm{diag}(1, 1, 1, -1)$$

and let $\varepsilon^1, \ldots, \varepsilon^4$ be the dual basis

$$\varepsilon^1 = e^{\frac{1}{2}\lambda}dr, \quad \varepsilon^2 = r\,d\theta, \quad \varepsilon^3 = r\sin\theta\,d\phi, \quad \varepsilon^4 = e^{\frac{1}{2}\nu}c\,dt = e^{\frac{1}{2}\nu}dx^4.$$

We will write Cartan's structural relations in terms of the 'lowered' connection forms $\omega_{\mu\nu}$ since, by Eq. (18.73), we have $\omega_{\mu\nu} = -\omega_{\nu\mu}$. Thus (18.64) can be written

$$d\varepsilon^\mu = -\omega^\mu{}_\nu \wedge \varepsilon^\nu = -\eta^{\mu\rho}\omega_{\rho\nu}\varepsilon^\nu$$

and setting successively $\mu = 1, 2, 3, 4$ we have, writing derivatives with respect to r by a prime ',

$$d\varepsilon^1 = \tfrac{1}{2}e^{\frac{1}{2}\lambda}\lambda'\,dr \wedge dr = 0 = -\omega_{12} \wedge \varepsilon^2 - \omega_{13} \wedge \varepsilon^3 - \omega_{14} \wedge \varepsilon^4, \quad (18.91)$$

$$d\varepsilon^2 = r^{-1}e^{-\frac{1}{2}\lambda}\varepsilon^1 \wedge \varepsilon^2 = \omega_{12} \wedge \varepsilon^1 - \omega_{23} \wedge \varepsilon^3 - \omega_{24} \wedge \varepsilon^4, \quad (18.92)$$

$$d\varepsilon^3 = r^{-1}e^{-\frac{1}{2}\lambda}\varepsilon^1 \wedge \varepsilon^3 + r^{-1}\cot\theta\varepsilon^2 \wedge \varepsilon^3 = \omega_{13} \wedge \varepsilon^1 + \omega_{23} \wedge \varepsilon^2 - \omega_{34} \wedge \varepsilon^4, \quad (18.93)$$

$$d\varepsilon^4 = \tfrac{1}{2}e^{-\frac{1}{2}\lambda}\nu'\varepsilon^1 \wedge \varepsilon^4 = -\omega_{14} \wedge \varepsilon^1 - \omega_{24} \wedge \varepsilon^2 - \omega_{34} \wedge \varepsilon^4. \quad (18.94)$$

From (18.93) it follows at once that $\omega_{34} = \Gamma_{344}\varepsilon^4$, and substituting in (18.94) we see that $\Gamma_{344} = 0$ since it is the sole coefficient of the 2-form basis element $\varepsilon^3 \wedge \varepsilon^4$. Similarly, from (18.94), $\omega_{24} = \Gamma_{242}\varepsilon^2$ and

$$\omega_{14} = \tfrac{1}{2}e^{-\frac{1}{2}\lambda}\nu'\varepsilon^4 + \Gamma_{141}\varepsilon^1.$$

Continuing in this way we find the following values for the connection 1-forms:

$$\omega_{12} = -r^{-1}e^{-\frac{1}{2}\lambda}\varepsilon^2, \quad \omega_{13} = -r^{-1}e^{-\frac{1}{2}\lambda}\varepsilon^3, \quad \omega_{23} = -r^{-1}\cot\theta\varepsilon^3,$$

$$\omega_{14} = \tfrac{1}{2}e^{-\frac{1}{2}\lambda}\nu'\varepsilon^4, \quad \omega_{24} = 0, \quad \omega_{34} = 0. \quad (18.95)$$

To obtain the curvature tensor it is now a simple matter of substituting these forms in the second Cartan structural equation (18.67), with indices lowered

$$\rho_{\mu\nu} = -\rho_{\nu\mu} = 2\,d\omega_{\mu\nu} + 2\omega_{\mu\rho} \wedge \omega_{\sigma\nu}\eta^{\rho\sigma}. \quad (18.96)$$

For example

$$\rho_{12} = 2\bigl(d\omega_{12} + \omega_{13} \wedge \omega_{32} + \omega_{14} \wedge \omega_{42}\bigr)$$

$$= 2\,d(-r^{-1}e^{-\frac{1}{2}\lambda}\varepsilon^2) \quad \text{since } \omega_{13} \propto \omega_{23} \text{ and } \omega_{42} = 0$$

$$= -2e^{-\frac{1}{2}\lambda}(r^{-1}e^{-\frac{1}{2}\lambda})'\varepsilon^1 \wedge \varepsilon^2 - 2r^{-1}e^{-\frac{1}{2}\lambda}\,d\varepsilon^2.$$

543

Connections and curvature

Substituting for $d\varepsilon^2$ using Eq. (18.92) we find
$$\rho_{12} = r^{-1}\lambda' e^{-\lambda} \varepsilon^1 \wedge \varepsilon^2.$$

Similarly,
$$\rho_{13} = r^{-1}\lambda' e^{-\lambda} \varepsilon^1 \wedge \varepsilon^3, \qquad \rho_{23} = 2r^{-2}(1 - e^{-\lambda})\varepsilon^2 \wedge \varepsilon^3,$$
$$\rho_{14} = e^{-\lambda}(v'' - \tfrac{1}{2}\lambda' v' + (v')^2)\varepsilon^1 \wedge \varepsilon^4,$$
$$\rho_{24} = r^{-1} v' e^{-\lambda} \varepsilon^2 \wedge \varepsilon^4, \qquad \rho_{34} = r^{-1} v' e^{-\lambda} \varepsilon^3 \wedge \varepsilon^4.$$

The components of the Riemann tensor in this basis are given by
$$R_{\mu\nu\rho\sigma} = \overline{R}(e_\mu, e_\nu, e_\rho, e_\sigma) = \rho_{\mu\nu}(e_\rho, e_\sigma).$$

The non-vanishing components are
$$R_{1212} = R_{1313} = \frac{\lambda'}{2r} e^{-\lambda},$$
$$R_{2323} = \frac{1 - e^{-\lambda}}{r^2},$$
$$R_{1414} = \frac{1}{4} e^{-\lambda}(2v'' - \lambda' v' + (v')^2),$$
$$R_{2424} = R_{3434} = \frac{v'}{2r} e^{-\lambda}. \tag{18.97}$$

The Ricci tensor components
$$R_{\mu\nu} = \eta^{\rho\sigma} R_{\rho\mu\sigma\nu} = \sum_{i=1}^{3} R_{i\mu i\nu} - R_{4\mu 4\nu}$$

are therefore
$$R_{11} = e^{-\lambda}\left(-\frac{v''}{2} + \frac{\lambda' v'}{4} - \frac{(v')^2}{4} + \frac{\lambda'}{r}\right), \tag{18.98}$$
$$R_{44} = e^{-\lambda}\left(\frac{v''}{2} - \frac{\lambda' v'}{4} + \frac{(v')^2}{4} + \frac{v'}{r}\right), \tag{18.99}$$
$$R_{22} = R_{33} = e^{-\lambda}\left(\frac{\lambda' - v'}{2r} - \frac{1}{r^2}\right) + \frac{1}{r^2}. \tag{18.100}$$

To solve Einstein's vacuum equations $R_{\mu\nu} = 0$, we see by adding (18.98) and (18.99) that $\lambda' + v' = 0$, whence
$$\lambda = -v + C \quad (C = \text{const.})$$

A rescaling of the time coordinate, $t \to t' = e^{C/2} t$, has the effect of making $C = 0$, which we now assume. By Eq. (18.99), $R_{44} = 0$ reduces the second-order differential equation to
$$v'' + (v')^2 + \frac{2v'}{r} = 0$$

and the substitution $\alpha = e^v$ results in
$$(r^2 \alpha')' = 0,$$

18.8 General relativity

whence

$$\alpha = e^\nu = A - \frac{2m}{r} \quad (m, A = \text{const.}).$$

If we substitute this into $R_{22} = 0$ we have, by (18.100), $(r\alpha)' = 1$ so that $A = 1$. The most general spherically symmetric solution of Einstein's vacuum equations is therefore

$$ds^2 = \frac{1}{1 - 2m/r} dr^2 + r^2(d\theta^2 + \sin^2\theta \, d\phi^2) - \left(1 - \frac{2m}{r}\right) c^2 \, dt^2, \tag{18.101}$$

known famously as the **Schwarzschild solution**. Converting the polar coordinates to equivalent cartesian coordinates x, y, z we have, as $r \to \infty$,

$$ds^2 \approx dx^2 + dy^2 + dz^2 - c^2 \, dt^2 + \frac{2m}{r}(c^2 \, dt^2 + \cdots)$$

and $g_{\mu\nu} \approx \eta_{\mu\nu} + h_{\mu\nu}$ where

$$h_{44} = \frac{2m}{r} \approx \frac{-2\phi}{c^2},$$

assuming the Newtonian approximation with potential ϕ is applicable in this limit. Since the potential of a Newtonian mass M is given by $\phi = GM/r$, it is reasonable to make the identification

$$m = \frac{GM}{c^2}.$$

The constant m has dimensions of length and $2m = 2GM/c^2$, where the metric (18.101) exhibits singular behaviour, is commonly known as the **Schwarzschild radius**. For a solar mass, $M_\odot = 2 \times 10^{33}$ g, its value is about 3 km. However the Sun would need to collapse to approximately this size before strong corrections to Newtonian theory apply.

When paths of particles (timelike geodesics) and photons (null geodesics) are calculated in this metric, the following deviations from Newtonian theory are found for the solar system:

1. There is a slowing of clocks at a lower gravitational potential. At the surface of the Earth this amounts to a redshift from a transmitter to a receiver at a height h above it of

$$z = \frac{GM_e h}{R_e^2 c^2}.$$

   This amounts to a redshift of about 10^{-15} m^{-1} and is measurable using the Mossbauer effect.

2. The perihelion of a planet in orbit around the Sun precesses by an amount

$$\delta\varphi = \frac{6\pi M_\odot G}{c^2 a(1 - e^2)} \text{ per revolution.}$$

   For Mercury this comes out to 43 seconds of arc per century.

Connections and curvature

3. A beam of light passing the Sun at a closest distance r_0 is deflected an amount

$$\delta\varphi = \frac{4GM_\odot}{r_0 c^2}.$$

For a beam grazing the rim of the Sun $r_0 = R_\odot$ the deflection is 1.75 seconds of arc.

The limit $r \to 2m$ is of particular interest. Although it appears that the metric (18.101) is singular in this limit, this is really only a feature of the coordinates, not of the space-time as such. A clue that this may be the case is found by calculating the curvature components (18.97) for the Schwarzschild solution,

$$R_{1212} = R_{1313} = R_{2424} = R_{3434} = -\frac{m}{r^3}, \qquad R_{2424} = -R_{1414} = \frac{2m}{r^3}$$

all of which approach finite values as $r \to 2m$.

Exercise: Verify these expressions for components of the Riemann tensor.

More specifically, let us make the coordinate transformation from t to $v = ct + r + 2m \ln(r - 2m)$, sometimes referred to as *advanced time* since it can be shown to be constant on inward directed null geodesics, while leaving the spatial coordinates r, θ, ϕ unchanged. In these **Eddington–Finkelstein coordinates** the metric becomes

$$ds^2 = -\left(1 - \frac{2m}{r}\right)dv^2 + 2dr\, dv + r^2(d\theta^2 + \sin^2\theta\, d\phi^2).$$

As $r \to 2m$ the metric shows no abnormality in these coordinates. Inward directed timelike geodesics in the region $r > 2m$ reach $r = 2m$ in finite v-time (and also in finite proper time). However, after the geodesic particle crosses $r = 2m$ no light signals can be sent out from it into $r > 2m$ (see Fig. 18.4). The surface $r = 2m$ acts as a one-way membrane for light signals, called an **event horizon**. Observers with $r > 2m$ can never see any events inside $r = 2m$, an effect commonly referred to as a **black hole**.

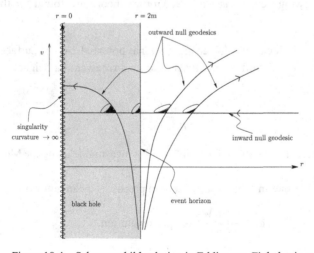

Figure 18.4   Schwarzschild solution in Eddington–Finkelstein coordinates

18.8 General relativity

Problems

Problem 18.20 A **linearized plane gravitational wave** is a solution of the linearized Einstein equations $\Box h_{\mu\nu} = 0$ of the form $h_{\mu\nu} = h_{\mu\nu}(u)$ where $u = x^3 - x^4 = z - ct$. Show that the harmonic gauge condition (18.89) implies that, up to undefined constants,

$$h_{14} + h_{13} = h_{24} + h_{23} = h_{11} + h_{22} = 0, \qquad h_{34} = -\tfrac{1}{2}(h_{33} + h_{44}).$$

Use the remaining gauge freedom $\xi_\mu = \xi_\mu(u)$ to show that it is possible to transform $h_{\mu\nu}$ to the form

$$[h_{\mu\nu}] = \begin{pmatrix} H & 0 \\ 0 & 0 \end{pmatrix} \quad \text{where} \quad H = \begin{pmatrix} h_{11} & h_{12} \\ h_{12} & -h_{11} \end{pmatrix}.$$

Setting $h_{11} = \alpha(u)$ and $h_{12} = \beta(u)$, show that the equation of geodesic deviation has the form

$$\ddot\eta_1 = \tfrac{\epsilon}{2}c^2(\alpha''\eta_1 + \beta''\eta_2), \qquad \ddot\eta_2 = \tfrac{\epsilon}{2}c^2(\beta''\eta_1 - \alpha''\eta_2)$$

and $\ddot\eta_3 = 0$. Make a sketch of the distribution of neighbouring accelerations of freely falling particles about a geodesic observer in the two cases $\beta = 0$ and $\alpha = 0$. These results are central to the observational search for gravity waves.

Problem 18.21 Show that every two-dimensional space-time metric (signature 0) can be expressed locally in *conformal coordinates*

$$ds^2 = e^{2\phi}(dx^2 - dt^2) \quad \text{where} \quad \phi = \phi(x, t).$$

Calculate the Riemann curvature tensor component R_{1212}, and write out the two-dimensional Einstein vacuum equations $R_{ij} = 0$. What is their general solution?

Problem 18.22 (a) For a perfect fluid in general relativity,

$$T_{\mu\nu} = (\rho c^2 + P)U_\mu U_\nu + Pg_{\mu\nu} \qquad (U^\mu U_\mu = -1)$$

show that the conservation identities $T^{\mu\nu}{}_{;\nu} = 0$ imply

$$\rho_{,\nu} U^\nu + (\rho c^2 + P) U^\nu{}_{;\nu},$$
$$(\rho c^2 + P) U^\mu{}_{;\nu} U^\nu + P_{,\nu}(g^{\mu\nu} + U^\mu U^\nu).$$

(b) For a pressure-free fluid show that the streamlines of the fluid (i.e. the curves $x^\mu(s)$ satisfying $dx^\mu/ds = U^\mu$) are geodesics, and ρU^μ is a covariant 4-current, $(\rho U^\mu)_{,\mu} = 0$.
(c) In the Newtonian approximation where

$$U_\mu = \left(\tfrac{v_i}{c}, -1\right) + O(\beta^2), \qquad P = O(\beta^2)\rho c^2, \qquad \left(\beta = \tfrac{v}{c}\right)$$

where $|\beta| \ll 1$ and $g_{\mu\nu} = \eta_{\mu\nu} + \epsilon h_{\mu\nu}$ with $\epsilon \ll 1$, show that

$$h_{44} \approx -\tfrac{2\phi}{c^2}, \qquad h_{ij} \approx -\tfrac{2\phi}{c^2}\delta_{ij} \quad \text{where} \quad \nabla^2\phi = 4\pi G\rho$$

and $h_{i4} = O(\beta)h_{44}$. Show in this approximation that the equations $T^{\mu\nu}{}_{;\nu} = 0$ approximate to

$$\frac{\partial\rho}{\partial t} + \nabla\cdot(\rho\mathbf{v}) = 0, \qquad \rho\frac{d\mathbf{v}}{dt} = -\nabla P - \rho\nabla\phi.$$

Problem 18.23 (a) Compute the components of the Ricci tensor $R_{\mu\nu}$ for a space-time that has a metric of the form

$$ds^2 = dx^2 + dy^2 - 2\,du\,dv + 2H\,dv^2 \qquad (H = H(x, y, u, v)).$$

547

(b) Show that the space-time is a vacuum if and only if $H = \alpha(x, y, v) + f(v)u$ where $f(v)$ is an arbitrary function and α satisfies the two-dimensional Laplace equation

$$\frac{\partial^2 \alpha}{\partial x^2} + \frac{\partial^2 \alpha}{\partial y^2} = 0,$$

and show that it is possible to set $f(v) = 0$ by a coordinate transformation $u' = ug(v)$, $v' = h(v)$.
(c) Show that $R_{i4j4} = -H_{,ij}$ for $i, j = 1, 2$.

Problem 18.24 Show that a coordinate transformation $r = h(r')$ can be found such that the Schwarzschild solution has the form

$$ds^2 = -e^{\mu(r')} dt^2 + e^{\nu(r')}(dr'^2 + r'^2(d\theta^2 + \sin^2\theta \, d\phi^2)).$$

Evaluate the functions e^μ and e^ν explicitly.

Problem 18.25 Consider an oscillator at $r = r_0$ emitting a pulse of light (null geodesic) at $t = t_0$. If this is received by an observer at $r = r_1$ at $t = t_1$, show that

$$t_1 = t_0 + \int_{r_0}^{r_1} \frac{dr}{c(1 - 2m/r)}.$$

By considering a signal emitted at $t_0 + \Delta t_0$, received at $t_1 + \Delta t_1$ (assuming the radial positions r_0 and r_1 to be constant), show that $\Delta t_0 = \Delta t_1$ and the **gravitational redshift** found by comparing *proper times* at emission and reception is given by

$$1 + z = \frac{\Delta \tau_1}{\Delta \tau_0} = \sqrt{\frac{1 - 2m/r_1}{1 - 2m/r_0}}.$$

Show that for two clocks at different heights h on the Earth's surface, this reduces to

$$z \approx \frac{2GM}{c^2} \frac{h}{R},$$

where M and R are the mass and radius of the Earth.

Problem 18.26 In the Schwarzschild solution show the only possible closed photon path is a circular orbit at $r = 3m$, and show that it is unstable.

Problem 18.27 (a) A particle falls radially inwards from rest at infinity in a Schwarzschild solution. Show that it will arrive at $r = 2m$ in a finite *proper time* after crossing some fixed reference position r_0, but that coordinate time $t \to \infty$ as $r \to 2m$.
(b) On an infalling extended body compute the tidal force in a radial direction, by parallel propagating a tetrad (only the radial spacelike unit vector need be considered) and calculating R_{1414}.
(c) Estimate the total tidal force on a person of height 1.8 m, weighing 70 kg, falling head-first into a solar mass black hole ($M_\odot = 2 \times 10^{30}$ kg), as he crosses $r = 2m$.

18.9 Cosmology

Cosmology is the study of the universe taken as a whole [18]. Generally it is assumed that on the broadest scale of observation the universe is homogeneous and isotropic – no particular positions or directions are singled out. Presuming that general relativity applies on this

18.9 Cosmology

overall scale, the metrics that have homogeneous and isotropic spatial sections are known as **flat Robertson–Walker models**. The simplest of these are the so-called **flat models**

$$ds^2 = a^2(t)(dx^2 + dy^2 + dz^2) - c^2 dt^2$$
$$= a^2(t)(dr^2 + r^2(d\theta^2 + \sin^2\theta\, d\phi^2)) - (dx^4)^2 \tag{18.102}$$

where the word 'flat' refers to the 3-surfaces $t = $ const., not to the entire metric. Setting

$$\varepsilon^1 = a(t)\, dr, \qquad \varepsilon^2 = a(t)r\, d\theta, \qquad \varepsilon^3 = a(t)r\sin\theta\, d\phi, \qquad \varepsilon^4 = c\, dt,$$

the first structural relations imply, much as in spherical symmetry,

$$\omega_{12} = -\frac{1}{ar}\varepsilon^2, \qquad \omega_{13} = -\frac{1}{ar}\varepsilon^3, \qquad \omega_{23} = -\frac{\cot\theta}{ar}\varepsilon^3$$

$$\omega_{14} = \frac{\dot{a}}{ca}\varepsilon^1, \qquad \omega_{24} = \frac{\dot{a}}{ca}\varepsilon^2, \qquad \omega_{34} = \frac{\dot{a}}{ca}\varepsilon^3$$

and substitution in the second structural relations gives, as in the Schwarzschild case,

$$\rho_{12} = \frac{2\dot{a}^2}{c^2 a^2}\varepsilon^1 \wedge \varepsilon^2, \qquad \rho_{13} = \frac{2\dot{a}^2}{c^2 a^2}\varepsilon^1 \wedge \varepsilon^3, \qquad \rho_{23} = \frac{2\dot{a}^2}{c^2 a^2}\varepsilon^2 \wedge \varepsilon^3$$

$$\rho_{14} = -\frac{2\ddot{a}}{c^2 a}\varepsilon^1 \wedge \varepsilon^4, \qquad \rho_{24} = -\frac{2\ddot{a}}{c^2 a}\varepsilon^2 \wedge \varepsilon^4, \qquad \rho_{34} = -\frac{2\ddot{a}}{c^2 a}\varepsilon^3 \wedge \varepsilon^4.$$

Hence the only non-vanishing curvature tensor components are

$$R_{1212} = R_{1313} = R_{2323} = \frac{\dot{a}^2}{c^2 a^2}, \qquad R_{1414} = R_{2424} = R_{3434} = -\frac{\ddot{a}}{c^2 a}.$$

The non-vanishing Ricci tensor components $R_{\mu\nu} = \sum_{i=1}^{3} R_{i\mu i\nu} - R_{4\mu 4\nu}$ are

$$R_{11} = R_{22} = R_{33} = \frac{1}{c^2}\left(\frac{\ddot{a}}{a} + \frac{2\dot{a}^2}{a^2}\right), \qquad R_{44} = -\frac{3\ddot{a}}{c^2 a},$$

and the Einstein tensor is

$$G_{11} = G_{22} = G_{33} = -\frac{1}{c^2}\left(\frac{2\ddot{a}}{a} + \frac{\dot{a}^2}{a^2}\right), \qquad G_{44} = \frac{3\dot{a}^2}{c^2 a^2}.$$

The **closed Robertson–Walker models** are defined in a similar way, but the spatial sections $t = $ const. are 3-spheres:

$$ds^2 = a^2(t)\big[d\chi^2 + \sin^2\chi(d\theta^2 + \sin^2\theta\, d\phi^2)\big] - c^2 dt^2. \tag{18.103}$$

Combining the analysis in Example 18.4 and that given above, we set

$$\varepsilon^1 = a\, d\chi, \qquad \varepsilon^2 = a\sin\chi\, d\theta, \qquad \varepsilon^3 = a\sin\chi\sin\theta\, d\phi, \qquad \varepsilon^4 = c\, dt.$$

Connections and curvature

The sections $t = \text{const.}$ are compact spaces having volume

$$V(t) = \int \varepsilon^1 \wedge \varepsilon^2 \wedge \varepsilon^3$$
$$= \int a^3(t) \sin^2 \chi \sin\theta \, d\chi \wedge d\theta \wedge d\phi$$
$$= a^3(t) \int_0^\pi \sin^2 \chi \, d\chi \int_0^\pi \sin\theta \, d\theta \int_0^{2\pi} d\phi$$
$$= 2\pi^2 a^3(t). \tag{18.104}$$

We find that ω_{i4} are as in the flat case, while

$$\omega_{12} = -\frac{\cot\chi}{a}\varepsilon^2, \qquad \omega_{13} = -\frac{\cot\chi}{a}\varepsilon^3, \qquad \omega_{23} = -\frac{\cot\theta}{a\sin\chi}\varepsilon^3.$$

The second structural relations result in the same ρ_{i4} as for the flat case, while an additional term $\dot{a}^2/c^2 a^2$ appears in the coefficients of the other curvature forms,

$$\rho_{12} = \left(\frac{\dot{a}^2}{c^2 a^2} + \frac{2\ddot{a}^2}{c^2 a^2}\right), \text{ etc.}$$

Finally, the so-called **open Robertson–Walker models**, having the form

$$ds^2 = a^2(t)\left[d\chi^2 + \sinh^2\chi(d\theta^2 + \sin^2\theta \, d\phi^2)\right] - c^2 \, dt^2, \tag{18.105}$$

give rise to similar expressions for ω_{ij} with hyperbolic functions replacing trigonometric, and

$$\rho_{12} = \left(\frac{\dot{a}^2}{c^2 a^2} - \frac{2\ddot{a}^2}{c^2 a^2}\right), \text{ etc.}$$

In summary, the non-vanishing curvature tensor components in the three models may be written

$$R_{1212} = R_{1313} = R_{2323} = \frac{\dot{a}^2}{c^2 a^2} + \frac{k}{a^2},$$

$$R_{1414} = R_{2424} = R_{3434} = -\frac{\ddot{a}}{c^2 a},$$

where $k = 0$ refers to the flat model, $k = 1$ the closed and $k = -1$ the open model. The Einstein tensor is thus

$$G_{11} = G_{22} = G_{33} = -\frac{1}{c^2}\left(\frac{2\ddot{a}}{a} + \frac{k\dot{a}^2}{a^2}\right), \qquad G_{44} = \frac{3\dot{a}^2}{c^2 a^2},$$

and Einstein's field equations $G_{\mu\nu} = \kappa T_{\mu\nu}$ imply that the energy–stress tensor is that of a perfect fluid $T = P(t)\sum_{i=1}^{3} \varepsilon^i \otimes \varepsilon^i + \rho(t)c^2 \varepsilon^4 \otimes \varepsilon^4$ (see Example 9.4), where

$$\frac{\dot{a}^2}{a^2} = \frac{8\pi G}{3}\rho - \frac{kc^2}{a^2}, \tag{18.106}$$

$$\frac{\ddot{a}}{a} = -\frac{4\pi G}{3}\left(\rho + 3\frac{P}{c^2}\right). \tag{18.107}$$

18.9 Cosmology

Taking the time derivative of (18.106) and substituting (18.107) gives

$$\dot{\rho} + \frac{3\dot{a}}{a}\left(\rho + \frac{P}{c^2}\right) = 0. \tag{18.108}$$

Exercise: Show that (18.108) is equivalent to the Bianchi identity $T^{4\nu}{}_{;\nu} = 0$.

If we set $P = 0$, a form of matter sometimes known as **dust**, then Eq. (18.108) implies that $d(\rho a^3)/dt = 0$, and the density has evolution

$$\rho(t) = \rho_0 a^{-3} \quad (\rho_0 = \text{const.}). \tag{18.109}$$

This shows, using (18.104), that for a closed universe the total mass of the universe $M = \rho(t)V(t) = \rho_0 2\pi^2$ is finite and constant. Substituting (18.109) into Eq. (18.108) we have the **Friedmann equation**

$$\dot{a}^2 = \frac{8\pi G \rho_0}{3} a^{-1} - kc^2. \tag{18.110}$$

It is convenient to define rescaled variables

$$\alpha = \frac{3c^2}{8\pi G \rho_0} a, \quad y = \frac{3c^3}{8\pi G \rho_0} t$$

and Eq. (18.110) becomes

$$\left(\frac{d\alpha}{dy}\right)^2 = \frac{1}{\alpha} - k. \tag{18.111}$$

The solutions are as follows.

$k = 0$: It is straightforward to verify that, up to an arbitrary origin of the time coordinate,

$$\alpha = \left(\frac{9}{4}\right)^{1/3} y^{2/3} \implies a(t) = (6\pi G \rho_0)^{1/3} t^{2/3}, \quad \rho = \frac{1}{6\pi G t^2}.$$

This solution is known as the **Einstein–de Sitter universe**.

$k = 1$: Equation (18.111) is best solved in parametric form

$$\alpha = \sin^2 \eta, \quad y = \eta - \sin \eta \cos \eta$$

a cycloid in the $\alpha - \eta$ plane, which starts at $\alpha = 0$, $\eta = 0$, rises to a maximum at $\eta = \pi/2$ then recollapses to zero at $\eta = \pi$. This behaviour is commonly referred to as an *oscillating universe*, but the term is not well chosen as there is no reason to expect that the universe can 'bounce' out of the singularity at $a = 0$ where the curvature and density are infinite.

$k = -1$: The solution is parametrically $\alpha = \sinh^2 \eta$, $y = \sinh \eta \cosh \eta - \eta$, which expands indefinitely as $\eta \to \infty$.

Collectively these models are known as **Friedmann models**, the Einstein–de Sitter model acting as a kind of critical case dividing closed from open models (see Fig. 18.5).

Connections and curvature

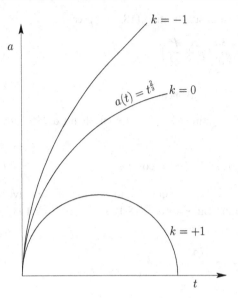

Figure 18.5   Friedmann cosmological models

Observational cosmology is still trying to decide which of these models is the closest representation of our actual universe, but most evidence favours the open model.

Problems

Problem 18.28   Show that for a closed Friedmann model of total mass M, the maximum radius is reached at $t = 2GM/3c^3$ where its value is $a_{max} = 4GM/3\pi c^2$.

Problem 18.29   Show that the *radiation filled universe*, $P = \frac{1}{3}\rho$ has $\rho \propto a^{-4}$ and the time evolution for $k = 0$ is given by $a \propto t^{1/2}$. Assuming the radiation is black body, $\rho = a_S T^4$, where $a_S = 7.55 \times 10^{-15}$ erg cm^{-3} K^{-4}, show that the temperature of the universe evolves with time as

$$T = \left(\frac{3c^2}{32\pi G a_S}\right)^{1/4} t^{-1/2} = \frac{1.52}{\sqrt{t}} \text{ K} \quad (t \text{ in seconds}).$$

Problem 18.30   Consider two radial light signals (null geodesics) received at the spatial origin of coordinates at times t_0 and $t_0 + \Delta t_0$, emitted from $\chi = \chi_1$ (or $r = r_1$ in the case of the flat models) at time $t = t_1 < t_0$. By comparing proper times between reception and emission show that the observer experiences a redshift in the case of an expanding universe ($a(t)$ increasing) given by

$$1 + z = \frac{\Delta t_0}{\Delta t_1} = \frac{a(t_0)}{a(t_1)}.$$

Problem 18.31   By considering light signals as in the previous problem, show that an observer at $r = 0$, in the Einstein–de Sitter universe can at time $t = t_0$ see no events having radial coordinate $r > r_H = 3ct_0$. Show that the mass contained within this radius, called the **particle horizon**, is given by $M_H = 6c^3 t_0 / G$.

18.10 Variation principles in space-time

We conclude this chapter with a description of the variation principle approach to field equations, including an appropriate variational derivation of Einstein's field equations [13, 19]. Recall from Chapter 17 that for integration over an orientable four-dimensional space-time (M, g) we need a non-vanishing 4-form Ω. Let $\varepsilon^1, \varepsilon^2, \varepsilon^3, \varepsilon^4$ be any o.n. basis of differential 1-forms, $g^{-1}(\varepsilon^\mu, \varepsilon^\nu) = \eta^{\mu\nu}$. The volume element $\Omega = \varepsilon^1 \wedge \varepsilon^2 \wedge \varepsilon^3 \wedge \varepsilon^4$ defined by this basis is independent of the choice of orthonormal basis, provided they are related by a proper Lorentz transformation. Following the discussion leading to Eq. (8.32), we have that the components of this 4-form in an arbitrary coordinate $(U, \phi; x^\mu)$ are

$$\Omega_{\mu\nu\rho\sigma} = \frac{\sqrt{|g|}}{4!} \epsilon_{\mu\nu\rho\sigma} \quad \text{where} \quad g = \det[g_{\mu\nu}],$$

and we can write

$$\Omega = \frac{\sqrt{|g|}}{4!} \epsilon_{\mu\nu\rho\sigma}\, dx^\mu \otimes dx^\nu \otimes dx^\rho \otimes dx^\sigma$$
$$= \sqrt{-g}\, dx^1 \wedge dx^2 \wedge dx^3 \wedge dx^4.$$

Every 4-form Λ can be written $\Lambda = f\Omega$ for some scalar function f on M, and for every regular domain D contained in the coordinate domain U,

$$\int_D \Lambda = \int_{\phi(D)} f \sqrt{-g}\, d^4x \quad (d^4x \equiv dx^1\, dx^2\, dx^3\, dx^4).$$

If $f\sqrt{-g} = A^\mu{}_{,\mu}$ for some set of functions A^μ on M, then

$$\Lambda = A^\mu{}_{,\mu}\, dx^1 \wedge dx^2 \wedge dx^3 \wedge dx^4$$

and a simple argument, such as suggested in Problem 17.8, leads to

$$\Lambda = d\alpha \quad \text{where} \quad \alpha = \frac{1}{3!}\epsilon_{\mu\nu\rho\sigma} A^\mu\, dx^\nu \wedge dx^\rho \wedge dx^\sigma$$

and by Stokes' theorem (17.3),

$$\int_D \Lambda = \int_D d\alpha = \int_{\partial D} \alpha. \tag{18.112}$$

Let $\Phi_A(x)$ be any set of fields on a neighbourhood V of regular domain $x \in D \subset M$, where the index $A = 1, \ldots, N$ refers to all possible components (scalar, vector, tensor, etc.) that may arise. By a **variation** of these fields is meant a one-parameter family of fields $\tilde{\Phi}_A(x, \lambda)$ on D such that

1. $\tilde{\Phi}_A(x, 0) = \Phi_A(x)$ for all $x \in D$.
2. $\tilde{\Phi}_A(x, \lambda) = \Phi_A(x)$ for all λ and all $x \in V - D$.

The second condition implies that the condition holds on the boundary ∂D and also all derivatives of the variation field components agree there, $\tilde{\Phi}_{A,\mu}(x, \lambda) = \Phi_{A,\mu}(x)$ for all $x \in \partial D$. We define the variational derivatives

$$\delta\Phi_A = \frac{\partial}{\partial \lambda} \tilde{\Phi}_A(x, \lambda)\bigg|_{\lambda=0}.$$

Connections and curvature

This vanishes on the boundary ∂D since $\tilde{\Phi}_A$ is independent of λ there. A **Lagrangian** is a function $L(\Phi_A, \Phi_{A,\mu})$ dependent on the fields and their derivatives. It defines a 4-form $\Lambda = L\Omega$ and an associated **action**

$$I = \int_D \Lambda = \int_{D\cap U} L\sqrt{-g}\, d^4x.$$

Field equations arise by requiring that the action be stationary,

$$\delta I = \frac{d}{d\lambda}\int_D \Lambda \bigg|_{\lambda=0} = \int_{D\cap U} \delta(L\sqrt{-g})\, d^4x,$$

called a **field action principle** that, as for path actions, can be evaluated by

$$0 = \int_{D\cap U} \frac{\partial L\sqrt{-g}}{\partial \Phi_A}\delta\Phi_A + \frac{\partial L\sqrt{-g}}{\partial \Phi_{A,\mu}}\delta\Phi_{A,\mu}\, d^4x$$

$$= \int_{D\cap U} \frac{\partial L\sqrt{-g}}{\partial \Phi_A}\delta\Phi_A + \frac{\partial}{\partial x^\mu}\left(\frac{\partial L\sqrt{-g}}{\partial \Phi_{A,\mu}}\delta\Phi_A\right) - \frac{\partial}{\partial x^\mu}\left(\frac{\partial L\sqrt{-g}}{\partial \Phi_{A,\mu}}\right)\delta\Phi_A\, d^4x.$$

Using the version of Stokes' theorem given in (18.112), the middle term can be converted to an integral over the boundary

$$\int_{\partial D} \frac{1}{3!}\epsilon_{\mu\nu\rho\sigma}\frac{\partial L}{\partial \Phi_{A,\mu}}\sqrt{-g}\delta\Phi_A\, dx^\nu \wedge dx^\rho \wedge dx^\sigma,$$

which vanishes since $\delta\Phi_A = 0$ on ∂D. Since $\delta\Phi_A$ are arbitrary functions (subject to this boundary constraint) on D, we deduce the Euler–Lagrange field equations

$$\frac{\delta L\sqrt{-g}}{\delta \Phi_A} \equiv \frac{\partial L\sqrt{-g}}{\partial \Phi_A} - \frac{\partial}{\partial x^\mu}\left(\frac{\partial L\sqrt{-g}}{\partial \Phi_{A,\mu}}\right) = 0. \tag{18.113}$$

It is best to include the term $\sqrt{-g}$ within the derivatives since, as we shall see, the fields may depend specifically on the metric tensor components $g_{\mu\nu}$.

Hilbert action

Einstein's field equations may be derived from a variation principle, by setting the Lagrangian to be the Ricci scalar, $L = R$. This is known as the **Hilbert Lagrangian**. For independent variables it is possible to take either the metric tensor components $g_{\mu\nu}$ or those of the inverse tensor $g^{\mu\nu}$. We will adopt the latter, as it is slightly more convenient (the reader may try to adapt the analysis that follows for the variables $\Phi_A = g_{\mu\nu}$. We cannot use the Euler–Lagrange equations (18.113) as they stand since R is dependent on $g^{\mu\nu}$ and its first *and second* derivatives. While the Euler–Lagrange analysis can be extended to include Lagrangians that depend on second derivatives of the fields (see Problem 18.32), this would be a prohibitively complicated calculation. Proceeding directly, we have

$$\delta I_G = \delta\int_D R\Omega = \int_{D\cap U} \delta(R_{\mu\nu}g^{\mu\nu}\sqrt{-g})\, d^4x,$$

18.10 Variation principles in space-time

whence

$$\delta I_G = \int_{D \cap U} (\delta R_{\mu\nu}\, g^{\mu\nu} \sqrt{-g} + R_{\mu\nu}\delta g^{\mu\nu} \sqrt{-g} + R\delta\sqrt{-g})d^4x. \tag{18.114}$$

We pause at this stage to analyse the last term, $\delta\sqrt{-g}$. Forgetting temporarily that $g_{\mu\nu}$ is a symmetric tensor, and assuming that all components are independent, we see that the determinant is a homogeneous function of degree in the components,

$$\delta g = \frac{\partial g}{\partial g_{\mu\nu}} \delta g_{\mu\nu} = G^{\mu\nu} \delta g_{\mu\nu}$$

where $G^{\mu\nu}$ is the cofactor of $g_{\mu\nu}$. We may therefore write

$$\delta g = g g^{\nu\mu} \delta g_{\mu\nu} = g g^{\mu\nu} \delta g_{\mu\nu} \tag{18.115}$$

since $g^{\mu\nu} = g^{\nu\mu}$. The symmetry of $g_{\mu\nu}$ may be imposed at this stage, without in any way altering this result. From $g^{\mu\nu} g_{\mu\nu} = \delta^\mu_\mu = 4$ it follows at once that we can write (18.115) as

$$\delta g = -g g_{\mu\nu} \delta g^{\mu\nu}. \tag{18.116}$$

Hence

$$\delta(\sqrt{-g}) = \frac{1}{2\sqrt{-g}} \delta(-g) = -\frac{\sqrt{-g}}{2} g_{\mu\nu} \delta g^{\mu\nu}. \tag{18.117}$$

A similar analysis gives

$$(\sqrt{-g})_{,\rho} = -\frac{\sqrt{-g}}{2} g_{\mu\nu} g^{\mu\nu}{}_{,\rho} \tag{18.118}$$

and from the formula (18.40) for Christoffel symbols,

$$\Gamma^\mu_{\rho\mu} = \frac{1}{2} g^{\mu\nu} g_{\nu\mu,\rho} = \frac{1}{\sqrt{-g}} (\sqrt{-g})_{,\rho}. \tag{18.119}$$

This identity is particularly useful in providing an equation for the covariant divergence of a vector field:

$$A^\mu{}_{;\mu} = A^\mu{}_{,\mu} + \Gamma^\mu_{\rho\mu} A^\rho = \frac{1}{\sqrt{-g}} (A^\mu \sqrt{-g})_{,\mu}. \tag{18.120}$$

We are now ready to continue with our evaluation of δI_G. Using (18.117) we can write Eq. (18.114) as

$$\delta I_G = \int_{D \cap U} (R_{\mu\nu} - \tfrac{1}{2} R) \delta g^{\mu\nu} \sqrt{-g} + \delta R_{\mu\nu}\, g^{\mu\nu} \sqrt{-g} d^4 x. \tag{18.121}$$

To evaluate the last term, we write out the Ricci tensor components

$$R_{\mu\nu} = \Gamma^\rho_{\mu\nu,\rho} - \Gamma^\rho_{\mu\rho,\nu} + \Gamma^\alpha_{\mu\nu} \Gamma^\rho_{\alpha\rho} - \Gamma^\alpha_{\mu\rho} \Gamma^\rho_{\alpha\nu}.$$

Since $\delta \Gamma^\rho_{\mu\nu}$ is the limit as $\lambda \to 0$ of a difference of two connections, it is a tensor field and we find

$$\delta R_{\mu\nu} = (\delta \Gamma^\rho_{\mu\nu})_{,\rho} - (\delta \Gamma^\rho_{\mu\rho})_{,\nu},$$

Connections and curvature

as may be checked either directly or, more simply, in geodesic normal coordinates. Hence, using $g^{\mu\nu}{}_{;\rho} = 0$,

$$\delta R_{\mu\nu} g^{\mu\nu} = W^\rho{}_{;\rho} \quad \text{where} \quad W^\rho = \delta\Gamma^\rho_{\mu\nu} g^{\mu\nu} - \delta\Gamma^\nu_{\mu\nu} g^{\mu\rho}$$

and from Eq. (18.120) we see that

$$\delta R_{\mu\nu} g^{\mu\nu} \sqrt{-g} = \left(W^\rho \sqrt{-g}\right)_{,\rho}.$$

Since W^ρ depends on $\delta g^{\mu\nu}$ and $\delta g^{\mu\nu}{}_{,\sigma}$, it vanishes on the boundary ∂D, and the last term in Eq. (18.121) is zero by Stokes' theorem. Since $\delta g^{\mu\nu}$ is assumed arbitrary on D, the Hilbert action gives rise to Einstein's vacuum field equations,

$$G_{\mu\nu} = 0 \Rightarrow R_{\mu\nu} = 0.$$

Energy–stress tensor of fields

With other fields Φ_A present we take the total Lagrangian to be

$$L = \frac{1}{2\kappa} + L_F(\Phi_A, \Phi_{A,\mu}, g^{\mu\nu}) \tag{18.122}$$

and we have

$$0 = \delta I = \delta \int_{D \cap U} \left(\frac{1}{2\kappa} R + L_F\right) \sqrt{-g}\, d^4x$$

$$= \int_{D \cap U} \left(\frac{1}{2\kappa} G_{\mu\nu} \delta g^{\mu\nu} \sqrt{-g} + \frac{\partial L_F \sqrt{-g}}{\partial g^{\mu\nu}} \delta g^{\mu\nu} + \frac{\delta L_F \sqrt{-g}}{\delta \Phi_A} \delta \Phi_A\right) d^4x.$$

Variations with respect to field variables Φ_A give rise to the Euler–Lagrange field equations (18.113), while the coefficients of $\delta g^{\mu\nu}$ lead to the full Einstein's field equations

$$G_{\mu\nu} = \kappa T_{\mu\nu} \quad \text{where} \quad T_{\mu\nu} = -\frac{2}{\sqrt{-g}} \frac{\partial L_F \sqrt{-g}}{\partial g^{\mu\nu}}. \tag{18.123}$$

Example 18.5 An interesting example of a variation principle is the **Einstein–Maxwell theory**, where the field variables are taken to be components of a covector field A_μ – essentially the electromagnetic 4-potential given in Eq. (9.47) – and the field Lagrangian is taken to be

$$L_F = -\frac{1}{16\pi} F_{\mu\nu} F^{\mu\nu} = \frac{1}{16\pi} F_{\mu\nu} F_{\rho\sigma} g^{\mu\rho} g^{\nu\sigma} \tag{18.124}$$

where $F_{\mu\nu} = A_{\nu,\mu} - A_{\mu,\nu}$. A straightforward way to compute the electromagnetic energy–stress tensor is to consider variations of $g^{\mu\nu}$,

$$T_{\mu\nu} \delta g^{\mu\nu} = -\frac{2}{\sqrt{-g}} \delta\left(L_F \sqrt{-g}\right)$$

$$= -\frac{2}{\sqrt{-g}} \left[\frac{-1}{8\pi} F_{\mu\nu} F_{\rho\sigma} \delta g^{\mu\rho} g^{\nu\sigma} \sqrt{-g} + L_F \delta(\sqrt{-g})\right]$$

$$= \tfrac{1}{4\pi}\left(F_{\mu\alpha} F_\nu{}^\alpha - \tfrac{1}{4} F_{\alpha\beta} F^{\alpha\beta} g_{\mu\nu}\right), \tag{18.125}$$

on using Eq. (18.117). This expression agrees with that proposed in Example 9.5, Eq. (9.59).

References

Variation of the field variables A_μ gives

$$\delta \int_D L_F \Omega = \frac{-1}{4\pi} \int_{D\cap U} \delta A_{\nu,\mu} F^{\mu\nu} \sqrt{-g}\, d^4x$$

$$= \frac{-1}{4\pi} \int_{D\cap U} \delta A_{\nu;\mu} F^{\mu\nu} \sqrt{-g}\, d^4x$$

$$= \frac{-1}{4\pi} \int_{D\cap U} \left(\delta A_\nu F^{\mu\nu}\right)_{;\mu} \sqrt{-g} - \delta A_\nu F^{\mu\nu}{}_{;\mu}\sqrt{-g}\, d^4x$$

$$= \frac{-1}{4\pi} \int_{D\cap U} \left(\delta A_\nu F^{\mu\nu}\sqrt{-g}\right)_{,\mu} - \delta A_\nu F^{\mu\nu}{}_{;\mu}\sqrt{-g}\, d^4x.$$

As the first term in the integrand is an ordinary divergence its integral vanishes, and we arrive at the charge-free covariant Maxwell equations

$$F^{\mu\nu}{}_{;\mu} = 0. \tag{18.126}$$

The source-free equations follow automatically from $F_{\mu\nu} = A_{\nu,\mu} - A_{\mu,\nu}$:

$$F_{\mu\nu,\rho} + F_{\nu\rho,\mu} + F_{\rho\mu,\nu} = 0 \iff F_{\mu\nu;\rho} + F_{\nu\rho;\mu} + F_{\rho\mu;\nu} = 0. \tag{18.127}$$

Problems

Problem 18.32 If a Lagrangian depends on second and higher order derivatives of the fields, $L = L(\Phi_A, \Phi_{A,\mu}, \Phi_{A,\mu\nu}, \ldots)$ derive the generalized Euler–Lagrange equations

$$\frac{\delta L\sqrt{-g}}{\delta \Phi_A} \equiv \frac{\partial L\sqrt{-g}}{\partial \Phi_A} - \frac{\partial}{\partial x^\mu}\left(\frac{\partial L\sqrt{-g}}{\partial \Phi_{A,\mu}}\right) + \frac{\partial^2}{\partial x^\mu \partial x^\nu}\left(\frac{\partial L\sqrt{-g}}{\partial \Phi_{A,\mu\nu}}\right) - \cdots = 0.$$

Problem 18.33 For a skew symmetric tensor $F^{\mu\nu}$ show that

$$F^{\mu\nu}{}_{;\nu} = \frac{1}{\sqrt{-g}}\left(\sqrt{-g}\, F^{\mu\nu}\right)_{,\nu}.$$

Problem 18.34 Compute the Euler–Lagrange equations and energy–stress tensor for a scalar field Lagrangian in general relativity given by

$$L_S = -\psi_{,\mu}\psi_{,\nu} g^{\mu\nu} - m^2\psi^2.$$

Verify $T^{\mu\nu}{}_{;\nu} = 0$.

Problem 18.35 Prove the implication given in Eq. (18.127). Show that this equation and Eq. (18.126) imply $T^{\mu\nu}{}_{;\nu} = 0$ for the electromagnetic energy–stress tensor given in Eqn. (18.125).

References

[1] C. de Witt-Morette, Y. Choquet-Bruhat and M. Dillard-Bleick. *Analysis, Manifolds and Physics*. Amsterdam, North-Holland, 1977.
[2] T. Frankel. *The Geometry of Physics*. New York, Cambridge University Press, 1997.
[3] M. Nakahara. *Geometry, Topology and Physics*. Bristol, Adam Hilger, 1990.

[4] W. H. Chen, S. S. Chern and K. S. Lam. *Lectures on Differential Geometry*. Singapore, World Scientific, 1999.

[5] S. Kobayashi and K. Nomizu. *Foundations of Differential Geometry*. New York, Interscience Publishers, 1963.

[6] E. Nelson. *Tensor Analysis*. Princeton, N.J., Princeton University Press, 1967.

[7] J. A. Schouten. *Ricci Calculus*. Berlin, Springer-Verlag, 1954.

[8] J. L. Synge and A. Schild. *Tensor Calculus*. Toronto, University of Toronto Press, 1959.

[9] W. Kopczyński and A. Trautman. *Spacetime and Gravitation*. Chichester, John Wiley & Sons, 1992.

[10] S. Hawking and W. Israel. *Three Hundred Years of Gravitation*. Cambridge, Cambridge University Press, 1987.

[11] R. d'Inverno. *Introducing Einstein's Relativity*. Oxford, Oxford University Press, 1993.

[12] R. K. Sachs and H. Wu. *General Relativity for Mathematicians*. New York, Springer-Verlag, 1977.

[13] E. Schrödinger. *Space-Time Structure*. Cambridge, Cambridge University Press, 1960.

[14] H. Stephani. *General Relativity*. Cambridge, Cambridge University Press, 1982.

[15] J. L. Synge. *Relativity: The General Theory*. Amsterdam, North-Holland, 1960.

[16] R. M. Wald. *General Relativity*. Chicago, The University of Chicago Press, 1984.

[17] P. Szekeres. The gravitational compass. *Journal of Mathematical Physics*, **6**:1387–91, 1965.

[18] J. V. Narlikar. *Introduction to Cosmology*. Cambridge, Cambridge University Press, 1993.

[19] W. Thirring. *A Course in Mathematical Physics, Vol. 2: Classical Field Theory*. New York, Springer-Verlag, 1979.

19 Lie groups and Lie algebras

19.1 Lie groups

In Section 10.8 we defined a topological group as a group G that is also a topological space such that the map $\psi : (g, h) \mapsto gh^{-1}$ is continuous. If G has the structure of a differentiable manifold and ψ is a smooth map it is said to be a **Lie group**. The arguments given in Section 10.8 to show that the maps $\phi : (g, h) \mapsto gh$ and $\tau : g \mapsto g^{-1}$ are both continuous are easily extended to show that both are differentiable in the case of a Lie group. The map τ is clearly a diffeomorphism of G since $\tau^{-1} = \tau$. Details of proofs in Lie group theory can sometimes become rather technical. We will often resort to outline proofs, when the full proof is not overly instructive. Details can be found in a number of texts, such as [1–7]. Warner [6] is particularly useful in this respect.

Example 19.1 The additive group $\mathbb{R}^n$, described in Example 10.20, is an n-dimensional abelian Lie group, as is the n-torus $T^n = \mathbb{R}^n/\mathbb{Z}^n$.

Example 19.2 The set $M_n(\mathbb{R})$ of $n \times n$ real matrices is a differentiable manifold, diffeomorphic to $\mathbb{R}^{n^2}$, with global coordinates $x^i{}_j(\mathbf{A}) = A^i{}_j$, where $\mathbf{A}$ is the matrix $[A^i{}_j]$. The general linear group $GL(n, \mathbb{R})$ is an n^2-dimensional Lie group, which is an open submanifold of $M_n(\mathbb{R})$ and a Lie group since the function ψ is given by

$$x^i{}_j(\psi(\mathbf{A}, \mathbf{B})) = x^i{}_k(\mathbf{A}) x^k{}_j(\mathbf{B}^{-1}),$$

which is differentiable since $x^k{}_j(\mathbf{B}^{-1})$ are rational polynomial functions of the components $B^i{}_j$ with non-vanishing denominator on $\det^{-1}(\mathbb{R})$. In a similar way, $GL(n, \mathbb{C})$ is a Lie group, since it is an open submanifold of $M_n(\mathbb{C}) \cong \mathbb{C}^{n^2} \cong \mathbb{R}^{2n^2}$.

Left-invariant vector fields

If G is a Lie group, the operation of **left translation** $L_g : G \to G$ defined by $L_g h \equiv L_g(h) = gh$ is a diffeomorphism of G onto itself. Similarly, the operation of **right translation** $R_g : G \to G$ defined by $R_g h = hg$ is a diffeomorphism.

Lie groups and Lie algebras

A left translation L_g induces a map on the module of vector fields, $L_{g*} : \mathcal{T}(M) \to \mathcal{T}(M)$ by setting

$$(L_{g*}X)_a = L_{g*}X_{g^{-1}a} \tag{19.1}$$

where X is any smooth vector field. On the right-hand side of Eq. (19.1) L_{g*} is the tangent map at the point $g^{-1}a$. A vector field X on G is said to be **left-invariant** if $L_{g*}X = X$ for all $g \in G$.

Given a tangent vector A at the identity, $A \in T_e(G)$, define the vector field X on G by $X_g = L_{g*}A$. This vector field is left-invariant for, by Eq. (15.17),

$$L_{g*}X_h = L_{g*} \circ L_{h*}A = (L_g \circ L_h)_*A = L_{gh*}A = X_{gh}.$$

It is clearly the unique left-invariant vector field on G such that $X_e = A$. We must show, however, that X is a differentiable vector field. In a local coordinate chart $(U, \varphi; x^i)$ at the identity e let the composition law be represented by n differentiable functions $\psi^i : \varphi(U) \times \varphi(U) \to \varphi(U)$,

$$x^i(gh^{-1}) = \psi^i(x^1(g), \ldots, x^n(g), x^1(h), \ldots, x^n(h)) \quad \text{where} \quad \psi^i = x^i \circ \psi.$$

For any smooth function $f : G \to \mathbb{R}$

$$Xf = (L_{g*}A)f = A(f \circ L_g)$$

$$= A^i \left. \frac{\partial f \circ L_g}{\partial x^i} \right|_{x^i(e)}$$

$$= A^i \left. \frac{\partial}{\partial y^i} \hat{f}(\psi^1(x^1(g), \ldots, x^n(g), \mathbf{y}), \ldots, \psi^n(x^1(g), \ldots, x^n(g), \mathbf{y})) \right|_{y^i = x^i(e)}$$

where $\hat{f} = f \circ \varphi^{-1}$. Hence Xf is differentiable at e since it is differentiable on the neighbourhood U. If g is an arbitrary point of G then gU is an open neighbourhood of g and every point $h \in gU$ can be written $h = gg'$ where $g' \in U$, so that

$$(Xf)(h) = A(f \circ L_h)$$
$$= A(f \circ L_g \circ L_{g'})$$
$$= X(f \circ L_g)(g')$$
$$= X(f \circ L_g) \circ L_{g^{-1}}(h).$$

Thus $Xf = X(f \circ L_g) \circ L_{g^{-1}}$ on gU, and it follows that Xf is differentiable at g.

Hence X is the unique differentiable left-invariant vector field everywhere on G such that $X_e = A$. Left-invariant vector fields on G are therefore in one-to-one correspondence with tangent vectors at e, and form a vector space of dimension n, denoted $\mathcal{G}$.

Lie algebra of a Lie group

Given a smooth map $\varphi : M \to N$ between manifolds M and M', we will say vector fields X on M and X' on M' are φ-**related** if $\varphi_*X_p = X'_{\varphi(p)}$ for every $p \in M$. In general there does not exist a vector field on M' that is φ-related to a given vector field X on M unless φ is a diffeomorphism (see Section 15.4).

19.1 Lie groups

Lemma 19.1 *If $\varphi : M \to M'$ is a smooth map and X and Y are two vector fields on M, φ-related respectively to X' and Y' on M', then their Lie brackets $[X, Y]$ and $[X', Y']$ are φ-related.*

Proof: If $f : M' \to \mathbb{R}$ is a smooth map on M' then for any $p \in M$

$$(X' f) \circ \varphi = X(f \circ \varphi)$$

since $X'_{\varphi(p)} f = (\varphi_* X_p) f = X_p(f \circ \varphi)$. Hence

$$\begin{aligned}[][X', Y']_{\varphi(p)} f &= X'_{\varphi(p)}(Y' f) - Y'_{\varphi(p)}(X' f) \\ &= X_p\big((Y' f) \circ \varphi\big) - Y_p\big((X' f) \circ \varphi\big) \\ &= X_p\big(Y(f \circ \varphi)\big) - Y_p\big(X(f \circ \varphi)\big) \\ &= [X, Y]_p(f \circ \varphi) \\ &= \varphi_*[X, Y]_p f,\end{aligned}$$

as required. ∎

Exercise: Show that if X is a left-invariant vector field then $(Xf) \circ L_g = X(f \circ L_g)$.

If X and Y are left-invariant vector fields on a Lie group G it follows from Lemma 19.1 that

$$L_{g*}[X, Y] = [X, Y]. \tag{19.2}$$

The vector space $\mathcal{G}$ therefore forms an n-dimensional Lie algebra called the **Lie algebra of the Lie group** G. Because of the one-to-one correspondence between $\mathcal{G}$ and $T_e(G)$ it is meaningful to write $[A, B]$ for any pair $A, B \in T_e(G)$, and the Lie algebra structure can be thought of as being imposed on the tangent space at the identity e.

Let $A_1, \ldots, A_n$ be a basis of the tangent space at the identity $T_e(G)$, and $X_1, \ldots, X_n$ the associated set of left-invariant vector fields forming a basis of $\mathcal{G}$. As in Section 6.5 define the structure constants $C^k_{ij} = -C^k_{ji}$ by

$$[X_i, X_j] = C^k_{ij} X_k.$$

Exercise: Show that the Jacobi identities (15.24) are equivalent to

$$C^k_{mi} C^m_{jl} + C^k_{mj} C^m_{li} + C^k_{ml} C^m_{ij} = 0. \tag{19.3}$$

Example 19.3 Let $\mathbb{R}^n$ be the additive abelian Lie group of Example 19.1. The vector field X generated by a tangent vector $A = A^i (\partial_{x^i})_e$ has components

$$X^i(g) = (Xx^i)(g) = (L_{g*} A) x^i = A(x^i \circ L_g).$$

Now $x^i \circ L_g = g^i + x^i$ where $g^i = x^i(g)$, whence

$$X^i(g) = A^j \frac{\partial}{\partial x^j}(g^i + x^i)\bigg|_{x^i=0} = A^j \delta^i_j = A^i.$$

If $X = A^i \partial_{x^i}$ and $Y = B^j \partial_{x^j}$ are left-invariant vector fields, then for any function f

$$[X, Y]f = A^i \frac{\partial}{\partial x^i}\left(B^j \frac{\partial f}{\partial x^j}\right) - B^j \frac{\partial}{\partial x^j}\left(A^i \frac{\partial f}{\partial x^i}\right)$$
$$= A^i B^j (f_{,ji} - f_{,ij}) = 0.$$

Hence $[X, Y] = 0$ for all left-invariant vector fields. The Lie algebra of the abelian Lie group $\mathbb{R}^n$ is commutative.

Example 19.4 Let A be a tangent vector at the identity element of $GL(n, \mathbb{R})$,

$$A = A^i{}_j \left(\frac{\partial}{\partial x^i{}_j}\right)_{X=I}.$$

The tangent space at $e = I$ is thus isomorphic with the vector space of $n \times n$ real matrices $M_n(\mathbb{R})$. The left-invariant vector field X generated by this tangent vector is

$$X = X^i{}_j \frac{\partial}{\partial x^i{}_j}$$

with components

$$X^i{}_j(\mathsf{G}) = (L_{\mathsf{G}*} A) x^i{}_j$$
$$= A(x^i{}_j \circ L_\mathsf{G})$$
$$= A^p{}_q \frac{\partial}{\partial x^p{}_q}(G^i{}_k x^k{}_j)\Big|_{X=I} \quad \text{where } G^i{}_j = x^i{}_j(\mathsf{G})$$
$$= A^p{}_q G^i{}_k \delta^k_p \delta^q_j$$
$$= x^i{}_k(\mathsf{G}) A^k{}_j.$$

Hence

$$X = x^i{}_k A^k{}_j \frac{\partial}{\partial x^i{}_j}.$$

If X and Y are left-invariant vector fields such that $X_e = A$ and $Y_e = B$, then their Lie bracket has components

$$[X, Y]^i{}_j = [X, Y] x^i{}_j = x^p_m A^m_q \frac{\partial}{\partial x^p_q}(x^i_k B^k_j) - x^p_m B^m_q \frac{\partial}{\partial x^p_q}(x^i_k A^k_j)$$
$$= x^i_m (A^m_k B^k_j - B^m_k A^k_j).$$

At the identity element $e = I$ the components of $[X, Y]$ are therefore formed by taking the matrix commutator product $\mathbf{AB} - \mathbf{BA}$ where $\mathbf{A} = [A^i{}_j]$ and $\mathbf{B} = [B^i{}_j]$, and the Lie algebra of $GL(n, \mathbb{R})$ is isomorphic to the Lie algebra formed by taking commutators of $n \times n$ matrices from $M_n(\mathbb{R})$, known as $\mathcal{GL}(n, \mathbb{R})$.

Maurer–Cartan relations

We say that a differential form α is **left-invariant** if $L_g^* \alpha = \alpha$ for all $g \in G$. Its exterior derivative $d\alpha$ is also left-invariant, for $L_g^* d\alpha = dL_g^* \alpha = d\alpha$. If ω is a left-invariant 1-form

19.1 Lie groups

and X a left-invariant vector field then $\langle \omega, X \rangle$ is constant over G, for

$$\langle \omega_g, X_g \rangle = \langle \omega_g, L_{g*} X_e \rangle = \langle L_g^* \omega_g, X_e \rangle = \langle \omega_e, X_e \rangle.$$

By the Cartan identity, Eq. (16.14), we therefore have

$$d\omega(X, Y) = \tfrac{1}{2}\big[X(\langle Y, \omega \rangle) - Y(\langle X, \omega \rangle) - \langle \omega, [X, Y] \rangle\big]$$
$$= -\tfrac{1}{2}\langle \omega, [X, Y] \rangle. \tag{19.4}$$

Let $E_1, \ldots, E_n$ be a left-invariant set of vector fields, forming a basis of the Lie algebra $\mathcal{G}$ and $\varepsilon^1, \ldots, \varepsilon^n$ the dual basis of differential 1-forms such that

$$\langle \varepsilon^i, E_i \rangle = \delta^i{}_j.$$

These 1-forms are left-invariant, for $L_g^* \varepsilon^i = \varepsilon^i$ for $i = 1, \ldots, n$ as

$$\langle L_g^* \varepsilon^i, E_j \rangle = \langle \varepsilon^i, L_{g*} E_j \rangle = \langle \varepsilon^i, E_j \rangle = \delta^i{}_j.$$

Hence, by (19.4),

$$d\varepsilon^i(E_j, E_k) = -\tfrac{1}{2}\langle \varepsilon^i, [E_j, E_k] \rangle = -\tfrac{1}{2}\langle \varepsilon^i, C^l_{jk} E_l \rangle = -\tfrac{1}{2} C^i_{jk},$$

from which we can deduce the **Maurer–Cartan relations**

$$d\varepsilon^i = -\tfrac{1}{2} C^i_{jk} \varepsilon^j \wedge \varepsilon^k. \tag{19.5}$$

Exercise: Show that the Jacobi identities (19.3) follow by taking the exterior derivative of the Maurer Cartan relations.

Theorem 19.2 *A Lie group G has vanishing structure constants if and only if it is isomorphic to the abelian group $\mathbb{R}^n$ in some neighbourhood of the identity.*

Proof: If there exists a coordinate neighbourhood of the identity such that $(gh)^i = g^i + h^i$ then from Example 19.3 $[X, Y] = 0$ throughout this neighbourhood. Thus $C^i_{jk} = 0$, since the Lie algebra structure is only required in a neighbourhood of e.

Conversely, if all structure constants vanish then the Maurer–Cartan relations imply $d\varepsilon^i = 0$. By the Poincaré lemma 17.5, there exist functions y^i in a neighbourhood of the identity such that $\varepsilon^i = dy^i$. Using these as local coordinates at e, we may assume that the identity is at the origin of these coordinates, $y^i(e) = 0$. For any a, g in the domain of these coordinates

$$L_g^*(\varepsilon^i)_{L_g a} = (\varepsilon^i)_a = da^i \quad \text{where } a^i = y^i(a),$$

and

$$L_g^*(\varepsilon^i)_{L_g a} = L_g^*(dy^i)_{ga} = \big(d(y^i \circ L_g)\big)_{ga}.$$

Writing $\phi(g, h) = gh$, we have for a fixed g

$$da^i = d\big(y^i \circ \phi(g, a)\big) = \frac{\partial \phi^i(g^1, \ldots, g^n, a^1, \ldots, a^n)}{\partial a^j} da^j$$

where $\phi^i = y^i \circ \phi$. Thus
$$\frac{\partial \phi^i}{\partial a^j} = \delta^i{}_j,$$
equations that are easily integrated to give
$$\phi^i(g^1, \ldots, g^n, a^1, \ldots, a^n) = f^i(g^1, \ldots, g^n) + a^i.$$
If $a = e$, so that $a^i = 0$, we have $\phi^i(g^1, \ldots, g^n, 0, \ldots, 0) = f^i(g^1, \ldots, g^n) = g^i$. The required local isomorphism with $\mathbb{R}^n$ follows immediately,
$$\phi^i(g^1, \ldots, g^n, a^1, \ldots, a^n) = g^i + a^i.$$
∎

Problems

Problem 19.1 Let $E^i{}_j$ be the matrix whose (i, j)th component is 1 and all other components vanish. Show that these matrices form a basis of $\mathcal{GL}(n, \mathbb{R})$, and have the commutator relations
$$[E^i{}_j, E^k{}_l] = \delta^i{}_l E^k{}_j - \delta^k{}_j E^i{}_l.$$
Write out the structure constants with respect to this algebra in this basis.

Problem 19.2 Let $E^i{}_j$ be the matrix defined as in the previous problem, and $F^i{}_j = iE^i{}_j$ where $i = \sqrt{-1}$. Show that these matrices form a basis of $\mathcal{GL}(n, \mathbb{C})$, and write all the commutator relations between these generators of $\mathcal{GL}(n, \mathbb{C})$.

Problem 19.3 Define the $T_e(G)$-valued 1-form θ on a Lie group G, by setting
$$\theta_g(X_g) = L_{g^{-1}*}X_g$$
for any vector field X on G (not necessarily left-invariant). Show that θ is left-invariant, $L_a^*\theta_g = \theta_{a^{-1}g}$ for all $a, g \in G$.

With respect to a basis E_i of left-invariant vector fields and its dual basis ε^i, show that
$$\theta = \sum_{i=1}^{n}(E_i)_e \varepsilon^i.$$

19.2 The exponential map

A **Lie group homomorphism** between two Lie groups G and H is a differentiable map $\varphi : G \to H$ that is a group homomorphism, $\varphi(gh) = \varphi(g)\varphi(h)$ for all $g, h \in G$. In the case where it is a diffeomorphism φ is said to be a **Lie group isomorphism**.

Theorem 19.3 *A Lie group homomorphism $\varphi : G \to H$ induces a Lie algebra homomorphism $\varphi_* : \mathcal{G} \to \mathcal{H}$. If φ is an isomorphism, then φ_* is a Lie algebra isomorphism.*

Proof: The tangent map at the origin, $\varphi_* : T_e(G) \to T_e(H)$, defines a map between Lie algebras $\mathcal{G}$ and $\mathcal{H}$. If X is a left-invariant vector field on G then the left-invariant vector

19.2 The exponential map

field $\varphi_* X$ on H is defined by

$$(\varphi_* X)_h = L_{h*} \varphi_* X_e.$$

Since φ is a homomorphism

$$\varphi \circ L_a(g) = \varphi(ag) = \varphi(a)\varphi(g) = L_{\varphi(a)} \circ \varphi(g)$$

for arbitrary $g \in G$. Thus the vector fields X and $\varphi_* X$ are φ-related,

$$\begin{aligned}\varphi_* X_a &= \varphi_* \circ L_{a*} X_e \\ &= L_{\varphi(a)*} \circ \varphi_* X_e \\ &= (\varphi_* X)_{\varphi(a)}.\end{aligned}$$

It follows by Theorem 19.1 that the Lie brackets $[X, Y]$ and $[\varphi_* X, \varphi_* Y]$ are φ-related,

$$[\varphi_* X, \varphi_* Y] = \varphi_*[X, Y],$$

so that φ_* is a Lie algebra homomorphism. In the case where φ is an isomorphism, φ_* is one-to-one and onto at G_e. ∎

A **one-parameter subgroup** of G is a homomorphism $\gamma : \mathbb{R} \to G$,

$$\gamma(t + s) = \gamma(t)\gamma(s).$$

Of necessity, $\gamma(0) = e$. The tangent vector at the origin generates a unique left-invariant vector field X such that $X_e = \dot\gamma(0)$, where we use the self-explanatory notation $\dot\gamma(t)$ for $\dot\gamma_{\gamma(t)}$. The vector field X is everywhere tangent to the curve, $X_{\gamma(t)} = \dot\gamma(t)$, for if $f : G \to \mathbb{R}$ is any differentiable function then

$$\begin{aligned}X_{\gamma(t)} f &= \left(L_{\gamma(t)*}\dot\gamma(0)\right)f \\ &= \dot\gamma(0)(f \circ L_{\gamma(t)}) \\ &= \frac{d}{du} f\bigl(\gamma(t)\gamma(u)\bigr)\Big|_{u=0} \\ &= \frac{d}{du} f\bigl(\gamma(t+u)\bigr)\Big|_{u=0} \\ &= \frac{d}{dt} f\bigl(\gamma(t)\bigr) = \dot\gamma(t) f.\end{aligned}$$

Conversely, given any left-invariant vector field X, there is a unique one-parameter group $\gamma : \mathbb{R} \to G$ that generates it. The proof uses the result of Theorem 15.2 that there exists a local one-parameter group of transformations σ_t on a neighbourhood U of the identity e such that $\sigma_{t+s} = \sigma_t \circ \sigma_s$ for $0 \le |t|, |s|, |t+s| < \epsilon$, and for all $h \in U$ and smooth functions f on U

$$X_h f = \frac{d f\bigl(\sigma_t(h)\bigr)}{dt}\bigg|_{t=0}.$$

For all $h, g \in U$ such that $g^{-1}h \in U$ we have

$$\begin{aligned} X_h f &= X_{L_g(g^{-1}h)} f \\ &= (L_{g*} X_{g^{-1}h}) f \\ &= X_{g^{-1}h} (f \circ L_g) \\ &= \left. \frac{df \circ L_g \circ \sigma_t(g^{-1}h)}{dt} \right|_{t=0} \\ &= \left. \frac{df(\sigma'_t(h))}{dt} \right|_{t=0} \end{aligned}$$

where

$$\sigma'_t = L_g \circ \sigma_t \circ L_{g^{-1}}.$$

The maps σ'_t form a local one-parameter group of transformations, since

$$\sigma'_t \circ \sigma'_s = L_g \circ \sigma_t \circ \sigma_s \circ L_{g^{-1}} = L_g \circ \sigma_{t+s} \circ L_{g^{-1}} = \sigma'_{t+s},$$

and generate the same vector field X as generated by σ_t. Hence $\sigma'_t = \sigma_t$ on a neighbourhood U of e, so that

$$\sigma_t \circ L_g = L_g \circ \sigma_t. \tag{19.6}$$

Setting $\gamma(t) = \sigma_t(e)$, we have

$$\sigma_t(g) = g\sigma_t(e) = g\gamma(t)$$

for all $g \in U$, and the one-parameter group property follows for γ for $0 \leq |t|, |s|, |t+s| < \epsilon$,

$$\begin{aligned} \gamma(t+s) = \sigma_{t+s}(e) = \sigma_t \circ \sigma_s(e) &= \sigma_t(\gamma(s)) \\ &= \gamma(s)\sigma_t(e) = \gamma(s)\gamma(t) = \gamma(t)\gamma(s). \end{aligned}$$

The local one-parameter group may be extended to all values of t and s by setting

$$\gamma(t) = (\gamma(t/n))^n$$

for n a positive integer chosen such that $|t/n| < \epsilon$. The group property follows for all t, s from

$$\gamma(t+s) = (\gamma((t+s)/n))^n = (\gamma(t/n)\gamma(s/n))^n = (\gamma(t/n))^n (\gamma(s/n))^n = \gamma(t)\gamma(s).$$

It is straightforward to verify that this one-parameter group is tangent to X for all values of t.

Exponential map

The **exponential map** $\exp : \mathcal{G} \to G$ is defined by

$$\exp X \equiv \exp(X) = \gamma(1)$$

19.2 The exponential map

where $\gamma : \mathbb{R} \to G$, the one-parameter subgroup generated by the left-invariant vector field X. Then

$$\gamma(s) = \exp sX. \tag{19.7}$$

For, let α be the one-parameter subgroup defined by $\alpha(t) = \gamma(st)$,

$$\alpha(t + t') = \gamma(s(t + t')) = \gamma(st)\gamma(st') = \alpha(t)\alpha(t').$$

If f is any smooth function on G, then

$$\dot{\alpha}(0)f = \frac{d}{dt}f(\gamma(st))\Big|_{t=0} = s\frac{d}{du}f(\gamma(u))\Big|_{u=0} = sX_e f.$$

Thus α is the one-parameter subgroup generated by the left-invariant vector field sX, and $\exp sX = \alpha(1) = \gamma(s)$.

We further have that

$$(Xf)(\exp tX) = X_{\gamma(t)}f = \dot{\gamma}(t)f = \frac{d}{dt}f(\gamma(t))$$

so that

$$(Xf)(\exp tX) = \frac{d}{dt}f(\exp tX). \tag{19.8}$$

The motivation for the name 'exponential map' lies in the identity

$$\exp(sX)\exp(tX) = \gamma(s)\gamma(t) = \gamma(s+t) = \exp(s+t)X. \tag{19.9}$$

Example 19.5 Let

$$X = x^i_k A^k_j \frac{\partial}{\partial x^i_j}$$

be a left-invariant vector field on the general linear group $GL(n, \mathbb{R})$. Setting $f = x^i_j$ we have

$$Xf = Xx^i_j = x^i_k A^k_j$$

and substitution in Eq. (19.8) gives

$$A^k_j x^i_k(\exp tX) = \frac{d}{dt}x^i_j(\exp tX).$$

If $y^i_k = x^i_k(\exp tX)$ are the components of the element $\exp(tX)$, and the matrix $Y = [y^i_k]$ satisfies the linear differential equation

$$\frac{dY}{dt} = YA$$

with the initial condition $Y(0) = I$, having unique solution

$$Y(t) = e^{tA} = I + tA + \frac{t^2 A^2}{2!} + \cdots,$$

then if X_e is identified with the matrix $A \in \mathcal{G}$, the matrix of components of the element $\exp(tX_e) \in GL(n, \mathbb{R})$ is e^{tA}.

For any left-invariant vector field X we have, by Eq. (19.8),

$$\frac{d}{dt} f(L_g(\exp tX)) = (X(f \circ L_g))(\exp tX)$$
$$= (L_{g*}X f)(\exp tX)$$
$$= (Xf)(g \exp tX)$$

since $L_{g*}X = X$. The curve $t \mapsto L_g \circ \exp tX$ is therefore an integral curve of X through g at $t = 0$. Since g is an arbitrary point of G, the maps $\sigma_t : g \to g \exp tX = R_{\exp tX} g$ form a one-parameter group of transformations that generate X, and we conclude that every left-invariant vector field is complete.

The exponential map is a diffeomorphism of a neighbourhood of the zero vector $0 \in \mathcal{G}$ onto a neighbourhood of the identity $e \in G$. The proof may be found in [6]. If $\varphi : H \to G$ is a Lie group homomorphism then

$$\varphi \circ \exp = \exp \circ \varphi_* \qquad (19.10)$$

where $\varphi_* : \mathcal{G} \to \mathcal{H}$ is the induced Lie group homomorphism of Theorem 19.3. For, let $\gamma : \mathbb{R} \to G$ be the curve defined by $\gamma(t) = \varphi(\exp tX)$. Since φ is a homomorphism, this curve is a one-parameter subgroup of G

$$\gamma(t)\gamma(s) = \varphi(\exp tX \exp sX) = \varphi(\exp(t+s)X) = \gamma(t+s).$$

Its tangent vector at $t = 0$ is $\varphi_* X_e$ since

$$\dot\gamma(0) f = \left.\frac{df \circ \varphi(\exp tX)}{dt}\right|_{t=0} = X_e(f \circ \varphi) = (\varphi_* X_e) f,$$

and γ is the one-parameter subgroup generated by the left-invariant vector field $\varphi_* X$. Hence $\varphi(\exp tX) = \exp t\varphi_* X$ and Eq. (19.10) follows on setting $t = 1$.

Problems

Problem 19.4 A function $f : G \to \mathbb{R}$ is said to be an analytic function on G if it can be expanded as a Taylor series at any point $g \in G$. Show that if X is a left-invariant vector field and f is an analytic function on G then

$$f(g \exp tX) = (e^{tX} f)(g)$$

where, for any vector field Y, we define

$$e^Y f = f + Yf + \frac{1}{2!} Y^2 f + \frac{1}{3!} Y^3 f + \cdots = \sum_{i=0}^{\infty} \frac{Y^n}{n!} f.$$

The operator Y^n is defined inductively by $Y^n f = Y(Y^{n-1} f)$.

Problem 19.5 Show that $\exp tX \exp tY = \exp t(X + Y) + O(t^2)$.

19.3 Lie subgroups

A **Lie subgroup** H of a Lie group G is a subgroup that is a Lie group, and such that the natural injection map $i : H \to G$ defined by $i(g) = g$ makes it into an embedded submanifold of G. It is called a **closed subgroup** if in addition H is a closed subset of G. In this case the embedding is regular and its topology is that induced by the topology of G (see Example 15.12). The injection i induces a Lie algebra homomorphism $i_* : \mathcal{H} \to (G)$, which is clearly an isomorphism of $\mathcal{H}$ with a Lie subalgebra of $\mathcal{G}$. We may therefore regard the Lie algebra of the Lie subgroup as being a Lie subalgebra of $\mathcal{G}$.

Example 19.6   Let $T^2 = S^1 \times S^1$ be the 2-torus, where S^1 is the one-dimensional Lie group where composition is addition modulo 1. This is evidently a Lie group whose elements can be written as pairs of complex numbers $(e^{i\theta}, e^{i\phi})$, where

$$(e^{i\theta}, e^{i\phi})(e^{i\theta'}, e^{i\phi'}) = (e^{i(\theta+\theta')}, e^{i(\phi+\phi')}).$$

The subset

$$H = \{(e^{iat}, e^{ibt}) \mid -\infty < t < \infty\}$$

is a Lie subgroup for arbitrary values of a and b. If a/b is rational it is isomorphic with S^1, the embedding is regular and it is a closed submanifold. If a/b is irrational then the subgroup winds around the torus an infinite number of times and is arbitrarily close to itself everywhere. In this case the embedding is not regular and the induced topology does not correspond to the submanifold topology. It is still referred to as a Lie subgroup. This is done so that all Lie subalgebras correspond to Lie subgroups.

The following theorem shows that there is a one-to-one correspondence between Lie subgroups of G and Lie subalgebras of $\mathcal{G}$. The details of the proof are a little technical and the interested reader is referred to the cited literature for a complete proof.

Theorem 19.4   *Let G be a Lie group with Lie algebra $\mathcal{G}$. For every Lie subalgebra $\mathcal{H}$ of $\mathcal{G}$, there exists a unique connected Lie subgroup $H \subseteq G$ with Lie algebra $\mathcal{H}$.*

Outline proof: The Lie subalgebra $\mathcal{H}$ defines a distribution D^k on G, by

$$D^k(g) = \{X_g \mid X \in \mathcal{H}\}.$$

Let the left-invariant vector fields $E_1, \ldots, E_k$ be a basis of the Lie algebra $\mathcal{H}$, so that a vector field X belongs to D^k if and only if it has the form $X = X^i E_i$ where X^i are real-valued functions on G. The distribution D^k is involutive, for if X and Y belong to D^k, then so does their Lie bracket $[X, Y]$:

$$[X, Y]_g = X^i(g) Y^j(g) C_{ij}^k E_k + X^i(g) E_i Y^j(g) E_j - Y^j(g) E_j X^i(g) E_i.$$

By the Frobenius theorem 15.4, every point $g \in G$ has an open neighbourhood U such that every $h \in U$ lies in an embedded submanifold N_h of G whose tangent space spans $\mathcal{H}$ at all points $h' \in N_h$. More specifically, it can be proved that through any point of G there exists a unique maximal connected integral submanifold – see [2, p. 92] or [6, p. 48]. Let H be the maximal connected integral submanifold through the identity $e \in G$. Since D^k

is invariant under left translations $g^{-1}H = L_{g^{-1}}H$ is also an integral submanifold of D^k. By maximality, we must have $g^{-1} \subseteq H$. Hence, if $g, h \in H$ then $g^{-1}h \in H$, so that H is a subgroup of G. It remains to show that $(g, h) \mapsto g^{-1}h$ is a smooth function with respect to the differentiable structure on H, and that H is the unique subgroup having $\mathcal{H}$ as its Lie algebra. Further details may be found in [6, p. 94]. ∎

If the Lie subalgebra is set to be $\mathcal{H} = \mathcal{G}$, namely the Lie algebra of G itself, then the unique Lie subgroup corresponding to $\mathcal{G}$ is the connected component of the identity, often denoted G_0.

Matrix Lie groups

All the groups discussed in Examples 2.10–2.15 of Section 2.3 are instances of **matrix Lie groups**; that is, they are all Lie subgroups of the general linear group $GL(n, \mathbb{R})$. Their Lie algebras were discussed heuristically in Section 6.5.

Example 19.7 As seen in Example 19.4 the Lie algebra of $GL(n, \mathbb{R})$ is isomorphic to the Lie algebra of all $n \times n$ matrices with respect to commutator products $[\mathsf{A}, \mathsf{B}] = \mathsf{AB} - \mathsf{BA}$. The set of all trace-free matrices $\mathcal{H} = \{\mathsf{A} \in M_n(\mathbb{R}) \mid \operatorname{tr} \mathsf{A} = 0\}$ is a Lie subalgebra of $\mathcal{GL}(n, \mathbb{R})$ since it is clearly a vector subspace of $M_n(\mathbb{R})$,

$$\operatorname{tr} \mathsf{A} = 0, \operatorname{tr} \mathsf{B} = 0 \implies \operatorname{tr}(\mathsf{A} + a\mathsf{B}) = 0,$$

and is closed with respect to taking commutators,

$$\operatorname{tr}[\mathsf{A}, \mathsf{B}] = \operatorname{tr}(\mathsf{AB}) - \operatorname{tr}(\mathsf{BA}) = 0.$$

It therefore generates a unique connected Lie subalgebra $H \subset GL(n, \mathbb{R})$ (see Theorem 19.4). To show that this Lie subgroup is the unimodular group $SL(n, \mathbb{R})$, we use the well-known identity

$$\det e^{\mathsf{A}} = e^{\operatorname{tr} \mathsf{A}}. \tag{19.11}$$

Thus, for all $\mathsf{A} \in \mathcal{H}$

$$\det e^{t\mathsf{A}} = e^{t \operatorname{tr} \mathsf{A}} = e^0 = 1,$$

and by Example 19.5 the entire one-parameter subgroup $\exp(t\mathsf{A})$ lies in the unimodular group $SL(n, \mathbb{R})$.

Since the map exp is a diffeomorphism from an open neighbourhood U of 0 in $\mathcal{GL}(n, \mathbb{R})$ onto $\exp(U)$, every non-singular matrix X in a connected neighbourhood of I is uniquely expressible as an exponential, $\mathsf{X} = e^{\mathsf{A}}$. Note the importance of connectedness here: the set of non-singular matrices has two connected components, being the inverse images of the two components of $\dot{\mathbb{R}}$ under the continuous map $\det : GL(n, \mathbb{R}) \to \mathbb{R}$. The matrices of negative determinant clearly cannot be connected to the identity matrix by a smooth curve in $GL(n, \mathbb{R})$ since the determinant would need to vanish somewhere along such a curve. In particular the subgroup $SL(n, \mathbb{R})$ is connected since it is the inverse image of the connected set $\{1\}$ under the determinant map. Every $\mathsf{X} \in SL(n, \mathbb{R})$ has $\det \mathsf{X} = 1$ and is therefore of the form $\mathsf{X} = e^{\mathsf{A}}$ where, by Eq. (19.11), $\mathsf{A} \in \mathcal{H}$. Let H be the unique connected

19.3 Lie subgroups

Lie subgroup H whose Lie algebra is $\mathcal{H}$, according to Theorem 19.4. In a neighbourhood of the identity every $X = e^A$ for some $A \in \mathcal{H}$, and every matrix of the form e^A belongs to H. Hence $SL(n, \mathbb{R}) = H$ is the connected Lie subgroup of $GL(n, \mathbb{R})$ with Lie algebra $\mathcal{SL}(n, \mathbb{R}) = \mathcal{H}$. Since a Lie group and its Lie algebra are of equal dimension, the Lie group $\dim SL(n, \mathbb{R}) = \dim \mathcal{H} = n^2 - 1$.

The Lie group $GL(n, \mathbb{C})$ has Lie algebra isomorphic with $M_n(\mathbb{C})$, with bracket $[A, B]$ again the commutator of the complex matrices A and B. As discussed in Chapter 6 this complex Lie algebra must be regarded as the complexification of a real Lie algebra by restricting the field of scalars to the real numbers. In this way any complex Lie algebra of dimension n can be considered as being a real Lie algebra of dimension $2n$. As a *real* Lie group $GL(n, \mathbb{C})$ has dimension $2n^2$, as does its Lie algebra $\mathcal{GL}(n, \mathbb{C})$. A similar discussion to that above can be used to show that the unimodular group $SL(n, \mathbb{C})$ of complex matrices of determinant 1 has Lie algebra consisting of trace-free complex $n \times n$ matrices. Both $SL(n, \mathbb{C})$ and $\mathcal{SL}(n, \mathbb{C})$ have (real) dimension $2n^2 - 2$.

Example 19.8 The orthogonal group $O(n)$ consists of real $n \times n$ matrices R such that

$$RR^T = I.$$

A one-parameter group of orthogonal transformations has the form $R(t) = \exp(tA) = e^{tA}$, whence

$$e^{tA}\left(e^{tA}\right)^T = e^{tA}e^{tA^T} = I.$$

Performing the derivative with respect to t of this matrix equation results in

$$Ae^{tA}e^{tA^T} + e^{tA}e^{tA^T}A^T = A + A^T = O,$$

so that A is a skew-symmetric matrix.

The set of skew-symmetric $n \times n$ matrices $\mathcal{O}(n)$ forms a Lie algebra since it is a vector subspace of $M_n(\mathbb{R})$ and is closed with respect to commutator products,

$$[A, B]^T = (AB - BA)^T$$
$$= B^T A^T - A^T B^T$$
$$= -[A^T, B^T] = -[A, B].$$

Since every matrix e^A is orthogonal for a skew-symmetric matrix A,

$$A = -A^T \implies e^A(e^A)^T = e^A e^{A^T} = e^A e^{-A} = I,$$

$\mathcal{O}(n)$ is the Lie algebra corresponding to the connected Lie subgroup $SO(n) = O(n) \cap SL(n, \mathbb{R})$. The dimensions of this Lie group and Lie algebra are clearly $\frac{1}{2}n(n - 1)$.

Similar arguments show that the unitary group $U(n)$ of complex matrices such that

$$UU^\dagger \equiv U\overline{U}^T = I$$

is a Lie group with Lie algebra $\mathcal{U}(n)$ consisting of skew-hermitian matrices, $A = -A^\dagger$. The dimensions of $U(n)$ and $\mathcal{U}(n)$ are both n^2. The group $SU(n) = U(n) \cap SL(n, \mathbb{C})$ has Lie algebra consisting of trace-free skew-hermitian matrices and has dimension $n^2 - 1$.

Problems

Problem 19.6 For any $n \times n$ matrix A, show that

$$\frac{d}{dt} \det e^{tA}\bigg|_{t=0} = \text{tr } A.$$

Problem 19.7 Prove Eq. (19.11). One method is to find a matrix S that transforms A to upper-triangular Jordan form by a similarity transformation as in Section 4.2, and use the fact that both determinant and trace are invariant under such transformations.

Problem 19.8 Show that $GL(n,\mathbb{C})$ and $SL(n,\mathbb{C})$ are connected Lie groups. Is $U(n)$ a connected group?

Problem 19.9 Show that the groups $SL(n,\mathbb{R})$ and $SO(n)$ are closed subgroups of $GL(N,\mathbb{R})$, and that $U(n)$ and $SU(n)$ are closed subgroups of $GL(n,\mathbb{C})$. Show furthermore that $SO(n)$ and $U(n)$ are compact Lie subgroups.

Problem 19.10 As in Example 2.13 let the symplectic group $Sp(n)$ consist of $2n \times 2n$ matrices S such that

$$S^T JS = J, \qquad J = \begin{pmatrix} O & I \\ -I & O \end{pmatrix},$$

where O is the $n \times n$ zero matrix and I is the $n \times n$ unit matrix. Show that the Lie algebra $\mathcal{S}_p(n)$ consists of matrices A satisfying

$$A^T J + JA = O.$$

Verify that these matrices form a Lie algebra and generate the symplectic group. What is the dimension of the symplectic group? Is it a closed subgroup of $GL(2n,\mathbb{R})$? Is it compact?

19.4 Lie groups of transformations

Let M be a differentiable manifold, and G a Lie group. By an **action** of G on M we mean a differentiable map $\phi : G \times M \to M$, often denoted $\phi(g, x) = gx$ such that

(i) $ex = x$ for all $x \in X$, where e is the identity element in G,
(ii) $(gh)x = g(hx)$.

This agrees with the conventions of a left action as defined in Section 2.6 and we refer to G as a **Lie group of transformations of** M. We may, of course, also have right actions $(g,x) \mapsto xg$ defined in the natural way.

Exercise: For any fixed $g \in G$ show that the map $\phi_g : M \to M$ defined by $\phi_g(x) = \phi(g,x) = gx$ is a diffeomorphism of M.

The action of G on M is said to be **effective** if e leaves every point $x \in M$ fixed,

$$gx = x \quad \text{for all } x \in M \implies g = e.$$

As in Section 2.6 the **orbit** Gx of a point $x \in M$ is the set $Gx = \{gx \mid g \in G\}$, and the action of G on M is said to be **transitive** if the whole of M is the orbit of some point in

19.4 Lie groups of transformations

M. In this case $M = Gy$ for all $y \in M$ and it is commonly said that M is a **homogeneous manifold** of G.

Example 19.9 Any Lie group G acts on itself by left translation $L_g : G \to G$, in which case the map $\phi : G \times G \to G$ is defined by $\phi(g, h) = gh = L_g h$. The action is both effective and transitive. Similarly G acts on itself to the right with right translations $R_g : G \to G$, where $R_g h = hg$.

Let H be a closed subgroup of a Lie group G, and $\pi : G \to G/H$ be the natural map sending each element of g to the left coset to which it belongs, $\pi(g) = gH$. As in Section 10.8 the factor space G/H is given the natural topology induced by π. Furthermore, G/H has a unique manifold structure such that π is C^∞ and G is a transitive Lie transformation group of G/H under the action

$$\phi(g, hH) = g(hH) = ghH.$$

A proof of this non-trivial theorem may be found in [6, p. 120]. The key result is the existence everywhere on G of *local sections*; every coset $gH \in G/H$ has a neighbourhood W and a smooth map $\alpha : W \to G$ with respect to the differentiable structure on G/H such that $\pi \circ \alpha = \mathrm{id}$ on $\alpha(W)$.

Every homogeneous manifold can be cast in the form of a left action on a space of cosets. Let G act transitively to the left on the manifold M and for any point $x \in M$ define the map $\phi_x : G \to M$ by $\phi_x(g) = gx$. This map is smooth, as it is the composition of two smooth maps $\phi_x = \phi \circ i_x$ where $i_x : G \to G \times M$ is the injection defined by $i_x(g) = (g, x)$. The isotropy group G_x of x, defined in Section 2.6 as $G_x = \{g \mid gx = x\}$, is therefore a closed subgroup of G since it is the inverse image of a closed singleton set, $G_x = \phi_x^{-1}(\{x\})$. Let the map $\rho : G/G_x \to M$ be defined by $\rho(gG_x) = gx$. This map is one-to-one, for

$$\rho(gG_x) = \rho(g'G_x) \Longrightarrow gx = g'x$$
$$\Longrightarrow g^{-1}g' \in G_x$$
$$\Longrightarrow g' \in gG_x$$
$$\Longrightarrow g'G_x = gG_x.$$

Furthermore, with respect to the differentiable structure induced on G/G_x, the map ρ is C^∞ since for any local section $\alpha : W \to G$ on a neighbourhood W of a given coset gG_x we can write $\rho = \pi \circ \alpha$, which is a composition of smooth functions. Since $\rho(gK) = g(\rho(K))$ for all cosets $K = hG_x \in G/G_x$, the group G has the 'same action' on G/G_x as it does on M.

Exercise: Show that ρ is a continuous map with respect to the factor space topology induced on G/G_x.

Example 19.10 The orthogonal group $O(n + 1)$ acts transitively on the unit n-sphere

$$S^n = \{\mathbf{x} \in \mathbb{R}^{n+1} \mid (x^1)^2 + (x^2)^2 + \cdots + (x^{n+1})^2 = 1\} \subset \mathbb{R}^{n+1},$$

since for any $\mathbf{x} \in S^n$ there exists an orthogonal transformation $\mathbf{A}$ such that $\mathbf{x} = \mathbf{A}\mathbf{e}$ where $\mathbf{e} = (0, 0, \ldots, 0, 1)$. In fact any orthogonal matrix with the last column having the same components as **x**, i.e. $A^i_{n+1} = x^i$, will do. Such an orthogonal matrix exists by a Schmidt orthonormalization in which **x** is transformed to the $(n+1)$th unit basis vector.

Let H be the isotropy group of the point **e**, consisting of all matrices of the form

$$\begin{pmatrix} & & & 0 \\ & [B^i_j] & & \vdots \\ & & & 0 \\ \cdots & & 0 & 1 \end{pmatrix} \quad \text{where } [B^i_j] \text{ is } n \times n \text{ orthogonal.}$$

Hence $H \cong O(n)$. The map $O(n+1)/H \to S^n$ defined by $\mathsf{A}H \mapsto \mathsf{A}\mathbf{e}$ is clearly one-to-one and continuous with respect to the induced topology on $O(n+1)/H$. Furthermore, $O(n+1)/H$ is compact since it is obtained by identification from an equivalence relation on the compact space $O(n+1)$ (see Example 10.16). Hence the map $O(n+1)/H \to S^n$ is a homeomorphism since it is a continuous map from a compact Hausdorff space onto a compact space (see Problem 10.17). Smoothness follows from the general results outlined above. Thus $O(n+1)/O(n)$ is diffeomorphic to S^n, and similarly it can be shown that $SO(n+1)/SO(n) \cong S^n$.

Example 19.11 The group of matrix transformations $\mathbf{x}' = \mathsf{L}\mathbf{x}$ leaving invariant the inner product $\mathbf{x} \cdot \mathbf{y} = x^1 y^1 + x^2 y^2 + x^3 y^3 - x^4 y^4$ of Minkowski space is the Lorentz group $O(3, 1)$. The group of all transformations leaving this form invariant including translations,

$$x'^i = L^i_j x^j + b^j,$$

is the Poincaré group $P(4)$ (see Example 2.30 and Chapter 9). The isotropy group of the origin is clearly $O(3, 1)$. The factor space $P_4/O(3, 1)$ is diffeomorphic to $\mathbb{R}^4$, for two Poincaré transformations P and P' belong to the same coset if and only if their translation parts are identical,

$$P^{-1} P' \in O(3, 1) \iff \mathsf{L}^{-1}(\mathsf{L}'\mathbf{x} + \mathbf{b}') - \mathsf{L}^{-1}\mathbf{b} = \mathsf{K}\mathbf{x} \quad \text{for } \mathsf{K} \in O(3, 1)$$
$$\iff \mathbf{b}' = \mathbf{b}.$$

Normal subgroups

Theorem 19.5 *Let H be a closed normal subgroup of a Lie group G. Then the factor group G/H is a Lie group.*

Proof: The map $\Psi : (aH, bH) \mapsto ab^{-1}H$ from $G/H \times G/H \to G/H$ is C^∞ with respect to the natural differentiable structure on G/H. For, if $(K_a, \alpha_a : K_a \to G)$ and $(K_b, \alpha_b : K_b \to G)$ are any pair of local sections at a and b then on $K_a \times K_b$

$$\Psi = \pi \circ \psi \circ (\alpha_a \times \alpha_b)$$

where $\psi : G \times G \to G$ is the map $\psi(a, b) = ab^{-1}$. Hence Ψ is everywhere locally a composition of smooth maps, and is therefore C^∞. ∎

Now suppose $\varphi : G \to H$ is any Lie group homomorphism, and let $N = \varphi^{-1}(\{e'\})$ be the kernel of the homomorphism, where e' is the identity of the Lie group H. It is clearly

19.4 Lie groups of transformations

a closed subgroup of G. The tangent map $\varphi_* : \mathcal{G} \to \mathcal{H}$ induced by the map φ is a Lie algebra homomorphism. Its kernel $\mathcal{N} = (\varphi_*)^{-1}(0)$ is an ideal of $\mathcal{G}$ and is the Lie algebra corresponding to the Lie subgroup N, since

$$\begin{aligned} Z \in \mathcal{N} &\iff \varphi_* Z = 0 \\ &\iff \exp t\varphi_* Z = e' \\ &\iff \varphi(\exp tZ) = e' \quad \text{by Eq. (19.10)} \\ &\iff \exp tZ \in N. \end{aligned}$$

Thus, if N is any closed normal subgroup of G then it is the kernel of the homomorphism $\pi : G \to H = G/N$, and its Lie algebra is the kernel of the Lie algebra homomorphism $\pi_* : \mathcal{G} \to \mathcal{H}$. That is, $\mathcal{H} \cong \mathcal{G}/\mathcal{N}$.

Example 19.12 Let G be the additive abelian group $G = \mathbb{R}^n$, and H the discrete subgroup consisting of all points with integral coordinates. Evidently H is a closed normal subgroup of G, and its factor group

$$T^n = \mathbb{R}^n/H$$

is the n-dimensional torus (see Example 10.14). In the torus group two vectors are identified if they differ by integral coordinates, $[\mathbf{x}] = [\mathbf{y}]$ in T^n if and only if $\mathbf{x} - \mathbf{y} = (k_1, k_2, \ldots, k_n)$ where k_i are integers. The one-dimensional torus $T = T^1$ is diffeomorphic to the unit circle in $\mathbb{R}^2$, and the n-dimensional torus group is the product of one-dimensional groups $T^n = T \times T \times \cdots \times T$. It is a compact group.

Induced vector fields

Let G be a Lie group of transformations of a manifold M with action to the right defined by a map $\rho : G \times M \to M$. We set $\rho(p, g) = pg$, with the stipulations $pe = p$ and $p(gh) = (pg)h$. Every left-invariant vector field X on G induces a vector field $\tilde{X}$ on M by setting

$$\tilde{X}_p f = \frac{d}{dt} f(p \exp tX) \Big|_{t=0} \quad (p \in M) \tag{19.12}$$

for any smooth function $f : M \to \mathbb{R}$. This is called the vector field **induced** by the left-invariant vector field X.

Exercise: Show that $\tilde{X}$ is a vector field on M by verifying linearity and the Leibnitz rule at each point $p \in M$, Eqs. (15.3) and (15.4).

Theorem 19.6 *Lie brackets of a pair of induced vector fields correspond to the Lie products of the corresponding left-invariant vector fields,*

$$\widetilde{[X, Y]} = [\tilde{X}, \tilde{Y}]. \tag{19.13}$$

Proof: Before proceeding with the main part of the proof, we need an expression for $[X, Y]_e$. Let σ_y be a local one-parameter group of transformations on G generated by the vector field X. By Eq. (19.6),

$$\sigma_t(g) = \sigma_t(ge) = g\sigma_t(e) = g \exp tX = R_{\exp tX}(g)$$

575

where the operation R_h is right translation by h. Hence

$$[X, Y]_e = \lim_{t \to 0} \frac{1}{t}\left[Y_e - (\sigma_{t*}Y)_e\right] = \lim_{t \to 0} \frac{1}{t}\left[Y_e - (R_{\exp(tX)*}Y)_e\right]. \tag{19.14}$$

Define the maps $\rho_p : G \to M$ and $\rho_g : M \to M$ for any $p \in M, g \in G$ by

$$\rho_p(g) = \rho_g(p) = \rho(p, g) = pg.$$

Then

$$\tilde{X}_p = \rho_{p*}X_e \tag{19.15}$$

for if f is any smooth function on M then, on making use of (19.8), we have

$$\rho_{p*}X_e f = X_e(f \circ \rho_p) = \frac{d}{dt}(f \circ \rho_p(\exp tX))\Big|_{t=0}$$
$$= \frac{d}{dt}(f(p \exp tX))\Big|_{t=0} = \tilde{X}_p f.$$

The maps $\tilde{\sigma}_t : M \to M$ defined by

$$\tilde{\sigma}_t(p) = p \exp tX = \rho_{\exp tX}(p)$$

form a one-parameter group of transformations of M since, using Eq. (19.9),

$$\tilde{\sigma}_{t+s}(p) = \tilde{\sigma}_t \circ \tilde{\sigma}_t(p)$$

for all t and s. By (19.12) they induce the vector field $\tilde{X}$, whence

$$[\tilde{X}, \tilde{Y}]_p = \lim_{t \to 0} \frac{1}{t}\left[\tilde{Y}_p - (\tilde{\sigma}_{t*}\tilde{Y})_p\right]. \tag{19.16}$$

Applying the definition of $\tilde{\sigma}_t$ we have

$$(\tilde{\sigma}_{t*}\tilde{Y})_p = (\rho_{\exp(tX)*}\tilde{Y})_p$$
$$= \rho_{\exp(tX)*}\rho_{p \exp(-tX)*}Y_e$$
$$= (\rho_{\exp tX} \circ \rho_{p \exp(-tX)})_* Y_e.$$

The map in the brackets can be written

$$\rho_{\exp tX} \circ \rho_{p \exp(-tX)} = \rho_p \circ R_{\exp tX} \circ L_{\exp(-tX)}$$

since

$$\rho_{\exp tX} \circ \rho_{p \exp(-tX)}(g) = p \exp(-tX)g \exp tX = \rho_p \circ R_{\exp tX} \circ L_{\exp(-tX)}(g).$$

Hence, since Y is left-invariant

$$(\tilde{\sigma}_{t*}\tilde{Y})_p = \rho_{p*} \circ R_{\exp(tX)*} \circ L_{\exp(-tX)*}Y_e$$
$$= \rho_{p*} \circ R_{\exp(tX)*}(Y_{\exp(-tX)})$$
$$= \rho_{p*}(R_{\exp(tX)*}Y_{\exp(-tX)})_e.$$

19.4 Lie groups of transformations

Since $\tilde{Y}_p = \rho_{p*} Y_e$ by Eq. (19.15), substitution in Eq. (19.16) and using Eq. (19.14) gives

$$[\tilde{X}, \tilde{Y}]_p = \rho_{p*} \left(\lim_{t \to 0} \frac{1}{t} \left[Y_e - \left(R_{\exp(tX)*} Y \right)_e \right] \right)$$
$$= \rho_{p*} [X, Y]_e = \widetilde{[X, Y]}_p,$$

which proves Eq. (19.13). ∎

Problems

Problem 19.11 Show that a group G acts effectively on G/H if and only if H contains no normal subgroup of G. [*Hint*: The set of elements leaving all points of G/H fixed is $\bigcap_{a \in G} a H a^{-1}$.]

Problem 19.12 Show that the special orthogonal group $SO(n)$, the pseudo-orthogonal groups $O(p, q)$ and the symplectic group $Sp(n)$ are all closed subgroups of $GL(n, \mathbb{R})$.

(a) Show that the complex groups $SL(n, \mathbb{C})$, $O(n, \mathbb{C})$, $U(n)$, $SU(n)$ are closed subgroups of $GL(n, \mathbb{C})$.
(b) Show that the unitary groups $U(n)$ and $SU(n)$ are compact groups.

Problem 19.13 Show that the centre Z of a Lie group G, consisting of all elements that commute with every element $g \in G$, is a closed normal subgroup of G.

Show that the general complex linear group $GL(n + 1, \mathbb{C})$ acts transitively but not effectively on complex projective n-space CP^n defined in Problem 15.4. Show that the centre of $GL(n + 1, \mathbb{C})$ is isomorphic to $GL(1, \mathbb{C})$ and $GL(n + 1, \mathbb{C})/GL(1, \mathbb{C})$ is a Lie group that acts effectively and transitively on CP^n.

Problem 19.14 Show that $SU(n + 1)$ acts transitively on CP^n and the isotropy group of a typical point, taken for convenience to be the point whose equivalence class contains $(0, 0, \ldots, 0, 1)$, is $U(n)$. Hence show that the factor space $SU(n + 1)/U(n)$ is homeomorphic to CP^n. Show similarly, that

(a) $SO(n + 1)/O(n)$ is homeomorphic to real projective space P^n.
(b) $U(n + 1)/U(n) \cong SU(n + 1)/SU(n)$ is homeomorphic to S^{2n+1}.

Problem 19.15 As in Problem 9.2 every Lorentz transformation $L = [L^i{}_j]$ has $\det L = \pm 1$ and either $L^4{}_4 \geq 1$ or $L^4{}_4 \leq -1$. Hence show that the Lorentz group $G = O(3, 1)$ has four connected components,

$$G_0 = G^{++} : \det L = 1, \ L^4{}_4 \geq 1 \qquad G^{+-} : \det L = 1, \ L^4{}_4 \leq -1$$
$$G^{-+} : \det L = -1, \ L^4{}_4 \geq 1 \qquad G^{--} : \det L = -1, \ L^4{}_4 \leq -1.$$

Show that the group of components G/G_0 is isomorphic with the discrete abelian group $Z_2 \times Z_2$.

Problem 19.16 Show that the component of the identity G_0 of a locally connected group G is generated by any connected neighbourhood of the identity e: that is, every element of G_0 can be written as a product of elements from such a neighbourhood.

Hence show that every discrete normal subgroup N of a connected group G is contained in the centre Z of G.

Find an example of a discrete normal subgroup of the disconnected group $O(3)$ that is not in the centre of $O(3)$.

Problem 19.17 Let $\mathcal{A}$ be a Lie algebra, and X any element of $\mathcal{A}$.

(a) Show that the linear operator $\mathrm{ad}_X : \mathcal{A} \to \mathcal{A}$ defined by $\mathrm{ad}_X(Y) = [X, Y]$ is a Lie algebra homomorphism of $\mathcal{A}$ into $\mathcal{GL}(\mathcal{A})$ (called the **adjoint representation**).

(b) For any Lie group G show that each inner automorphism $C_g : G \to G$ defined by $C_g(a) = gag^{-1}$ (see Section 2.4) is a Lie group automorphism, and the map $\mathrm{Ad} : G \to GL(\mathcal{G})$ defined by $\mathrm{Ad}(g) = C_{g*}$ is a Lie group homomorphism.
(c) Show that $\mathrm{Ad}_* = \mathrm{ad}$.

Problem 19.18 (a) Show that the group of all Lie algebra automorphisms of a Lie algebra $\mathcal{A}$ form a Lie subgroup of $\mathrm{Aut}(\mathcal{A}) \subseteq GL(\mathcal{A})$.
(b) A linear operator $D : \mathcal{A} \to \mathcal{A}$ is called a *derivation* on $\mathcal{A}$ if $D[X, Y] = [DX, Y] + [X, DY]$. Prove that the set of all derivations of $\mathcal{A}$ form a Lie algebra, $\partial(\mathcal{A})$, which is the Lie algebra of $\mathrm{Aut}(\mathcal{A})$.

19.5 Groups of isometries

Let (M, g) be a pseudo-Riemannian manifold. An **isometry** of M is a transformation $\varphi : M \to M$ such that $\widetilde{\varphi} g = g$, where $\widetilde{\varphi}$ is the map induced on tensor fields as defined in Section 15.5. This condition amounts to requiring

$$g_{\varphi(p)}(\varphi_* X_p, \varphi_* Y_p) = g_p(X_p, Y_p)$$

for all $X_p, Y_p \in T_p(M)$.

Let G be a Lie group of isometries of (M, g), and $\mathcal{G}$ its Lie algebra of left-invariant vector fields. If $A \in \mathcal{G}$ is a left-invariant vector field then the induced vector field $X = \widetilde{A}$ is called a **Killing vector** on M. If σ_t is the one-parameter group of isometries generated by A then, by Eq. (15.33), we have

$$\mathcal{L}_X g = \lim_{t \to 0} \frac{1}{t}(g - \widetilde{\sigma}_t g) = 0.$$

In any coordinate chart $(U; x^i)$ let $X = \xi^i \partial_{x^i}$ and, by Eq. (15.39), this equation becomes

$$\mathcal{L}_X g_{ij} = g_{ij,k} \xi^k + \xi^k_{,i} g_{kj} + \xi^k_{,j} g_{ik} = 0, \qquad (19.17)$$

known as **Killing's equations**. In a local chart such that $\xi^i = (1, 0, \ldots, 0)$ (see Theorem 15.3), Eq. (19.17) reads

$$g_{ij,1} = \frac{\partial g_{ij}}{\partial x^1} = 0$$

and the components of g are independent of the coordinate x^1, $g_{ij} = g_{ij}(x^2, \ldots, x^n)$. By direct computation from the Christoffel symbols or by considering the equation in geodesic coordinates, ordinary derivatives may be replaced by covariant derivatives in Eq. (19.17)

$$g_{ij;k} \xi^k + \xi^k_{;i} g_{kj} + \xi^k_{;j} g_{ik} = 0,$$

and since $g_{ij;k} = 0$ Killing's equations may be written in the *covariant form*:

$$\xi_{i;j} + \xi_{j;i} = 0. \qquad (19.18)$$

By Theorem 19.6, if $X = \widetilde{A}$ and $Y = \widetilde{B}$ then $[X, Y] = \widetilde{[A, B]}$. We also conclude from Problem 15.18 that if X and Y satisfy Killing's equations then so does $[X, Y]$. In fact, there

19.5 Groups of isometries

can be at most a finite number of linearly independent Killing vectors. For, from (19.18) and the Ricci identities (18.29), $\xi_{k;ij} - \xi_{k;ji} = \xi_a R^a{}_{kij}$ (no torsion), we have

$$\xi_{k;ij} + \xi_{j;ki} = \xi_a R^a{}_{kij}.$$

From the cyclic first Bianchi identity (18.26), $R^i{}_{jkl} + R^i{}_{klj} + R^i{}_{ljk} = 0$, we have $\xi_{i;jk} + \xi_{j;ki} + \xi_{k;ij} = 0$, whence

$$\xi_{i;jk} = -\xi_{j;ki} - \xi_{k;ij} = -\xi_a R^a{}_{kij} = \xi_a R^a{}_{kji}. \tag{19.19}$$

Thus if we know the components ξ_i and $\xi_{i;j}$ in a given pseudo-Riemannian space, all covariant derivatives of second order of ξ_i may be calculated from Eq. (19.19). All higher orders may then be found by successively forming higher order covariant derivatives of this equation. Assuming that ξ_i can be expanded in a power series in a neighbourhood of any point of M (this is not actually an additional assumption as it turns out), we only need to know ξ_i and $\xi_{i;j} = -\xi_{j;i}$ at a specified point p to define the entire Killing vector field in a neighbourhood of p. As there are $n + \binom{n}{2} = n(n+1)/2$ linearly independent initial values at p, the maximum number of linearly independent Killing vectors in any neighbourhood of M is $n(n+1)/2$. In general of course there are fewer than these, say r, and the general Killing vector is expressible as a linear combination of r Killing vectors $X_1, \ldots, X_r$,

$$X = \sum_{i=1}^{r} a^i X_i, \quad (a^i = \text{const.})$$

generating a Lie algebra of dimension r with structure constants $C^k_{ij} = -C^k_{ji}$,

$$[X_i, X_j] = C^k_{ij} X_k.$$

Maximal symmetries and cosmology

A pseudo-Riemannian space is said to have **maximal symmetry** if it has the maximum number $n(n+1)/2$ of Killing vectors. Taking a covariant derivative of Eq. (19.19),

$$\xi_{i;jkl} = \xi_{a;l} R^a{}_{kji} + \xi_a R^a{}_{kji;l},$$

and using the generalized Ricci identities given in Problem 18.9,

$$\xi_{i;jkl} - \xi_{i;jlk} = \xi_{a;j} R^a{}_{ikl} + \xi_{i;a} R^a{}_{jkl}$$
$$= \xi_{a;j} R^a{}_{ikl} - \xi_{a;i} R^a{}_{jkl},$$

we have

$$\xi_a \left(R^a{}_{kji;l} - R^a{}_{lji;k} \right) = \xi_{a;b} \left(R^a{}_{ikl} \delta^b_j - R^a{}_{jkl} \delta^b_i - R^a{}_{kji} \delta^b_l + R^a{}_{lji} \delta^b_k \right).$$

Since for maximal symmetry ξ_a and $\xi_{a;b} = -\xi_{b;a}$ are arbitrary at any point, the antisymmetric part with respect to a and b of the term in parentheses on the right-hand side vanishes,

$$R^a{}_{ikl} \delta^b_j - R^a{}_{jkl} \delta^b_i - R^a{}_{kji} \delta^b_l + R^a{}_{lji} \delta^b_k = R^b{}_{ikl} \delta^a_j - R^b{}_{jkl} \delta^a_i - R^b{}_{kji} \delta^a_l + R^b{}_{lji} \delta^a_k.$$

Contracting this equation with respect to indices b and l, we find on using the cyclic symmetry (18.26),

$$(n-1)R^a{}_{kji} = R_{ik}\delta^a_j - R_{jk}\delta^a_i.$$

Another contraction with respect to k and i gives $nR_{.j}^a = R\delta^a_j$ and substituting back in the expression for $R^a{}_{kji}$, we find on lowering the index and making a simple permutation of index symbols

$$R_{ijkl} = \frac{R}{n(n-1)}(g_{ik}g_{jl} - g_{il}g_{jk}). \tag{19.20}$$

The contracted Bianchi identity (18.60) implies that the Ricci scalar is constant for $n > 2$ since

$$R^a{}_{j;a} = \tfrac{1}{2}R_{,j} \implies nR_{,j} = \tfrac{1}{2}R_{,j} \implies R_{,j} = 0.$$

Spaces whose Riemann tensor has this form are known as spaces of **constant curvature**. Example 18.4 provides another motivation for this nomenclature and shows that the 3-sphere of radius a is a space of constant curvature, with $R = 6/a^2$. The converse is in fact true – every space of constant curvature has maximal symmetry. We give a few instances of this statement in the following examples.

Example 19.13 Euclidean 3-space $ds^2 = \delta_{ij}\,dx^i\,dx^j$ of constant curvature zero. To find its Killing vectors, we must find all solutions of Killing's equations (19.18),

$$\xi_{i,j} + \xi_{j,i} = 0.$$

Since this implies $\xi_{1,1} = \xi_{2,2} = \xi_{3,3} = 0$ we have

$$\xi_{i,11} = -\xi_{1,1i} = 0, \qquad \xi_{i,22} = \xi_{i,33} = 0,$$

whence there exist constants $a_{ij} = -a_{ji}$ and b_i such that

$$\xi_i = a_{ij}x^j + b_i.$$

Setting $a_{ij} = -\epsilon_{ijk}a^k$ and $b^k = b_k$ we can express the general Killing vector in the form

$$X = a^1 X_1 + a^2 X_2 + a^3 X_3 + b^1 Y_1 + b^2 Y_2 + b^3 Y_3$$

where $X_1 = x^2\partial_3 - x^2\partial_3$, $X_2 = x^3\partial_1 - x^1\partial_3$, $X_3 = x^1\partial_2 - x^2\partial_1$ and $Y_1 = \partial_1$, $Y_2 = \partial_2$, $Y_3 = \partial_3$. As these are six independent Killing vectors, the space has maximal symmetry. Their Lie algebra commutators are

$$[X_1, X_2] = -X_3, \qquad [X_2, X_3] = -X_1, \qquad [X_3, X_1] = -X_2,$$
$$[Y_1, Y_2] = [Y_1, Y_3] = [Y_2, Y_3] = 0,$$

$$[X_1, Y_1] = 0, \qquad [X_2, Y_1] = Y_3, \qquad [X_3, Y_1] = -Y_2,$$
$$[X_1, Y_2] = -Y_3, \qquad [X_2, Y_2] = 0, \qquad [X_3, Y_2] = Y_1,$$
$$[X_1, Y_3] = Y_2, \qquad [X_2, Y_3] = -Y_1, \qquad [X_3, Y_3] = 0.$$

This is known as the Lie algebra of the **Euclidean group**.

19.5 Groups of isometries

Example 19.14 The 3-sphere of Example 18.4,

$$ds^2 = a^2(d\chi^2 + \sin^2 \chi (d\theta^2 + \sin^2 \theta \, d\phi^2)),$$

has Killing's equations

$$\xi_{1,1} = 0, \tag{19.21}$$
$$\xi_{1,2} + \xi_{2,1} - 2\cot\chi\, \xi_2 = 0, \tag{19.22}$$
$$\xi_{1,3} + \xi_{3,1} - 2\cot\chi\, \xi_3 = 0, \tag{19.23}$$
$$\xi_{2,2} + \sin\chi \cos\chi\, \xi_1 = 0, \tag{19.24}$$
$$\xi_{2,3} + \xi_{3,2} - 2\cot\chi\, \xi_3 = 0, \tag{19.25}$$
$$\xi_{3,3} + \sin\chi \cos\chi \sin^2\theta\, \xi_1 + \sin\theta \cos\theta\, \xi_2 = 0. \tag{19.26}$$

From (19.21) we have $\xi_1 = F(\theta, \phi)$ and differentiating (19.22) with respect to $x^1 = \chi$ we have a differential equation for ξ_2,

$$\xi_{2,11} - 2\cot\chi\, \xi_{2,1} + 2\csc^2\chi\, \xi_2 = 0.$$

The general solution of this linear differential equation is not hard to find:

$$\xi_2 = -\sin\chi \cos\chi\, f(\theta, \phi) + \sin^2\chi\, G(\theta, \phi).$$

Substituting back into (19.22) we find $f = -F_{,2}$ where $x^2 = \theta$. Similarly,

$$\xi_3 = F_{,3} \sin\chi \cos\chi + H(\theta, \phi) \sin^2\chi.$$

Substituting these expressions in the remaining equations results in the following general solution of Killing's equations dependent on six arbitrary constants $a^1, a^2, a^3, b^1, b^2, b^3$:

$$X = \xi^i \partial_i = a^i X_i + b^j Y_j$$

where $\xi^1 = \xi_1, \xi^2 = \xi_2/\sin^2\chi, \xi^3 = \xi_3/\sin^2\chi \sin^2\theta$, and

$$X_1 = \cos\phi\, \partial_\theta - \cot\theta \sin\phi\, \partial_\phi,$$
$$X_2 = \sin\phi\, \partial_\theta + \cot\theta \cos\phi\, \partial_\phi,$$
$$X_3 = \partial_\phi,$$
$$Y_1 = \sin\theta \sin\phi\, \partial_\chi + \cot\chi \cos\theta \sin\phi\, \partial_\theta + \frac{\cot\chi}{\sin\theta} \cos\phi\, \partial_\phi,$$
$$Y_2 = -\sin\theta \cos\phi\, \partial_\chi - \cot\chi \cos\theta \cos\phi\, \partial_\theta + \frac{\cot\chi}{\sin\theta} \sin\phi\, \partial_\phi,$$
$$Y_3 = \cos\theta\, \partial_\chi - \sin\theta \cot\chi\, \partial_\theta.$$

The Lie algebra brackets are tedious to calculate compared with those in the previous example, but the results have similarities to those of the Euclidean group:

$$[X_1, X_2] = -X_3, \quad [X_2, X_3] = -X_1, \quad [X_3, X_1] = -X_2,$$
$$[Y_1, Y_2] = -X_3, \quad [Y_2, Y_3] = -X_1, \quad [Y_3, Y_1] = -X_2,$$
$$[X_1, Y_1] = 0, \quad [X_1, Y_2] = -Y_3, \quad [X_1, Y_3] = Y_2,$$
$$[X_2, Y_1] = Y_3, \quad [X_2, Y_2] = 0, \quad [X_2, Y_3] = -Y_1,$$
$$[X_3, Y_1] = -Y_2, \quad [X_3, Y_2] = Y_1, \quad [X_3, Y_3] = 0.$$

Not surprisingly this Lie algebra is isomorphic to the Lie algebra of the four-dimensional rotation group, $SO(4)$.

The Robertson–Walker cosmologies of Section 18.9 all have maximally symmetric spatial sections. The sections $t = $ const. of the open model (18.105) are 3-spaces of constant negative curvature, called *pseudo-spheres*. These models are called homogeneous and isotropic. It is not hard to see that these space-times have the same number of independent Killing vectors as their spatial sections. In general they have six Killing vectors, but some special cases may have more. Of particular interest is the **de Sitter universe**, which is a maximally symmetric space-time, having 10 independent Killing vectors:

$$ds^2 = -dt^2 + a^2 \cosh^2(t/a)\big[d\chi^2 + \sin^2\chi(d\theta^2 + \sin^2\theta\, d\phi^2)\big].$$

This is a space-time of constant curvature, which may be thought of as a hyperboloid embedded in five-dimensional space,

$$x^2 + y^2 + z^2 + w^2 - v^2 = a^2.$$

Since it is a space of constant curvature, $R_{\mu\nu} = \frac{1}{4} R g_{\mu\nu}$, the Einstein tensor is

$$G_{\mu\nu} = R_{\mu\nu} - \frac{1}{2} R g_{\mu\nu} = -\frac{1}{4} R g_{\mu\nu} = \frac{3}{a^2} g_{\mu\nu}.$$

This can be thought of in two ways. It can be interpreted as a solution of Einstein's field equations $G_{\mu\nu} = \kappa T_{\mu\nu}$ with a perfect fluid $T_{\mu\nu} \propto g_{\mu\nu}$ having negative pressure $P = -\rho c^2$. However it is more common to interpret it as a *vacuum* solution $T_{\mu\nu} = 0$ of the modified Einstein field equations with **cosmological constant** Λ,

$$\mathbb{G}_{\mu\nu} = \kappa T_{\mu\nu} - \Lambda g_{\mu\nu} \quad (\Lambda = 3a^{-2}).$$

This model is currently popular with advocates of the *inflationary cosmology*. Interesting aspects of its geometry are described in [8, 9].

Sometimes cosmologists focus on cosmologies having fewer symmetries. A common technique is to look for homogeneous models that are not necessarily isotropic, equivalent to relaxing the Lie algebra of Killing vectors from six to three, and assuming the orbits are three-dimensional subspaces of space-time. All three-dimensional Lie algebras may be categorized into one of nine **Bianchi types**, usually labelled by Roman numerals. A detailed discussion may be found in [10]. The Robertson–Walker models all fall into this classification, the flat model being of Bianchi type I, the closed model of type IX, and the open model of type V. To see how such a relaxation of symmetry gives rise to more general

19.5 Groups of isometries

models, consider type I, which is the commutative Lie algebra

$$[X_1, X_2] = [X_2, X_3] = [X_1, X_3] = 0.$$

It is not hard to show locally that a metric having these symmetries must have the form

$$ds^2 = e^{2\alpha_1(t)}(dx^1)^2 + e^{2\alpha_2(t)}(dx^2)^2 + e^{2\alpha_3(t)}(dx^3)^2 - c^2 dt^2.$$

The vacuum solutions of this metric are (see [11])

$$\alpha_i(t) = a_i \ln t \quad (a_1 + a_2 + a_3 = a_1^2 + a_2^2 + a_3^2 = 1),$$

called **Kasner solutions**. The pressure-free dust cosmologies of this type are called **Heckmann–Schücking solutions** (see the article by E. Heckmann and O. Schücking in [12]) and have the form

$$\alpha_i = a_i \ln(t - t_1) + b_i \ln(t - t_2) \quad (b_i = \tfrac{2}{3} - a_i, \ \textstyle\sum_{i=1}^3 a_i = \sum_{i=1}^3 (a_i)^2 = 1).$$

It is not hard to show that $\sum_{i=1}^3 b_i = \sum_{i=1}^3 (b_i)^2 = 1$. The density in these solutions evolves as

$$\rho = \frac{1}{6\pi G(t - t_1)(t - t_2)}.$$

The flat Friedmann model arises as the limit $t_1 = t_2$ of this model.

Spherical symmetry

A space-time is said to be **spherically symmetric** if it has three spacelike Killing vectors X_1, X_2, X_3 such that they span a Lie algebra isomorphic with $\mathcal{SO}(3)$,

$$[X_i, X_j] = -\epsilon_{ijk} X_k \quad (i, j, k \in \{1, 2, 3\})$$

and such that the orbits of all points are two-dimensional surfaces, or possibly isolated points. The idea is that the orbits generated by the group of transformations are in general 2-spheres that could be represented as $r = \text{const.}$ in appropriate coordinates. There should therefore be coordinates $x = r$, $x^2 = \theta$, $x^3 = \phi$ such that the X_i are spanned by ∂_θ and ∂_ϕ, and using Theorem 15.3 it should be locally possible to choose these coordinates such that

$$X_3 = \partial_\phi, \quad X_1 = \xi^1 \partial_\theta + \xi^2 \partial_\phi, \quad X_2 = \eta^1 \partial_\theta + \eta^2 \partial_\phi.$$

We then have

$$[X_3, X_1] = -X_2 \implies \eta^1 = -\xi^1_{,\phi}, \ \eta^2 = -\xi^2_{,\phi}$$
$$[X_3, X_2] = X_1 \implies \xi^1 = \eta^1_{,\phi}, \ \xi^2 = \eta^2_{,\phi}$$

whence $\xi^i_{,\phi\phi} = -\xi^i \ (i = 1, 2)$, so that

$$\xi^1 = f \sin\phi + g \cos\phi, \qquad \xi^2 = h \sin\phi + k \cos\phi$$
$$\eta^1 = -f \cos\phi + g \sin\phi, \qquad \eta^2 = -h \cos\phi + k \sin\phi$$

where the functions f, g, h, k are arbitrary functions of θ, r and t. The remaining commutation relation $[X_1, X_2] = -X_3$ implies, after some simplification,

$$fg_\theta - gf_\theta + gk + fh = 0 \qquad (19.27)$$
$$fk_\theta - gh_\theta + h^2 + k^2 = -1 \qquad (19.28)$$

where $g_\theta \equiv \partial g/\partial \theta$, etc. A coordinate transformation $\phi' = \phi + F(\theta, r, t), \theta = G(\theta, r, t)$ has the effect

$$\partial_\phi = \partial_{\phi'}, \qquad \partial_\theta = F_\theta \, \partial_{\phi'} + G_\theta \, \partial_{\theta'}$$

and therefore

$$X_1 = \xi^1 G_\theta \, \partial_{\theta'} + \left(\xi^1 F_\theta + \xi^2\right) \partial_{\phi'}.$$

Hence, using addition of angle identities for the functions sin and cos,

$$(\xi^1)' = \xi^1 G_\theta = \left(f \sin(\phi' - F) + g \cos(\phi' - F)\right) G_\theta$$
$$= (f \cos F + g \sin F) G_\theta \sin \phi' + (-f \sin F + g \cos F) G_\theta \cos \phi'.$$

Choosing

$$\tan F = -\frac{f}{g} \quad \text{and} \quad G_\theta = \frac{1}{g \cos F - f \sin F}$$

we have $(\xi^1)' = \cos \phi'$. We have thus arrived at the possibility of selecting coordinates θ and ϕ such that $f = 0$, $g = 1$. Substituting in Eqs. (19.27) and (19.28) gives $k = 0$ and $h = -\cot(\theta - \theta_0(r, t))$. Making a final coordinate transformation $\theta \to \theta - \theta_0(r, t)$, which has no effect on ξ^1, we have

$$X_1 = \cos \phi \, \partial_\theta - \cot \theta \sin \phi \, \partial_\phi, \qquad X_2 = \sin \phi \, \partial_\theta + \cot \theta \cos \phi \, \partial_\phi, \qquad X_3 = \partial_\phi.$$

From Killing's equations (19.17) with $X = X_3$ we have $g_{\mu\nu} = g_{\mu\nu}(r, \theta, t)$ and for $X = X_2, X_3$, we find that these equations have the form

$$\xi^2 \partial_\theta g_{\mu\nu} + \xi^2_{,\mu} g_{2\nu} + \xi^3_{,\mu} g_{3\nu} + \xi^2_{,\nu} g_{2\mu} + \xi^3_{,\nu} g_{3\mu} = 0$$

and successively setting $\mu\nu = 11, 12, \ldots$ we obtain

$$g_{11} = g_{11}(r, t), \qquad g_{14} = g_{14}(r, t), \qquad g_{44} = g_{44}(r, t),$$
$$g_{12} = g_{13} = g_{42} = g_{43} = g_{23} = 0,$$
$$g_{22} = f(r, t), \qquad g_{33} = f(r, t) \sin^2 \theta.$$

As there is still an arbitrary coordinate freedom in the radial and time coordinate,

$$r' = F(r, t), \qquad t' = G(r, t)$$

it is possible to choose the new radial coordinate to be such that $f = r'^2$ and the time coordinate may then be found so that $g'_{14} = 0$. The resulting form of the metric is that postulated in Eq. (18.90),

$$ds^2 = g_{11}(r,t)\,dr^2 + r^2(d\theta^2 + \sin^2\theta\,d\phi^2) - |g_{44}(r,t)|c^2\,dt^2.$$

If g_{11} and g_{44} are independent of the time coordinate then the vector $X = \partial_t$ is a Killing vector. Any space-time having a timelike Killing vector is called **stationary**. For the case considered here the Killing vector has the special property that it is orthogonal to the 3-surfaces $t = $ const., and is called a **static** space-time. The condition for a space-time to be static is that the covariant version of the Killing vector be proportional to a gradient $\xi_\mu = g_{\mu\nu}\xi^\nu = \lambda f_{,\mu}$ for some functions λ and f. Equivalently, if ξ is the 1-form $\xi = \xi_\mu\,dx^\mu$, then $\xi = \lambda\,df$ which, by the Frobenius theorem 16.4, can hold if and only if $d\xi \wedge \xi = 0$. For the spherically symmetric metric above, $\xi = g_{44}c\,dt$ and $d\xi \wedge \xi = dg_{44} \wedge c\,dt \wedge g_{44}c\,dt = 0$, as required. An important example of a metric that is stationary but not static is the Kerr solution, representing a rotating body in general relativity. More details can be found in [8, 13].

Problem

Problem 19.19 Show that the non-translational Killing vectors of pseudo-Euclidean space with metric tensor $g_{ij} = \eta_{ij}$ are of the form

$$X = A_j^{\,k} x^j \partial_{x^k} \quad \text{where} \quad A_{kl} = A_k^{\,j}\eta_{jl} = -A_{lk}.$$

Hence, with reference to Example 19.3, show that the Lie algebra of $SO(p,q)$ is generated by matrices I_{ij} with $i < j$, having matrix elements $(I_{ij})_a^{\,b} = \eta_{ia}\delta_j^b - \delta_i^b\eta_{ja}$. Show that the commutators of these generators can be written (setting $I_{ij} = -I_{ji}$ if $i > j$)

$$[I_{ij}, I_{kl}] = I_{il}\eta_{jk} + I_{jk}\eta_{il} - I_{ik}\eta_{jl} - I_{jl}\eta_{ik}.$$

References

[1] L. Auslander and R. E. MacKenzie. *Introduction to Differentiable Manifolds*. New York, McGraw-Hill, 1963.

[2] C. Chevalley. *Theory of Lie Groups*. Princeton, N.J., Princeton University Press, 1946.

[3] T. Frankel. *The Geometry of Physics*. New York, Cambridge University Press, 1997.

[4] S. Helgason. *Differential Geometry and Symmetric Spaces*. New York, Academic Press, 1962.

[5] W. H. Chen, S. S. Chern and K. S. Lam. *Lectures on Differential Geometry*. Singapore, World Scientific, 1999.

[6] F. W. Warner. *Foundations of Differential Manifolds and Lie Groups*. New York, Springer-Verlag, 1983.

[7] C. de Witt-Morette, Y. Choquet-Bruhat and M. Dillard-Bleick. *Analysis, Manifolds and Physics*. Amsterdam, North-Holland, 1977.

[8] S. Hawking and G. F. R. Ellis. *The Large-Scale Structure of Space-Time*. Cambridge, Cambridge University Press, 1973.

[9] E. Schrödinger. *Expanding Universes*. Cambridge, Cambridge University Press, 1956.

[10] M. P. Ryan and L. C. Shepley. *Homogeneous Relativistic Cosmologies*. Princeton, N.J., Princeton University Press, 1975.

[11] L. D. Landau and E. M. Lifshitz. *The Classical Theory of Fields*. Reading, Mass., Addison-Wesley, 1971.

[12] L. Witten (ed.). *Gravitation: An Introduction to Current Research*. New York, John Wiley & Sons, 1962.

[13] R. d'Inverno. *Introducing Einstein's Relativity*. Oxford, Oxford University Press, 1993.

Bibliography

The following is a list of books that the reader may find of general interest. None covers the entire contents of this book, but all relate to significant portions of the book, and some go significantly beyond.

V. I. Arnold. *Mathematical Methods of Classical Mechanics*. New York, Springer-Verlag, 1978.

N. Boccara. *Functional Analysis*. San Diego, Academic Press, 1990.

R. Courant and D. Hilbert. *Methods of Mathematical Physics, Vols. 1 and 2*. New York, Interscience, 1953.

R. W. R. Darling. *Differential Forms and Connections*. New York, Cambridge University Press, 1994.

L. Debnath and P. Mikusiński. *Introduction to Hilbert Spaces with Applications*. San Diego, Academic Press, 1990.

H. Flanders. *Differential Forms*. New York, Dover Publications, 1989.

T. Frankel. *The Geometry of Physics*. New York, Cambridge University Press, 1997.

R. Geroch. *Mathematical Physics*. Chicago, The University of Chicago Press, 1985.

S. Hassani. *Foundations of Mathematical Physics*. Boston, Allyn and Bacon, 1991.

S. Hawking and G. F. R. Ellis. *The Large-Scale Structure of Space–Time*. Cambridge, Cambridge University Press, 1973.

F. P. Hildebrand. *Methods of Applied Mathematics*. Englewood Cliffs, N.J., Prentice-Hall, 1965.

J. M. Jauch. *Foundations of Quantum Mechanics*. Reading, Mass., Addison-Wesley, 1968.

S. Lang. *Algebra*. Reading, Mass., Addison-Wesley, 1965.

L. H. Loomis and S. Sternberg. *Advanced Calculus*. Reading, Mass., Addison-Wesley, 1968.

M. Nakahara. *Geometry, Topology and Physics*. Bristol, Adam Hilger, 1990.

C. Nash and S. Sen. *Topology and Geometry for Physicists*. London, Academic Press, 1983.

I. M. Singer and J. A. Thorpe. *Lecture Notes on Elementary Topology and Geometry*. Glenview, Ill., Scott Foresman, 1967.

Bibliography

M. Spivak. *Differential Geometry, Vols. 1–5*. Boston, Publish or Perish Inc., 1979.

W. H. Chen, S. S. Chern and K. S. Lam. *Lectures on Differential Geometry*. Singapore, World Scientific, 1999.

F. W. Warner. *Foundations of Differential Manifolds and Lie Groups*. New York, Springer-Verlag, 1983.

Y. Choquet-Bruhat, C. de Witt-Morette and M. Dillard-Bleick. *Analysis, Manifolds and Physics*. Amsterdam, North-Holland, 1977.

Index

ϵ-symbol, 215
σ-algebra, 287
   generated, 288
φ-related vector fields, 560
k-cell, 486
   support of, 486
k-dimensional distribution, 440
   integral manifold, 440
   involutive, 441
   smooth, 440
   vector field belongs to, 440
   vector field lies in, 440
n-ary function, 11
   arguments, 11
n-ary relation on a set, 7
nth power of a linear operator, 101
p-boundary on a manifold, 496
p-chain, 494
p-chain on a manifold, 496
   boundary of, 496
p-cycle on a manifold, 496
r-forms, 204
   simple, 212
r-vector, 162, 184, 204
   simple, 162, 211
1-form, 90, 420
   components of, 421
   exact, 428
2-torus, 9
2-vector, 161
   simple, 161
3-force, 242
3-sphere, 530
4-acceleration, 241
4-current, 246
4-force, 242
4-momentum, 242
4-scalar, 232
4-tensor, 228
   4-covector, 232
   of type (r, s), 232
4-tensor field, 244
4-vector, 232
   future-pointing, 233
   magnitude of, 233
   null, 233
   past-pointing, 233
   spacelike, 233
   timelike, 233
4-velocity, 241

aberration of light, 238
absolute continuity, 356
absolute entropy, 463
absolute gas temperature, 458
absolute temperature, 463
absolute zero of temperature, 463
accumulation point, 260
action, 465, 554
action of Lie group on a manifold, 572
action principle, 554
addition modulo an integer, 29
addition of velocities
   Newtonian, 229
   relativistic, 238
additive group of integers modulo an integer, 29
adiabatic enclosure, 458
adiabatic processes, 458
adjoint representation of a Lie algebra, 577
advanced time, 546
affine connection, *see* connection
affine space, 231
   coordinates, 231
   difference of points in, 231
   origin, 231
affine transformation, 54, 58
algebra, 149
   associative, 149
   commutative, 149
algebra homomorphism, 150
algebra isomorphism, 151
almost everywhere, 135, 299
alternating group, 33
angular 4-momentum, 252

Index

angular momentum, 378
    orbital, 385
    spin, 385
angular momentum operator, 392
annihilator of a subset, 95
annihilator subspace, 454
antiderivation, 214, 448
anti-homomorphism, 50
antilinear, 134
antilinear transformation, 388
antisymmetrical state, 397
antisymmetrization operator, 205, 405
anti-unitary transformation, 388
associative law, 11, 27
atlas, 412
    maximal, 412
automorphism
    inner, 278
    of groups, 42
autonomous system, 118
axiom of extensionality, 4

Banach space, 282
basis, 73
Bayes' formula, 294
Bessel's inequality, 336
Betti numbers, 497
Bianchi identities, *see* Bianchi identity, second
Bianchi identity
    first, 514, 529
    second, 515, 526, 529
Bianchi types, 582
bijection, 11
bijective map, 11
bilinear, 127, 181
binary relation, 7
bivector, 161
block diagonal form, 107
boost, 236
Borel sets, 288
Bose–Einstein statistics, 397
boson, 397
boundary, 496
boundary map, 487
boundary of a p-chain, 495
boundary of a set, 260
bounded above, 17
bounded linear map, 283
bounded set, 273
bra, 369
bra-ket notation, 342, 369
bracket product, 167

canonical 1-form, 478
canonical commutation relations, 372
canonical map, 12

Cantor set, 15, 299
Cantor's continuum hypothesis, 16
Cantor's diagonal argument, 15
Carathéodory, 460
cardinality of a set, 13
Cartan formalism, 527–34
Cartan's first structural equation, 528
Cartan's lemma, 213
Cartan's second structural equation, 528
cartesian product, 266
cartesian product of two sets, 7
cartesian tensors, 201, 228
category, 23
category of sets, 23
Cauchy sequence, 265
Cauchy–Schwartz inequality, 136, 336
Cayley–Hamilton theorem, 105
cellular automaton, 22
centre of a group, 46
certainty event, 293
character of a representation, 147
characteristic equation, 102
characteristic function, 12
characteristic polynomial, 102
charge density, 245
    proper, 245
charge flux density, 245
charts having same orientation, 481
chemical potential, 407
Christoffel symbols, 519
Clifford algebra, 158
closed set, 259
closed subgroup, 277
closed subgroup of a Lie group, 569
closure
    of vector subspace, 335
closure of a set, 260
closure property, 27
coarser topology, 261
coboundary, 497
cochains, 497
cocycle, 497
codimension, 79
cohomologous cocycles, 498
collection of sets, 4
commutator, 167
    of observables, 371
    of vector fields, 432
compact support, 309
compatible charts, 412
complementary subspaces, 67
complete orthonormal set, *see* orthonormal basis
completely integrable, 455
complex number, 152
    conjugate of, 153

Index

inverse of, 153
modulus of, 153
complex structure on a vector space, 155
complexification of a vector space, 154
component of the identity, 278
components of a linear operator, 76
components of a vector, 75
composition of maps, 11
conditional probability, 293
configuration space, 469
conjugacy class, 43
conjugation by group element, 43
connected component, 275
connected set, 273
connection, 507
    Riemannian, 518
    symmetric, *see* connection, torsion-free
    torsion-free, 513
connection 1-forms, 527
connection vector field, 522
conservation of charge, 252
conservation of total 4-momentum, 244
conservative system, 468
conserved quantity, 393
constant of the motion, 393
constants, 2
constrained system, 468
constraint, 468
contact 1-form, 479
contact transformation
    homogeneous, 479
continuous function, 257, 262
continuous spectrum, 363
contracted Bianchi identities, 527
contravariant degree, 186
contravariant transformation law of components, 84
convergence to order m, 310
convergent sequence, 255, 256, 260, 280
Conway's game of life, 23
coordinate chart, 411
coordinate functions, 411
coordinate map, 411
coordinate neighbourhood, 411
coordinate system at a point, 411
coordinates of point in a manifold, 411
correspondence between classical and quantum mechanics, 382
correspondence principle, 375
coset
    left, 45
    right, 46
coset of a vector subspace, 69
cosmological constant, 582
cosmology, 548
cotangent bundle, 424
cotangent space, 420

countable set, 13
countably infinite set, 13
countably subadditive, 295
covariant degree, 186
covariant derivative, 507
    components of, 508
    of a tensor field, 510
    of a vector field, 507
    of a vector field along a curve, 508
covariant vector transformation law of components, 94
covector at a point, 420
covector field, 423
covectors, 90
covering, 271
    open, 271
critical point, 118
cubical k-chain, 486
current density, 245
curvature 2-forms, 528
curvature tensor, 513, 516
    physical interpretation, 537–539
    Riemann curvature tensor, 524
        number of independent components, 526
        symmetries of, 524
curve
    directional derivative of a function, 417
    passes through a point, 416
    smooth parametrized, 416
curves, one parameter family of, 522
cycle, 496

d'Alembertian, 248
de Rham cohomology spaces (groups), 498
de Rham's theorem, 499
de Sitter universe, 582
delta distribution, 312
    derivative of, 316
dense set, 261
dense set in $\mathbb{R}$, 14
density operator, 398
diathermic wall, 458
diffeomorphic, 416
diffeomorphism, 416
difference of two sets, 6
differentiable 1-form, 423
differentiable manifold, 412
differentiable structure, 412
differentiable vector field, 422
differential r-form, 447
    closed, 497
    exact, 497
differential exterior algebra, 447
differential of a function, 423
differential of a function at a point, 421
dimension of vector space, 72

Index

Dirac delta function, 308, 353
   change of variable, 318
   Fourier transform, 320
Dirac measure, 293
Dirac string, 504
direct product
   of groups, 48
direct sum of vector spaces, 67
discrete dynamical structure, 22
discrete dynamical system, 18
discrete symmetries, 393
discrete topology, 261
disjoint sets, 6
displacement of the origin, 390
distribution, 311–5
   density, 312
   derivative of, 315
   Fourier transform of, 321
   inverse Fourier transform of, 321
   of order m, 311
   regular, 311
   singular, 312
   tempered, 321
distributive law, 59, 149
divergence-free 4-vector field, 245
division algebra, 153
domain
   of a chart, 411
   of an operator, 357
   of a mapping, 10
dominated convergence, 306
double pendulum, 471
dual electromagnetic tensor, 247
dual Maxwell 2-form, 502
dual space, 90, 283
dummy indices, 81
dust, 551
dynamical variable, 475

Eddington–Finkelstein coordinates, 546
effective action, 572
eigenvalue, 100, 351
   multiplicity, 102
eigenvector, 100, 351
Einstein elevator, 535
Einstein tensor, 527
Einstein's field equations, 537
Einstein's gravitational constant, 537
Einstein's principle of relativity, 229
Einstein–Cartan theory, 518
electric field, 246
electrodynamics, 246
electromagnetc field, 246
electromagnetic 4-potential, 247
electron spin, 373
elementary divisors, 105

embedded submanifold, 429
embedding, 429
empirical entropy, 461
empirical temperature, 458
empty set, 5
energy operator, 379
energy–stress tensor, 252
   of electromagnetic field, 253
ensemble, 401
   canonical, 401
   grand canonical, 406
   microcanonical, 401
entropy, 403
Eötvös experiment, 535
equal operators, 357
equality of sets, 4
equation of continuity, 245
equivalence class, 8
equivalence relation, 8
Euclidean geometry, 19
Euclidean group, 580
Euclidean space, 53
Euclidean transformation, 54
Euclidean vector space, 127
Euler characteristic, 497
Euler–Lagrange equations, 467
even parity, 394
event, 54, 230, 536
event horizon, 546
exponential map, 566
exponential operator, 350
extended real line, 287
exterior algebra, 164, 209
exterior derivative, 448
exterior product, 163, 208
external direct sum, 67
extremal curve, 467

factor group, 47
factor space, 8
family of sets, 4
Fermi–Dirac statistics, 397
fermion, 397
field, 60
finer topology, 261
finite set, 4, 13
flow, 433
focus
   stable, 118
   unstable, 118
four-dimensional Gauss theorem, 251
Fourier series, 338
Fourier transform, 320
   inverse, 320
Fourier's integral theorem, 320
free action, 50

Index

free associative algebra, 182
free indices, 81
free vector space, 178
Friedmann equation, 551
Frobenius theorem, 441
   expressed in differential forms, 455
Fubini's theorem, 306
function, 10
   analytic, 410
   differentiable, 410
   real differentiable, 415
fundamental n-chain, 489

Galilean group, 54
Galilean space, 54
Galilean transformation, 55, 229
gauge transformation, 248, 542
Gauss theorem, 491
general linear group on a vector space, 65
general linear groups, 37
   complex, 39
general relativity, 536–552
generalized coordinates, 468
generalized momentum, 473
generalized thermal force, 460
geodesic, 508
   affine parameter along, 509
   null, 536
   timelike, 536
geodesic coordinates, 520–22
geodesic deviation, equation of, 522–24
geodesics
   curves of stationary length, 519
graded algebra, 164
gradient
   of 4-tensor field, 244
Gram–Schmidt orthonormalization, 129
graph of a map, 269
Grassmann algebra, 160, 166
gravitational redshift, 548
Green's theorem, 491
group, 27
   abelian, 28
   cyclic, 29
      generator of, 29
   finite, 29
   order of, 29
   simple, 46
group of components, 279

Hénon map, 22
Hamilton's equations, 477
Hamilton's principle, 469
Hamilton–Jacobi equation, 480
Hamiltonian, 475
Hamiltonian operator, 379

Hamiltonian symmetry, 393
Hamiltonian vector field, 475
harmonic gauge, 542
harmonic oscillator
   quantum, 384, 402
heat 1-form, 459
heat added to a system, 459
Heaviside step function, 316
Heckmann–Schücking solutions, 583
Heisenberg picture, 380
Heisenberg uncertainty relation, 372
hermite polynomials, 338
hermitian operator
   complete, 352
   eigenvalues, 351
   eigenvectors, 351
   spectrum, 355
Hilbert Lagrangian, 554
Hilbert space, 330–34
   finite dimensional, 134
   of states, 369
   separable, 332, 335
Hodge dual, 220–27
Hodge star operator, 223
homeomorphic, 263
homeomorphism, 263
homogeneous manifold, 573
homologous cycles, 497
homology spaces (groups), 497
homomorphism
   of groups, 40

ideal
   left, 152
   right, 152
   two-sided, 152
ideal gas, 458
idempotent, 205
identity element, 27
identity map, 12
ignorable coordinate, 473
image of a linear map, 70
immersion, 429
impossible event, 293
inclusion map, 12
independent events, 293
independent set of points in $\mathbb{R}^n$, 493
indexing set, 4
indiscrete topology, 261
indistinguishable particles, 395
induced, 267
induced topology, 259
induced vector field, 575
inertial frame, 228, 230, 232
infinite set, 13
infinitesimal generator, 170, 390

Index

injection, 11, 267
injective map, 11
inner automorphism, 43
inner measure, 300
inner product, 330
   complex
      components of, 137
   components of, 128
   Euclidean, 127
   index of inner product, 131
   Minkowskian, 131
   non-singular, 127
   of p-vectors, 221
   on complex spaces, 133
   positive definite, 127
   real, 126
inner product space
   complex, 134
instantaneous rest frame, 240
integral curve of a vector field, 433
integral of n-form with compact support, 484
integral of a 1-form on the curve, 428
interior of a set, 260
interior product, 213, 452
internal energy, 459
intersection of sets, 6
intertwining operator, 121
invariance group of a set of functions, 53
invariance of function under group action, 52
invariance of Lagrangian under local flow, 474
invariant subspace, 99, 121
invariants of electromagnetic field, 247
inverse element, 27
inverse image of a set under a mapping, 10
inverse map, 11
is a member of, 3
isentropic, 461
isometry, 578
isomorphic
   groups, 42
isomorphic algebras, 151
isomorphic vector spaces, 64
isomorphism
   of groups, 42
isotherm, 458
isotropic space, 534
isotropy group, 50

Jacobi identity, 167, 432
Jordan canonical form, 113

Kasner solutions, 583
kernel index notation, 196
kernel of a linear map, 70
kernel of an algebra homomorphism, 152
kernel of homomorphism, 47

ket, 369
Killing vector, 578
Killing's equations, 578
kinetic energy, 468
Kronecker delta, 36, 83

Lagrange's equations, 469
Lagrangian, 469, 554
Lagrangian function, 465
Lagrangian mechanical system, 469
law of composition, 18
   associative, 18
   commutative, 18
Lebesgue integrable function, 304
Lebesgue integral
   non-negative measurable function, 301
   over measurable set, 302
   simple functions, 301
Lebesgue measure, 295–300
   non-measurable set, 299
Lebesgue–Stieltjes inetgral, 356
left action of group on a set, 49
left translation, 51, 277, 559
left-invariant differential form, 562
left-invariant vector field, 560
Leibnitz rule, 418
Levi–Civita symbols, 215
Lie algebra, 166–77
   commutative, 167
   factor algebra, 168
   ideal, 168
Lie algebra of a Lie group, 561
Lie bracket, 432
Lie derivative, 507
   components of, 439
   of a tensor field, 438
   of a vector field, 436
Lie group, 559
Lie group homomorphism, 564
Lie group isomorphism, 564
Lie group of transformations, 572
Lie subgroup, 569
lift of a curve
   to tangent bundle, 424, 465
light cone, 230
light cone at a point, 233
lim inf, 291
lim sup, 291
limit, 255, 256, 280
limit point, 260
linear connection, *see* connection
linear functional, 88
   components of, 91
   on Banach space, 283
linear map, 63
linear mapping, 35
   matrix of, 36

Index

linear operator, 65
  bounded, 344
  continuous, 344
linear ordinary differential equations, 116
linear transformation, 36, 65
linearized approximation, 539
linearly dependent set of vectors, 73
linearly independent vectors, 73
local basis of vector fields, 422
local flow, 434
local one-parameter group of transformations, 434
  generated by a vector field, 434
locally Euclidean space, 411
  dimension, 411
locally flat space, 532–34
locally integrable function, 311
logical connectives, 3
logical propositions, 3
logistic map, 22
Lorentz force equation, 247
Lorentz gauge, 248
  gauge freedom, 248
Lorentz group, 56
Lorentz transformation, 56
  improper, 230
  proper, 230
Lorentz–Fitzgerald contraction, 237
lowering an index, 200

magnetic field, 246
magnitude of a vector, 127
map, 10
  differentiable at a point, 416
  differentiable between manifolds, 415
mapping, 10
mass of particle, 467
matrix
  adjoint, 40
  components of, 36
  improper orthogonal, 38
  non-singular, 36
  orthogonal, 38
  proper orthogonal, 38
  symplectic, 39
  unimodular, 37
  unitary, 40
matrix element of an operator, 347
matrix group, 37
matrix Lie groups, 169–72, 570
matrix of components, 98
matrix of linear operator, 76
Maurer–Cartan relations, 563
maximal element, 17
maximal symmetry, 579
Maxwell 2-form, 502
Maxwell equations, 246
  source-free, 246

measurable function, 289–92
measurable set, 287
measurable space, 288
measure, 292
  complete, 300
measure space, 287, 292
metric, 264
metric space, 264
  complete, 265
metric tensor, 189, 469, 516
  inverse, 516
metric topology, 264
Michelson–Morley experiment, 229
minimal annihilating polynomial, 105
Minkowski space, 55, 230, 231, 535
  inner product, 233
  interval between events, 232
Minkowski space–time, 230
mixed state, 399
  stationary, 400
module, 63
momentum 1-form conjugate to generalized velocity, 472
momentum operator, 361, 375
monotone convergence theorem, 302
morphism, 23
  composition of, 23
  epimorphism, 24
  identity morphism, 23
  isomorphism, 25
  monomorphism, 24
multilinear map, 186
multiplication operator, 345, 347, 350, 353
multiplicities of representation, 147
multivector, 183, 208
multivectors, 163

natural numbers, 4
negatively oriented, 217
neighbourhood, 255, 256, 262
Newtonian tidal equation, 540
nilpotent matrix, 110
nilpotent operator, 110
node
  stable, 118
  unstable, 118
Noether's theorem, 473
non-degenerate 2-form, 474
norm
  bounded linear operator, 350
  of linear operator, 344
norm of a vector, 135
normal coordinates, *see* geodesic coordinates
normed space
  complete, 282
null cone, 233
null vector, 127

Index

nullity of a linear operator, 80
number of degrees of freedom of constrained system, 468

objects, 23
observable, 370
  compatible, 372
  complementary, 372
  complete, 370
  expectation value of, 371
  root mean square deviation of, 371
occupation numbers, 405
octonians, 158
odd parity, 394
one-parameter group of transformations, 433
one-parameter group of unitary transformations, 390
one-parameter subgroup, 171
one-parameter subgroup of a Lie group, 565
one-to-one correspondence between sets, 11
one-to-one map, 11
onto map, 11
open ball, 256, 264
open interval, 255
open map, 278
open neighbourhood, 262
open set, 256, 257
open submanifold, 413, 430
operator
  adjoint, 346–8
  closed, 358
  densely defined, 357
  extension of, 357
  hermitian, 348–349
  idempotent, 348
  in Hilbert space, 357
  invertible, 345
  isometric, 349
  normal, 351
  projection, 348
  self-adjoint, 360
  symmetric, 360
  unbounded, 357
  unitary, 349
opposite orientation on a manifold, 481
oppositely oriented, 217
orbit, 50
orbit of a point under a Lie group action, 572
orbit of point
  under a flow, 433
ordered n-tuple, 7
ordered p-simplex, 493
  vertices, 493
ordered pair, 7
orientable manifold, 481
orientation, 217
orientation on a manifold, 481

oriented manifold, 481
oriented vector space, 217
orthogonal 4-vectors, 233
orthogonal complement, 341–42
orthogonal complement of a vector subspace, 142
orthogonal groups, 38
  complex, 39
  proper, 38
orthogonal projection, 341
orthogonal vectors, 127, 137, 341
orthonormal basis, 129, 137, 335, 370
outer measure, 295

paracompact topological space, 482
parallel transport, 509
parallelogram law, 140, 330
parametrized curve
  in Minkowski space, 239
  length of, 517
  null, 239
  spacelike, 239
  timelike, 239
parity observable, 394
Parseval's identity, 339
partial order, 9
partially ordered set, 9
particle horizon, 552
particle number operator, 359
partition function
  canonical, 402
  grand canonical, 406
  one-particle, 406
partition of a set, 8
partition of unity subordinate to the covering, 482
Pauli exclusion principle, 397
Pauli matrices, 173
Peano's axioms, 4
perfect fluid, 253
  pressure, 253
periods of an r-form, 499
permutation, 30
  cycle, 31
  cyclic, 31
  cyclic notation, 31
  even, 32
  interchange, 32
  odd, 32
  order of, 34
  parity, 33
  sign of, 33
permutation group, 30
permutation operator, 395
Pfaffian system of equations, 455
  first integral, 455
  integral submanifolds, 456

Index

phase space, 478
    extended, 479
photon, 242, 367
    direction of propagation, 242
plane gravitational waves, 547
plane pendulum, 470
plane waves, 366
Poincaré group, 56
Poincaré transformation, 56, 230
point spectrum, 363
Poisson bracket, 383, 475
Poisson's equation, 323, 534
    Green's function, 323–25
polarization
    circular, 367
    elliptical, 367
    linear, 367
poset, 9
position operator, 360, 375
positively oriented, 217
potential energy, 468
power set of a set, 5
Poynting vector, 253
pre-Hilbert space, 330
principal pressures, 253
principle of equivalence, 534–37
probability measure, 293
probability of an event, 293
probability space, 293
product, 18
    of elements of a group, 27
    of manifolds, 414
    of vectors, 149
product of linear maps, 65
product topology, 266
projection map, 11
    tangent bundle, 424
projection operator, 352
projective representation, 389
proper time, 241, 536
pseudo-orthogonal groups, 39
pseudo-orthogonal transformations, 132
pseudo-Riemannian manifold, 516–22, 529
    hyperbolic manifold, 517
    Minkowskian manifold, 517
    Riemannian manifold, 516
pseudo-sphere, 582
pullback, 426
pure state, 399

quantifiers, 3
quasi-static process, 458
quaternion, 157
    conjugate, 157
    inverse, 158
    magnitude, 158
    pure, 157
    scalar part, 157
    vector part, 157
quaternions, 157–58
quotient vector space, 69

raising an index, 200
range
    of an operator, 357
    of a mapping, 10
rank of a linear operator, 80
rational numbers, 14
ray, 369
ray representation, *see* projective representation
real inner product space, 127
real projective n-space, 269
real projective plane, 269
real structure of a complex vector space, 155
rectilinear motion, 54
refinement of an open covering, 482
    locally finite, 482
reflexive relation, 8
regular k-domain, 490
regular domain, 489
regular embedding, 430
regular value, 353, 362
relative topology, 259
representation, 120
    completely reducible, 122
    degree of, 120
    equivalent, 120
    faithful, 120
    irreducible, 121
    unitary, 141
representation of a group, 50
    complex, 50
residue classes modulo an integer, 8
resolvent operator, 362
rest mass, 242
rest-energy, 242
restriction of map to a subset, 12
reversible process, 458
Ricci identities, 514
    general, 515
Ricci scalar, 526
Ricci tensor, 526
Riemann tensor, *see* curvature tensor, Riemann curvature tensor
Riemannian manifold, 469
Riesz representation theorem, 342, 369
Riesz–Fischer theorem, 333
right action, 50
right translation, 51, 278, 559
rigid body, 472
ring, 59

Index

Robertson–Walker models, 549, 582
  closed, 549
  flat, 549
  open, 550
rotation, 53
rotation group, 38, 53
  spinor representation, 174
rotation operator, 392

saddle point, 118
scalar, 60
scalar multiplication, 61
scalar potential, 248
scalar product, 133
Schmidt orthonormalization, 138, 335
Schrödinger equation, 379
  time-independent, 383
Schrödinger picture, 380
Schur's lemma, 124
Schwarzschild radius, 545
Schwarzschild solution, 545
sectional curvature, 532
Segré characteristics, 114
self-adjoint, 348
self-adjoint operator, 141
semi-direct product, 57
semigroup, 18
  homomorphism, 19
  identity element, 18
  isomorphism, 19
separation axioms, 269
sequence, 13
set, 2
set theory, 3
shift operators, 344, 347, 353
similarity transformation, 43, 86
simple function, 290
simultaneity, 236
simultaneous events, 54
singleton, 4
singular p-simplex, 496
singular chains, 489
smooth vector field, 422
  along a parametrized curve, 424
  on open set, 424
source field, 246
space of constant curvature, 532, 534, 580
space–time, 536
spacelike 3-surface, 251
spatial inversion, 393
special linear group, 37
specific heat, 464
spectral theorem
  hermitian operators, 356
  self-adoint operaotrs, 363
spectral theory
  unbounded operators, 362–64

spectrum
  bounded operator, 353–57
  continuous spectrum, 353
  point spectrum, 353
  unbounded operator, 362
spherical pendulum, 471
spherical symmetry, 542, 583
spin-statistics theorem, 397
standard n-simplex, 494
state, 369
  dispersion-free, 371
static metric, 585
stationary metric, 585
step function, 290
stereographic projection, 413
Stern–Gerlach experiment, 368
Stokes' theorem, 487, 491
strange attractors, 22
stronger topology, 261
structure constants, 150
structured set, 9
subalgebra, 151
subcovering, 271
subgroup, 28
  conjugate, 43
  generated by a subset, 278
  normal, 46
  trivial, 28
subrepresentation, 121
subset, 5
subspace
  dense, 357
  generated by a subset, 72, 335
  Hilbert, 335
  spanned by a subset, 72
sum of vector subspaces, 67
summation convention, 81
superposition
  of polarization states, 367
superset, 5
support of a function, 309
support of an n-form, 484
surface of revolution, 534
surjection, 11
surjective map, 11
Sylvester's theorem, 130
symmetric group, 30
symmetric relation, 8
symmetrical state, 397
symmetrization operator, 405
symmetry group of a set of functions, 53
symmetry group of Lagrangian system, 474
symmetry transformation
  between observers, 387
symplectic groups, 39
symplectic manifold, 475
symplectic structure, 475

Index

tangent 4-vector, 239
tangent bundle, 424
tangent map, 426
tangent space at a point, 418
tangent vector
   components, 420
   to a curve at a point, 420
tangent vector at point in a manifold, 418
temperature, 404
tensor, 180
   antisymmetric, 201, 204
   antisymmetric part, 205
   components of, 188, 190, 194
   contraction, 198
   contravariant of degree 2, 181, 190
   covariant of degree 2, 181, 187
   mixed, 192
   symmetric, 189, 201
tensor of type (r, s), 186
   at a point, 421
tensor product, 194
   of covectors, 187
   of vectors, 190
tensor product of dual spaces, 186
tensor product of two vector spaces, 179
tensor product of two vectors, 179
tensor product of vector spaces, 186
test functions, 309
   of order m, 309
test particle, 536
thermal contact, 458
thermal equilibrium, 458
thermal variable assumption, 459
thermodynamic system
   equilibrium states of, 457
   internal variables, 457
   number of degrees of freedom, 457
   thermal variables, 457
thermodynamic variables, 457
tidal forces, 536, 539
time dilatation, 237
time translation, 392
time-reversal operator, 394
topological group, 277
   discrete, 277
topological invariant, 263
topological manifold, 411
topological product, 266
topological space, 258
   compact, 271
   connected, 273
   disconnected, 273
   first countable, 262
   Hausdorff, 269
   locally connected, 278
   normal, 265
   second countable, 262
   separable, 262
topological subgroup, 277
topological subspace, 259
topological vector space, 279
topologically equivalent, 263
topology, 21, 257
   by identification, 268
   generated by a collection of sets, 261
   induced, 265, 266
torsion 2-forms, 527
torsion map, 512
torsion tensor, 512–13
torus
   n-torus, 415
total angular momentum, 385
total order, 9
trace, 107
transformation
   of a manifold, 433
   of a set, 12
transformation group, 30
transformation of velocities
   Newtonian, 229
   relativistic, 237
transition functions, 411
transitive action, 50, 572
transitive relation, 8
translation operators, 391
transmission probability, 367
transpose, 37
transpose of a linear operator, 96
triangle inequality, 137, 264, 330
trivial topology, 261

unary relation, 7
uncountable set, 14
unimodular group, 37
   complex, 39
union of sets, 5
unit matrix, 36
unitary group, 40
   special, 40
unitary operator
   eigenvalues, 352
   eigenvectors, 352
unitary transformation, 139

variables, 3
variation
   of a curve, 465
   of fields, 553
variation field, 466
variational derivative, 553
vector, 60

Index

vector addition, 61
vector field
  complete, 434
  parallel along a curve, 508
vector field on a manifold, 422
vector product, 220
vector space, 60
  finite dimensional, 72
  infinite dimensional, 72
vector space homomorphism, 63
vector space isomorphism, 63
vector subspace, 66
volume element, 215
volume element on a manifold, 481
vortex point, 118

wave equation
  Green's function, 326–28
  inhomogeneous, 326
wave operator, 248
weak field approximation, 537
weaker topology, 261
wedge product, 208
Weierstrass approximation theorem, 337
Wigner's theorem, 388
work done by system, 458
work done on system, 458
work form, 459
world-line, 239, 536

zero vector, 61
zeroth law of thermodynamics, 458

Printed in the United States
By Bookmasters